Handbook of Cucurbits

Growth,Cultural Practices, and Physiology

Handbook of Cucurbits

Growth, Cultural Practices, and Physiology

Mohammad Pessarakli

The University of Arizona
School of Plant Sciences
Tucson, Arizona, USA

CRC Press
Taylor & Francis Group
Boca Raton London New York

CRC Press is an imprint of the
Taylor & Francis Group, an **informa** business

CRC Press
Taylor & Francis Group
6000 Broken Sound Parkway NW, Suite 300
Boca Raton, FL 33487-2742

First issued in paperback 2021

© 2016 by Taylor & Francis Group, LLC
CRC Press is an imprint of Taylor & Francis Group, an Informa business

No claim to original U.S. Government works

ISBN 13: 978-1-03-209806-7 (pbk)
ISBN 13: 978-1-4822-3458-9 (hbk)

Visit the Taylor & Francis Web site at
http://www.taylorandfrancis.com

and the CRC Press Web site at
http://www.crcpress.com

To the memory of my beloved parents, Fatemeh and Vahab, who regretfully, did not live to see this work and my other works, which, in no small part, resulted from their gift of many years of unconditional love to me.

Contents

SECTION I   Introductory Chapters

SECTION II   Cucurbit Physiological Stages
of Growth and Development I

SECTION III   Cultural Practices of Cucurbits

SECTION IV   Cucurbits Physiological Stages of Growth and Development II

SECTION V   Genetics, Genomics, and Breeding of Cucurbits

SECTION VI   Cucurbit Grafting

SECTION VII   Cucurbit Pathology and Diseases

SECTION VIII   Weed Control, Pest Control, and Insects of Cucurbits

SECTION IX   Therapeutic and Medicinal Values of Cucurbits

SECTION X   Growth Responses of Cucurbits under Stressful Conditions (Abiotic and Biotic Stresses)

Preface

Cucurbits are part of the daily diet of people around the world. Therefore, cucurbit products are important and need special attention in their cultural practices, physiology, and production. There are numerous books and articles available on cucurbits, but these all exist relatively in isolation of each other, covering only one or a few specific topics on cucurbits. Therefore, the information on these important plants is scattered. I felt the need for a single unique comprehensive source of information that includes as many factors as possible on cucurbits, and this resulted in the *Handbook of Cucurbits*. It is a complete collection of the factors on cucurbits.

This comprehensive source is an up-to-date reference book effectively addressing issues and concerns related to cucurbit growth, physiology, cultural practices, diseases, and production. These aspects of cucurbits have efficiently and effectively been addressed in this unique handbook.

While previous authors have indeed competently covered relevant areas separately in various publications, the areas are, nonetheless, interrelated and should be covered comprehensively in a single text, which is the purpose of this book.

The *Handbook of Cucurbits* has been prepared by many competent and knowledgeable scientists, specialists, and researchers in agriculture and horticulture from several countries. It is intended to serve as a resource for both lectures and independent purposes. Scientists, agriculture researchers, agriculture practitioners, and students will benefit from this unique comprehensive guide, which covers issues related to cucurbits from planting to production.

As with other fields, accessibility of knowledge is among the most critical of factors involved with cucurbit production. Without due consideration of all the elements contributing to cucurbit crop production, it is unlikely that a successful production system will be achieved. For this reason, as many factors as possible are included in this handbook. To further facilitate the accessibility of the desired information in various areas covered in this collection, the book is divided into 11 sections: Introductory Chapters; Cucurbit Physiological Stages of Growth and Development I; Cultural Practices of Cucurbits; Cucurbit Physiological Stages of Growth and Development II; Genetics, Genomics, and Breeding of Cucurbits; Cucurbit Grafting; Cucurbit Pathology and Diseases; Weed Control, Pest Control, and Insects of Cucurbits; Therapeutic and Medicinal Values of Cucurbits; Growth Responses of Cucurbits under Stressful Conditions (Abiotic and Biotic Stresses); and Examples of Cucurbit Crop Plants Growth and Development and Cultural Practices. Each of these sections consists of one or more chapters to discuss, independently, as many aspects of cucurbits as possible in that specific topic.

Section I consists of two chapters, including one that provides basic and general introductory information on cucurbit plants/crops.

Section II contains one chapter, which provides details on cucurbit carbohydrate metabolism.

Section III includes five chapters, each of which presents in-depth information on their topic.

Section IV contains four chapters that provide detailed information on various physiological stages of cucurbit growth and development.

Section V consists of two chapters that provide information on muskmelon and ash gourd, respectively.

Section VI includes three chapters that discuss improvements in cucurbit productions and stress tolerance responses of cucurbit plants/crops

Section VII contains one chapter that presents important diseases of cucurbits and their proper management strategies.

Section VIII consists of three chapters that present up-to-date, detailed information on the control and management practices of weeds, pests, and insects of cucurbit plants/crops.

Section IX contains one chapter devoted to the health benefits of cucurbit plants/crops.

Section X consists of three chapters that discuss in detail the responses of cucurbit plants/crops under stressful conditions.

Finally, Section XI, consisting of six chapters, presents detailed information on various cucurbit plants/crops.

Numerous figures and tables are included in the handbook to facilitate comprehension of the presented material. Hundreds of index words are also included to further increase accessibility to desired information.

Mohammad Pessarakli, PhD
Professor
University of Arizona
Tucson, Arizona

Acknowledgments

I express my appreciation for the assistance I received from the secretarial and administrative staff of the School of Plant Sciences, College of Agriculture and Life Sciences, University of Arizona. The encouraging words of several of my colleagues, which are always greatly appreciated, have certainly been a driving force for the successful completion of this project.

In addition, I sincerely acknowledge Randy Brehm (acquiring editor, Taylor & Francis Group, CRC Press) whose professionalism, patience, hard work, and proactive methods helped in the completion of this project and my previous book projects. This job would not have been completed as smoothly and rapidly without her valuable support and efforts.

I am indebted to Jill Jurgensen (senior project coordinator, Taylor & Francis Group, CRC Press) for her professional and careful handling of this book and my previous publications. I also acknowledge the eye for detail, sincere efforts, and the hard work put in by the copy editor and the project editor, Rachael Panthier.

The collective efforts and invaluable contributions of several experts on cucurbits plants/crops made it possible to produce this unique resource that presents comprehensive information on this subject. Each and every one of these contributors and their contributions are greatly appreciated.

Last, but not least, I thank my wife, Vinca, a high school science teacher, and my son, Dr. Mahdi Pessarakli, MD, who supported me during the course of this work.

Editor

Mohammad Pessarakli, PhD, is a professor in the School of Plant Sciences, College of Agriculture and Life Sciences, University of Arizona, Tucson, Arizona. His work at the University of Arizona includes research and extension services and teaching courses in turfgrass science, management, and stress physiology. He is the editor of the *Handbook of Plant and Crop Stress* and the *Handbook of Plant and Crop Physiology* (both titles published by Taylor & Francis Group, CRC Press [formerly Marcel Dekker, Inc.]), and the *Handbook of Photosynthesis*, and *Handbook of Turfgrass Management and Physiology*. Dr. Pessarakli has written 20 book chapters; has been an editorial board member of the *Journal of Plant Nutrition*, *Communications in Soil Science and Plant Analysis*, *Advances in Plants and Agriculture Research Journal* and the *Journal of Agricultural Technology*; a member of the Book Review Committee of the Crop Science Society of America, and a reviewer of the *Crop Science, Agronomy, Soil Science Society of America* and the *HortScience* journals. He is author or coauthor of 185 journal articles and 55 trade magazine articles. Dr. Pessarakli is an active member of the Agronomy Society of America, the Crop Science Society of America, and the Soil Science Society of America, among others. He is an executive board member of the American Association of University Professors (AAUP), Arizona Chapter.

Dr. Pessarakli is a well-known, internationally recognized scientist and scholar and an esteemed member (invited) of *Sterling Who's Who*, *Marques Who's Who*, *Strathmore Who's Who*, *Madison Who's Who*, and *Continental Who's Who*, as well as numerous other honor societies (i.e., Phi Kappa Phi, Gamma Sigma Delta, Pi Lambda Theta, Alpha Alpha Chapter). He is a certified professional agronomist and a certified professional soil scientist (CPAg/SS), designated by the American Registry of the Certified Professionals in Agronomy, Crop Science, and Soil Science. Dr. Pessarakli is a United Nations consultant in agriculture for underdeveloped countries. He earned his BS (1977) in environmental resources in agriculture and MS (1978) in soil management and crop production from The Arizona State University, Tempe, and earned his PhD (1981) in soil and water science from the University of Arizona, Tucson. Dr. Pessarakli's environmental stress research work and expertise on plants and crops are internationally recognized.

For more information about Dr. Pessarakli, please visit http://ag.arizona.edu/pls/faculty/pessarakli.htm, http://cals.arizona.edu/spls/people/faculty

Contributors

Lord Abbey
Department of Plant and Animal Sciences
Dalhousie University
Nova Scotia, Canada

Fernando Antonio Souza de Aragão
Laboratory of Plant Breeding and Genetic
Resources
Embrapa Tropical Agroindustry
Fortaleza, Brazil

T.K. Behera
Division of Vegetable Science
Indian Agricultural Research Institute
New Delhi, India

Meni Ben-Hur
Institute of Soil, Water and Environmental
Sciences
Agricultural Research Organization
Bet Dagan, Israel

L.K. Bharati
Central Horticultural Experimentation Station
Indian Institute of Horticultural Research
Orissa, India

Isac Gabriel Abrahão Bomfim
Department of Animal Sciences
Federal University of Ceará
Fortaleza, Brazil

Prem Chand
India Council of Agricultural Research
Jawaharlal Nehru Krishi Vishwa Vidyalaya
Madhya Pradesh, India

B.R. Choudhary
Division of Crop Improvement
Central Institute for Arid Horticulture
Rajasthan, India

Rakesh Kr. Dubey
Department of Vegetable Science
Central Agricultural University
Arunachal Pradesh, India

Menahem Edelstein
Department of Vegetable Crops
Agricultural Research Organization
Ramat Yishay, Israel

Breno Magalhães Freitas
Department of Animal Sciences
Federal University of Ceará
Fortaleza, Brazil

Ruparao T. Gahukar
Arag Biotech Pvt. Ltd.
Maharashtra, India

Wenjing Guan
Department of Horticulture and Landscape
Architecture
Southwest Purdue Agricultural Program
Purdue University
Vincennes, Indiana

Maryam Haghighi
Department of Horticulture
Isfahan University of Technology
Isfahan, Iran

Donald J. Huber
Horticultural Sciences Department
University of Florida
Gainesville, Florida

Jamal Javanmardi
Department of Horticultural Sciences
College of Agriculture
Shiraz University
Shiraz, Iran

Suhas G. Karkute
ICAR-Indian Institute of Vegetable Research
Uttar Pradesh, India

Viveka Katoch
Department of Vegetable Science and
Floriculture
Himachal Pradesh Agricultural University
Himachal Pradesh, India

A.V.V. Koundinya
Department of Vegetable Crops
Bidhan Chandra Krishi Viswavidyalaya
West Bengal, India

Sanjeev Kumar
Division of Vegetable Science and Floriculture
Sher-e-Kashmir University of Agricultural
    Sciences and Technology
Jammu and Kashmir, India

S. Ramesh Kumar
Department of Horticulture
Tamil Nadu Agricultural University
Tamil Nadu, India

Meenu Kumari
Central Horticultural Experimentation Station
Indian Institute of Horticultural Research
Orissa, India

Hong Li
Environment and Plant Protection Institute
Chinese Academy of Tropical Agricultural Science
Haikou, Hainan, China

and

Nova Scotia Institute of Agrologists
Nova Scotia, Canada

Paul J. McLeod
Department of Entomology
University of Arkansas
Fayetteville, Arkansas

Minmin Miao
School of Horticulture and Plant Protection
Yangzhou University
Wenhui, Yangzhou, People's Republic of China

Reena Nair
Department of Horticulture
Jawaharlal Nehru Krishi Vishwa Vidyalaya
Madhya Pradesh, India

Satya S. Narina
Department of Agriculture
Agricultural Research Station
Virginia State University
Petersburg, Virginia

David O. Ojo
National Horticultural Research Institute
University of Ibadan
Ibadan, Nigeria

A.K. Pandey
College of Horticulture and Forestry
Central Agricultural University
Arunachal Pradesh, India

Sudhakar Pandey
Division of Crop Improvement
ICAR-Indian Institute of Vegetable Research
Uttar Pradesh, India

M.K. Pandit
Department of Vegetable Crops
Bidhan Chandra Krishi Viswavidyalaya
West Bengal, India

Mohammad Pessarakli
School of Plant Sciences
The University of Arizona
Tucson, Arizona

Hari Har Ram
Department of Vegetable Science
GB Pant University Agriculture & Technology
Uttrakhand, India

Chanchal Rana
Department of Vegetable Science and
    Floriculture
Himachal Pradesh Agricultural University
Himachal Pradesh, India

Tahir Rashid
Department of Agriculture
Alcorn State University
Mound Bayou, Mississippi

Puja Rattan
Department of Agriculture
DAV University
Punjab, India

Amir Hossein Saeidnejad
Department of Agriculture
Payame Noor University
Tehran, Iran

R.K. Samnotra
Division of Vegetable Science and
    Floriculture
Sher-e-Kashmir University of Agricultural
    Sciences and Technology
Jammu and Kashmir, India

Akhilesh Sharma
Department of Vegetable Science and
    Floriculture
Himachal Pradesh Agricultural University
Himachal Pradesh, India

Atena Sheibanirad
Department of Horticulture
Isfahan University of Technology
Isfahan, Iran

Kumari Shiwani
Department of Vegetable Science and
    Floriculture
Himachal Pradesh Agricultural University
Himachal Pradesh, India

Beena Singh
SG College of Agriculture and Research Station
Indira Gandhi Agricultural University
Chhattisgarh, India

H.S. Singh
Central Horticultural Experimentation Station
Indian Institute of Horticultural Research
Orissa, India

Krishan Pal Singh
College of Horticulture and Research Station
Indira Gandhi Agricultural University
Chhattisgarh, India

Saurabh Singh
Department of Vegetable Science and Floriculture
Himachal Pradesh Agricultural University
Himachal Pradesh, India

Shrawan Singh
Division of Horticulture and Forestry
Central Island Agricultural Research Institute
Andaman and Nicobar Islands, India

Vikas Singh
Department of Vegetable Science
Central Agricultural University
Arunachal Pradesh, India

Garima Upadhyay
Department of Vegetable Science
Central Agricultural University
Arunachal Pradesh, India

Stuart Alan Walters
Department of Plant, Soil and Agricultural
    Systems
Southern Illinois University
Carbondale, Illinois

Zhiping Zhang
School of Horticulture and Plant Protection
Yangzhou University
Wenhui, Yangzhou, People's Republic of China

Xin Zhao
Horticultural Sciences Department
University of Florida
Gainesville, Florida

Section I

Introductory Chapters

1 Cucurbits
History, Nomenclature, Taxonomy, and Reproductive Growth

S. Ramesh Kumar

CONTENTS

1.1 INTRODUCTION

Cucurbits are vegetable crops belonging to the family Cucurbitaceae, which primarily comprise of species consumed as food worldwide. The family consists of about 118 genera and 825 species. Although most of them originated in the Old World, many species originated in the New World and at least seven genera in both hemispheres. There is tremendous genetic diversity within the family, and the range of adaptation for cucurbit species includes tropical and subtropical regions, arid deserts, and temperate regions (Rai et al., 2008). The genetic diversity in cucurbits extends to both vegetative and reproductive characteristics and considerably to the monoploid (x) chromosome number, including 7 (*Cucumis sativus*), 11 (*Citrullus* spp., *Momordica* spp., *Lagenaria* spp., *Sechium* spp., and *Trichosanthes* spp.), 12 (*Benincasa hispida*, *Coccinia cordifolia*, *Cucumis* spp. other than *C. sativus*, and *Praecitrullus fistulosus*), 13 (*Luffa* spp.), and 20 (*Cucurbita* spp.). Chakravarty (1982) estimated 36 genera and 100 species in India. Cucurbits are consumed in various forms: as salads (cucumber, gherkin, long melon), in sweets (ash gourd, pointed gourd), and in pickles (gherkins), desserts (melons), as well as used for other culinary purposes. Some of them (e.g., bitter gourd) are well known for their unique medicinal properties. In recent years, abortifacient proteins with ribosome-inhibiting properties have been isolated from several cucurbit species, which include momordicin (from *Momordica charantia*), trichosanthin (from *Trichosanthes kirilowii*), and beta-trichosanthin (from *Trichosanthes cucumeroides*).

1.2 GENERAL FEATURES

Cucurbit vegetables have the following common features (Goplakrishnan, 2007):

1. *Long tap root system*: Tap root may grow up to 175–180 cm, and laterals are confined to the top 60 cm. Hence, crops like bottle gourd, ash gourd, and pointed gourd are largely part of river bed cultivation.
2. *Branched stem*: The stem is 3–8 branched, prostrate/climbing, and spread up to 9–10 m in *Cucurbita* and *Lagenaria*. Crops like *Cucurbita pepo* have short internodes and are bushy. Nodes usually produce roots upon contact with soil.
3. Leaves are simple and mostly 3–5 lobed, usually palmate, and rarely pinnately lobed (*Citrullus* spp.).
4. *Leaves*: Tendrils on axils of leaves are simple in *Cucumis*, simple or bifid in others, and absent in bush types.
5. *Pollination*: Cucurbits are highly cross-pollinated, and pollination is done by honey bees and bumble bees. Flowers are born in axils of leaves and are solitary or in racemose clusters. Individual flowers are unisexual, large, and showy.
6. *Fruit*: The fruit is essentially an inferior berry and is called "pepo" due to its hard rind when mature. Fruits can be stored for a long period in ash gourd, pumpkin, oriental pickling melon, etc., while keeping quality is less in cucumber, snake gourd, bitter gourd, etc. The fruits of all cucurbits except chayote have many seeds.
7. *Seeds*: Seeds are borne in parietal placentation. Placenta is the edible portion in watermelon, while in ash gourd, ridge gourd, and smooth gourd, it is endocarp. In muskmelon, the edible portion is mostly pericarp with a little mesocarp.
8. *Propagation*: Cucurbits are mostly seed propagated. A few are vegetative propagated like parwal and coccinia.
9. *Life cycle*: Most cucurbits are annuals except chow chow and coccinia, both being perennial.
10. *Cucurbitacins*: A majority of cucurbits are characterized by the presence of bitter principles, cucurbitacins, at some parts of the plant and at some stages of development. Cucurbitacins are tetracyclic triterpenes having extensive oxidation levels. Its highest concentration is in fruits and roots and is less in leaves. Pollen grain also carries a fairly good

amount of bitter principles. This is a common problem in oriental pickling melon, cucumber, and bottle gourd and is rarely noticed in ridge gourd and snake gourd. The consumers usually remove fruit tips during conception to avoid the possibility of bitterness in fruits.

11. *Sex forms*: A wide range of sex forms like monoecious, andromonoecious, gynomonoecious, and dioecious are noticed in the family.

   a. *Hermaphrodite form*: This is the most primitive form, and plants produce only bisexual flowers. This is noticed in the Satputia variety of ridge gourd and in a few lines of cucumber and muskmelon.

   b. *Monoecious form*: This is the advanced form, and plants produce both male and female flowers in a plant. A majority of the cucurbits exhibit monoecious conditions.

   c. *Andromonoecious form*: Muskmelon and some cultivars of watermelon produce both male and bisexual flowers in a plant. However, nondessert forms like oriental pickling melon, phoot under *Cucumis melo* are monoecious.

   d. *Gynomonoecious form*: This is noticed in cucumber, and the plants produce female and bisexual flowers.

   e. *Gynoecious form*: Lines producing female flowers alone are rarely noticed in cucumber and have got a great potential for commercial F_1 production.

   f. *Trimonoecious form*: This is a condition wherein the male, female, and bisexual flowers are produced in a single plant.

   g. *Dioecious form*: Male and female flowers are produced on separate plants in parwal, coccinia, and teasle gourd.

12. *Flowering*: A majority of cucurbits start flowering 30–45 days after sowing and follow a definite sequence.

13. *Cultural requirements*: The cultivation practices of different cucurbits are similar.

1.3 HISTORY

Cucurbits or the gourd family is essentially a group of tropical plants, and it is believed to be a relatively old one. There had been a long and intimate association of humankind with this group. Jeffrey (1980a) estimated the relationship as the longest standing (ca. 15,000 years BP) and of great economic importance. This family is notable for a comparatively large number of species of cultivated plants. Even in prehistoric civilization of the Negritos and then the proto-australoids (called "Nisada" in Sanskrit literature), a cucurbit called "alabu," probably the gourd of "kalinga" identified as watermelon, has been noted by philological studies of Jeen Przyluski, Jules Bosch, and Sylvan Levi (Om Prakash, 1961). Some cucurbits like *Cucumis* spp., melon, colocynth, etc., are important seeds obtained from archaeological sites of the Indian subcontinent (Kajale, 1991).

Sanskrit prose, scriptures, and poetical works like "Vedas," "Upanishads," epics like "Ramayana" and "Mahabharata," "Brahmanas" (ritual texts), "Aranayakas" (text for forest dwellers), dating back to ages before the Christian era, mention several kinds of cucurbits. These suggest many kinds of association of humans, like some plants or trees visualized as being of sacred origin, some plants or parts thereof liked by gods and goddesses, such as fruits of *Momordica dioica* mentioned in "Yogini Tantrum" et (Sensarma, 1998). The "Puranas" (dated between the ages of Vedas and classical literature) contained detailed ethnobotanical information as in "Matsya Purana" on "alabu" (bottle gourd) and "trapusa" (cucumber). Brief historical notes of some of the cultivated cucurbits are given in the following discussion (Seshadri and More, 2004).

1.3.1 CUCUMBER

Cucumber has a very long ancient history of more than 5000 years and originated in India. It was mentioned in Rigveda (ca. 2000–1500 BC). From India, it spreads eastward to China and westward to Asia Minor, North Africa, and later to other parts of Europe. The Greeks and Romans grew cucumber in 300 BC. Columbus introduced cucumber into the New World and planted it in Haiti in 1494 from where it moved to the United States by 1539.

1.3.2 PUMPKIN AND SQUASHES

The summer squash, *C. pepo*, was the oldest to be domesticated in southern Mexico to southwestern United States, as early as 8000 BC. The pumpkin, *C. moschata*, was cultivated in southern Mexico in 5000 BC and in Peru in 3000 BC. *Cucurbita maxima* was domesticated in southern South America, as evidenced by the excavated seeds that dated to AD 1200 (Vishnu Swarup, 2006).

1.3.3 MELONS

Melon was cultivated in Egypt during 2400 BC. It was introduced into the United States by Columbus in 1494. It is now grown in both the Old World and the New World. Colocynth was known as "indravaruni" in medical treatise "Carka Samhita" and "Puranas." Watermelon was known as "kalinga" to the Ptotoaustraloids of prehistoric times and had grown during Indus Valley Civilization (Kajale, 1991).

1.3.4 GOURDS

Bitter gourd has domesticated from eastern India and southern China. It has a long history of cultivation. It was introduced from the Old World in Brazil. It widely distributed in India, China, Malaysia, and tropical Africa (Miniraj et al., 1993; Singh, 1990). Bitter gourd has been used for centuries in the ancient traditional medicine of India, China, Africa, and Latin America. Bitter gourd fruits also possess antioxidant, antimicrobial, antiviral, and antidiabetic activities (Raman and Lau, 1996; Welihinda et al., 1986). Based on historical literature (Chakravarty, 1990; Miniraj et al., 1993; Walters and Decker-Walters, 1988a) and recent random amplified polymorphic DNA (RAPD) (Dey et al., 2006), inter simple sequence repeat (ISSR) (Singh et al., 2007) and amplified fragment length polymorphism (AFLP) (Gaikwad et al., 2008) molecular analyses, eastern India may be considered as a probable primary center of diversity of bitter gourd.

Although apparently native to Africa, the bottle gourd had reached Asia and the Americas 9000–8000 years ago, possibly as a wild species whose fruits had floated across the sea (Whitaker and Carter, 1954). The bottle gourd had a broad New World distribution 8000 years ago. Independent domestications from wild populations are believed to have occurred in both the Old World and New World (a variety of plants and animals were independently domesticated in multiple parts of the world between 5,000 and 10,000 years ago). A range of data suggest that the bottle gourd was present in the Americas as a domesticated plant by 10,000 BP, which would make it among the earliest domesticated species in the New World. Comparisons of DNA sequences from archaeological bottle gourd specimens and modern Asian and African landraces identify Asia as the source of its introduction to the New World.

The genus *Luffa*, including another cultivated species ridge gourd (*L. acutangula* [L.] Roxb.) and a few wild species, viz., *L. graveolens* Roxb. (var. *longistyla*), *L. echinata* Roxb., *L. tuberosa* Roxb., and *L. umbellata* Roem, is considered to be an essentially Old World genus (Seshadri and More, 2009). *L. operculata* L. (Cogn.), *L. quinquefida* (Hook. and Arn.) Seem and *L. astorii* Svens are confined to the New World (Seshadri and More, 2009). Whitaker and Davis (1962) stated that "loofah" gourd is either cultivated or grows as an "escape" in practically all of the tropical regions of the world and it is very difficult or even impossible to point out with accuracy the indigenous area of the species. They further stated that pending the emergence of convincing evidence, we can assume with confidence that *L. cylidrica* is indigenous to tropical Asia, probably India. The name "loofah" and "luffa" is of Arabic origin because of the sponge characteristic in Egyptian writing. Kosataki and dharmaragava are the equivalent names in Sanskrit for luffa.

1.4 TAXONOMY OF CUCURBITS (TABLE 1.1)

1.4.1 TRIBE-1—MELOTHRIEAE

Cucumis L.: It is the most economically important genus with two major cultivated crops *Cucumis sativus* and *C. melo.* The two minor crops are *C. anguria,* the West Indian gherkin confined to Africa and Central America, and *C. metuliferus,* the African horned cucumber confined to Africa only. It can be subdivided into five cross-sterile species groups under two subgenera (Jeffrey, 1980a).

1. *Subgenus: Cucumis*: Monoeciuous and adromonoecious species, including *C. sativus* and *C. hystrix* (*C. muriculatus*). There is a record of *C. sativus* var. *sikkimensis* having 7–9 lobed leaves and, five carpelled ovaries (Jeffrey, 1980b). Chakravarty (1982) distinguished two botanical varieties, var. *sativus* and var. *sikkimensis,* on the *basis of higher* number of leaves and higher number of placentae in the latter.
2. *Subgenus: Melo*: There are about 25 species mostly in tropical and South Africa.

TABLE 1.1

Classification of Cucurbitaceae

I. Subfamily	Zanonioideae	n = 8 (18 genera and 80 spp.)
II. Subfamily	Cucurbitoideae	(100 genera, 745 spp.)
	Tribe-1	Melothrieae (34 genera, 250 spp., n = 12 (13, 11, 7)
		Melothria, Zeheneria, Cucumeropsis, Pasadaca, Melancium, Cucumis
	Subtribe	Dendrosicynthinae
		Apodanthera, Kedrostis, Corallocarpus, Iberivilliea
	Subtribe	Guraniinae
		Gurania, Psiguria
	Tribe-2	Schizopeponeae (1 genus, 8 spp., n = 10)
		Schizopepon
	Tribe-3	Jollifeae (5 genera, 76 spp., n = 14, 11, 9)
		Momordica, Thladiantha, Telfairia
	Tribe-4	Trichosantheae (10 genera, 75 spp., n = 11)
		Trichosanthes, Hodgsonia, Ampelosicyos
	Tribe-5	Benincaseae (17 genera, 85 spp., n = 12 (11, 13, 10)
		Subtribe Benincaseae
		Coccinia, Benincasa, Lagenaria, Citrullus, Acanthosicyos, Praecitrullus
		Subtribe Luffinae
		Luffa
	Tribe-6	Cucurbitae (12 genera, 110 spp., n = 20)
		Cucurbita, Sicana, Cayaponia
	Tribe-7	Cyclantherae (12 genera, 75 spp., n = 8)
		Cyclanthera, Marah, Elateriopsis, Rytidostylis
	Tribe-8	Sicyoeae (19 genera, 140 spp., n = 12)
		Sechium, Sicyos

Source: Jeffrey, C., An outline of the Cucurbitaceae, in: *Biology and Utilization of Cucurbitaceae,* Bates, D.M., Robinson, R.W., and Jeffrey, C. (eds.), Cornell University Press, Ithaca, NY, 1990, pp. 449–463.

1. *Metuliferus group*: Monoecious annuals with red spiny fruits
2. *Anguria group*: Dioecious, monoecious, or adromonoecious perennials or monoecious or adromonoecious annuals with yellowish or brown stripe fruits; about 20 species including *C. prophetarum*, *C. myriocarpus*, *C. anguria*, and *C. sacluexii*
3. *Melo group*: Monoecious or adromonoecious perennials or annuals with smooth fruits, four species *C. melo*, *C. trigonus*, *C. sagittatus*, and *C. humifructus*
4. *Hirsutus group*: Dioecious perennials with smooth orange fruits, one species *C. hirsutus*

There is another classification proposed by Kirkbride (1993) which consists of two subgenera, viz., *Cucumis* that includes *C. sativus* var. *hardwickii*, var. *sativus*, *xishuangbannaensis*, and *C. hystrix*. The subgenus *melo* comprises of six series (grouping 30 species).

1.4.2 TRIBE-2—SCHIZOPEPONEAE

There are no cultivated taxa and is a small tribe with a single genus.

1.4.3 TRIBE-3—JOLLIFEAE

Some species of *Thladiantha, Momordica, and Telfairia* are economically important.

Momordica L.: It consists of a large number of species including *M. charantia*, the commonly cultivated vegetable—bitter gourd. Among the 65 species, only 7 occur in India and 23 in Africa. There is some confusion between *M. dioica and M. cochinchinensis* Spreng. The oil glands on the outer margin of the leaf lamina and petioles are distinctly present only in *M. cochinchinensis*, and both are dioecious and tuberous rooted perennials cultivated in eastern and northeastern India. In the flora of tropical East Africa, Jeffrey (1967) has recognized 23 species under *Momordica*, viz., *sessifolia, spinosapetri, cissoides, multiflora, foetida, leiocarpa, angiosantha, pterocarpa, friesionium, pycnantha, glabra, boionii, cymbalaria, kirkii, trifoliate, rostrata, cardiospermoides*, and *calantha*, besides cultivated species, *charantia* and *balsamina*. There is new *M. littorea* with dioecious habit, with succulent foliage leaves. But, Meeuse (1962) identified seven species of *Momordica*, viz., *clematida, welwitschii, repens*, besides *charantia and balsamina* in tropical South Africa. Chakravarty (1982) recorded abundant calcium oxalate and calcium carbonate crystals in the leaves in the form of cystoliths in *M. charantia*.

1.4.4 TRIBE-4—TRICHOSANTHEAE

Trichosanthes is the largest genus and monotypic *Hodgsonia*, an Indo-Malayan genus, is grown for its large edible seed rich in oil and protein in China and Northeast India.

Trichosanthes: *Trichosanthes cucumerina* var. *anguina* L., snake gourd, and *T. dioica* Roxb., parwal or pointed gourd, are the two important vegetable crops belonging to this genus. *Trichosanthes* consists of 44 species of which 22 occur in India. Chakravarty (1982) listed the 22 species of *Trichosanthes* occurring in India. They are *cuspidate, anguina, cucumerina, lobata, horsefieldii, villosa, perrottetiana, truncata, ovata, cordata, wallichiana, majuscula, bracteata, lepiniana, anamalaiensis, himalensis, ovigera, dicaelosperma, dioica, tomentosa, and listeria*.

1.4.5 TRIBE-5—BENINCASEAE

Besides *Citrullus, Lagenaria, Luffa*, and *Coccinia* there are two monotypic genera, viz., *Benincasa* and *Praecitrullus* which have their species cultivated as vegetables.

1. *Citrullus*: Taxonomic classification of Citrullus consisted of the following four species (Rehm et al., 1957; Shimotsuma, 1965):
   a. *Citrullus lanatus* var. *lanatus*, cultivating watermelon, and *C. lanatus* var. *citroides*, preserving melon

b. *Citrullus colocynthis* compact growth, deeply lobed leaves, and small and hard fruits found in Namibia

c. *C. ecirrhosus* Cogn. Xerophytic perennial, roughly haired vines, tendrils absent, with striped firm rinded, white-fleshed fruits

d. *C. naudinianus* (Sond.) Dioecioue perennial, ligneous xerophyte, roughly and deeply divided leaves, confined to South-West Africa

A fifth species has been identified by Fursa (1985), *C. mucospermus* Fursa, confined to Nigeria and Ghana. It is cultivated for its seeds. Another species *C. rehmii* De Winter from Namibia has been described by De Winter (1990).

2. *Praecitrullus* Pang.: This is a new genus of Russian botanist Pangalo to include Indian squash or "tinda." *Praecitrullus fistulosus* Pang. is the only species recognized, mostly confined to northwestern India and Pakistan. Even though erroneously called round melon or squash melon, it is not related to *Cucumis melo* (melon) or *Citrullus lanatus* (watermelon), and the previous botanical names *Citrullus lanatus* var. *fistulosus* Chakra or *Citrullus fistulosus* stocks are no longer accepted.

3. *Benincasa* Savi: It is of Indo-Malaysian distribution, and the cultivated vegetable crop is *Benincasa hispida* called wax or ash or white gourd. Two botanical forms have been recognized in Japan. One is called typical which is characterized by a velvety testa and marginal band around the seeds, while this characteristic is absent in the other form called *F. marginata*. Decker-Walters and Walters (1989) reevaluated the cultigen, which is morphologically diverse, and their allosyme studies suggest that the species is relatively uniform. Four major cultivar groups were recognized by Herklots (1972), which were redefined by Decker-Walters and Walters (1989).

4. *Coccinia* Wight and Arn: It is reported that there are 30 species, mostly confined to Africa. Only one species *Coccinia grandis* L. (Voigt), called ivy gourd, "kundsru" or "tondli," is cultivated in India as a vegetable from very early times. It is a dioecious perennial having XY chromosome differentiated sex mechanism. Variability in leaf lobation made Cogniaux (Chakravarty, 1982) differentiate three varieties: (1) *genuina*, (2) *wightiaana*, and (3) *Alceafolia*. Meeuse (1962) identified seven species of *Coccinia* in South Africa, viz., *palamata, sessifolia, quirqueloba, varifolia, hirtella, rehmanii*, and *adoensis*. Calcium carbonate punctuations on the upper surface of the leaves, especially when mature, in similar white scale-like deposition on the lower main veins in *C. indica* are also characteristic (Chakravarty, 1982).

5. *Luffa* Mill.: It is a genus distributed both in Old and New Worlds. *Luffa* includes seven species, three of which are native in the New World and four in the Old World. The Old World species are: (1) *Luffa cylindrica* L. Roem., sponge or smooth gourd, and (2) *L. acutangula* L. Roxb., the ribbed or ridge gourd, usually 10 angled. The two feral species are *L. graveolens* Roxb. and *L. echinata* Roxb. (dioecious perennial). The immature fruits of domesticated species (first two) as well as those of nonbitter selections of the Central and South American *L. operculata* are cooked and eaten also as vegetables. In India, Chakravarty (1982) has recorded *L. tuberosa* Roxb. and *L. umbellata* Roem., the later being endemic in Kerala (southwest India). *Luffa cylindrica* is known as wild populations *Luffa cylindrica* var. *leiocarpa* having small fruits deeply furrowed and bitter fruits, distributed from Myanmar to the Philippines, Australia, and Tahiti.

1.4.6 TRIBE-6—CUCURBITAE

The cultivated five species of *Cucurbita* and a genus of the New World, *Sicana odorifera* (Vell) Naud. are listed.

Cucurbita L.: Under the genus *Cucurbita*, the five cultivated species are *Cucurbita moschata* Duch. (Lam) Poir. (Pumpkin), *Cucurbita maxima* Duch ex. Lam. (winter squash), *Cucurbita pepo* L.

(summer squash—cooked immature), *Cucurbita mixta* Pang. now named *C. argyrosperma* Huber and *C. ficifolia* Bouche, the latter two principally grown in Latin America. The first three species, especially the first one, is extensively cultivated in India, while the fifth one was later introduced into Meghalaya (Northeast India)

Although as many as 27 species have been described in the New World by Whitaker and Bemis (1975), recent observations of synonyms and taxanomic recognition of subspecies and botanical varieties have reduced to nearly 15. Some of the synonyms are *C. argyrosperma* subsp. *sororia* and *C. argyrosperma* var. *palmeri*; *C. pepo* and *C. pepo* var. *texana*; *C. maxima*; *C. martinezii*, etc.

1.4.7　TRIBE-7—CYCLANTHERAE

The cultivated taxon *Cyclantherae pedata* L. (Schrad) is grown in South America and has been introduced in west Himalaya and Nepal. There is one more cultivated species *C. brachystachia* Cogn.

1.4.8　TRIBE-8—SICYOEAE

Essentially of the New World and Hawaii, the chow chow or chayote, *Sechium edule* (Jacq) SW, is a vigorous perennial introduced in Northeast India, suited to human situations.

1.5　CHEMOTAXONOMY

Besides cucurbitacins, modern chemotaxonomy treatments include several compounds like protein and enzymes, estimated by starch gel electrophoresis (Mulcahy, 1980). Isoelectric focusing (IEF) electrophoretic seed alubumins (Pichl, 1980), nonprotein amino acids or free amino acids (Fowden, 1990), and fatty acids of seed oils (Hopkins, 1990) are important. Cucurbitacins are the bitter principles of cucurbits and are oxygenated tetracyclic triterpenes (Enslin and Rehm, 1958; Kupchan et al., 1978) having taxonomic significance. Of the known 19 cucurbitacins (A to S), Cucurbitacins B and E constitute the primary components. Most of the African species of *Cucumis* have Cucurbitacins like B and D and traces of G and H. Cucurbitacin C is found in *Cucucmis hardwickii* and *Cucumis sativus,* likewise Rhem et al. (1957) and Rhem (1960) reported that cucurbitacin E was the main bitter principle in *Cucumis vulgaris, C. colocynthis,* and *C. ecirrhosus. Praecitrullus fistulosus,* the Indian squash or tinda, has only Cucurbitacin B significantly.

1.6　KARYOTAXONOMY

Karyologically, the cucurbits are difficult to study. Varghese (1973) and Roy and his associates (Sinha et al., 1983) have devised some good methods. Basic chromosome numbers for the different taxa vary widely—8, 10, 11, 12, 20 and other numbers 7, 13, and 14 are also known (Jeffrey, 1980). Ayyangar (1967) suggested a karyotaxonomic explanation on the phylogenetic genealogy. The haploid number of 12 was found to have high frequency. Next in order of frequency are 11 and 22. Cytological observations on secondary association of bivalents during meiosis have shown that 3 and 5 are possible primary basic numbers, while 6, 10, and 11 may be considered as secondary basic numbers. All the other numbers might have been evolved through autoploidy, alloploidy, aneuploidy, and secondary polyploids.

1.7　FLOWER DEVELOPMENT

Flowers develop in leaf axils. The flower type varies depending on genetics and other factors. Plants may be monoecious (separate male and female flowers), andromonoecious (separate male and perfect flowers), gynoecious (female flowers only), or hermaphroditic (perfect flowers). The most

TABLE 1.2

Commonly Grown Major Cucurbits in the World

English Name	Scientific Name	Origin	Chromosome Number
Cucumber	*Cucumis sativus*	India	14
Bitter gourd	*Momordica charantia*	Indo-Burma	22
Bottle gourd	*Lagenaria siceraria*	Ethiopia	22
Watermelon	*Citrullus lanatus*	Tropical Africa	22
Melon	*Cucumis melo*	Tropical Africa	24
Long/serpentine melon	*Cucumis melo* var. *flexuosus*	India	24
Melon	*Cucumis melo* var. *momordica*	India	
Ridge gourd	*Luffa acutangula*	India	26
Sponge gourd	*Luffa cylindrica*	India	26
Pumpkin	*Cucurbita moschata*	Peru and Mexico	40
Summer squash	*Cucurbita pepo*	Peru and Mexico	40
Winter squash	*Cucurbita maxima*	Peru and Mexico	40
Ash gourd	*Benincasa hispida*	Southeast Asia	24
Pointed gourd	*Trichosanthes dioica*	India	22
Ivy or scarlet gourd	*Coccinia cordifolia* (syn. *C. indica*)	India	24
Round melon	*Praecitrullus fistulosuos*	Indo-Burma	24
Sweet gourd	*Momordica cochinchinensis*	Southeast Asia	28

Source: Rai, M. et al., IIVR: Fifteen years of accomplishments, Indian Institute of Vegetable Research Publication, Varanasi, India, 2008, pp. 1–45.

common forms for various species are listed in Table 1.2. In monoecious and andromonoecious plants, several male flowers usually open before any pistillate (female or perfect) flowers open. At any one time, there are usually several times more male flowers open than female flowers. Most typically, a stem develops a series of nodes with male flowers, one node with a pistillate flower, another series of nodes with male flowers, a second pistillate flower, and so on. Generally, as the plant develops, the proportion of nodes with female flowers increases. In older plants of cucumbers and summer squash, the distal portion of the stem may have pistillate flowers at every node (Loy, 2004; Wien, 1997). In muskmelons, the pistillate flowers form on short lateral branches either on the main stem or on one of several large basal branches (McGlasson and Pratt, 1963). The first pistillate flower to open on muskmelon is usually on a short lateral of a branch near the base of the plant.

Whether a particular node initiates a male or pistillate flower and whether that flower develops fully to bloom is determined by genetics and the environment. Cool temperatures promote development of pistillate flowers in cucumber, squash, and pumpkin. Under these conditions, the first pistillate flower forms at a node closer to the base of plant, and the ratio of male to pistillate flowers is reduced. For some summer squash, this means the first pistillate flower opens before any male flowers are open, and pollination and hence fruit set does not occur. High temperatures promote male flowers and delay female flower development. For instance, in pumpkins, temperatures of 90°F day/70°F night lead to abortion of female flower buds. Light levels are also important. High light levels promote female flower production, and shade can reduce the number of female flowers. Photoperiod does not appear to play a major role in field production, but under controlled environments, some cucumber cultivars produce more pistillate flowers under short days. Plant nutritional status also plays a role; high nitrogen fertilization can delay production of pistillate flowers (Loy, 2004; Wien, 1997). Sex expression in cucurbits is influenced by hormones produced within the plant as well as synthetic growth regulators applied to the plant. Gibberellins promote male flower development in *Cucumis* and *Cucurbita*. In these genera, ethylene promotes pistillate flower development and suppresses male flowers. Natural and synthetic auxins promote pistillate flower development in cucumber. Very little work on watermelon has been done in this area, and whether it responds similarly to other genera is not known (Wien, 1997).

1.8 SEX MODIFICATION

A majority of cucurbits are monoecious, and the sex ratio (male:female) ranges from 25–30:1 to 15:1. The sex ratio is influenced by environmental factors. High nitrogen content in the soil, long days, and high temperature favor maleness. Besides environmental factors, endogenous levels of auxins, gibberellins, ethylene, and abscisic acid also determine the sex ratio and sequence of flowering. A primordium can form either a female or a male flower, and it can be manipulated by addition or deletion of auxins. Exogenous application of plant growth regulators can alter sex form, if applied at 2–4 leaf stage. A high ethylene level induces female sex and is suggested to increase female flowers in cucumber, muskmelon, summer squash, and pumpkin. In cucumber, maleic hydrazide (50–100 ppm) GA 3 (5–10 ppm), Ethrel (150–200 ppm), TIBA (25–50 ppm), and boron (3 ppm) also induce female flowers.

Gibberellins promote maleness and are antagonistic to the action of ethylene and abscisic acid. In fact, a gynoecious line of cucumber is maintained by inducing male flowers through spray of GA 3 (1500–2000 ppm). Silver nitrate (300–400 ppm) also induces maleness.

1.9 FLORAL BIOLOGY

Bitter gourd: Bitter gourd is a monoecious annual climber with a duration of 100–120 days. Leaves are palmately 5–9 lobed. Flowers are axillary with long pedicel and are yellow in color. Stamens are five in number, with free filaments and united anthers. Stigma is divided. The fruit is pendulous, fusiform, ribbed with numerous tubercles. The bitterness of the fruit is due to the presence of an alkaloid, Momordicin. Anthesis is from 4:00 to 7:00 a.m. Anther dehiscence takes place between 5:00 and 7:30 a.m. Stigma is receptive 24 h before and after anthesis.

Snake gourd: T. anguina is a monoecious annual climber with small white flowers. Female flowers are solitary, sessile with long narrow ovary and fimbriate corolla and are 5-partite. Male flowers appear in clusters on long densely pubescent stalks of 10–20 cm long. Plants flower at early hours of night. Anthesis begins at 5:15 a.m. and continues up to 9:30 a.m. Anther dehiscence occurs before flower opening and is completed by 1 h. Pollen grains remain viable 10 h after anthesis. Stigma is receptive 7 h before and 51 h after opening of flower.

Pumpkin: Pumpkin produces largest flower among cultivated cucurbits with large yellow corolla and large ovary with variable green color. Anthesis occurs between 4:30 and 4:50 a.m. Distinctly visible variation in shape, size, and color of various floral parts like calyx, corolla, anther, stigma, and ovary are noticed. Calyx is always green, but shape and size differ among genotypes. The opened flowers are companulate in shape, measuring 15–20 cm in length. The diameter of the opened flower at the distal end varies from 15 to 25 cm. Pistillate flowers are larger than staminate flowers. Corolla color is yellow, but a tinge of variation in shades of yellow color may be seen among the genotypes. The crop is invariably monoecious, where axillary staminate and pistillate flowers are found on separate nodes of the same plant. Both staminate and pistillate flowers are strictly solitary. In a very few genotypes, in a particular set of climatic conditions at the same location, a transitory trimonoecious (gynomonoecious) condition is also encountered, where a few abnormal, deformed hermaphrodite flowers are formed in some plants of the genotype. Both staminate as well as pistillate flowers have a tendency to convert into abnormal hermaphrodite flowers. The frequency of pistillate flowers turning into hermaphrodite flowers is rather low. Staminate flowers converting into hermaphrodite flowers have superior ovary—an unusual phenomenon in cucurbits. Pistillate flowers converting into hermaphrodite flowers have an usual inferior ovary. The hermaphrodite flowers formed out of either staminate or pistillate flowers never turn into normal, effective fruits.

Ash gourd: This monoecious crop produces large male flowers with long pedicels and female flowers with densely haired ovary and short peduncle on the same plant. Corolla is yellow in colour and large in size. The ratio of staminate to pistillate flowers is 34:1. Anthesis takes place between 4:30 and 7:30 a.m. and anther dehiscence is between 3:00 and 5:00 a.m. Stigma is receptive from 8 h before to 18 h after anthesis.

Bottle gourd: Bottle gourd is commonly monoecious in nature. However, an andromonoecious isolate Andromon-6 has been reported by Singh et al. (1990). Transitory unstable trimonoecious sex form also appears in both monoecious and andromonoecious forms. Generally, solitary staminate, pistillate, or hermaphrodite flowers are present in the leaf axils of separate nodes. However, occasionally, two staminate, one staminate and one pistillate, or even two pistillate flowers are present on the same node in certain genotypes. Very rarely both of the pistillate flowers on a node turn into effective fruits.

Usually, in common genotypes, first staminate flower anthesis takes place earlier at a lower node number, and first pistillate flower anthesis takes place 1–10 days later at a higher node of the vine, but the sequence changes in early/prolific genotypes, where first pistillate flower anthesis occurs 1–2 days earlier at a lower node as compared to first staminate flower anthesis. Anthesis in bottle gourd takes place between 4:30 and 7:00 p.m. In early genotypes, anthesis occurs relatively earlier as compared to late genotypes. In all the genotypes, pistillate flowers open about an hour earlier than staminate flowers on the same plant. In narrow petalled genotypes, opening of petals/corolla of pistillate flower begins early in the day, between 12:00 noon and 2:00 p.m., and staminate flowers also open a bit earlier as compared to normal genotypes. Dehiscence in all the genotypes takes place between 1:00 and 2:00 p.m. Anthesis and dehiscence time is moderately influenced by humidity and temperature conditions.

Cucumber: Cucumber is monoecious, that is, male and female flowers present on the same plant. Flowers are bracteate, pedicellate, unisexual, actinomorphic, pentamerous, and epigynous. The whole developmental process from the initial bud stage to the stage when the flower is detached from the pedicel is divided into eight stages. The opening and closing of the male flowers is mainly influenced by the sunrise and sunset, that is, by light and the time of the day.

Pollination and fruit set: Flowers open at temperatures above 50°F (pumpkins and squash), 60°F (cucumber and watermelon), or 65°F (muskmelon). They remain open for a day in the case of watermelons, muskmelons, and cucumbers or half a day or less for *Cucurbita* spp. In most cases, fruit set requires the activity of pollinators, such as honeybees or native squash bees. Enough viable pollen must be delivered to the stigma so that there is one grain of pollen for each developing seed in the fruit. Once pollen is on the stigma, fertilization is still not guaranteed. The pollen must germinate and grow a pollen tube down the stigma to deliver sperm to the ovule. If there is not enough pollen or conditions are not suitable for pollen tube growth, only the ovules closest to the blossom may be successfully fertilized. Seeds developing close to the blossom end stimulate growth in that part of the fruit, but the rest of the fruit remains small. The result is a misshapen fruit. In the case of triploid watermelon, pollination with viable pollen is necessary to stimulate fruit growth even though seeds do not develop. Triploid plants do not produce enough viable pollen themselves, so a pollenizer variety must be planted nearby. Traditionally, seeded varieties that produce fruit visually different from fruit of the seedless variety are used as pollenizers. More recently, some varieties have been developed that are marketed solely as pollenizers; they do not produce marketable fruit. For successful fruit set in triploid watermelons, it is critical that pollenizer varieties produce viable pollen at the time female flowers are open on the triploid plant. Some gynoecious cucumbers are parthenocarpic (able to set fruit without fertilization of ovules) and so do not require pollinators. Natural parthenocarpy is known to occur in other cucurbits as well, notably summer squash, but is generally not relied upon for commercial production (Wien, 1997). Environmental conditions and the condition of the plant can interfere with pollination and fruit set. Weather conditions influence pollinator activity. For example, honeybees are less active when it is hot and dry. Pesticide applications or residues can kill or deter bees. Fruit already developing on the plant hinder successful fruit set in younger flowers, especially those on the same branch or stem.

Fruit development: Cucurbit fruit grows exponentially for a period after fruit set and then the growth rate slows. The increase in fruit size after pollination is largely a result of cell expansion rather than an increase in the number of cells. Cucurbits can be divided into two major groups based on whether the fruit are harvested when immature—summer squash and cucumbers—or mature—all types of melons, winter squash, pumpkins, gourds. Cucumbers and summer squash are harvested during the period of rapid growth. They may be ready for harvest as early as 3 days after pollination, depending on the market requirements. In the other crops, fruits typically reach their full size about 2–3 weeks after pollination and take another 3 or more weeks to mature to a harvestable stage. During this time, seeds develop to maturity and sweetness, flavor, and color develop in the fruit. The rind toughens, becomes less permeable to water, and in the case of muskmelon, develops corky netting. A color change may occur, either subtle, as in the change from pale green to yellowish in muskmelon; or just on the portion of the fruit near the ground, as in a yellow ground spot of a watermelon; or across the entire fruit surface, as in pumpkin. Watermelons and muskmelons typically mature 42–46 days after pollination, while winter squash and pumpkins take 50–90 days to reach harvest maturity. Developing fruit places heavy demands on the plant, reducing the growth of new leaves, roots, and any other fruit developing at the same time (Loy, 2004; Wien, 1997). Indicators of harvest maturity vary depending on the crop. Cucumbers and summer squash are usually harvested based on size. Muskmelons form an abscission layer between the peduncle and fruit, so they "slip" from the vine when fully ripe; commercial harvest occurs after the layer begins to form but before the melon falls off the vine. Watermelon harvest maturity is identified by yellowing of the ground spot and wilting of tendrils near the place of fruit attachment. Sugar levels do not increase in muskmelon or watermelon after harvest (Wien, 1997). In winter squash and pumpkin, hardening of the rind and change in rind color at the ground spot or over the entire rind indicate the earliest readiness for harvest. At this stage, seed development may not be complete, and leaving the fruit on the vine for another 10–20 days may improve postharvest quality (Loy, 2004). The size of the mature fruit is influenced by genetics, the environment, and plant conditions during development of the pistillate flower and fruit. Conditions that reduce the amount of assimilate available tend to decrease the size of individual fruit. Increased plant density, greater numbers of fruit per plant, and reduced water supply tend to decrease fruit size. In muskmelon and watermelon, the soluble solids content of the fruit is an important measure of quality. Like fruit size, soluble solids tend to be lower under conditions that reduce assimilate level. High night temperatures, reduced leaf area, increased numbers of fruit per plant, and increased plant density can all reduce soluble solids (Wien, 1997). In contrast to its affect on fruit size, reduced water supply can increase fruit-soluble solids (Bhalla, 1971) (Table 1.3).

1.10　HARVESTING AND POSTHARVESTING

Ash gourd: Ash gourds are mature when the stems connecting the fruit to the vine begin to shrivel. Cut fruits from the vines carefully using pruning shears or a sharp knife leaving 34 in. of stem attached. Snapping the stems from the vines results in many broken or missing "handles." The fruits can be harvested at different stages depending on the purpose for which it will be used. Normally, green fruits are ready for harvest within 45–60 days; matured ones coated with powdery substance are harvested between 80 and 90 days after sowing. The fruit yield can vary depending on variety and crop management. Average marketable yields are 20–25 t/ha. The harvested fruits can be stored for several weeks in ambient conditions. It can be kept for 23 months in temperatures from 10°C to 12°C and 50–75% relative humidity. Avoid cutting and bruising the ash gourds when handling them.

Watermelon: Watermelons do not slip from the vine or emit an odor when ripening, unlike muskmelons. Indicators for picking watermelons include color change (the most reliable), blossom-end conditions, rind roughness, and drying of the nearest tendril to the fruit (less reliable). A sharp

TABLE 1.3

Scientific Names and Typical Sex Expression of Common Cucurbit Vegetables

Crop	Scientific Name	Sex Expression
Cucumber	*Cucumis sativus*	Monoecious[a], gynoecious, hermaphroditic, andromonoecious
Muskmelon	*Cucumis melo*	Andromonoecious[a], monoecious
Pumpkin (most jack-o-lantern and fresh market types)	*Cucurbita pepo*	Monoecious
Pumpkin (processing)	*Cucurbita moschata* or *Cucurbita maxima*	Monoecious
Squash, summer	*Cucurbita pepo*	Monoecious
Squash, winter		
Acorn, sweet dumpling	*Cucurbita pepo*	Monoecious
Buttercup, Kabocha	*Cucurbita maxima*	Monoecious
Butternut	*Cucurbita moschata*	Monoecious
Watermelon	*Citrullus lanatus*	Monoecious[a], andromonoecious

Source: Loy, J.B., *Crit. Rev. Plant Sci.*, 23(4), 337, 2004; Wien, H.C., The cucurbits: Cucumber, melon, squash and pumpkin, in: *The Physiology of Vegetable Crops*, Wien, H.C. (ed.), CAB International, New York, 1997, pp. 345–386.

[a] Most common.

knife should be used to cut melons from the vines; melons pulled from the vine may crack open. Harvested fruit is windrowed to nearby roadways, often located 10 beds apart. A pitching crew follows the cutters and pitches the melons from hand to hand, then loads them in trucks to be transported to a shed. Melons should never be stacked on the blossom end as excessive breakage may occur. Loss of foliage covering the melons can increase sunburn. Exposed melons should be covered with vines, straw, or excelsior as they start to mature to prevent sunburn. Each time the field is harvested, the exposed melons must be re-covered. Most fields are picked at least twice. Some fields may be harvested a third or fourth time depending upon field condition and market prices.

1.10.1 HARVEST INDEXES COULD BE USED

- Tapping—A dull or hollow sound is an indication of maturity.
- Color—The fruit part resting on the ground becomes a distinct yellow patch as in sugar baby.
- Tendril right behind each fruit is dried down up to the base.

1.10.2 SORTING AND GRADING

Seeded melons are sorted and packed in large, sturdy, "tri-wall" fiberboard containers. The melons are sorted according to grade: number 1, 6.4–11.8 kg, and number 2, 3.6–6.4 kg. Inferior melons may be sold at nearby markets; rejects (discolored, misshapen, sugar-cracked, blossom-end rot, and insect-damaged fruit) are discarded. Containers that hold 60–80 melons and weigh 500–545 kg are shipped on flatbed trucks to terminal markets or wholesale receivers. The containers are covered to prevent sunburn in transit. Seedless melons are sorted according to size and packed in cartons containing 3, 4, 5, 6, or 8 fruits. "Fours" and "fives" are preferred to sizes; "sixes" and "eights" are common later in the season after the crown-set melons have been removed from the vine. The rough gross weight of a carton is 18–23 kg. Seedless melons may also be sold in large, bulk containers. Personal, seedless watermelons are sorted by size and packed in single layer boxes containing 6, 8, 9, or 11 fruits. Shipping boxes roughly weigh 15 kg and are arranged 50 boxes/pallet.

1.10.3 PACKAGING

The seeded melons are sorted and packed in large, sturdy, "tri-wall" fiberboard containers. The melons are sorted according to grade and number. Bins hold 60–80 melons and will weigh 500–550 kg. Two-third bins hold about 360 kg of melons. Discolored, misshapen, sugar-cracked, blossom-end rot, and insect-damaged fruits are regarded as culls but still may be sold to nearby markets.

The containers are loaded on flatbed, 18-wheel trucks destined for terminal market resale. The tops of the containers should be covered to prevent sunburn in transit. Watermelon sales usually are based upon a 1%–2% shrink because of breakage. The buyer is responsible for supplying bins and lids or the shipper will send a bill for the cost of these items. Seedless watermelons are sorted according to size and packed in cartons containing 4, 5, 6, or 8 fruits. "Fours" and "fives" are preferred sizes. "Sixes" and "eights" are common later in the season after the crown-set melons are removed from the vine. The rough weight of a carton is 18–23 kg. Some bins and cartons have high-resolution graphics for logos that may increase overall cost.

1.10.4 STORAGE

Watermelons are not adapted to long-term storage. Normally, the upper limit of suitable storage is about 3 weeks. However, this will vary from variety to variety. Storage for more than 2 weeks triggers a loss in flesh crispness. Storing melons for several weeks at room temperature will result in poor flavor. However, when a fruit is held just a few days at warmer temperatures, the flesh color tends to intensify. Sugar content does not change after harvest. Watermelons' flesh tends to lose its red color if held too long at temperatures below 10°C.

Watermelons may lose crispness and color in prolonged storage. They should be held at 10°C–15°C and 90% relative humidity. Sugar content does not change after harvest, but flavor may be improved because of a drop in acidity of slightly immature melons. Chilling damage will occur after several days below 5°C. The resulting pits in the rind will be invaded by decay-causing organisms.

1.10.5 MARKET PREPARATION

Watermelon: Watermelons usually are sold by the hundred weight at harvest time. The bulk of the commercial crop is shipped out. Many are sold from smaller plantings through temporary or permanent roadside stands or at farmers' markets. Some growers sell their fields to shippers or brokers as harvest time approaches. An important consideration in successful marketing is to have adequate facilities for transporting the crop to market outlets. Although earliness usually results in higher prices, quality and maturity should be of prime importance in marketing watermelons.

Pumpkin: Pumpkins are manually harvested when they have reached maturity. Pumpkins should be picked only when the fruit surface is completely dry. The fruit should be carefully clipped off the vine, leaving about a 2.5 cm (1 in.) stem attached to the fruit. A pair of sharp pruning shears is needed to sever the stem and create an attractive, smooth, clean cut. Do not pick up the pumpkin by the stem as it may separate from the fruit and provide an easy access for decay organisms. A short length of stem should always remain attached to the fruit. Once removed from the vine, the pumpkins should be put in wooden or strong plastic field crates for transport to the collection site or packinghouse. Outgrading is required in the field to remove pumpkins affected by disease, insects, or physical damage. During harvesting, handling, and field transport, every effort should be made to avoid bruising or puncturing the rind. Also, harvested pumpkins should not be exposed to direct sunlight or rainfall. Ideally, pumpkins should not be stacked on top of each other. Stacking is a sure way to create bruises. Padding material, such as grain straw, should be used liberally if fruits have to be stacked during harvest. Spread out a layer of dry straw on the ground and set the pumpkins on this. Keep the fruit dry at all times and never store pumpkins on moist bare ground. If the pumpkins must be stacked for transport, the pile should not be more than 1 m (3 ft) deep.

1.10.6 SORTING/GRADING

Pumpkin fruit are quite variable in size, shape, and color; therefore, it is difficult to obtain consistent uniformity of product from a single harvest. However, grading for uniformity of appearance is important to meet market requirements. There are three established size categories (small, medium, and large) for domestic marketing of pumpkins, based on fruit weight. Small-sized pumpkins weigh between 1.4 and 3.2 kg (3 and 7 lb), medium-sized pumpkins weigh between 3.3 and 5.5 kg (7 and 12 lb), and large-sized pumpkins weigh 5.6 kg (12 lb) or more. Export markets accept a range in fruit size, although large-sized fruits weighing between 5.6 and 8 kg (12 and 18 lb) are preferred. Fruit shape may vary from round, to oval, to slightly flat. Similarly, rind color ranges from green, to blue-green, to tan. The striping pattern or mottling of the rind also varies, although the striations are typically white or cream colored. The rind may be smooth or sutured. Domestic consumers and importers prefer uniformly regular-shaped fruits that have a smooth, tough rind.

All fruit should be examined for external maturity characteristics, and only mature pumpkins should be packed. The fruit should be free of noticeable skin blemishes. The rind should not be discolored or have any surface mould growth. Fruit should be free of insect or mechanical damage, and any partially decayed fruit should be discarded. The fruit must have a closed blossom end and be free of cracking in order to avoid serious decay problems. The flesh should be thick and dark orange, since many pumpkins are sold as cut fruit in the market. Randomly selected fruit should occasionally be cut open for assessment of internal color.

1.10.7 PACKING

Packages used to market pumpkins vary depending on market destination. Fruit sold in the domestic market and nearby Caribbean export destinations is usually packed in mesh sacks. The sacks typically contain from three to seven fruits and weigh around 23 kg (50 lb). However, mesh sacks provide little or no protection against bruising and physical injury. Variability in fruit size will also cause bulging problems of the mesh sack. Smaller sized pumpkins intended for more distant export markets should be packed in strong, well-ventilated fiberboard cartons containing 19 kg (42 lb) of fruit. The cartons should have a minimal bursting strength of 275 psi and internal dividers should be used to separate and protect the fruit. Large wooden bulk bins holding from 360 to 410 kg (800–900 lb) of fruit may be used for marine transport to export market destinations. Pumpkins packed in cartons and transported by marine containers should include an additional 5% weight to account for moisture and respiratory weight loss that will occur during transport.

1.10.8 TEMPERATURE MANAGEMENT

Pumpkin: Pumpkins not intended for immediate sale should be held in a cool, dry, well-ventilated area. The optimum temperature for pumpkin storage is 12°C (54°F). Sound fruit can be stored for up to 3 months at this temperature without a significant loss in quality. Storage at ambient temperature will result in excessive weight loss, loss of surface color intensity, and a decline in culinary quality. Green-skinned cultivars will gradually turn yellow at high temperature, and the flesh will become dry and stringy. The storage life of pumpkins at ambient temperatures is limited to several weeks. On the other hand, the fruit should not be stored at cold temperatures. Pumpkins are susceptible to chilling injury (CI) and should never be stored below 10°C (50°F).

Bitter gourd: Harvesting starts 55–60 days after sowing. Picking is done when fruits are fully grown but still young and tender. Seeds should not be hard at the time of harvest. From a good crop, 15–20 harvests are possible, and harvesting is done twice a week. If fruits are allowed to ripen on vines, further bearing is adversely affected. Fruits after harvest are packed in thin gunny bags or directly packed in trucks and marketed. Since keeping quality of fruits is less, fruits should be marketed without any delay to nearby markets on the same day itself. Otherwise, tubercles will be dropped and freshness and appearance of fruits will be adversely affected.

Snake gourd: To obtain straight fruits in long and slender-fruited varieties, there is a practice of hanging a weight from the bottom of developing fruits. Fully mature fruits will be fibrous and hard. Hence, fruits are harvested at tender stage. Usually, harvesting is twice a week, and it continues for 2 months. The productivity of snake gourd is more than that of bitter gourd, and it varies with soil and season. The average yield in a well-managed field will be about 30–35 t/ha in Kerala. Fruits, immediately after harvest, are tied into bundles of 15–20 kg and covered with arecanut sheath and marketed to nearby places. Picking should be done once in 2 days. The fruits are to be handled carefully to avoid damage. The fruits are then graded and packed in bamboo baskets or cartons without causing damage. Covering baskets with moistened jute bags will reduce the rate of physiological activities.

Sponge gourd: The crop is ready for harvest in about 60 days after sowing. Both crops are picked at immature tender stage. Fruits attain marketable maturity 5–7 days after anthesis. Overmature fruits will be fibrous and are unfit for consumption. To avoid overmaturity, picking is done at 3–4 days interval. Harvested fruits are packed in baskets to avoid injury and can be kept for 3–4 days in a cool atmosphere.

Bottle gourd: Fruits are harvested at tender stage when they grow to one-third to half. Fruits attain edible maturity 10–12 days after anthesis and are judged by pressing the fruit skin and noting pubescence persisting on the skin. At edible maturity, seeds are soft. Seeds become hard and the flesh turns coarse and dry during aging. Tender fruits with a cylindrical shape are preferred in market. Harvesting starts 55–60 days after sowing and is done at 3–4 days intervals. While harvesting, care should be taken to avoid injury to vines as well as to fruits. Plucking of individual fruits is done with sharp knives by keeping a small part of fruit stalk along with fruit. Average yield is 20–25 t/ha for open-pollinated varieties and 40–50 t/ha for F_1 hybrids.

Fruits can be stored for 3–5 days under cool and moist condition. For export purpose, fruits are packed in polythene bags, and bags are kept in boxes of 50–100 kg capacity.

Muskmelon: Fruits maturing on the vine, without becoming overripe, are superior in quality to those harvested immature or left on the vine after they have become mature. Harvesting at the proper stage is of major importance in marketing good quality produce. The fruits for distant markets should be harvested as they reach the half-slip maturity to avoid losses from the over-ripeness and decay. For local markets, harvesting should be done at "full-slip" stage. Hara Madhu however, never reaches the full-slip stage and color of the rind can be taken as the criteria of maturity. Flavor and texture of the flesh improve for a few days after harvesting and attain highest quality if the fruits are harvested when they have developed their maximum sugar content. Fruits harvested when immature never attain these desirable, characteristics. The soluble solids are considerably affected by the environment, the incidence of diseases and the vigor of the plant.

Round gourd/Tinda: It takes 60–90 days from sowing to first fruit picking, depending upon the cultivar and season. Fruits reach edible maturity 6–8 days after fruit set. Picking should be done when these are still immature and small in size. Large-sized fruits are not liked in the market even if immature and soft. Therefore, picking should be done at every third to fourth day. The first formed one or two fruits at basal nodes should be harvested/nipped early to allow better vine growth so that the plant can bear more number of fruits later.

A good crop can give 80–120 q/ha green, unripe fruits. Plant growth-promoting chemicals also influence the yield. Maleic Hydrazide (50 ppm) aqueous solution sprayed at 2 and 4 leaf stage stimulates vine growth, giving more femaleness, and enhance female flowering at lower nodes. All these factors improve yield by 50%–60%.

Keeping the fruits in a cool environment or under the shade with frequent watering of covering/packing material without causing bruises to their surface could help in storing fruits for 3–4 days.

All deformed and damaged fruits should be sorted out and rejected. Healthy fruits should be graded according to their size. Graded fruits fetch a higher price. The produce is then packed in baskets with some filler, preferable leaves with soft texture and low moisture content. For distant markets, even perforated cardboard boxes with fillers are used. For local market, jumble packing in baskets or gunny bags is done, and water is frequently sprinkled to keep the packing cover wet and cool. Since fruits respire and liberate heat, there should be enough aeration between the fruits and the warm air should go out, otherwise the fruits turn pale and become unmarketable. For transporting, a rack system should be preferred rather than dumping in a truck or heaping in the carriage. Due to high water content, the fruits are likely to get spoiled early, therefore, fast transportation and quick disposal/consumption should be kept in mind.

Cucumber: Fruits of cucumber attain edible maturity within a week from anthesis of female flowers, though variation for edible maturity exists among its varieties. Picking of fruits at the right edible stage depends upon individual varieties and marketing requirement. In salad or slicing cucumber, dark green skin color should not turn brownish-yellow or russet. White spine color is also a useful indication for their edible maturity. Further, overmature fruits show carpel separation in transverse section of the fruits. For commercial purposes, cucumbers are harvested at immature condition 5–7 days after pollination, depending upon the cultivar. If cucumber vines are trained vertically, their fruits reach harvestable size a day or two early. The cucumber should be picked at 2 days intervals. However, their seeds mature 25–30 days after pollination. For seed purpose, pale-yellow and golden-yellow (mature fruit color) cucumbers should be harvested in white and black spine varieties respectively. Its yield varies according to the system of cultivation, cultivar, season, and other factors. Generally, cucumber yields are about 80–120 q/ha.

Cucumber is packed in baskets and transported to markets. The river-bed farmers sell their produce to transport contractors and mandi agents who advance funds to them for cultivation.

Gherkin: The crop is ready for harvest in 30–35 days. As the tender immature fruits are meant for canning, the price of the produce is decided by the stage of maturity. The smallest fruit (stage 1), which will weigh approximately 4.0 g (250 fruits/kg), will fetch the maximum price followed by stage 2 and stage 3. To maintain the grade, the harvesting of fruits should be done every day; a day's break would result in outsized or overgrown gherkin, therefore a loss to the farmer. Avoid sharp sun and high temperature while harvesting. For this, picking of fruits must be done in the very early morning or late evening. Harvest the fruits by retaining the stalk on the plant. Harvested fruits must be collected under the shade. The flower head has to be removed from the fruit. Water should not be sprinkled on harvested fruits at any stage. Even if there is surface water during harvest, it should be dried by aeration. For collection of fruits, jute bags alone have to be used and plastic bags should be totally avoided. The harvested produce should be transported to the factory on the same day before dusk. Leaving the gherkin unprocessed overnight would result in poor quality produce.

REFERENCES

Ayyangar, K.R. 1967. Taxonomy of cucurbitaceae. *Bull. Nat. Inst. Sci. India*, 34:380–396.

Bhalla, P.R. 1971. Gibberellin like substances in developing watermelon seeds. *Physiol. Plant.*, 24:106–111.

Chakravarty, H.L. 1982. Cucurbitaceae. In: *Fascicles of Flora of India—Fascicle-11* (ed. Thothathri, K.), Botanical Survey of India, Calcutta, India.

Chakravarty, H.L. 1990. Cucurbits of India and their role in the development of vegetable crops. In: *Biology and Utilization of Cucurbitaceae* (eds. Bates, D.M., Robinson, R.W., and Jeffrey, C.), Cornell University Press, Ithaca, NY, pp. 325–334.

Decker-Walters, D.S. and Walters, T.W. 1989. Allozyme studies in Benincaseae. *Cucurbit Genet. Coop. Rpt.*, 12:89–90.

Dey, S.S., Singh, A.K., Chandel, D., and Behera, T.K. 2006. Genetic diversity of bitter gourd (*Momordica charantia* L.) genotypes revealed by RAPD markers and agronomic traits. *Sci. Horticult.*, 109:21–28.

Enslin, P.R. and Rehm, S. 1958. The distribution and biogenesis of cucurbitacins in relation to the taxonomy of Cucurbitaceae. In: *Proceedings of the Linnean Society of London*, London, U.K., 69th Section, Part 3, pp. 230–236.

Fowden, L. 1990. Amino acids as chemotaxonomic indices. In: *Biology and Utilization of the Cucurbitaceae* (eds. Bates, D.M., Robinson, R.W., and Jeffrey, C.), Cornell University Press, Ithaca, NY, pp. 29–37.

Fursa, T.B. 1985. A new species of watermelon—*Citrullus mucosospermus*. *Trudy po Prikladnoi Botanike, Genetike-i-Selektsii*, 81:108–112.

Gaikwad, A.B., Behera, T.K., Singh, A.K., Chandel, D., Karihaloo, J.L., and Staub, J.E. 2008. AFLP analysis provides strategies for improvement of bitter gourd (*Momordica charantia*). *Hort. Sci.*, 43:127–133.

Gopalakrishnan, T.R. 2007. *Vegetable Crops* (ed. Peter, K.V.). New India Publishing Agency, New Delhi, India, pp. 104–105.

Herklots, G.A.C. 1972. *Vegetables in South Asia*. George Allen and Unwin Ltd., London, U.K.

Hopkins, C.Y. 1990. Fatty acids of Cucurbitaceae seed oil in relation to taxonomy. In: *Biology and Utilization of the Cucurbitaceae*. Cornell University, Press, New York, pp. 38–50.

Jeffrey, C. 1967. Cucurbitaceae. In: *Flora of Tropical East Africa* (eds. Milne Redhead, E. and Polhills, R.M.), London Crown Agents, London, U.K.

Jeffrey, C. 1980a. A review of Cucurbitaceae. *Bot. J. Linn. Soc.*, 81(3):233–247.

Jeffrey, C. 1980b. Further notes on Cucurbitaceae. The Cucurbitaceae of the Indian sub-continent. *Kew Bull.*, 34(4):789–809.

Jeffrey, C. 1990. An outline of the Cucurbitaceae. In: *Biology and Utilization of Cucurbitaceae* (eds. Bates, D.M., Robinson, R.W., and Jeffrey, C.), Cornell University Press, Ithaca, NY, pp. 449–463.

Kajale, M.D. 1991. Current status on Indian paleoethanobotany. In: *New Light on Early Farming: Recent Developments in Palaeonomy* (ed. Renfrew, J.M.), Edinburgh University Press, Edinburgh, U.K., pp. 155–189.

Kirkbride, J.H. 1993. *Biosystematic Monograph of the Genus Cucumis: Botanical Identifications for Cucumbers and Melons*. Parkway Publishers, Boone, NC, 159pp.

Kupchan, S., Morris, H.M., Haim, M., and Sneden, A.T. 1978. New cucurbitacins from *Phormium tenax* and *Marah oreganus*. *Phytochemistry*, 17:767–770.

Loy, J.B. 2004. Morpho-physiological aspects of productivity and quality in squash and pumpkins (*Cucurbita* spp.). *Crit. Rev. Plant Sci.*, 23(4):337–363.

McGlasson, W.B. and Pratt, H.K. 1963. Fruit-set patterns and fruit growth in cantaloupe. *Proceedings of American Society for Horticultural Sciences*, 83:495–505.

Meeuse, A.D.J. 1962. The Cucurbitaceae of South Africa. *Bothalia*, 8:1–11.

Miniraj, N., Prasanna, K.P., and Peter, K.V. 1993. Bitter gourd *Momordica* spp. In: *Genetic Improvement of Vegetable Plants* (eds. Kalloo, G. and Bergh, B.O.), Pergamon Press, Oxford, U.K., pp. 239–246.

Mulchy, D.L. 1980. Electrophoresis of proteins and single pollen grains. In: *Conference of the Biology and Chemistry of the Cucurbitaceae*, Cornell University, Ithaca, NY, August 3–6, 1980, p. 9.

Om Prakash. 1961. *Foods and Drinks in Ancient India*. Munshiram Manohar Lal Oriental Booksellers and Publishers, Delhi, India.

Pichl, I. 1980. Electrophoretic profile of water soluble proteins isolated from seeds of various species of Cucurbitaceae. In: *Conference of the Biology and Chemistry of the Cucurbitaceae*, Cornell University, Ithaca, NY, August 3–6, 1980, p. 9.

Rai, M., Singh, M., Pandey, S., Singh, B., Yadav, D.S., Rai, A.B., Singh, J. et al. 2008. IIVR: Fifteen years of accomplishments. Indian Institute of Vegetable Research Publication, Varanasi, India, pp. 1–45.

Raman, A. and Lau, C. 1996. Anti-diabetic properties and phytochemistry of *Momordica charantia* L. (Cucurbitaceae). *Phytomedicine*, 2:349–362.

Rhem, S. 1960. Die bitterstoffe der cucurbitaceen. *Ergebnisse der Biologia.*, 22:108–136.

Rehm, S., Enslin, P.R., Meeuse, A.D.J., and Wessels, J.H. 1957. Bitter principles of Cucurbitaceae. VII. The distribution of bitter principles in this plant family. *J. Sci. Food. Agric.*, 8:679–686.

Sensarama, P. 1998. Conservation in ancient India. Centre for Indigenous Knowledge on Indian Bio Resources, Lucknow, India.

Seshadri, V.S. and More, T.A. 2004. History and antiquity of cucurbits in India. In: *Progress in Cucurbit Genetics and Breeding Research. Proceedings of the Cucurbitaceae 2004, Eighth Eucarpa Meeting on Cucurbit Genetics and Breeding*, Palacky University, Olomouc, Czech Republic, July 12–17, 2004, pp. 81–90.

Seshadri, V.S. and More, T.A. 2009. *Cucurbit Vegetables-Biology, Production and Utilization*. Studium Press (India) Pvt. Ltd., New Delhi, India.

Shimotsuma. 1965. Chromosome studies of some *Cucumis*. *Seiken Ziho*, 17:11–16.

Singh, A.K. 1990. Cytogenetics and evolution in the Cucurbitaceae. In: *Biology and Utilization of Cucurbitaceae* (eds. Bates, D.M., Robinson, R.W., and Jeffrey, C.), Cornell University Press, Ithaca, NY, pp. 10–28.

Singh, S.P., Singh, V., Singh, N.K., Pathak, S.P., and Kumar, P. (2008). Inheritance of powdery mildew resistance in F1 hybrid of bottle gourd [*Lagenaria siceraria* (Mol.) Standl.]. *Indian Phytopathol*, 61:400.

Sinha, U., Saran, S., and Dutt, B. 1983. Cytogenetics of *Cucurbits*. In: *Cytogenetics of Crop Plants*. Macmillan, Madras, India, pp. 555–582.

Vargheese, B.M. 1973. Studies on cytology and evolution of the South Indian Cucurbitaceae. PhD thesis, Kerala University, Trivandrum, India.

Vishnu, S. 2006. *Vegetable Science and Technology in India*. Kalyani Publishers, New Delhi, India.

Walters, T.W. and Decker-Walters, D.S. 1988a. Origin, evolution and systematics of *Cucurbita pepo*. *Econ. Bot.*, 42:4–15.

Walters, T.W. and Decker-Walters, D.S. 1988b. Balsam pear (*Momordica charantia*, Cucurbitaceae). *Econ. Bot.*, 42:286–288.

Welihinda, A.A., Tirasophon, W., and Kaufman, R.J. 2000. The transcriptional co-activator ADA5 is required for HAC1 mRNA processing in vivo. *J. Biol. Chem.*, 275(5):3377–3381.

Whitaker, T.W. and Bemis, W.P. 1975. Origin and evaluation of the cultivated Cucurbita. *Bull. Torrey Bot. Club.*, 102:362–368.

Whitaker, T.W. and Cutler, H.C. 1971. Pre-historic cucurbits from the Valley of Oaxaca, *Econ. Bot.*, 25:123–127.

Whitaker, T.W. and Davis, G.N. 1962. *Cucurbits—Botany, Cultivation and Utilization*. Leonard Hill (Book), Ltd./London and Interscience Publisher Inc., New York/London, U.K.

Wien, H.C. 1997. The cucurbits: Cucumber, melon, squash and pumpkin. In: *The Physiology of Vegetable Crops* (ed. Wien, H.C.), CAB International, New York, pp. 345–386.

2 Cucurbits
Importance, Botany, Uses, Cultivation, Nutrition, Genetic Resources, Diseases, and Pests

David O. Ojo

CONTENTS

2.1 *CUCUMEROPSIS MANNII* NAUDIN

2.1.1 INTRODUCTION

Synonym is *Cucumeropsis edulis* (Hook.f.) Cogn. (1881). Common names are Egusi-itoo (Yoruba), white seed melon, dark egusi (English). Egousi-itoo, égousi, gousi (French). Lipupu (Portuguese) (Schaefer and Renner 2011; Achigan-Dako et al. 2015).

Egusi-itoo occurs wild from Guinea Bissau, east to southern Sudan and Uganda, and south to Angola. It is mostly cultivated in West Africa, especially in Nigeria, but occasionally also elsewhere, for example, in Côte d'Ivoire, Cameroon, and the Central African Republic. Egusi-itoo was very important as a seed vegetable in West Africa and parts of Central Africa at a time when there was plenty of forest to practice shifting cultivation. Now, it is rapidly declining and is replaced by egusi melon (*Citrullus lanatus* (Thunb.) Matsum. & Nakai) (Schaefer and Renner 2011; Achigan-Dako et al. 2015).

2.1.2 USES

Egusi-itoo is mainly grown for its oily seed. The seeds are prepared for consumption by parching and pounding to free the kernels of the seed coat (Sarwar et al. 2013). The kernels are milled into a whitish paste, which is used in soups and stews. The seeds (including seed coat) are also roasted and served as a snack. They resemble groundnut in flavor (Ogunbusola et al. 2012; Onawola et al. 2012).

An expensive semi-drying oil is extracted from the kernel, and the residue is fed to animals or used in the preparation of local snacks. The oil is suitable for cooking, soap making, and, less commonly, illumination. It can readily be refined into superior products for table use. It is of better quality and higher value than cottonseed oil. The flesh of the fruit, though edible, is not commonly eaten (Kapseu and Parmentier 1997; Kamda et al. 2015). In Ghana, the fruit juice mixed with other ingredients is applied to the navel of newborn babies to accelerate the healing process until the cord-relics drop off. Macerated leaves are used in Gabon for purging constipated suckling babies. In Sierra Leone, cattle boys traditionally use the dried fruit-shell of an egusi-itoo type with small elongated fruits as a warning horn (Burkill 1985; Ajuru and Okoli 2013; Kamda et al. 2014).

2.1.3 PRODUCTION AND TRADE

Egusi-itoo is regarded as the original indigenous egusi melon in West and Central Africa, and the seed can be found in most markets in the region. In Nigeria, the demand for the seeds, particularly in the towns, led to large-scale planting. Although its production is declining, egusi-itoo still is a common commodity in the markets. The trade is mostly local. Export occurs from Côte d'Ivoire to Nigeria, but the quantities involved are not reported.

2.1.4 NUTRITIONAL PROPERTIES

The nutritional composition of egusi-itoo seed per 100 g is water 8.3 g, energy 2282 kJ (545 kcal), protein 26.2 g, fat 47.3 g, carbohydrate 14.2 g, fiber 4.0 g, and Ca 86 mg. The seeds are rich in niacin (14.3 mg/100 g). The oil content of the kernel is 44% by weight. A sample of egusi-itoo seed oil from Côte d'Ivoire consisted of linoleic acid 64.9%, oleic acid 12.4%, stearic acid 11.8%, and palmitic acid 10.9% (Anhwange et al. 2010; John et al. 2014; Kwiri et al. 2014). Egusi-itoo is replaced in many regions by egusi melon (*Citrullus lanatus*) (Burkill 1985).

2.1.5 BOTANICAL DESCRIPTION

It is a monoecious, scandent herb 5–10 m long, climbing by simple tendrils; stem is angular, sparsely hairy. Leaves alternate, simple; stipules absent; petiole 2–15 cm long, initially hairy but glabrescent; blade broadly ovate in outline, 6–21 cm × 7–21 cm, deeply cordate at base, pentagonal to palmately 3–5-lobed with triangular to ovate lobes, margin sinuate-toothed, sparsely hairy on the veins, scabrid-punctate, palmately veined. Flowers unisexual, regular, five-merous, yellow; calyx campanulate, lobes up to 6 mm × 1.5 mm; corolla with lobes shortly united at base; male flowers in an axillary raceme, often umbel-like, pedicel up to 2 cm long, corolla lobes up to 7 mm × 5 mm, with three free stamens almost lacking filaments; female flowers solitary in leaf axils, pedicel up to 5 cm long, corolla lobes up to 11 mm × 6 mm, with inferior, fusiform, one-celled ovary, style columnar, stigmas three, two-lobed. Fruit, an ellipsoid to obovoid berry 17–25 cm × 8–18 cm, green to pale yellow or creamy white, mottled, glossy, flesh white, many-seeded. Seeds obovate, flattened, 1–2 cm × 0.5–1 cm, smooth, white. Seedling with epigeal germination; cotyledons leafy, elliptical. *Cucumeropsis* comprises a single species. It belongs to the tribe *Melothrieae*, together with *Cucumis* (Schaefer and Renner 2011).

2.1.6 GROWTH AND DEVELOPMENT

In West Africa, egusi-itoo is usually planted March–May at the start of the rainy season and harvested 6–8 months later (September–December). The crop requires support and is commonly found at the edge of gardens, climbing into shrubs or trees. When grown in shifting cultivation, debris left after burning serves as support. Egusi-itoo does not do well in the open or on flat land.

2.1.7 ECOLOGY

Egusi-itoo grows in forest, often at the margin or in openings, but also in swamp forest, more humid savanna, and abandoned fields, up to 1150 m altitude.

2.1.8 CULTIVATION AND MANAGEMENT

Egusi-itoo is still mainly collected from wild stands, which are often retained when clearing fields. In cultivation, it requires a soil rich in manure or partially decomposed organic matter. Application of N and K fertilizer can increase yields considerably, but P fertilizer has shown little effect.

At the beginning of the rainy season, three to four seeds per hole are sown. The 1000-seed weight is 150–250 g. Seedlings usually appear within 6–8 days. Egusi-itoo is often grown between other crops, growing on stakes along with yam, or supported by a strong trellis of at least 1 m tall.

2.1.9 DISEASES AND PESTS

In Nigeria, a severe damping-off disease caused by *Macrophomina phaseolina* has been reported. The fruits are sometimes attacked by the fruit fly *Dacus punctifrons*. The larvae develop in the fruit and eventually cause rot. Fruit flies attack fruits at every stage of development and can severely affect production. The pupae are found in the soil, and it is therefore advised not to plant in the same field the following year. The aphid-like flea hopper *Halticus tibialis* may suck sap from the leaves; young leaves become wrinkled, older ones become swollen around the sucking holes and later die. Several other pests that attack cucurbits are also found on egusi-itoo.

Seeds of egusi-itoo stored in open jars may be seriously damaged by beetles within a few weeks of storage; these have been identified as *Triboleum castaneum* and *Lasioderma serricorne*, and are also found in dried okra (*Abelmoschus* spp.) and roselle (*Hibiscus sabdariffa* L.) fruits (Adekunle and Uma 2012).

2.1.10 HARVESTING AND YIELD

Fruits are collected when the stems have dried and fruits have changed color from green to creamy white or yellow.

Under extensive management, where egusi-itoo is planted around the remaining trunks of trees, seed yield is about 300 kg/ha. In more intensive cropping systems, where land has been cleared and burnt before cultivation, it may reach 900 kg/ha. A plant usually produces two to five fruits; each fruit weighs 0.8–1.8 kg and contains 90–400 seeds (up to 100 g) (Nantoume et al. 2012; Yao et al. 2014).

2.1.11 POSTHARVEST HANDLING

After collection, fruits are cracked or split open; they are then placed in a heap or pit and are left for 14–20 days to let the fruit pulp rot. During this period, a strong pungent smell is produced, and this explains why seed extraction takes place at a distance from the homestead. Then, the seeds are removed and thoroughly washed to remove thick mucilage covering them; next, they are covered with sand or ash to prevent sticking, which would make hulling difficult. The seeds are dried to about 10% moisture content before packing. Packaging must be thorough and packs must be stored away from moisture, as seeds otherwise may germinate. Hulling is facilitated by heating to 60°C. The weight of the decorticated seed is about 60% of the whole dry seed. The kernels are milled and used as a vegetable or for producing vegetable oil for domestic use. Processing the seed of egusi-itoo is time-consuming and labor intensive; this is one of the reasons why it has been partly replaced by egusi melon (Olanrewaju and Moriyike 2013; Touré et al. 2015).

2.1.12 GENETIC RESOURCES

Germplasm of several *Cucurbitaceae* species used as seed vegetable, including *C. mannii*, is being maintained at the genebank of the National Centre for Genetic Resources and Biodiversity (NACGRAB), Ibadan, Nigeria (Ladipo et al. 1999).

2.1.13 PROSPECTS

Unless the seed yield of egusi-itoo can be increased and its crop management and seed processing can be simplified, it seems likely that its replacement by cultivars of egusi melon will continue, although specialty markets may develop (Pius et al. 2014; Touré et al. 2015).

2.2 *CUCUMIS ACUTANGULUS* L. (1753)/*LUFFA ACUTANGULA* (L.) ROXB.

2.2.1 INTRODUCTION

Synonym is *Cucumis acutangulus* L. (1753). Common names are ridged gourd, angled loofah, ribbed gourd, Chinese okra, silk squash (En). Papengaye, liane torchon (Fr). *Lufa riscada* (Po). Mdodoki (Sw).

C. *acutangulus* is believed to have originated in India, where wild types still occur, but has now spread pantropically to all areas with a high rainfall. It is cultivated and locally naturalized in West Africa, from Sierra Leone to Nigeria. It is cultivated from the coastal areas to the semi-dry savanna, for example, in Sierra Leone, Côte d'Ivoire, Ghana, Benin, and Nigeria. In East Africa, ridged gourd is grown on a small scale near the big cities as an exotic vegetable for consumers of Asian origin, and it is also locally cultivated and naturalized in Madagascar, Réunion, and Mauritius. In southern and eastern Asia, it is a widely cultivated vegetable (Robinson and Decker-Walters 1997).

2.2.2 USES

Immature fruits of less-bitter cultivars of *Luffa acutangula* are used as a vegetable. They are cooked or fried and used in soups and sauces. Occasionally, the stem tops with young leaves and flower buds are used as a leafy vegetable. In Southeast Asia, ridged gourd is a popular vegetable because of the mildly bitter flavor, the slightly spongy texture, and sweet juiciness. Young fruits of sweet cultivars are also eaten raw and small fruits are sometimes pickled. The seeds yield an edible oil that is, however, sometimes bitter and toxic (Burkill 1985; Lim 2012a,b).

In some parts of West Africa, a leaf extract of ridged gourd is applied on sores caused by guinea worms to kill the parasite. Leaf sap is also used as an eyewash to cure conjunctivitis. The fruits and seeds are used in herbal preparations for the treatment of venereal diseases, particularly gonorrhoea. In Mauritius, the seeds are eaten to expel intestinal worms and the leaf juice is applied to skin affections such as eczema. The plant including the seed is insecticidal. Mature fruits, when harvested dry, are processed into sponges and used for scrubbing the body while bathing or for domestic purposes, such as washing of cooking utensils, and as filters for local drinks such as palm wine. Industrial use is made of these fibers for making hats. However, the sponge gourd (*Luffa cylindrica* (L.) M. Roem., synonym: *Luffa aegyptiaca* Mill.) is preferred for making sponges because its fiber is easier to extract (Adebisi and Ladipo 2000). The trailing stem is used as temporary tying rope for firewood and crops to be carried home. The plant is occasionally used as an ornamental climber for enclosures.

2.2.3 PRODUCTION AND TRADE

Ridged gourd is mainly produced as a home garden crop. Thailand exports ridged gourd to western Europe as a vegetable for the Asian communities. Japan and Brazil are the main exporters of loofah sponges, mostly to the United States, but these are mainly from sponge gourd. In West Africa, mature fruits of ridged gourd or sponge gourd are sold as sponges in street markets and supermarkets (Huyskens et al. 1993).

2.2.4 NUTRITIONAL PROPERTIES

The composition of ridged gourd fruits per 100 g edible portion (tough skin removed, edible portion 62%) is water 94.2 g, energy 70 kJ (17 kcal), protein 0.8 g, fat 0.1 g, carbohydrate 3.3 g, fiber 1.7 g, Ca 12 mg, P 32 mg, Fe 0.3 mg, carotene 26 µg, thiamin 0.07 mg, riboflavin 0.02 mg, niacin 0.4 mg, folate 37 µg, and ascorbic acid 3 mg. The composition of young *Luffa* leaves per 100 g edible portion is water 89 g, protein 5.1 g, carbohydrate 4 g, fiber 1.5 g, Ca 56 mg, Fe 11.5 mg, β-carotene 9.2 mg, and ascorbic acid 95 mg (Holland et al. 1991). The oil content in the seeds is 26%; the fatty acid composition is linoleic acid 34%, oleic acid 24%, palmitic acid 23%, and stearic acid 10% (Uriostegui Arias 2015).

Two trypsin inhibitors and a ribosome-inactivating peptide (luffangulin) have been isolated from ridged gourd seeds. The glycoprotein luffaculin, also isolated from the seeds, exhibits abortifacient, antitumor, ribosome-inactivating and immunomodulatory activities (Fernando and Grün 2001).

Young fruits of sponge gourd (*L. cylindrica*) are used as a substitute for ridged gourd as a vegetable, although much less popular.

2.2.5 BOTANICAL DESCRIPTION

Monoecious, annual, climbing, or trailing herb, with acutely five-angled stem; tendrils up to six-fid, hairy. Leaves alternate, simple; stipules absent; petiole up to 15 cm long; blade broadly ovate to kidney-shaped in outline, 10–25 cm × 10–25 cm, shallowly palmately five to seven-lobed with broadly triangular to broadly rounded lobes, cordate at base, shallowly sinuate-dentate, pale green, scabrous, palmately veined. Male inflorescence racemose with 15–35 cm long peduncle. Flowers unisexual, regular, five-merous, 5–9 cm in diameter; receptacle tube obconic below, expanded above, c. 0.5 cm long, lobes triangular, 1–1.5 cm long; petals free, pale yellow; male flowers with three free stamens inserted on the receptacle tube, connectives broad; female flowers solitary, on pedicels 2–15 cm long, with inferior, densely pubescent, longitudinally ridged ovary, stigma three-lobed.

Fruit a club-shaped, dry and fibrous capsule 15–50 cm × 5–10 cm, acutely 10-ribbed, brownish, dehiscent by an apical operculum, many-seeded. Seeds broadly elliptical in outline, compressed, up to 1.5 cm long, smooth, dull black (Renner and Pandey 2013).

Luffa comprises seven species, four of these native to the Old World tropics and three somewhat more distantly related species indigenous to South America (Heiser and Schilling 1990). In *L. acutangula*, three varieties have been distinguished: var. *acutangula*, the large-fruited cultivated types; var. *amara* (Roxb.) C.B.Clarke, a wild or feral type with extremely bitter fruits and confined to India; and var. *forskalii* (Harms) Heiser & E.E.Schill., confined to Yemen, where it occurs wild or possibly as an escape. *L. acutangula* cultivars grown as vegetables have larger fruits and are less bitter than the wild types. In West Africa, local cultivars are used as vegetables, whereas in East Africa commercial growers use improved cultivars imported from Asian countries for the Asian customers (Robinson and Decker-Walters 1997).

2.2.6 GROWTH AND DEVELOPMENT

Spontaneous growth of plants commences with the beginning of the rainy season. Flowering and fruiting take place throughout the rainy season, while fruits mature and seed dispersal commences as the whole plants become dry at the peak of the dry season. In cultivation, seedlings emerge 4–7 days after sowing after soaking the seeds in cold water overnight to soften the hard seed coat. Ridged gourd tends to be day-neutral. Flowering starts 6–10 weeks after sowing. Initially male flowers are produced, later female ones at a ratio of male to female flowers of about 40:1. This ratio can be changed by chemical treatment. The flowers open in the evening and the stigmas have been found to remain receptive from a few hours before to 36–60 h after anthesis. The flowers are cross-pollinated by many insects, including bees, butterflies, and moths (Robinson and Decker-Walters 1997).

2.2.7 ECOLOGY

Ridged gourd may be common as a spontaneous plant on abandoned land and as a fallow crop on garbage heaps. Unlike many other cucurbits, it grows well in tropical lowlands. It prefers seasonal climates because dry-season planting is more successful than wet-season planting. In Africa, it thrives in the dry forest or moist savanna area, around 8°N–10°N. Outside these latitudes, too much rain or excessive dryness often affect the development of the fruits. In humid areas, growth is directed toward the production of leaf biomass, whereas under dry conditions the energy is directed toward abundant flowering. Too much heavy rainfall during flowering and fruiting leads to fruit rot. Frost is not tolerated. Ridged gourd prefers a well-drained soil with a high organic matter content and a pH of 6.5–7.5 (Heiser and Schilling 1990; Robinson and Decker-Walters 1997).

2.2.8 CULTIVATION MANAGEMENT

In commercial cultivation, the crop needs good care. Planting on raised beds assures good drainage in the rainy season. Irrigation is required during dry conditions at regular intervals, particularly before the flowering period. NPK fertilizer is applied to enhance growth, flowering and fruit formation. A basal dressing of NPK (e.g., 14–14–14) at the rate of 25 g/hill can be given, followed by side dressings of 20 g/hill of urea or NPK at 2-week intervals. Lateral stems are pruned if they grow too abundantly. Some top and leaf pruning may promote flower and fruit development, resulting in a higher yield. For optimal production, the number of fruits per stem may be limited to 20–25. For the spontaneous plants of abandoned farmland or on refuse dumps, hardly any management care is given.

Ridged gourd is normally grown on supports or trellises up to 3 m high. During the dry season, it may also be allowed to trail on the ground, but this practice lowers the yield and quality. The seeds are sown on mounds or ridges, two to three seeds per hill, 50–60 cm apart in the row, and 200 cm between the rows in a trellized system. Without support, 300 cm between the rows can be practiced, or about one hole per m each way. Alternatively, seedlings may be raised in containers and transplanted. The 1000-seed weight is around 90 g. For direct sowing, 2–3 kg seed is needed per ha for transplanting 1–1.5 kg. In the Philippines, a planting distance of 2 m × 2 m is practiced for a superior F_1 hybrid, with a seed requirement of only 500 g/ha (Huyskens et al. 1993).

2.2.9 DISEASES AND PESTS

Ridged gourd is not very susceptible to diseases and pests. Powdery mildew (*Erysiphe cichoracearum*) and downy mildew (*Pseudoperonospora cubensis*) are reported. Fruits rot easily in contact with wet soil. In Southeast Asia, the larvae of fruit flies (*Dacus* spp.) may damage young fruits; a high infection of thrips may cause stunted growth, and also caterpillars, leaf miners, and aphids are reported as pests.

2.2.10 HARVESTING AND YIELD

Young immature fruits of 300–400 g are picked 12–15 days after fruit set. Fruits can be picked every 3 days throughout the fruiting season, by hand or with a knife. Individual plants may produce 15–20 fruits; yield declines after 8–13 weeks of harvesting. For sponge production, the fruits are left for 2 months on the vines till they turn brown. For seed production, the seeds are shaken out of the completely dry fruits.

Landraces produce 10–15 t/ha. An average yield of 27 t/ha of young fruits is reported for hybrid cultivars in the Philippines under good management.

2.2.11 POSTHARVEST HANDLING

Immature fruits of ridged gourd are easily damaged. For long-distance transport, the fruits have to be carefully packed. The fruits can be stored for 2–3 weeks at 12°C–16°C. The processing of sponges from the ripe fruits involves immersing the fruit in running water until the rind disintegrates and disappears, then the pulp and seeds are washed out, and the sponges are bleached with hydrogen peroxide and dried in the sun.

2.2.12 GENETIC RESOURCES AND BREEDING

Germplasm collections of *L. acutangula* are kept at genebanks in India and Taiwan, at the Institute for Plant Breeding in the Philippines, and in Nigeria at the National Centre for Genetic Resources and Biotechnology (NACGRAB) at Ibadan (Renner and Pandey 2013).

Many local cultivars are found in the Asian countries and improved cultivars are available from several seed companies. Populations are very variable. F_1 hybrid cultivars are used in several Asian countries. East-West Seed Company in Thailand developed F_1 hybrids for tropical lowland with good market quality, for example, pale or dark green fruits, short (35 cm) to long (50 cm) fruits. Malika F_1 is a hybrid with high disease tolerance and especially suited for the rainy season (Talano et al. 2012).

2.2.13 PROSPECTS

Ridged gourd is a high-yielding and easy-to-cultivate vegetable. Breeding and production technology research combined with market development might give it a chance to develop into a market vegetable of importance in Africa, as in Asian countries (Renner and Pandey 2013). The use of fiber from the mature fruits and the use in agroforestry as a plant for soil rehabilitation with a heavy production of leaf biomass might be investigated.

2.3 *CUCUMIS AFRICANUS* (WILD WATERMELON)

2.3.1 INTRODUCTION

Cucumis africanus belongs to the family Cucurbitaceae with the chromosome number 2n = 24. Its origin and geographic distribution occurs in Angola, Namibia, Botswana, and South Africa. It is also found in Madagascar, where it was introduced. *C. africanus* leaves are eaten as a cooked vegetable by many tribes in its area of origin. Its nonbitter fruit types serve as a source of water and are eaten as a vegetable (Kull et al. 2015). The leaves contain per 100 g: water 92.2 g, protein 1.3 g, fat 0.3 g, carbohydrate 3.4 g, fiber 1.2 g, Ca 216 mg, Mg 175 mg, P 11 mg, Fe 12 mg, thiamin 0.02 mg, riboflavin 0.11 mg, niacin 0.34 mg, and ascorbic acid 81 mg. The fruits contain per 100 g: water 88.2 g, protein 2.8 g, fat 1.6 g, carbohydrate 3.3 g, fiber 2.9 g, Ca 13 mg, Mg 29 mg, P 20 mg, Fe 1.1 mg, thiamin 0.2 mg, riboflavin 0.03 mg, niacin 0.84 mg, and ascorbic acid 13 mg (Arnold et al. 1985; Nkgapele and Mphosi 2014, 2015; Mphosi 2015).

C. *africanus* types with nonbitter, large, and oblong fruits occur wild in Angola, Namibia, and South Africa. The smaller, ellipsoid fruit types found in other *C. africanus* are bitter, could be poisonous, and not used for consumption. A third type, intermediate in taste and shape, seems to exist as well but is not well documented. Medically, the fruit of *C. africanus* contains considerable amounts of cucurbitacin A, B, and D and traces of cucurbitacin G and H. Cucurbitacins, which are known from many *Cucurbitaceae* and various other plant species, exhibit cytotoxicity (including antitumor activity), anti-inflammatory and analgesic activities (Jeffrey 1980).

2.3.2 BOTANICAL DESCRIPTION

C. africanus is an annual, monoecious, prostrate or scandent herb, sometimes with woody, thickened roots, stems up to 1 m long; tendrils simple. Leaves alternate, simple; stipules absent; petiole 1–1.5 cm long; blade ovate, deeply palmately (three to five) lobed, 1.6–8.2 cm × 1.8–7 cm, cordate at base, lobes elliptical, and broadly elliptical to ovate-elliptical. Flowers are unisexual, regular, five-merous; receptacle 3–5 mm long; sepals 1.5–3 mm long; petals bright yellow, 5–11 mm long; male flowers one to five together in small fascicles, with pedicel up to 1 cm long, stamens three; female flowers solitary, with pedicel 1–4 cm long, ovary inferior, densely softly spiny. Fruit an ellipsoid to oblong-ellipsoid berry 3–9 cm × 2–4.5 cm, when ripe strongly longitudinally striped pale greenish-white and purplish-brown, with spines 3–6 mm long; fruit stalk 2–4.5 cm long, slender, not expanded upward. Seeds ellipsoid, compressed, 4–7 mm × 2–3.8 mm × 1–1.2 mm (Kirkbride 1993; Regassa et al. 2015).

The genus *Cucumis* includes about 30 species, 4 of which are economically important: cucumber (*Cucumis sativus* L.), melon and snake cucumber (*Cucumis melo* L.), West Indian gherkin (*Cucumis anguria* L.), and horned melon (*Cucumis metuliferus* Naudin) (Pelinganga et al. 2013). *C. africanus* is placed in the "*anguria*" group of the subgenus *melo*. *C. africanus* flowers from January to June and occurs in dry bushland areas close to habitation (Jeffrey 1980; Kull et al. 2015).

2.3.3 GENETIC RESOURCES AND BREEDING

C. africanus is common in its area of origin, thus is not threatened with genetic erosion or extinction. *C. africanus* germplasm is stored in the United States, the United Kingdom, the Czech Republic, and Spain. Within the "*anguria*" group of about 16 spiny-fruited *Cucumis* species to which *C. anguria* belongs as well, there seem to be no major barriers to gene exchange. Several interspecific crosses have been made in this group. An intermediate response to downy mildew (*P. cubensis*) has been reported for *C. africanus*. In southern Africa, *C. africanus* is considered to have potential for domestication. The variation within the species will allow successful breeding and selection. Breeders' interest will focus on disease resistance within the scope of gene transfer to the economically important *Cucumis* species (Schippers 2000; van Wyk and Gericke 2000; Ojo et al. 2013; Kull et al. 2015).

2.4   *CUCUMIS ANGURIA* L. (GHERKIN)

2.4.1   INTRODUCTION

Synonym is *Cucumis longipes* Hook.f. (1871). *C. anguria* vernacular names are gooseberry gourd (English), Concombre antillais, ti-concombre, macissis (French), Pepino das Antilhas, cornichão das Antillas, machiche, maxixé (Portuguese).

   C. anguria is synonymous to *C. longipes* Hook.f. (1871) of African origin, and it occurs wild in East and southern Africa. It has bitter fruits, but occasionally nonbitter types occur. Seeds were taken to the Americas with the slave trade, where the cultivated West Indian gherkin was developed. This edible, nonbitter type spread through the Caribbean, parts of Latin America, and the southern United States. It can now be found in a semiwild state as an escape from cultivation, and in some cases, it appears to be an element of the indigenous flora. It is an invasive weed in parts of North America and in Australia and a serious weed in peanut fields of the southern United States. The nonbitter edible form was reintroduced into Africa (e.g., Cape Verde, Senegal, Sierra Leone, DR Congo, Réunion, Madagascar, South Africa), where it is grown for its fruits. In Madagascar, *C. anguria* is probably not originally wild but naturalized because it is localized around human habitations.

2.4.2   USES

The leaves of bitter forms of *C. anguria* are cooked and eaten in the same manner as pumpkin leaves (*Cucurbita* spp.). In Ruwangwe, Zimbabwe, it is known as "mubvororo" and used to prepare a special dish for the father of the household. In Namibia, it is one of a range of edible wild greens, which are dried into cakes and stored for use during the dry season. Elsewhere in Africa, the nonbitter form is cultivated for its fruits. It is recorded near Thiès (Senegal), where the immature fruits are pickled green. In South Africa, the fruits are eaten both fresh and dried.

   In the New World, West Indian gherkin refers to the cultivated nonbitter form, a favorite pickle since the seventeenth century and sometimes eaten fresh. Fruits are also relatively common as a table vegetable, and they are used in soups and stews. In Brazil, the mature fruits are cooked as the main ingredient of a traditional soup called "maxixada." Immature fruit are used as fresh cucumbers.

   Bitter forms of *C. anguria* are sometimes used in Zimbabwe as a natural pesticide in stored crops. The juice of the fruit is reportedly used as an antifeedant in granaries. In Matabeleland (Zimbabwe), the fruit is used as a lure in rock and stick traps. Medicinal uses are reported from Tanzania where an enema of the wild plant is used to treat stomach pain. In Zimbabwe, traditional medical practitioners consider the bitter fruit as poisonous and the juice of the fruit is used to treat septic wounds in livestock. In America, medicinal uses are varied, including root decoctions as a remedy for stomach trouble in Mexico, and to reduce oedema in Cuba. The fruit is eaten to treat jaundice in Curaçao, and leaf juice preparations are applied to freckles in Cuba. Kidney problems are treated with a decoction in Colombia, where it is believed that the fruits eaten raw dissolve kidney stones. The fruit is applied to hemorrhoids in Cuba, and the leaves after being steeped in vinegar are used against ringworm (Bates et al. 1990).

2.4.3   PRODUCTION AND TRADE

C. anguria as a leafy vegetable is collected from the wild or grown on a small scale in southern Africa, but no data on its production or trade are available. The cultivation of *C. anguria* for nonbitter young fruits is also practiced on a small scale only. In the New World, where it is always cultivated for its immature fruits, it is also of minor importance and in statistics is combined with pickling cucumber (*C. sativus* L.) (Baird and Thieret 1988).

2.4.4 Nutritional Properties

The nutrient composition of the fresh fruit of West Indian gherkin per 100 g edible portion is water 93 g, energy 71 kJ (17 kcal), protein 1.4 g, fat 0.1–0.5 g, total sugar 1.9–2.5 g, starch 0.3–0.4 g, Ca 25–27 mg, P 33–34 mg, Fe 0.6 mg, vitamin A 200–325 IU, thiamin 0.05–0.15 mg, riboflavin 0.40 mg, niacin 0.3–0.5 mg, and ascorbic acid 48–54 mg (Whitaker and Davis 1962). No data are reported on the composition of the leaves, but this is probably similar to other East African dark green leafy vegetables. The seed oil of fruits of the wild bitter form is composed of palmitic, stearic, oleic, linoleic, and linolenic acids.

Many cucurbits have both bitter and nonbitter forms within the same species. In the bitter forms of *C. anguria*, the bitterness increases considerably as the fruit ripens. The bitter principles, known as cucurbitacins, are tetracyclic triterpenoids. Cucurbitacins are among the most bitter substances known and are extremely toxic to mammals. In *C. anguria*, the main bitter principle is cucurbitacin B ($C_{32}H_{48}O_8$) with a much smaller amount of cucurbitacin D ($C_{30}H_{46}O_7$) and traces of cucurbitacins G and H. Toxicity studies showed the juice of the fruits to be highly toxic to rats (LD_{50} 1.6 mg/kg). The toxicity is reported to be reduced more than 100-fold if the juice is first boiled. Studies on the larvicidal activity of aqueous, ethanolic and citric acid extracts from *C. anguria* on *Aedes aegypti*, the yellow fever and dengue fever mosquito, showed that concentrations of 0.5 mg/mL after 24 h exposure caused larval mortalities of up to 40% (Petrus 2014).

2.4.5 Botanical Description

Annual, monoecious herb with trailing or scandent stems, having solitary, simple, setose tendrils 3–6 cm long; stems grooved, with bristle-like hairs. Leaves alternate, simple; stipules absent; petiole (2–)6–13 cm long, hispid to setose; blade broadly ovate in outline, 3–12 cm × 2–12 cm, shallowly to deeply palmately 3–5(–7)-lobed, with punctate to hispidulous hairs on both surfaces. Flowers unisexual, regular, five-merous; sepals narrowly triangular, 1–3 mm long; petals united at base, 4–8 mm long, yellow; male flowers in 2–10-flowered fascicles, with pedicel 0.5–3 cm long, stamens three; female flowers solitary, with pedicel 2–10 cm long, ovary inferior, ellipsoid, 7–9 mm long, softly spiny, stigma three-lobed. Fruit an ellipsoid to subglobose berry 3–4.5 cm × 2–3.5 cm, on a stalk 2.5–21 cm long, beset with soft, thin spines with transparent tips, green, ripening yellow, many-seeded. Seeds ellipsoid, 5–6 mm long, compressed with rounded margins and smooth (Bates et al. 1990; Chen 2011).

The approximately 30 *Cucumis* species are native to Africa, except the cucumber (*C. sativus* L.), which probably originates from India. Wild and cultivated types of *C. anguria* differ in bitterness of the fruits but also in the length of fruit spines (longer in wild forms). Wild types have been distinguished as var. *longipes* (Hook.f.) A. Meeuse or var. *longaculeatus* J.H. Kirkbr, cultivated ones as var. *anguria*. However, plants with short-spined fruits are often naturalized in tropical America and rarely in Africa (Kirkbride 1993).

2.4.6 Growth and Development

In its native habitat in southern Africa, *C. anguria* germinates in a few days during the summer rains when night temperatures are above 12°C and the soil is sufficiently wet. Early growth is upright; the primary stem may reach a height of 20 cm and does not produce flowers. This is followed quickly by several trailing procumbent stems, which branch off from the base, reaching a length of 2–3 m. Male flowers appear first, followed by female ones. Plants are self-fertile and cross-pollination is by insects. Day length plays an important role in flowering. Longer days combined with high temperatures tend to keep plants in the male-flowering phase of development, whereas lower temperatures and shorter days encourage development of female flowers. Fruits may be produced within 60 days from time of planting. They continue to be produced and to ripen over the hot season, giving up to 50 fruits/stem. Fruits remain attached to the withered annual stems long after these have died back at the end of the growing season (Sonnewald 2013).

2.4.7 Ecology and Cultivation Management

Wild *C. anguria* is a common inhabitant of semi-deciduous and deciduous woodland, tree and shrub savanna, grassland and semi-desert, up to 1500 m altitude. Wild and semi-domesticated forms can be found growing near compounds, in woodland and grassland, often on abandoned cultivated land, near cattle kraals, or occasionally as a weed in cultivation.

Plants tolerate a wide range of soil types, including Kalahari sands (regosols), red clays (fersial-litics), and black cotton soils (vertisols). In its southern African habitat, rainfall occurs in summer and varies from less than 400 to over 1000 mm. Temperatures during the growing season range from 15°C to 35°C. *C. anguria* is intolerant of frosts and cold temperatures.

The culture and agronomic requirements are similar to those of the common garden cucumber. In cultivation, the plants should be trailed. The application of organic manure and NPK fertilizer is beneficial. Irrigation can be given in periods of drought. In South Africa, the first fruit of a plant is tasted and if it is bitter, the whole plant is discarded.

2.4.8 Propagation and Planting

West Indian gherkin is propagated by seed, which requires light for germination. Seeds are sown in pockets of three to four at a spacing of 30 cm in the row and 100–150 cm between rows. The seed requirement is 2.5–4.5 kg/ha. In the growing season, the period from seeding to first harvest is 2–2.5 months. Plants continue to flower and set fruit for several months. For leaf production, the same cultural practices can be followed (Fernandes 2011).

2.4.9 Diseases and Pests

West Indian gherkin is quite resistant to pests and diseases. It displays varying degrees of natural resistance to pathogens and insects, such as the cucumber green mottle mosaic virus, root-knot nematodes, powdery mildew, and greenhouse whitefly. The fruits are seldom parasitized by fruit fly larvae, which attack most other cucurbit species in southern Africa.

2.4.10 Harvesting and Yield

As the fruits are preferred for pickling, they are harvested in the immature stage, while still green. If grown for leaves, these can be picked many times during several months. A single plant can produce 50 or more fruits. No statistics on fruit or leaf yield are reported. The yield potential is probably higher than for pickling cucumbers. The fruits can be kept for a few days at room temperature; the leaves should be consumed or marketed within a day.

2.4.11 Genetic Resources and Breeding

C. anguria is not in danger of extinction in its native habitat. The National Plant Germplasm System of the U.S. Department of Agriculture maintains numerous accessions of cultivated types of *C. anguria* at its regional plant introduction station in Ames, IA. Another collection is maintained at the Centro Agronómico Tropical de Investigación y Enseñanza (CATIE), Turrialba, Costa Rica. Various Western seed companies offer seed of West Indian gherkin, including "African Heirloom," and "West Indian Burr Gherkin" (Sepasal 2003).

C. anguria and related species have been the focus of investigations by plant scientists to identify resistances to the many pests (viruses, bacteria, fungi, insects) attacking cucumber and melon, which might be genetically transferred. *C. anguria* proved to be totally immune to cucumber green mottle mosaic virus (CGMV). Resistance also occurred to root-knot nematodes and powdery mildew. In a study in South Africa, where fungal diseases and fruit parasitization by trypetid larvae is usually severe in Cucurbitaceae, *C. anguria* showed a high resistance to both fungi and trypetids. Research efforts to transfer resistances into cucumber and melon have been undertaken. Repeated attempts to hybridize different *Cucumis* species have not been entirely successful; some species have never been successfully crossed to produce a fertile F_1 generation, whereas other species have been crossed to a limited extent.

The possibility of using *C. anguria* as a rootstock has been suggested, where scions of desirable crop species are grafted to it. In populations of some cucurbit species that normally produce bitter or toxic fruits, individuals may occasionally arise spontaneously that produce nonbitter, edible fruit. These variants are genetically stable when removed from the bitter gene pool. In *C. anguria*, a single gene distinguishes the bitter from the nonbitter type, the gene producing bitterness being dominant. Multiple factors appear to be involved in controlling bitterness, including various physiological conditions (Schippers 2000; Esteras et al. 2011; Manamohan and Chandra 2011).

2.4.12 PROSPECTS

C. anguria, both as a semi-wild leafy vegetable and as the West Indian gherkin, merits more attention from plant breeders and agronomists. It is an attractive alternative to the common garden cucumber for use as a pickle, with fewer pest and disease problems and a larger fruit production (Sepasal 2003; Fabricante et al. 2015).

2.5 *CUCUMIS HIRSUTUS* SOND. (WILD HIRSUTUS)

2.5.1 INTRODUCTION

Synonym is *Cocculus hirsutus* (Linn.). *C. hirsutus* is distributed from Cameroon to Sudan and southwards to South Africa (Cape Province) as well as in Madagascar.

2.5.2 USES

In Malawi, the leaves are eaten in the same way as pumpkin leaves, that is, sliced and cooked. The raw fruits are eaten as well but are not much appreciated. In South Africa, *C. hirsutus* is considered a poisonous plant. A decoction of the root is used by the Zulu tribe to treat chronic cough (Fand and Suroshe 2015).

2.5.3 NUTRITIONAL PROPERTIES

There is no information on nutritional values, but the leaf composition is probably comparable to other dark green leaf vegetables and that of the fruits to cucumber. Several cucurbitacins have been isolated from the roots of *C. hirsutus*. Cucurbitacins, which are known from many Cucurbitaceae and various other plant species, exhibit cytotoxicity (including antitumor activity), anti-inflammatory and analgesic activities.

2.5.4 BOTANICAL DESCRIPTION

Dioecious, perennial, prostrate or scandent herb, with simple tendrils; roots fibrous, woody; stems up to 2.5 m long, thickened and woody at base. Leaves alternate, simple; stipules absent; petiole 0.5–5.5 cm long; blade broadly ovate, ovate-triangular or narrowly ovate, 2–15 cm × 1–10 cm, slightly cordate at base, unlobed or variously palmately three to five-lobed, lobes ovate-triangular to linear. Flowers unisexual, regular, five-merous; receptacle 3–9 mm long; sepals 1–9 mm long; petals white, cream or yellow; male flowers 1–12 together in fascicles, pedicel 0.5–7.5 cm long, petals up to 2 cm long; female flowers solitary or paired, pedicel 0.5–2.5 cm long, petals up to

3 cm long, ovary inferior, densely appressed or patent hairy. Fruit a globose to oblong-ellipsoid berry 2.5–7 cm × 1.5–6 cm, brownish-orange when ripe, smooth; fruit stalk 2–6 cm long, slender, not expanded upward. Seeds ovoid, compressed, 6.5–9 mm × 5–6.5 mm × 2–3 mm, white, smooth.

The genus *Cucumis* includes about 30 species, 4 of which are economically important: cucumber (*C. sativus* L.), melon and snake cucumber (*C. melo* L.), West Indian gherkin (*C. anguria* L.), and horned melon (*C. metuliferus* Naudin). *C. hirsutus* is the only species in the "*hirsutus*" group of the subgenus *melo*. In Malawi, the leaves are eaten at the end of the dry and beginning of the rainy season (October–November) (Kirkbride 1993).

2.5.5 Ecology

C. hirsutus is found in woodland, wooded grassland and grassland, and as a weed on formerly cultivated ground, up to 2500 m altitude.

2.5.6 Management

C. hirsutus is exclusively collected from the wild.

2.5.7 Genetic Resources and Breeding

Since *C. hirsutus* is widespread, there is no serious risk of genetic erosion. Only in the United States are a few accessions registered, all originating from South Africa. Breeders' interest in *C. hirsutus* is limited as transfer of genes by conventional breeding techniques to economically important *Cucumis* species is not possible (Sarvalingam et al. 2014; Garad et al. 2015).

2.5.8 Prospects

It is likely that *C. hirsutus* will remain a vegetable of local interest only.

2.6 *CUCUMIS MELO* L. (MUSKELON)

2.6.1 Introduction

Vernacular names are melon, muskmelon, cantaloupe (En). Melon (Fr). Melão (Po). Mtango, mtango mungunyana, mmumunye (Sw).

Melon probably originated in East Africa, where wild populations still occur, for example, in Sudan, Ethiopia, Eritrea, Somalia, Uganda, and Tanzania. Possibly, it also occurs wild in southern Africa, but the exact distribution of wild *C. melo* is unclear because of the regular occurrence of plants escaped from cultivation. Melon was domesticated in the eastern Mediterranean region and West Asia at least 4000 years ago and subsequently spread into Asia. During the long period of cultivation, many types developed with many fruit shapes and with either sweet or nonsweet flesh. Important centers of genetic diversity of cultivated melon developed in Iran, Uzbekistan, Afghanistan, China, and India. In Africa, important variations occur in Sudan and Egypt. The name "cantaloupe" derives from a fifteenth-century introduction of melon from Turkish Armenia to the papal residence at Cantalupi near Rome. Melon is now grown worldwide. It is a typical fruit vegetable of subtropical and warm temperate areas.

Melon occurs throughout the warm and dry areas of Africa, where it is grown either for its fruit or for its seeds.

Several nonsweet types of *C. melo* are grown traditionally. The most important one is snake melon, called "ajjur," "faqqus," or "qatta" in Arabic. It is found in many parts of Asia, from Turkey to Japan, and locally in Europe (Italy) and the United States. In Africa, it seems to be restricted to Sudan and North Africa (Egypt, Morocco, Tunisia), where it is quite important. In Sudan, the immature fruits of a melon type locally known as "tibish" are used in the same way as snake melon. Some other types grown in Africa have bitter flesh and are grown for their edible seeds (Robinson and Decker-Walters 1997; Nonaka and Ezura 2015).

2.6.2 USES

Mature fruits of sweet melon cultivars are usually consumed fresh for the sweet and juicy pulp. The pulp is also mixed with water and sugar, or sometimes with milk, and served as a refreshing drink or made into ice cream. Immature fruits of nonsweet types, including snake melon, are used as a fresh, cooked, or pickled vegetable; they are also stuffed with meat, rice, and spices and fried in oil. Snake melon is often confused with cucumber and used as such. The seeds are eaten after roasting; they contain edible oil. The Hausa people in Nigeria grind the kernels to a paste and make it into fermented cakes. The young leaves are occasionally consumed as a potherb and in soups. The leafy stems and also the fruit provide good forage for all livestock. In Réunion and Mauritius, a decoction of seeds and roots is used as a diuretic and vermifuge (Robinson and Decker-Walters 1997).

2.6.3 PRODUCTION AND TRADE

Annual world production of melon has increased from 9 million t (700,000 ha) in 1992 to 22 million t (1.2 million ha) in 2002. Major producing countries are China with 400,000 ha, West Asia (Turkey, Iran, Iraq) 200,000 ha, the Americas (the United States, Mexico, Central and South American countries) 165,000 ha, northern Africa (Egypt, Morocco, Tunisia) 110,000 ha, southern Asia (India, Pakistan, Bangladesh) 100,000 ha, European Union (Spain, Italy, France, Greece, Portugal) 95,000 ha, Romania 50,000 ha, Japan 13,000 ha, and Korea 11,000 ha.

Each country has its own specific melon cultivars, and most of the crop is sold in local markets. Production for export has developed in the Mediterranean region, the United States, Mexico, Australia, Taiwan, and Japan, using F_1 hybrid cultivars with good shipping and storage characteristics.

In Africa, sweet melon is a luxury crop for urban markets, grown in drier regions and in highlands. Statistics on production are not available for most countries, except Cameroon (3500 ha) and Sudan (1200 ha). Senegal and surrounding countries export melon during the winter to Europe. Snake melon is important in Sudan, where it is grown for home use and local markets. The area grown is about 4000 ha with an annual production of 80,000 t. It is not exported (Jones et al. 2001).

2.6.4 NUTRITIONAL PROPERTIES

The edible portion of a mature melon fruit is 45%–80%. Fruits (raw, peeled) contain per 100 g edible portion: water 90.2 g, energy 142 kJ (34 kcal), protein 0.8 g, fat 0.2 g, carbohydrate 8.2 g, fiber 0.9 g, Ca 9 mg, Mg 12 mg, P 15 mg, Fe 0.2 mg, Zn 0.2 mg, vitamin A 3382 IU, thiamin 0.04 mg, riboflavin 0.02 mg, niacin 0.7 mg, folate 21 μg, and ascorbic acid 37 mg (USDA 2002).

The nutritional composition of snake melon per 100 g edible portion is water 94.5 g, energy 75 kJ (18 kcal), protein 0.6 g, fat 0.1 g, carbohydrate 4.4 g, fiber 0.3 g, and ascorbic acid 13 mg (Polacchi et al. 1982).

Sugar content and aroma are important factors determining the quality of sweet melon. Esters derived from amino acids are important components of the characteristic flavor; sulfur-containing compounds also play a role. Several C-9 alcohols and aldehydes, including Z-non-6-enal, are characteristic of the melon aroma. To get the best aroma, fruits should be harvested only 2–3 days before they are fully ripe. The edible seed kernel contains approximately 46% of yellow oil and 36% protein (USDA 2014).

2.6.5 Uses

Snake melon for use in salads can be replaced by cucumber (*C. sativus* L.). As a fruit, sweet melon can be replaced by papaya (Wyllie et al. 1995; Falah et al. 2015).

2.6.6 Botanical Description

Monoecious, climbing, creeping, or trailing, annual herb, having simple tendrils; root system large, mostly distributed in the top 30–40 cm of the soil, a few roots descending to 1 m depth; stem up to 3 m long, ridged or striate, hairy. Leaves alternate, simple; stipules absent; petiole 4–10 cm long; blade orbicular or ovate to reniform, 3–20 cm in diameter, angular or shallowly palmately five to seven-lobed, cordate at base, shallowly sinuate-toothed, surfaces hairy. Flowers axillary, unisexual or bisexual, regular, five-merous; pedicel 0.5–3 cm long; sepals linear, 6–8 mm long; corolla campanulate, lobes almost orbicular, up to 2 cm long, yellow; male flowers in two to four-flowered fascicles, with three free stamens; female or bisexual flowers solitary, with inferior, ellipsoid ovary, stigma three-lobed. Fruit a globose, ovoid or oblongoid berry weighing 0.4–2.2 kg, smooth or furrowed, rind smooth to rough and reticulate, white, green, yellowish-green, yellow, yellowish-brown, speckled yellow or orange with green or yellow background, flesh yellow, pink, orange, green or white, many-seeded. Seeds compressed ellipsoid, 5–12 mm × 2–7 mm × 1–1.5 mm, whitish or buff, smooth. Seedling with epigeal germination (Robinson and Decker-Walters 1997).

Most of the about 30 *Cucumis* species are native to Africa. They all have a chromosome number of 2n = 24, except *C. sativus* L. (cucumber) with 2n = 14; probably this species originated from Asia. *C. sativus* fruits are beset with spinous tubercles and warts when young, whereas ovaries of *C. melo* are hairy without tubercles and warts.

C. melo is polymorphic. Wild and weedy plants are often distinguished as subsp. *agrestis* (Naudin) Pangalo, having shortly pubescent ovaries and comparatively small flowers and fruits, whereas the cultivated plants (subsp. *melo*) have villous ovaries and generally larger flowers and fruits.

Cultivated plants belong to many different cultivar groups, of which the most important with sweet fruits for modern market gardening are as follows:

1. *Reticulatus group (muskmelon or netted melon)*: Fruit globular (1–1.8 kg), rind strongly reticulate, sometimes furrowed, yellowish-green with orange flesh (Italo-American) or finely reticulate to smooth, yellowish-green with pale green flesh (Japanese, Mediterranean, e.g., "Galia"), sugar content high (13%–15%), aromatic, shelf life medium.
2. *Cantaloupe group (cantaloupe or muskmelon)*: Fruit flattish to globular and often ribbed (1.2–1.8 kg), rind smooth or reticulate, flesh usually orange, carotene and sugar content high, flavor rich, shelf life short, mainly grown in southwestern Europe (e.g., "Charentais") and the Americas.
3. *Inodorus group (winter melon)*: Fruit ovoid (1.5–2.5 kg), late maturing, rind smooth, wrinkled or slightly reticulate, often striped or splashed, grey, green or yellow, flesh firm, white or pale green, sugar content high but little flavor, shelf life long, mainly grown in Iran, Afghanistan and China, but also in Spain, the United States and Japan; important cultivars are "Casaba," "Honeydew," "Piel de Sapo," "Jaune Canari," and "Chinese Hami."

Examples of groups with nonsweet fruits used as a vegetable in Africa are as follows:

1. *Flexuosus group (snake melon or "snake cucumber")*: Fruit up to 2 m long, more than six times as long as wide, rind pale green or striped pale and dark green, ribbed or wrinkled, flesh white.
2. *Tibish group*: Fruit small, ovoid to oblate, without ribs, rind smooth, dark green with pale green stripes, flesh firm, white, particularly important in Sudan; a similar type named "seinat" in Sudan is grown for its seed.

Sweet melon grown for the urban markets in tropical Africa comprises nowadays mostly of F_1 hybrid cultivars of reticulatus group (e.g., "Galia") and cantaloupe group (e.g., "Charentais").

2.6.7 GROWTH AND DEVELOPMENT

Melon seed will remain viable for at least 6 years when stored dry (moisture content 6%) at temperatures below 18°C. Priming may improve germination after long storage. Seedlings appear 2–14 days after sowing. Numerous horizontal lateral roots develop rapidly from the taproot. The roots grow mostly at a depth of 30–40 cm. The first true leaf appears 5–6 days after unfolding of the cotyledons. The first two to four axillary buds on the main stem produce vigorous primary branches, which check the growth of the main stem. In most types, the first clusters of male flowers appear on the 5th–12th node of primary branches, while bisexual or female flowers appear on tertiary branches, formed from the 14th node of primary branches onward. Flowers are open for 1 day only and insects, mostly bees, effect pollination. The fruit is a heavy sink for assimilates and minerals and per plant usually only 3–6 fruits will develop out of 30–100 female/bisexual flowers. The fruit development curve is sigmoid, with maximum growth at 10–40 days after flowering; maturation with little further expansion occurs during the last 10 days when sugars accumulate in the fruit flesh and the net tissue on the fruit surface develops. Fruits mature 75 days (early cultivars of reticulatus group and cantaloupe group) to 120 days (inodorus group) after sowing. In the fruits of reticulatus group and cantaloupe group, ethylene plays an essential role in the ripening process (climacteric), for example, for flesh softening, yellowing of the rind, and abscission from the pedicel. Shelf life of these fruits is short (<1 week for "Charentais") to medium (2–3 weeks for "Galia"). Fruits of cultivars of inodorus group do not produce ethylene during ripening (non-climacteric) and consequently have a long 3 months for "Piel de Sapo"). Ethylene-independent ripening processes are flesh coloration and accumulation of sugars and organic acids (Robinson and Decker-Walters 1997).

2.6.8 ECOLOGY

Wild *C. melo* plants occur in open woodland, especially along rivers, and as a weed in fields and waste places, up to 1200 m altitude.

Melon requires warm and dry weather with plenty of sunshine for growth and production. The optimum temperature range is 18°C–28°C, growth being severely retarded below 12°C. Melon easily withstands several hours per day of very high temperatures, up to 40°C. Plants are killed instantly by frost. In snake melon, stem elongation was found to be greater under short 8 h days than under 16 h days. High humidity will reduce growth, adversely affect fruit quality, and encourage leaf diseases. Melon grows best on deep, well-drained, and thoroughly cultivated fertile loamy soils with pH 6–7. It does not tolerate very acid soils or waterlogged soil (Robinson and Decker-Walters 1997).

2.6.9 CULTIVATION AND MANAGEMENT

Melon can be grown in normal upland conditions, provided that it is rotated with noncucurbit crops to avoid soilborne diseases and nematodes. The soil should be ploughed, harrowed, and rotavated to attain a well-pulverized and well-leveled soil. Furrow or drip irrigation is common since plants have a high demand for water until the fruits have reached maturity. Melon is less dependent on daily irrigation than cucumber or pumpkin. During the dry season, 1 L water per planting hole may be given. In Sudan, an established snake melon crop is irrigated every 10–12 days during the hot rainy season (March–October) and every 14–18 days during the cool season.

Fertilizer requirements depend on crop performance and nutrient status of the soil. Removal of nutrients in a harvest of 20 t/ha of fruits is N 60–120 kg, P 9–18 kg, K 100–120 kg, Ca 70–100 kg, and Mg 10–30 kg. Melon responds well to organic manures applied at 25–30 t/ha. A complete fertilizer should be applied before sowing or planting, followed by a N topdressing when the stems are 20–30 cm long. Melon is particularly sensitive to Ca deficiency, which causes glassiness or watercore in the fruits. It is also sensitive to molybdenum deficiency occurring in ferralitic soils. Melon is very sensitive to a number of herbicides, including Atrazine, and may even be damaged by herbicide residues from preceeding crops.

Mulching is a well-established practice in the production of melon. In subtropical areas, black, transparent, or silver-painted polythene sheets are commonly used not only to control weeds but also to raise or lower the temperature of the soil. In the tropics, common mulching materials are rice straw or grass and increasingly plastic foil. In areas where mulching materials are not available, weeding is necessary until the plants start producing long stems. Hand hoeing or pulling of large weeds is often practiced. Various methods of pruning primary and secondary branches are applied to regulate vegetative growth and fruit set (three to five fruits/plant). Snake melon is sometimes grown along a trellis to get straight fruits (Rubatzky and Yamaguchi 1997).

Melon is usually direct-seeded: two to three seeds per hole, sown 2–4 cm deep on mounds or ridges, later thinned to one plant. Spacing is 50–60 cm within and 120–200 cm between the rows, giving a density of 8,000–16,000 plants/ha. Alternatively, seedlings are raised in polythene pots or in soil blocks and transplanted carefully to the field when 4 weeks old, taking care not to damage the root system. The weight of 1000 seeds is (8–) 25–35 g. Seed rates per ha are 1.5–2 kg for direct-seeded melon and 0.5 kg for the transplant method. In Sudan, snake melon is sown directly on raised flats of 2 m wide, in holes on both sides of the flats (Huang et al. 2012).

2.6.10 DISEASES AND PESTS

Several diseases may affect melon. Gummy stem blight (*Didymella bryoniae*) causes stem and fruit stalk canker, fruit rot, and plant wilting. Is a serious disease in humid and hot conditions. It is controlled by the use of disease-free seed, seed disinfection, crop rotation, spraying fungicides, and especially planting resistant cultivars. Powdery mildew (*Sphaerotheca fuliginea* and *E. cichoracearum*) can be controlled by fungicides, but modern F_1 hybrids have high tolerance to most types. Downy mildew (*P. cubensis*) is important in hot and humid climates and can be controlled by fungicides; polygenically controlled resistance is available in certain Indian accessions. Anthracnose (*Colletotrichum lagenarium*) can be controlled by seed treatment, crop rotation, and fungicides. Damping-off (*Pythium* sp. and *Rhizoctonia* sp.) has to be prevented by treating seed with fungicides. Bacterial soft rot or bacterial wilt (*Erwinia tracheiphila*) is controlled by removing affected plants and by eliminating the vector (striped and spotted cucumber beetle) with insecticide sprays. Angular leaf-spot (*Pseudomonas lachrymans*) is primarily a cucumber disease but occurs occasionally also in melon. *Fusarium* wilt (*Fusarium oxysporum* f.sp. *melonis*) can be effectively prevented only by resistant cultivars. In snake melon, it is also a serious problem. It is most aggressive at lower temperatures (18°C–20°C) and does not occur in the lowlands. Sudden wilt caused by the soilborne fungus *Monosporascus cannonballus* has become a serious problem in areas with a subtropical climate, neutral or alkaline soils, high (30°C–35°C) soil temperatures, and use of plastic mulch; it has not yet been recorded in tropical Africa but is a potential menace.

The most frequent virus in tropical conditions is the aphid-transmitted papaya ringspot virus (PRSV-W, formerly WMV-1), for which good resistance is available. Cucumber mosaic virus (CMV), watermelon mosaic virus (WMV-2), and zucchini yellow mosaic virus (ZYMV), all three transmitted by aphids, particularly *Aphis gossypii*, affect melon and predominate in subtropical conditions; there are various sources of resistance to these three viruses and also to the vector. Other virus diseases in melon are muskmelon necrotic spot virus (MNSV), transmitted by the soil fungus *Olpidium* sp., the soil and seedborne cucumber green mottle mosaic virus (CGMMV), and beet curly top virus (BCTV) transmitted by leafhoppers.

Root-knot nematodes (*Meloidogyne* spp.) can be a serious problem when melons are grown without proper crop rotation; control can be done by soil solarization or by wide-spectrum soil fumigants, but the latter are expensive and hazardous to the environment.

Pests in melon are thrips (*Thrips palmi* and *Frankliniella* spp.), spider mite (*Tetranychus urticae*), aphids (*A. gossypii*), melon fruit fly (*Dacus cucurbitae*), cucumber beetles (*Diabrotica* spp.), leaf folder (*Diaphania indica*), leaf feeder (*Aulacophora similis*), and the fly *Bactrocera cucurbitae*, which is especially active in the humid tropics and causes young fruits to drop by tunnelling in the pedicel. Farmers usually control these pests with insecticides. However, indiscriminate use of insecticides only aggravates pest problems by destroying useful parasitic insects (Nonaka and Ezura 2015).

2.6.11 HARVESTING AND YIELD

Cantaloupe and muskmelon tend to separate from the pedicel at the base of the fruit at maturity due to the formation of an abscission layer. This is called "full slip." Harvesting occurs usually at the "half slip" stage. Fruits of inodorus group do not form an abscission layer, and maturity is indicated by color change, for example, from green to yellow. Tender, immature, pale green fruits of snake melon are harvested starting 45–60 days from sowing. Harvested fruits are about 20 cm long with a diameter of 3 cm, weighing 90–100 g. If the fruits are left for seed, they are harvested when fully ripe.

On average, yields of fresh melon fruits reach 18 t/ha, ranging from 5 to 40 t/ha depending on cultivar and cultural practices. Seed yields are about 300–500 kg/ha for open-pollinated and 100–200 kg/ha for hybrid cultivars. The yield of a snake melon crop in Sudan is on average 20 t/ha.

Muskmelon fruits for storage should be cooled to 10°C–15°C immediately after harvesting to retard ripening. Storage for 10–15 days at 3°C–4°C (90% relative humidity) is possible, but lower temperatures can cause chilling injury. "Honeydew" and other winter melon fruits can be stored at 10°C–15°C for longer periods, some cultivars up to 90 days. Heavily netted melon fruits (e.g., the Mediterranean "Galia" and the American "Western Shipper") are relatively resistant to handling and transport. The fruits of snake melon are packed in plastic or jute sacs for transport to nearby town markets. They are treated in the same way as cucumber fruits (Rubatzky and Yamaguchi 1997; Simsek and Comlekcioglu 2013).

2.6.12 GENETIC RESOURCES AND BREEDING

The genetic diversity within *C. melo* is fairly well preserved in germplasm collections. The most extensive collections are maintained in the United States (North Central Regional Plant Introduction Station; Ames, IA), the Russian Federation (N.I. Vavilov All-Russian Scientific Research Institute of Plant Industry, St. Petersburg, Russia), and China (Institute of Crop Germplasm Resources [CAAS], Beijing, China), but many other countries hold significant collections. In Africa, important collections are held at National Horticultural Research Institute, Ibadan, Nigeria, and Agricultural Research Corporation, Wad Medani, Sudan. The melon germplasm collections preserved in Wad Medani include about 70 accesssions of snake melon and 45 of the local vegetable melon "tibish." The germplasm collections could be complemented by further collection of germplasm in the secondary centers of genetic diversity in Afghanistan, India, China, Pakistan, and Sudan. Some cultivar groups with small fruits close to wild types appear to be particularly good sources of host resistance to major melon diseases (Tindall 1983; Andres 2003).

Much of the cultivar improvement in melon is based on mass and line selection in open-pollinated populations. However, these are now rapidly giving way to F_1 hybrid cultivars, especially in Europe, the United States, Japan, and Taiwan. Pure-line development in melon is easy as there is practically no inbreeding depression after repeated selfing. On the other hand, there is also little hybrid vigor in hybrids between inbred lines. The main advantages of F_1 hybrids are, however, uniformity of plant and fruit type and recombination of favorable characteristics of different melon types in one genotype: fruit quality (round shape, good flavor, high sugar content, small seed cavity), long shelf life, adaptation to more humid climates, and especially resistance to diseases and pests.

Most cultivars have male as well as bisexual flowers on the same plant, and F_1 hybrid seed production requires emasculation of the bisexual flowers followed by hand pollination. Monoecious plant types would enable hybrid seed production with bee pollination, as the female parent line can be temporarily induced to produce female flowers only by sprays with ethrel. However, the change to monoecious F_1 hybrids is slowed down by the fact that monoecy in melon is linked to elongated shape and large size of the fruits, while the aim for cantaloupe and muskmelon is round and compact fruits. However, monoecious F_1 hybrids are now becoming increasingly common.

The main breeding objectives for snake melon are disease and insect resistance and better fruit quality. Breeding for resistance is a top priority because the fruit is often consumed fresh, making it hazardous to use chemicals, especially where restrictions are not observed. Breeding for quality aims at producing cultivars with slender and tender fruits for the fresh consumption in green salads.

In a recent evaluation study of melon germplasm collected from Sudan, several snake melon accessions were found, including resistant plants to some types of Fusarium wilt and *S. fuliginea*. Crosses with normal melon are easy and occur spontaneously. Resistance to Fusarium wilt was also detected in some accessions of tibish group and wild populations of *C. melo*, known locally in Sudan as "humaid." Resistance to some viral diseases, especially ZYMV, was also detected in some accessions of "humaid" (Ezura et al. 2002; Garcia-Mas et al. 2012; Pitrat 2013).

2.6.13 PROSPECTS

Melon is well liked by most people, and the importance of this crop will further increase through better adaptation to hot and humid growing conditions. A factor limiting melon production is the multitude of diseases (viruses in particular) and pests. New techniques from cellular (protoplast fusion) and molecular (genetic transformation, DNA markers) biology are now within reach of the melon breeders. This has opened up prospects for exploiting germplasm from other *Cucumis* species for disease and pest resistance and other characteristics not available through conventional interspecific hybridization.

Snake melon is adapted to hot and dry conditions, which makes it interesting for further expansion to some African regions outside Sudan. Germplasm collection and breeding need more attention (Stepansky et al. 1999; Pitrat 2013).

2.7 *CUCUMIS METULIFERUS* E.MEY. EX NAUDIN

2.7.1 INTRODUCTION

Vernacular names are horned melon, African cucumber, horned cucumber, kiwano (En). Concombre cornu, métulon, kiwano (Fr). Maxije (Po).

C. metuliferus occurs naturally throughout the tropical and subtropical sub-Saharan regions of Africa, from Senegal to Somalia and South Africa. It has also been recorded in Yemen. In Kenya, New Zealand, France, and Israel, the fruits of improved cultivars are commercially grown for export. *C. metuliferus* has become naturalized in Australia and is reported as adventive in Croatia.

2.7.2 USES

The fruits of horned melon are mainly eaten, and in some parts of Africa, the leaves are also used as a vegetable. The fruits are peeled and eaten in either the immature or the mature stages. Fruits in the unripe stages have the appearance and taste of cucumber. Mature fruits may have a sweet dessert-fruit flavor. Mature fruits may also be split open and dried in the sun for later use. In Botswana, the

Kalahari San people prepare the fruits by roasting. In Zimbabwe, young leaves are stripped from the stems, washed and boiled as spinach, in the same way as musk pumpkin leaves (*Cucurbita moschata* Duchesne), adding peanut butter prior to serving.

Fruits from wild-growing plants are often bitter and inedible. Traditional medical practitioners in Zimbabwe consider the bitter wild fruits as poisonous if taken by mouth. The root is used in the Mutare area (Zimbabwe) for the relief of pain following childbirth. In Benin, the fruit is said to possess medico-magical properties and is used to treat eruptive fevers in "Sakpata voodoo" rituals. The decorticated fruit macerated in distilled palm wine or lemon juice is used to treat smallpox and skin rashes.

In Western countries, *C. metuliferus* is currently mostly marketed as an ornament for its decorative fruit, with a unique appearance and extended keeping qualities (Metcalf and Rhodes 1990; Offiah et al. 2011; Lim 2012).

2.7.3 PRODUCTION AND TRADE

In southern Africa, horned melon is considered a traditional fruit vegetable. Cultivation has been on a small scale, for example, in Zimbabwe, it is cultivated in rural and peri-urban areas for sale in traditional markets and by street vendors. The development of the African horned melon into an international crop started in New Zealand, where it has been commercially grown and exported since the 1980s. There it was given the name "kiwano" in an attempt to promote the new fruit crop in Japan and the United States. Since then, it has also been grown commercially to a limited extent in California for the U.S. market and in Israel and Kenya from where the fruits are exported to markets in Europe. Recent efforts to grow the crop during the summer in southern France for the European market have been successful (Robinson and Decker-Walters 1997).

2.7.4 NUTRITIONAL PROPERTIES

The nutrient content for fresh fruits of *C. metuliferus* (per 100 g edible portion) is water 91.0 g, energy 134 kJ (32 kcal), protein 1.1 g, fat 0.7 g, carbohydrate 5.2 g, crude fiber 1.1 g, Ca 11.9 mg, Mg 22.3 mg, P 25.5 mg, Fe 0.53 mg, thiamin 0.04 mg, riboflavin 0.02 mg, niacine 0.55 mg, and ascorbic acid 19 mg (Wehmeyer 1986). Some values may vary depending on fruit maturity as the fruits are eaten at both immature and mature stages.

The leaf composition is approximately the same as other dark green leafy vegetables.

In wild *C. metuliferus*, plants with bitter and nonbitter fruits occur, and the two types are morphologically indistinguishable. A significant proportion of wild-growing horned melon plants encountered in southern Africa are bitter-fruited and have caused poisoning. Bitter, mature fruits may remain completely intact on the plants, as neither baboons nor other wildlife eat them. The amount of bitterness varies in immature and mature fruits on the same plant, with younger fruits having a less bitter taste. Bitterness is due primarily to the presence of cucurbitacins, bitter and toxic compounds occurring in Cucurbitaceae. Cucurbitacins can cause severe illness and death due to their potent action as purgatives and laxatives. *C. metuliferus* contains cucurbitacin B, a triterpene known to exhibit cytotoxic, antitumor, and anti-inflammatory activities (Lim 2012).

2.7.5 BOTANICAL DESCRIPTION

Vigorous annual herb with climbing or prostrate stems, having solitary, simple tendrils 4–10.5 cm long; root system strong, fibrous; stems reaching several meter in length, grooved, with long stiff spreading hairs. Leaves alternate, simple; stipules absent; petiole 3–12 cm long, setose; blade ovate or pentagonal in outline, 3.5–14 cm × 3.5–13.5 cm, shallowly palmately three to five-lobed, hispid setulose especially on veins below, becoming scabrid-punctate. Flowers unisexual, regular, five-merous; sepals filiform, 2–4 mm long; petals united at base, 0.5–1.5 cm long, yellow; male flowers in 1–10-flowered fascicles, with pedicel 2–18 mm long, stamens three; female flowers solitary, with pedicel 5–35 mm long, ovary inferior, ellipsoid, 1–2.5 cm long, covered with large soft spines, stigma three-lobed. Fruit an oblong-cylindrical berry 6–16 cm × 3–9 cm, on a stalk 2–7 cm long, rounded at both ends and beset with stout, broad-based, spiny protuberances 1–1.5 cm long, dark mottled green, ripening through yellow to bright orange, many-seeded. Seeds narrowly ovoid, 5–8 mm long, compressed with rounded margins, sericeous hairy.

Most of the approximately 30 *Cucumis* species are native to Africa, but the cucumber (*C. sativus* L.) probably originates from India. *C. metuliferus* with its "horned fruits" and hairy seeds is genetically more closely related to melon (*C. melo* L.) than cucumber but has proven to be cross-incompatible with other species. Based on meiotic and crossing studies, flavonoid patterns, chloroplast DNA data, and isozyme analyses, *C. metuliferus* is isolated from the other species in the genus. Specimens with nonbitter fruits, totally lacking spiny protuberances, have been observed both in the wild and under semi-cultivated conditions near Bulawayo (Zimbabwe) (Kirkbride 1993).

2.7.6 GROWTH AND DEVELOPMENT

At optimum temperatures of 20°C–35°C, seed germination takes place in 3–8 days. Below 8°C germination is completely inhibited. Vegetative stem growth, either climbing or sprawling, exhibits typical cucurbit exuberance, and the plants are capable of smothering nearby plant growth. Flowering starts about 8 weeks after sowing, with male flowers appearing first, followed after several days by female flowers. Pollination is by insects. In experiments in Israel, maximum fruit weight (on average 200 g) was reached 30–40 days after pollination, and the main period of fruit ripening on the plant in terms of changes in fruit constituents and color (from green to yellow) occurred 37–51 days after pollination. Under field conditions, time from sowing to harvest was 3.5 months.

In Zimbabwe, sweet-fruited plants reseed themselves with little management and protection. Fruits continue to be sweet, barring the occasional pollination by bitter-fruited wild plants. Fruits of horned melon in southern Africa continue ripening after the cessation of the rainy season, long after the stems have died back.

2.7.7 ECOLOGY

The natural habitat of horned melon ranges from low-altitude riverine semi-evergreen forest to semi-arid highlands and Kalahari sands. Horned melon is a warm-season grower in tropical to subtropical regions and does not tolerate cold conditions. It occurs at altitudes from near sea level to 1800 m. In southern Africa, seeds germinate with the summer rains when night temperatures are above 12°C. A semi-arid climate with a warm-season rainfall regime appears to enhance the fruit ripening stage, allowing fruits to develop their full flavor. Plants tolerate a wide range of soil types throughout their natural distribution area (Robinson and Decker-Walters 1997).

2.7.8 CULTIVATION AND MANAGEMENT

Horned melon may be grown under field conditions similar to cantaloupe melon or cucumber. In Spain, field-grown plants using supports did not produce as satisfactorily as those grown without, and this was thought to be due to climatic factors, principally the wind. Greenhouse planting is an option in which case pollinators must be introduced at the time of flowering. According to local soil conditions and soil test recommendations, compost, manure, or inorganic fertilizers can be incorporated (Benzioni et al. 1993).

2.7.9 PROPAGATION AND PLANTING

C. metuliferus is propagated by seed. The weight of 1000 seeds is about 14 g. Seeds may be sown directly, or seedlings are transplanted when they have two true leaves. The optimum time of planting is when soil and air temperatures are above 14°C. A planting density of 10,000 plants/ha produced good yields in Israel. In Spain, direct planting of seed retarded production; the use of transplants with well-developed root systems was recommended.

2.7.10 DISEASES AND PESTS

In southern Africa, horned melon plants are seldom affected by diseases or pests in their natural habitat. *C. metuliferus* is susceptible to cucumber mosaic virus, tobacco ringspot virus, tomato ringspot virus, watermelon mosaic virus 2, and a severe strain of bean yellow mosaic virus. Some accessions are susceptible to Fusarium wilt (*F. oxysporum*). Plantings in Israel were affected by powdery mildew (*S. fuliginea*) and squash mosaic virus. Greenhouse plantings in Spain, with high temperatures and humidity, were affected by powdery mildew (*E. cichoracearum*) and the greenhouse white fly (*Trialeurodes vaporariorum*), but field plantings were unaffected. African horned melon is resistant to the musk melon yellow virus. Some accessions are highly resistant to watermelon mosaic virus 1 due to a single completely dominant gene and hypersensitive-resistant to squash mosaic virus.

An orange and black cucurbit beetle, *Sonchia pectoralis*, has been observed damaging the leaves of young plants but not to the extent of harming overall growth. The ubiquitous pumpkin fly, which ravages other cucurbit crops in southern Africa, does not attack horned melon.

Horned melon is highly resistant to root-knot nematodes (*Meloidogyne* spp.). Resistance to powdery mildew, melon aphid (*A. gossypii*), greenhouse white fly, and Fusarium wilt has been recorded in several accessions (Marsh 1993; Walters et al. 1993).

2.7.11 HARVESTING AND YIELD

Stems of horned melon die at the end of the growing season, while the fruits remain attached and continue ripening to a bright orange color. They may be harvested over successive months. Immature fruits may be harvested at any time during the growing period. Care is needed during picking because the stiff sharp hairs on the stems and the spiny "horns" on the fruits can easily puncture the skin; it is recommended that gloves be worn for harvesting. For home consumption, leaves are picked from plantings or are collected from wild plants.

There are no records on yield from Africa. In New Zealand, growers harvest up to 20 t/ha of horned melon fruits, in California about 8 t/ha. Growers in Israel harvested approximately 230,000 fruits/ha, fruits weighing on average 200 g, totaling 46 t/ha. Results of experiments conducted in Spain showed that each plant produced on average 66 fruits weighing 15 kg.

Fruits should not be stacked without protective covering; the sharp spines easily puncture the skin of other fruits, causing a dark-orange to reddish discharge from the wounds. The spines may be rendered less harmful by use of sandpaper or a file. Fruits have an exceptionally long keeping quality at room temperature and may be kept for many months without losing their decorative appeal. Fruits picked at the onset of ripening and kept at 24°C were undamaged after three months of storage, though they failed to develop the desired orange color. Ethylene treatment resulted in fruit color changing from green to yellow in three days but had no effect on total soluble solids levels. There was a rise in reducing sugars during storage, unrelated to ethylene treatment. Fruit ripened on the stem showed higher values for total soluble solids and reducing sugars. At 20°C, 30% spoilage among stored fruits occurred by day 37, and at storage temperatures below 12°C, all ripe fruits spoiled within 55 days. Chilling symptoms in the form of opaque rind spots appeared on the fruit surface when fruits were stored at 4°C; cold storage is, therefore, not recommended (Robinson and Decker-Walters 1997).

2.7.12 GENETIC RESOURCES AND BREEDING

Germplasm of horned melon is held at the National Plant Genetic Resources Centre in Windhoek, Namibia, and in the United States (Department of Agriculture, North Carolina Plant Introduction Station). In the wild, *C. metuliferus* is widespread and occurs in a variety of habitats, so there is no reason to consider it liable to genetic erosion.

Fruit quality in horned melon is measured by color, size, taste, acidity, and aroma. Original cultigens tested were found somewhat lacking in the taste factor; however, these lines were being tested for plant vigor and pest and disease resistances. There is a need to identify sweet-fruited cultivars. More recent studies with germplasm from Botswana and Zimbabwe showed promising results for increased size and improved taste. Within the germplasm being grown in Zimbabwe and South Africa, cultivars are found with large (up to 18 cm) fruits; they are orange when ripe and have pleasantly tasting flesh, used as an attractive ingredient in fruit salads. *C. metuliferus* possesses important genes for disease and pest resistance that would be of benefit to the gene pool of other commercially important *Cucumis* species, that is, musk melon and cucumber, if they could be transferred. However, many attempts by various research groups to introduce these genes using traditional sexual hybridization methods have not been successful due to strong incompatibility barriers. Neither embryo culture nor somatic hybridization by protoplast fusion has produced successful results to date (Fassuliotis and Nelson 1988; Kirkbride 1993; Schippers 2000; Roy et al. 2012).

2.7.13 PROSPECTS

The prospects for horned melon as a subsistence leafy vegetable are rather poor, but as a cultivated cucumber-like vegetable or dessert fruit, it has a bright future. The immature green fruits are highly prized in Zimbabwe. Because of the somewhat insipid taste of the cultivars being grown, it has not caught on as a dessert fruit or cucumber substitute in the United States and Europe to the extent that marketers had hoped. Until the flavor can be improved to the satisfaction of the consumer, the present marketing technique for horned melon is as a decorative ornamental fruit. Given its unique form and appearance, together with its appealing orange-ripe color plus an extended shelf life, *C. metuliferus* rightfully deserves centerpiece fruit-bowl attention. The preferred climate pattern of warm-season rain followed by a dry cool season could be an important consideration in any future successful commercial growing venture (Schippers 2000; Offiah et al. 2011; Quiroz and van Andel 2015).

2.8 *CUCUMIS MYRIOCARPUS* NAUDIN

2.8.1 INTRODUCTION

Synonym is *Cucumis leptodermis* Schweick. (1933). Common names are gooseberry cucumber and, prickly paddy cucumber (English).

Cucumis myriocarpus originates from Zambia, Botswana, Zimbabwe, Mozambique, South Africa, and Lesotho. It has been introduced in several other regions and has become naturalized in southern Europe, California, and Australia. Locally it is considered a weed; in California, it has been declared noxious.

2.8.2 USES

Leaves of *C. myriocarpus* are collected in the wild for use as a cooked vegetable. The fruit pulp is widely used as an emetic and purgative. Overdosing and/or inclusion of seeds in the pulp have been blamed for fatal cases of poisoning. *C. myriocarpus* is suspected of causing photosensitization in sheep and blindness in cattle. However, reports on medicinal use and toxicity of *C. myriocarpus* are suspected of wrong identification and may refer to other species (Sola et al. 2014).

2.8.3 NUTRITIONAL PROPERTIES

The nutritive value of the leaves is unknown, but it is probably comparable to *C. africanus* L.f. leaves and other dark green leaf vegetables. Fresh fruits of *C. myriocarpus* contain cucurbitacins A and D. Cucurbitacins, which are known from many Cucurbitaceae and various other plant species, exhibit cytotoxicity (including antitumor activity), anti-inflammatory and analgesic activities.

2.8.4 BOTANICAL DESCRIPTION

Annual, monoecious, prostrate, or scandent herb, with simple tendrils; stems up to 2 m long. Leaves simple, alternate; petiole 1.5–10 cm long; blade broadly ovate, very deeply palmately five-lobed, 2.5–10 cm × 2–7.5 cm, shallowly cordate at base, lobes elliptical, each again rather deeply three to five-lobed. Flowers unisexual, regular, five-merous; receptacle 2–5 mm long; sepals 1–3 mm long; petals 3–6(–10) mm long, pale yellow; male flowers one to two (rarely more) together, pedicel 0.5–2.5 cm long, stamens three; female flowers solitary, together with male, pedicel 1.5–4.5 cm long, ovary shortly and softly spiny. Fruit a globose to ellipsoid berry 3–6(–9) cm × 1.5–2.5 cm, with spines up to 2 mm long; fruit stalk 4.5–7 cm long, slender. Seeds ellipsoid, 5–6 mm × 2.5–3 mm × 1–1.5 mm.

The genus *Cucumis* includes about 30 species, 4 of which are economically important: cucumber (*C. sativus* L.), melon and snake cucumber (*C. melo* L.), West Indian gherkin (*C. anguria* L.), and horned melon (*C. metuliferus* Naudin). *C. myriocarpus* is placed in the "anguria" group of the subgenus *melo*. Two subspecies are distinguished based on fruit characteristics: subsp. *leptodermis* (Schweick.) C. Jeffrey & P. Halliday is confined to South Africa, subsp. *myriocarpus* has a much wider distribution and differs from the former by its fruits having a more distinct pattern of stripes (dark green, pale green, white, purplish, brown and rusty orange) and being densely covered with spines (Mafeo 2014).

2.8.5 ECOLOGY

C. myriocarpus is found in open localities in grassland and wooded grassland, especially on sandy soils at altitudes of 350–2000 m. It is frequently found as a weed of cultivated land and irrigation furrows (Nkgapele and Mphosi 2012).

2.8.6 CULTIVATION AND MANAGEMENT

In southern Africa, *C. myriocarpus* is an important weed in maize fields. Poisoning of cattle grazing the stubble of weedy fields is a serious risk. As a weed, it can easily be controlled mechanically (Pofu et al. 2013; Hallett et al. 2014).

2.8.7 GENETIC RESOURCES AND BREEDING

As *C. myriocarpus* is widespread and common, it is not threatened by genetic erosion. There are genebank accessions recorded in Germany, the Czech Republic, Spain, the United Kingdom, and the United States.

Within the "*anguria*" group of about 16 spiny-fruited *Cucumis* species to which *C. anguria* belongs as well, there are no major barriers to gene exchange. Several interspecific crosses have been made in this group. Individual plants of *C. myriocarpus* have been found with intermediate resistance to downy mildew (*P. cubensis*) and resistance to scab (*Cladosporium cucumerinum*) (Kirkbride 1993; Telford et al. 2011).

2.8.8 PROSPECTS

As a vegetable, *C. myriocarpus* will remain only locally of some importance and its spread is to be discouraged. It is promising as a source of disease and pest resistance for economically important *Cucumis* species (Schippers 2000).

2.9 *CUCUMIS NAUDINIANUS* SOND. (1862)/ *ACANTHOSICYOS NAUDINIANUS*

2.9.1 INTRODUCTION

A. naudinianus belongs to the family Cucurbitaceae with the chromosome number n = 11. *A. naudinianus* has as its synonyms: *Cucumis naudinianus* Sond. (1862), *Citrullus naudinianus* (Sond.) Hook.f. (1871), *Colocynthis naudianus* (Sond.) Kuntze (1891); it has the vernacular names: herero cucumber, gemsbok cucumber, wild melon. It is native to Zambia, Angola, Namibia, Botswana, Zimbabwe, Mozambique, and South Africa.

The mature fruits of *A. naudinianus* are eaten raw or roasted; unripe fruits cause a burning sensation of the tongue and lips when eaten raw. The fruit also provides an important source of water. The fruit skin and the seeds are roasted and pounded to make a meal. The tuberous roots are considered inedible or even poisonous, and in Zambia, they have been reportedly used for homicidal purposes. The preparation and use of arrow poison made from the roots of *A. naudinianus* is widespread among bushmen tribes in Angola, Namibia, and Botswana.

Some *A. naudinianus* plants produce bitter fruits. The bitter taste is attributed to cucurbitacin B (c. 0.001%). Fruits contain per 100 g: water 90.6 g, energy 111 kJ (27 kcal), protein 1.3 g, fat 0.2 g, carbohydrate 4.8 g, fiber 2.1 g, Ca 21 mg, Mg 23 mg, P 25 mg, Fe 0.5 mg, thiamin 0.09 mg, riboflavin 0.03 mg, niacin 0.98 mg, and ascorbic acid 35 mg (Arnold et al. 1985). The seed kernel yields approximately 15% thin, yellow nondrying oil, and the residue contains approximately 20% protein. In the older roots, the total content of cucurbitacins amounts to 1.4%. Cucurbitacins, which are also known from other *Cucurbitaceae* and various other plant species, exhibit cytotoxicity (including antitumor activity) and anti-inflammatory and analgesic activities (Azimova and Glushenkova 2012).

2.9.2 BOTANICAL DESCRIPTION

A. naudinianus is a perennial, dioecious, scandent herb with solitary, spiniform tendrils; root tuberous, up to 1 m long; stem annual, up to 6 m long, rooting at the nodes, glabrescent. Leaves alternate, simple; stipules absent; petiole 0.7–7.5 cm long; blade ovate to broadly ovate in outline, usually deeply palmately five-lobed, 3–18 cm × 2.5–14 cm. Flowers solitary, unisexual, five-merous; petals yellow to white, 1.4–2.5 cm × 0.9–1.3 cm; male flowers with pedicel up to 2 cm long, receptacle campanulate, up to 6 mm long, pale green, sepals up to 6 mm × 1.5 mm, stamens three or five; female flowers with pedicel up to 8 cm long, receptacle cylindrical, 3 mm long, sepals 3–4 mm long, three small staminodes, ovary inferior, spiny. Fruit an ellipsoid or subglobose berry 6–12 cm × 4–8 cm, weight c. 250 g, fleshy, covered with seta-tipped fleshy spines, many-seeded. Seeds ellipsoid, slightly compressed, 7.5–10 mm × 4–6 mm.

Acanthosicyos comprises two species and is placed in the tribe *Benincaseae* together with important genera such as *Benincasa, Coccinia, Citrullus, Lagenaria,* and *Praecitrullus*. The better known nara melon (*Acanthosicyos horridus*) differs notably from *A. naudinianus* by its shrubby habit and leafless, spiny stems and is restricted to Angola, Namibia, and South Africa (Neuwinger 1996; Ojo et al. 2010).

2.9.3 ECOLOGY

A. naudinianus is a typical Kalahari species that prefers deep sandy soils. It occurs in woodland, wooded grassland, and grassland at altitudes of 900–1350 m. It is not frost tolerant but tolerates a saline subsoil. The fruits of *A. naudinianus* are exclusively collected from the wild (Sepasal 2013).

2.9.4 GENETIC RESOURCES AND BREEDING

There is no indication that *A. naudinianus* is threatened. As in many other cucurbits, there is considerable variation in the bitterness of the fruits. This will allow for selection and breeding of more palatable lines. There are four documented accessions held in the United States and two at the Royal Botanic Gardens Kew (United Kingdom). In view of the increasing demands for edible oil and protein in arid lands, *A. naudinianus* is a candidate for development as a high-yielding, dry country crop. It yields a crop quickly; harvesting the fruits is easy; it has a wide ecological adaptation; it is easily propagated and handled; and fruits store well. As such, it compares favorably with *A. horridus* as a candidate for domestication (Chang et al. 2014; Chomicki and Renner 2015).

2.10 *CUCUMIS SATIVUS* L.

2.10.1 INTRODUCTION

Vernacular names are cucumber, gherkin (En). Concombre, cornichon (Fr). Pepino (Po). Tango (Sw).

C. *sativus* is believed to have originated in the southern Himalayan foothills region of Asia. The wild *C. sativus* var. *hardwickii* (Royle) Gabaev (synonym: *Cucumis hardwickii* Royle), which is seen as the possible progenitor, can still be found there. It has small, very bitter, spiny fruits and is fully compatible with *C. sativus*. An alternative view, however, suggests that var. *hardwickii* is a derivative that escaped from cultivation.

Cucumber is said to have been cultivated in India for at least 3000 years and in eastern Iran and China probably for 2000 years. China is considered a secondary center of genetic diversification. Cucumber was carried to Europe before our era and was introduced into the New World by early travellers and explorers. In tropical Africa, it probably arrived first in the west with the Portuguese. Cucumber is now cultivated worldwide. In tropical Africa, it can be found in all city markets and is common in most supermarkets.

2.10.2 USES

The main use of cucumber is as the immature fruit in salads either with the skin or peeled. Fruits are sliced or cut into pieces and served with vinegar or a dressing, on their own or mixed with other vegetables. Young fruits of special small-fruited cultivars called "gherkin" are pickled in vinegar. Young or ripe cucumber fruits are occasionally used as cooked vegetables or made into chutney. In Asia, types with large, white or yellow fruits are boiled and eaten as an ingredient of stews, young shoots are consumed as a leafy vegetable, and seeds are consumed or used to extract an edible oil, but these uses have not been recorded for Africa. In tropical Africa, cucumber is considered an exotic or Western vegetable of relatively recent introduction, mostly used by city consumers. However, it is rapidly gaining popularity in the African kitchen; in East Africa, it is regularly used in "kachumbari," a kind of African coleslaw.

Ripe raw cucumber fruits are said to cure sprue, and in Indo-China, cooked immature fruits are given to children to treat dysentery. The seed has some anthelmintic property. Cucumber extract is known to have cleansing, soothing, and softening properties; it is used as an ingredient in a variety of health and beauty products for the skin. Cucumber peel, when eaten by cockroaches, is reported to kill them after several nights. Nonfood uses of cucumber are not common in Africa (Bates et al. 1990).

2.10.3 PRODUCTION AND TRADE

In 2002, the world area under *C. sativus* was estimated at about 2 million ha, with a total production of 36 million t. Asia is the world leader, with China alone accounting for over 60%. Cucumber is grown in all countries of tropical Africa but nowhere on a large scale. In 2002, Africa produced

507,000 t on 25,000 ha, accounting for just under 1.5% of production. Egypt is the largest African producer with 360,000 t. Detailed data on countries of tropical Africa are lacking. International trade in 2002 amounted to 1.5 million t, with Mexico, the Netherlands, and Spain as the main exporters; international trade from African countries is modest and unrecorded (FAO 2013).

2.10.4 NUTRITION PROPERTIES

The nutritional composition of cucumber per 100 g edible portion (ends trimmed, not peeled, edible part 97%) is water 96.4 g, energy 42 kJ (10 kcal), protein 0.7 g, fat 0.1 g, carbohydrate 1.5 g, dietary fiber 0.6 g, Ca 18 mg, Mg 8 mg, P 49 mg, Fe 0.3 mg, Zn 0.1 mg, carotene 60 µg, thiamin 0.03 mg, riboflavin 0.01 mg, niacin 0.2 mg, folate 9 µg, and ascorbic acid 2 mg (Holland et al. 1991). The edible portion is about 85% when peeled. Seed kernels contain approximately 42% oil and 42% protein.

The bitter principle cucurbitacin C occurs in *C. sativus*. Cucurbitacins are terpene components in the foliage and fruits, the evolutionary role being to protect the plant against herbivore attack. As a result of breeding, modern cultivars are not bitter. The presence of a saponin and the slightly poisonous alkaloid hypoxanthine might explain the anthelmintic property of the seed.

The fruits of *C. anguria* L., the West Indian gherkin, may replace those of *C. sativus* for pickling, and the fruits of snake melon (*C. melo* L.) for pickling and fresh use.

2.10.5 BOTANICAL DESCRIPTION

Annual, monoecious herb with trailing or scandent stems up to 5 m long, having simple tendrils up to 30 cm long; stem four to five-angled, sparingly branched, with bristle-like hairs; root system extensive and largely superficial. Leaves alternate, simple; stipules absent; petiole 5–20 cm long; blade triangular-ovate in outline, 7–20 cm × 7–15 cm, palmately three to seven-lobed, deeply cordate at base, acute at apex, toothed, bristly hairy. Flowers unisexual, regular, five-merous; sepals narrowly triangular, 0.5–1 cm long; corolla widely campanulate, lobes up to 2 cm long, yellow; male flowers in three to seven-flowered fascicles, with pedicel 0.5–2 cm long, stamens three; female flowers solitary, with pedicel short and thick up to 0.5 cm long, lengthening in fruit up to 5 cm, ovary inferior, ellipsoid, 2–5 cm long, prickly hairy or warty, stigma three-lobed. Fruit a pendulous, globose to cylindrical berry up to over 30 cm long, often slightly curved, beset with spinous tubercles and warts when young, skin usually green, but in some cultivars white, yellow or brown, flesh pale green, many-seeded. Seeds ovate-oblong in outline, 8–10 mm × 3–5 mm, compressed, white, smooth. Seedling with epigeal germination (Nicodemo et al. 2012).

With the chromosome number of 2n = 14, cucumber and its wild relative are different from all other members of the genus, which have 2n = 24. It is also the only *Cucumis* species thought to have originated in Asia; the other species are indigenous to Africa. *C. anguria* is often confused with the small cucumber types that are used for pickles, since both are commonly called "gherkin."

A satisfactory classification of the cultivated cucumber does not exist. A large variation of fruit shapes, sizes, colors, and rind characteristics can be found in different combinations, and numerous cultivars have been developed all over the world. Commonly cultivated types include

- *American slicer*: Fruits dark green, smooth-skinned but quite spiny, medium-sized; popular open-pollinated cultivars grown worldwide are "Marketmore 76," "Poinsett 76," "Ashley," and the hybrids "Cyclone" and the gynoecious "Dasher II"; the hybrid "Kande" was specifically developed for tropical climates by East-West Seed Company; popular open-pollinated cultivars as well as hybrids such as "Tokyo" and "Olympic" are distributed in Africa by Technisem.
- *European greenhouse cucumber*: Fruits very long, slim, nearly spineless but with a rough skin, grown in greenhouses; cultivars are all hybrids, gynoecious, and with parthenocarpic fruit, for example, "Mystica" and "Sabrina."

- *Beit Alpha*: Mainly grown in and around the Middle East; fruits medium-sized, with a somewhat ribbed though spineless skin; often gynoecious and/or parthenocarpic, for example, the hybrids "Basma" and "Excel," distributed in Africa by Technisem.
- *Pickling cucumber*: Usually a bit smaller than slicers, around 15 cm or less, often with prominent warts, used for production of pickled fruits (gherkins); common and popular cultivars are "Calypso" and "Eureka," bush types such as "Little Leaf" require less space and set fruits simultaneously.
- *White cucumber*: Grown in India, Sri Lanka, and other Asian countries; fruits with white smooth skin, medium- to large-sized; a popular cultivar is "Long White," hybrids are also available, for example, "Keisha" from East-West Seed Company and "Shivneri" from Seminis.
- *Asian or mottled cucumber*: Popular in many southern and eastern Asian countries; many hybrids are available in fruit sizes ranging from mini (around 7 cm), for example, "Kiros" (East-West Seed Company), to medium-sized, for example, "Ninja" (Chia Tai) and "Kasinda" (East-West Seed Company).
- *Chinese and Japanese (oriental) cucumber*: Fruits relatively long, slim, rather spiny; in southern China sometimes with black spines, in Japan perfect size 22 cm × 2–3 cm, white-spined; Chinese cultivars include "Beijing Dachi" and "Ganfeng 3" from GAAS, popular Japanese hybrid cultivars are "Sharp 1" and "Nao-Yoshi" from Saitama Gensyu Ikuseikai.

In tropical Africa, mainly slicing cucumber is grown, for which mostly the open-pollinated cultivars "Ashley" and "Poinsett" are used. Beit Alpha types are also grown, especially in northern Africa. Pickling cucumber is planted as well, but this may often be *C. anguria* instead of *C. sativus* (Bates et al. 1990).

2.10.6 GROWTH AND DEVELOPMENT

Germination takes three days at optimum temperatures. Flowering normally starts 40–45 days after sowing, but early cultivars such as "Kiros" can start flowering within 30 days. The female flowers develop later than the more numerous male flowers. The ratio male/female flowers largely depends on daylength, temperature, and cultivar. Long days and high temperatures tend to keep the plants in the male phase or change the ratio to a higher male proportion.

Several growth regulators can be used to influence sex expression; spraying of ethephon induces female flowering. Many modern cucumber cultivars are gynoecious (having only female flowers). To increase seeds of a gynoecious line, or to use it as a male parent, spraying with silver nitrate, silver thiosulfate, or gibberallic acid will induce male flowering. Concentration and duration of spraying depend on the genotype and the intended result; usually spraying can start at the two to three true leaf stage and can be repeated every two days for up to five times.

For gynoecious or highly female cultivars that are not parthenocarpic, commercial seed is usually mixed with 10%–15% of a highly male line. Bees are the main pollinating agents and should be sufficiently available for good fruit development. Poor pollination results in deformed or curved fruits. However, the European parthenocarpic greenhouse cucumber should not be pollinated since this will result in unwanted seeded fruits and fruit deformation. Greenhouses are, therefore, kept insect free to prevent pollination.

Fruits are harvested 1–2 weeks after flowering, depending on the genotype, usually before they are physiologically mature. Frequent harvesting of immature, marketable fruits will result in a continuation of new fruit set and a longer life cycle of the crop. Large, maturing fruits that are left on the plant inhibit the development of additional fruits. Very early, field-grown cultivars can senesce quickly and may die after only 2–3 months, especially when diseases start to affect the plants during fruit setting stage. The crop cycle of cucumber grown in glasshouses in Europe can be extended to around six months under specific conditions.

2.10.7 ECOLOGY

Cucumber requires a warm climate. In cool, temperate countries, it is grown in greenhouses; only during hot summers can it be grown in the open. The optimum temperature for growth is about 30°C and the optimum night temperature 18°C–21°C; the minimum temperature for good development is 15°C. Pickling cultivars are usually more adapted to low temperatures. Sensitivity to daylength differs per cultivar; short daylengths usually promote vegetative growth and female flower production. High light intensity is needed for optimum yields. Cucumber needs a fair amount of water but it cannot stand waterlogging. Low relative humidity results in high plant evaporation due to the large leaf area, and sufficient irrigation is then very important. High relative humidity facilitates the occurrence of downy mildew. The soil should be fertile, well drained, with a pH of 6.0–7.0. In tropical Africa, elevations up to 2000 m appear to be suitable for cucumber cultivation (Robinson and Decker-Walters 1997).

2.10.8 CULTIVATION AND MANAGEMENT

Planting on raised beds will improve drainage, which is especially important during the rainy season, and can support good root development. The use of plastic mulch makes weed control and water management easier and can help in reducing insect populations at an early stage. Weed control is necessary until the plants cover the soil entirely. Support (stakes) can be provided, which will generally improve fruit quality, reduce disease incidence through better air circulation in the crop, and make it easier to pick the fruits. Irrigation is required at short intervals; a high level of soil moisture should be maintained throughout the growing period. The use of drip irrigation is highly recommended for an optimum and uniform use of available water.

Fertilizers can be included in the drip system. Cucumber responds well to fertilizers. In addition to the initial organic manure, a general recommendation is 700 kg/ha of an NPK mixture, followed by N fertilizer every 2–3 weeks until the fruits form. However, it is always best to base fertilizer gifts on a soil analysis before planting. Micronutrients are also essential for a good development; shortages can result in strong deficiency symptoms in both plants and fruits, leading to low quality yields.

The tip of the main stem may be nipped off to encourage branching; in plants with very strong vegetative growth, lateral shoots may be pruned after the first fruits have formed to limit leaf and flower production. Excessive use of N promotes stem growth and the production of male flowers (Rubatzky and Yamaguchi 1997).

Cucumber is propagated by seeds. The 1000-seed weight ranges from 20 to 35 g. During soil preparation, generous incorporation of organic manure (about 25–35 t/ha) is required. About 1–3 kg of seed is needed per ha depending on the method of sowing. Direct sowing, which is still a common practice especially in open fields, requires larger amounts of seed. The use of transplants will result in a more uniform crop stand, if done properly. In open fields in the tropics, seedlings can already be transplanted after around 7 days or at the two-true-leaf stage, but in cooler areas or for greenhouse production, much older transplants of up to 33 days are used. When directly sown, cucumbers are planted on hills, 90–120 cm apart, with several seeds per hill and thinned to two to three plants, or they are sown in rows 1–2 m apart and thinned to 30 cm between plants. When planted as a ground crop, the wider distances are used, whereas for trellised crops, closer planting can be applied. Cucumber cultivated for pickles is planted closer, up to 250,000 plants/ha (Nwofia et al. 2015).

2.10.9 DISEASES AND PESTS

Many diseases and pests can affect cucumber in all stages of development. Leaf diseases that can result in serious damage are the fungal diseases downy mildew (*P. cubensis*), powdery mildew (*E. cichoracearum* and *S. fuliginea*), anthracnose (*C. lagenarium*), target leaf spot (*Corynespora cassiicola*), and gummy stem blight (*D. bryoniae*), as well as the bacterial disease angular leaf spot (*P. lachrymans*). Anthracnose also causes symptoms on fruits. Good air circulation, for example, through trellizing, reduces the incidence of these diseases to some extent. Spraying of systemic fungicides such as benomyl (Benlate) or metalaxyl (Ridomil) can reduce spread; they can be alternated with broad spectrum fungicides such as copper oxychloride (Vitigran Blue) or mancozeb (Dithane). Other wilting in cucumber may be caused by soilborne Fusarium wilt (*F. oxysporum* f. sp. *cucumerinum*), or bacterial wilt (*E. tracheiphila*), which is spread by cucumber beetles. In protected cultivation in temperate countries, especially in Japan and Korea, grafting of cucumber on *Cucurbita ficifolia* Bouché or *Cucurbita maxima* Duchesne × *C. moschata* Duchesne rootstock is often practiced to avoid soilborne diseases such as root rot caused by the fungi *Phomopsis sclerotioides* and *F. oxysporum*; no experience with grafting in tropical Africa has been reported. Cucumber is susceptible to damping off, resulting in seedling death soon after emergence; it occurs more often when the soil is poorly drained and can be caused by several fungi, for example, *Pythium* spp. or *Phytophthora* spp., some of which can also cause root rot in older plants.

Fruit damage can be caused by scab (*C. cucumerinum*), a fungus that also attacks the leaves of susceptible cultivars, by bacterial soft rot (*Erwinia*), phytophthora fruit rot (*Phytophthora capsici*), and belly rot (*Rhizoctonia solani*). Resistance to scab is widely available; to prevent soft rot and other rotting of fruits, fruits should be handled with care especially during harvest to prevent damage as much as possible. Wounds on fruits are often the starting points of infection. Belly rot is soilborne and infects the fruits at the place where they touch the soil; preventing contact, for example, through the use of mulch or trellis systems can prevent this disease. Other fungal diseases observed in tropical Africa (Côte d'Ivoire) are *Alternaria* sp., *Cercospora citrullina*, *Choanephora cucurbitarum*, *Myrothecium roridum*, *Oidium tabaci*, and *Sclerotium rolfsii*.

Commonly found viruses that can cause considerable yield losses in cucumber in the tropics are the aphid-borne cucumber mosaic virus (CMV), zucchini yellow mosaic virus (ZYMV), papaya ringspot virus (PRSV), and a range of whitefly transmitted viruses that cause yellowing, such as cucumber vein yellowing virus (CVYV) and cucumber yellows virus (CYV). Another important virus is cucumber green mottle mosaic virus (CGMMV), which is highly seed transmitted. Special care must be taken not to grow seeds produced on infected plants; CGMMV can be easily spread mechanically, but it is unknown whether there is an insect or other vector. Root-knot nematodes (*Meloidogyne* spp.) can severely affect plant growth, resulting in stunting or wilting, thereby reducing yields.

A general recommendation is to grow cucumber only on sites where no other cucurbits have been grown for a number of years, to prevent soilborne diseases. Several of the diseases mentioned, such as angular leaf spot, scab, anthracnose, and phytophthora, can be seedborne. Use of disease-free seed or seed treated with chemicals can prevent early disease infection or insect attack and will reduce risk levels considerably. For most leaf diseases and viruses, resistant cultivars are available; using the relevant ones is a good way to minimize problems.

Aphids, whitefly, and thrips are insects that can cause major problems, mainly because they act as vectors for viruses or diseases. General insect damage may be caused by beetles, leaf miners, and leaf hoppers. The melon worm or pyralid moth (*Diaphania hyalinata*) and red spider mites (*Tetranychus* spp.) may cause much damage to the leaves, and fruit flies (*D. cucurbitae*) may cause fruit rotting. Nonbitter cultivars are more susceptible to damage by spider mites than bitter ones. The use of natural insect enemies is a more environmentally friendly method than spraying chemicals against pests, but until now, it has mainly or exclusively been practiced in protected cultivation of cucumber (Gismervik et al. 2015).

2.10.10 HARVESTING AND YIELD

Cucumber fruits for fresh consumption are harvested before they are fully mature; depending on the type, this can be 1–2 weeks after flowering. The moment of first harvest is 40–60 days after sowing, depending on climate and cultivar. Harvesting is done every other day to every few days. For pickling types, immature fruits of several stages are harvested. Cucumber fruits for the fresh market are harvested by hand, but pickling types are also harvested mechanically all in a single harvest, especially on the large fields in the United States. For seed production, the fruits are allowed to mature on the plant.

In 2002, the average world yield for cucumber reached 18 t/ha, but the range is very wide. For Africa, few data are available; estimates for DR Congo and Ghana are 4 and 10 t/ha respectively. In tropical Asia, countries such as Thailand, Indonesia, and India have an estimated average yield of just below 10 t/ha. Hybrid cultivars in Thailand yield over 100 t/ha. The European Union as a whole produces an average of 90 t/ha, but under protected conditions in greenhouses this can be even higher, mainly because the crop's life cycle is extended considerably (Choi et al. 2015).

2.10.11 Postharvest Handling

Cucumber fruits should be treated with care as they are sensitive to transportation damage. The maximum storage period is about 14 days at 13°C with a relative humidity of 95%. Below 10°C, chilling injury may occur, and above 16°C, fruits rapidly become yellow. Waxing or packaging in plastic film reduces moisture loss. In tropical countries, fruits will usually keep an acceptable marketable quality for around 5 days unless they are stored under cool conditions. After that, they become soft and lose their crispy texture, and they can become yellowish (Sarwar et al. 2011).

2.10.12 Genetic Resources and Breeding

Important germplasm collections are available in the Czech Republic (Breeding Station, Kvetoslavov), Germany (Institute for Plant Cultivation and Plant Breeding, Braunschweig), India (Kerala Agricultural University, Trichur), the Netherlands (Centre for Genetic Resources, Wageningen), the Philippines (Institute for Plant Breeding, Los Baños), Turkey (AARIR, Menemen, Izmir), the Russian Federation (N.I. Vavilov Institute of Plant Industry, St. Petersburg), and the United States (NCRPIS, Iowa State University, Ames, IA; NSSL, USDA-ARS Colorado State University, Fort Collins, CO).

Genetically, cucumber is one of the best-known vegetable species and a great deal of work on breeding has been done. The first hybrid, the slicer "Burpee Hybrid," was released in 1945. Gynoecious sex expression was found in a Korean cultivar. Since then, the development of hybrid, gynoecious, and parthenocarpic cultivars has led to extremely high yields, especially in cucumber for fresh consumption cultivated in greenhouses. Breeding of disease- and pest-resistant cultivars, combined with better cultivation practices, has led to more than threefold increases in the yield of pickling cucumber over the past 60 years.

For tropical Africa, breeding work should aim at producing suitable cultivars for hot and humid conditions, with the necessary disease tolerances. The French company Technisem focuses on breeding for the tropics, mainly for West Africa, with experimental stations in, Senegal, Mali, Benin, and Cameroon under the name Tropicasem. For cucumber, the common open-pollinated slicer cultivars are sold, as well as improved hybrids in slicer, Beit Alpha, and pickling cucumber types. Depending on type preference, climate, and disease problems, various options are available. The F_1 hybrid "Tokyo" (slicer) is popular in West Africa. It tolerates hot and humid conditions and has tolerance to downy mildew and CMV. Another, also heat-tolerant, gynoecious hybrid is "Olympic," which in addition to downy mildew and CMV has tolerance to powdery mildew and angular leaf spot. The pickling cucumber hybrid "Antilla" is tolerant to heat, as well as to anthracnose, downy and powdery mildew. "Arizona F_1" is a less heat-tolerant pickler but adds angular leaf spot and CMV tolerance. The Beit Alpha type "F_1 Basma" has tolerance to both mildews and to CMV and WMV viruses. The hybrid "Excel" has a less vigorous plant and is parthenocarpic; it is especially suited for greenhouse growing.

Cucumber cultivars from East-West Seed Company, developed in Southeast Asia, are also adapted to hot and humid conditions. Available cultivars for Africa in the slicer type are the F_1 hybrids "Kande" and "Kosey," which have strong vigor and tolerance to downy mildew; "Kosey" has tolerance to ZYMV and PRSV.

Studies have shown that the genetic base within cultivated cucumber is rather narrow. Genebank accessions from various locations are, however, found to be genetically diverse, and often different from the commercial germplasm. More use of this material, as well as hybridization with related wild taxa, might be promising to obtain new desirable characteristics. Combinations with *C. sativus* var. *hardwickii* are made to increase branching and fruit set, especially in pickling cucumber cultivars. Interspecific crosses with *Cucumis hystrix* Chakrav. (2n = 24) have been attempted for quite some time. *C. hystrix* is rather similar to cucumber and especially interesting because of its resistance to nematodes. After embryo rescue techniques and chromosome doubling, an amphidiploid (2n = 38) was obtained, which is now being further developed to try to obtain lines directly crossable with *C. sativus* (Staub and Ivandic 2000; Sun et al. 2014; Afroz et al. 2015).

2.10.13 Prospects

Cucumber is quite important in tropical Africa and is becoming increasingly so because it is an easy-to-prepare vegetable and suited for sale in supermarkets and the big city markets. Moreover, cultivars are becoming available that are more adapted to the climatic conditions prevailing in tropical Africa and that also include the relevant disease tolerances (Sun et al. 2014).

2.11 *CUCUMIS ZEYHERI* SOND.

2.11.1 Introduction

Synonym is *Cucumis prophetarum* L. subsp. *zeyheri* (Sond.) C. Jeffrey (1962). Common name is wild cucumber (English).

C. zeyheri is restricted to the eastern part of southern Africa: Zambia, Zimbabwe, Mozambique, and South Africa (Jeffrey 1978).

2.11.2 Uses

The nonbitter fruits are eaten raw or pickled. The bitter fruits are used as a (drastic) purgative (Jeffrey 1978).

2.11.3 Nutritional Properties

There is no information on the nutritive value of the fruits, but it is probably comparable to cucumber. The fruits of *C. zeyheri* contain cucurbitacins B and D. Cucurbitacins, which are known from many *Cucurbitaceae* and various other plant species, exhibit cytotoxicity (including antitumor activity), anti-inflammatory and analgesic activities (Decker-Walters 2011).

2.11.4 Botanical Description

Perennial, monoecious, prostrate, or rarely scandent herb, with simple tendrils; roots fibrous, woody; stems up to 2 m long. Leaves alternate, simple; petiole 0.5–4 cm long; blade ovate (rarely ovate-triangular), 2.5–9 cm × 2–6 cm, deeply palmately three to five-lobed, rarely unlobed, slightly cordate at base, lobes variable but mostly elliptical or narrowly elliptical, each three to five-lobed. Flowers solitary, unisexual, regular, five-merous; receptacle 2.5–4.5 mm long; sepals 1–4 mm long;

petals 4.5–9 mm long, yellow; male flowers with pedicel 3–12(–22) mm long, stamens three; female flowers with pedicel 3–12 mm long, ovary inferior, densely softly hairy. Fruit an oblong, ellipsoid or obovoid-ellipsoid berry, 3.5–5 cm × 2.5–3.5 cm, concolorous yellow when ripe, spines slender, up to 1.5 cm long; fruit stalk 2–5 cm long. Seeds elliptical in outline, 5–8.5 mm × 2–4.5 mm × 1.5–2 mm, whitish.

The genus *Cucumis* includes about 30 species, 4 of which are economically important: cucumber (*C. sativus* L.), melon and snake cucumber (*C. melo* L.), West Indian gherkin (*C. anguria* L.), and horned melon (*C. metuliferus* Naudin). *C. zeyheri* belongs to the "*anguria*" group of the subgenus *melo* (Chen 2011).

2.11.5 ECOLOGY

C. zeyheri occurs in open woodland, in grassland, and as a weed on cultivated ground at 300–1650 m altitude (Decker-Walters 2011).

2.11.6 CULTIVATION AND MANAGEMENT

Fruits are collected from wild plants.

2.11.7 GENETIC RESOURCES AND BREEDING

As *C. zeyheri* is common, it is not threatened by genetic erosion. There are genebank accessions recorded in the Czech Republic, Spain, France, and the United States. Fully fertile hybrids can be obtained from crosses of *C. zeyheri* with *C. anguria*, *C. africanus* L.f., and *C. myriocarpus* Naudin. Resistance to downy mildew (*P. cubensis*) and scab (*C. cucumerinum*), both economically important pathogens of cucumber, has been found in *C. zeyheri* (Kirkbride 1993; Esteras et al. 2011; Navrátilová et al. 2011; Zhang et al. 2012).

2.11.8 PROSPECTS

As a vegetable, *C. zeyheri* will remain of some importance locally. As a source of disease resistance, it may make an important contribution to breeding, especially of cucumber (Decker-Walters 2011; Mabona and Van Vuuren 2013; Liu et al. 2015).

2.12 *BENINCASA HISPIDA* (THUNB. EX MURRAY) COGN./*CUCURBITA HISPIDA* THUNB. EX MURRAY (1784)/*BENINCASA CERIFERA* SAVI (1818)

2.12.1 INTRODUCTION

Synonyms are *Cucurbita hispida* Thunb. ex Murray (1784), *Benincasa cerifera* Savi (1818). Common names are wax gourd, Chinese winter melon, white gourd, ash gourd, fuzzy melon (English). Courge cireuse, bidao, courgette velue (French). Abóbora d'agua, comalenge (Portuguese).

Wax gourd is a cultigen probably originating from Indo-China. It is not found in the wild and no related species are known. It has been grown since ancient times in southern China, Japan, and southern and southeastern Asia. Wax gourd is now widely cultivated throughout tropical Asia and is also rather popular in the Caribbean and the United States. In Africa, it is a vegetable crop of limited importance, grown mainly in East and southern Africa. In Madagascar and Mauritius, it was formerly cultivated, but it seems to have vanished at present (Robinson and Decker-Walters 1997).

2.12.2 Uses

Wax gourd is grown both for its immature and mature fruits. The immature fruits, called fuzzy melons, have a delicate taste and flavor and are prepared in the same way as summer squash from *Cucurbita pepo* L. In India, they are used extensively in curries. The ripe fruits have juicy greenish-white flesh with a flat taste. They are especially popular among people of Asian descent, but they are also liked by many Africans. The skin is peeled or scraped off, seeds and pith are removed, and the flesh is cooked in soups. In China, the fruits are often stuffed with meat, shrimps, and vegetables, and then steamed in a pot. The firm flesh of the older fruits is also candied with sugar and can be dried for later use. Young shoots, leaves, and flowers are occasionally eaten too. The seeds are prepared as a snack by frying. In India, the wax of the fruits was formerly scraped off to make candles. Wax gourd is sometimes used as a rootstock for melon (*C. melo* L.). The fruits are valued for their medicinal properties. In India and China, they are used as an anthelmintic, antiperiodic, aphrodisiac; for lowering blood sugar; against epilepsy, insanity, and other nervous diseases, haemophysis and haemorrhage; and as a diuretic, laxative, and bitter tonic. They are recommended in Ayurvedic medicine for the management of peptic ulcers. The seeds are used as a vermifuge (Pagare et al. 2011; Zaini and Saari 2011; Aromal and Philip 2012; Mukhtar et al. 2012; Han et al. 2013).

2.12.3 Production and Trade

Wax gourd is a rather important market vegetable in subtropical and tropical Asia, and the immature fruits are increasingly popular in city markets. In Africa, wax gourd is grown occasionally for local and city markets, in peri-urban gardens more for the young fruits, in rural areas more for the mature ones. There is no known statistical data (Robinson and Decker-Walters 1997; Rubatzky and Yamaguchi 1997).

2.12.4 Nutritional Properties

The edible portion of wax gourd is about 70% of the total fruit weight. The nutritional composition of mature wax gourd fruits per 100 g edible portion is water 96.1 g, energy 54 kJ (13 kcal), protein 0.4 g, fat 0.2 g, carbohydrate 3.0 g, dietary fiber 2.9 g, Ca 19 mg, Mg 10 mg, P 19 mg, Fe 0.4 mg, Zn 0.6 mg, carotene absent, thiamin 0.04 mg, riboflavin 0.11 mg, niacin 0.40 mg, folate 5 μg, and ascorbic acid 13 mg (USDA 2002). The content of micronutrients in young fruits is probably somewhat higher, notably of ascorbic acid (USDA 2012).

Fruit extracts of wax gourd showed antiulcer activity in tests with mice and rats. Fruit juice also showed significant activity against symptoms of morphine withdrawal in tests with mice. Histamine-release inhibitors were isolated from wax gourd fruits; the triterpenes alnusenol and multiflorenol were the most active inhibitors. An immuno-potentiator has been isolated from the seeds, markedly stimulating the proliferation and differentiation of murine B cells. Wax gourd fruits contain cucurbitacin B, which is known to exhibit cytotoxic and anti-inflammatory activities.

In dishes, wax gourd can be replaced by young fruits of bottle gourd (*Lagenaria siceraria* (Molina) Standl.) (Han et al. 2013; Samad et al. 2013; Tahir et al. 2013; Savant 2015).

2.12.5 Botanical Description

Usually monoecious, annual herb climbing by two to three-fid tendrils up to 35 cm long; stem up to 5 m long, thick, terete, longitudinally furrowed, whitish-green with scattered rough hairs. Leaves distichously alternate, simple; stipules absent; petiole 5–20 cm long; blade broadly ovate in outline, 10–25 cm × 10–20 cm, deeply cordate at base, apex acuminate, margin more or less deeply and irregularly 5–11-angular or -lobed and irregularly undulate-crenate or toothed, densely patently hispid on both sides, 5–7-veined from the base. Flowers solitary in leaf axils, unisexual, regular, five-merous, 6–12 cm in diameter; calyx campanulate, densely silky; petals almost free, yellow; male flowers with pedicel 5–15 cm long and three stamens; female flowers with pedicel 2–4 cm long, an inferior, ovoid or cylindrical, densely villose ovary, and a short style with three curved stigmas. Fruit an ovoid-oblong, ellipsoid or globose berry 20–200 cm × 10–25 cm, dark green to speckled pale green or glaucous, hispid when immature, thinly hispid or subglabrous when ripe, covered with a chalk-white, easily removable wax layer; flesh greenish white, juicy, slightly fragrant, spongy in the middle, containing many seeds. Seeds ovate-elliptical, flattened, 1–1.5 cm long, yellow-brown, sometimes prominently ridged (Walters and Decker-Walters 1989).

Benincasa comprises a single species. Numerous types are distinguished, mainly differing from each other in fruit size and shape, color, hairiness, and amount of wax present. A classification into 16 cultivar groups has been proposed, but the main distinction in improved cultivars is between types suitable for harvest of the young fruits, the "fuzzy melon" type, and cultivars grown for the mature gourds, the true "wax gourd" type. From some landraces in India, both the young and mature fruits are used (Robinson and Decker-Walters 1997; Rubatzky and Yamaguchi 1997).

2.12.6 GROWTH AND DEVELOPMENT

Wax gourd is a vigorous grower, but it needs a long growing season of 4–5 months. Flowering starts about 45 days after sowing for early types and up to over 100 days after sowing for late ones. The flowers are insect-pollinated. Sex ratio is 1 female flower to 20–33 male ones in primitive types, but modern selections and hybrids are predominantly female. The ratio of female to male flowers increases with lower temperatures and with shorter days. Young fruits are harvestable eight days after anthesis or later, depending on the size wanted by the market. The fruits need 1–2 months from anthesis until full maturity, 50–72 days in modern improved cultivars. A regular harvest of young fruits prolongs the flowering period and the crop duration. Fruits contain 15–45 g of seed (Robinson and Decker-Walters 1997).

2.12.7 ECOLOGY

Wax gourd is best suited for moderately dry areas in the tropics. It is relatively drought tolerant. It grows well at temperatures above 25°C, the optimum temperature for growth ranging from 23°C to 28°C (24 h average). It is suited to tropical lowland conditions and elevations up to 1000 m altitude. It prefers a well-drained light soil with pH 6.0–7.0 (Robinson and Decker-Walters 1997; Rubatzky and Yamaguchi 1997).

2.12.8 CULTIVATION AND MANAGEMENT

Wax gourd grown for mature fruits is planted on flat land, whereas the fuzzy melon type is mostly grown upright, for example, against trellises. It needs a fertile soil and responds well to much organic matter, for example, 30 t manure/ha. An application of NPK fertilizer is recommended before sowing and a nitrogenous fertilizer as a side-dressing at regular intervals until flowering. Ample irrigation is needed during dry periods. In the rainy season, fruits of trailing plants can be protected from rotting by putting them on some straw. Pruning of stem tips and flowers is sometimes carried out to achieve better growth of fruits (Karmakar et al. 2015).

Both direct-seeding and sowing in pots and transplanting is practiced if grown for immature fruits, whereas for the production of mature wax gourd fruits, only direct sowing is practiced. Direct sowing is done in trenches or planting holes filled with manure or compost. When grown trellised for young fruits, plants are spaced at 50–70 cm in the row, with the rows 1.5–2.0 m apart or about 10,000 plants/ha, when grown for mature fruits and stems allowed to trail there are about 5,000 plants/ha. In intensive growing systems for immature fruits, the seed requirement is 400–500 g/ha if transplanting is practiced and 800–1000 g/ha for direct sowing. One gram contains 12–25 seeds. Farmers who produce farm-saved seed usually use up to 2 kg/ha. For wax gourd in India, farmers use a seed rate of about 5 kg/ha (Sharma and Tarsem 1998; Huong et al. 2013; Gunasena and Pushpakumara 2015).

2.12.9 Diseases and Pests

Wax gourd is moderately susceptible to anthracnose (*C. lagenarium*) and gummy stem blight (*D. bryoniae*), for which no tolerance has been identified yet. It is also moderately susceptible to fruit rot (*Fusarium solani*) and cavity rot (*Verticillium dahliae*) but rather resistant to leaf fungi, such as downy mildew (*P. cubensis*) and powdery mildew (*E. cichoracearum and S. fuliginea*), that are devastating on other cucurbits. Wax gourd is susceptible to watermelon mosaic virus (WMV) transmitted by aphids. Among the insect pests are squash beetle (*Aulacophora foveicollis*), aphids (*A. gossypii*), and fruit flies (*Dacus* spp.). These pests and diseases are seldom serious enough to justify chemical sprays.

Western seed companies have selected special wax gourd cultivars as disease-resistant rootstock for other cucurbits (cucumber, water melon, melon) because of growth vigor and resistance to Fusarium wilt and root-knot nematodes. Wax gourd combines best with melon, but in practice, growers prefer rootstocks from *Cucurbita* species (hybrids of *C. moschata* Duchesne and *C. maxima* Duchesne) and bottle gourd (*L. siceraria*) because wax gourd is susceptible to *P. sclerotioides*, a root disease of melon and cucumber in protected cultivation (Rubatzky and Yamaguchi 1997; Gouda 2011; Tahir et al. 2013).

2.12.10 Harvesting and Yield

Immature fruits are harvested at weekly intervals when they weigh 300–1000 g. Mature fruits are harvested from 100 to 160 days after sowing; depending on the cultivar, the harvestable fruit weight varies from 3 to 40 kg but is commonly around 10 kg.

Yields of young fruits of more than 30 t/ha, harvested 60–100 days after sowing, have been reported. Mature wax gourd in India yields about 20 t/ha. A seed yield of 100–150 kg/ha is recorded from India; 200–300 kg/ha from Thailand (Huong et al. 2013; Reuzeau 2015).

2.12.11 Postharvest Handling

Young fruits are rather perishable. Mature fruits can be stored for over a year if kept at 13°C–15°C and 75% relative humidity due to the waxy layer that protects them from attack by micro-organisms (Robinson and Decker-Walters 1997; Rubatzky and Yamaguchi 1997).

2.12.12 Genetic Resources and Breeding

With the increased popularity of improved cultivars in Asian countries, local cultivars are disappearing in Asia. No germplasm collections are known from Africa, but germplasm collections are available at horticultural institutes in other tropical regions, mainly in the Philippines (Institute of Plant Breeding), India (Kerala Agricultural University), Russia (N.I. Vavilov Institute of Plant Industry, St. Petersburg), and the United States (Southern Regional Plant Introduction Station, Georgia; Cornell University, New York) (National Academy of Sciences 1975; Sharma and Tarsem 1998).

Several seed companies in India, Thailand, Taiwan, China, and Japan have carried out selection work on local cultivars. Selection criteria are fruit quality, few seeds, high yield, earliness, and resistance to diseases. Wax gourds of both types are offered in seed catalogues. Seed companies in India, Thailand, Taiwan, China, Japan, and the United States offer improved cultivars of fuzzy melons.

Several seed companies including East-West Seed Company have developed vigorous F_1 cultivars of fuzzy melon with superior yield and fruit quality. Examples are "Pearl F_1," a pale green midlate cultivar, harvestable 75 days after sowing, and "Jade F_1," an early cultivar, harvestable 55 days from sowing. Wax gourd cultivars were developed at Coimbatore in South India, for example, "Co-2," a selection with small fruits (3 kg) harvestable 120 days after sowing (Esquinas-Alcazar and Gulick 1983; Afroze et al. 2013).

2.12.13 PROSPECTS

Although at present of little importance in tropical Africa, wax gourd merits attention as an easy-to-grow vegetable suitable for home gardens and market production, with immature fruits having a good taste and mature fruits having excellent keeping quality (Walters and Decker-Walters 1989; Rubatzky and Yamaguchi 1997; Huong et al. 2013; Gunasena and Pushpakumara 2015).

REFERENCES

Achigan-Dako, E. G., Avohou, E. S., Linsoussi, C., Ahanchede, A., Vodouhe, R. S., and Blattner, F. R. (2015). Phenetic characterization of *Citrullus* spp. (Cucurbitaceae) and differentiation of egusi-type (*C. mucosospermus*). *Genetic Resources and Crop Evolution*, 62(8): 1–2.

Adebisi, A. A. and Ladipo, D. O. (2000). Evaluation of the utilization spectrum of some Cucurbits in South West Nigeria. CENRAD Development Series 07. CENRAD, Ibadan, Nigeria, 14pp.

Adekunle, A. and Uma, N. U. (2012). Management of insects pests in storage for preservation of *Cucumeropsis mannii* seeds. *Indian Phytopathology*, 57: 280.

Afroz, S., Noman, M. S., Hossain, M. S., Mamun, A. A., Howlader, N., and Ara, S. (2015). Multivariate analysis approach to select parents for hybridization aiming at yield improvement in cucumber (*Cucumis sativus* L.). *Journal of Environmental Science and Natural Resources*, 6(1): 33–36.

Afroze, F., Rasul, M. G., Islam, A. A., Mian, M. A. K., and Hossain, T. (2013). Genetic divergence in ash gourd (*Benincasa hispida* Thunb.). *Bangladesh Journal of Plant Breeding and Genetics*, 20(1): 19–24.

Ajuru, M. G. and Okoli, B. E. (2013). The morphological characterization of the melon species in the family Cucurbitaceae Juss., and their utilization in Nigeria. *International Journal of Modern Botany*, 3(2): 15–19.

Andres, T. C. 2003. Web site for the plant family Cucurbitaceae and home of the curcubit network. http://www.curcubit.org/family.html. Accessed on May 2014.

Anhwange, B. A., Ikyenge, B. A., Nyiatagher, D. T., and Ageh, J. T. (2010). Chemical analysis of *Citrullus lanatus* (Thunb.), *Cucumeropsis mannii* (Naud.) and *Telfairia occidentalis* (Hook F.) seeds oils. *Journal of Applied Sciences Research*, 6(3): 265–268.

Arnold, T. H., Wells, M. J., and Wehmeyer, A. S. (1985). Khosian food plants: taxa with potential for future economic exploitation. In: Wickens, G.E., Goodin, J.R., and Field, D.V. (eds.). *Plants for Arid Lands*. Allen and Unwin, London, U.K., pp. 69–86.

Aromal, S. A. and Philip, D. (2012). *Benincasa hispida* seed mediated green synthesis of gold nanoparticles and its optical nonlinearity. *Physica E: Low-Dimensional Systems and Nanostructures*, 44(7): 1329–1334.

Azimova, S. S. and Glushenkova, A. I. (2012). *Citrullus maxima* L. In: *Lipids, Lipophilic Components and Essential Oils from Plant Sources*. Springer, London, U.K., pp. 291–297.

Baird, J. R. and Thieret, J. W. (1988). The bur gherkin (*Cucumis anguria* var. *anguria*, Cucurbitaceae). *Economic Botany*, 42: 447–451.

Bates, D. M., Robinson, R. W., and Jeffrey, C. (1990). *Biology and Utilization of the Cucurbitaceae*. Cornell University Press, New York, 485pp.

Benzioni, A., Mendlinger, S., Ventura, M., and Huyskens, S. (1993). Germination, fruit development, yield and postharvest characteristics of *Cucumis metuliferus*. In: Janick, J. and Simon, J.E. (eds.). *New Crops*. John Wiley & Sons, New York, pp. 553–557.

Burkill, H. M. (1985). *The Useful Plants of West Tropical Africa*, 2nd edn., Vol. 1: *Families A–D*. Royal Botanic Gardens, Kew, U.K., 960pp.

Chang, K., Cheng, X., Wang, J., Li, Z., and Han, Y. (2014). Comparative analysis of rDNA loci in *Citrullus* species and *Acanthosicyos naudinianus* by fluorescence in situ hybridization. *Cucurbitaceae 2014 Proceedings*, Bay Harbor, MI, p. 73.

Chen, J. F. (2011). Basic botany of the genus *Cucumis*. Taxonomy. In: *Wild Crop Relatives: Genomic and Breeding Resources: Vegetables*, Springer, Berlin, Heidelberg, p. 67, Sections 6.2 and 6.2.1.

Choi, J. W., Park, M. H., Lee, J. H., Do, K. R., Choi, H. J., and Kim, J. G. (2015). Changes of postharvest quality in "Bagdadagi" cucumber (*Cucumis sativus* L.) by storage temperature. *Journal of Food and Nutrition Sciences*, 3(1–2): 143–147.

Chomicki, G. and Renner, S. S. (2015). Watermelon origin solved with molecular phylogenetics including Linnaean material: Another example of museomics. *New Phytologist*, 205(2): 526–532.

Decker-Walters, D. (2011). *Cucumis zeyheri* (Cultivated). Texas A&M University, TX, 1(2011): 98976. ·Available electronically from http://hdl. handle. net/1969.

Esquinas-Alcazar, J. T. and Gulick, P. J. (1983). Genetic resources of Cucurbitaceae. IBPGR, Rome, Italy, 101pp.

Esteras, C., Nuez, F., Picó, B., YiHong, W., Behera, T. K., and Kole, C. (2011). Genetic diversity studies in Cucurbits using molecular tools. *Genetics, Genomics and Breeding of Cucurbits*, 2011: 140–198.

Ezura, H., Akashi, Y., Kato, K., and Kuzuya, M. (2002). Genetic characterization of long shelf-life in honeydew (*Cucumis melo* var. *inodorus*) melon. *Acta Horticulturae*, 588: 369–372.

Fabricante, J. R., Ziller, S. R., de Araújo, K. C. T., Furtado, M. D. D. G., and de Arantes Basso, F. (2015). Non-native and invasive alien plants on fluvial islands in the São Francisco River, Northeastern Brazil. *Check List*, 11(1): 1535.

Falah, M. A. F., Nadine, M. D., and Suryandono, A. (2015). Effects of storage conditions on quality and shelf-life of fresh-cut melon (*Cucumis melo* L.) and Papaya (*Carica papaya* L.). *Procedia Food Science*, 3: 313–322.

Fand, B. B. and Suroshe, S. S. (2015). The invasive mealybug *Phenacoccus solenopsis* Tinsley, a threat to tropical and subtropical agricultural and horticultural production systems—A review. *Crop Protection*, 69: 34–43.

FAO. (2013). FAOSTAT agriculture data. http://apps.fao.org/page/collections?subset=agriculture. Accessed on 2014.

Fassuliotis, G. and Nelson, B. V. 1988. Interspecific hybrids of *Cucumis metuliferus* × *C. anguria* obtained through embryo culture and somatic embryogenesis. *Euphytica*, 37: 53–60.

Fernandes, C. A. (2011). Evaluation of seed sources and cultural practices of maxixe (*Cucumis anguria* L.) for production in Massachusetts.

Fernando, L. N. and Grün, I. U. (2001). Headspace-SPME analysis of volatiles of the ridge gourd (*Luffa acutangula*) and bitter gourd (*Momordica charantia*) flowers. *Flavour and Fragrance Journal*, 16(4): 289–293.

Garad, K. U., Gore, R. D., and Gaikwad, S. P. (2015). A synoptic account of flora of Solapur District, Maharashtra (India). *Biodiversity Data Journal*, 3: e4282. DOI: 10.3897/BDJ.3.e4282.

Garcia-Mas, J., Benjak, A., Sanseverino, W., Bourgeois, M., Mir, G., González, V. M., and Puigdomènech, P. (2012). The genome of melon (*Cucumis melo* L.). *Proceedings of the National Academy of Sciences of the United States of America*, 109(29): 11872–11877.

Gismervik, K., Aspholm, M., Rørvik, L. M., Bruheim, T., Andersen, A., and Skaar, I. (2015). Invading slugs (*Arion vulgaris*) can be vectors for *Listeria monocytogenes*. *Journal of Applied Microbiology*, 118(4): 809–816.

Gouda, S. (2011). Anthelmintic activity of root of *Benincasa hispida* (Petha). *American Journal of Pharm Research*, 1(1): 1–6.

Gunasena, H. P. and Pushpakumara, D. K. N. G. (2015). Chena cultivation in Sri Lanka. Shifting cultivation and environmental change. *Indigenous People, Agriculture and Forest Conservation*, Springer Netherlands, 199pp.

Hallett, L. M., Standish, R. J., Jonson, J., and Hobbs, R. J. (2014). Seedling emergence and summer survival after direct seeding for woodland restoration on old fields in south-western Australia. *Ecological Management and Restoration*, 15(2): 140–146.

Han, X. N., Liu, C. Y., Liu, Y. L., Xu, Q. M., Li, X. R., and Yang, S. L. (2013). New triterpenoids and other constituents from the fruits of *Benincasa hispida* (Thunb.) Cogn. *Journal of Agricultural and Food Chemistry*, 61(51): 12692–12699.

Heiser, C. B. and Schilling, E. E. (1990). The genus *Luffa*: A problem in phytogeography. In: Bates, D.M., Robinson, R.W., and Jeffrey, C. (eds.). *Biology and Utilization of the Cucurbitaceae*. Comstock (Cornell University Press), Ithaca, NY, pp. 120–133.

Holland, B., Unwin, I. D., and Buss, D. H. (1991). *Vegetables, Herbs and Spices: The Fifth Supplement to McCance & Widdowson's The Composition of Foods*, 4th ed., 163 Seiten. Royal Society of Chemistry, Cambridge, U.K.

Huang, C. H., Zong, L., Buonanno, M., Xue, X., Wang, T., and Tedeschi, A. (2012). Impact of saline water irrigation on yield and quality of melon (*Cucumis melo* cv. Huanghemi) in Northwest China. *European Journal of Agronomy*, 43: 68–76.

Huong, P. T. T., Everaarts, A. P., Neeteson, J. J., and Struik, P. C. (2013). Vegetable production in the Red River Delta of Vietnam. II. Profitability, labour requirement and pesticide use. *NJAS—Wageningen Journal of Life Sciences*, 67: 37–46.

Huyskens, S., Mendlinger, S., Benzioni, A., and Ventura, M. (1993). Optimization of agrotechniques in the cultivation of *Luffa acutangula*. *Journal of Horticultural Science*, 68(6): 989–994.

Jeffrey, C. (1962). Vernacular names Wild cucumber (En). Origin and geographic distribution *Cucumis zeyheri*. *Kew Bull*. 15(3): 351.

Jeffrey, C. (1978). Cucurbitaceae. In: Launert, E. (ed.). *Flora Zambesiaca*, Vol. 4. Flora Zambesiaca Managing Committee, London, U.K., pp. 414–499.

Jeffrey, C. (1980). A review of the Cucurbitaceae. *Botanical Journal of the Linnean Society*, 81: 233–247.

John, I., Olugbenga, O. S., and Kayode, Q. (2014). Determination of thermal conductivity and thermal diffusivity of three varieties of melon. *Journal of Emerging Trends in Engineering and Applied Sciences*, 5(2): 123–128.

Jones, B., Pech, J. C., Bouzayen, M., Lelièvre, J. M., Guis, M., Romajaro, F., and Latché, A. (2001). Ethylene and developmentally-regulated processes in ripening climacteric fruit. *Acta Horticulturae*, 553: 133–138.

Kamda, A. G. S., Fokou, E., Loh, M. B. A., Raducanu, D., Kansci, G., Ifrim, I., and Lazar, I. (2014). Protective effect of edible cucurbitaceae seed extract from cameroon against oxidative stress. *Environmental Engineering and Management Journal*, 13(7): 1721–1727.

Kamda, A. G. S., Ramos, C. L., Fokou, E., Duarte, W. F., Mercy, A., Germain, K., and Schwan, R. F. (2015). In vitro determination of volatile compound development during starter culture-controlled fermentation of Cucurbitaceae cotyledons. *International Journal of Food Microbiology*, 192: 58–65.

Kapseu, C. and Parmentier, M. (1997). Composition en acides gras de quelques huiles vegetales du Cameroun. *Sciences des Aliments*, 17: 325–331.

Karmakar, P., Pandey, S., and Bhardwaj, D. R. (2015). Improved production technology for cucurbits. *E-Manual on Improved Production Technologies in Vegetable Crops*, 61–71. DOI: 10.13140/2.1.2105.6805.

Kirkbride Jr., J. H. (1993). *Biosystematic Monograph of the Genus Cucumis (Cucurbitaceae): Botanical Identification of Cucumbers and Melons*. Parkway Publishers, Boone, NC, 159pp.

Kull, C. A., Alpers, E. A., and Tassin, J. (2015). Marooned plants: Vernacular naming practices in the Mascarene Islands. *Environment and History*, 21(1): 43–75.

Kwiri, R., Winini, C., Musengi, A., Mudyiwa, M., Nyambi, C., Muredzi, P., and Malunga, A. (2014). Proximate composition of pumpkin gourd (*Cucurbita pepo*) seeds from Zimbabwe. *International Journal of Nutrition and Food Sciences*, 3(4): 279.

Ladipo, D. O., Sarumi, M. B., Adewusi, H. G., and Adebisi, A. A. (1999). "Egusi" diversity in Nigeria. A commissioned survey report submitted to IPGRI-SSA. IPGRI, Rome, Italy.

Lim, T. K. (2012a). *Cucumis metuliferus*. In: *Edible Medicinal and Non-Medicinal Plants*. Springer, Dordrecht, the Netherlands, pp. 235–238.

Lim, T. K. (2012b). Sechium edule. In: *Edible Medicinal and Non-Medicinal Plants*. Springer, Dordrecht, the Netherlands, pp. 384–391.

Liu, B., Ren, J., Zhang, Y., An, J., Chen, M., Chen, H., and Ren, H. (2015). A new grafted rootstock against root-knot nematode for cucumber, melon, and watermelon. *Agronomy for Sustainable Development*, 35(1): 251–259.

Mabona, U. and Van Vuuren, S. F. (2013). Southern African medicinal plants used to treat skin diseases. *South African Journal of Botany*, 87: 175–193.

Mafeo, T. P. (2014). Pre-sowing temperature treatment effect on emergence of *Cucumis myriocarpus* seedlings. *African Journal of Agricultural Research*, 9(26): 2028–2030.

Manamohan, M. and Chandra. G. S. 2011. Cucurbits: Advances in Horticulture Biotechnology—Gene Cloning & Transgenics. *Advances in Horticulture Biotechnology. Gene Cloning & Transgenics* 5: 227–259.

Marsh, D. B. (1993). Evaluation of *Cucumis metuliferus* as a speciality crop for Missouri. In: Janick, J. and Simon, J.E. (eds.). *New Crops*. Wiley, New York, pp. 558–559.

Metcalf, R. L. and Rhodes, A. M. (1990). Coevolution of the Cucurbitaceae and Luperini (Coleoptera: Chrysomelidae): Basic and applied aspects. In: Bates, D.M., Robinson, R.W., and Jeffrey, C. (eds.). *Biology and Utilization of the Cucurbitaceae*. Cornell University Press, New York, pp. 167–182.

Mphosi, M. S. (2015). Biomass yield and partitioning of greenhouse-grown wild watermelon *Cucumis africanus* in response to different irrigation intervals and NPK fertilizer levels. *African Journal of Agricultural Research*, 10(9): 933–937.

Mukhtar, H. M., Vashishth, D., Ali, B., and Kaur, R. (2012). A study for quality assessment of the dried seeds of *Benincasa hispida*. *Research Journal of Pharmacognosy and Phytochemistry*, 4(4): 201–204.

Nantoume, A. D., Christiansen, J. L., Andersen, S. B., and Jensen, B. D. (2012). On-farm yield potential of local seed watermelon landraces under heat-and drought-prone conditions in Mali. *The Journal of Agricultural Science*, 150(6): 665–674.

National Academy of Sciences. (1975). *Underexploited Tropical Plants with Promising Economic Value*. National Academy of Sciences, Washington, DC, 188pp.

Navrátilová, B., Skálová, D., Ondrej, V., Kitner, M., and Lebeda, A. (2011). Biotechnological methods utilized in *Cucumis* research—A review. *Horticultural Science*, 38: 150–158.

Neuwinger, H. D. (1996). *African Ethnobotany: Poisons and Drugs*. Chapman & Hall, London, U.K., 941pp.

Nicodemo, D., Malheiros, E. B., Jong, D. D., and Couto, R. H. N. (2012). Floral biology of greenhouse-grown of Japanese cucumber plants (*Cucumis sativus* L.). *Científica (Jaboticabal)*, 40(1): 35–40.

Nkgapele, R. J. and Mphosi, M. S. (2012). Biomass yield and partitioning of greenhouse grown indigenous crop *Cucumis myriocarpus* in response to irrigation frequency and NPK fertilizer application rate. *Sustainable Irrigation and Drainage IV: Management, Technologies and Policies*, 168: 339.

Nkgapele, R. J. and Mphosi, M. S. (2014). The yield and yield-character variability of traditional leafy-crop *Cucumis africanus* in response to variation in irrigation intervals and NPK fertilizer rates. *Sustainable Irrigation and Drainage V: Management, Technologies and Policies*, 185: 67.

Nkgapele, R. J. and Mphosi, M. S. (2015). Response of vegetative yield characters and yield of biomass fractions of wild-watermelon *Cucumis africanus* to irrigation interval and NPK fertilizer. *African Journal of Agricultural Research*, 10(9): 938–943.

Nonaka, S. and Ezura, H. (2015). Melon (*Cucumis melo*). In: *Agrobacterium Protocols*. Springer, New York, pp. 195–203.

Nwofia, G. E., Amajuoyi, A. N., and Mbah, E. U. (2015). Response of three cucumber varieties (*Cucumis sativus* L.) to planting season and NPK fertilizer rates in lowland humid tropics: Sex expression, yield and inter-relationships between yield and associated traits. *International Journal of Agriculture and Forestry*, 5(1): 30–37.

Offiah, N. V., Makama, S., Elisha, I. L., Makoshi, M. S., Gotep, J. G., Dawurung, C. J., and Shamaki, D. (2011). Ethnobotanical survey of medicinal plants used in the treatment of animal diarrhoea in Plateau State, Nigeria. *BMC Veterinary Research*, 7(1): 36.

Ogunbusola, E. M., Fagbemi, T. N., and Osundahunsi, O. F. (2012). Chemical and functional properties of full fat and defatted white melon (*Cucumeropsis mannii*) seed flours. *Journal of Food Science and Engineering*, 2: 691–696.

Ojo, D., Taiwo, S., Aiyelaagbe, I., and Martin, C. (2013). Watermelon (*Citrullus lanatus*) as live mulch for climate change adaptation in African humid tropics' cropping systems. A paper presented and abstracted at the *First Biennial Conference on Agricultural Research and Extension* held at Kigali Senera Hotel, Kigali, Rwanda, August 21–23, 2013, Abstracted pp. 149.

Ojo, O. D., Akinfasoye, J. A., Famosipe, O. O., and Ngere, P. I. (2010). Effect of organo-mineral soil amendments on cucumber production. A paper presented and abstracted at the *28th Annual Conference of the Horticultural Society of Nigeria (HORTSON)* held at the Federal University of Technology, Minna, Nigeria, November 7–11, 2010.

Olanrewaju, A. S. and Moriyike, O. E. (2013). Physicochemical characteristics and the effect of packaging materials on the storage stability of selected cucurbits oils. *American Journal of Food and Nutrition*, 1(3): 34–37.

Onawola, O., Asagbra, A., and Faderin, M. (2012). Comparative soluble nutrient value of ogiri obtained from dehulled and undehulled boiled melon seeds (*Cucumeropsis mannii*). *Food Science and Quality Management*, 4: 10–15.

Pagare, M. S., Patil, L., and Kadam, V. J. (2011). *Benincasa hispida*: A natural medicine. *Research Journal of Pharmacy and Technology*, 4(12): 1941–1944.

Pelinganga, O., Mashela, P., Mphosi, M., and Nzanza, B. (2013). Optimizing application frequency of diluted (3%) fermented *Cucumis africanus* fruit in tomato production and nematode management. *Acta Agriculturae Scandinavica, Section B–Soil and Plant Science*, 63(3): 278–282.

Petrus, A. J. A. (2014). An approach to the chemosystematics of the genus *Cucumis* L. *Oriental Journal of Chemistry*, 30(1): 149–154.

Pitrat, M. (2013). Phenotypic diversity in wild and cultivated melons (*Cucumis melo*). *Plant Biotechnology*, 30(3): 273–278.

Pius, A., Ekebafe, L., Ugbesia, S., and Pius, R. 2014. Modification of adhesive using cellulose micro-fiber (CMF) from melon seed shell. *American Journal of Polymer Science*, 4(4): 101–106.

Pofu, K. M., Mashela, P. W., and Mphosi, M. S. (2013). Management of *Meloidogyne incognita* in nematode-susceptible watermelon cultivars using nematoderesistant *Cucumis africanus* and *Cucumis myriocarpus* rootstocks. *African Journal of Biotechnology*, 10(44): 8790–8793.

Polacchi, W., MacHargue, J. S., and Perloff, B. P. (1982). Food composition tables for use in the Middle East. Food and Agriculture Organization, Rome, Italy, 265pp.

Quiroz, D. and van Andel, T. (2015). Evidence of a link between taboos and sacrifices and resource scarcity of ritual plants. *Journal of Ethnobiology and Ethnomedicine*, 11(1): 5.

Regassa, T., Kelbessa, E., and Asfaw, Z. (2015). Ethnobotany of wild and semi-wild edible plants of Chelia District, West-Central Ethiopia. *Science, Technology and Arts Research Journal*, 3(4): 122–134.

Renner, S. S. and Pandey, A. K. (2013). The Cucurbitaceae of India: Accepted names, synonyms, geographic distribution, and information on images and DNA sequences. *PhytoKeys*, 20: 53.

Reuzeau, C. (2015). Plants having enhanced yield-related traits and method for making the same. U.S. Patent No. 20,150,007,367, January 1, 2015.

Robinson, R. W. and Decker-Walters, D. S. (1997). *Cucurbits*. CAB International, Wallingford, U.K., 226pp.

Roy, A., Bal, S. S., Fergany, M., Kaur, S., Singh, H., Malik, A. A., and Dhillon, N. P. S. (2012). Wild melon diversity in India (Punjab State). *Genetic Resources and Crop Evolution*, 59(5): 755–767.

Rubatzky, V. E. and Yamaguchi, M. (1997). *World Vegetables: Principles, Production and Nutritive Values*, 2nd edn. Chapman & Hall, New York, 843pp.

Samad, N. B., Debnath, T., Jin, H. L., Lee, B. R., Park, P. J., Lee, S. Y., and Lim, B. O. (2013). Antioxidant activity of *Benincasa hispida* seeds. *Journal of Food Biochemistry*, 37(4): 388–395.

Sarvalingam, A., Rajendran, A., and Sivalingam, R. (2014). Wild edible plant resources used by the Irulas of the Maruthamalai Hills, Southern Western Ghats, Coimbatore, Tamil Nadu. *Indian Journal of Natural Products and Resources*, 5(2): 198–201.

Sarwar, M., Xu, X., Wang, E., and Wu, K. (2011). The potential of four mite species (Acari: Phytoseiidae) as predators of sucking pests on protected cucumber (*Cucumis sativus* L.) crop. *African Journal of Agricultural Research*, 6(1): 73–78.

Sarwar, M. F., Sarwar, M. H., Sarwar, M., Qadri, N. A., and Moghal, S. (2013). The role of oilseeds nutrition in human health: A critical review. *Journal of Cereals and Oilseeds*, 4(8): 97–100.

Savant, C. (2015). Acute toxicity study of *Benincasa hispida*. *Journal of AYUSH: Ayurveda, Yoga, Unani, Siddha and Homeopathy*, 3(3): 28–30.

Schaefer, H. and Renner, S. S. (2011). Cucurbitaceae. In: *Flowering Plants: Eudicots*. Springer, Berlin, Germany, pp. 112–174.

Schippers, R. R. 2000. *African Indigenous Vegetables: An Overview of the Cultivated Species*. Natural Resources Institute/ACP-EU Technical Centre for Agricultural and Rural Cooperation, Chatham, U.K., 214pp.

Sepasal. (2003). *Cucumis anguria* vars. *anguria* and *longipes*. Survey of economic plants for arid and semi-arid lands (SEPASAL) database. Royal Botanic Gardens, Kew, U.K. http://www.rbgkew.org.uk/ceb/sepasal/internet/. Accessed on June 2013.

Sepasal. (2013). *Acanthosicyos naudinianus*. Survey of economic plants for arid and semi-arid lands (SEPASAL) database. Royal Botanic Gardens, Kew, U.K. http://www.rbgkew.org.uk/ceb/sepasal/acantho.htm. Accessed on October 2013.

Sharma, B. R. and Tarsem, L. (1998). Improvement and cultivation: *Cucurbita* and *Benincasa*. In: Nayar, N.M. and More, T.A. (eds.). *Cucurbits*. Science Publishers Inc., Enfield, NH, pp. 155–168.

Simsek, M. and Comlekcioglu, N. (2013). Effects of different irrigation regimes and nitrogen levels on yield and quality of melon (*Cucumis melo* L.). *African Journal of Biotechnology*, 10(49): 10009–10018.

Sola, P., Mvumi, B. M., Ogendo, J. O., Mponda, O., Kamanula, J. F., Nyirenda, S. P., and Stevenson, P. C. (2014). Botanical pesticide production, trade and regulatory mechanisms in sub-Saharan Africa: Making a case for plant-based pesticidal products. *Food Security*, 6(3): 369–384.

Sonnewald, U. (2013). Physiology of development. In: *Strasburger's Plant Sciences*. Springer, Berlin, Germany, pp. 411–530.

Staub, J. E. and Ivandic, V. (2000). Genetic assessment of the United States national cucumber collection. *Acta Horticulturae*, 510: 113–121.

Stepansky, A., Kovalski, I., and Perl-Treves, R. 1999. Intraspecific classification of melons (*Cucumis melo* L.) in view of their phenotypic and molecular variation. *Plant Systematics and Evolution*, 217: 313–333.

Sun, J., Sui, X., Wang, S., Wei, Y., Huang, H., Hu, L., and Zhang, Z. (2014). The response of rbcL, rbcS and rca genes in cucumber (*Cucumis sativus* L.) to growth and induction light intensity. *Acta Physiologiae Plantarum*, 36(10): 2779–2791.

Tahir, L., Chand, B., and Rahman, S. (2013). Antibacterial studies on *Benincasa hispida* and *Nigella sativa* oil. *International Research Journal of Pharmacy*, 4(4): 121–122.

Talano, A. M., Laura Wevar Oller, A., Gonzalez, S. P., and Agostini, E. (2012). Hairy roots, their multiple applications and recent patents. *Recent Patents on Biotechnology*, 6(2), 115–133.

Telford, I. R., Sebastian, P., Bruhl, J. J., and Renner, S. S. (2011). *Cucumis* (Cucurbitaceae) in Australia and Eastern Malesia, including newly recognized species and the sister species to *C. melo*. *Systematic Botany*, 36(2): 376–389.

Tindall, H. D. (1983). *Vegetables in the Tropics*. Macmillan Press, London, U.K., 533pp.

Touré, A. I., Loukou, A. L., Koffi, K. K., Mouaragadja, I., Bohoua, G. L., Mbatchi, B., and Zoro, B. I. A. (2015). Feedstuffs potential of harvest by-products from two oleaginous curcurbits. *African Journal of Biotechnology*, 14(6): 459–465.

Uriostegui Arias, M. T. (2015). Phytochemical analysis and antiproliferative effect of genotypes S. edule (Jacq.) Sw. On breast cancer. A Thesis (MSc, specialist in Botany), University of Mexico.

US Department of Agriculture (USDA). (2002). Vegetables 2001 summary. USDA National Agricultural Statistics Service, Washington, DC.

USDA (United States Department of Agriculture). (2012). USDA nutrient database for standard reference, release 15. U.S. Department of Agriculture, Beltsville Human Nutrition Research Center, Beltsville, MD. http://www.nal.usda.gov/fnic/foodcomp. Accessed on 2013.

USDA. 2014. USDA nutrient database for standard reference, release 15. U.S. Department of Agriculture, Beltsville Human Nutrition Research Center, Beltsville, MD. http://www.nal.usda.gov/fnic/foodcomp. Accessed on June 2015.

van Wyk, B. E. and Gericke, N. (2000). *People's Plants: A Guide to Useful Plants of Southern Africa*. Briza Publications, Pretoria, South Africa, 351pp.

Walters, S. A., Wehner, T. C., and Barker, K. R. (1993). Root-knot nematode resistance in cucumber and horned cucumber. *HortScience*, 28(2): 151–154.

Walters, T. W. and Decker-Walters, D. S. (1989). Systematic re-evaluation of *Benincasa hispida* (Cucurbitaceae). Notes on economic plants. *Economic Botany*, 43(2): 274–278.

Wehmeyer, A. S. (1986). Edible wild plants of Southern Africa: Data on the nutrient contents of over 300 species. NFRI Report of Council for Scientific and Industrial Research (CSIR), National Food Research Institute, Pretoria, South Africa.

Whitaker, T. W. and Davis, G. N. (1962). *Cucurbits: Botany, Cultivation, and Utilization*. The University of Michigan, Ann Arbor, MI, 250pp.

Wyllie, S. G., Leach, D. N., Wang, Y.-M., and Shewfelt, R. L. (1995). Key aroma compounds in melons: Their development and cultivar dependence. In: Rousseff, R.L. and Leahy, M.M. (eds.). *Fruit Flavors: Biogenesis, Characterization and Authentification*. American Chemical Society, Washington, DC, pp. 248–257.

Yao, K. A., Koffi, K. K., Ondo-Azi, S. A., Baudoin, J. P., and Zoro, B. I. (2014). Seed yield component identification and analysis for exploiting recombinative heterosis in bottle gourd. *International Journal of Vegetable Science*. DOI:10.1080/19315260.2014.895791.

Zaini, N. A. M., Anwar, F., Hamid, A. A., and Saari, N. (2011). Kundur [*Benincasa hispida* (Thunb.) Cogn.]: A potential source for valuable nutrients and functional foods. *Food Research International*, 44(7): 2368–2376.

Zhang, W. W., Pan, J. S., He, H. L., Zhang, C., Li, Z., Zhao, J. L., and Cai, R. (2012). Construction of a high density integrated genetic map for cucumber (*Cucumis sativus* L.). *Theoretical and Applied Genetics*, 124(2): 249–259.

Section II

Cucurbit Physiological Stages of
Growth and Development I

3 Carbohydrate Metabolism of Cucurbits

Minmin Miao and Zhiping Zhang

CONTENTS

3.1 INTRODUCTION

Carbohydrate metabolism and its regulation are of central importance to the growth and development of plants. Cucurbits have long been considered ideal materials for carbohydrates metabolism studies for a number of reasons. First, the cucurbits synthesize and translocate assimilates mainly in the form of raffinose family oligosaccharides (RFOs). As a result, the loading, translocation, and unloading mechanisms in cucurbits are largely different from those in sucrose-translocating plants. Second, cucurbit fruits are relatively fast growing. Some types of pumpkins show maximum dry weight gain of 1.71 g h^{-1} during growth (Crafts and Lorenz, 1944). Recently (October 2014), a Swiss farmer claimed a new world record with a pumpkin weighing 953.5 kg. These evidences indicate that the mass transfer rate in cucurbit phloem is quite fast. Third, the long petioles and stem internodes and relatively abundant phloem sap exudation from the incised stem facilitate radiolabeling studies and sample collection and analysis.

In cucurbits, RFOs are not only primary translocated assimilates but also abundant soluble sugars throughout all plant parts, except immature seeds and fleshy fruits (Schaffer et al., 1996). The pathway of synthesis and catabolism of these galactosyl-sucrose (mainly stachyose and raffinose) is well established (Gross and Pharr, 1982; Handley et al., 1983a; Schaffer et al., 1987; Keller and Pharr, 1996; Dai et al., 2006), as shown in Figures 3.1 and 3.2.

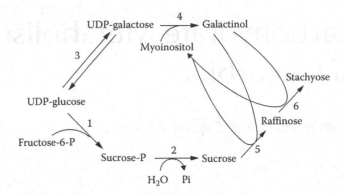

FIGURE 3.1 Pathway of biosynthesis of stachyose and raffinose in cucurbits. 1, sucrose phosphate synthase (EC.2.4.1.14); 2, sucrose phosphate phosphatase (EC 3.1.3.17); 3, UDP-glucose epimerase (EC 5.1.3.2); 4, galactinol synthase (EC 2.4.1.123); 5, raffinose synthase (EC 2.4.1.82); 6, stachyose synthase (EC.2.4.1.67).

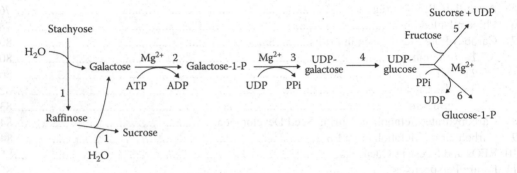

FIGURE 3.2 Pathway of catabolism of stachyose and raffinose in cucurbits. 1, α-galactosidase (EC 3.2.1.22); 2, galactokinase (EC 2.7.1.6); 3, UDP-galactose pyrophosphorylase (EC 2.7.7.10); 4, UDP-glucose epimerase (EC 5.1.3.2); 5, sucrose synthase (EC 2.4.1.13); 6, UDP-glucose pyrophosphorylase (EC 2.7.7.9).

3.2 CARBOHYDRATE METABOLISM DURING SEED GERMINATION

Cucurbits accumulate lipids, proteins, starch, and soluble sugars in mature seeds as sources of nutrition for germination. Lipids are the primary component in cucurbit seeds; in mature cucumber (*Cucumis sativus*) seeds, about 38% of dry weight is that of lipids (Pharr and Motomura, 1989). The lipid level could reach to approximately 50% of dry weight in the mature seeds of watermelon (*Citrullus vulgaris*) and pumpkin (*Cucurbita pepo*) (El-Adawy and Taha, 2001). Mature cucurbit seeds are also rich in protein, and the protein level in mature seeds of watermelon and pumpkin is about 35% (EI-Adawy and Taha, 2001). There is a small amount of starch in cucurbit seeds, the content of which is less than 1 mg g⁻¹ in cucumber seeds (Handley et al., 1983b). Furthermore, cucurbit seeds also contain a certain amount of glucose, sucrose, raffinose, stachyose, and verbascose (Handley et al., 1983a,b; Botha and Small, 1985; EI-Adawy and Taha, 2001).

Stachyose is the primary soluble sugar in mature cucumber seeds. Stachyose and raffinose are rapidly consumed during the first 36 h after imbibition, accompanied by an increase in the concentration of their enzymatic degradation product sucrose (Handley et al., 1983b). α-Galactosidases catalyze the first step of RFO hydrolysis (Figure 3.1). There are six α-galactosidase genes in the cucumber genome, three of them are acid and the other three are alkaline, based on their activity response to pH (Yang, 2012). The author found that mRNA expression of all six genes gradually increased during the 5 days after seed imbibition. Thomas and Webb (1978) also found that the activities of two acid α-galactosidases rose and RFO levels declined rapidly within 48 h of pumpkin seed germination. Blöchl et al. (2008) reported that the RFOs were broken down by both acid α-galactosidase and alkaline α-galactosidase together in pea (*Pisum sativum*) seeds. An acid α-galactosidase is predominately expressed during seed maturation and hydrolyzed RFOs at the early stage of germination, while alkaline α-galactosidase is only expressed in the latter stage of seed germination. It is not clear whether RFO catabolism in cucurbit seeds is similar to that in pea seeds.

After 36 h imbibition, lipids began to break down into cucumber seeds (Pharr and Motomura, 1989). Lipid degradation in watermelon seeds was processed even later than that in cucumber seeds (Botha and Small, 1985), indicating the importance of soluble sugars as an initial energy substrate during the early germination stage. Bhardwaj et al. (2012) reported that there were significant activities of protease and β-amylase after 12 h of germination in cucumber seeds, suggesting that the decomposition of proteins and starch may be earlier than lipids. Pharr and Motomura (1989) observed that there was starch synthesis in the cucumber seeds after 36 h of germination.

Galactinol synthase is a key enzyme in the RFO biosynthesis pathway (Figure 3.1). Its activity was undetectable in cucumber seeds during the first 36 h after imbibition, which then increased rapidly between 36 to 60 h. RFO levels maintained at a stable low level during the latter stage, indicating that RFOs may be resynthesized during the catabolism of lipids and other large biological molecules (Handley et al., 1983b; Pharr and Motomura, 1989). At this stage, vascular bundles in cotyledons to the roots and growing seedling axis gradually formed, and the resynthesized RFOs may be transported in the phloem from the cotyledons (source) to the roots or seedling axis (sink) (Schaffer et al., 1996; Savage et al., 2013).

Enzymes in lipid metabolism pathway are more sensitive to low oxygen stress than those in starch metabolism pathway. As a result, seeds storing lipids as the major food reserve (such as cucurbit seeds) were more sensitive to anaerobic stress than those containing starch predominantly (Al-Ani et al., 1985). Thus, utilization of starch is important for seed germination under low oxygen conditions. The inhibition of lipid degradation and improvement of the utilization of readily metabolizable carbohydrates under anaerobic stress have been reported by several research groups (Pharr and Motomura, 1989). Todaka et al. (2000) also found that β-amylase activity and sucrose level in cucumber cotyledons increased after water stress. The increased soluble sugar level may act as both energy sources and protective agents. It is suggested that selecting genotypes with a high seed starch reserve may be a good idea for breeding new cultivars with high germination rate under abiotic-stressed conditions (Schaffer et al., 1996).

3.3 CARBOHYDRATE METABOLISM IN LEAVES

3.3.1 SINK-TO-SOURCE TRANSITION

Young leaves of cucurbits are heterotrophic organs whose growth is maintained by imported soluble carbohydrates from mature leaves. Turgeon and Webb (1975) indicated that the leaf of *C. pepo* was first capable of net CO_2 fixation when 8% expanded, began to export excess photosynthate when the phase of rapid decrease in relative growth rate was almost complete at about 45% expansion, and reached the maximum net photosynthesis rate at 70% expansion. Similar to other dicotyledonous plants, the maturation (transition from sink to source) of the cucurbit leaf develops in a basipetal direction (Figure 3.3) (Turgeon and Webb, 1973; Savage et al., 2013). Assimilates import into the lamina tip of *C. pepo* leaves stops when the blade is 10% expanded, while the leaf base still needs assimilate supply at this time. As a result, sugars exported from the leaf tip are redistributed to the less-matured leaf base during this period, delaying export from the lamina until the blade is 35% expanded (Turgeon and Webb, 1973). During the sink-to-source transition, there may be a short period in which assimilates travel toward the leaf in the adaxial

FIGURE 3.3 Transition of immature leaves from sink to source tissue. The perfusion of the phloem-mobile tracer (CF; green) declines with transport into the leaf, indicating reduced dependence on external carbon. Cucumber leaves were imaged with a fluorescent scope during a 5-day period to visualize their source–sink transition. Bar = 5 mm. (Copied from Savage, J.A. et al., *Plant Physiol.*, 163, 1409, 2013. With permission.)

phloem and away from the leaf in the abaxial phloem simultaneously (Peterson and Currier, 1969). In addition, the loss of import capacity of the petiole is also basipetal and dorsoventral (Turgeon and Webb, 1973).

Transition of cucurbit leaves from sink to source involves complex anatomical changes. First, intercellular air spaces increase as the leaf expands, which greatly enhances the CO_2 diffusion and fixation. Second, as the cucurbit leaf grows, maturation of veins develops progressively from the largest toward the smallest elements, and commencement of sugar export is coincident with maturation of the abaxial phloem of the minor veins (Turgeon and Webb, 1976).

Changes in soluble carbohydrate levels were observed during leaf development. In cucumber, sucrose and raffinose concentrations decrease, while galactinol and stachyose concentrations increase during leaf expansion (Pharr and Sox, 1984). In mature cucumber leaves, stachyose is the most abundant soluble carbohydrate (Pharr and Sox, 1984). Sucrose is synthesized at all development stages, whereas the first detectable synthesis of raffinose and stachyose coincides with the beginning of assimilate export from leaf tip (Turgeon and Webb, 1975). The raffinose pool in sink leaves is from the degradation of imported stachyose, rather than from *in situ* synthesis (Turgeon and Webb, 1975).

Activity changes of enzymes involved in stachyose degradation and synthesis play an important role during leaf sink-to-source transition. The function of acid α-galactosidase during leaf development is not clear. Thomas and Webb (1978) and Smart and Pharr (1980) indicated that there was no significant correlation between the acid α-galactosidase activity and the leaf development of cucumbers and pumpkins. However, Pharr and Sox (1984) found that acid α-galactosidase activity declined during leaf development. More evidences support that alkaline α-galactosidase is involved in the catabolism of RFOs imported from the phloem in sink leaves (Gaudreault and Webb, 1982, 1986; Pharr and Sox, 1984). They found that the activity of alkaline α-galactosidase is high in young leaves and declines as leaves expand. Yang (2012) reported that the mRNA level of cucumber *alkaline α-galactosidase 1* (*AGA1*, Genbank accession number DQ157703) was positively correlated with the activity of alkaline α-galactosidase during leaf development, indicating *AGA1* may play a role in RFO unloading in young leaves of cucumber. On the other hand, galactinol synthase activity increases during leaf growth, coinciding with the increasing accumulation of stachyose in maturing pumpkin leaves (Pharr and Sox, 1984).

3.3.2 Phloem Loading

The vascular anatomy of the cucurbit leaf affects assimilate loading profoundly, and the following description is based primarily on the studies of Turgeon et al. (1975) and Schmitz et al. (1987) (Figure 3.4). In cucurbits, veins in mature leaves are always classified into seven orders, among which veins of orders 1–3 are defined as major veins and veins of orders 4–7 as minor veins, according to whether they contain intermediary cells. Photosynthate loading of mature leaves occurs at the

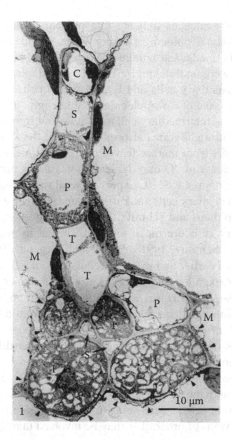

FIGURE 3.4 Transverse section through a minor vein of *Cucumis melo*. S, sieve element; I, intermediary cell; P, parenchyma cell; C, companion cell; T, tracheid. Note the large number of plasmodesmata (▲) in the common wall between intermediary cells and mesophyll cells (M). (Copied from Schmitz, K. et al., *Planta*, 171, 19, 1987. With permission.)

minor veins. The veins in cucurbit leaves are bicollateral, even including veins of order 7. In these smallest veins, the adaxial phloem and abaxial phloem are separated by one to two tracheid cells and a parenchyma cell. The whole minor vein is bounded by a single layer of bundle sheath cells. The areoles are delimited by small veins, yet the distance between any mesophyll cell and the next vein is no more than three to four cell diameters.

The adaxial phloem in the smallest vein is composed of one sieve element and one companion cell, and the adaxial sieve elements are approximately equal in diameter to their companion cells. These companion cells have typical characteristics of common companion cells, that is, a central vacuole surrounded by a dense layer of cytoplasm, which are termed as ordinary companion cells. In the adaxial phloem, plasmodesmata connection is observed between the companion cell and the sieve element but not between the companion cell or sieve element and the bundle sheath cells.

The abaxial phloem in the smallest veins consists of one to two sieve elements and two to four companion cells. The abaxial companion cell is specialized, has dense cytoplasm, and contains numerous mitochondria and small vesicles, which are termed as the intermediary cell. The adjacent cell walls of intermediary and bundle sheath cells are traversed by numerous plasmodesmata, which occur in clusters in large pit fields. The plasmodesmata are highly branched, more so on the intermediary cell side than on the bundle sheath side. In the abaxial phloem, the intermediary cell is much larger than the sieve element, which is smaller than the sieve element in the adaxial element. There is also extensive symplastic contact between intermediary cells and sieve elements; these plasmodesmata are also branched on the intermediary cell side.

A considerable amount of evidences indicate that phloem loading of cucurbits occurs by a symplastic pathway at the abaxial phloem of minor veins. There are abundant plasmodesmata connecting intermediary cells to adjacent bundle sheath cells and intermediary cells to sieve elements. Dye-coupling studies were undertaken by Turgeon and Hepler (1989) to determine whether plasmodesmata between intermediary cells and bundle sheath cells in the minor veins of mature *C. pepo* leaves are open to passage of low-molecular-weight compounds, and the results indicated that the dye can spread from one intermediary cell to another and from intermediary cells to bundle sheath and mesophyll cells. Autoradiograms also indicated that the abaxial phloem of minor veins is responsible for carbon export from mature leaves of *Cucumis melo* (Schmitz et al., 1987). The authors also found that the release of ^{14}C into the leaf apoplast was minor and PCMBS (a sucrose transporter inhibitor) only slightly reduced ^{14}C export. Several studies have suggested that stachyose synthase is located in the intermediary cells and raffinose and stachyose for long-distance translocation are synthesized there (Holthaus and Schmitz, 1991a,b). In addition, concentrations of raffinose and stachyose in intermediary cells are much higher than that in mesophyll cells (Schmitz and Holthaus, 1986; Holthaus and Schmitz, 1991b; Beebe and Turgeon, 1992; Haritatos et al., 1996). Based on the results mentioned earlier and the research data from other RFO-transporting species, Turgeon (1991) suggested a phloem loading process known as polymer trapping. According to this model, sucrose from the mesophyll diffuses to the bundle sheath and then into the intermediary cells and is converted to raffinose and stachyose. These RFOs are too large to diffuse back to the bundle sheath and mesophyll through the intermediary cell plasmodesmata. As a result, they accumulate to high concentrations in the intermediary cell. They further diffuse into the sieve element for long-distance translocation.

In lower-order (larger) veins, the number of SE–CC complex increases. Only those companion cells that have direct contact with the bundle sheath cells are specialized as intermediary cells; others are still ordinary companion cells. These veins (6–4) are considered to be responsible for both loading and transport, while veins of orders 1–3 may be involved mainly in long-distance transport (Schmitz et al., 1987).

Three pathways of phloem loading exist in the plant kingdom, symplastic and passive, symplastic followed by polymer trapping, and transporter driven via the apoplast (Rennie and Turgeon, 2009). Individual species always employ multiple loading strategies, as we have seen in cucurbits (Slewinski et al., 2013). If compartmentation is taken into account (cytosol made up only 5% of the mesophyll cell volume), sucrose level is higher in mesophyll cells than in intermediary cells (Haritatos et al., 1996). Thus, sucrose could be loaded through a symplastic and passive way. In addition, sucrose could be loaded at the adaxial phloem of minor veins through an apoplastic way. This may explain why the sucrose level in the phloem sap is higher than that in the sieve element–intermediary cell complex (Haritatos et al., 1996). Experimental data from CMV-infected *C. melo* support this view (Shalitin and Wolf, 2000; Gil et al., 2011, 2012). The authors found that CMV infection increases the sucrose–stachyose ratio in phloem sap of *C. melo*. Further research indicated that the enhancement expression and activity of a sucrose transporter was responsible for this increase. Furthermore, the phloem loading was inhibited by the sucrose transporter inhibitor PCMBS in this case. The physiological significance of CMV-induced quantitative shift from symplastic to apoplastic phloem loading is not clear. It is suggested that the plant partly closed the symplastic loading pathway to prevent CMV entering the phloem and improved the apoplastic loading pathway to ensure enough assimilate supply.

3.4 LONG-DISTANCE PHLOEM TRANSPORT

There are two distinct types of phloem in cucurbits, fascicular phloem (FP) and extrafascicular phloem (EFP). FP is bicollateral and presents in the vascular bundles on both sides of the xylem (so-called internal FP and external FP, respectively). EFP exists outside the bundles, which is composed of peripheral sieve tubes located at the margin of vascular bundles, entocyclic sieve tubes

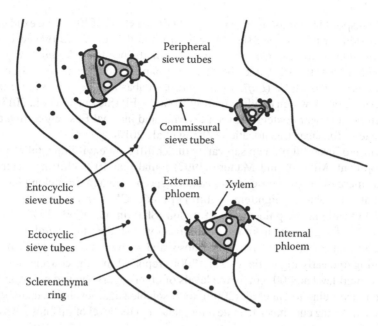

FIGURE 3.5 Schematic diagram of a transverse section of a pumpkin stem. (Adapted from Crafts, A.S., *Plant Physiol.*, 7, 183, 1932.)

just inside the sclerenchyma ring, and laterally oriented commissural sieve tubes linking entocyclic sieve tubes, peripheral sieve tubes, and fascicular sieve tubes to each other. EFP also includes the longitudinal ectocyclic sieve tubes located outside the sclerenchyma ring (Figure 3.5) (Crafts, 1932; Zhang et al., 2012).

After loading at the minor vein, assimilates enter the phloem for long-distance translocation. Fruits of cucurbits usually grow fast (some giant pumpkins probably have the fastest-growing fruit in the plant kingdom), indicating the mass flow rate in the phloem is quite fast in these plants. Data from both fruit weight increase measurement and ^{14}C labeling experiment suggest that assimilate translocation velocity is between 55 and 160 cm h^{-1} in cucurbits (Crafts and Lorenz, 1944; Schaffer et al., 1996). Environmental factors, including light, temperature, and CO_2 concentration, affect assimilate translocation rate profoundly, either by altering the quantity of dry matter available for transport or by directly affecting the flow velocity in the phloem. Several reports have focused on the negative effect of low temperature on the assimilate translocation in cucurbit phloem (Murakami and Inayama, 1974; Toki et al., 1978; Kanahamam and Hori, 1980). In a typical Chinese energy-saving sunlight greenhouse, the temperature at daytime could reach 30°C or higher and would drop to 12°C or even lower at night. Thus, low translocation rate of photoassimilates from source to sink under low night temperature becomes one of the key limitations of cucumber production in these greenhouses in winter, and cultivars with relatively high assimilate translocation rate under low night temperature have been the target of Chinese cucumber breeders (Miao et al., 2009). On the other hand, Matsumoto et al. (2012) reported that heating bearing shoots near fruits promote sugar accumulation in melon fruit.

Stachyose, sucrose, and raffinose are major sugars translocated in the phloem of cucurbits (Weidner, 1964; Pharr et al., 1977; Richardson et al., 1982, 1984; Mitchell et al., 1992). It is difficult to sample "pure" phloem exudate from cucurbit plants and measure its exact sugar concentration since the fluid is always contaminated by adjacent parenchyma cells and diluted by the xylem sap (Zhang et al., 2012). Microdissecting FP tissues from freeze-dried stem is a more accurate method than stem cutting for phloem sap sampling. Using this technique, Zhang et al. (2010) found that the total sugar content is around 1 M in FP, which is consistent with other species and the measurements in cucurbit leaf cells. However, sugar concentration in EFP is much lower than that in FP, indicating that FP is largely responsible for assimilate transport (Zhang et al., 2010). If the exudate is collected from the incised stem or petiole, one should keep in mind that there are two types of cucurbits; the first group includes pumpkin (*Cucurbita maxima*) and Zucchini (*C. pepo*), whose exudate is primarily originated from the EFP, and the second group includes cucumber, watermelon, bitter apple (*Citrullus colocynthis*), luffa (*Luffa acutangula*), calabash (*Lagenaria siceraria*), and winter melon (*Benincasa hispida*), which exudate mainly from the FP (Slewinski et al., 2013). There is no significant difference of sugar content between external and internal FPs, suggesting that they may have similar capacity for sugar translocation (Zhang et al., 2010).

The composition of cucurbit phloem sap varies under different environmental conditions or during plant development. Mitchell and Madore (1992) found that 10°C chilling treatment for 72 h caused a general increase of stachyose, raffinose, and sucrose content in melon phloem sap. It has been reported that some abiotic (high temperature) or biotic (CMV infection) stresses can increase the sucrose to RFO ratio in the phloem sap of melon (Shalitin and Wolf, 2000; Gil et al., 2012). The diurnal rhythm of sugar content in the phloem sap was studied in several cucurbits. Mitchell et al. (1992) found that levels of stachyose, raffinose, and sucrose in melon phloem sap increased during the morning and early night. Jiang et al. (2006) reported that the concentration of stachyose achieved the maximal level at 4:00 p.m. while levels of other sugars showed no distinct diurnal patterns. In addition, according to Hu et al. (2009), total soluble sugar levels in cucumber phloem sap gradually increased during cucumber fruit development, and the level of raffinose increased smaller than those of stachyose and sucrose.

Two mechanisms may be involved in the composition change of cucurbit phloem sap. First, some environmental factors may alter the loading pathways, as mentioned in Section 3.2. A typical example is CMV infection caused by sucrose transporter activation in melon minor veins, which led to a quantitative shift from symplastic loading to apoplastic loading and an increase of sucrose to stachyose ratio in the phloem sap (Gil et al., 2011; Slewinski et al., 2013). Second, solutes are exchanged between the phloem and surrounding tissues, and the unbalanced exchange of some solutes may cause composition variation in the phloem sap during long-distance translocation. Richardson et al. (1982) suggested that monosaccharide concentration is higher in the phloem sap sampled near the sink tissue than that near the leaves. Ayre et al. (2003) proposed that in *Coleus blumei* (RFO-transporting species), sucrose leaks from the phloem symplast and is retrieved by transporters at similar rates. As a result, the concentration of sucrose remains unchanged during transport. Galactinol also leaks from the phloem but is not retrieved, and it is eventually depleted from the translocation stream. RFOs leak minimally and can be efficiently transported to sink tissues. The authors indicated that one advantage of transporting RFOs is it reduces solute leakage during long-distance transport and hence improves translocating efficiency.

3.5 ASSIMILATE PARTITIONING

In cucurbit plants, the older fruit gets higher priority in obtaining assimilates over the younger fruit, which is known as "first-fruit inhibition" (McCollum, 1934; Schaffer et al., 1996). This inhibitory effect may contribute to the mechanism of cucumber plants to obtain mature fruits and seeds in a shorter period when the total assimilate supply is limited. However, in cucurbit production, this limited simultaneous fruit set pattern greatly reduces the yield, especially in once-over harvest systems (Ramirez and Wehner, 1984).

Assimilate limitation is an obvious explanation for this inhibitory effect. In general, a fruit is a strong sink for assimilates. Several leaves act as principal assimilate suppliers for individual fruits. When assimilate supply is limited, fruit competition results in fruit drop, particularly among the later-set fruits (Ho, 1992). Ells (1983) observed that about 9–10 leaf nodes typically intervene between successfully growing fruits in picking cucumber plants, suggesting that assimilates from about 10 leaves may be required simultaneously for the growth of an individual fruit. During rapid growth period, about 3300–3400 mg assimilates are required to be transported to an individual fruit everyday (Schapendonk and Challa, 1980; Pharr et al., 1985), while the leaves can provide no more than 225 mg of assimilate dm^{-2} during an 11 h photoperiod (Pharr et al., 1985). According to the average area of leaf blade, a balanced source–sink relationship between approximately 9–10 leaves and a single fruit can be calculated. In melon plants, assimilates from six leaves were needed for the growth of a single fruit (Hughes et al., 1983). In general, fruits tend to assimilate photosynthate from closer leaves (Hughes et al., 1983; Shishido et al., 1992). However, Zhang et al. (2015) found that in cucumber plants setting two fruits, assimilates from ^{14}C-labeled leaves tended to move to the first fruit, although these leaves were closer to the second fruit. In watermelon plants, upper leaves of fruit set node were more important for fruit growth than lower leaves of fruit set node (Lee et al., 2000).

Another possible factor involved in the limited simultaneous fruit setting of cucumber plants is the level of plant hormones. It was reported that application of some plant growth regulators, such as chlorflurenol or 6-benzyl aminopurine (BA), could overcome the inhibition effect caused by the early fruit set (Cantliffe, 1976; Baniel et al., 2008; Zhang et al., 2015). Schapendonk and Brouwer (1984) also pointed out that limiting assimilate supply by defoliation caused reduced fruit growth rate rather than abortion, indicating other factors may be involved in the growth competition between fruits. Zhang et al. (2015) revealed that in cucumber plant with two fruits, sucrose and trehalose-6-phosphate (T6P) levels in the peduncle of the first fruit are higher than those of the second fruit, while sucrose nonfermenting 1–related kinase 1 [SnRK1] activity is lower in the peduncle of the first fruit than that of the second fruit from 0 to 8 days after anthesis. The growth rate and sink activity of the second fruit were enhanced by removing the first fruit or by treating it with BA, in comparison with an increase of sucrose and T6P levels and a decrease of SnRK1 activity in its peduncle. Furthermore, the mRNA level of *CsAGA1* in cucumber calli was upregulated by exogenous trehalose treatment. These results suggested that T6P- and SnRK1-mediated signaling is involved in the regulation of cucumber first-fruit inhibition.

Growth competition exists not only between fruits on the same plant but also between fruits and vegetative tissues and between different vegetative tissues in plants without fruit. In fruiting plants, the fruit is the strongest sink. Compared with vegetative plants, the growth of young leaves, petioles, stems, and roots was significantly suppressed in fruiting plants (Pharr et al., 1985). The influence of fruit setting on the growth of root was greater than that of aboveground vegetative parts in cucumber (Marcelis, 1994). In vegetative plants, photoassimilates tended to translocate to root at first in melon (Schmitz et al., 1987), while in pumpkin, young leaves and seedling shoots were the regions most actively importing assimilates (Webb and Gorham, 1964). In cucumber plants with no fruit removed, the partitioning rates of $^{14}CO_2$ in fruits, stems, roots, and leaves were 90%, 5%, 3%, and 2%, respectively; in vegetative cucumber plants, $^{14}CO_2$ was mostly distributed into stems (Choi et al., 1997). The effects of restricted root growth on the leaf carbohydrate metabolism were studied by Robbins and Pharr (1988). The results showed that the reduced sink demand, induced by restricted root growth, may led to increased starch concentrations and a reduction in stachyose biosynthesis (indicated by a decrease of galactinol synthase activity) in cucumber source leaves.

Compared with the vegetative plants, the carbon exchange rate was higher in plants bearing fruits (Marcelis, 1991). Pharr et al. (1985) suggested that carbon exchange rate, carbon export rate, and starch accumulation rate in the day and starch degradation rate at night were higher in the plants with fruits than those without fruits. Higher carbon export rate is often accompanied by higher leaf stachyose synthase activity but has no relevance with sucrose phosphate synthase (SPS) and galactinol synthase (Pharr et al., 1985; Holthaus and Schmitz, 1991b; Miao et al., 2003), while the concentration of galactinol in the vegetative plants is significantly higher than fruiting plants (Pharr et al., 1985; Miao et al., 2003). Additional fruit loading did not increase the photosynthetic rate and assimilate partitioning rate to the fruits, which further confirmed that there is competition for assimilates among fruits (Marcelis, 1991). Girdling of the petiole had a similar effect on the leaf photosynthesis by removing fruits (Mayoral et al., 1985).

The diurnal fluctuation of the assimilate export rate of cucumber has been studied by several research groups. Some researchers suggested that the assimilate export rate is higher during daytime (Verkleij and Baan Hofman-Eijer, 1988; Verkleij and Challa, 1988). However, Toki et al. (1978) reported that photoassimilates were mainly exported from leaves during early night. According to these data, the authors proposed that higher temperature should be maintained during early night to facilitate carbon export, while relatively lower temperature during later night is beneficial to reduce respiration. Miao et al. (2009) found that under 28°C/22°C (day/night) condition, cucumber fruits grew fast during afternoon and early night and slowly during late night and morning, while under 28°C/12°C condition, the fruit growth during the night reduced significantly, indicating that carbon export from leaves was inhibited by low temperature. Verkleij and Baan Hofman-Eijer (1988) reported that under a short-day condition, the accumulation of dry matter and the volume expansion of cucumber fruit were out of phase during a diurnal cycle. Most assimilates were exported to the fruit during the 8 h light period, while about 70% fruit growth occurred during the 16 h dark period. This conclusion was confirmed in our study under a similar condition (Miao et al., 2009).

Branch removing, topping, and fruit thinning are common practices to manipulate the source–sink relationship during cucurbit cultivation. Fruit thinning is usually carried out in the production of cucurbits with large fruits, such as melon, watermelon, and pumpkin, since strong competition exists among fruits in these plants (Long et al., 2004). Besides fruits, the seedling shoot is another strong sink in the cucurbit plant. Thus, lower parts of topping plants always accumulate more assimilates than those of nontopping plants (Lee et al., 2000). Park et al. (2002) demonstrated that compared with one-stem plants, no more than four extra side shoots provided more assimilates to roots, stimulating the root activity and root growth of hydroponically grown cucumber plants. In addition, it has been reported that plants tend to accumulate more assimilates into roots under moderate nutrient deficiency conditions (Ciereszk et al., 1996; Hermans et al., 2006). Although little research is carried out on cucurbit plants in this area, we did observe similar phenomenon during cucurbits cultivation.

3.6 PHLOEM UNLOADING

An understanding of the vascular anatomy of the pedicel and fruit is beneficial to unveil the unloading mechanism in cucurbits. The vascular network of cucumber and melon were described by several research groups (Barber, 1909; Judson, 1929; Kanahama and Saito, 1987; Masuko et al., 1989). Basically, there are 10 main bundles in the pedicel. These bundles extend into the fruit, further branch and anastomose and extend inward to the placenta and outward to the epidermis. In addition to the main vascular bundles there is also an anastomosing minor vascular system in the fruit.

α-Galactosidase is the initial enzyme in the metabolic pathway of stachyose and raffinose (Figure 3.2). Plant α-galactosidase can be divided into two groups, acid and alkaline, based on their activity response to pH. There is a negative correlation between the activity of alkaline α-galactosidase and stachyose levels in cucumber pedicels, indicating that the alkaline activity was responsible for stachyose hydrolysis (Pharr and Hubbard, 1994) rather than the acid activity. Irving et al. (1997) also found that alkaline α-galactosidase activity was higher than acid activity in *Cucurbita maxima* fruit at anthesis. Two alkaline α-galactosidase genes were identified in the melon genome, namely, *CmAGA1* (AY114164) and *CmAGA2* (AY114165) (Carmi et al., 2003). *CmAGA2* was relatively specific for stachyose, whereas *CmAGA1* showed significant activity with both stachyose and raffinose. *CmAGA1* enzyme activity increased during the early stages of melon ovary development and fruit set, while *CmAGA2* declined in enzyme activity during this period. From these data, they supposed that *CmAGA1* may play a key role in melon assimilate unloading (Gao et al., 1999; Gao and Schaffer, 1999; Carmi et al., 2003). However, the data mentioned here cannot rule out the role of acid α-galactosidase in RFO catabolism during sink unloading in cucurbit plants.

There are two opinions about where RFOs are metabolized to sucrose when they reach the fruit sink. Some researchers thought that the catabolism of RFOs to sucrose takes place in the pedicel. There are several evidences supporting this view. (1) Little RFOs were detected in the fruits (Pharr et al., 1977; Gross and Pharr, 1982; Handley et al., 1983b; Hughes and Yamaguchi, 1983). (2) Significant activity of enzymes responsible for RFO catabolism, such as α-galactosidase, sucrose synthase (SS, synthetic activity), SPS, and UDP-galactose pyrophosphorylase, existed in the pedicel (Smart and Pharr, 1981; Gross and Pharr, 1982; Burger and Schaffer, 1991). (3) The ratio of RFO to sucrose in the pedicel was lower than that in the petiole and stem, indicating that RFOs were hydrolyzed to sucrose in the pedicel (Ohkawa et al., 2010).

However, other researchers suggested that RFOs are translocated into the fruit and rapidly metabolized to sucrose then. The evidences supporting this view are as follows: (1) RFOs indeed exist in the cucurbit fruit, although the level is quite low (Hubbard et al., 1989; Chrost and Schmitz, 1997; Irving et al., 1997; Kim et al., 2007; Hu et al., 2009). In addition, exudate from the fruit's main bundles contained stachyose as the major sugar (Pharr and Hubbard, 1994; Schaffer et al., 1996). (2) Significant activities of alkaline α-galactosidase and other RFO metabolic enzymes were detected in cucurbit fruits (Pharr and Hubbard, 1994; Schaffer et al., 1996; Chrost and Schmitz, 1997; Irving et al., 1997; Gao and Schaffer, 1999; Carmi et al., 2003; Dai et al., 2006). (3) The sugar composition of phloem exudate collected from peduncles cut near the stem or fruit of the melon plant was identical, indicating RFOs are translocated into cucurbit fruits (Chrost and Schmitz, 1997).

Recently, we have studied vascular exudate from cucumber pedicels and fruits. We cut the pedicels and fruits from near the stem end to the fruit end and collected and analyzed exudate from each cut surface (from plant side). The results showed that RFO levels in the exudate gradually decreased from the stem end to the fruit end, indicating RFOs were hydrolyzed throughout the main vasculature along the pedicel and fruit (unpublished date). In addition, in melon pedicels, sucrose and hexoses were prevalent until day 18 after anthesis, while from the 25th day after anthesis, the RFO increased until fruit maturity (Chrost and Schmitz, 1997), suggesting the location of RFO catabolism may change from pedicel to fruit during fruit development.

Hu et al. (2011) reported that the sieve element-companion cell (SE–CC) complex of the main bundles in cucumber fruit is apparently symplasmically restricted and there are cell-wall invertase and sucrose transporter located at the SE–CC complex. The authors concluded that phloem unloading pathway in cucumber fruit is apoplasmic. In addition, the RFOs were absent in the tissue immediately adjacent to the main bundles, indicating the location of RFO metabolism is restricted in the main vasculature area (Schaffer et al., 1996). Cheng et al. (2015) cloned three cucumber hexose transporter genes, among which the protein of *CsHT3* is located at the plasma membrane, and its expression level increased in peduncles and fruit tissues along with cucumber fruit enlargement, suggesting that *CsHT3* probably plays an important role in apoplastic phloem unloading of cucumber fruit. If the apoplasmic unloading pathway is true, acid cell-wall α-galactosidases should be responsible for the RFO hydrolysis when they are transferred to the free space around the SE–CC complex. However, there are abundant evidences indicating that alkaline α-galactosidase is important during RFO unloading (Pharr and Hubbard, 1994; Irving et al., 1997; Gao et al., 1999; Gao and Schaffer, 1999; Carmi et al., 2003). Another mystery is why RFOs cannot symplasmically diffuse from main vasculars to those anastomosing minor veins throughout the fruit. In summary, much remains to be studied to elucidate the full mechanism of assimilate unloading in cucurbits.

Little is known about how assimilates are unloaded in seedling shoots, young leaves, and roots of cucurbits. Activity of alkaline α-galactosidase is higher than that of acid α-galactosidase in immature cucumber leaves (Pharr and Sox, 1984). Furthermore, the activity of alkaline α-galactosidases is higher in immature leaves and roots than in mature leaves (Gaudreault and Webb, 1982, 1986). These results indicated that RFO may be translocated to immature leaves or roots symplasmically and then hydrolyzed by alkaline α-galactosidases.

3.7 CARBOHYDRATE METABOLISM IN FRUITS

Fruit size varies significantly among cucurbit plants. The final fruit size depends on how fast and how long it grows. However, Sinnott (1945) thought the size of cucurbit fruit is mainly dependent on the duration of growth; growth rate plays a minor role. The growth pattern of most cucurbit fruits generally follows an S-curve (Sinnott, 1945). Among those that belong in the Cucurbitaceae family, cucumber, melon, watermelon, and pumpkin are especially important for their economic value. As a result, most researches about fruit carbohydrate metabolism of cucurbits are focused on these four species.

3.7.1 CUCUMBER

Glucose and fructose are main soluble sugars in cucumber fruits (6–12 mg gFW^{-1}), and the level of sucrose is low (about 0.5 mg gFW^{-1}) (Handley et al., 1983b; Schaffer et al., 1987). Although a few RFOs were found in young fruits, little RFOs were detected in mature cucumber fruits (Pharr et al., 1977; Handley et al., 1983b). However, Hu et al. (2009) found that RFOs can be detected in cucumber fruits 20 days after anthesis, which make up 2.51% of total soluble sugars. Similar results were observed by Wang (2014). The different results may be due to the different cultivars they used or different fruit parts they sampled. Handley et al. (1983b) used a pickling cucumber variety, while Hu et al. (2009) and Wang (2014) used North China–type cucumber varieties. Furthermore, fruit tissue containing major veins may have relatively higher RFO levels, as mentioned in Section 3.6. In addition, starch grains in the cucumber mesocarp were observed by Patchareeya et al. (2011). The content of starch of cucumber fruit ranged from 1 to 7 mg gFW^{-1} (Schaffer et al., 1987; Hu et al., 2009).

The fluctuation patterns of carbohydrate levels during cucumber fruit development have been studied by several groups. Handley et al. (1983b) and Schaffer et al. (1987) suggested that the contents of glucose, fructose, sucrose, and starch in cucumber fruit almost remained unchanged during the development. However, Hu et al. (2009) reported that contents of glucose and fructose increased in the cucumber fruit from anthesis to 20 days after anthesis, while levels of sucrose, stachyose, raffinose, and starch decreased during this period. The different results may be due to the different varieties they used or different development stages they observed.

Little is known about the diurnal pattern of carbohydrates in cucumber fruits. Jiang (2006) reported that levels of glucose and fructose were low during 10:00–13:00 and had a peak value at 16:00, while levels of sucrose and starch remained stable during a photoperiod. The fluctuation of the hexose level should be caused by the diurnal changes of both assimilate tanslocation into the fruits and the carbon consumption by fruit respiration and growth and should be affected remarkably by the environment.

Significant acid invertase and SS (cleavage direction) were observed during cucumber fruit development, and the activities of sucrose degrading enzymes were always higher than that of sucrose-synthesizing enzymes in mesocarp tissues (Schaffer et al., 1987; Hu et al., 2009). These data may explain why levels of glucose and fructose are much higher than that of sucrose in cucumber fruits. Activities of raffinose synthase and stachyose synthase were also detected in cucumber fruits (Hu et al., 2009; Sui et al., 2012), indicating there may be basic RFO synthesis in the mesocarp tissues.

The contribution of green cucumber fruit photosynthesis to its own dry matter accumulation cannot be ignored. The quantity of fixed carbon by cucumber fruit was about equal to its respiration consumption (Todd et al., 1961). Marcelis and Baan Hofman-Eijer (1994) reported that photosynthetic contribution of cucumber fruits to their carbon accumulation was about 1%–5%. Obviously, chlorophyll level of the fruit epidermis, fruit shape, and fruit age are major factors impacting the importance of cucumber fruit photosynthesis.

3.7.2 MELON

Melon shows abundant genetic variation in fruit characteristics. Considerable variation in the sugar content and composition in mature fruits was observed (Stepansky et al., 1999). In general, the most dominant soluble sugar in the fruit of sweet cultivars is sucrose, while glucose and fructose are primary sugars in nonsweet cultivars. One exception is the Soviet cultivar "Kuvsinka," which belongs to the cantalupensis type, containing 82 mg gFW^{-1} soluble sugars in the fruit, but the contents of sucrose, glucose, and fructose are 24, 19, and 28 mg gFW^{-1}, respectively (Stepansky et al., 1999). In most types, the content of glucose is always similar to that of fructose. There are also some exceptions. Stepansky et al. (1999) reported that an Indian cultivar that belongs to chito type contains 19.6 mg gFW^{-1} glucose and 30.2 mg gFW^{-1} fructose in the fruit. Hubbard et al. (1989) reported that about 0.15 ± 0.06 mg gFW^{-1} starch was detected in melon fruit. The existence of a small amount of starch in the melon fruit was reported by a few other researches (Schaffer et al., 1987; Combrink et al., 2001). In addition, low concentrations of the RFO and galactose were also detected in melon fruits (Hubbard et al., 1989; Chrost and Schmitz, 1997; Kim et al., 2007). RFOs in the ovary will disappear after development, similar to what happens during cucumber fruit development (Hughes and Yamaguchi, 1983).

In general, soluble sugars were present in greater concentrations in the inner tissues than in the outer tissues in melon fruits (Lingle and Dunlap, 1987; Hubbard et al., 1989; Combrink et al., 2001). The difference in sugar concentration between inner and outer tissues is mainly caused by the concentration difference of sucros rather than those of hexose (Hubbard et al., 1989). The distribution pattern of RFO is similar to that of sucrose (Hubbard et al., 1989). In the other direction, sucrose and total sugar gradients were observed ascending from mesocarp adjacent to pedicle to mesocarp adjacent to umbilicus (Zhang and Li, 2005).

Generally, there are two types of melon: the high sucrose accumulation type and low sucrose accumulation type (Stepansky et al., 1999; Burger et al., 2009). For high sucrose accumulation type, in the early fruit development stage, the concentration of sucrose remains low, while glucose and fructose levels are relatively high. During the later stage, sucrose accumulates quickly, and its level achieves a similar level as hexose in some genotypes, or much higher than hexose levels in other varieties. For low sucrose accumulation type, the concentration of sucrose is maintained at a low level throughout the fruit development (Lingle and Dunlap, 1987; Schaffer et al., 1987; Hubbard et al., 1989; Gao et al., 1999; Zhang and Li, 2005). The gradient of sucrose concentration between the inner and outer fruit tissues is due to the faster rate or longer duration of sucrose accumulation in inner fruit tissues (Lingle and Dunlap, 1987; Hubbard et al., 1989; Wang et al., 2014). For high sucrose accumulation type, the exponential stage of fruit growth is always earlier than sucrose quick accumulation stage, and when the fruit growth slows down in the late stage of the S-curve, sucrose still accumulates rapidly; the time lag between the two periods varies with different varieties (Lingle and Dunlap, 1987; Gao et al., 1999; Villanueva et al., 2004). In some varieties, hexose and starch content remained little changed or slightly increased during fruit development (Lingle and Dunlap, 1987; Schaffer et al., 1987; McCollum et al., 1988, Hubbard et al., 1989; Zhang and Li, 2005). However, in other varieties, their levels gradually decreased with the accumulation of sucrose, suggesting the conversion from hexose and starch to sucrose during this period.

Acid invertase and SPS are two key enzymes regulating sucrose accumulation during melon fruit development (Lingle and Dunlap, 1987; Schaffer et al., 1987; Hubbard et al., 1989; Ranwala et al., 1991; Lee et al., 1997; Lester et al., 2001; Burger and Schaffer, 2007). Sucrose accumulation is always accompanied by a decrease in acid invertase activity and an increase in SPS activity. In addition, the inner tissue has higher SPS activity and low acid invertase activity than the outer tissue in melon fruit, further proving the important role of the two enzymes in determining the sucrose level (Lingle and Dunlap, 1987; Wang et al., 2014). The role of the SS and neutral invertase in sucrose accumulation may be different among varieties and varies according to the state of the plant growth. Hubbard et al. (1989) showed that the activities of SS and neutral invertase are low throughout the fruit growth, indicating minor roles of these two enzymes in sucrose accumulation. Wang et al. (2014) also discovered that there is no significant difference in the two enzyme activities between inner and outer fruit tissues. However, Ranwala et al. (1991) and Lee et al. (1997) found that sucrose accumulation was accompanied by the sharp decline of neutral invertase activity. Schaffer et al. (1987) suggested that SS activities in both synthesis and cleavage direction increased during sucrose accumulation and proposed that SS is related to sucrose accumulation. Lingle and Dunlap (1987) found SS activity (cleavage) was higher than that of SPS in the period of sucrose accumulation and the activity in the outer tissue was higher than in the inner tissue. They speculated that the SS may provide the substrate to glycolysis for cell growth. Burger and Schaffer (2007) studied sucrose accumulation in seven melon genotypes and found that SS and neutral invertase activities were positively correlated with sucrose accumulation. They proposed that final sucrose content was determined by the duration of "sucrose accumulation metabolism," which was characterized by acid invertase activity less than threshold values, together with SPS, SS, and neutral invertase activities higher than threshold level. In addition, cell-wall acid invertase was also detected in the melon fruit (Schaffer et al., 1987; Ranwala et al., 1991). As mentioned in Section 3.6, there may be an apoplastic unloading pathway in the melon fruit. The cell-wall acid invertase may be responsible for the unloading of sucrose and the RFO.

Dai et al. (2011) analyzed the expression levels of 42 carbohydrate metabolism–related genes in melon fruits at different development stages by 454 pyrosequencing technology. They found that the expression levels of *CmAIN2* (ICuGI contig ID: MU22596), *CmSUS1* (MU21164), *CmNIN3* (MU26813), *CmHK1* (MU29719), *CmFK3* (MU22877), *CmAAG1* (MU20940), and *CmAAG2* (MU39965) were high in the early stage of fruit development, while significant expressions of *CmSPS1* (MU23155), *CmSPP1* (MU23296), and *CmSUS3* (MU31768) were observed in sucrose accumulation stage. The authors cannot correlate these expression patterns with the specific metabolic processes since most enzymes are encoded by more than one gene, so it is difficult to identify the contribution of each gene to the final activity. In addition, the functions of genes are also regulated at the posttranscription, translation, and posttranslation level, as well as in the presence of inhibitors. There may not be significant correlations between transcription levels and enzyme activities. Therefore, a lot of work needs to be done to clarify the function of these genes in fruit growth and carbohydrate metabolism in melon fruits.

3.7.3 WATERMELON

Glucose, fructose, and sucrose are important sugars in watermelon fruits (Kano, 1991; Yativ et al., 2010; Zhang, 2010; Liu et al., 2013). Broad variations in total sugar concentration and the ratio among the three sugars were observed in different genotypes. The total sugar concentration was as high as 100 mg gFW^{-1} in sweet genotypes but was as low as 10 mg gFW^{-1} in nonsweet genotypes (Liu et al., 2013). Yativ et al. (2010) suggested that glucose and sucrose levels ranged from 20% to 40%, whereas the fructose levels changed within 30%–50% in the commercial watermelon varieties. Liu et al. (2013) showed that the levels of sucrose and fructose were similar at 35 days after anthesis in a high sugar variety, while Zhang (2010) suggested that sucrose content was higher than hexoses in ripening fruit of an inbred line. Generally, fructose level is always higher than glucose level in mature watermelon fruits (Kano, 1991; Yativ et al., 2010; Zhang, 2010; Liu et al., 2013). Unlike melon, there is no significant correlation between the total sugar concentration and the sucrose concentration in watermelon fruits among different genotypes (Yativ et al., 2010).

The change pattern of sugar levels in watermelon fruits during development was extensively studied in different genotypes. Soluble sugar levels are very low within the first 10 days after anthesis and then fructose and glucose begin to accumulate, while the sucrose level begins to increase rapidly at 18–35 days after anthesis, depending on the cultivars and planting conditions (Elmstrom and Davis, 1981; Kano, 1991; Yativ et al., 2010; Zhang; 2010; Liu et al., 2013). In the latter period of the sucrose accumulation, glucose and fructose levels slightly increase or remain at a stable level in some genotypes (Elmstrom and Davis, 1981; Brown and Summers, 1985; Zhang, 2010) or decrease with the increase of the sucrose level in other genotypes (Elmstrom and Davis, 1981; Kano, 1991; Yativ et al., 2010). For nonsweet genotypes, three sugars are maintained at very low levels during the entire growth period (Liu et al., 2013).

There is an upward gradient of sugar levels from outer tissue to inner tissue in watermelon fruits, both longitudinally and equatorially (Kano, 1991). In the early stage of fruit development (before the sucrose accumulation), the sugar gradient was mainly caused by the different distributions of glucose and fructose, while in mature fruit, it was primarily due to the higher sucrose concentration in the inner tissues. There were no concentration gradients of glucose and fructose between the inner and outer tissues at this time, indicating that more hexoses were converted into sucrose in the inner tissue than in the outer tissue. In addition, sugar content was high around the seeds (Kano, 1991).

Comparing with the low-sucrose-accumulation genotypes, lower soluble invertase activity and higher SPS and SS activities were observed in high-sucrose-accumulation genotypes, indicating the important roles of these enzymes in the sucrose accumulation of watermelon fruits (Yativ et al., 2010). The role of SPS in sucrose accumulation of watermelon fruit has also been demonstrated in other studies (Zhang, 2010; Liu et al., 2013). As in the melon, the functions of SS and neutral invertase in sugar accumulation may depend on the variety. In low sugar accumulation varieties, activities of all enzymes were maintained at low levels (Liu et al., 2013). The role of cell-wall invertase in the watermelon fruit development is controversial. Yativ et al. (2010) found that insoluble invertase activity was high and constant throughout fruit development in genotypes accumulating low levels of sucrose, while in genotypes accumulating high levels of sucrose, the activity declined sharply 4 weeks after pollination. However, Liu et al. (2013) showed that insoluble invertase was the only enzyme whose activity was positively correlated to sucrose accumulation during fruit development. Insoluble invertase may be responsive for both sucrose and RFO unloading (Section 3.6; Roitsch and Gonzalez, 2004) and sucrose hydrolysis in the fruit. Thus, the relationship between insoluble invertase and sucrose accumulation depends on how much hexoses from sucrose and RFO apoplastic unloading are transferred into the cell and used to resynthesize sucrose.

3.7.4 PUMPKIN

Plants with edible fruits in genus *Cucurbita* are collectively called pumpkin or squash, mainly including *C. pepo, C. maxima, C. moschata*, and *C. argyrosperma*. Generally, there are two types of squash, summer squash and winter squash. Summer squash is harvested in the immature fruit stage, while winter squash is harvested in the mature fruit stage. Squash fruits contain glucose, fructose, sucrose, RFO, galactose, and starch; the content of these components changes among different types (Terazawa et al., 2011). Glucose and fructose are major soluble sugars in zucchini fruit, whereas galactose, RFO, and sucrose show very low concentrations. Irving et al. (1997) divided the developmental growth pattern of buttercup squash (*C. maxima*) into three phases: (1) early growth, from flowering up to 30 days after flowering; (2) maturation, from 30 days until 60 days after flowering (or harvest); and (3) ripening, from 60 days (or harvest) until about 100 days after flowering. During the early stage, dry matter and starch accumulation are largely completed. Fruits contain significant glucose, fructose, a little RFO, and very low sucrose. During the maturation stage, levels of dry matter and starch remain unchanged, while sucrose begins to accumulate. During ripening, starch is degraded and sucrose continues to accumulate. A similar pattern was observed during fruit development of other varieties of *C. maxima* and *C. moschata* (Tateishi et al., 2004; Sun et al., 2008). In addition, squash fruits contain a number of functional components such as polysaccharides and galactinol (Yang et al., 2008).

The content and property of starch largely determine the edible, storage, and processing quality of squash fruit. The fruit dry matter starch content ranges from 3% to 60% among different types (Stevenson et al., 2005), and the level of amylopectin is usually higher than that of amylose (Nakkanong et al., 2012). Structures and physicochemical properties of the starch from squash fruits were studied by Stevenson et al. (2005) and Zhou et al. (2013). They found that squash starch exhibited the B-type x-ray diffraction pattern with granules at the size of 1–15 µm. Scanning electron micrograph of pumpkin starch revealed that most of the granules of *Cucurbita moschata* starch were polygon-shaped, while those of *C. maxima* starch were oval-shaped (Yin, 2012). In addition, the gelatinization temperature and viscosity also vary among different types (Stevenson et al., 2005; Zhou et al., 2013).

As mentioned earlier, the sucrose concentration in the early development stage of squash fruit is very low. SS and acid invertase are two major enzymes responsible for the catabolism of sucrose translocated into the fruits. Irving et al. (1997) reported that SS activity was higher than that of acid invertase activity in young squash fruit, indicating that SS played a major role in sucrolysis during this period. However, data from Sun et al. (2008) showed that acid invertase was the major contributor toward sucrose decomposition at this stage. These results suggested that sucrose metabolism are different among various types; maybe in genotypes accumulating high concentration of starch, SS is more important, while in genotypes with abundant hexose, acid invertase plays a more significant role. On the other hand, high SPS activity in the late fruit development stage indicates that this enzyme is primarily responsible for sucrose accumulation during this period, although the import from mature leaf and the synthesis by SS in the synthesis direction cannot be ruled out (Irving et al., 1997; Tateishi et al., 2007). Nakkanonga et al. (2012) analyzed the expression levels of genes associated with starch biosynthesis of three genotypes and found that *AGPaseL* gene (encoding ADP-glucose pyrophosphorylase) expressed at early development stage and its transcript level was positively correlated with the starch content among different genotypes. A similar relationship was observed between transcript level of *GBSSI* (responsible for amylose biosynthesis) and the content of amylose. However, there was no significant correlation between *SSII*, *SBEII*, and *ISAI* (all involved in amylopectin biosynthesis) expression and the content of amylopectin. Data from Irving et al. (1999) suggested that starch breakdown during the fruit's late development stage or storage was primarily catalyzed by α-amylase in buttercup squash.

3.8 CARBOHYDRATE METABOLISM DURING SEED DEVELOPMENT

In the early stage of development, cucumber seeds contain relatively high levels of glucose, fructose, and sucrose, while the levels of RFO are quite low (Widders and Kwantes, 1995). As the cucumber seed developed, the contents of glucose and fructose decreased, and sucrose concentration first decreased and then increased at the late maturation stage. RFO concentrations increased throughout the fruit development (Handley et al., 1983b). Widders and Kwantes (1995) found that there was also a small amount of arabinose in cucumber seeds. In addition, the dry weight of seeds was positively related to fruit diameter in the cucumber (Widders and Kwantes, 1995), indicating a continuing assimilate input to cucumber seeds during fruit development. Differing from cucumber seeds, the main storage sugar in mature melon seeds is sucrose (Chrost and Schmitz, 1997).

Handley et al. (1983b) found that during the accumulation of RFO in cucumber seeds, galactinol synthase activity gradually increased. Funiculi from maturing fruit contained a high level of sucrose but little raffinose and stachyose. Holthaus and Schmitz (1991b) also reported that there are significant activities of raffinose synthase and stachyose synthase in the melon seeds. These results indicated that sucrose is transported to the seeds through funiculi and RFOs are biosynthesized *in situ* in the seeds. However, these evidences cannot exclude the possibility that RFOs are directly transported into the seeds through funiculi (Widders and Kwantes, 1995). In addition, cucumber seeds also contain a certain amount of starch (Handley et al., 1983b), but little is known about how starch is synthesized and accumulated during seed development.

3.9 CARBOHYDRATE METABOLISM IN ROOTS

The roots of Cucurbitaceae plants contain glucose, fructose, sucrose, and RFO (Du and Tachibana, 1994; Su et al., 1998). Du and Tachibana (1994) found that high temperature led to a significant increase of RFO in cucumber root. Further studies showed that this increase was not due to the inhibition of α-galactosidase activity. Su et al. (1998) suggested that soluble sugar significantly increased in roots of *Luffa cylindrica* and *Momordica charantia* under high temperature and waterlogging stress. In addition, some cucurbits, such as *Apodanthera biflora*, *Cucurbita foetidissima*, and *Cucurbita digitata*, store a large amount of starch in fresh roots (Berry et al., 1975; Bemis et al., 1978; Clark et al., 2012).

3.10 RFOs AND STRESS IN CUCURBITS

The roles of RFOs in plant response to abiotic stress have been well established. It is suggested that RFOs act as desiccation protectant with glassy state during seed maturation, osmoprotectant under dehydration stress, and antioxidant to scavenging reactive oxygen species (ElSayed et al., 2014). However, the relationship of RFOs and stress tolerance in RFO-transporting species, such as in cucurbits, has not been well characterized. Meng et al. (2008) reported that sucrose, glucose, fructose, galactose, raffinose, and stachyose were all upregulated by low-temperature treatment in cucumber seedlings. Stachyose synthase activity was higher in the cold-tolerant genotype than that in the cold-sensitive genotype and can be enhanced by abscisic acid (ABA) treatment. In addition, the expression of raffinose synthase gene could be induced by cold stress and ABA. As a result, raffinose synthase activity and raffinose content increased in the cucumber leaves under low temperature (Sui et al., 2012). Dong et al. (2011) also reported the levels of glucose, fructose, raffinose, and stachyose were increased in both leaves and roots by the treatment of salicylic acid, a stress-related phytohormone. These data indicated that RFO may also play a role in enhancing abiotic stress tolerance of cucurbits. In fact, there was a controversy about the site of RFO synthesis during the 1970s and 1980s. Most researchers think that the minor vein in the leaf is the location of RFO synthesis for transport outside, as mentioned in Section 3.3.2. However, a few RFOs were found to be synthesized in mesophyll cells (Madore and Webb, 1982). Sprenger and Keller (2000) cloned two galactinol synthase genes from *Ajuga reptans*, a RFO-transporting species, and found that one gene is responsible for synthesis of transport RFO and another for storage RFO. There are four galactinol synthase genes in the cucumber genome; we have studied the expression of these genes under normal and different stress conditions recently, and a similar phenomenon was observed (unpublished). Based on these data, we proposed that in cucurbits, basic synthesis of RFO in mesophyll cells under normal growth conditions is less. The synthesis of RFO would be enhanced when plants experience stress conditions.

3.11 FUTURE PERSPECTIVES

Recent studies have revealed that sugars act as not only carbon sources but also critical signaling molecules affecting most processes in the plant life, including carbohydrate metabolism itself (Smeekens and Hellmann, 2014). Important sugar signaling systems identified recently are hexokinase (HXK) glucose sensor, T6P signal, target of rapamycin kinase system, SNF1-related protein kinase 1 (SnRK1), and C/S1 bZIP transcription factor network. Some key enzymes involved in sucrose and starch metabolism, such as SPS, SS, UDP-glucose pyrophosphorylase, and α-amylase, were proven to be regulated by sugar signals mentioned earlier in sucrose-transporting plants (Rolland et al., 2006; Lastdrager et al., 2014). However, till date little is known about the sugar signal systems in RFO-transporting species, including cucurbits. Further work needs to be done to elucidate if RFO metabolism-related enzymes are also regulated by these signal systems or if there are other molecular mechanisms to control carbohydrate partitioning that are specific to RFO-transporting species.

Compared with loading mechanism, phloem unloading and the role of vasculature within the fruit have attracted less attention. As mentioned in Section 3.6, the exact location of RFO catabolism still remains unclear. Hu et al. (2011) supposed that the RFO transporters should be located at the plasma membrane of companion cells in the cucumber fruit if RFOs are translocated into the fruit and unloaded through an apoplasmic pathway. Greuter and Keller (1993) confirmed the existence of an active stachyose transporter on the tonoplast of artichoke (*Stachys sieboldii*) tuber vacuoles. Schneider and Keller (2009) suggested that raffinose was indeed transported across the chloroplast envelope by a raffinose transporter in the cold-treated common bugle (*Ajuga reptans*) leaf. However, no RFO transporters have been identified at the molecular level to date in any of the higher plants. Looking for these mysterious transporters is very important to fully clarify RFO metabolism and regulation in cucurbits.

Six putative α-galactosidase genes and four putative galactinol synthase genes were identified in the cucumber genome. It is hard to elucidate the exact function of each form of these enzymes since some forms may play important roles in multiple physiological processes. One example is a cell-wall acid α-galactosidase that may be responsible for both cell-wall polysaccharides hydrolysis and assimilate apoplasmic unloading. Thus, traditional gene knockout technology, such as constitutive RNAi, may cause serious deformity or even plant lethality. Microdissecting technology (Zhang et al., 2010) and chemical-induced RNAi (Guo et al., 2003) may enable us to enhance the research in this area.

We are lucky to live in a postgenomic era. The genome of cucumber, melon, and watermelon were sequenced in recent years (Huang et al., 2009; Garcia-Mas et al., 2012; Guo et al., 2013). The next-generation sequencing technology will accelerate the process of sequencing other cucurbit genomes and resequencing different types of cucumber, melon, and watermelon. The sharing of these massive data sets around the world enables us to develop high-throughput technologies such as transcriptome sequencing and microarray. These powerful genomic tools will provide a global view on the transcript dynamics of source–sink interaction and identify novel regulatory components and target genes and then have major impacts on the production and breeding of cucurbit crops.

REFERENCES

Al-Ani, A., Bruzau, F., Raymond, P., Saint-Ges, V., Leblanc, J.M., Pradet, A. 1985. Germination, respiration, and adenylate energy charge of seeds at various oxygen partial pressures. *Plant Physiol.* 79: 885–890.

Ayre, B.G., Keller, F., Turgeon, R. 2003. Symplastic continuity between companion cells and the translocation stream: Long-distance transport is controlled by retention and retrieval mechanisms in the phloem. *Plant Physiol.* 131: 1518–1528.

Baniel, A., Saraf-Levy, T., Perl-Treves, R. 2008. How does the first fruit inhibit younger fruit set in cucurbits? In *Proceedings of the IXth EUCARPIA Meeting on Genetics and Breeding of Cucurbitaceae*, ed. M. Pitrat, pp. 597–601, Avignon, France.

Barber, K.G. 1909. Comparative histology of fruits and seeds of certain species of cucurbitaceae. *Bot. Gaz.* 47: 263–310.

Beebe, D.U., Turgeon, R. 1992. Localization of galactinol, raffinose, and stachyose synthesis in *Cucurbita pepo* leaves. *Planta* 188: 354–361.

Bemis, W.P., Berry, J.M., Weber, C.W., Whitaker, T.W. 1978. The buffalo gourd: A new potential crop. *HortScience* 13: 225–240.

Berry, J.W., Bemis, W.P., Weber, C.W., Philip, T. 1975. Cucurbit root starches: Isolation and some properties of starches from *Cucurbita foetidissima* hbk and *Cucurbita digitata* gray. *J. Agric. Food Chem.* 23: 825–826.

Bhardwaj, J., Anand, A., Nagarajan, S. 2012. Biochemical and biophysical changes associated with magnetopriming in germinating cucumber seeds. *Plant Physiol. Biochem.* 57: 67–73.

Blöchl, A., Peterbauer, T., Hofmann, J. 2008. Enzymatic breakdown of raffinose oligosaccharides in pea seeds. *Planta* 228: 99–110.

Botha, F.C., Small, J.G.C. 1985. Effect of water stress on the carbohydrate metabolism of *Citrullus lanatus* seeds during germination. *Plant Physiol.* 77: 79–82.

Burger, Y., Paris, H.S., Cohen, R., Katzir, N., Tadmor, Y., Lewinsohn, E., Schaffer, A.A. 2009. Genetic diversity of *Cucumis melo. Hortic. Rev.* 36: 165–198.

Burger, Y., Schaffer, A.A. 1991. Sucrose metabolism in mature fruit peduncles of *Cucumis melo* and *Cucumis sativus*. In *Recent Advances in Phloem Transport and Assimilate Compartmentation*, eds. J.L. Bonnemain, S. Delrot, W.J. Lucas, J. Dainty, pp. 244–247, Ouest Editions, Nantes, France.

Burger, Y., Schaffer, A.A. 2007. The contribution of sucrose metabolism enzymes to sucrose accumulation in *Cucumis melo. J. Am. Soc. Hortic. Sci.* 132: 704–712.

Cantliffe, D.J. 1976. Improvement fruit set on cucumbers by plant growth regulator sprays. *Proc. Fla. State Hortic. Soc.* 89: 94–96.

Carmi, N., Zhang, G., Petreikov, M., Gao, Z., Eyal, Y., Granot, D., Schaffer, A.A. 2003. Cloning and functional expression of alkaline α-galactosidase from melon fruit: Similarity to plant SIP proteins uncovers a novel family of plant glycosyl hydrolases. *Plant J.* 33: 97–106.

Cheng, H.T., Li, X., Yao, F.Z., Shan, N., Li, Y.H., Zhang, Z.X., Sui, X.L. (2015). Functional characterization and expression analysis of cucumber (*Cucumis sativus* L.) hexose transporters, involving carbohydrate partitioning and phloem unloading in sink tissues. *Plant Sci.* 237: 46–56.

Choi, Y.H., Rhee, H.C., Kweon, G.B., Cheong, J.W., Hong, Y.P. 1997. Effects of fruit removal and pinching on the translocation and partition of photoassimilates in the cucumber (*Cucumis sativus* L.). *RDA J. Hortic. Sci.* 39: 1–7.

Chrost, B., Schmitz, K. 1997. Changes in soluble sugar and activity of α-galactosidases and acid invertase during muskmelon (*Cucumis melo* L.) fruit development. *J. Plant Physiol.* 151: 41–50.

Ciereszk, I., Gniazdowska, A., Mikulska, M., Rychter, A.M. 1996. Assimilate translocation in bean plants (*Phaseolus vulgaris* L.) during phosphate deficiency. *J. Plant Physiol.* 149: 343–348.

Clark, D., Tupa, M., Bazán, A., Chang, L., Gonzáles, W.L. 2012. Chemical composition of *Apodanthera biflora*, a cucurbit of the dry forest in northwestern Peru. *Rev. Peru Biol.* 19(2): 199–203.

Combrink, N.J.J., Agenbag, G.A., Langenhoven, P., Jacobs, G., Marais, E.M. 2001. Anatomical and compositional changes during fruit development of "Galia" melons. *S. Afr. J. Plant Soil* 18: 7–14.

Crafts, A.S. 1932. Phloem anatomy, exudation, and transport of organic nutrients in cucurbits. *Plant Physiol.* 7: 183–225.

Crafts, A.S., Lorenz, O.A. 1944. Fruit growth and food transport in cucurbits. *Plant Physiol.* 19: 131–138.

Dai, N., Petreikov, M., Portnoy, V., Katzir, N., Pharr, D.M., Schaffer, A.A. 2006. Cloning and expression analysis of a UDP-galactose/glucose pyrophosphorylase from melon fruit provides evidence for the major metabolic pathway of galactose metabolism in raffinose oligosaccharide metabolizing plants. *Plant Physiol.* 142: 294–304.

Dong, C., Wang, X., Shang, Q. 2011. Salicylic acid regulates sugar metabolism that confers tolerance to salinity stress in cucumber seedlings. *Sci. Hortic.* 129: 629–636.

Du, Y.C., Tachibana, S. 1994. Effect of supraoptimal root temperature on the growth, root respiration and sugar content of cucumber plants. *Sci. Hortic.* 58: 289–301.

El-Adawy, T.A., Taha, K.M. 2001. Characteristics and composition of watermelon, pumpkin, and paprika seed oils and flours. *J. Agric. Food Chem.* 49: 1253–1259.

Ells, J.E. 1983. Chlorflurenol as a fruit-setting hormone for gynoecious pickling cucumber production under open-field conditions. *J. Am. Soc. Hortic. Sci.* 108: 164–168.

Elmstrom, G.W. and Davis, P.L. 1981. Sugars in developing and mature fruits of several watermelon cultivars. *J. Am. Soc. Hortic. Sci.* 106: 330–333.

ElSayed, A.I., Rafudeen, M.S., Golldack, D. 2014. Physiological aspects of raffinose family oligosaccharides in plants: Protection against abiotic stress. *Plant Biol.* 16: 1–8.

Gao, Z., Petreikov, M., Zamski, E., Schaffer, A.A. 1999. Carbohydrate metabolism during early fruit development of sweet melon (*Cucumis melo*). *Physiol. Plant.* 106: 1–8.

Gao, Z., Schaffer, A.A. 1999. A novel alkaline α-galactosidase from melon fruit with a substrate preference for raffinose. *Plant Physiol.* 119: 979–987.

Garcia-Mas, J., Benjak, A., Sanseverino, W. et al. 2012. The genome of melon (*Cucumis melo* L.). *Proc. Natl. Acad. Sci. USA* 109: 11872–11877.

Gaudreault, P.R., Webb, J.A. 1982. Alkaline α-galactosidase in leaves of *Cucurbita pepo*. *Plant Sci. Lett.* 24: 281–288.

Gaudreault, P.R., Webb, J.A. 1986. Alkaline α-galactosidase activity and galactose metabolism in the family Cucurbitaceae. *Plant Sci.* 45: 71–75.

Gil, L., Ben-Arib, J., Turgeon, R., Wolf, S. 2012. Effect of CMV infection and high temperatures on the enzymes involved in raffinose family oligosaccharide biosynthesis in melon plants. *J. Plant Physiol.* 169: 965–970.

Gil, L., Yaron, I., Shalitin, D., Sauer, N., Turgeon, R., Wolf, S. 2011. Sucrose transporter plays a role in phloem loading in CMV-infected melon plants that are defined as symplastic loaders. *Plant J.* 66: 366–374.

Greutert, H., Keller, F. 1993. Further evidence for stachyose and sucrose/H+ antiporters on the tonoplast of Japanese artichoke (*Stachys sieboldii*) tubers. *Plant Physiol.* 101: 1317–1322.

Gross, K.C., Pharr, D.M. 1982. A potential pathway for galactose metabolism in *Cucumis sativus* L., a stachyose transporting species. *Plant Physiol.* 69: 117–121.

Guo, H.S., Fei, J.F., Xie, Q., Chua, N.H. 2003. A chemical-regulated inducible RNAi system in plants. *Plant J.* 34: 383–392.

Guo, S., Zhang, J., Sun, H. et al. 2013.The draft genome of watermelon (*Citrullus lanatus*) and resequencing of 20 diverse accessions. *Nat. Genet.* 45: 51–58.

Handley, L.W., Pharr, D.M., Mcfeeters, R.F. 1983a. Relationship between galactinol synthase activity and sugar composition of leaves and seeds of several crop species. *J. Am. Soc. Hortic. Sci.* 108: 600–605.

Handley, L.W., Pharr, D.M., McFeeters, R.F. 1983b. Carbohydrate changes during maturation of cucumber fruit. *Plant Physiol.* 72: 498–502.

Haritatos, E., Keller, F., Turgeon, R. 1996. Raffinose oligosaccharide concentrations measured in individual cell and tissue types in *Cucumis melo* L. leaves: Implications for phloem loading. *Planta* 198: 614–622.

Hermans, C., Hammond, J.P., White, P.J., Verbruggen, N. 2006. How do plants respond to nutrient shortage by biomass allocation? *Trends Plant Sci.* 11: 610–617.

Ho, L.C. 1992. Fruit growth and sink strength. In *Fruit and Seed Production: Aspects of Development, Environmental Physiology and Ecology*, eds. C. Marshall, and J. Grace, pp. 101–124. Cambridge University Press, Cambridge, U.K.

Holthaus, U., Schmitz, K. 1991a. Stachyose synthesis in mature leaves of *Cucumis melo*. Purification and characterization of stachyose synthase (EC. 2.4.1.67). *Planta* 184: 525–531.

Holthaus, U., Schmitz, K. 1991b. Distribution and immunolocalization of stachyose synthase in *Cucumis melo* L. *Planta* 188: 479–486.

Hu, L., Sun, H., Li, R., Zhang, L., Wang, S., Sui, X., Zhang, Z. 2011. Phloem unloading follows an extensive apoplasmic pathway in cucumber (*Cucumis sativus* L.) fruit from anthesis to marketable maturing stage. *Plant Cell Environ.* 34: 1835–1848.

Hu, L.P., Meng, F.Z., Wang, S.H., Sui, X.L., Li, W., Wei, Y.X., Sun, J.L., Zhang, Z.X. 2009. Changes in carbohydrate levels and their metabolic enzymes in leaves, phloem sap and mesocarp during cucumber (*Cucumis sativus* L.) fruit development. *Sci. Hortic.* 121: 131–137.

Huang, S., Li, R., Zhang, Z. et al. 2009. The genome of the cucumber, *Cucumis sativus* L. *Nat. Genet.* 41: 1275–1281.

Hubbard, N.L., Huber, S.C., Pharr, D.M. 1989. Sucrose phosphate synthase and acid invertase as determinants of sucrose concentration in developing muskmelon (*Cucumis melo* L.) fruits. *Plant Physiol.* 91: 1527–1534.

Hughes, D.L., Bosland, J., Yamaguchi, M. 1983. Movement of photosynthates in muskmelon plants. *J. Am. Soc. Hortic. Sci.* 108: 189–192.

Irving, D.E., Hurst, P.L., Ragg, J.S. 1997. Changes in carbohydrates and carbohydrate metabolizing enzymes during the development, maturation, and ripening of buttercup squash (*Cucurbita maxima* D. "Delica"). *J. Am. Soc. Hortic. Sci.* 122: 310–314.

Irving, D.E., Shingleton, G.J., Hurst, P.L. 1999. Starch degradation in buttercup squash (*Cucurbita maxima*). *J. Am. Soc. Hortic. Sci.* 124: 587–590.

Jiang, Y.H., Miao M.M., Chen, X.H., Cao, P.S. 2006. Diurnal changes of non-structural carbohydrates and relative enzymes in leaves and phloem sap of cucumber. *Acta Hortic. Sin.* 33: 529–533.

Judson, J.E. 1929. The morphology and vascular anatomy of the pistillate flower of the cucumber. *Am. J. Bot.* 16: 69–86.

Kanahama, K., Hori, Y. 1980. Time course of export of ^{14}C-assimilates and their distribution pattern as affected by feeding time and night temperature in cucumber plants. *Tohoku J. Agric. Res.* 30: 142–152.

Kanahama, K., Saito, T. 1987. Vascular system and carpel arrangement in the fruits of melon, cucumber and *Luffa acutangula* Roxb. *J. Jpn. Soc. Hortic. Sci.* 55: 476–483.

Kano, Y. 1991. Changes of sugar kind and its content in the fruit of watermelon during its development and after harvest. *Environ. Control Biol.* 29(4): 159–166.

Kim, Y., Hwang, B., Kim, J. 2007. Changes in soluble and transported sugars content and activity of their hydrolytic enzymes in muskmelon (*Cucumis melo* L.) fruit during development and senescence. *Kor. J. Hortic. Sci.* 25: 89–96.

Lastdrager, J., Hanson, J., Smeekens, S. 2014. Sugar signals and the control of plant growth and development. *J. Exp. Bot.* 65: 799–807.

Lee, S.G., Seong, K.C., Kang, K.K., Ko, K.D., Lee, Y.B. 2000. Characteristics of translocation and distribution of photo-assimilates in leaves at different nodes in watermelon. *J. Kor. Soc. Hortic. Sci.* 41: 257–260.

Lee, T.I.K., Jeong, C.S., Yeoung, Y.R. 1997. Sucrose accumulation and changes of sucrose synthase, sucrose phosphate synthase and invertase activities during the development of muskmelon fruits. *J. Kor. Soc. Hortic. Sci.* 38: 1–5.

Lester, G.E., Arias, L.S., Lim, M.G. 2001. Muskmelon fruit soluble acid invertase and sucrose phosphate synthase activity and polypeptide profiles during growth and maturation. *J. Am. Soc. Hortic. Sci.* 126: 33–36.

Lingle, S.E., Dunlap, J.R. 1987. Sucrose metabolism in netted muskmelon fruit during development. *Plant Physiol.* 84: 386–389.

Liu, J., Guo, S., He, H., Zhang, H., Gong, G., Ren, Y., Xu, Y. 2013. Dynamic characteristics of sugar accumulation and related enzyme activities in sweet and non-sweet watermelon fruits. *Acta Physiol. Plant* 35: 3213–3222.

Long, R.L., Walsh, K.B., Rogers, G., Midmore, D.J. 2004. Source-sink manipulation to increase melon (*Cucumis melo* L.) fruit biomass and soluble sugar content. *Aust. J. Agric. Res.* 55: 1241–1251.

Madore, M.A., Webb, J.A. 1982. Stachyose synthesis in isolated mesophyll cells of *Cucurbita pepo*. *Can. J. Bot.* 60: 126–130.

Marcelis, L.F.M. 1991. Effects of sink demand on photosynthesis in cucumber. *J. Exp. Bot.* 42: 1387–1392.

Marcelis, L.F.M. 1994a. Simulation model for dry matter partitioning in cucumber. *Ann. Bot.* 74: 43–52.

Marcelis, L.F.M. 1994b. Fruit growth and dry matter partitioning in cucumber. PhD dissertation, Research Institute for Agrobiology and Soil Fertility, Wageningen, the Netherlands.

Masuko, A., Kozai, T., Kanahama, K. 1989. Three-dimensional vascular system model of Cucurbitaceae plants on a color video display unit of a microcomputer. *J. Jpn. Soc. Hortic. Sci.* 58: 685–690.

Matsumoto, J., Kano, Y., Madachi, T., Aoki, Y. 2012. Heating bearing shoots near fruits promoters sugar accumulation in melon fruit. *Sci. Hortic.* 133: 18–22.

Mayoral, M.L., Plaut, Z., Reinhold, L. 1985. Effect of translocation-hindering procedures on source leaf photosynthesis in cucumber. *Plant Physiol.* 77: 712–717.

McCollum, J.P. 1934. *Vegetative and Reproductive Responses Associated with Fruit Development in Cucumber*, 163pp. Cornell University Agriculture Experiment Station, Ithaca, NY.

McCollum, T.G., Huber, D.J., Cantliffe, D.J. 1988. Soluble sugar accumulation and activity of related enzymes during muskmelon fruit development. *J. Am. Soc. Hortic. Sci.* 113: 399–403.

Meng, F., Hu, L., Wang, S. 2008. Effects of exogenous abscisic acid (ABA) on cucumber seedling leaf carbohydrate metabolism under low temperature. *Plant Growth Regul.* 56: 233–244.

Miao, M., Zhang, Z., Xu, X., Wang, K., Cheng, H., Cao, B. 2009. Different mechanisms to obtain higher fruit growth rate in two cold-tolerant cucumber (*Cucumis sativus* L.) lines under low night temperature. *Sci. Hortic.* 119: 357–361.

Miao, M.M., Cao, P.S., Xue, L.B., Xu, Q. 2003. Effect of pistillate flower removing on sugar content and relative enzyme activities in mature leaves of *Cucumis melo*. *Plant Physiol. Commun.* 39(2): 131–133.

Mitchell, D.E., Gadus, M.V., Madore, M.A. 1992. Patterns of assimilate production and translocation in muskmelon (*Cucumis melo* L.). I. Diurnal patterns. *Plant Physiol.* 99: 959–965.

Mitchell, D.E., Madore, M.A. 1992. Patterns of assimilate production and translocation in muskmelon (*Cucumis melo* L.). II. Low temperature effects. *Plant Physiol.* 99: 966–971.

Murakami, T., Inayama, M. 1974. The effect of temperature on the translocation of photosynthates in cucumber plants. II. Effect of the night temperature on the translocation of ^{14}C-photosynthates in cucumber seedlings. *J. Jpn. Soc. Hortic. Sci.* 43: 43–54.

Nakkanong, K., Yang, J.H., Zhang, M.F. 2012. Starch accumulation and starch related genes expression in novel inter-specific inbred squash line and their parents during fruit development. *Sci. Hortic.* 136: 1–8.

Ohkawa, W., Kanayama, Y., Daibo N., Sato T., Nishiyama M., Kanahama K. 2010. Metabolic process of the ^{14}C-sugars on the translocation pathways of cucumber plants. *Sci. Hortic.* 124: 46–50.

Park, K.H., Chun, Y.T., Kim, W.S., Chung, S.J. 2002. Effects of extra side shoot on the growth and fruit yield of hydroponically grown cucumber plants. *J. Kor. Soc. Hortic. Sci.* 43: 549–552.

Patchareeya, B., Akio, T., Shoko, H., Yoko, M., Nobuo, S. 2011. Effects of fruit load on fruit growth, mesocarp starch grain appearance and sucrose-catalysing enzyme activity in a gynoecious cucumber fruit. *Environ. Control Biol.* 49: 119–125.

Peterson, C.A., Currier, H.B. 1969. An investigation of bidirectional translocation in the phloem. *Physiol. Plant.* 22: 1238–1250.

Pharr, D.M., Hubbard, N.L. 1994. Melons: Biochemical and physiological control of sugar accumulation. In *Encyclopedia of Agricultural Sciences*, Vol. 3, pp. 25–37.

Pharr, D.M., Huber, S.C., Sox, H.N. 1985. Leaf carbohydrate status and enzymes of translocate synthesis in fruiting and vegetative plants of *Cucumis sativus* L. *Plant Physiol.* 77: 104–108.

Pharr, D.M., Motomura, Y. 1989. Anaerobiosis and carbohydrate status of the embryonic axis of germinating cucumber seeds. *HortScience* 24: 120–122.

Pharr, D.M., Sox, H.N. 1984. Changes in carbohydrate and enzyme levels during the sink to source transition of leaves of *Cucumis sativus* L., a stachyose translocator. *Plant Sci. Lett.* 35: 187–193.

Pharr, D.M., Sox, H.N., Smart, E.L., Lower, R.L., Fleming, H.P. 1977. Identification and distribution of soluble saccharides in pickling cucumber plants and their fate in fermentation. *J. Am. Soc. Hortic. Sci.* 102: 406–409.

Ramirez, D.R., Wehner, T.C. 1984. Growth analysis of three cucumber lines differing in plant habit and yield. *Cucurbit Genet. Coop. Rep.* 7: 17–18.

Ranwala, A.P., Iwanami, S.S., Masuda, H. 1991. Acid and neutral invertases in the mesocarp of developing muskmelon (*Cucumis melo* L. cv. Prince) fruit. *Plant Physiol.* 96: 881–886.

Rennie, E.A., Turgeon, R. 2009. A comprehensive picture of phloem loading strategies. *Proc. Natl. Acad. Sci. USA* 106: 14162–14167.

Richardson, P.T., Baker, D.A., Ho, L.C. 1982. The chemical composition of cucurbit vascular exudates. *J. Exp. Bot.* 33: 1239–1247.

Richardson, P.T., Baker, D.A., Ho, L.C. 1984. Assimilate transport in cucurbits. *J. Exp. Bot.* 35: 1575–1581.

Robbins, N.S., Pharr, D.M. 1988. Effect of restricted root growth on carbohydrate metabolism and whole plant growth of *Cucumis sativus* L. *Plant Physiol.* 87: 409–413.

Roitsch, T., Gonzalez, M.V. 2004. Function and regulation of plant invertases: Sweet sensations. *Trends Plant Sci.* 9: 606–613.

Rolland, F., Baena-Gonzalez, E., Sheen, J. 2006. Sugar sensing and signaling in plants: Conserved and novel mechanisms. *Annu. Rev. Plant Biol.* 57: 675–709.

Savage, J.A., Zwieniecki, M.A., Holbrook, N.M. 2013. Phloem transport velocity varies over time and among vascular bundles during early cucumber seedling development. *Plant Physiol.* 163: 1409–1418.

Schaffer, A.A., Aloni, B.A., Fogelman, E. 1987. Sucrose metabolism and accumulation in developing fruit of Cucumis. *Phytochemistry* 26: 1883–1887.

Schaffer, A.A., Pharr, D.M., Madore, M.A. 1996. Cucurbits. In *Photoassimilate Distribution in Plants and Crops*, eds. E. Zamski, A.A. Schaffer, pp. 729–757. Marcel Dekker, New York.

Schapendonk, A.H.C.M., Brouwer, P. 1984. Fruit growth of cucumber in relation to assimilate supply and sink activity. *Sci. Hortic.* 23: 21–33.

Schapendonk, A.H.C.M., Challa, H. 1980. Assimilate requirements for growth and maintenance of the cucumber fruit. *Acta Hortic.* 118: 73–82.

Schmitz, K., Cuypers, B., Moll, M. 1987. Pathway of assimilate transfer between mesophyll cells and minor veins in leaves of *Cucumis melo* L. *Planta* 171: 19–29.

Schmitz, K., Holthaus, U. 1986. Are sucrosyl-oligosaccharides synthesized in mesophyll protoplasts of mature leaves of *Cucumis melo*. *Planta* 169: 529–535.

Schneider, T., Keller, F. 2009. Raffinose in chloroplasts is synthesized in the cytosol and transported across the chloroplast envelope. *Plant Cell Physiol.* 50: 2174–2182.

Shalitin, D., Wolf, S. 2000. Cucumber mosaic virus infection affects sugar transport in melon plants. *Plant Physiol.* 123: 597–604.

Shishido, Y., Yuhashi, T., Seyama, N., Imada, S. 1992. Effects of leaf position and water management on translocation and distribution of ^{14}C-assimilates in fruiting muskmelon. *J. Jpn. Soc. Hortic. Sci.* 60: 897–903.

Sinnott, E.W. 1945. The relation of growth to size in cucurbit fruits. *Am. J. Bot.* 32: 439–446.

Slewinski, T.L., Zhang, C., Turgeon, R. 2013. Structural and functional heterogeneity in phloem loading and transport. *Front. Plant Sci.* 4: 244.

Smart, E.L., Pharr, D.M. 1980. Characterization of α-galactosidase from cucumber leaves. *Plant Physiol.* 66: 731–734.

Smart, E.L., Pharr, D.M. 1981. Separation and characteristics of galactose-1-phosphate and glucose-1-P uridyltransferase from fruit peduncles of cucumber. *Planta* 153: 370–375.

Smeekens, S., Hellmann, H.A. 2014. Sugar sensing and signaling in plants. *Front. Plant Sci.* 5: 113.

Sprenger, N., Keller, F. 2000. Allocation of raffinose family oligosaccharides to transport and storage pools in *Ajuga reptans*: The roles of two distinct galactinol synthases. *Plant J.* 21: 249–258.

Stepansky, A., Kovalski, I., Schaffer, A.A., Perl-Treves, R. 1999. Variation in sugar levels and invertase activity in mature fruit representing a broad spectrum of *Cucumis melo* genotypes. *Genet. Resour. Crop Evol.* 46: 53–62.

Stevenson, D.G., Yoo, S.H., Hurst, P.L., Jane, J.L. 2005. Structural and physicochemical characteristics of winter squash (*Cucurbita maxima* D.) fruit starches at harvest. *Carbohydr. Polym.* 59: 153–163.

Su, P.H., Wu, T.H., Lin, C.H. 1998. Root sugar level in *Luffa* and bitter melon is not referential to their flooding tolerance. *Bot. Bull. Acad. Sin. (Taipei)* 39: 175–179.

Sui, X.L., Meng, F.Z., Wang, H.Y. et al. 2012. Molecular cloning, characteristics and low temperature response of raffinose synthase gene in *Cucumis sativus* L. *J. Plant Physiol.* 169: 1883–1891.

Sun, S.R., Yang, Z.Q., Zhang, J.P., Zhai, Q.H., Chen, J.F., Gong, Z.H. 2008. Changes of sugar metabolism and related enzyme activities in the development of pumpkin fruit. *J. Northwest A&F Univ. (Nat. Sci. Ed.)* 36: 159–164.

Tateishi, A., Nakayama, T., Isobe, K., Nomura, K., Watanabe, K., Inoue, H. 2004. Changes in sugar metabolism enzyme activities in cultivars of two pumpkin species (*Cucurbita maxima* and *C. moschata*) during fruit development. *J. Jpn. Soc. Hortic. Sci.* 73: 57–59.

Terazawa, Y., Ito, K., Masuda, R., Yoshida, K. 2001. Changes in carbohydrate composition in pumpkins [*Cucurbita maxima*] (kabocha) during fruit growth. *J. Jpn. Soc. Hortic. Sci.* 70: 656–658.

Thomas, B., Webb, J.A. 1978. Distribution of α-galactosidase in *Cucurbita pepo*. *Plant Physiol*. 62: 713–717.

Todaka, D., Matsushima, H., Morohashi, Y. 2000. Water stress enhances β-amylase activity in cucumber cotyledons. *J. Exp. Bot*. 51: 739–745.

Todd, G.W., Bean, R.C., Propst, B. 1961. Photosynthesis and respiration of developing fruits. II. Comparative rates at various stages of development. *Plant Physiol*. 36: 69–73.

Toki, T., Ogiwara, H., Aoki, H. 1978. Effect of varying night temperature on the growth and yields in cucumber. *Acta Hortic*. 87: 233–237.

Turgeon, R. 1991. Symplastic phloem loading and the sink-source transition in leaves: A model. In *Recent Advances in Phloem Transport and Assimilate Compartmentation*, eds. J.L. Bonnemain, S. Delrot, W.J. Lucas, J. Dainty, pp. 18–22. Ouest Editions, Nantes, France.

Turgeon, R., Hepler, P.K. 1989. Symplastic continuity between mesophyll and companion cells in minor veins of mature *Cucurbita pepo* L. leaves. *Planta* 179: 24–31.

Turgeon, R., Webb, J.A. 1973. Leaf development and phloem transport in *Cucurbita pepo*: Transition from import to export. *Planta* 113: 179–191.

Turgeon, R., Webb, J.A. 1975. Leaf development and phloem transport in *Cucurbita pepo*: Carbon economy. *Planta* 123: 53–62.

Turgeon, R., Webb, J.A. 1976. Leaf development and phloem transport in *Cucurbita pepo*: Maturation of the minor veins. *Planta* 129: 265–269.

Turgeon, R., Webb, J.A., Evert, R.F. 1975. Ultrastructure of minor veins in *Cucurbita pepo* leaves. *Protoplasma* 83: 217–232.

Verkleij, F.N., Baan Hofman-Eijer, L.B. 1988. Diurnal export of carbon and fruit growth in cucumber. *J. Plant Physiol*. 33: 345–348.

Verkleij, F.N., Challa, H. 1988. Diurnal export and carbon economy in an expanding source leaf of cucumber at contrasting source and sink temperature. *Physiol. Plant*. 74: 284–293.

Villanueva, M.J., Tenorio, M.D., Esteban, M.A., Mendoza, M.C. 2004. Compositional changes during ripening of two cultivars of muskmelon fruits. *Food Chem*. 87: 179–185.

Wang, L.J., Shi, X.Y., Liu, Y., Li, J.W. 2014. Study on gradient distribution of sugar components in developing melon fruit. *J. Fruit Sci*. 31(3): 430–437.

Wang, X.L. 2014. A study on RFOs catabolize location and relationship with acidic, alkaline α-galactosidases in fruit of cucumber (*Cucumis sativus* L.). MS dissertation, Yangzhou University, Yangzhou, China.

Webb, J.A., Gorham, P.R. 1964. Stachyose transport in plants. *Plant Physiol*. 39: 973–977.

Weidner, T.M. 1964. Translocation of photosynthetically labeled ^{14}C compounds in bean, cucumber and white ash. PhD dissertation, Ohio State University, Columbus, OH.

Widders, I.E., Kwantes, M. 1995. Ontogenic changes in seed weight and carbohydrate composition as related to growth of cucumber (*Cucumis sativus* L.) fruit. *Sci. Hortic*. 63: 155–165.

Yang, H., Song, R., Ma, K., Gu, W., Tang, Q. 2008. Dynamic changes of inositol, polysaccharide and reductive sugars in pumpkin during fruit development. *Acta Hortic. Sin*. 35(1): 127–130.

Yang, X.G. 2012. A preliminary study on the physiological functions of six α-galactosidases in cucumber (*Cucumis sativus* L.). MS dissertation, Yangzhou University, Yangzhou, China.

Yativ, M., Harary, I., Wolf, S. 2010. Sucrose accumulation in watermelon fruits: Genetic variation and biochemical analysis. *J. Plant Physiol*. 167: 589–596.

Yin, L. 2012. Sensory assess and starch properties analysis of pumpkin. MS dissertation, Chinese Academy of Agricultural Sciences, Beijing, China.

Zhang, B., Tolstikov, V., Turnbull, C., Hicks, L.M., Fiehn, O. 2010. Divergent metabolome and proteome suggest functional independence of dual phloem transport systems in cucurbits. *Proc. Natl. Acad. Sci. USA* 107: 13532–13537.

Zhang, C., Yu, X., Ayre, B.G., Turgeon, R. 2012. The origin and composition of cucurbit "phloem" exudate. *Plant Physiol*. 158: 1873–1882.

Zhang, L. 2010. Inheritance of sugar content and the relationship between sugar accumulation and enzymes related to sucrose metabolism in fruits of watermelon. MS dissertation, Gansu Agriculture University, Lanzhou, China.

Zhang, M.F., Li, Z.L. 2005. A comparison of sugar-accumulating patterns and relative compositions in developing fruits of two oriental melon varieties as determined by HPLC. *Food Chem*. 90: 785–790.

Zhang, Z.P., Deng, Y.K., Song, X.X., Miao, M.M. 2015. Trehalose-6-phosphate and SNF1-related protein kinase 1 are involved in the first-fruit inhibition of cucumber. *J. Plant Physiol*. doi:10.1016/j.jplph.2014.09.009.

Zhou, A.M., Yang, H., Yang, L., Liu, X.J., Liu, X., Yang, G.M., Chen, Y.Q. 2013. Physicochemical properties of pumpkin starches from different cultivars. *Mod. Food Sci. Technol*. 29(8): 1784–1790.

Section III

Cultural Practices of Cucurbits

4 Cultivation and Bioprospecting of Perennial Cucurbits

Shrawan Singh and L.K. Bharati

CONTENTS

4.1   INTRODUCTION

Nowadays, perennial cucurbits are gaining importance because of their ecological consideration and nutritional composition. The important perennial cucurbits are gac (*Momordica cochinchinensis* Spreng), spine gourd (*Momordica dioica* Roxb.), teasel gourd (*Momordica subangulata* ssp. *renigera* [G. Don] de Wilde), pointed gourd (*Trichosanthes dioica* Roxb.), ivy gourd (*Coccinia cordifolia* [Voigt] L.), and chow-chow (*Sechium edule* [Jack] Sw). The fruits of these cucurbits are consumed either raw or cooked and in pickled form and are cultivated in small scale and, to some extent, in commercial scale in tropical regions; however, they are still considered cucurbits of secondary status. *M. cochinchinensis* and *C. cordifolia* are still collected from wild habitats in Andaman and Nicobar Islands and northeast India for their value in local culinary preparations. They are commonly used at the green stage only for dietary preparations, while in some places, the use of ripe fruits for the extraction of natural color used in food preparation, particularly in rice, is also reported. These crops have regenerating capacity and hardiness to overcome different biotic stresses, particularly damage by cyclones and heavy rains. Besides, these are important source of functional constituents such as lycopene and carotenoids and suitable for processing and value-added products. Lycopene has strong free radical quenching capacity, and aril fraction of ripe fruits of *M. cochinchinensis* has been reported as the richest source of (70 times higher than tomato) lycopene. This chapter highlights the significance of perennial vegetables in the food industry and their potential in tropical regions with special reference to Andaman and Nicobar Islands, India.

The family Cucurbitaceae consists of 117 genera and 825 species, which are mainly distributed in warmer regions of both hemispheres. Around 36 genera and 100 species of cucurbits were reported in India (Chakravarty, 1982). Though there are many perennial cucurbits, this chapter will cover the commonly known perennial cucurbits such as gac (*Momordica cochinchinensis* Spreng), spine gourd (*Momordica dioica* Roxb.), teasel gourd (*Momordica subangulata* ssp. *renigera* [G. Don] de Wilde), pointed gourd (*Trichosanthes dioica* Roxb.), ivy gourd (*Coccinia cordifolia* [Voigt] L.), and chow-chow (*Sechium edule* [Jack] Sw). Among them, *T. dioica* is common throughout India, while *M. dioica* is cultivated in West Bengal, Bihar, Orissa, Maharashtra, and northeast India; *M. subangulata* ssp. *renigera* is cultivated in Maharashtra, Chhattisgarh, northeast India, and Andaman and Nicobar Islands; and *C. cordifolia* is abundant as cultivated and wild spontaneous plants in Andaman and Nicobar Islands, India.

The investigations on plants and their parts searching fhealth-benefiting compounds such as natural antioxidant have shown some interesting results, and many of them were used in the preparation of functional foods. It is now an established fact that the "antioxidant" acts as a defending agent against free radicals, which otherwise damage the immune system of a body and cause degenerative diseases. The important antioxidants are carotenoids, ascorbic acid, tocopherols, anthocyanins, phenolics, tannins, and flavonoids (Rahman, 2007). The kind and concentration of such antioxidants varies with species (Singh et al., 2011; Singh and Singh, 2012), genotypes (Singh et al., 2014), and stage and kind of the plant tissue (Kubola and Siriamornpun, 2011). The variation is mainly attributed to genetic makeup and pressure of the associated environmental factors on the plant or its parts. The variation in uptake of micronutrients between species and genotypes and the nature of movement and storage of micronutrients in different parts in plant systems also determine the variation in kind and concentration of antinutrients because these micronutrients (particularly Fe, Ca, Si, Cu, and Mn) have active role as cofactors in free radical scavenging activity (Evans and Halliwell, 2001). *M. subangulata* ssp. *renigera* and *M. cochinchinensis* are consumed as delicious vegetables (Bharathi et al., 2011), while their ripe fruits are rich in carotenoids and lycopene, which has great potential in the functional food industry.

4.2 PERENNIAL CUCURBITS: USES AND PREFERENCES

Perennial cucurbits were considered less important than annual cucurbits, and *M. cochinchinensis* and *C. cordifolia* are still growing wild. They are used as vegetables and in the preparation of value-added products like pickles and dry flecks (Table 4.1). A survey in Andaman and Nicobar Islands on indigenous vegetables revealed that the household consumption of perennial cucurbitaceous vegetables in the islands was observed to be 445.0 g/consumption day (*T. dioica*) to 665.6 g/consumption day (*M. subangulata* ssp. *renigera*). Although the average consumption frequency (days/months) of cucurbits is influenced by the season, quantity of harvest from wild habitats, and relative market price, these vegetables are consumed 2.2–6.6 days in a month. Though these cucurbits are partially cultivated and still underutilized, they have a great potential in changing the climatic scenario and shifting agriculture toward less-fertile or culturable wastelands.

TABLE 4.1

Common Perennial Cucurbits and Their Habitats in Andaman and Nicobar Islands

Local Name	Scientific Name	Center of Origin/Diversity	Chromosome No. (2n)	Habitat	Season	Habit	Edible Portion	Culinary Uses
Teasel gourd or kakrol	*Momordica subangulata* ssp. *renigera* (G. Don) de Wilde	Indo-Burma	56	C/W/H	D	P/V	Fruits	Vegetables, pickles, dry flecks
Ivy gourd or kundru	*Coccinia grandis* (L.) J. Voigt or *Coccinia cordifolia* (Voigt) L.	India	24	C/W	R	P/V	Fruits, leaves	Vegetables, pickles, dry flecks
Pointed gourd or parwal	*Trichosanthes dioica* Roxb. (L.)	India	22	C	R/D	P/V	Fruits, leaves	Vegetables, pickles
Gac or sweet gourd	*Momordica cochinchinensis* (Lour.) Spreng	Cochinchina	28	W	R	P/V	Fruits	Vegetables, pickles
Chow-chow or chayote	*Sechium edule* (Jack) Sw.	Central America	22–28	C	R	P/V	Fruits	Vegetables

Notes: C, cultivated; W, wild; R, rainy season; D, dry season; A, annual; P, perennial; V, vine.

4.3 CULTIVATION PRACTICES FOR PERENNIAL CUCURBITS

4.3.1 Soil and Climate

In general, all perennial cucurbits (pointed gourd, *T. dioica*; ivy gourd, *Coccinia grandis*; Bankunari, *Solena amplexicaulis* (Lam.) Gandhi.; teasel gourd, *M. subangulata* subsp. *renigera*; sweet gourd, *M. cochinchinensis*; spine gourd, *M. dioica*; mountain spine gourd, *Momordica sahyadrica* sp. nov.; Athalakkai, *Momordica cymbalaria* Hook f.; chayote, *S. edule*) require warm and humid areas where rains are abundant, though without waterlogging; however, spine gourd can adapt equally well to arid and semiarid tracts. Senescence of aerial parts and dormancy of tubers for 3–4 months are reported during winter months. They can be grown successfully in sandy loam to loam soil and should not be grown in heavy soils. Medium acidic soils of pH 6–8 and temperature range of 25°C–30°C and 1500–2500 mm rainfall are ideal for teasel gourd. Sweet gourd comes very well in monsoon climate with a relatively warm growing period of 240 days. It does not tolerate freezing temperature. It grows well in soils of shallow to medium depth (50–150 cm) and prefers acidic to slightly alkaline soils with a pH range of 6–7.5.

4.3.2 Propagation Techniques

Seed propagation is slightly difficult in these perennial cucurbits due to high seed dormancy. However, plants developed from such vine cuttings do not perennate beyond one season in spine gourd and mountain spine gourd. Germination of seeds of spine gourd and mountain spine gourd may be enhanced through careful decortication just before sowing of 4-month-old seeds. Teasel gourd can be propagated either by tuber cuttings or rooted vine cuttings; however, higher yields can be realized in tuber-raised plants. An average plant produces 20–25 adventitious tubers of 60–80 g each. Large tubers can be randomly cut into pieces and used as propagule. The cut tubers should be treated with 2% Dithane M-45 solution and shade-dried for wound healing before planting. But only longitudinal cuttings with a portion of intact apical bud give rise to new plants in the case of spine gourd and mountain spine gourd. A better and reliable method for sweet gourd is to use rooted vine cuttings as propagules. Mid- and top-level cuttings root well with Seradix treatment. Closed media sachet technique may be adopted for home gardens. Basal two-third portion of the cutting should be planted in polybags containing rooting medium (soil + sand + farm yard manure (FYM)) after treatment with hormone during the month of October–November and be kept under partial shade. In sweet gourd, wedge grafting of the growing tips on vigorous sprouts give nearly 100% establishment, compact vine growth, and higher yield. Use of tubers has been advocated for commercial cultivation of *M. cymbalaria*, and the tuber weight of 60 g and above recorded significantly higher fruit yield per plant (120 g) over the rest of the tubers (Reddy et al., 2007). Pointed gourd is usually propagated through vine cuttings and root suckers. Seeds are generally not used for commercial propagation owing to poor germination and unpredictable sex. The vine cuttings can be planted during October–November or February–March by following any one of the following methods.

4.3.3 Planting Methods

4.3.3.1 *Lunda* or *Lachhi* Method

In this system, young vines of 1–1.5 cm long with 8–10 nodes per cutting are taken and folded into a figure of eight (8) commonly known as lunda or lachhi. This lachhi should be placed flat in the pit and pressed 3–5 cm deep into the pits filled with FYM and soil. Fresh cow dung may be applied over the central part of the pit to enhance the sprouting in moisture-deficient conditions.

4.3.3.2 Moist Lump Method

In this method, 60 cm long vine is encircled over a lump of moist soil leaving both ends 15 cm free. Such lumps are buried 10 cm deep into the well-prepared pits, leaving the ends of vine above the soil. Underground vine develops into root and the exposed ends give sprouts.

4.3.3.3 Straight Vine Method

In this method, the vine cuttings are planted end to end horizontally 15 cm deep into furrows filled with a mixture of farmyard manure and soil.

4.3.3.4 Ring Method

In this method, the vine cutting is coiled into a spiral or ring shape and planted directly on the mounds, covering one-half to two-thirds of the ring under the soil.

4.3.3.5 Vine Cutting Method

The cuttings from the mature vines (3–4 nodes) are treated with Rootex No. 1 and planted in sand beds. After 15 days, the rooted vine cuttings are shifted to polybags filled with FYM, sand, and garden soil. Cuttings will be ready for planting between 15 and 20 days after shifting in polybags. Success rate will be high during rainy season.

Pointed gourd can also be propagated through tuberous roots that are uprooted and planted on the mounds. The propagation through this method is easier and faster and gives assured success. Care should be taken to discriminate between tubers collected from male and female vines. Ivy gourd and *Solena* is propagated through stem cuttings. The stem cuttings from older shoots can be directly planted in the main field. In the case of young shoots, cuttings can be planted in polybags and transplanted after rooting. Chayote is usually propagated by sprouted fruits, although basal shoots are sometimes used as well.

4.3.4 MICROPROPAGATION TECHNIQUES

Shekhawat et al. (2011) developed an *in vitro* propagation method for female plants of *M. dioica* (Roxb.) using nodal segments on Murashige and Skoog's (MS) agar-gelled medium + 2.0 mg L^{-1} 6-benzylaminopurine (BAP) + 0.1 mg L^{-1} indole-3 acetic acid (IAA). The cultures were amplified on MS medium with 1.0 mg L^{-1} BAP + 0.1 mg L^{-1} IAA and shoots multiplied by subculturing of shoot clump on MS medium + 0.5 mg L^{-1} BAP + 0.1 mg L^{-1} IAA. Root formation was done on half-strength MS medium + 2.0 mg L^{-1} indole-3 butyric acid (IBA) (89% success rate) and greenhouse hardened plantlets transferred to the field. Aileni et al. (2009) attempted a protocol for *in vitro* propagation of *Momordica tuberosa* (Cogn) Roxb. using nodal segments and shoot apices on MS medium and reported the highest regeneration efficiency when MS medium is supplemented with 4.40 µM BA combined with 4.60 µM Kn. The BA at 13.30 µM induced regeneration from shoot apex cultures. Microshoots were rooted onto MS medium supplemented with 4.90 µM IBA and regenerated plants established with 90% survival rate. Park et al. (2012) used IBA for treating vine cuttings and observed better responses. However, tissue culture seems to be the most effective option for large-scale multiplication of true to type male and female plants, but there is still need to have robust and efficient protocols through research efforts on these crops.

4.3.5 SPACING AND TRANSPLANTING

Planting density varies in species depending upon the area covered by each species. About 7400 plants/ha for pointed gourd, 2500–3000 plants/ha for ivy gourd and *Solena*, 6000 plants/ha for spine gourd and mountain spine gourd, 4500 plants/ha for teasel gourd, and 1500–2000 plants/ha for sweet gourd are generally recommended. A few male plants to the extent of 10% on staggered sowing may be retained on field borders as pollen source. In the case of ivy gourd, a male plant is not necessary as it produces fruits through vegetative parthenocarpy. A spacing of 2.4 × 1.8 m in chayote, 0.9 × 1.5 m in pointed gourd, 2–2.5 × 2–2.5 m in ivy gourd, 1 × 1 m in spine gourd and mountain spine gourd, 1.5 × 1.5 m in teasel gourd, and 3 × 3 m in sweet gourd may be ideal. Sowing time varies across the country depending upon the availability of premonsoon rains. The field should be well ploughed and harrowed to remove weeds and plant debris. Pits of 30–45 cm^2 for spine gourd, pointed gourd, and ivy gourd and 60 cm^2 for sweet gourd are dug and filled with the recommended quantity of FYM and topsoil and raised to mounds of 20–30 cm height, with provision for water drainage as none of the three species can tolerate waterlogging. In tropical regions, frequent waterlogging in low and medium lands, where raised beds of 15–20 cm height and 1 m width should be made at 50–75 cm (furrows) distance, is found. The furrows between two beds allow drainage of excess water and prevent the crop from short-period waterlogging. This technique was found to be most suitable for teasel gourd in the tropical climate of Andaman Islands.

4.3.6 TRAINING AND PRUNING

Perennial cucurbits require permanent frame to allow proper spread of vines for better yield and quality fruits. There are several methods of trellising, namely, bower type, lean-to type, arch type, and single-row vertical trellis. The single-row vertical trellis is found to be beneficial for all these perennial gourds. For the single-row vertical trellis, the trellis should be about 6 ft high, constructed from stakes 2 m apart, with a system of vertical strings between horizontal wires. The bottom horizontal line is fixed at 3 ft high and the subsequent lines at 1 ft interval. Train the vines on the vertical trellis regularly by tying the vines to the trellis. Where the drainage is a problem or soil is wet continuously for 7–10 days in any period of the year, the plants should be trained in trellis.

Nonproductive side branches should be removed up to first 10 nodes. Annual pruning after the cessation of active growth period helps to keep a healthy and compact canopy in sweet gourd. In the case of pointed gourd, spine gourd, and sweet gourd aerial parts dry up naturally with the end of the southwest monsoon and hence do not need any pruning. The most important in cultivation of ivy gourd is the time of pruning the vines. Repeated pruning is recommended as the newly developing vines produce more flowers and yield more. Pruning the vines must be done every 3–4 months to maximize yield. When the vines start to look a bit weak and there is a change in color of the leaves to yellow, pruning is done.

4.3.7 WEED AND WATER MANAGEMENT

In tropical ecology, weeds are a menace for crop production because they compete with crop plants for soil moisture, nutrients, and sunlight and sometimes cause allelopathic exudates to harm crop plants. Many of the weeds are alternate host for pathogens and insect pests, which multiply on host crop in its season and take a heavy toll. During rainy season, the weeds in between the rows can be controlled by manual weeding or using power tiller, and within rows, manual weeding has to be done in the plant basins. Plastic mulching or organic mulching with paddy straw, banana leaves, sugarcane trash, or dried grass can effectively be used as mulching to control the weeds and conserve the moisture in plant basin.

Chayote requires large quantities of water and should be copiously irrigated in regions of low rainfall and during periods of drought. Ivy gourd, *Solena*, sweet gourd, spine gourd, and teasel gourd do not require irrigation if regular rainfall is received during active crop season, though sweet gourd require life-saving irrigation during summer months. In case of uneven rains, light irrigation at 3–5 days interval will be advantageous. However, irrigation at 8–10 days interval during winter and 4–5 days interval during summer is essential. Waterlogging should be avoided to save the roots from rotting and infection by pathogenic fungi *Fusarium*. Pressurized irrigation system like drip irrigation may be used to economize water-use and labor-use efficiencies.

4.3.8 NUTRITION MANAGEMENT

The kind and quantity of fertilizer needed depend on the soil type and amount of nutrients already available in the soil; a blanket fertilizer dose is difficult to recommend. However, Cucurbits require low nitrogen and high potassium and phosphorous for good fruit development. Excess nitrogen application should be avoided as it encourages vine growth and retards fruiting. However, the vines grow to several meters and produce many offshoots; fertilizer management has to be monitored regularly. Prepare a solution of fresh cow dung (10 kg), MOP (5 kg), neem cake (2 kg), and vermicompost (5 kg) in 200 L water and apply 1 L to one plant at 30 days after planting. It is highly recommended to apply two baskets of fresh organic manure to every pit, every month, along with 100–200 g NPK. During fertilizer application, the earthing up of the soil is done to break the clods and crust formation.

4.3.9 POLLINATION MANAGEMENT

Spine gourd is pollinated naturally by moths and other nocturnal visitors. Pointed gourd and sweet gourd are also pollinated by native bees, but hand pollination increases fruit set. However, teasel gourd requires hand pollination for assured fruit set, especially outside its home range where its natural pollinators are absent. Using a camel hairbrush, the pollen can be dusted on receptive stigma preferably in the morning hours for higher fruit setting.

4.3.10 PEST AND DISEASE MANAGEMENT

Fusarium wilt-affected leaves wilt suddenly and vascular bundles in the collar region become yellow or brown. It is difficult to control the disease since the fungus persists in the soil. However, it can be controlled to some extent by raising soil temperature using plastic mulch. Soil application of a fungicidal mixture of carbendazim (1 g/L) and captan (2 g/L) or topsin-M (2 g/L) to the root zones of the plant is recommended. Care should be taken to avoid excess soil moisture in the root zone as it aggravates the situation. The anthracnose appears as small yellowish spots on the leaves as water-soaked areas, which enlarge in size, coalesce and turn brown to black in color. Repeated spray of Dithane M-45 at 5–7 days interval controls the disease. In powdery mildew, initially white or fluffy growth appears in circular patches or spots on the leaves. Severely infected leaves become brown and shriveled and defoliation may occur. Spray of sulfur fortnightly minimizes the disease. Downy mildew symptoms appear as irregular shape yellow to brown angular spots that become evident on the upper sides of the leaves with corresponding mycelia growth on the downside of the leaves. Leaves die due to necrotic spots that increase in size and cause severe defoliation. Moist condition favors the development of the disease. It causes heavy defoliation. Spraying of Ridomil MZ 72 (2.5 g/L) has been found more effective to control the disease. Besides, mosaic also occurs in which infected plants should be removed and destroyed as soon as the disease appears in the field. Spray Malathion mixed with 2 mL/L of water at 5-day interval to control aphid that transmits the disease.

Among many pests, adult red pumpkin beetles eat the leaves, resulting in holes in the foliage. The insect attacks at seedling stage. Spray a mixture of Sevin and 2 g/L of water at 6–7-day interval to control insect attacks. In some regions in mainland India, the node borer grubs were observed as pest of these cucurbits, which bore into the main vines of the plants and produced a swelling of the stem. The feeding tunnel is usually directed toward nodes and is filled with glutinous waste material. Other than scar tissue at the site where the larvae entered the stem, infested plants usually show no conspicuous symptoms. Under very severe infestations, young plants may die, but older plants often live to produce fruit with reduced yields. Spraying of endosulfan, chlorpyrifos, or fenvalerate, which are found to be effective in managing the disease, contributes to 70%–85% increase in production yield. The red spider mite nymphs and adults cause damage in plants by web formation on the leaves in large number, giving an unhealthy appearance. Usually, these mites feed on the underside of the leaves by sucking cell sap. Gradually, chlorophyll degradation takes place and leaves dry. Spray neem seed kernel extract (NSKE) (5%) at weekly interval. Spray dicofol (mixed in 2.75 mL/L of water) or vertmac (mixed in 0.8 mL/L of water) at 15–20-day interval. Adequate supply of water in summer season also reduces the incidence of mites. The root knot nematode-infested plants are heavily knotted, the growth of vines badly arrested, leaves appear yellow, leaf size reduces, and the majority of the veins become dry and wither. Vine dipping in monocrotophos 1000 ppm for 6 h before planting and application of neem cake (500 g) and Bio-Nematon (10 g) per pit as basal application with FYM is effective to control the infestation by nematode. A second dose of Bio-Nematon (10 g/pit) at 40 DAP may also reduce infestation. The aphids can cause curling, yellowing, and distortion of leaves and stunting of shoots. They can also produce large quantities of a sticky exudate known as honeydew, which often turns black with the growth of a sooty mold fungus. Aphids feed on the flower and affect fruit development. Dimethoate (mixed with 2 mL/L of water) or imidacloprid (mixed with 1 mL/L of water) effectively control this problem.

The leaf miner makes serpentine mines in the leaves. Severe infestation of these causes drying and dropping of leaves. Collect and destroy mined leaves. Spray NSKE (3%). A cucumber moth has young bright green larva with a pair of white mid-dorsal lines, which initially scrapes the chlorophyll content. Later it folds and webs the leaves and feeds within. It also feeds on flowers and bores into the developing fruits. The early-stage caterpillars should be collected and destroyed. In case of heavy infestation, spray with Malathion 50 EC mixed in 2 mL/L of water or Bt mixed in 2 mL/L of water. The fruit fly maggots feed on the pulp of the fruits, which causes oozing of resinous fluid from fruits, resulting in distorted and malformed fruits. The infested and fallen fruits should be collected and dump in deep pits. To manage the pest, expose the pupae by ploughing and turning over soil after harvest. For trapping the adult sex pheromone, blended plywood blocks may be used. For this purpose, a mixture of ethanol, Cue-Lure, Malathion in the ratio of 6:4:1 may be prepared. Plywood blocks of 5 × 5 × 1.2 cm are prepared and soaked in this solution for 48 h. These can be used in bottle traps at 10 blocks/ha. Under severe infestation, 100 g jiggery may be mixed in 1 L of water with 2 mL carbaryl and sprayed on foliage in spots at the distance of 7 m.

4.3.11 HARVESTING

Fruits are ideally harvested at tender stage, while the outer skin is still green. Fruits that are 10–12 days old fetch a premium price in the market, and beyond 15 days, fruits develop gradual reddening/yellowing of the outer skin and seeds get hardened. Pointed gourd, ivy gourd, teasel gourd, and spine gourd are handpicked, while sweet gourd needs to be harvested using a sharp knife with a portion of the fruit stalk. The immature fruits of chayote should be harvested 100–120 days after planting, that is, prior to seed development. Sweet gourd for "gac" (as commonly known in Vietnam) products should be harvested at dead ripe stage (90–110 days) when the rind becomes scarlet red and soft upon pressing.

4.4 CROP IMPROVEMENT IN PERENNIAL VEGETABLES

Information on floral biology is basically needed before setting up a breeding program. Spine gourd, mountain spine gourd, teasel gourd, sweet gourd, pointed gourd, *S. amplexicaulis*, and ivy gourd are dioecious, and *M. cymbalaria* is monoecious. However, natural occurrence of hermaphrodite flowers in *M. dioica* (Jha and Roy, 1989) and *M. subangulata* subsp. *renigera* (Bharathi and Joseph, 2013) has been reported. *M. dioica* starts flowering 30–40 days after planting, depending upon prevailing weather conditions (Dod et al., 2007), teasel gourd starts flowering 50–60 days after planting (Vijay et al., 1977), and *M. cochinchinensis* begins flowering very late, that is, 90–100 days after planting (Bharathi and Joseph, 2013). Anthesis takes place in the early morning hours in most of the perennial cucurbits. However, anthesis takes place in later of the day in chayote (12–12.5 h),

M. cymbalaria (10:30–11:30 a.m.), and *S. amplexicaulis* (12–13:00 h) and in the late evening hours as in spine gourd (6:30–8:30 p.m.) (Bharathi and Joseph, 2013). Spine gourd is pollinated by moths and teasel gourd and sweet gourd by oil bees (Schaefer and Renner, 2010), and stingless bees are reported to be common and efficient pollinators of chayote flowers.

Seshadri and Parthasarathi (2002) considered the differentiation of sex in *M. dioica* to be entirely genic or genetical without any cytological evidence of heterogamety. However, Jha (1990) reported that sexual mechanism in *M. dioica* as an incipient type of sexual dimorphism (an intermediate stage toward X/Y chromosome basis), in which a pair of autosomes is responsible for sexual dimorphism. In the case of ivy gourd, there is one pair of heteromorphic sex chromosome; the female plant and male plant carry XX and XY chromosome, respectively. In the case of pointed gourd, OPC 07_{567} marker was reported as a molecular marker for sex expression (Singh et al., 2002).

In the case of spine gourd and teasel gourd, intergenotypic crosses were made possible through the induction bisexual flowers (Ali et al., 1991; Hossain et al., 1996) to facilitate recombination of desirable characters of parents in homosexual hybrid. Selection of high-yielding clones from such homosexual hybrids may lead to establishing a variety of species in a short period as the dioecious species are vegetatively propagated. Foliar sprays with $AgNO_3$ (400 ppm) at preflowering stage could induce 70%–90% hermaphrodite flowers in *M. dioica* (Rajput et al., 1994). Application of 500 mg/L $AgNO_3$ on female plants produced the maximum proportion of induced hermaphrodite flowers in *M. cochinchinensis*, and the pollen viability was similar to that of a normal male plant (Sanwal et al., 2011).

Seedlessness is appreciated by consumers and can contribute to increasing the quality of the fruits when seeds are hard or have a unpalatable taste. Furthermore, the shelf life of seedless fruit is expected to be longer than seeded fruit because seeds produce hormones that trigger senescence. This effect has been observed in watermelons, in which seeds hasten fruit deterioration. In addition, if the seeds and their cavities are replaced with edible fruit tissue, this is more attractive to the consumer. In the case of sweet gourd and spine gourd, the young and developing seed coat is whitish, soft, and delicate, subsequently turning ash color to black and becoming hard, which is a major bottleneck in consumer acceptance, and hence it is obvious that the development of parthenocarpic fruit will greatly enhance its food value and consumer acceptability (Handique, 1988; Chowdhury et al., 2007). Hormonal induction of parthenocarpy had been reported in teasel gourd (Vijay and Jalikop, 1980; Handique, 1988) and spine gourd (Chowdhury et al., 2007; Rasul et al., 2008). Seedlessness was also induced in spine gourd and pointed gourd (Singh, 1978) with pollen of related taxa (*Momordica charantia* and *Lagenaria leucantha*) and a mixture of the pollens from these two species. A high parthenocarpic fruit set (>90%) was also observed in an interspecific hybrid between *M. dioica* and *M. cochinchinensis* when the F_1 was pollinated with pollen from *M. cochinchinensis* (Bharathi and Joseph, 2013). Natural parthenocarpy has been observed in pointed gourd and ivy gourd.

Very little work has been attempted toward breeding though there is wide diversity available in the local germplasm. Selection of high-yielding clones is an important method in developing high-yielding cultivars in perennial cucurbits as most of them are dioecious and clonally propagated. In chayote, selection based on the number of fruits per plant and fruit weight with preferable physical appearance is useful in improving the crop, for example, two high-yielding varieties of ivy gourd (Arka Neelachal Kunki, Arka Neelachal Sabuja), spine gourd (Arka Neelachal Sree, Indira Kankad, Visal Small Baby Doll), teasel gourd (Arka Neelachal Gaurav), and pointed gourd (Arka Neelachal Kirti, Swarna Alaukik, Swarna Rekha, Faizabad Parwal 1, Faizabad Parwal 3, Faizabad Parwal 4, Rajendra Parwal 1, Rajendra Parwal 2, Shankolia, Konkan Harita). In chayote, no named varieties and available types are known based on fruit shape and color. Round white, long white, pointed green, broad green, and Germany green types of chayote are grown. However, there are a few reports of interspecific hybridization between spine gourd and teasel gourd (Bharathi et al., 2011) and spine gourd and sweet gourd (Mohanty et al., 1994; Mondal et al., 2006). Though many F_1 hybrids among spine gourd, teasel gourd, and sweet gourd have been developed without much difficulty, most of the efforts could not be sustained due to high sterility associated with meiotic abnormalities, which hindered the utilization of these hybrids. However, recently, two fertile interspecific hybrids were developed through backcrossing (Bharathi et al., 2014b) and ploidy manipulation (Bharathi et al., 2014a).

Polyploidy breeding has not been promising in pointed gourd because of lack of fertility (Hazra and Ghosh, 2001). However, the development of a fertile interspecific hybrid in *Momordica*, an auto- or allopolyploid (*M. subotica*), between *M. subangulata* subsp. *renigera* (tetraploid, 2n = 2x = 56) and *M. dioica* (diploid, 2n = 2x = 28) via chromosome doubling (of *M. dioica*) and hybridization represents an important step in the genus *Momordica*. The hybrid is naturally fertile and has the superior agronomic traits of both parents, making it a good choice as a new vegetable crop, and is expected to revolutionize the cultivation of *M. dioica* and *M. subangulata* subsp. *renigera* by saving labor and producing better propagation efficiency. Such successful efforts open up new opportunities for future *Momordica* breeding program.

Further, lycopene is an important functional compound in cucurbits, particularly *Momordica*, and watermelon. Though limited information exists on lycopene genetics in cucurbits (Grassi et al., 2013), it has been investigated to some extent in other plant models such as papaya (Blas et al., 2010; Devitt et al., 2010) and tomato (Rosati et al., 2000; Sun et al., 2012). The transfer of gene for phytoene synthase from *Erwinia uredovora* to tomato genetic background by Fraser et al. (2000) showed that the total fruit carotenoids of primary transformants were 2–4-fold higher than the controls, whereas phytoene, lycopene, β-carotene, and lutein levels were increased 2.4-, 1.8-, and 2.2-fold, respectively, by reducing the regulatory effect over the carotenoid pathway and without affecting activities of other enzymes in the pathway. Thus, the study showed the way for genetic manipulation to increase the lycopene content in tomato and other sources. For this, *Momordica* spp. can be investigated to understand the genetics and biochemistry of the tissue-specific expression of the gene (in aril portion). The *Momordica* spp. may have different genes or may be different promoters or other unknown factors like genetic background factors or tissue-specific factors, which may have great potential in transforming the conventional lycopene sources and enriching the lycopene trait in different crop plants.

4.5 GENETIC STUDIES IN PERENNIAL CUCURBITS

The literature contains very little information on genetics and cytogenetics of perennial cucurbits. The large field space requirements for cucurbits and the need for laborious hand pollinations for selfing and crossing cucurbits have been the constraints for genetic investigations (Robinson and Walters, 1997). These perennial cucurbits were considered less important earlier; consequently, very little work of this nature has been reported. Crossing experiments indicated that sex is controlled by a single factor, with heterozygous males and homozygous recessive females in *M. dioica* (Hossain et al., 1996) and *M. cochinchinensis* (Sanwal et al., 2011).

Selection for characters, such as number of fruits per plant, individual fruit weight, and fruit volume, is more important for the yield improvement of *M. dioica* (Bharathi et al., 2006). Characters such as fruit weight, fruit length, fruit diameter, fruit pulp content, fruit volume, and number of primary branches, leaves, and fruits per plant are the major yield contributing characters in pointed gourd (Singh et al., 1993, 2007; Sarkar et al., 1999; Hazra et al., 2003). Additive gene action for traits like fruit length, fruit volume, yield, primary branches, and number of fruits per plant was reported (Singh et al., 1985, 1986, 1992; Sarkar, 1989). Significant positive correlation had been reported between late flowering and fruit yield (Sarkar and Datta, 1987; Prasad and Singh, 1990), number of fruits per plant and fruit yield (Singh et al., 1986; Sarkar, 1989; Singh and Prasad, 1989; Dora et al., 2003), fruit weight and yield (Singh et al., 1986; Sarkar, 1989; Singh and Prasad, 1989; Sarkar et al., 1999), and fruit length and yield (Singh et al., 1987). In pointed gourd, stem and leaf pubescence, tendril branching, and coiling were reported to be governed by a single gene (Kumar et al., 2008).

4.6 BIOPROSPECTING OF PERENNIAL CUCURBITS

Although cucurbits are widely used for vegetable preparations, their use in traditional health systems is also not uncommon. The bioprospecting of specific health-benefiting properties or compounds from perennial cucurbits could contribute in food and health industry. In perennial cucurbits, the gac is used for extraction of lycopene-rich aril fraction for use in the food and pharmaceutical industry (Voung et al., 2005). The understanding of indigenous traditional knowledge is useful for speeding up the process of discovery and commercialization of new products.

4.7 HEALTH PERCEPTIONS

Traditionally, cucurbits are used in treating various health problems like cancer, HIV, diabetes, and other maladies. Researchers reported a large number of health benefits of perennial cucurbits (Bawara et al., 2010; Kumar et al., 2010; Gupta et al., 2011; Dhiman et al., 2012; Hasan and Khatoon, 2012). The traditional health system of uses of *M. cochinchinensis* leaves and fruits was reported by Rahman (2013) from Rajshahi division in Bangladesh for curing lumbago, ulceration, and fracture of bones. The seeds are used as aperients and in treating ulcers, sores, and obstructions of the liver and spleen. The roots are used in treating rheumatism with swelling of the lower limbs. He also reported that the seed of *M. dioica* is used as a cardiac tonic and as an astringent. The root is alterative, cholagogue, diuretic, and galactagogue. He also reported that the fruit of *M. charantia* are considered tonic, stomachic, febrifuge, carminative, and cooling; they are used in rheumatism, gout, and diseases of liver and spleen. The seeds are used in the production of anthelmintics. An alcoholic extract of the whole plant is used in treating stomachic against colic and fever, while the plant juice is used in administering diabetes, piles, leprosy, and jaundice and as vermifuge. Sato et al. (2011) reported that dietary intake of *M. dioica* flesh inhibits triacylglycerol absorption and lowers the risk for development of fatty liver in rats. Rahman et al. (2013) also reported that *M. charantia* reduces serum sialic acid in type 2 diabetic patients, which evidence the delay of the process of atherosclerosis. The antidiabetic effects of *M. charantia* (bitter melon) and its medicinal potency were also reported by Joseph and Jini (2013).

Health supplements are now common in markets, and the public is well aware of their uses and benefits. These products contain antioxidants or health supportive compounds, which work against free radicals in the body. The free radicals are responsible for extensive damage to deoxyribonucleic acid (DNA), protein, and lipid and even associated with degenerative diseases like cardiovascular diseases, cancer, immunity-associated problems, brain dysfunction, and cataracts. The antioxidants like lycopene and carotenoids play key role in scavenging free radicals supporting the *in-built* homeostasis mechanism. Lycopene is at the top of the antioxidants due to its strong capacity to quench singlet oxygen. It is two times more effective than β-carotene and 100 times more effective than α-tocopherol. It has been considered safe and nontoxic, and its consumption is usually without side effects so that even lycopene from food sources is suggested during pregnancy (Heber and Lu, 2002). Lycopene has the capacity to prevent free radical damage in cells caused by reactive oxygen species. It has shown potential antioxidant activity during *in vitro* and in human studies, reducing the susceptibility of lymphocyte DNA to oxidative damage, inactivating hydrogen peroxide and nitrogen dioxide, and protecting lymphocytes from nitrogen oxide-induced membrane damage and cell death twice as efficiently as β-carotene. The researchers indicated other mechanisms of action for lycopene, including modulation of intercellular gap junction communication, and anticancer mechanism by interfering with growth factor receptor signaling and cell cycle progression, specifically in prostate cancer cells (Rao and Agarwal, 2000; Heber and Lu, 2002).

4.8 FUNCTIONAL FOODS

The markets are now flooded with various kinds of processed food as well as plant-derived functional foods. These items have a great potential particularly in the health sector due to their wide acceptance and easy bioavailability among different age groups along with multiple benefits. The functional food items have one or more phytochemicals as their major ingredients, which have been reported with strong biological activity. Similarly, lycopene and carotenoids are two important phytochemicals for which the aril and pulp fractions are used for the preparation of various food items. The gac aril portion is most commonly used as a source of lycopene for the preparation of various health supplements, particularly in Western countries and East Asian countries. But the use of this plant is nil or very rare in India, and here, tomato is the only source of lycopene, which is in fact a poorer source than these plant species (Ishida et al., 2004). Some of the products like lycopene rich capsules, lycopene-rich drinks, snacks, sauce, powder for fortification, and food color. The aril powder, paste, concentrated juice, or other forms have great potential as natural color agent in vegetarian foods without adding tomato taste. It can be used for preparing food items for different age groups and even in producing candies, toffees, or ice creams. Due to its natural source, the acceptance of red color from gac will certainly be high among people. Besides, a number of products and formulations can be made for different groups of people, for example, various kinds of attractive snacks, food products, and beverages for children and functionally rich beverages and easily digestible foods for aged people. However, systematic consumer survey and product profiling may help in designing lycopene-rich new food items or food supplements. The gac is a newly domesticated crop having good adaptability to Indian soil and climate; therefore, the supply side of the raw material is quite good. Further research is needed for genetic improvement, planting material production, and understanding the biochemistry of the bioactives in *Momordica* species, which will be helpful for their industrial projects.

4.8.1 RECOVERY OF ARIL FRACTION FROM RIPE FRUITS

Anatomical dissection of ripe fruits of *M. cochinchinensis* and *M. subangulata* ssp. *renigera* revealed that the aril fraction was 24.8% and 19.6%, respectively (Table 4.2). The recovery of dry aril portion was 39.2 g/fruit of *M. cochinchinensis*, while it was only 4.2 g/fruit in the case of *M. subangulata* ssp. *renigera*. Also, the recovery of dry aril powder from sun drying process was higher in *M. subangulata* ssp. *renigera* (28.9%) than *M. cochinchinensis* (22.4%), which was similar to the observations made in gac by Ishida et al. (2004).

4.8.2 CUCURBITS AS A SOURCE OF LYCOPENE

Lycopene is one among the more than 600 carotenoids in photosynthetic organisms. It is a tetraterpene hydrocarbon, contains 40 carbon atoms, and has 56 hydrogen atoms with a molecular mass of 536. Though tomato is the most common source of lycopene and is used regularly in different food

TABLE 4.2

Recovery of Aril Fraction from the Ripe Fruits of *Momordica cochinchinensis* and *Momordica subangulata* ssp. *renigera*

Parameters	Gac (*M. cochinchinensis*)	Teasel Gourd (*M. subangulata* ssp. *renigera*)
Fruit weight (g)	718.5 ± 20.4	73.8 ± 18.9
No. of seeds/fruit	12.0 ± 0.7	28.0 ± 2.8
Aril (g)	178.3 ± 9.5 (24.8%)	14.5 ± 5.8 (19.6%)
Dry aril portion (g)	39.2 ± 1.5	4.2 ± 1.5

TABLE 4.3

Lycopene Content in Different Plant Sources

Fruit and Vegetable	Lycopene	Reference
Watermelon	49 µg/g	Holden et al. (1999)
	23–72 µg/g	Rao and Rao (2007)
Gac aril	1000–2000 µg/g	Ishida et al. (2004)
	2000–2300 µg/g	Rao and Rao (2007)
	1903 µg/g	Ishida et al. (2004)
	380 µg/g	Aoki et al. (2002)
	7020 µg/g	Kubola and Siriamornpun (2011)
Gac pulp	1800–6200 µg/g	Kubola and Siriamornpun (2011)
	408 µg/g	Voung et al. (2005)
	2073 µg/g	Ishida et al. (2004)
Gac pulp (green)	1690 µg/g	Kubola and Siriamornpun (2011)
Gac peel	1600–3400 µg/g	Kubola and Siriamornpun (2011)
Bitter melon	411 µg/g dw	Tran and Raymundo (1999)
Range	3.6–7020 µg/g	

items, the seed aril of gac (*M. cochinchinensis*) fruits is the richest source of lycopene and reported to contain even 70 times more lycopene than tomato (Ishida et al., 2004; Rao and Rao, 2007). It is reported to have anticancer activities, particularly those against prostate cancer. This trait attracts the industry for the commercial utilization of the gac in its goal of developing various lycopene-rich health supplements. Further, the bioavailability of the lycopene from aril fraction is also high due to its oil content because lycopene is a carotenoid and soluble in oil (Voung et al., 2005). A systematic review of the previous reports on lycopene determination by Singh et al. (2014) in various cucurbits indicates that the gac (*M. cochinchinensis*) is the richest source of lycopene, followed by *M. subangulata* ssp. *renigera* and *M. charantia* (Table 4.3). These sources can be used for lycopene extraction at commercial scale.

4.8.3 PHYTOCHEMICALS CHANGES IN *MOMORDICA*

The composition analysis of *M. subangulata* ssp. *renigera* fruits with methanol solvent at stage one showed that the tannins (495.4 mg/100 g; 492.4 mg/100 g) and flavonoids (396.0 mg/100 g; 389.4 mg/100 g) were predominant phytochemical aril and pulp fractions. At this stage, the aril fraction had a significantly ($p < 0.05$) higher content of lycopene (57.0 µg/g) and total content of carotenoid (149.8 µg/g) than pulp fraction, which had values of 51.7 µg/g and 68.9 µg/g, respectively. β-carotene (36.7 µg/g) and chlorophyll (56.1 µg/g) were significantly higher at stage one than later stages of fruit maturity. Total carotenoids and lycopene in aril fraction were increased from mature green to fully ripe stage by 205.0% and 303.7%, respectively, while in pulp fraction, the changes were 633.2% and 1113.2%, respectively. On the contrary, flavonoid (18.3%, 16.5%), tannin (24.4%, 26.8%), phenolic (20.2%, 19.5%), β-carotene (87.4%, 70.6%), and chlorophyll (55.1%, 82.0%) contents declined in aril and pulp fractions (Figure 4.1).

The 2,2′-azino-bis (3-ethylbenzthiazoline-6-sulphonic acid and 2,2-diphenyl-1-picrylhydrazyl activities of aril and pulp fractions reduced from 75.3% to 40.2% and 80.7 to 44.0% and from 68.3% to 39.6% and 77.0% to 41.0%, respectively, with fruit ripening stages and showed strong correlation with antioxidants (Figure 4.1). The information on changes in phytochemicals during fruit stages suggests for further investigations on the biochemical processes.

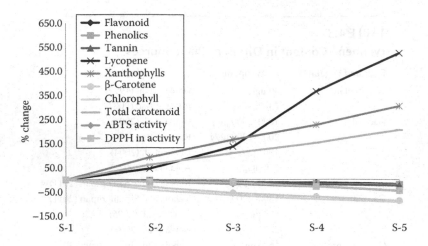

FIGURE 4.1   Percentage change in phytochemicals and antioxidant activity in aril fraction during fruit development in *Momordica subangulata* ssp. *renigera*.

4.9   CONCLUSIONS

Perennial cucurbits have a better tolerance to climatic vagaries and are the preferred food in rural and tribal communities. On the one side, many of them are still in forest or grow wild and their habitat degradation through natural and anthropogenic factors is a serious challenge for their existence on earth. On the other side, initial research efforts on certain aspects like antidiabetic juice of bitter gourd and lycopene-rich fraction from *Momordica* species showed great prospects for this group of cucurbits. Of course, all cucurbits will not have lycopene but there are certain other bioactive compounds in fruits or seeds of perennial cucurbits, which attracts the pharmaceutical and herbal industry. Therefore, *M. subangulata* ssp. *renigera*, *M. charantia*, and *M. cochinchinensis* can be used for the extraction of the lycopene-rich aril fraction. This will diversify the raw materials and ensure continuous supply of the raw material for sustaining the industrial firm. Besides, a preliminary investigation showed that the pulp fraction is also rich in carotenoids and phytochemicals that can also be used for the development of various products. The seeds of gac have showed potential for the extraction of oil and fatty acids, while gac oil is in great demand in international markets. Thus, these crops have great prospects in the herbal and nutraceutical industry for the development of cost-effective health supplements.

REFERENCES

Aileni M, Kota SR, Kokkirala VR, Umate P, Abbagani S (2009). Efficient in vitro regeneration and micropropagation of medicinal plant *Momordica tuberosa* Roxb. *Journal of Herbs, Spices & Medicinal Plants*, 15(2): 141–148.

Ali M, Okubo H, Fujii T, Fujiedan K (1991). Techniques for propagation and breeding of kakrol (*Momordica dioica* Roxb.). *Scientia Horticulturae* 47: 335–343.

Aoki H, Kieu MTN, Kuze N, Tomisaka K, Chuyen VN (2002). Carotenoid pigments in gac fruit (*Momordica cochinchinensis* Spreng). *Bioscience Biotechnology and Biochemistry* 66(11): 2479–2482.

Bawara B, Dixit M, Chauhan NS, Dixit VK, Saraf DK (2010). Phyto-pharmacology of *Momordica dioica* Roxb. ex Willd: A review. *International Journal of Phytomedicine* 2: 01–09.

Bharathi LK, Joseph JK (2013). *Momordica Genus in Asia: An Overview*, 147pp. Springer, New Delhi, India.

Bharathi LK, Munshi AD, Vinod CS, Behera TK, Das AB, John KJ, Nath V (2011). Cytotaxonomical analysis of *Momordica* L. (Cucurbitaceae) species of Indian occurrence. *Journal of Genetics* 90(1): 21–30.

Bharathi LK, Naik G, Dora DK (2006). Studies on genetic variability in spine gourd. *Indian Journal of Horticulture* 63: 96–97.

Bharathi LK, Singh HS, John, JK (2014a). A novel synthetic species of *Momordica* (*M.* × *subotica* Bharathi) with potential as a new vegetable crop. *Genetic Resources and Crop Evolution* 61: 979–999. doi 10.1007/s10722-014-0092-7.

Bharathi LK, Singh HS, Shivashankar S, Ganeshamurthy AN (2014b). Characterization of a fertile backcross progeny derived from inter-specific hybrid of *Momordica dioica* and *M. subangulata* subsp. *renigera* and its implications on improvement of dioecious *Momordica* spp. *Scientia Horticulturae* 172: 143–148.

Blas AL, Ming R, Liu Z, Veatch OJ, Paull RE, Moore PH, Yu Q (2010). Cloning of the papaya chromoplast-specific lycopene beta-cyclase, CpCYC-b, controlling fruit flesh color reveals conserved microsynteny and a recombination hot spot. *Plant Physiology* 152(4): 2013–2022.

Chakravarty HL (1982). *Fascicle Flora of India—Fascicle II Cucurbitaceae*. Botanical Survey of India, Calcutta, India.

Chowdhury RN, Rasul MG, Islam AKMA, Mian MAK, Ahmed JU (2007). Effect of plant growth regulators for induction of parthenocarpic fruit in kakrol (*Momordica dioica* Roxb.). *Bangladesh Journal of Plant Breeding and Genetics* 20(2): 17–22.

Devitt LC, Fanning K, Dietzgen RG, Holton TA (2010). Isolation and functional characterization of a lycopene β-cyclase gene that controls fruit colour of papaya (*Carica papaya* L.). *Journal of Experimental Botany* 61(1): 33–39.

Dhiman K, Gupta A, Sharma DK, Gill NS, Goyal A (2012). A review on the medicinally important plants of the family Cucurbitaceae. *Asian Journal of Clinical Nutrition* 4(1): 16–26.

Dod VN, Mahorkar VK, Peshattiwar PD (2007). Floral biology of underutilized vegetable: Kartoli. *The Asian Journal of Horticulture* 2: 285–287.

Dora DK, Behera TK, Acharya GC, Mohapatra P, Mishra B (2003). Genetic variability and character association in pointed gourd (*Trichosanthes dioica* Roxb.). *Indian Journal of Horticulture* 60(2): 163–166.

Evans P, Helliwell B (2001) Micronutrients: Oxidant/antioxidant status. *British Journal of Nutrition* 85: 567–574.

Fraser PD, Romer S, Shipton CA, Mills PB, Kiano JW, Misawa N et al. (2000). Evaluation of transgenic tomato plants expressing an additional phytoene synthase in a fruit-specific manner. *Nature Biotechnology* 99(2): 1092–1097.

Grassi S, Piro G, Lee JM, Zheng Y, Fei Z, Dalessandro G, Giovannoni JJ, Lenucci MS (2013). Comparative genomics reveals candidate carotenoid pathway regulators of ripening watermelon fruit. *BMC Genomics* 14: 781. doi: 10.1186/1471-2164-14-781.

Gupta R, Mathur M, Bajaj VK, Katariya P, Yadav S, Kamal R, Gupta RS (2011). Antidiabetic and renoprotective activity of *Momordica dioica* in diabetic rats. *Diabetologia Croatica* 40(3): 81–88.

Handique AK (1988). Hormonal induction of parthenocarpy in *Momordica cochinchinensis* Spreng. *Current Science* 57: 896–898.

Hasan I, Khatoon S (2012). Effect of *Momordica charantia* (bitter gourd) tablets in diabetes mellitus: Type 1 and Type 2. *Prime Research on Medicine* 2(2): 72–74.

Hazra P, Ghosh R (2001). Induced polyploidy as a breeding approach in pointed gourd. *SABRAO Journal of Breeding and Genetics* 33(1): 47–49.

Hazra P, Ghosh R, Nath S (2003). Identification of important yield components in pointed gourd (*Trichosanthes dioica* Roxb.) clones. *Journal of Vegetation Science* 25(2): 162–165.

Heber D, Lu QY (2002). Overview of mechanisms of action of lycopene. *Experimental Biology and Medicine* (*Maywood*) 227: 920–923.

Holden JM, Eldridge AL, Beecher GR, Buzzard IM, Bhagwat SA, Davis CS, Douglass LW, Gebhardt SE, Hayowitz DB, Shakel S (1999). Carotenoid content of U.S. foods: An update of the database. *Journal of Food Composition and Analysis* 12: 169–196.

Hossain MA, Islam M, Ali M (1996). Sexual crossing between two genetically female plants and sex genetics of kakrol (*Momordica dioica* Roxb.). *Euphytica* 90: 121–125.

Ishida KB, Turner C, Chapman HM, Mckeon AT (2004). Fatty acid and carotenoid composition of Gac (*Momordica cochinchinensis* Spreng.) fruit. *Journal of Agricultural and Food Chemistry* 52: 274–279.

Jha UC (1990). Autosomal chromosomes carrying sex genes in *Momordica dioica* Roxb. *Current Science* 59: 606–607.

Jha UC, Roy RP (1989). Hermaphrodite flowers in dioecious *Momordica dioica* Roxb. *Current Science* 58: 1249–1250.

Joseph B, Jini D (2013). Antidiabetic effects of *Momordica charantia* (bitter melon) and its medicinal potency. *Clinical Intervention in Aging* 2(2): 219–236.

Kubola J, Siriamornpun S (2011). Phytochemicals and antioxidant activity of different fruit fractions (peel, pulp, aril and seed) of Thai gac (*Momordica cochinchinensis* Spreng). *Food Chemistry* 127: 1138–1145.

Kumar DS, Vamshi Sharathnath K, Yogeswaran P, Harani A, Sudhakar K, Sudha P, Banji D (2010). A medicinal potency of *Momordica charantia*. *International Journal of Pharmaceutical Sciences Review and Research* 1(2): 95–100.

Kumar S, Singh BD, Pandey S, Ram S (2008). Inheritance of stem and leaf morphological traits in pointed gourd (*Trichosanthes dioica* Roxb.). *Cucurbit Genetics Cooperative Report* 26: 74–75.

Mohanty CR, Maharana T, Tripathy P, Senapati N (1994). Interspecific hybridization in *Momordica* species. *Mysore Journal of Agricultural Sciences* 28: 151–156.

Mondal A, Ghosh GP, Zuberi MI (2006). Phylogenetic relationship of different kakrol collections of Bangladesh. *Pakistan Journal of Biological Sciences* 9: 1516–1524.

Parks SE, Murray CT, Gale DL, et al. (2012). Propagation and production of gac (*Momordica Cochinchinensis* Spreng.), a Greenhouse Case Study. Exp Agr, 49, 234–243.

Prasad VSRK, Singh DP (1990). Studies on morphological and agronomical components of pointed gourd (*Trichosanthes dioica* Roxb.). *Indian Journal of Horticulture* 47: 337.

Rahman AHMM (2013). Ethno-medico-botanical studies on cucurbits of Rajshahi division, Bangladesh. *Journal of Medicinal Plants Studies* 1(3): 118–125.

Rahman K (2007). Studies on free radicals, antioxidants, and co-factors. *Clinical Interventions in Aging* 2(2): 219–236.

Rajput JC, Parulekar YR, Sawant SS, Jamadagni BM (1994). Sex modification in kartoli (*Momordica dioica* Roxb.) by foliar sprays of silver nitrate (AgNO$_3$). *Current Science* 66: 779.

Rao AV, Agarwal S (2000). Role of antioxidant lycopene in cancer and heart disease. *Journal of the American College of Nutrition* 19: 563–569.

Rao AV, Rao LG (2007). Carotenoids and human health. *Pharmacological Research* 55(3): 207–216.

Rasul MG, Mian MAK, Cho Y, Ozaki Y, Okubo H (2008). Application of plant growth regulators on the parthenocarpic fruit development in teasle gourd (kakrol, *Momordica dioica* Roxb.). *Journal of the Faculty of Agriculture, Kyushu University* 53(1): 39–42.

Reddy VSK, Subbaiah YPV, Reddy MGDM (2007). Exploitation of kasara kaya (*Momordica tuberosa*) for diversification of vegetables. *Acta Horticulture* (ISHS) 752: 577–580.

Robinson RW, Decker-Walters DS (1997). *Cucurbits*. CABI Publishers, Oxford, U.K., pp. 97–101.

Rosati C, Aquilani R, Dharmapuri S, Pallara P, Marusic C, Tavazza R et al. (2000). Metabolic engineering of b-carotene and lycopene content in tomato fruit. *The Plant Cell* 24(3): 413–419.

Sanwal SK, Kozak M, Kumar S, Singh B, Deka BC (2011) Yield improvement through female homosexual hybrids and sex genetics of sweet gourd (*Momordica cochinchinensis* Spreng.). *Acta Physiologiae Plantarum* 33: 1991–1996.

Sarkar DD, Datta KB (1987). Cytomorphology of some wild and cultivated members of *Trichosanthes* L. *Cytologia* 52: 419–423.

Sarkar SK 1989. Studies on genetic improvement of pointed gourd (*Trichosanthes dioica* Roxb.). PhD dissertation, Bidan Chandra Krishi Viswa Vidyalaya, Mohanpur, India.

Sarkar SK, Maity TK, Som MG (1999) Correlation and path-coefficient studies in pointed gourd (*Trichosanthes dioica* Roxb.). *Indian Journal of Horticulture* 56(3): 252–255.

Sato M, Ueda T, Nagata K, Shiratake S, Tomoyori H, Kawakami M et al. (2011). Dietary teasel gourd (*Momordica dioica* Roxb.). flesh inhibits triacylglycerol absorption and lowers the risk for development of fatty liver in rats. *Experimental Biology and Medicine* (*Maywood*) 236(10): 1139–1146.

Schaefer H, Renner SS (2010). A three-genome phylogeny of *Momordica* (Cucurbitaceae) suggests seven returns from dioecy to monoecy and recent long-distance dispersal to Asia. *Molecular Phylogenetics and Evolution* 54: 553–560.

Seshadri VS, Parthasarathi VA (2002). Cucurbits. In: Bose TK, Som MG (eds.), *Vegetable Crops in India*. Naya Prakash, Calcutta, India, pp. 91–164.

Shekhawat MS, Shekhawat NS, Harish, Ram K, Phulwaria M, Gupta AK (2011). High frequency plantlet regeneration from nodal segment culture of female *Momordica dioica* (Roxb.), *J Crop Sci Biotechnol.* 14: 133–137.

Singh AK, Singh RD, Singh JP (1993). Correlation and path coefficient analysis in pointed gourd. *Indian Journal of Horticulture* 50(1): 68–72.

Singh BR, Mishra GM, Niha R (1987). Inter-relationship between yield and its components in parwal. *South Indian Horticulture* 35: 245–246.

Singh DP, Prasad VSRK (1989). Variability and correlation studies on pointed gourd. *Indian Journal of Horticulture* 46(2): 204–209.

Singh DP, Sachan SCP, Mehta ML (1992). A note on occurrence of hermaphrodite flowers on pistillate plants of *Trichosanthes dioica* Roxb. *Haryana Journal of Horticulture Science* 21(1–2): 115–116.

Singh H (1978). Parthenocarpy in *Trichosanthes dioica* Roxb. and *Momordica dioica* Roxb. *Current Science* 47: 735.

Singh KP, Jha RN, Mohan K, Hague M (2007). Correlation and path co-efficient analysis in pointed gourd (*Trichosanthes dioica* Roxb.). *Asian Journal of Horticulture* 2(1): 9–11.

Singh M, Kumar S, Singh AK, Ram D, Kalloo G (2002). Female sex associated RAPD marker in pointed gourd (*Trichosanthes dioica* Roxb.). *Current Science* 82(2): 131–132.

Singh RR, Mishra GM, Jha RN (1985). Studies on variability and scope for improvement in pointed gourd (*Trichosanthes dioica* Roxb.). *South Indian Horticulture* 33: 257.

Singh S, Singh DR (2012). Bioprospecting of lycopene trait in native *Momordica* spp. of Andaman Islands. In: Singh DR et al. (eds.), *Abstract Book of National Seminar on Innovative Technologies for Conservation and Sustainable Utilization of Island Biodiversity.* CIARI, Port Blair, India, pp. 68–67 (Abstract).

Singh S, Singh DR, Banu VS, Avinash N (2014). Functional constituents (micronutrients and phytochemicals) and antioxidant activity of *Centella asiatica* (L.) Urban leaves. *Industrial Crops and Products* 61: 115–119.

Singh S, Singh DR, Salim KM, Singh LB, Srivastava A, Srivastava RC (2011). Estimation of proximate composition, micronutrients and phytochemical compounds in traditional vegetables from Andaman and Nicobar Islands. *International Journal of Food Science and Nutrition* 62(7): 765–773.

Singh VP, Singh K, Jaiswal RC (1986). Genetic variability and correlation studies in pointed gourd (*Trichosanthes dioica* Roxb.). *Narendra Deva Journal of Agricultural Research* 1: 120.

Sun YD, Liang Y, Wu JM, Li YZ, Cui X, Qin L (2012). Dynamic QTL analysis for fruit lycopene content and total soluble solid content in a *Solanum lycopersicum* × *S. pimpinellifolium* cross. *Genetics and Molecular Research* 11(4): 3696–3710.

Tran TLH, Raymundo LC (1999). Biosynthesis of carotenoids in bittermelon at high temperature. *Phytochemistry* 52: 275–280.

Vijay OP, Jalikop SH (1980). Production of parthenocarpic fruit with growth regulators in kakrol (*Momordica cochinchinensis* Spreng). *Indian Journal of Horticulture* 37: 167–169.

Vijay OP, Jalikop SH, Nath P (1977). Studies on floral biology in kakrol (*Momordica cochinchinensis* Spreng.). *Indian Journal of Horticulture* 34: 284–288.

Vuong LT, Franke AA, Custer LJ, Murphy SP (2005). *Momordica cochinchinensis* Spreng. (Gac) fruit carotenoids reevaluated. *Journal of Food Composition and Analysis* 19(6–7): 664–668.

5 Watermelon (*Citrullus lanatus*) Live Mulch Climatic Adaptation Capabilities in Humid Tropics Cropping System

David O. Ojo

CONTENTS

5.1 INTRODUCTION

In situ live mulch crops such as watermelon (*Citrullus lanatus*) can suppress weed populations, resulting in reduced reliance on herbicides, reduced soil temperature, improved soil moisture, and additional income from sale of produce harvested in mixed cropping systems in sub-Saharan Africa. Previous studies indicate that cover crops could suppress weed populations and reduce reliance on herbicides (Duppong, 2004; Saini et al., 2008; Pelosi et al., 2009; Anugroho and Kitou, 2011; Muleke et al., 2012; Hertwig et al., 2013; Isah et al., 2014) and that in situ live mulch cover crops give short-term cash supplements from produce harvested, thereby making adoption by farmers more attractive, easier, and quicker (Ali, 1999; Anyszka et al., 2006; Jodaugienė, 2006; Willard et al., 2008; Elzaki et al., 2013). It is suspected that changes in crop production practices might be due to weed control, temperature, moisture, and flexibility in African humid tropics cropping systems. Grain amaranth (*Amaranthus cruentus* L.) is important for its higher protein content compared to other staple crops, such as rice, maize, sorghum, and millet (Bressani, 1988) in sub-Saharan Africa where meeting daily dietary requirements is challenging. It is drought tolerant and highly adaptable to the tropics as a potential crop, thereby contributing to food self-security in the region. Watermelon commands higher prices than the local nonexotic crops especially during dry seasons. The present investigations, therefore, seek to quantify the impact of various densities of in situ watermelon live mulch on weed control and yield potential and climatic change adaptation capability strategy in the humid tropics amaranth production system.

5.2 MATERIALS AND METHODS

An experiment was conducted at the National Institute of Horticultural Research (NIHORT) headquarters, Ibadan, Nigeria (3°54′E, 7°30′N, 213 m above sea level) from October 2009 to June 2010 using supplemental irrigation facilities during low rainfall periods. Treatments comprised three sowing densities of watermelon: 1.5 × 0.45 m; 1.5 × 0.90 m; 1.5 × 1.50 m. Grain amaranth as an intercrop was transplanted at 0.75 × 0.75 m spacing. There was a control plot left bare without crop-ping and a check plot with only grain amaranth forming five treatments in each of five replicates in a randomized complete block (RCB) design trial. Weeding was done by hand at 3-week intervals, commencing 3 weeks after planting (WAP) of watermelon. Naturally occurring weed population was used, that is, no supplemental weed planting was used. Net plot size was six data rows, 15 m long, 0.75 m apart (i.e., 15 m × 4.5 m), plus a border row on each side to minimize interference of the weed control treatments. Treatments, design, and agronomic practices used were based on experimental results at the National Horticultural Research Institute (NIHORT) from 1985 to 1999 (NIHORT 1985–1999) and on the results obtained by David (1997). Soil maximum and minimum temperature data were collected by inserting two maximum–minimum dial thermometers per plot at 0.10 m depth for 12 weeks, commencing 3 weeks after transplanting (WAT) amaranth. Readings were made at 3 WAT amaranth, 50% amaranth anthesis, and harvest of amaranth. Readings made by instruments in each plot were averaged for use in data analyses. Soil moisture content in the sur-face 0.15 m was determined gravimetrically at 50% anthesis of amaranth. Four soil samples were taken randomly from each plot and combined in a sealed plastic bag for drying as soon as practi-cable. Soil was dried at 120°C for 3 days, and soil moisture content was expressed as percentage weight of moisture lost during drying relative to soil dry weight. Transmitted and incident photosyn-thetic photon flux density (PPFD) was measured using AccuPAR PAR 80 Model within 1 h of solar noon during clear weather at 50% amaranth anthesis between the center of amaranth and water-melon rows. Percent transmittance was determined from readings under and above the vegetation after adjustment for measured differences among sensors in unobstructed sunlight readings. Weeds within four 0.50 m² quadrates per plot were sampled at 3 WAT amaranth, 50% amaranth anthesis, and amaranth harvest. Weeds within quadrats were cut at ground level and oven dried at 80°C until constant weight to determine total weed biomass. All data were subjected to analysis of variance using procedures of SAS software. Interaction effects are only reported when significant at P = 0.05.

5.3   RESULTS AND DISCUSSION

Averaged over the two consecutive croppings, amaranth grain and watermelon fruit yields were highest at 1.5 × 0.90 m watermelon plant spacing (Table 5.1). These values are within the range reported by David (1997).

Watermelon live mulch treatment consistently reduced weed density and biomass relative to the control in both croppings (Table 5.2).

Lowest mean light penetration of live mulch was 0.29% and 0.22% in the first and second crop-pings, respectively. All mulch densities transmitted less PPFD relative to the control (Table 5.3). Light levels of less than 0.1% transmittance are required to activate phytochrome-mediated

TABLE 5.1
Yield of Watermelon and Grain Amaranth with Watermelon as
Green Mulch and Amaranth as an Intercrop

Mulch Treatment[a]	Watermelon (kg/m²)	Grain Amaranth (g/m²)
Check	—	88
1.5 × 0.45 m	8.2	82
1.5 × 0.90 m	8.6	93
1.5 × 1.5 m	9.4	90
Lsd (P = 0.05)	0.2	3.8

[a] Photosynthetic photon flux density is % ratio of PPFD under mulch divided by unobstructed PPFD made above mulch.

TABLE 5.2
Weed Density and Biomass with Watermelon as a Green Mulch and Grain Amaranth as an Intercrop

Mulch Treatment[a]	Weed Density (No/m²)	Weed Biomass (g/m²)
Check	111	177
Control	1344	880
1.5 × 0.45 m	832	182
1.5 × 0.90 m	876	230
1.5 × 1.5 m	964	500
Lsd (5%)	40.2	30.8
	Interaction (P = 0.05)	
Mulch/control	[b]	[b]

[a] Photosynthetic photon flux density is % ratio of PPFD under mulch divided by unobstructed PPFD made above mulch.
[b] Significance of interaction.

TABLE 5.3
Light Transmittance with Watermelon as a Green Mulch and Amaranth as an Intercrop in Two Successive Crops

Mulch Treatment	Light Transmittance (% PPFD)[a]		
	First Cropping	Second Cropping	Mean
Check	0.11	0.77	0.56
Control	0.88	0.65	0.77
1.5 × 0.45 m	0.29	0.22	0.26
1.5 × 0.90 m	0.35	0.24	0.30
1.5 × 1.5 m	0.38	0.25	0.32
Lsd (5%)	0.05	0.03	0.04

[a] Photosynthetic photon flux density is % ratio of PPFD under mulch divided by unobstructed PPFD made above mulch.

germination and even lower levels are required for the very low fluence rate response (Kronenberg et al., 1986). This finding is supported by previous research that live mulch reduces light penetration a quarter to a half times compared to bare natural weed infested soil (Teasdale et al., 1993).

Maximum soil temperature was reduced by amaranth and melon in each cropping (Table 5.4), and there was a trend toward a significant increase in minimum soil temperature under live mulch. Britow (1988) first observed this under mulch for modifying soil environmental temperature, but the implication in adapting to climate change needed to be clarified (Challinor et al., 2007). The practical implication is that live mulch could regulate environmental and soil temperature by reducing temperature in a hotter environment and increasing it in a colder environment and during cool periods of the day. Melon may have a greater influence on weed germination by affecting the daily amplitude between maximum and minimum temperatures (Table 5.4). A diurnal temperature change of approximately 10°C is required for germination of some weed seeds (Taylorson, 1987), but it is unknown whether the species (predominantly elephant grass) in this experiment had a temperature amplitude requirement.

Soil moisture content was significantly greater in the live mulch treatments compared to the bare soil (Table 5.5).

TABLE 5.4
Soil Maximum and Minimum Temperature under Live Mulch

Mulch Treatment[a]	Soil Temperature (°C)[a]		
	Maximum	Minimum	Amplitude
Check	40.1	32.7	7.40
Control	44.3	38.0	6.30
1.5 × 0.45 m	34.0	32.1	1.90
1.5 × 0.90 m	36.1	31.8	4.30
1.5 × 1.5 m	39.8	30.3	8.50
Lsd (5%)	0.88	1.04	0.42

[a] Averaged temperature recorded daily/weekly from 50% anthesis till final harvest of amaranth.

TABLE 5.5
Soil Moisture Content under Live Mulch

Mulch Treatment	Moisture Content (%)		
	First Cropping	Second Cropping	Mean
Check	14.0	12.8	13.4
Control	15.0	13.0	14.0
1.5 × 0.45 m	20.8	20.5	20.7
1.5 × 0.90 m	18.1	18.0	18.1
1.5 × 1.5 m	16.3	15.9	16.1
Lsd (5%)	0.89	0.91	0.64

These research findings demonstrated that live mulch suppressed weeds, had greater light extinction, and had lower diurnal soil temperature amplitude that accounted for usefulness of live mulch in cropping systems, and we believe it offers opportunities in adaptation of agricultural production systems under climate change scenarios.

5.4 SUMMARY AND CONCLUSIONS

In situ live mulch crops such as watermelon (*C. lanatus*) can suppress weed populations resulting in reduced reliance on herbicides, reduced soil temperature, improved soil moisture, and additional income from sale of produce harvested in mixed cropping systems in sub-Saharan Africa. The present investigations, therefore, seek to quantify the impact of various densities of in situ watermelon live mulch on weed control and yield potential and climatic change adaptation capability strategy in the humid tropics amaranth production system. Treatments comprised three sowing densities of watermelon: 1.5 × 0.45 m; 1.5 × 0.90 m; and 1.5 × 1.50 m. Grain amaranth as an intercrop was transplanted at 0.75 × 0.75 m spacing. There was a control plot left bare without cropping and a check plot with only grain amaranth forming five treatments in each of five replicates in an randomized complete block design (RCBD) design. Weeding was at 3-week intervals, commencing 3 WAP of watermelon. Naturally occurring weed population was used. Averaged over the two consecutive croppings, amaranth grain and watermelon fruit yields were highest at 1.5 × 0.90 m watermelon plant spacing. All mulch densities transmitted less PPFD relative to the control. Soil moisture content was significantly greater in the live mulch treatments compared to the bare soil. Our research demonstrated that live mulch suppressed weeds, had greater light extinction, and had lower diurnal soil temperature amplitude that accounted for usefulness of live mulch in cropping systems, and we believe it offers opportunities in adaptation of agricultural production systems under climate change scenarios.

REFERENCES

Ali, M. 1999. Evaluation of green manure technology in tropical lowland rice systems. *Field Crops Research* 61(1): 61–78.

Anugroho, F. and Kitou, M. 2011. Effect of live hairy vetch and its incorporation on weed growth in a subtropical region. *Weed Biology and Management* 11.1(2011): 1–6.

Anyszka, Z. and Dobrzanski, A. 2006. Impact of cover crops and herbicides usage on weed infestation, growth and yield of transplanted leek. *Zeitschrift Fur Pflanzenkrankheiten Und Pflanzenschutz-Sonderheft* 20: 733.

Bittencourt, H.V.H., Lovato, P.E., Comin, J.J., Lana, M.A., Altieri, M.A., Costa, M.D., and Gomes, J.C. 2013. Effect of winter cover crop biomass on summer weed emergence and biomass production. *Journal of Plant Protection Research* 53(3): 248–252.

Bresani, R. 1988. Amaranth: The nutritive value and potential uses. *Food and Nutrition Bulletin* 10: 49–59.

Bristow, K.L. 1988. Mulch and its architecture in modifying soil temperature. *Australian Journal of Soil Research* 26: 269–280.

Challinor, A., Wheeler, T., Garforth, C., Craufurd, P., Kassam, A., Challinor, A., Wheeler, T., and Garforth, C. 2007. Assessing the vulnerability of food crop systems in Africa to climate change. *Climatic Change* 83(3): 381–388.

David, O. 1997. Effect of weeding frequencies on grain amaranth growth and yield. *Crop Protection* 16(5): 463–466.

Duppong, L.M. 2004.The effect of natural mulches on crop performance, weed suppression and biochemical constituents of catnip and St. John's wort. *Crop Science* 44(3): 861–869.

Elzaki, R.M., Elbushra, A.A., Eissa, A.M., and Ahmed, S.E.H.A. 2013. Crop biodiversity: Potential of sustainability indicators and poverty reduction in farming systems in Sudan. *American Journal of Agriculture and Forestry* 1(4): 55–62.

Isah, A.S., Amans, E.B., Odion, E.C., and Yusuf, A.A. 2014. Growth rate and yield of two tomato varieties (*Lycopersicon esculentum* Mill) under green manure and NPK fertilizer rate samaru northern guinea savanna. *International Journal of Agronomy* 2014: 1–8.

Jodaugienė, D. 2006. The impact of different types of organic mulches on weed emergence. *Agronomy Research* 4: 197–201.

Kronenberg, G.H.M. and Kendrick, R.E. 1986. Phytochrome. The physiology of action. In R.E. Kendrick and G.H.M. kronenberg, eds. *Photo Morphogenesis in Plants*. Martinus Nijhoff Publishers, Dordrecht, the Netherlands, pp. 17–33.

Muleke, E.M., Saidi, M., Itulya, F.M., Martin, T., and Ngouajio, M. 2012. The assessment of the use of eco-friendly nets to ensure sustainable cabbage seedling production in Africa. *Agronomy* 3(1): 1–12.

NIHORT (National Horticultural Research Institute). 1985–1999. Grain amaranth agronomic characterization, evaluation, yield. NIHORT Annual Reports 1985–1999, NIHORT, Idi Ishin, Jericho GRA, Ibadan, Nigeria.

Pelosi, C., Bertrand, M., Roger-Estrade, J., and Pelosi, C. 2009. Earthworm community in conventional, organic and direct seeding with living mulch cropping systems. *Agronomy for Sustainable Development* 29(2): 287–295.

Saini, M., Price, A.J., Van Santen, E., Arriaga, F.J., Balkcom, K.S., and Raper, R.L. 2008. Planting and termination dates affect winter cover crop biomass in a conservation-tillage corn-cotton rotation: Implications for weed control and yield. In *Proceedings of the 30th Southern Conservation Agricultural Systems Conference*, Tifton, GA: Southern Conservation Agricultural Systems, pp. 137–141.

Taylorson, R.B. 1987. Environment and chemical manipulation of weed seed dormancy. *Reviews of Weed Science* 3: 135–154.

Teasdale, J.R. and Daughtry, C.S.T. 1993. Weed suppression by live and desiccated hairy vetch (*Vicia villosa*). *Weed Science* 41: 207–212.

Willard, D. and Valenti, H.H. 2008. Juneberry growth is affected by weed control methods. *HortTechnology* 18(1): 75–79.

6 Cultural Practices of Persian Melons

Jamal Javanmardi

CONTENTS

6.1 INTRODUCTION

Persian melons are one of the most important cucurbits in Iran with a wide variety of types. Due to their high economical and nutritional values, they are cultivated in most cucurbit cultivation areas in Iran and some surrounding countries. It is believed that melon domestication began in Iran. Historical records and archeological remains show the place of origin of melons as Egypt and Iran (Robinson and Decker-Walters, 1997). Persian melons have a considerable tolerance to drought and saline condition and need a long warm growing season (Javanmardi, 1998). The best conditions for growing Persian melons are around the central desert in Iran, where other plants rarely could be grown. Farmers in those areas have grown Persian melons for centuries. Since Persian melons are endemic to Iran, internationally published literature on their research is rare and most of the information is local or based on centuries of farmers' experiences.

6.2 BOTANY

According to Naudin's classification in 1882 for melons detailed by Robinson and Decker-Walters (1997), most of the Persian melons do not meet the criteria described for *Cucumis melo* var. *inodorous* (Javanmardi, 2010). The definition for *inodorous* group or winter melons is that the fruits are usually larger, mature later, and keep longer than those of the *cantalupensis* group. Also they have smooth or wrinkled rind surface but not netted, typically white or green flesh and no musky odor from which their name of *inodorus* is derived. Fruits do not detach from the peduncle when mature, plants are usually andromonoecious (Robinson and Decker-Walters, 1997; Pitrat, 2008). The *reticulatus* group is defined as fruits with round to oval shape, with a typical netted skin, and with or without ribs. The flesh is usually orange, aromatic, and sweet (Pitrat, 2008). In fact, Persian melons are not smooth but have netted rind with or without ribs. The skin color can be white, yellow, gray, dark green, and even black, uniform or with spots. They are round, oval, or oblong fruits with a musky odor. The flesh is usually from white to cream, light green, or orange in color. Fruits in most cultivars do not detach from peduncle at maturity (Javanmardi, 2010). According to the definition for cultigroup (cultivar group) by Pitrat (2008), it seems Persian melons can be defined as a cultigroup within the aforementioned botanical groups.

Cross-pollination among different cultivars of Persian melons can occur freely. Therefore, there is a wide diversity of cultivars of intermediate types. Each cultivar has been adapted to local environmental conditions. They are locally selected, cultivated, and named by generations according to the fruit type and quality; resistance to environmental stresses, pests, and diseases; and ease of cultivation. It should be noticed that there is diminution of the number of cultigroups (such as "Doody Sabz" and "Ghandi Zard") and an increase of cultivars (such as "Susky Sabz" and "Susky Zard") within some cultigroups. Some of the common cultivars in Iran are Abatar, Abbas Shuri, Akbar Abadi, Alam Gargar, Atashin, Bahar Hamedan, Bakharman, Balkhi, Ebrahimkhani, Eivanaki, Garmsari, Ghalamghash, Gorgab, Haj-Mashallahi, Jabbari, Jafar abadi, Jafari, Jarju, Kadkhoda Hosseini, Khaghani or Shakhteh Mashad, Kharcheh Zanjan, Khatuni, Majidi, Marand, Mashhadi, Mohajeran Hamedan, Mohamad Abadi, Rostagh, Sabuni, Shafi Abadi, Shah abadi, Shahsavari, Shirazi, Shomal, Suski Sabz, Susky Zard, Tashkandi, Tokhm Mohamad, Zarand Saveh, Zarcheh Esfahan, Zard Garmsari or Aliabadi, Zard Jalal, and Zard Tabriz. It is possible that some similar cultivars are known by different names in different regions (Javanmardi, 2010).

6.3 SOME CULTIVARS' DESCRIPTIONS

- *Abbas Shuri*: Vigorous, late crop (120–150 days), suitable for semiarid hot and dry and semisaline condition, oval-shaped fruits (20–45 cm L, 12–25 cm W), green grayish fine netted rind, firm and suitable for transportation and storage, white greenish flesh with 10%–15% sugar. It was widely cultivated in Garmsar, Varamin, Eivanaki, Tehran, and the northwest of the central desert in Iran. It had been the best suited cultivar for dry farming before Suski Sabz and Suski Zard were developed.
- *Khaghani Mashad*: Vigorous, semiearly crop (90–130 days), cylindrical fruit, green yellowish thin rind, netted with deep smooth ribs, green to orange flesh, 10%–15% sugar, big fruit (40–45 cm L and 15–18 cm W) with an average weight of 3.5–4 kg. It is the most cultivated cultivar in the northeast of Iran, which is distributed to other parts of Iran.
- *Zard Karaj*: Semilate crop (100–120 days), oval fruit (20–30 cm L, 17–25 cm W), and 1.5–2 kg, deep yellow netted rind, green whitish firm flesh, 8%–12% sugar.
- *Tashkandi* (*Dar Gazi*): Vigorous plant, semiearly crop (90–100 days), cylindrical fruit (20–40 cm L, 12–17 cm W), average weight of 4–5 kg, full slip at ripening, creamy to yellowish with brown strips on rind, white and soft flesh with 10%–14% sugar, low transportability and storability.
- *Jabbari*: Vigorous, medium crop (100–110 days), oval fruit, netted with 8–9 ribs, yellowgrayish rind, 11%–14.5% sugar, suitable for semiarid regions and saline soils.
- *Khatuni*: Vigorous, early crop (90–100 days), several fruits on plant, oval fruit, netted with 6–8 yellow ribs, yellow-greenish rind, 11%–14.5% sugar, suitable for semiarid regions and saline soils (Kashi, 2000) (Figure 6.1).

6.4 CLIMATIC CONDITIONS

Persian melons require a long growing season (90–180 days) depending on the cultivar and climatic conditions. They favor sunshine and hot and dry climatic condition to have the best quality (sugar and flavor). Although the plant grows well in warm and humid regions, foliage diseases are especially serious and fruits do not develop to the best quality. In Iran, a large part of the crop is grown in arid and semiarid regions under irrigation; however, in some parts, it is grown under dry farming cultivation.

Suski Sabz

Eivanaki

Tashkandi

FIGURE 6.1   **(See color insert.)** Some Persian melon cultivars.

6.5   SOILS AND SOIL PREPARATION

Melons are grown on many classes of soil from sands and sandy loams to silt loams and clay loams. Where earliness is important or growing season is short, a sandy loam is considered excellent. In fact, this class of soil is almost ideal in most producing regions. Any friable, well-drained soil is satisfactory, provided other conditions are favorable to melon growing. In any kind of soil, organic material and water-holding capacity are important for production. For dry farming, high water-holding capacity and soil structure determine success in production.

Melons grow on neutral or alkaline soils, but blossoms can drop off if grown in acid soils. The soil should be well prepared as for other cultivated crops. In some areas, the land is prepared for planting on ridges or beds to ensure good surface drainage. Where the crop is grown under irrigation, the land is bedded to conform to the irrigation practices.

Land preparation starts from autumn by plowing 30 cm deep to improve rainfall absorption, lower disease inoculums, and give time to decay previous crop residues. This is important especially when heavy soils are used. Application of manure and chemical phosphorus and potash fertilizers (based on soil analyses) is recommended at this time. Fertilizers are broadcast before beds are formed or banded into the beds between the seed rows and the irrigation furrow or added to the irrigation water. Disking and leveling the surface is achieved before planting during spring.

In most irrigated areas, the soil is shaped to furrow (50 cm deep) and ridge (2–4 m wide) depending on the soil type, amount of available water, and the cultivar's vigor. Raised beds improve drainage, modify temperature, and increase the depth of the rooting zone. This also helps plants against root, crown, stem, leaves, and fruit rot caused by *Phytophthora*. Before laser land leveling became widespread in agriculture, the farmers put small barriers (called "Pateh") along the furrows every few meters, depending on the land slope and furrow length, to raise water level during irrigation. This holds water in furrows for a longer time and improves water infiltration into side ridge soil where the plant roots are growing. The height of the Pateh is determined in such a way that water does not override the ridge. The top of the Pateh is usually covered with plastic to prevent erosion.

FIGURE 6.2   Installed Pateh in the furrows.

The first irrigation is before seed sowing to determine the final level of watermark and adjust the height of Patehs (if applicable). Currently, irrigation time is prolonged until water reaches to where the seeds are sown (Figure 6.2).

Land preparation for dry farming Persian melons is different. Usually, in the late winter, a heavy deep soil land with high water-holding capacity is chosen, and after making big basins, it is heavily irrigated. It is plowed, disked, and thoroughly leveled after the soil reaches field capacity. Leveling helps to lower water loss due to evaporation. Fertilization with chemicals is not usual since the land should be already fertile. Selecting suitable cultivar adapted to dry farming, proper seed sowing date, and practices decreasing water loss during the growth are initial necessities. Nowadays, due to the possibility of using underground water for irrigation, the method is under semidry farming through supplemental irrigation at fruit set stage and its subsequent fruit growth. This method is very beneficial for increasing fruit size and total yield; however, fruit sugar content and flavor will decrease (Kashi, 2000).

6.6   PLANTATION

Persian melon plants are very tender, and the seeds will not germinate at temperatures less than 15°C; hence, planting in the field should be delayed until all danger of chilling or frost is over and the soil becomes warm. In the central parts of Iran, unheated plastic tunnels have become popular for commercial production of Persian melons to promote earliness. Polyethylene plastic is typically suspended by wire hoops (every 1.5 m) over ridges, covering seed rows on both sides. The cover is secured by soil to keep it in place. Tunnels increase temperature during the day, protect young plants from the wind, and exclude harmful insects and insect-transmitted diseases if they are around since at that time, the field is still cold and insects have not started their activity. This covering provides only slight protection against frost. In any case, covers must be removed once female flowers open; otherwise, the flowers cannot be pollinated (Figure 6.3).

A large part of the commercial crop is grown from direct seeding in the field by the hill method on the flat, on ridges, or on raised beds. The choice of system usually follows the local custom that has developed according to the irrigation systems used and the soil's drainage characteristics. Although using transplants are not usual due to susceptibility of seedlings to the transplanting shock, it is possible to use 3–4-week-old transplants if only they were grown in the plug system (Javanmardi, 2009).

FIGURE 6.3 Plastic tunnels covered the furrows.

The seeds are usually soaked for days at 20°C–25°C before planting until a radicle tip appears on the seeds. Traditionally, several seeds are sown by hand in each hole to help in emerging through the soil (Postchi, 1965). The hills where the seeds are sown are usually 10 cm away from the watermark on ridges to prevent damping off in seedlings.

Plant spacing depends on the vigor of the cultivar as well as the irrigation system. Seeds are sown in rows at 50–75 cm spacing for irrigated farming and 100 cm spacing for dry farming. Rows are usually from 1.25 to 2 m apart. In this spacing, 3–4 kg/ha of seeds are needed. The plants are thinned carefully to one or two when they have two to three leaves and are well established. The remaining plants should not be disturbed during thinning. When row cover tunnels are used, a small sharp knife with a long handle is used to thin the extra seedlings. This is done by piercing through the plastic cover without rupture (Figure 6.4).

In the use of plastic tunnels, care should be taken to prevent damage to the plants during warm sunny days. Excess heat can build up under the plastic, and high temperatures can injure plants unless ventilation is provided. To do this, farmers slit the plastic on the sides of the tunnel. Some

FIGURE 6.4 Cutting extra seedlings with long handle knife from the outside of the plastic tunnel.

growers just open the end sides of tunnel on warm days. After the danger of cold weather is over, the cover is gradually torn out during one month. The plastic tunnels are gathered afterward. At this time, plants are turned toward the top of ridges and secured by dry soil to prevent twisting by wind and falling into furrows. Care should be taken not to break the stem during this practice (Figures 6.5 through 6.7).

FIGURE 6.5   Gradual tearing out of plastic tunnel above the seedlings.

FIGURE 6.6   Torn out of plastic tunnel after a month.

FIGURE 6.7   Turning vine toward the ridge and securing with dry soil.

6.7   IRRIGATION

Although Persian melons are adapted to dry and hot climates and need less water compared to other cucurbits, they need a high amount of water due to high temperatures during their long growing period.

The farm is irrigated right after seeding. The second irrigation is delayed until the seedlings show temporary wilting. This helps the seedlings to develop roots deeper and more distributed in the subsoil than in the surface soil and increase tolerance to drought and harsh conditions. Previously, traditional farmers used a 14-day-irrigation interval. They also intentionally missed another irrigation usually after plant and/or early-formed flower–fruit pruning to induce fruit setting. The irrigation interval is reduced to 6–8 days during fruit growth and prior to harvest. Excess water during fruit maturation reduces sweetness and storage life.

6.8   PRUNING

Plant, flower, side branch, and fruit pruning are important practices for Persian melon production. Plant pruning is done after seedlings are at the stage of five to six leaves when the first side branches have a 5 cm length. It comprises of removing the main stem's apical meristem while keeping two side branches. These branches will form the final plant structure. The main reason for this pruning is so that the main stem bears male flowers and delays side branching that bears male and female flowers. In the dry farming method, two side branches are spread in opposite directions in order to cover the ground, reducing water evaporation and weed growth. In conventional farming, two side branches are directed over the ridge.

The more the plant is ramified, the more female flowers are produced. But not every flower gives a desirable fruit as there is strong competition between young fruits. Therefore, side-branch and early-formed flower–fruit pruning (locally called "Tarash") are achieved when side branches grow at least 50 cm in length. Secondary side branches, all male and female flowers, and fruits that exist before the sixth and after the eighth leaves are then removed. Fruits on sixth to eighth leaves are kept and make the final yield. The final number of fruits on the plant depends on the plant vigor. Usually, two fruits are kept on each plant. Fruits before the sixth leaf and distal end fruits on branches in most Persian melons are usually smaller than those the in middle part of the branch. Intact plants will produce side branches and fruits. All fruits, especially those set earlier, will be small. In fact, Tarash synchronizes plant vegetative and reproductive development toward big and quality fruits. This is more important for those cultivars with large fruits than those with small fruits. Tarash for small-sized fruit cultivars is not necessary (Kashi, 2000).

Flower pruning is another cultural practice for Persian melons that consists of selecting, keeping one or two healthy, pest-free fruits, and removing all remaining flowers and fruits. Meanwhile, there were indigenous practices (before the widespread use of chemical pesticides) to protect the chosen fruit against melon fruit fly (*Myiopardalis pardalina*). Fruits of about 2 cm length were wrapped in a leaf of the same plant and buried under the soil. An alternate method in small-scale farms was to moisten the fruit with spit and cover it with soil. As the melon grows, repeatedly cover with soil until the rind gets thicker, preventing melon fruit fly from penetrating its ovipositor into the fruit and laying eggs (Amiri-Ardakani and Emadi, 2006).

6.9   FRUIT TURNING

Fruits on the plant do not develop at a same rate. Fruits are monitored carefully for their growth, health, defection, infection, etc., during growth and development. Turning the fruits during the growth is done in order to prevent uneven fruit rind coloration or soilborne infections on the part of the fruit in contact with the ground. Meanwhile, unnecessary side branches, fruits, flowers, and weeds are removed.

6.10 HARVESTING

Persian melons should be left on the vines until they are fully mature, are of high quality, and have highest sugar content, but are still solid. If it remains on the vine, the sugar will not increase but its flesh gets soft and losses its tenderness. Fruits on the same plant do not ripe at the same time.

There are certain changes that take place, which may be used as an index, but these should be correlated with the internal condition including texture, flavor, and sweetness. As fruits approach maturity, the netting becomes fully rounded out, while its color gets dull. In immature fruits, the flatten netting has a slight crease along the top. As ripening advances, small cracks develop around the peduncle at the base of the fruit. The fruit's blossom end gets soft at this point and its aroma increases. Persian melons, except some cultivars such as Tashkandi, do not develop abscission layer called "full slip" when they are mature. The sugar (soluble solids) content in the juice of the edible portion is one of the most important factors that are determined by refractometer (Javanmardi, 2010). The fruits are graded after harvest based on size, shape, color, and general appearance (Figure 6.8).

6.11 TRANSPORTATION AND STORAGE

Persian melons are transported to local and distant markets usually by trucks. Some wheat straws are usually put around the fruits when loading in the trucks. This prevents fruit bruising and keeps the fruits dry. Persian melons ordinarily are not stored except for late season types, which are usually consumed during mid-winter. They can be kept at 7°C–10°C and a relative humidity of 85%–90% (Javanmardi, 2010) (Figures 6.9 and 6.10).

6.12 SEED PRODUCTION

The different groups of melons within *C. melo* are all cross compatible. They are insect pollinated; therefore, the minimum isolation distance for commercial hybrid seed production should be 1000 m, although some seed regulatory authorities specify distances of only 500 m (Raymond, 2009). The heterosis is not very important when the parents belong to the same cultigroup; however, it is clearly observed in F1 hybrids between two different parental lines (Pitrat, 2008).

Conditions for high-quality melon seed production usually are the same for good-quality fruit production. Although the most important selection criteria for melon growers are fruit size and fruit quality including sugar content, aroma, flesh texture, and color (Javanmardi, 2010); the roguing for seed production is achieved along production processes starting before flowering (vegetative habit, characters of foliage, and undeveloped fruit), after flowering (fruit type and morphology), during

FIGURE 6.8   (See color insert.) Harvested melons.

FIGURE 6.9 **(See color insert.)** Truck loading the Persian melon "Khaghani" with wheat straw.

FIGURE 6.10 **(See color insert.)** Tashkandi (left) and Khaghani (right) Persian melons on display.

fruit set (fruit type, morphology, and external characters), and harvesting period (fruit type, morphology, external characters, fruit quality factors) (Raymond, 2009).

The melon seeds are usually extracted by hand in Iran. The fully ripe selected melons are cut in halves and seeds with placentas are scraped into a container. The mixture is left for a few days to be fermented in water before washing. The "good seeds" sink at the bottom of the tank, while placentas and empty seeds float. Seeds are then washed in running water. Some small-scale growers do not follow the fermentation process. They wash the mixture in suitable-sized sieves to extract seeds. The seeds are then usually sun-dried. Keeping dry seeds in the dark at 5°C and low humidity will secure viability for 10 years or more.

In addition to using seeds for the next year's plantation, some seeds are consumed as roasted nut (Javanmardi, 2010) because they are rich in lipids and proteins that represent a great amount of energy (Pitrat, 2008).

REFERENCES

Amiri-Ardakani, M. and M.H. Emadi. 2006. *An Overview of Indigenous Knowledge of Iranian Farmers on Plant Protection.* Nosouh Press, Isfahan, Iran.

Javanmardi, J. 1998. Effects of salinity on plant growth, absorption and transportation of mineral elements in five Iranian native Persian melon cultivars. MSc thesis. Tehran University, Tehran, Iran (in Farsi).

Javanmardi, J. 2009. Scientific and applied basis for vegetable transplant production. Mashad University Press, Mashad, Iran (in Farsi).

Javanmardi, J. 2010. *Growing Organic Vegetables.* Mashad University Press, Mashad, Iran (in Farsi).

Kashi, A. 2000. *Complementary to Vegetable Crop Production*. Tehran University Press, Tehran, Iran (in Farsi).

Pitrat, M. 2008. Melon. In: Prohens, J. and Nuez, F. (eds.). *Handbook of Plant Breeding—Vegetables I*. Springer, New York, pp. 283–316.

Postchi, E. 1965. *Cucurbits and Cucurbits Production*. Shiraz University Press, Shiraz, Iran (in Farsi).

Raymond, A.T.G. 2009. *Vegetable Seed Production*, 3rd ed. CAB International, Wallingford, Oxford, UK.

Robinson, R.W. and D. Decker-Walters. 1997. *Cucurbits*. CAB International, Wallingford, Oxford, UK.

7 No-Tillage Production Systems for Cucurbit Vegetables

Stuart Alan Walters

CONTENTS

7.1 INTRODUCTION

Many cucurbit vegetable growers are interested in or are converting at least a portion of their production schemes to incorporate conservation tillage practices. No-tillage (NT) is a conservation tillage practice that has been implemented by some cucurbit growers, especially those that grow pumpkins, although it has not been implemented in other cucurbit crops to the same extent. Cucumber and squash are the two other cucurbit crops in which NT has been extensively evaluated, with most research results indicating that yields for these two crops are similar between NT and conventional tillage (CT). Most NT research studies conducted in melons and watermelons either are inconclusive or show a negative influence of NT on their overall productivity. This review will focus on providing an overview of NT production systems for cucurbit vegetables, as well as complementary management practices that are often utilized in association with this type of production system.

NT is a type of conservation tillage practice that provides an undisturbed soil before and after planting, with crops generally directly planted into the residue remaining on the soil surface (Morse, 1999). Vegetable crops provide little organic matter to soils after harvest is completed and tend to provide little surface residue to protect soils from wind and water erosion (Bellinder and Gaffney, 1995). Although most vegetables are grown using CT practices, NT is becoming more prominent in vegetable production due to growers gaining more understanding of the economic and ecological benefits associated with the NT production system. NT production practices provide effective and low cost methods to reduce soil erosion, increase soil organic matter, and improving both the water-holding capacity and the nutrient-holding capacity of soils (Johnson and Hoyt, 1999). NT systems will also significantly increase the amount of surface residue and reduce soil erosion compared to other tillage systems (Harrelson et al., 2007). NT cultivation with cover crops is an effective technique for halting soil erosion and making food production truly sustainable (Jelonkiewicz and Borowy, 2009).

Although NT practices provide various environmental benefits (Johnson and Hoyt, 1999; Morse, 1999), many growers prefer this production system for cucurbit crops since cleaner fruit often results from residing on crop residues (Walters and Young, 2008) or from reduced soil splashing onto the fruit surface. Besides providing a cleaner fruit than CT, NT can also help in reducing disease incidence on fruits. Osmond et al. (2011) indicated that reduced tillage methods providing mulch residues on the soil surface can reduce the risk of cucumber belly rot (*Rhizoctonia solani*) since less fruit to soil contact results from this production system.

Vegetable growers, oftentimes, have difficulty transitioning from CT to NT, since NT typically requires more planning and management compared to CT production systems. Management problems include integration of cover crops into the system, use of crop residues as mulches, fertility management, stand establishment, and lack of adequate weed control. Thus, many vegetable growers do not implement NT production systems due to the extra management required and the potential problems that may be encountered.

7.2 COVER CROPS

Cover crops are often integrated into NT production systems. A cover crop is an alternative crop that is planted with the primary cash crop(s) or soon afterward and is usually suppressed or killed before the crop is planted (Walters, 2011). Numerous cover crops are used in NT production systems with the specific species or mixture of species chosen depending on their overall purpose (Table 7.1).

TABLE 7.1
Some Cover Crops for No-Tillage Cucurbit Vegetable Production

Cover Crop	Primary Purpose	Potential Utilization in NT Production Scheme
Winter grass cover crops such as annual ryegrass, oats, wheat, and winter rye	Biomass production	Build soil organic matter
	Companion plants for winter legume cover crops	Recycle soil nutrients
	Organic mulch	Organic mulch over soil
		Soil erosion control
		Weed suppression
Winter legume crops including hairy vetch and various clovers	Companion plants for winter grass cover crops	Add some organic matter
	Organic mulch	Organic mulch over soil
	Nitrogen-fixing capability	Provide nutrients to system
	Some biomass	Soil erosion control
		Weed suppression
Summer grass cover crops, mostly sorghum–sudangrass	Biomass production	Alleviate soil compaction
	Breakup subsoil	Build soil organic matter
	Organic mulch	Organic mulch over soil
		Recycle soil nutrients
		Weed suppression
Summer legumes cover crops such as sweet clover	Nitrogen-fixing capability	Add some organic matter
	Organic mulch	Organic mulch over soil
	Some biomass	Provide nutrients to system
		Weed suppression
Summer broadleaf cover crops such as buckwheat	Loosen topsoil	Add some organic matter
	Nutrient assimilation	Organic mulch over soil
	Organic mulch	Provide nutrients to system
	Some biomass	Soil erosion control
		Weed suppression

Some produce high amounts of biomass (e.g., wheat or winter rye) that can produce consistent thick mulches from the resides left behind to cover the soil surface after being either harvested or killed in some manner, while others (such as the various legumes that include hairy vetch and clovers) are used to improve soil nutrient levels. In most cucurbit production systems, grass cover crops are used to increase crop residue biomass levels and thus soil organic matter content, while legumes are grown for their soil nutrient improvement properties. In most NT cucurbit production systems, autumn cover crop plantings are most often used to obtain some plant growth before the onset of winter. Cover crop seeds are planted either by drilling or by broadcasting with light incorporation into shallow tillage. Higher germination rates will occur when seeds are drilled. It is important that the cover crop attain maximum growth in the spring before it is suppressed or killed in some manner, and this is just before flowering and seed development. This is important to maximize the development of lignin in grass cover crops that will allow the thickest mulch possible for covering the soil surface. Legumes, such as hairy vetch, should be killed when the first flowers form and definitely before seeds develop. When flowers are seen on hairy vetch, vegetative growth has stopped and it must be killed in some manner; otherwise, it can become a serious weed if allowed to go to seed. Additionally, nitrogen fertilizer applications in late winter are often used to accelerate spring growth. Many cover crops, such as wheat or winter rye, must be killed or suppressed in some manner to minimize competition with the cucurbit crop (Zandstra et al., 1998). Cover crops used for NT plantings are often killed with a nonselective herbicide in spring (typically either glyphosate or paraquat) and mowed or chopped in some manner before seeding or transplanting the cucurbit crop.

Cover crops serve multiple purposes in NT production systems as they conserve soil moisture, increase soil organic matter content, provide multiple soil conservation advantages (Johnson and Hoyt, 1999), and improve overall weed control (Teasdale et al., 1991). Cover crops or previous crop residues play an important role in adding organic matter to the soil to build soil structure and other soil physical properties, such as cation exchange capacity and soil porosity. Furthermore, the use of a legume cover crop has the added benefit of adding nitrogen (N) back into the cropping system which can decrease the amount of N fertilizer that must be applied (Table 7.1). Cover crops will also reduce wind and water erosion that can occur after removal of the primary cash crop as little residue is often left on the soil after harvest operations are complete.

7.2.1 Mulches

Cover crop residues (or organic mulches) on the soil surface can provide multiple benefits to NT cucurbit production. The decayed root channels of cover crop residues allow more percolation into the soil profile from either irrigation or rain water. Additionally, crop residues in contact with the soil surface reduce runoff and allow more rain water or irrigation water to infiltrate into the soil. Additionally, this residue slows or prohibits the flow of water through a field, thus decreasing erosion of soil and reducing fertilizer/agrochemical runoff. Cover crop residue in NT fields also allows equipment to enter fields under wet conditions, which cannot usually be done in CT fields. Furthermore, the timing of spray applications or harvests may be delayed in CT fields until soils become dry, while in NT fields, these operations can occur due to the residues on the soil surface.

Cover crop mulch residues can improve water-use efficiency in NT production fields. The mulch residue on the soil surface often acts as a buffer between the air and soil to decrease evaporation of water from the soil surface. In comparison, tilled soils in CT systems will often dry quickly from higher evaporation rates. Thus, cucurbit crops growing under drought conditions in NT systems will most likely grow better and produce greater yields compared to those grown under similar conditions in CT systems. However, irrigation may be still required to produce adequate yields if the lack of rainfall persists for a long period of time.

Organic mulches on the soil surface produced from cover crop or crop residues will insulate the soil and generally slow early plant growth and delay crop maturity (Morse, 1999). Walters and

Young (2008) indicated that soil covered with winter rye was generally about 5°C–6°C cooler compared with bare soil during the spring. These lower soil temperatures will often delay the development of cucurbit vegetables (e.g., cucumber and squash), resulting in later fruit production (Walters et al., 2005, 2007). Earliness of crop production is often extremely important in the marketing and profitability of many vegetables (Morse, 1999). Thus, the loss of crop earliness is a major concern that many growers have when utilizing NT production systems early in the growing season for certain short-season cucurbits.

NT cucurbit vegetable systems, when combined with residues left from small grains, have generally yielded similar to those grown in CT systems (Harrelson et al., 2007; Walters et al., 2007; Walters and Young, 2010, 2011). The residues that remain on the soil surface in NT most likely improve vegetable yields by increasing soil moisture compared with CT systems (Johnson and Hoyt, 1999), and those tillage systems that leave 30% or more residue on the soil surface after planting will generally increase soil moisture during the growing season due to increased infiltration and decreased evaporation (Johnson and Hoyt, 1999). Thus, NT production can be especially beneficial for reducing potential water stress associated with nonirrigated vegetable crops (Hoyt, 1999).

7.2.2 FERTILITY MANAGEMENT

In cucurbit vegetable production systems, including both CT and NT, fertilizers are often overapplied to ensure the highest yields possible. Many cucurbit vegetables have high nutrient demands, including cucumbers and squash, due to fruit removal from multiple harvests in a growing season. Furthermore, most cucurbits are inefficient users of nutrients and high fertilizer rates are often used that exceed crop demand to insure high yields since fertilizers are a relatively low-cost input for growers. These excess nutrients (especially N) are prone to be lost through leaching, immobilization, or volatilization (Shennan, 1992). Agricultural practices remain a major source of nitrate pollution of groundwater, and excessive N fertilization accompanied by poor soil and crop management practices have increased nitrate pollution of groundwater supplies (Hallberg, 1989). Thus, agricultural practices that conserve and manage soil nutrients are essential in maintaining soil and water quality, as well as sustaining vegetable crop productivity.

Excess nitrogen fertilization in cucurbit vegetable crops will often cause extra vegetative growth at the expense of fruit set, which can reduce or delay fruit set, resulting in reduced yields. Although fertility management in NT systems is important to maximize productivity, there is little available information on fertilizer recommendations for vegetable crops (including cucurbits) that are grown in this type of production system. Hoyt et al. (1994) indicated that fertilizer rates and application methods must be adjusted when NT systems are used in vegetable production. Although grass cover crops (such as wheat or winter rye) are often used in cucurbit vegetable NT production systems to reduce soil erosion and to minimize nutrient leaching from soil, their residues tend to tie up soil nitrogen in the spring and have a slow rate of decomposition during the summer (Morse and Seward, 1986). Grass cover crops may create N deficiency for the succeeding vegetable crop through N assimilation and/or immobilization (Skarphol et al., 1987; Vyn et al., 1999). Cover crops that provide mulch residues with high C/N ratios may reduce the yield of the following vegetable crops due to N immobilization (Francis et al., 1998; Schonbeck et al., 1993), and net N immobilization is more likely to occur if the cover crop mulch residues have a C/N ratio above 25 (Paul and Clark, 1989). Thus, breakdown of the cover crop residues and subsequent N release is in part regulated by the C/N ratio. During the spring, a herbicide-killed small grain will oftentimes induce immobilization of mineral N due to the high C/N ratio of the straw mulch residue. However, grass cover crops can increase nitrogen availability by preventing N losses due to leaching during the winter months (Knavel and Herron, 1986). Grass cover crops such as winter rye, wheat, oats, barley, and annual ryegrass can take up large amounts of residual soil nitrogen in the autumn months, whereas legumes tend to establish slower in the autumn and are poor nitrogen scavengers since they fix much of their own nitrogen.

TABLE 7.2

Suggested Nitrogen Fertilization Rate Adjustments Based on Previous Field Cropping History prior to Planting No-Tillage Cucurbit Vegetables

Previous Field Cropping History	Nitrogen Adjustment Rate (kg/ha)
Grass cover crop	Add 25
Legume cover crop, legume vegetables, or soybean	Subtract 40
Corn or wheat grain as previous crop	Add 50
Nonlegume vegetables	Add 30

Although cover crops primarily affect the soil structure rather than the fertility through increases in plant material biomass, it is generally accepted that cucurbit vegetables grown in NT systems with grass or cereal mulch residues on the soil surface require increased amounts of fertilizers compared to those grown in CT systems due to the immobilization of N in this system. Thus, the field cropping history plays an important role in nutrient management. Table 7.2 provides suggested nitrogen fertilization rate adjustments for cucurbit crops based on previous field cropping history. Besides the N requirements typically recommended for a specific cucurbit vegetable crop, an additional 25 kg N/ha should be supplemented to those fields in which a grass cover crop was grown immediately prior to the cucurbit crop, and if a cereal or corn grain crop was harvested prior to establishment of the NT cucurbit crop, the field should be supplemented with 50 kg N/ha. However, if a legume cover crop or vegetable or soybean is grown in the field prior to the cucurbit crop, then a small nitrogen credit can be given, with about 40 kg N/ha removed from the specific cucurbit crop's nitrogen fertilization rate.

Another factor to consider is soil organic matter content. Soil organic matter breaks down very rapidly in soils used for conventional vegetable production due primarily to intensive cultivation; however, NT vegetable production systems improve and build soil organic matter content compared to tillage. Franzluebbers (2010) found that reduced tillage is most effective at enhancing soil organic matter when combined with rotations or cover crops that include deep-rooting and high-residue-producing plants, such as alfalfa. The inclusion of grass or legume cover crops in a crop production scheme helps to maintain organic matter levels in the soil, as well as providing improved soil structure components, such as improved nutrient retention capabilities (or cation exchange capacity). Although legumes release nitrogen and other nutrients quickly back into the soil for plants to use, they are not as effective as grass cover crops at building up the organic matter levels in the soil. In comparison, plants that tend to have more lignin and be more fibrous (e.g., grass cover crops) release nutrients much more slowly into the soil system but will improve organic matter levels in the soil.

Cucurbits can effectively use nitrogen that is released from organic materials in the soil. Soils may release up to 100 kg N/ha depending on the soil organic matter content and the amount and type of organic materials that have been incorporated into the soil. However, this amount is more likely to be around 40 kg N/ha. It is also important to consider a field's previous cropping history, as a legume cover crop may supply upward of 40–50 kg N/ha. Again, Table 7.2 provides guidelines for adjusting N rate based on the previous crop or cover crop.

High-organic-matter soils have the ability to hold more nutrients compared to low-organic-matter soils, which can definitely affect the fertility of a given soil. Thus, the accumulation of soil organic matter can often lead to enhanced soil fertility through the sequestration of plant nutrients, especially N; the amounts of fertilizers added to soils are often adjusted due to the amount of organic matter present in soils. Soil organic matter is a huge reserve of potentially available nitrogen, phosphorus, sulfur, and other nutrients. It is thought that about 1% organic matter in a soil will release on average about 11 kg available N/ha over the course of a growing season. Since most cucurbit crops require large amounts of nitrogen to maximize productivity, growers do not adjust nitrogen fertility

TABLE 7.3

Suggested Nitrogen Fertilization Rates in No-Tillage for Selected Cucurbit Vegetables

Cucurbit Crop	Nitrogen Application Rates for No-Tillage Production Systems (kg/ha)
Cucumber—pickling	135–180 (Osmond et al., 2011)
Cucumber—slicing	155–235 (Rich, 2013)
Pumpkin	≥120 (Harrelson et al., 2008)
Zucchini squash	125–135 (Rich, 2013)

rates based on soil organic matter levels, and taking into account that the estimated N release rate of soil organic matter can be quite variable and most soils contain less than 3% soil organic matter (which provides about 33 kg N/ha), organic matter only provides a small portion of the nitrogen that is required for cucurbit crop growth.

There is little available information available for NT cucurbit crops regarding optimum N fertility rates. Although the use of conservation tillage combined with cover cropping is a management scheme that is often considered by vegetable growers, not much is known about optimum N fertility rates in this type of production system for cucurbit crops. A few nitrogen fertilization studies have been conducted for NT cucurbit vegetable production, with results generally indicating a need for extra nitrogen compared to CT systems (Table 7.3). Harrelson et al. (2008) indicated that greater amounts of N fertilization are needed in NT pumpkins compared to CT systems and suggested that optimum yields may occur with rates ≥120 kg/ha. Although nitrogen rates did not differ between CT, NT, or strip-tillage (ST) systems for pickling cucumber productivity, Osmond et al. (2011) indicated that the optimum N rate was influenced by the year, and there was a linear response of yield with N rate up to 200 kg N/ha in one year, while during the other, yield was maximized at 90 kg N/ha. Rich (2013) indicated that the nitrogen fertilizer rate required to obtain the greatest amounts of zucchini squash and slicing cucumber yield in NT was 125–135 and 155–235 kg N/acre, respectively; and in both NT cucumbers and zucchini squash, high nitrogen fertilizer rates stimulated vegetative growth, with resulting fruit yields significantly reduced. It appears that if extra nitrogen is available in the soil through overapplication, these plants will suffer from luxury nitrogen consumption, resulting in extra vegetative growth with less overall fruit production.

7.2.3 STAND ESTABLISHMENT

Seeded vegetable crops generally require a clean, freshly tilled seed bed to obtain high germination rates, and seeding vegetable crops into NT fields that have surface residues can oftentimes be a challenge. However, large seeded vegetable crops, like cucurbits, can be successfully established in NT production systems either through direct seeding or using transplants (Morse, 1999). Supplemental irrigation is oftentimes essential to improve stand establishment for both seeded and transplanted NT cucurbit crops, especially if planted under drought conditions or in drought-prone areas.

When directly seeding cucurbits in a NT field, proper seed-to-soil contact is required for adequate seed germination and seedling establishment. When using a NT seeder, it is important that proper seeding depth placement, row closure, and adequate seed-to-soil contact occur. Sometimes, the seed will not make it down into the soil and can be observed as only pressed onto or into the organic mulch residues, which will drastically reduce germination rates to the point where there is oftentimes no germination. To ensure adequate stand establishment, the seeding rate may need to be increased so that adequate amounts of seeds germinate, but this can be costly. Another problem that often occurs is rodent (primarily mice) feeding on the recently planted cucurbit seed. Mulch residues in NT provide a great habitat for mice, and large-sized cucurbit seeds (e.g., pumpkins and squash) provide a great source of food. If mice are a problem, some growers have had success with lightly spraying seed with diesel fuel and then allowing the seed to dry before planting.

Compared to direct seeding, cucurbit vegetables grown from transplants provide several advantages to the grower. First, those cucurbits grown from transplants can be typically harvested earlier than those grown from seed. Growers who use cucurbit transplants (regardless of the production system) often target earlier markets to improve overall revenue generation. Second, transplants are normally grown in greenhouses, which can serve to protect developing seedlings from various environmental stresses, animal predations, disease pathogens, and insect pests. Third, the use of transplants can actually reduce establishment costs due to the possibility of poor germination rates in the field or excess plant thinning that may be required to obtain proper plant spacing. Fourth, transplants allow growers to obtain nearly perfect stands of plants that are of the same age and size. However, it is important to note that the use of transplants can be capital intensive and requires specialized equipment for planting.

Cucurbits can actually have high survival rates when planted into the field if proper transplanting techniques are utilized. After seeding cucurbits in a greenhouse, they require about 3–4 weeks of growth to reach the two- to three-leaf stage that will normally produce plants of sufficient size for transplanting; transplanting cucurbits at an early age will reduce transplant shock. Furthermore, transplant establishment is improved when water is applied to the soil–root environment at transplanting. Applying adequate moisture and using the proper planting depth for a cucurbit transplant are both critical to reduce plant loss in the field to drying, regardless of the type of production system. In NT systems, cucurbit growers have adapted their production techniques to use either direct seeding or transplanting based on what has worked best for them in the past regarding plant field establishment.

7.2.4 WEED CONTROL

The adoption of NT by commercial cucurbit vegetable growers has been limited primarily due to the lack of effective weed management (Walters, 2011). Broadleaf weed control is a major impediment to the success of NT vegetable production systems because there are relatively few herbicides available to control these weeds in most vegetable crops, and the number of herbicides available is dependent on the specific vegetable that is grown. Weeds often become the primary problem in NT production systems since between-row tillage is often used in CT systems to reduce weed populations that preemergence (PRE) herbicides fail to control (Walters, 2011).

The establishment of a dense, uniformly distributed mulch residue on the soil surface is critical for enhancing weed control in NT systems (Morse, 1999). The more crop residue left on the soil surface, the greater the likelihood for smothering germinating weeds at the soil surface and preventing them from reaching light. Most suggest establishment of NT cucurbits into a weed-free field, but many times this is not possible. PRE herbicides can be used, but it is important to understand that they may not be as successful in a NT field covered with crop residues compared to a bare soil situation, as the mulch residues may prevent herbicide residues from reaching the soil (Walters et al., 2007). Various PRE herbicides are available for use on some cucurbits including bensulide, clomazone, ethalfluralin, halosulfuron, metolachlor, and trifluralin; there is also the widely used premix product of clomazone + ethalfluralin that is used as PRE in many situations. This may seem like a good arsenal of herbicides to control weeds, but many provide limited control to only a few weed species, and the residual activity in the soil of these herbicides is only about 30 days, which limits their effectiveness.

The main problem for NT weed control, however, is the lack of sufficient postemergence (POST) herbicides for use in cucurbits once the PRE herbicides lose their effectiveness. There are very few herbicides labeled POST for cucurbit vegetable crops. Halosulfuron is labeled for POST use in some cucurbit crops to control certain broadleaf weeds and sedges. This herbicide is specifically used for common cocklebur (*Xanthium strumarium* L.), pigweed (*Amaranthus* spp.), and yellow nutsedge (*Cyperus esculentus* L.) control. Halosulfuron must be applied when the cucurbit crop has at least some growth (at least two to five true leaves) but before the plants begin to flower or spread out over the soil surface. Gramoxone and glyphosate are both labeled for postemergent, nonselective control of weeds but must be applied with a spray shield to prevent drift onto the crop. These herbicides are also often used as a nonselective application to kill all weeds in the mulch residues prior to planting. To control annual grasses and some perennial grasses, sethoxydim or clethodim can be applied POST at labeled rates without damaging the cucurbit crop. It is important to note that once the cucurbit plant is established and spreads out over the soil surface, plant competition provided by the crop will definitely hinder weed growth.

7.3 CUCURBIT CROPS FOR NO-TILLAGE

Cucumbers, squash, and pumpkin are cucurbit crops well suited to NT production practices. Although there have been studies conducted in other cucurbit crops, such as melon and watermelon, most are either inconclusive or show a negative influence of NT on their overall productivity.

7.3.1 CUCUMBERS

Cucumbers can be effectively managed in NT culture (Lonsbary et al., 2004; Osmond et al., 2011; Walters et al., 2007; Wang and Ngouajio, 2008; Weston, 1990). Although cucumber plant growth was reduced by NT, no reduction in total yield was observed compared to CT (Lonsbary et al., 2004; Wang and Ngouajio, 2008), and the reduced vegetative growth that occurs in NT may actually be an advantage for mechanical harvesting procedures (Lonsbary et al., 2004). Furthermore, Ogutu and Caldwell (1999) found that the use of cucumber transplants provided more biomass accumulation at three weeks after planting in NT, which resulted in higher early yields due to earlier flowering and fruit set than those that were directly seeded. However, many studies have indicated that when cucumber is grown under adverse growing conditions in NT, fruit yields can be lower than in CT (Jelonkiewicz and Borowy, 2009; Walters et al., 2007; Zandstra et al., 1998). Lastly, since weed control is a problem in NT cucumber production, it is important to note that adequate weed control systems must be developed in some manner before NT will be widely utilized for this crop (Walters et al., 2007).

7.3.2 SQUASH

Research has suggested that it is possible to effectively produce summer squash using NT with no yield reductions compared to CT (NeSmith et al., 1994; Walters and Kindhart, 2002). Since early season squash yields may be reduced when using NT, this production system would not generally be beneficial to those growers seeking early fruit production due to the likelihood of lower revenue generation with later harvests (NeSmith et al., 1994). Although early season squash yields may be reduced, the lack of weed control is really the most important limiting factor to widespread implementation of NT squash production (Walters et al., 2004, 2005; Walters and Young, 2011). Therefore, the success of NT for squash production ultimately rests on the effectiveness of sufficient weed control through various cropping systems (e.g., cover crops) and/or herbicides.

7.3.3 PUMPKINS

Of all cucurbit crops, pumpkins are probably the best suited for NT culture. Several studies have all indicated that NT and CT produce comparable pumpkin yields when sufficient weed control is achieved in NT production systems (Galloway and Weston, 1996; Rapp et al., 2004; Walters et al., 2008). Additionally, pumpkin vegetation will provide some soil shading and weed suppression once vines form across the soil surface, but other weed control measures are generally necessary to achieve adequate weed control in NT pumpkin production. The use of herbicides and cover crops often plays an important role in the management of weeds in NT pumpkin production, and the use of a thick, evenly distributed mulch residue over the soil surface can enhance weed suppression in NT pumpkins, which will often reduce or even eliminate the need for PRE herbicides.

7.4 CONCLUSIONS

The use of conservation tillage practices, including NT, is gaining attention and wide acceptance by many vegetable growers, since research has generally indicated that these production systems are effective in maintaining yields and profitability while promoting sustainability. Research has shown that many cucurbit crops produced in NT (including cucumbers, pumpkins, and squash) yield similar to those grown in CT systems. Furthermore, in some years, cucurbits grown in alternative tillage systems have performed better than those grown in tilled fields, and this is often associated with a more consistent supply of soil moisture throughout the season in fields with cover crop surface residues. However, it is important to emphasize that NT systems for vegetables are more management intensive than CT production systems. Cover crops, mulching, fertility management, stand establishment, and weed control are few of the factors that must be considered to maximize yields in NT cucurbit vegetable production systems.

REFERENCES

Bellinder, R.R. and F.B. Gaffney. 1995. Erosion control in crops that produce sparse residue, pp. 68–79. In: R.L. Blevins and W.C. Moldenhauer (eds.). *Crop Residue Management to Reduce Erosion and Improve Soil Quality.* Appalachian and Northeast USDA-ARS Conservation Research Report #41, Washington, D.C.

Francis, G.S., K.M. Bartley, and F.J. Tabley. 1998. The effect of winter cover crop management on nitrate leaching losses and crop growth. *J. Agric. Sci.* 131:299–308.

Franzluebbers, A.J. 2010. Achieving soil organic carbon sequestration with conservation agricultural systems in the southeastern United States. *Soil Sci. Soc. Am. J.* 74:347–357.

Galloway, B.A. and L.A. Weston. 1996. Influence of cover crop and herbicide treatment on weed control and yield in no-till sweet corn (*Zea mays* L.) and pumpkin (*Cucurbita maxima* Duch.). *Weed Technol.* 10:341–346.

Hallberg, G.R. 1989. Nitrate in groundwater in the United States, pp. 35–74. In: R.F. Follett (ed.). *Nitrogen Management and Groundwater Protection.* Elsevier, Amsterdam, the Netherlands.

Harrelson, E.R., G.D. Hoyt, J.L. Havlin, and D.W. Monks. 2007. Effect of winter cover crop residue on no-till pumpkin yield. *HortScience* 42:1568–1574.

Harrelson, E.R., G.D. Hoyt, J.L. Havlin, and D.W. Monks. 2008. Effect of planting date and nitrogen fertilization rates on no-till pumpkins. *HortScience* 43:857–861.

Hoyt, G.D. 1999. Tillage and cover residue affects on vegetable yields. *HortTechnology* 9(3):351–358.

Hoyt, G.D., D.W. Monks, and T.J. Monaco. 1994. Conservation tillage for vegetable production. *HortTechnology* 4(2):129–135.

Jelonkiewicz, M. and A. Borowy. 2009. Growth and yield of cucumber under no-tillage cultivation using rye as a cover crop. *Acta Agrobot.* 62(1):147–153.

Johnson, A.M. and G.D. Hoyt. 1999. Changes to the soil environment under conservation tillage. *HortTechnology* 9(3):380–393.

Knavel, D.E. and J.W. Herron. 1986. Response of vegetable crops to nitrogen rates in tillage systems with and without vetch and ryegrass. *J. Am. Soc. Hortic. Sci.* 111:502–507.

Lonsbary, S.K., J. O'Sullivan, and C.J. Swanton. 2004. Reduced tillage alternative for machine-harvested cucumbers. *HortScience* 39:991–995.

Morse, R.D. 1999. No-till vegetable production—Its time is now. *HortTechnology* 9(3):373–379.

Morse, R.D. and D.L. Seward. 1986. No-tillage production of broccoli and cabbage. *Appl. Agric. Res.* 1(2):96–99.

NeSmith, D.S., G.G. Hoogenboom, and D.V. McCracken. 1994. Summer squash production using conservation tillage. *HortScience* 29(1):28–30.

Ogutu, M.O. and J.S. Caldwell. 1999. Stand differences in no-till and plasticulture direct seeded and transplanted cucumbers (*Cucumis sativus* L.). *Acta Hortic. (ISHS)* 504:129–134.

Osmond, D.L., S.L. Cahill, J.R. Schultheis, G.J. Holmes, and W.R. Jester. 2011. Tillage practices and nitrogen rates on pickling cucumber production. *Int. J. Veg. Sci.* 17(1):13–25.

Paul, E.A. and F.E. Clark. 1989. *Soil Microbiology and Biochemistry.* Academic Press Inc., Toronto, Ontario, Canada.

Rapp, H.S., R.R. Bellinder, H.C. Wien, and F.M. Vermeylen. 2004. Reduced tillage, rye residues, and herbicides influence weed suppression and yield of pumpkins. *Weed Technol.* 18:953–961.

Rich, H.N. 2013. Nitrogen fertility management in no-tillage cucumbers and squash. MS thesis, Southern Illinois University, Carbondale, IL.

Schonbeck, M., S. Herbert, R. DeGregorio, F. Mangan, K. Guillard, E. Sideman, J. Herbst, and R. Jaye. 1993. Cover cropping system in the northeastern United States: I. Cover crop and vegetable yields, nutrients, and soil conditions. *J. Sustain. Agric.* 3:105–132.

Shennan, C. 1992. Cover crops, nitrogen cycling, and soil properties in semi-irrigated vegetable production systems. *HortScience* 27:749–754.

Skarphol, B.J., K.A. Corey, and J.J. Meisinger. 1987. Response of snap beans to tillage and cover crop combinations. *J. Am. Soc. Hortic. Sci.* 112:936–941.

Teasdale, J.R., C.E. Beste, and W.E. Potts. 1991. Response of weeds to tillage and cover crop residue. *Weed Sci.* 39:195–199.

Vyn, T.J., K.L. Janovicek, M.H. Miller, and E.G. Beauchamp. 1999. Soil nitrate accumulation and corn response to preceding small-grain fertilization and cover crops. *Agron. J.* 91:17–24.

Walters, S.A. 2011. Weed management systems for no-tillage vegetable production, pp. 17–40. In: S. Soloneski and M. Larramendy (eds.). *Herbicides: Theory and Applications.* InTech, Rijeka, Croatia.

Walters, S.A. and J.D. Kindhart. 2002. Reduced tillage practices for summer squash production in southern Illinois. *HortTechnology* 12(1):114–117.

Walters, S.A., S.A. Nolte, and B.G. Young. 2005. Influence of winter rye and pre-emergence herbicides on weed control in no-tillage zucchini squash production. *HortTechnology* 15(2):238–243.

Walters, S.A. and B.G. Young. 2008. Utility of winter rye living mulch for weed management in zucchini squash production. *Weed Technol.* 22(4):724–728.

Walters, S.A. and B.G. Young. 2010. Effect of herbicide and cover crop on weed control in no-tillage jack-o-lantern pumpkin production. *Crop Prot.* 29(1):30–33.

Walters, S.A. and B.G. Young. 2011. Weed management in no-till zucchini squash. *Int. J. Veg. Sci.* 17:383–392.

Walters, S.A., B.G. Young, and J.D. Kindhart. 2004. Weed control in no-tillage zucchini squash production. *Acta Hortic.* 638:201–207.

Walters, S.A., B.G. Young, and S.A. Nolte. 2007. Cover crop and pre-emergence herbicide combinations in no-tillage fresh market cucumber production. *J. Sustain. Agric.* 30(3):5–19.

Walters, S.A., B.G. Young, and R.F. Krausz. 2008. Influence of tillage, cover crop, and preemergence herbicides on weed control and pumpkin yield. *Int. J. Veg. Sci.* 14(2):148–161.

Wang, G. and M. Ngouajio. 2008. Integration of cover crop, conservation tillage, and low herbicide rate for machine-harvested pickling cucumbers. *HortScience* 43:1770–1774.

Weston, L.A. 1990. Cover crop and herbicide influence on row crop seedling establishment in no-tillage culture. *WeedScience* 38:166–171.

Zandstra, B.H., W.R. Chase, and J.G. Masabni. 1998. Interplanted small grain cover crops in pickling cucumbers. *HortTechnology* 8:356–360.

8 Botany and Crop Rotation Management of New Specialty Japanese Melon in Humid, Tropical Climate Zones

Hong Li

CONTENTS

8.1 INTRODUCTION: BOTANY OF A NEW SPECIALTY JAPANESE MELON

High-value *Cucumis melo* crops are susceptible to high temperatures, water saturation, and soil acidic environments. The winter season is the right time to grow *C. melo* crops in the tropical zone for high production returns when fruit supply is short in the spring. However, it is a challenge to cultivate high-value melons when the winter turns hot with water saturation, flooding, nitrogen deficiency, and excess Fe ions in the soil. This chapter is focused on the relationships between the new specialty Japanese melon (*C. melo* L.) and environmental soil stressors and on melon crop rotation with N-fixing yardlong beans (*Vigna unguiculata* subsp. *sesquipedalis*) in the humid, acidic, tropical environment. The results show that crop rotation with N-fixation legume is a useful farming practice that helps Japanese melon plants to cope with soil acidic stress and N deficiency. The introduction of Japanese melon, a fist-sized, fast-growing, and tasteful fruit, as a new specialty melon, enhanced by improved crop rotation management, can lead to a healthy, profitable melon production in the winter season in humid, tropical areas.

Cucumis melo L. in the cucurbit (Cucurbitaceae) family has a tremendous diversity of melon species of cultivated plants of great economic importance. The plants are high-value crops, consisting of many popular, useful species, and the three most common forms of *C. melo* are cantaloupes, muskmelons, and honeydew melons (Elmstrom and Maynard, 1992; Elamin and Wilcox, 2013). Melon, the generic name applied to annual plants of the *C. melo* species, is a tropical species that is believed to have origins in Africa (Kerje and Grum, 2000). The primary center for diversity in melons is southwestern and central Asia (McCreight et al., 2004). These delicious fruit crops and other specialty melons have been bred by cultures worldwide and come in all shapes, sizes, and colors (Nerson et al., 1988; Gibson, 2009). Melons are grown in many countries worldwide for fresh markets or for processing juices. Each of the melon crops has a unique flavor, texture, and appearance.

Cucumis melo crops are tropical annual vines with long, spreading stems. Melons require full sunlight. The crops are monoecious, that is, male (staminate) and female (pistillate) flowers on a single plant (Nerson et al., 1988; Gibson, 2009). The crops are observed to have several conspicuous features: (1) The vines are usually annuals with five-lobed or palmately divided leaves, having long petioles. (2) The leaves are alternately arranged on the stem. (3) The flowers have five fused petals and five stamens (male) or an inferior ovary (female). (4) The fruits are large and fleshy, usually with a hard outer covering (a special type of berry termed as "pepo"). (5) The seeds are attached to the ovary wall (parietal placentation) and not to the center (Gibson, 2009).

These groups of melons require moderately warm temperature, fertile deep soil with good drainage, and a neutral soil pH (Elamin and Wilcox, 2013). The crops are traditionally primarily grown with nearly all significant commercial production in more favorable conditions of a dry, warm climate during summer for fruit supplies in the fall and winter seasons (Simon et al., 1993; Elamin and Wilcox, 2013). Yet, fruits are in short supply during spring. Therefore, cucurbit crop production during winter in tropical areas becomes a cropping option that is profitable. Since muskmelons, cantaloupes, and honeydew melons are large, handling and storage remain their marketing issues.

Japanese melon (*C. melo* L.), a green, fist-sized, tasteful fruit, is a new specialty melon being introduced in the tropical areas in China in recent years to produce fresh fruits in the winter seasons for fruit supplies in the spring. Unlike the usual muskmelon, cantaloupe, and honeydew, Japanese melons are small, weighing only 300–450 g each, only enough for one person to consume so that a storage for leftovers is no longer needed. This makes the Japanese melon a consumer preference for easy handling and marketing. In addition, since the fruit is small, it grows faster in the winter in the tropical areas for early harvesting in the spring, wherein melons can be of high price because there would be shortage of fresh fruits in the local and national markets. Because of these three reasons (excellent taste, easy consumption, and high price), Japanese melons are the most widely grown melons in the aforesaid areas in recent years.

The Japanese melon is a type of melon that diversifies the fruit production in warm winter climates. Like the other melon crops, the Japanese melon plants are vulnerable to high heat radiation and soil–water saturation. The physiological aspects of Japanese melons have some differences compared to other common melons described earlier. The plants are monoecious with separate male and female flowers on each vine. The crops are produced on nonclimbing vines that are prostrate on the soil. Vines (stems) with tendrils grow rapidly with many-lobed, palmately divided leaves that have long petioles (Figure 8.1a). For healthy plants, the canopy has large, lobed, generally heart-shaped leaves that are alternately arranged on the stems. Flowers are bright yellow, funnel-shaped, and grown at each petiole node on the stems (Figure 8.1b). The flowers are produced when the vines reach maturity. Each flower has six to eight fused petals (Figure 8.1c). The female flowers have an inferior ovary. The pollen-producing male flowers open first, followed by the female flowers. Each female flower must be pollinated by bees for fruit set.

Japanese melons are short-season crops, with earlier-maturing cultivars, especially those that can mature within 3 months from seeding to harvest. Popular cultivars of Japanese melons cultivated in the local tropics are hybrid cv. "Gulaba", "Gaolic", "Elizabeth", "West Lotto", etc. The fruits are attractive with a smooth, green rind turning pale as the fruits ripen (Figure 8.2a). Its shape is generally round in different kinds ranging 6–7 cm size (Figure 8.2b). The fruits are small and fleshy with a hard outer covering. Like the other types, Japanese melon seeds are also attached to the ovary wall, not to the center of the fruit (Figure 8.2c). Japanese melons are very tasteful containing a total dissolved solid content of 14% on an average (Hong Li, unpublished data). These melons are sold quickly nationwide as specialty goods.

8.2 JAPANESE MELON CROPPING AND ENVIRONMENTAL ABIOTIC STRESSORS IN HUMID, TROPICAL CLIMATE ZONES

In crop production, generally, high temperatures and soil acidity, water saturation, and flooding are among the most important abiotic environmental stress factors faced by plants (Simpson and Daft, 1990; Kocsy et al., 2004; Li et al., 2006b, 2007b; Johnson et al., 2010). Plant stress is defined as

FIGURE 8.1 **(See color insert.)** The experimental trial of Japanese melon (*Cucumis melo* L.) crops at the vegetative–flowering stage, grown in the mulched beds in mid-April 2013 in the tropical Hainan Island in southern China. The 6-week-old blooming Japanese melon plants (a), leaf and flower positions on vines (b), and fully opened flowers (c).

an external factor that exerts a disadvantageous influence on a plant (Mittler, 2006; Laughlin and Abella, 2007; Li, 2009). Plants show some degree of stress when exposed to unfavorable environments, and these stresses can collectively affect plant nutrient assimilation and crop productivity (Taiz and Zeiger, 2006; Li et al., 2010, 2014a). It is estimated that up to 82% of potential crop yields are lost due to abiotic stress annually (Taiz and Zeiger, 2006).

Humid, tropical climate is characterized by excess rainfall and high temperatures. Within the tropics there are different vegetation zones, classified based on the availability of water for plants to grow (Peel et al., 2012; Li et al., 2014b). In the tropical Hainan Island, southern China, the climate is a humid, tropical monsoon with hot summers and warm winters. The annual temperature averages 23.8°C and the annual rainfall totals 2430 mm. The growing degree-days for plants are up to 6200°C. The winter temperatures in this tropical island are high (18°C ± 6°C daily), and the monthly rainfall is still high (102 ± 23 mm) for small fruit and vegetable production (Li, 2015). High temperatures can cause intense soil surface heating that can injure plants (Li et al., 2008, 2010), and excess rainfall can lead to water saturation and flooding, resulting in soil nutrient (i.e., NO_3) losses through leaching and runoff process (Li et al., 2003; Taiz and Zeiger, 2006). Also, excess rainfall can aggravate soil acidity problems (Yoo and James, 2003; Li, 2009).

Iron (Fe) ions and active H^+ ions are dominant elements in humid areas (Kidd and Proctor, 2001; Li et al., 2007a). In the paddy soils where Japanese melons are grown in the humid, acidic

FIGURE 8.2   (See color insert.) Japanese melon (*C. melo* L.) crops at the fruit bulking stage and maturity stage in mid-June 2013 in the tropical Hainan Island in southern China. Japanese melon fruit setting stage (a), the mature fruits (b), and the cross section of a Japanese melon showing the seeds attached to the ovary wall (c).

environments of Hainan Island, the soils in the rooting depth (0.20 cm) are acidic (pH 5.51–5.70) and poor in organic matter (10.4–11.4 g kg^{-1}) (Table 8.1). Nitrate concentrations are especially low and variable (2.04–2.77 mg kg^{-1}, Table 8.1). Most of the soil nutrients including Fe, P, K, Mn, Cu, and Zn at the three melon field sites are significantly different (Table 8.1). Soil acidity and nitrate deficiency should be due to H$^+$ ions from high rainfall that also causes nitrate loss through leaching.

Soil Fe excess (279.2–370.3 mg kg^{-1}, Table 8.1) in Japanese melon fields was also found. Iron is an essential micronutrient for plant growth, but excess Fe holding within cells can be toxic to plants (Larbi et al., 2006). High soil Fe concentrations were associated with citrus tree decline in

TABLE 8.1

Comparisons of the Means of Soil Chemical Variables and Nutrient Concentrations in the Trials for Japanese Melons Using Least Significant Difference Test

Variables	pH	OC[a]	NO$_3$-N[a]	P[a]	K[a]	Fe[a]	Mn[a]	Zn[a]	Cu[a]
Site 1	5.70 a	11.4 a	2.04 b	22.5 ab	86.4 c	370.3 a	8.80 c	1.67 b	0.68 ab
Site 2	5.52 ab	11.1 a	2.66 a	25.6 a	112.5 b	279.2 c	18.2 a	2.02 a	0.77 a
Site 3	5.51 ab	10.4 ab	2.77 a	20.4 b	143.2 a	344.4 b	12.9 b	1.90 ab	0.72 a

The means with the same letters are not significantly different at α < 0.05.

[a] OC (organic carbon) in g kg^{-1} and NO$_3$–N, NH$_4$–N, P, K, Fe, Mn, Zn, and Cu in mg kg^{-1}.

(a)   (b)

FIGURE 8.3 The experimental trial of Japanese melon (*C. melo* L.) crops at the early vegetative stage, grown in the mulched beds in mid-March 2013 in the tropical Hainan Island in southern China. Flooding after a heavy rain (a) and the injury produced can quickly show up on seemingly healthy, vigorous plants from early April (b).

subtropical Florida humid, acidic orchards (Li et al., 2007a), and reduced yard-long bean yield in humid, tropical Hainan (Li, 2015). In contrast, soil Mn concentrations are low (Table 8.1) compared to the Mn levels shown by Elamin and Wilcox (2013). The differences in soil Fe and Mn concentrations would be due to their relations with soil acidity.

As new specialty fruit crops, Japanese melons cultivated on acidic soils in humid, tropical climates deal with some growth and production problems that include poor germination, early plant damage, low yields, and irregular quality. These problems are associated with periodic heat and flooding events. As the melon plants are susceptible to heat wave and heavy rainfall, often, there is a general lack of vigorous plant growth that is obvious from time of transplanting when winter changes to hot, rainy, and flooding times (Figure 8.3a). The injury can quickly show up on likely healthy plants from early April through late May following a few days of high heat or heavy rainfall (Figure 8.3b). The melon plants can break down with the death of older leaves when the plants are carrying a load of fruits (Figure 8.2a). As the soil Mn levels are low, Mn toxicity signs are not visible on melon leaves, different than that shown in Elamin and Wilcox (2013).

The problems of soil acidity and too much Fe ions in the soils can be treated with liming to increase soil pH levels while lowering Fe concentrations in the soils. Lime materials are relatively less expensive. Liming has been the common farming practice to increase soil pH and maintain crop yields associated with soil acidity (Li et al., 2007b; Elamin and Wilcox, 2013). Although organic matter can be used to seal the soils to reduce nutrient NO_3^- ion lost from leaching and runoff processes in the humid areas (Li et al., 2004; Kong et al., 2011), chemical N fertilizers can also be applied for meeting the crop's N requirements (Li et al., 2003, 2006a), yet, chemical N fertilizers are expensive, nonrenewable resources. With increased high prices on chemical amendments associated with energy crisis in recent years, continued inputs of high priced chemical N fertilizers are not sustainable.

8.3 CROP ROTATION AS A TOOL FOR JAPANESE MELON PLANTS TO COPE WITH ENVIRONMENTAL LIMITATION

Reduced N availability and low organic matter level in the soil, among the most important plant unfavorable stressors such as high temperatures, soil acidity, water saturation, and flooding, can affect plant physiological developments and crop productivity (Li et al., 2004, 2006a, 2008, 2011; Kong et al., 2011). Nitrogen is the most required primary macronutrient for plant growth (Li et al., 2003, 2013; Lea and Azevedo, 2006; Swarbreck et al., 2011; Bloom et al., 2012). Nitrogen nutrition is one of the key processes involved in determining crop development, which has marked effects on crop productivity and quality components (Hirel et al., 2007; Foyer et al., 2011; Li et al., 2013). Plant N assimilation is a fundamental biological process, and N determines the synthesis of amino acids and therefore proteins and vitamins in plants (Gastal and Lemaire, 2002; Nestby et al., 2005; Reich et al., 2006; Bloom et al., 2012).

Nitrogen is found in all parts of plants in substantial amounts, and N nutrient is absorbed in large quantities, representing 2%–5% of plant tissues on a dry matter basis (Li et al., 2003; Taiz and Zeiger, 2006; Dordas, 2012). In humid areas, crop N use efficiency is related to soil water holding, N availability and plant physiological aspects (Li et al., 2003, 2013; Hirel et al., 2007). Plant vigor is associated with water and nutrient use efficiency (Lawlor and Tezara, 2009; Li and Lascano, 2011). Japanese melon plant stress from N limitation was shown by reduced leaf photosynthesis, diminished plant leaf macronutrient NPK retention, and yield loss (Hong Li, unpublished data).

Crop rotation is the farming practice tool to help Japanese melon plants to cope with environmental soil acidic stress, water saturation, N deficiency, and Fe excess in tropical Hainan Island. It is the farming practice of growing different types of crops in the same areas in sequential seasons (Anderson, 2010). Crop rotation can improve soil environmental conditions to promote plant macronutrient uptake, which helps enhance plant capacity to cope with limited nutrient supply (Li et al., 2014a). Crop rotation has been used as one of the most efficient measures to increase plant nutrient use efficiency and crop productivity (Six et al., 2006; Li et al., 2014a). This farming practice has also been used to insure soil health by balancing soil nutrients and preventing disease and pest incidents through the concepts of crop rotation, that is, different crops have different nutrient requirements and crops in different botanical families tend to avoid the same pest and disease problems (Stanger and Lauer, 2008; Murugan and Kumar, 2013). Roots and top parts of the previous crops can contribute nutrients and organic matter to the soil after they are tilled under the soils (Anderson, 2010; Kong et al., 2011).

For the cultivation management of Japanese melons in humid, acidic environments, the crop rotation practices in the experimental trials have been used in the production of the vegetable yard-long bean (*V. unguiculata* subsp. *sesquipedalis*), followed by the short-season small fruit crop, Japanese melon (*C. melo* L.). The yard-long bean plant, a vigorous climbing annual vine vegetable, is a true legume. Yard-long bean plants can interact with soil rhizobial bacteria to form root nodules for N fixation from the atmosphere. Nitrogen requirements by yard-long bean crops are derived from soil N and symbiotically fixed atmospheric N (Li, 2015). The return of the N-rich nodules, roots, and crop residues of yard-long beans to the soils could contribute 25–34 kg N ha^{-1} to the soil for the followed crop of Japanese melon plants (Hong Li, unpublished data).

The Japanese melon plant leaf N holding could increase from 2.2% to 2.9% on a dry matter basis and the melon yields improved from 29 to 38 Mg ha^{-1}, which was related to the increase in yard-long bean whole plant biomass (Hong Li, unpublished data). This result could mean that legume crop residues of previous crop could be beneficial for the Japanese melon plants. Soil pH also increased by 0.3 units, probably because soil H$^+$ ions were used in the N-fixation process by the yard-long bean plants. In other small fruit crop production, it was reported that different types of rotation could help the strawberry plant cope with the limited primary macronutrient NPK supplies (Li et al., 2013, 2014a).

Crop rotation practices have been among the most useful measures to insure successful melon crop production. Yet, lengthy rotations may need to build healthier soils because of the times necessary for roots and top residues of the previous crops to correct soil acidity problem and to balance soil macro- and micronutrients in the humid, hot areas. It would be useful to further understand how crop rotation and other farming practices (i.e., organic amendment inputs, planting schedules) can help this new specialty crop cope with environmental stress (i.e., high-temperature radiations, water saturation, and toxicity effects from excess Fe concentrations) by improving soil conditions or planting times to maintain or even improve the melon crop productivity.

8.4 CONCLUSIONS

Crop rotation with N-fixation legume is a useful farming practice tool to help Japanese melon plants cope with environmental soil acidic stress and N deficiency in humid, tropical climate zones. The introduction of Japanese melon as a new specialty melon, combined with the adaptation of improved crop rotation management, can be useful for profitable melon production in the winter in humid, tropical areas. The results have the implication in improving cultivation management for profitable production of this high-value, new specialty Japanese melon crop.

ACKNOWLEDGMENTS

The author would like to thank the support given by the Fundamental Research Funds for Chinese Academy of Tropical Agricultural Science (#Hzs1202, #1630032015005) and Hainan Soil Improvement Key Techniques Research and Demonstration Program (#HNGDg12015). The author would like also to greatly express his thanks to Xutong Wang for her assistance in conducting the field trials and Dr. Qinghuo Lin for his assistance in the soil chemical analysis.

REFERENCES

Anderson, R.L. 2010. A rotation design to reduce weed density in organic farming. *Renewable Agriculture and Food Systems* 25:189–195.

Bloom, A.J., L. Randall, and A.R. Taylor. 2012. Deposition of ammonium and nitrate in the roots of maize seedlings supplied with different nitrogen salts. *Journal of Experimental Botany* 63:1997–2006.

Dordas, C. 2012. Variation in dry matter and nitrogen accumulation and remobilization in barley as affected by fertilization, cultivar, and source-sink relations. *European Journal of Agronomy* 37:31–42.

Elamin, O. and G.E. Wilcox. 2013. Muskmelon problems on acid sandy soils. Manganese toxicity and magnesium deficiency: Diagnosis and correction. Paper HO-191. Purdue University Cooperative Extension Service, West Lafayette, IN. https://www.extension.purdue.edu/extmedia/HO/HO-191.html.

Elmstrom, G.W. and D.N. Maynard. 1992. Exotic melons for commercial production in humid regions. *Second International Symposium on Specialty and Exotic Vegetable Crops*, vol. 318, pp. 117–124. DOI:10.17660/ActaHortic.1992.318.14, http://dx.doi.org/10.17660/ActaHortic.1992.318.14.

Foyer, C.H., G. Noctor, and M. Hodges. 2011. Respiration and nitrogen assimilation: Targeting mitochondria-associated metabolism as a means to enhance nitrogen use efficiency. *Journal of Experimental Botany* 62:1467–1482.

Gastal, F. and G. Lemaire. 2002. N uptake and distribution in crops: An agronomical and ecophysiological perspective. *Journal of Experimental Botany* 53:789–799.

Gibson, A.C. 2009. Cucurbitaceae—Fruit for peons, pilgrims, and pharaohs, Cucurbitaceae, Gourd Family. In: *Writeups and Illustrations of Economically Important Plants*. UCLA, Los Angeles, CA. http://www.botgard.ucla.edu/html/botanytextbooks/economicbotany/Cucurbita/. p. 2. Accessed date 15 March 2015.

Hirel, B., F. Chardon, and J. Durand. 2007. The contribution of molecular physiology to the improvement of nitrogen use efficiency in crops. *Journal of Crop Science and Biotechnology* 10:129–136.

Johnson, N.C., G.W.T. Wilson, M.A. Bowker, J.A. Wilson, and R.M. Miller. 2010. Resource limitation is a driver of local adaptation in mycorrhizal symbioses. *Proceedings of the Natural Academy of Sciences of the United States of America* 107:2093–2098.

Kerje, T. and M. Grum. 2000. The origin of melon, Cucumis melo: A review of the literature. *Acta Horticulturea* 510:37–44.

Kidd, P.S. and J. Proctor. 2001. Why plants grow poorly on very acid soils: Are ecologists missing the obvious? *Journal of Experimental Botany* 52:791–799.

Kocsy, G., K. Kobrehel, G. Szalai, M.-P. Duveau, Z. Buzas, and G. Galiba. 2004. Abiotic stress-induced changes in glutathione and thioredoxin h levels in maize. *Environmental and Experimental Botany* 52:101–112.

Kong, A.Y., K.M. Scow, A.L. Cordova-Kreylos, W.E. Holmes, and J. Six. 2011. Microbial community composition and carbon cycling within soil microenvironments of conventional, low-input, and organic cropping systems. *Soil Biology and Biochemistry* 43:20–30.

Larbi, A., A. Abadia, J. Abadia, and F. Morales. 2006. Down co-regulation of light absorption, photochemistry, and carboxylation in Fe-deficient plants growing in different environments. *Photosynthesis Research* 89:113–126.

Laughlin, D.C. and S.R. Abella. 2007. Abiotic and biotic factors explain independent gradients of plant community composition in ponderosa pine forests. *Ecological Modeling* 205:231–240.

Lawlor, D.W. and W. Tezara. 2009. Causes of decreased photosynthetic rate and metabolic capacity in water-deficient leaf cells: A critical evaluation of mechanisms and integration of processes. *Annals of Botany* 103:561–579.

Lea, P.J. and R.A. Azevedo. 2006. Nitrogen use efficiency. I: Uptake of nitrogen from the soil. *Annals of Applied Biology* 149:243–247.

Li, H. 2009. Citrus tree abiotic and biotic stress and implication of simulation and modeling tools in tree management. *Tree and Forestry Science and Biotechnology* 3:66–78.

Li, H., S.H. Futch, R.J. Stuart, J.P. Syvertsen, and C.W. McCoy. 2007a. Associations of soil iron with citrus tree decline and variability of sand, soil water, pH, magnesium and *Diaprepes abbreviatus* root weevil: Two-site study. *Environmental and Experimental Botany* 59:321–333.

Li, H., H.M. Jiang, and T. Li. 2011. Broccoli plant nitrogen, phosphorus and water relations at field scale and in various growth media. *International Journal of Vegetable Science* 17:1–21.

Li, H. and R.J. Lascano. 2011. Deficit irrigation for enhancing sustainable water use: Comparison of cotton nitrogen uptake and prediction of lint yield in a multivariate autoregressive state-space model. *Environmental and Experimental Botany* 71:224–231.

Li, H., T. Li, R.J. Gordon, S. Asiedu, and K. Hu. 2010. Strawberry plant fruiting efficiency and its correlation with solar irradiance, temperature and reflectance water index variation. *Environmental and Experimental Botany* 68:165–174.

Li, H., T.X. Li, G. Fu, and K. Hu. 2014a. How strawberry plants cope with limited phosphorus supply: Nursery-crop formation and phosphorus and nitrogen uptake dynamics. *Journal of Plant Nutrition and Soil Science* 177:260–270.

Li, H., T.X. Li, G. Fu, and P. Katulanda. 2013. Induced leaf intercellular CO_2, photosynthesis, potassium and nitrate retention and strawberry early fruit formation under macronutrient limitation. *Photosynthesis Research* 115:101–114.

Li, H., L.E. Parent, and A. Karam. 2006a. Simulation modeling of soil and plant nitrogen use in a potato cropping system in the humid and cool environment. *Agriculture, Ecosystem and Environment* 115:248–260.

Li, H., L.E. Parent, A. Karam, and C. Tremblay. 2003. Efficiency of soil and fertilizer nitrogen of a sod-potato system in the humid, acid and cool environment. *Plant and Soil* 251:23–36.

Li, H., L.E. Parent, A. Karam, and C. Tremblay. 2004. Potential of *Sphagnum* peat for improving soil organic matter, water holding capacity, bulk density and potato yield in a sandy soil. *Plant and Soil* 265:355–365.

Li, H., W.A. Payne, C. Bielder, and T.X. Li. 2014b. Spatial characterization of scaled soil hydraulic conductivity functions in the internal drainage process leading to tropical semiarid environment. *Journal of Arid Environments* 105:64–74.

Li, H., W.A. Payne, G.J. Michels, and C.M. Rush. 2008. Reducing plant abiotic and biotic stress: Drought and attacks of greenbugs, corn leaf aphids and virus disease in dryland sorghum. *Environmental Experimental Botany* 63:305–316.

Li, H., J.P. Syvertsen, and C.W. McCoy. 2007b. Controlling factors of environmental flooding, soil pH and *Diaprepes abbreviatus* (L.) root weevil feeding in citrus: Larval survival and larval growth. *Applied Soil Ecology* 35:553–565.

Li, H., J.P. Syvertsen, R.J. Stuart, C.W. McCoy, and A. Schumann. 2006b. Water stress and root injury from simulated flooding and *Diaprepes* root weevil feeding in citrus. *Soil Science* 171:138–151.

Li, H. 2015. Co-limitation of soil iron in root nodulation and chlorophyll-bean formation of yardlong-bean plants in tropical, humid environment. *International Journal of Bioscience, Biochemistry and Bioinformatics* 5:232–240.

McCreight, J.D., J.E. Staub, A. Lopez-Sese, and S.-M. Chung. 2004. Isozyme variation in India and Chinese melon (*Cucumis melo* L.). *Journal of American Society for Horticultural Science* 129:811–818.

Mittler, R. 2006. Abiotic stress, the field environment and stress combination. *Trends in Plant Science* 11:15–19.

Murugan, R. and S. Kumar. 2013. Influence of long-term fertilisation and crop rotation on changes in fungal and bacterial residues in a tropical rice-field soil. *Biology and Fertility of Soils* 49:847–856.

Nerson, H., H.S. Paris, M. Edelstein, Y. Burger, and Z. Karchi. 1988. Breeding pickling melons for a concentrated yield. *HortScience* 23:136–138.

Nestby, R., F. Lieten, D. Pivot, L.C. Raynal, and M. Tagliavini. 2005. Influence of mineral nutrients on strawberry fruit quality and their accumulation in plant organs: A review. *International Journal of Fruit Science* 5:139–156.

Peel, M.C., B.L. Finlayson, and T.A. McMahon. 2012. World map of Köppen-Geiger climate classification. The University of Melbourne, Parkville, Victoria, Australia.

Reich, P.B., S.E. Hobbie, T. Lee, D.S. Ellsworth, J.B. West, J.M.H. Knops, S. Naeem, and J. Trost. 2006. Nitrogen limitation constrains sustainability of ecosystem response to CO_2. *Nature* 440:922–925.

Simon, J.E., M.R. Morales, and D.J. Charles. 1993. Specialty melons for the fresh market, pp. 547–553. In: J. Janick and J.E. Simon (eds.), *New Crops*. Wiley, New York.

Simpson, D. and M.J. Daft. 1990. Interactions between water-stress and different mycorrhizal inocula on plant growth and mycorrhizal development in maize and sorghum. *Plant Soil* 121:179–186.

Six, J., S.D. Frey, R.K. Thiet, and K.M. Batten. 2006. Bacterial and fungal contributions to carbon sequestration in agroecosystems. *Soil Science Society of America Journal* 70:555–569.

Stanger, T.F. and J. Lauer. 2008. Corn grain yield response to crop rotation and nitrogen over 35 years. *Agronomy Journal* 100:643–650.

Swarbreck, S.M., M. Defoin-Platel, M. Hindle, M. Saqi, and D.Z. Habash. 2011. New perspectives on glutamine synthetase in grasses. *Journal of Experimental Botany* 62:1511–1522.

Taiz, L. and E. Zeiger. 2006. *Plant Physiology*, 4th edn. Sinauer Associates, Sunderland, MA, 764pp.

Yoo, M.S. and B.R. James. 2003. Zinc extractability and plant uptake in flooded, or organic waste-amended soils. *Soil Science* 168:686–698.

Section IV

Cucurbits Physiological Stages of Growth and Development II

9 Cucurbits Physiological Stages of Growth

Lord Abbey

CONTENTS

9.1 REPRODUCTIVE GROWTH AND DEVELOPMENT: FLOWERING AND FRUIT SET

9.1.1 FLORAL HABIT

Cucurbitaceae is a family of angiosperms that is made up of approximately 120 genera and 850 species (Gerrath et al., 2008; Welbaum, 2014). It is the most commonly used plant species for food, utensils, medicine, and artifacts throughout the world (Okoli, 1984). Plants belonging to the Cucurbitaceae family are botanically identical with similar flower anatomy. However, members of this plant family can easily be identified by the differences in growth habits, physical features, and size of their leaves, tendrils, vines (stem), and fruits (Rubatzky and Yamaguchi, 1997). The different cucurbitaceous fruits also have varied concentrations of the tetracyclic triterpene compound, cucurbitacin, which is also known as the "bitter principle." This chemical compound characterizes the bitter taste experienced when eating cucurbitaceous fruits such as cucumbers and melons (Welbaum, 2014).

The potential yield performance of angiosperms is dependent on flower formation and development, the ratio of male to female flowers, and the success of pollination and fruit set (Maynard and Hochmuth, 2007; Westerfield, 2014). The sex of a cucurbitaceous flower can easily be identified before the bud develops into full flower or before the flower opens (Lerner and Dana, 2014). Figure 9.1 shows the differences in the female (pistillate, P) and male (staminate, S) flowers of some cucurbitaceous plants. For a female flower, the petals (corolla) enclose the female parts: stigma, style, and ovary. The female flowers are attached to a swollen base referred to as the embryonic fruit (Figure 9.1). These organs are small and undeveloped fruits. The male flower comprises the male parts: filament and anther, which are also enclosed in petals. The male flowers are attached to an ordinary flower stalk without the embryonic fruit. This is common among all members of the

FIGURE 9.1 (**See color insert.**) Different types of squash plants showing female (pistillate, P) flowers with embryonic fruits and male (staminate, S) flowers without embryonic fruits at the base.

Cucurbitaceae family. Typically, the female and male organs are found in separate flowers on the vines of some plants like cucumber, squash, and pumpkin. This manner of flower arrangement is botanically referred to as monoecious (Loy, 2004; Maynard and Hochmuth, 2007). Some varieties of muskmelon and watermelon may demonstrate different floral habits known as andromonoecious under the influence of varying genetic and environmental factors, which will be explained later. In andromonoecy, each flower that is formed on the vine has both male and female sex organs and is termed as a complete or hermaphrodite flower. Additionally, andromonoecious plants have separate male flowers on their vines in addition to complete flowers. Some species may also have female flowers alone referred to as gynoecious or carpillate flowers. Even though it is rare to find only male plants, also known as androecious plants, among family members, traces were reported in cucumber and muskmelon plants (reviewed by Saito et al., 2007). Table 9.1 shows examples of cucurbitaceous plants with their typical floral characteristics. Some of them exhibit more than one floral habit.

9.1.2 DEVELOPMENT OF AXILLARY BUD COMPLEX AND FLOWERS

Plants of the Cucurbitaceae family are herbaceous and have annual or perennial vines. Cucurbitaceae, like other plant families of the angiosperms, develops specialized structural organs on the vines that resemble threadlike projections (Maynard and Hochmuth, 2007; Gerrath et al., 2008). These projections are called tendrils (Figure 9.2). The functions of tendrils are mainly mechanical support and "movement" when they come into close contact with physical objects (Putz and Holbrook, 1991).

TABLE 9.1

Floral Habits of Some Common Cucurbitaceous Plants

Scientific Name	Common Name	Floral Habit
Cucumis sativus L.	Cucumber	Monoecious, andromonoecious, gynoecious, hermaphrodite
Cucumis melo L.	Muskmelon, cantaloupe	Monoecious, andromonoecious
Cucurbita pepo L.	Pumpkin, buttercup, kabocha	Monoecious
C. moschata Duchesne		
C. maxima Duchesne		
C. pepo L.	Squash (summer and winter), acorn	Monoecious
Citrullus lanatus Thunb.	Watermelon	Monoecious, andromonoecious
Momordica charantia L.	Bitter melon	Monoecious
Citrullus colocynthis L.	Egusi melon	Monoecious
Luffa aegyptiaca Mill.	Sponge gourd	Monoecious, dioecious
Lagenaria siceraria (Mol.) Standl.	Bottle gourd	
Coccinia grandis L.	Ivy gourd	

(a)   (b)

FIGURE 9.2 (**See color insert.**) Watermelon (a) and squash (b) plants showing the origin and placement of the tendril, leaf, vine (stem), and flower in the axillary bud complex of cucurbitaceous plants.

This phenomenon in plants is known as thigmotropism. The mode of support and movement using these modified organs include creeping, trailing, climbing, clinging, or twinning around physical objects. Another important but indirect physiological function of the tendril is to help spread out the foliage and floral parts of the sprawling cucurbitaceous plant. This increases canopy light penetration and energy-use efficiency. Additionally, air circulation and overall humidity within the vegetation are enhanced while exposing the flowers to agents of pollination.

The exact anatomical origin of the tendril in Cucurbitaceae is not clearly understood. However, it is accepted that the tendril does not have the same anatomical features as the leaf, and it is therefore not a modified leaf (Gerrath et al., 2008). It was also thought by earlier workers that the tendril in Cucurbitaceae is a modified flower but recent reports refuted this assertion. It is now generally acknowledged that the tendril is structurally a part of an axillary bud complex (ABC). The ABC is made up of the tendril, leaf, and flower formed at meristematic points on the vine (Figure 9.2). The development and emergence of the tendril, leaf, and flower occur in the leaf axil or at the extra-axillary point on the vine depending on the species. To further elucidate this, it was reported that the initiation and development of tendrils in the ABC of wild cucumber (*Echinocystis lobata* Michaux) occur earlier in the growth stage (Gerrath et al., 2008), and it is structurally different from the floral primordia. The tendrils and floral primordia are also physically separated and emerge from

FIGURE 9.3   A cucumber plant with a mix of female–male flowers on the vine.

different locations within the ABC. These findings strongly suggest that tendrils are not modified flowers as thought by previous workers, but are likely to be modified stem practices in Cucurbitaceae.

Developing flower buds are located between the leaf and the tendril at each nodal point in mature cucurbitaceous plants. This flower bud is formed from a lateral bud and comprises primordia of sepals, petals, stamens, and pistils (Saito et al., 2007; Gerrath et al., 2008). Nonetheless, prevailing intrinsic and extrinsic conditions can cause selective arrested development of the stamen or pistil to determine the sex of the opened flower (Saito et al., 2007; Shiber et al., 2008; Grumet and Taft, 2012). In cucumber plants, for instance, the lateral (flower) buds can be found after the third or fourth node downstream from the apical meristem. Development of stamen in cucumber can be inhibited when the pistil and stamen primordia appeared at the same level leading to predominantly female flowers. Generally, most of the male flowers open earlier than their female counterparts during growth even though a previous report indicated early opening of the female flowers in cucumber (Yang et al., 2000). Similarly, the development of male flowers in Egusi melon (*Citrullus colocynthis* L.) plants is early and faster than that of the female flowers. This early opening of the male flowers guarantees adequate pollen availability during anthesis for successful pollination and early fruit set. On the vines of cucurbitaceous plants such as cucumber, squash, muskmelon, and watermelon, the first series of male flowers are produced (Figure 9.3). This is then followed by a mixed phase of female–male flowers in that order. This trend is often terminated by a phase of female flowers. Sex ratio for most of the Cucurbitaceous plants may range from 30:1 to 15:1 in favor of male flowers in monoecious plants (Ghani et al., 2013). Such a high male/female sex ratio has both production and economic benefits since more female flowers per plant are likely to be successfully pollinated to produce fruits. The proportion of nodes with female flowers at the distal portion of the vine increases as the plant grows older (Wang and Zeng, 1996; Yang et al., 2000; Wien, 2007). For instance, the distal portions of the vines of cucumber and summer squash plants were found to have more female flowers at each nodal point as the plant ages. Older muskmelon plants produce female flowers on short lateral branches.

9.1.3   Genetic and Molecular Basis of Sex Expression

The sex of cucurbitaceous plants can be identified early in their vegetative growth stage. The relative numbers of male and female flowers determines the sex of the plant (Dellaporta and Calderon, 1994; Saito et al., 2007). The two major genes that regulate phenotypic sex expression in Cucurbitaceae are the *F* and *M* genes (Saito et al., 2007; Shiber et al., 2008). Genotypically, gynoecious lines

(*M-F-*) produce only females; monoecious lines (*M-ff*) produce males and females; andromonoecious lines (*mmff*) produce males and bisexuals; and hermaphrodite lines (*mmF-*) produce only bisexual flowers. The dominant *F* gene regulates the formation of female flowers, while the interaction between the dominant *M* and *F* genes confirm the production of female flowers. The recessive *m* alleles (*mm*) produce bisexual flowers. Androecious phenotypes that produce only male flowers have been reported in cucumber and muskmelon plants. The phenotypic expression of androecious lines can be ascribed to the presence of the recessive *A* and *F* genes in the form of *aaff*.

Monoecious lines have a single copy of the *CS-ACS1* gene that encodes the enzyme 1-aminocyclopropane-1-carboxylic acid synthase (ACS) (Saito et al., 2007; Shiber et al., 2008). This enzyme is required for ethylene biosynthesis. Further work suggested that the *CS-ACS1* gene does not only act on growth and development of cucumber plants but is also involved in sex determination by virtue of hormonal influence. Gynoecious lines have additional copy of this gene denoted by *CS-ACS1G*, which is strongly linked to the female *F* locus. Gynoecy can, therefore, be attributed to the suppression of stamen primordia development by the *CS-ACS1G* gene. Additionally, high levels of the ethylene-inducible gene *CS-ACS2* can be found in flower buds of cucurbitaceous plants destined to develop into female flowers. This gene is specifically found in tissues that surround the ovules and placenta in the ovary. The expression of the *CS-ACS2* gene in flower buds occurs at the bisexual stage of development after the expression of the *CS-ACS1G* gene. These are genetic traits and as such phenotypic expression of these genes at any sex determining loci can be influenced by several external factors. These factors include plant growth hormones, agroecological factors, and environmental variables. These external variables affect the survival, potential growth, and development of the flowers and ultimately fruit production and yield performance.

9.1.4 OTHER FACTORS INFLUENCING FLOWER SEX RATIO

There are many other external and agronomic factors that affect flower sex expression in cucurbitaceous plants apart from the genotypic variables explained earlier in Section 9.1.3. These factors interact with the ACC synthase genes to modify the sex of the plant (Kamachi, 1997). Some of these factors can be explained as follows:

9.1.4.1 Plant Growth Regulators

Plant growth regulators are groups of (agro) chemicals that are needed in small concentrations to change the growth and developmental habits of plants. These may include changes in seed germination, plant form and height, flowering behavior, sex ratio, fruit size, and fruit ripeness and senescence (Susila et al., 2010; Ghani et al., 2013). The sequence of flower formation and sex ratio in cucurbitaceous plants can be altered by both exogenous and endogenous plant hormones, namely, auxins, gibberellins, cytokinins, ethylene, and abscisic acid (Sedghi et al., 2008; Arabsalmanik et al., 2012; Ghani et al., 2013). Endogenous plant hormones effects on sex ratio are manifested at the time and onset of ontogeny. On the other hand, exogenous plant hormones function at the two- to four-true-leaf stages, which is the critical stage for inherent sex determination in most, if not all, cucurbitaceous species. These plant hormones thus interact with the relevant genes to modify genotypic expression at critical growth and development stages. For instance, it is reported that gibberellic acid has masculinization effect on gynoecious *Cucumis* and *Cucurbita* species through the reduction of the expression of the *CS-ACS1G* gene. On the contrary, auxins and ethylene promote femaleness and suppress the development of male flowers. These variations in plant response may be due to the stage of plant development, climatic conditions, and concentration and time of application of the chemical hormone. For example, although the development of female flowers in cantaloupe (*Cucumis melo* L.) can be increased by ethephon (a source of ethylene) application, the concentration and time of application is critical. Further studies will be required to elucidate plant growth hormone response in other cucurbitaceous plants since different genera of Cucurbitaceae respond differently under varied or unknown conditions, as shown in Table 9.2.

TABLE 9.2

Summary of Plant Growth Hormones and Their Effects on Flower Sex Determination in Some Cucurbitaceous Plants

Hormone	Maleness or Femaleness	Critical	Sources
Gibberellic acid	Maleness in bitter gourd, *Cucumis* and *Cucurbita* spp.; femaleness in bitter melon	Concentration and application time	Wien (1997), Sedghi et al. (2008), Arabsalmani et al. (2012), and Ghani et al. (2013)
Auxins	Femaleness in *Cucumis* and *Cucurbita* spp.	Concentration	Sulochanamma (2001)
Ethephon	Femaleness	Concentration	Arabsalmani et al. (2012)
α-/γ-Naphthalene acetic acid	Femaleness in bitter gourd and pumpkin	Concentration	Sedghi et al. (2008) and Ghani et al. (2013)
Brassinosteroid	Femaleness in watermelon	Concentration and stage of plant growth	Susila et al. (2010)
Ethrel	Maleness in watermelon; femaleness in gherkin	Concentration	Devaraju et al. (2002), Susila et al. (2010), and Ghani et al. (2013)
Paclobutrazol	Femaleness in watermelon	Concentration	Susila et al. (2010)
β-Indoleacetic acid	Femaleness in pumpkin	Not applicable	Ntui et al. (2007)
Cycocel	Maleness in bitter gourd; femaleness in gherkin	Species and concentration	Wang and Zeng (1996) and Devaraju et al. (2002)
Mepiquat chloride	Femaleness in gherkin	Concentration	Devaraju et al. (2002)

9.1.4.2 Temperature

Maximum temperatures of 21°C in the day and 16°C at night increase retention of female flowers and fruit set in cucumber, squash, and pumpkin (Maynard and Hochmuth, 2007). Development of female flowers in cucurbitaceous plants is reduced when the plants are subjected to high diurnal and night temperatures above 30°C and 18°C, respectively. Abortion of female flowers and young fruits is also high when the plants are subjected to temperature extremities. Total fruit production and yield are reduced as a result of high abortion of the female flower and fruit drop.

9.1.4.3 Light

Angiosperms are usually classified as long-day, short-day, or day-neutral plants based on their inherent light requirements for flower induction and development (Welbaum, 2014). A day length of at least 16 h (long-day) promotes male flower development in cucumbers, whereas a short-day length of at most 9 h at moderate light intensity promotes female flower development. Oriental-type cucumbers are often grouped into two groups, namely, day-neutral North China group and short-day South China group. Exposure of the short-day South China varieties to short days leads to an increase in the production of female flowers, whereas male flowers are increased under long-day conditions. This was also evident in gherkin cucumbers. The effect of the photoperiod seems to be more pronounced in monoecious plants than in andromonoecious plants. It was reported that under short-day conditions, evolution of endogenous ethylene in shoot apices increased in monoecious cucumber plants. A rise in endogenous ethylene correlated with an increase in female flower development in monoecious cucumber plants but not in andromonoecious plants.

9.1.4.4 Nutrition

Nitrogen fertilization is important for vegetative growth of plants and needs to be managed properly to achieve a balance between generative and vegetative growth. High nitrogen fertilization leads to excessive vegetative growth and reduction in reproductive growth. It also delays and/or reduces female flower development in cucurbitaceous plants (Abd El-Fattah and Sorial, 2000; Agbaje et al., 2012). Potassium fertilization and the application of biofertilizers can increase female flower production in cucumber and squash plants (Abd El-Fattah and Sorial, 2000). In sub-Saharan Africa, female flower formation in early season pumpkin is enhanced by the application of 15–15–15 compound fertilizer.

9.1.5 POLLINATION, FERTILIZATION, AND FRUIT SET

The life cycle of cucurbitaceous plants is comprised of a diploid ($2n$) sporophyte (i.e., roots, vines, leaves, tendrils, flowers, and fruits) and a haploid (n) gametophyte (i.e., male and female sex cells or gametes). The male and female gametes are produced by gametogenesis of pollen mother cells in the anthers and the embryo sac mother cells in the ovary, respectively. Gametogenesis in the anthers begins with mitotic cell division of cells in the anthers to form the diploid ($2n$) pollen mother cells (Biology Reference, 2014). These pollen mother cells then divide by meiosis to produce four groups of haploid (n) microspores called tetrads. The tetrads mitotically divide once or twice to form two or three haploid nuclei (n) in the mature pollen grains, but only two become functional sperm nuclei. Gametogenesis in the ovary begins at the flower bud stage. Cells in the ovule called archespores (diploid, $2n$) divide by meiosis to form four haploid (n) megaspores. However, only one of the megaspores divides thrice by mitosis to yield eight haploid cells called embryonic sac cells. These cells then differentiate into two synergids, three antipodal cells, two fused endosperm nuclei, and one egg cell in the ovule.

Flower opening is followed by pollination, a process of pollen grain transfer at anthesis from the anther of the male organ to the stigma of the female organ (Biology Reference, 2014). Pollination is a sexual means of reproduction that leads to fertilization, fruit set, and seed development. The condition of the stigma is important in pollination (Edlund et al., 2004; Westerfield, 2014). In cucurbitaceous plants, stigma receptiveness of pollen grains may last for about 2–3 days to coincide with anthesis. Nonetheless, stigma receptivity is largely dependent on the pollen–stigma interface, pollen viability, genetic compatibility, and environmental variables (Peng et al., 2004; Westerfield, 2014). Adequate pollen grain coverage of the stigma lobes is necessary to ensure good set and development of fruits. Pollen viability of newly opened cucurbitaceous flowers is about 96% but reduces gradually during the day to as little as 6%–8% the next day. Pollen grain germination and fertilization are facilitated by stigma exudates secreted from papilla cells (Dumas et al., 1978). The precursors of these exudates are derived from photoassimilate. Hydration is important in the polarization of the pollen grains, which is then followed by various cellular activities. Important cellular processes include activation of pollen grain nuclei and enzymes, germination of pollen grain, and formation of pollen tube in readiness for fertilization.

At pollination, the vegetative pollen grains adhered to a receptive stigma. Viable pollen grains germinate and develop pollen tubes (Figure 9.4). The pollen tubes are made of sheaths of pectin that are filled with water, nutrients and other solutes, and the sperm nuclei. Each pollen tube grows down the style of the pistil into the ovary (Biology Reference, 2014). The pollen tube reaches the ovule in the ovary through the placenta (an attachment to the ovary wall) by chemical signals.

 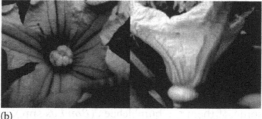

(a) (b)

FIGURE 9.4 (**See color insert.**) Cross-sectional diagram of a male flower (a) at anthesis and a receptive female flower (b). (Courtesy of Chris Rasel, Horticultural Production, ACC, Brandon, Manitoba, Canada.)

FIGURE 9.5   Withering of remnant cucumber flower parts after fertilization and fruit set.

The pollen tube penetrates the synergids and breaks open to release the two male sex nuclei. The synergids facilitate the process of bringing the egg cell and the male sex nuclei closer for fertilization to take place. One of the two male sex nuclei fuses with the egg cell in the embryo sac. This leads to the formation of a single-cell zygote that develops by mitosis into a multicellular diploid (2n) embryo of the sporophyte. The other male sex nucleus fuses with the diploid endosperm nucleus to form a triploid (3n) endosperm. The endosperm is a nutritive tissue in the seed that sustains the embryo and seedling.

The ovary forms the fruit, and the ovules in the ovary form the seeds (Agbagwa et al., 2007; Westerfield, 2014). The young fruit grows exponentially by cell expansion before slowing down when getting to its potential or full size. The other parts of the flower then wither and fall off the young developing fruits (Figure 9.5). Female flowers that are not pollinated produce distinctively unmarketable smaller, shriveled, misshapen, and seedless fruits (Lerner and Dana, 2014).

The opening of flowers and anthesis requires optimum temperature and relative humidity (Agbagwa et al., 2007; Welbaum, 2014). Most of the cucurbitaceous flowers will open at temperatures >10°C. For instance, flowers of pumpkins and squashes will open at >10°C, cucumber and watermelon at approximately 16°C, and muskmelon and cantaloupe at approximately 18°C. The optimum temperature for anthesis and pollen dehiscence is between 13°C and 18°C in pumpkin, cucumber, muskmelon, and watermelon. Usually, fruit set in cucurbitaceous plants occurs early in the morning but there are exceptions. Gourd plants such as the snake gourd and pointed gourd set fruits in the morning, whereas bottle gourd and ridge gourd set fruits at midday when temperatures are high.

Cucurbitaceous plants are generally cross-pollinated due to the nature of their floral habit. Cross-pollination is about 60%–80%, and it is aided by the large, showy flowers and high proportion of male/female flower ratio. Cross-pollination can successfully occur between plants belonging to the same genus. For example, watermelon and citron belong to the genus *Citrullus* and can successfully cross-pollinate. But cucumber (*Cucumis* spp.) and watermelon may fail to cross-pollinate because they belong to different genera. Pollination is largely entomophilous, and the main pollination agents are bees. There are many agronomically important bee pollinators; some of them are bumblebees (*Bombus* spp.), honeybees (*Apis* spp.), and squash bees (*Peponapis* spp. and *Xenoglossa* spp.; Mussen and Thorp, 2014). These insects visit flowers to collect nectar and pollen for food. Among the insect pollinators, bees are the most efficient. Their efficiency is attributable to their body forms, hairs, agility, ability to visit many flowers in a single trip, and their selective feeding on nectar and pollen grain. It is recommended that growers place colonies of bees in their fields to help improve pollination and productivity. One colony of bees per hectare was found to be adequate in Atlantic Canada for gynoecious hybrid varieties of cucumber when placed 6 days just after beginning of bloom.

The activity of bees is highest when flowers are open shortly after sunrise up to late afternoon to early evening. Therefore, any condition affecting the activities of bees, flower production, and flower opening may affect pollination and fruit development of the Cucurbitaceae family of plants (Agbagwa et al., 2007; Mussen and Thorp, 2014). Growers and environmentalists have shown concern regarding the recent decline in bee populations. It is important to strategize agronomic activities to attract bees to fields. Artificial pollination techniques such as hand pollination are sometimes used to increase pollination efficiency and number of fertilized ovules. This technique increases the number of seeds and fruit size, a phenomenon known as xenia. However, some workers reported that there is no evidence of pollen effects on fruit or seed characters in cucurbits, a phenomenon known as metaxenia (Rai and Rai, 2006). This also means that crossing of inbred lines may not have any influence on hybrid vigor, which is in sharp contrast to responses in some other plant species. Some hybrid cucurbitaceous plants do not require pollination prior to fruiting but are able to produce seedless fruits known as parthenocarpic fruits. Unlike seeded fruits, parthenocarpic fruit-producing plants have comparatively higher fruit quality (Welbaum, 2014). The main reason is that parthenocarpy has no limitation on subsequent fruit production on the vine of the same plant like seeded plants do. The immature fruits of seeded plants need to be harvested frequently to maintain quality of subsequent fruits. This is because at maximum fruit-carrying capacity, physiological balance is attained in seeded plants beyond which fruit quality and yield are compromised.

Nutritionally, adequate molybdenum (Mo) and nitrogen (N) fertilization is required to increase flower numbers and to improve pollen production and fruit set. Potassium (K), phosphorus (P), and boron (B) fertilization affects the filling, sizing, cellular sugar production and transport, and fruit color. Excessive or late nitrogen fertilizer application promotes excessive vegetative growth and reduces fruit yield and quality.

9.1.6 MATURITY AND HARVESTING

The stage of development at which a cucurbitaceous fruit is said to be mature varies among the different species and varieties (Lerner and Dana, 2014; Welbaum, 2014). Principally, maturity is determined by a minimum acceptable quality defined by the final consumer. Maturity of horticultural crops can be classified as physiological maturity or edible (marketable) maturity. Physiological maturity is when the fruit is allowed to complete its natural growth and development cycles, whereas edible maturity is strictly determined by the consumer and market forces. Edible maturity can be at any stage in the growth and development phases of the fruit. Based on market-accepted criteria, some zucchinis are harvested at an early stage during flowering; cucumber is harvested at the midstage; and sponge gourd is harvested when fully mature at a late stage of development (Table 9.3). The time to harvest any of the cucurbitaceous plants is largely dependent on a variety of quality indices. These indices can be assessed using objective and subjective methods that are often based on external and internal quality characteristics (Reid, 2002). The quality parameters may include flavor, acidity, specific gravity, texture, firmness, and water and nutrient contents of the fruits. These parameters are also critical for the assessment of storage ability and marketable life of the fresh produce.

Other important characteristics are shape and fruit size. The shapes of cucurbitaceous fruits can vary depending on species, varieties, and environmental factors. Fruit shapes may be columnar, spherical, blocky, oblong, or rounded. Some of the gourds have wide posterior and long, narrow anterior ends. The color, texture, luster, and the number and size of spiny growth structures on the surfaces of cucurbitaceous fruits also vary with species and variety, growth stage, and

TABLE 9.3
Harvesting, Curing, and Storage Conditions for Some Selected Cucurbitaceous Plants

Crop Types	Examples of Types/Varieties	Harvesting, Cooling, and Curing	Storage
Cucumber	*Pickling types:* Cross Country, Patio Pickles *Novelty types:* White Wonder	Young, immature, and small fruits harvested before first frost; stem should be cut 5 cm long to prevent infection; length of fruit $\leq$12.5 cm; remove field heat by R, FA, and FA-EC[a].	Good air circulation using appropriate bulk container; store at 10°C–13°C and 90%–95% relative humidity (RH) to prevent shriveling; up to 100% for pickling type; highly sensitive (10–100 μL/kg h) to exogenous ethylene; low production of endogenous ethylene (1 μL/kg h); waxed or shrink-wrapped in polyethylene film to reduce physical damage, desiccation, disease, and pests attack; 3%–5% O_2, 3%–5% CO_2 for controlled atmosphere (CA) storage.
Field cucumber	*Slicing types:* English Telegraph, Fanfare, Stonewall, Sweet Slender, Beit Alpha	Mature stage or small fruits; remove field heat by R, FA, and FA-EC[a] before curing and storing.	
	Other oriental types: North and South China cultivars	Mature stage of fruits; remove field heat by R, FA, and FA-EC[a] before curing and storing.	

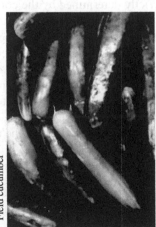

Field cucumber

Fuzzy cucumber (Mogwa)

(Continued)

TABLE 9.3 (*Continued*)
Harvesting, Curing, and Storage Conditions for Some Selected Cucurbitaceous Plants

Crop Types	Examples of Types/Varieties	Harvesting, Cooling, and Curing	Storage
English cucumber 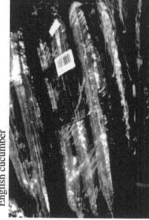	*Armenia types*: snake or serpent cucumbers *Heirloom varieties*: lemon or apple cucumbers		
Squash	*Summer squash*: zucchini, pick yellow types, patty pan or scallop	Immature stage: usually, ready for harvest 4–8 days after anthesis; fruit size 180 mm in length and 50 mm wide; remove field heat by R, FA, and FA-EC[a] before curing and storing.	*Summer squash*: 5°C–10°C at 95% RH for 1–2 weeks; CA 3%–5% O_2 and 5%–10% CO_2.
Zucchini 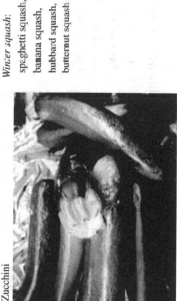	*Winter squash*: spaghetti squash, banana squash, hubbard squash, butternut squash	Shriveling and grayish vines; dull green and waxy fruits; also golden yellow for Spaghetti types and golden orange for Banana types; remove field heat by [1]R before curing for about 2 weeks at 27°C–30°C at 80% relative humidity except acorn-type squashes.	*Winter squash*: 13°C–15°C at 50%–70% RH; storage potential varies (2–6 months); varieties respond differently to CA; endogenous ethylene production is low and sensitivity to exogenous ethylene is medium (1–10 μL/kg h) for all.

(Continued)

TABLE 9.3 (Continued)
Harvesting, Curing, and Storage Conditions for Some Selected Cucurbitaceous Plants

Crop Types	Examples of Types/Varieties	Harvesting, Cooling, and Curing	Storage
Butternut squash	*Acorn types:* Sugar Loaf, Delicata, Sweet Dumpling, Thelma Sanders		
Muskmelons and allied crops	*Muskmelon:* Alaska, Earlidew, Earlisweet, Fastbreak, Seneca Bender *Cantaloupe:* Hearts of Gold, Ambrosia, Athens, Honey Bun Hybrids, Hales Best Jumbo, Sweet 'N Early Hybrid *Honeydew:* Gourmet, Passport	*Mature stage:* split stem at point of attachment to the fruit (1/4 or 1/2 slip); harvest preslip for distant market; tanned or yellow between netting of rind; cantaloupe cultivars have heavy netting, thick flesh and small dry seed cavity; remove field heat by HC, FA, and PI[a] before curing and storing.	7°C–10°C at 85%–90% RH for casaba, crenshaw, honeydew (low 5°C), and Persian melons; ripe cantaloupe can be stored at 2°C–5°C at ≥95% RH for 2–3 weeks; high ethylene production but medium ethylene sensitivity; CA 3%–5% O_2 and 10%–15% CO_2.

(Continued)

(Continued)

TABLE 9.3 *(Continued)*
Harvesting, Curing, and Storage Conditions for Some Selected Cucurbitaceous Plants

Crop Types	Examples of Types/Varieties	Harvesting, Cooling, and Curing	Storage
Honeydew melon 	*Casaba:* Casaba Golden Beauty, Casaba Sungold	*Mature stage:* fruit does not slip; pale to yellowish or creamy white in color and soft blossom end; matured casaba is greenish yellow with thick and hard rind; field-heat removal same as mentioned earlier.	
Cantaloupe 	Persian melon	*Mature stage:* do not slip; rind turns purplish; soft blossom end and slightly orange flesh; green rind with slight tan cracks or netting; sweet fruity aroma; field-heat removal same as mentioned earlier.	Storage temperature is 13°C–16°C at 80% RH for 2–3 weeks.

TABLE 9.3 (Continued)
Harvesting, Curing, and Storage Conditions for Some Selected Cucurbitaceous Plants

Crop Types	Examples of Types/Varieties	Harvesting, Cooling, and Curing	Storage
Watermelon	Sugar Baby, Gypsy, New Queen, Yellow Doll, Tom Watson	*Mature stage:* smooth rind; color varies from very dark green to yellow with solid, striped, or mottled coloring; pale or white patch where the fruit rests on the ground; rind color fades from glossy to dull green; remove field heat by FA and HC[a].	Store at 10°C–15°C at 70%–90% RH for 2–3 weeks; tolerates ≥–0.4°C; very low endogenous ethylene production but high sensitivity to exogenous ethylene.
Pumpkin	*Small sizes for cooking and pies* (4–6 lb): Kumi Kumi, Winter Luxury Pie Medium to large (15–25 lb) for cooking; *Jack-o'-lanterns:* Lady Godiva, Streaker, Triple Treat, Eat All *Jumbo size* (≥50 lb) *for exhibitions:* Atlantic giant	*Mature stage:* vines die back; stems should be cut 5–10 cm long; fruits must have deep reddish to orange color with hard rind; cure for about 2 weeks at 27°C–30°C and 80% relative humidity.	12°C–15°C, 50%–70% RH for 2–3 weeks; tolerates –0.8°C; low internal ethylene production and medium sensitivity to exogenous ethylene.

(Continued)

TABLE 9.3 (Continued)
Harvesting, Curing, and Storage Conditions for Some Selected Cucurbitaceous Plants

Crop Types	Examples of Types/Varieties	Harvesting, Cooling, and Curing	Storage
Gourds	*Luffa or sponge gourd*: Chinese okra	*Mature stage*: ripened on vine; ready when vines are killed by frost; shell of sponge gourd turns yellow or brown.	Store in 10°C–12°C at 90%–95% RH; 2–3 weeks shelf life; sponge gourds—warm, dry, and well-ventilated area to encourage rapid drying, maintain rind color, and prevent dampness and mold growth.
	Lagenaria: bottle or dipper gourd, snake gourd (edible), bushel basket gourd, New Guinea penis sheath gourd	*Mature stage*: thin and hard shell; stems die back and turn brown; harvest before first frost; sun cure for 2–4 weeks. *Immature and edible stage*: harvest before first frost; green rind; picked 1 week after flowering or at the edible maturity stage; leave short stem attached to fruit; remove field heat by R and FA[a].	
Chinese okra			

Source: Reid, M.S. 2002. Maturation and maturity indices. In: A.A. Kader (ed.). *Postharvest Technology of Horticultural Crops*, pp. 55–65. University of California Agriculture and Natural Resources, Oakland, CA, Publication 3311; Thompson, J.F., F.G. Mitchell, and R.F. Kasmire. 2002. Cooling horticultural commodities. In: A.A. Kader (ed.). *Postharvest Technology of Horticultural Crops*, pp. 97–112. University of California Agriculture and Natural Resources, Oakland, CA, Publication 3311; Thompson, A.K. 2003. *Fruit and Vegetables: Harvesting, Handling and Storage*. Blackwell Publishing Ltd., Oxford, U.K.; Rajan, S. and B.L. Markose. 2007. *Propagation of Horticultural Crops*. Vol. 6: Horticultural Science Series. New India Publishing Agency, New Delhi, India; Lerner, B.R. and M.N. Dana. 2014. Growing cucumbers, melons, squash, pumpkins and gourds. Purdue University Cooperative Extension Service, West Lafayette, IN, Revised 4/01. http://www.hort.purdue.edu/ext/ho-8.pdf. Accessed on August 2, 2014; Welbaum, G.E. 2014. Family Cucurbitaceae. In: G.E. Welbaum (ed.). *Vegetable Production and Practices*, pp. 10–49. CABI, Boston, MA. http://www.Cabi.org/Uploads/CABI/OpenResources/45346/Welbaum.Chapter-10.pdf. Accessed on September 28, 2014.

[a] R, room cooling; FA, forced-air cooling; FA-EC, forced-air-evaporative cooling; HC, hydrocooling; PI, package icing.

environmental conditions under which they were grown. The spiny growth structures are usually more prominent on younger fruits of some cucurbitaceous plants like cucumber fruits.

Computational analysis based on duration between successful pollination and edible maturity can vary from 4 to 6 days for pickling cucumbers, 60 to 70 days for squash, and 80 to 120 days for pumpkins. The species of cucurbitaceous plants such as summer squash and pickling cucumber types whose fruits are eaten fresh, immature, and young are preferably harvested regularly from the vines before the fruits and seeds become too developed, fibrous, and hard. Late harvest may cause the fruit to exceed the edible maturity stage, which can also reduce the value of the crop, lower the processing quality, and limit postharvest storage life. For squash, the best harvesting period is before the first frost. Exposure to frost can cause chilling injury leading to external and internal tissue breakdown. One important advantage of regular harvesting is the stimulation of production of new fruits as the sink strength is continuously reduced. Generally, members of the gourds such as sponge gourd (*Luffa cylindrica*) and bottle gourd (*Lagenaria siceraria*) are harvested when they are fully matured and their skin become thick and tough, whereas the edible types are harvested young and immature with a soft rind.

Cucurbitaceous fruits are prone to easy damage during harvesting, handling, and transportation by virtue of the nature of their rinds and the succulence of their flesh. Damages such as cuts, bruises, scraping, and compression are common. Consequently, handpicking or mechanical harvesting must protect the intact fruits from damage. Damages can lead to rapid deterioration or serve as entry points for insect pests and disease pathogens.

9.1.7 Curing, Postharvest Storage, and Senescence

Like other fresh horticultural produce, almost all harvested cucurbitaceous fruits are perishables. The exception to this is those fruits that are harvested mature with low water content and dried rind such as the *Luffa* types used for utensils and sponge. The high perishability can be attributed to the high water content, high postharvest respiration rate, responsiveness to climatic conditions, and sensitivity to exogenous gases, especially ethylene. Therefore, the ability to control the internal temperature and to reduce the amount of time the harvested produce is exposed to suboptimal temperature is critical to lowering respiration rate and slowing down senescence. Table 9.3 shows some examples of cucurbitaceous fruits and their required harvesting criteria, field-heat removal, and storage conditions.

Prior to storage, field heat must be removed and the effects of exogenous gases such as ethylene must be prevented as much as possible. Field-heat removal can be done simply by adopting any of the following methods or their combinations: harvest at night; move the harvested produce quickly to a cooler; use a container whose color reflects light and heat; cover the container with a lid or tarp; temporaryily store the produce in the shade; use a refrigerated truck; and cool the produce as soon as they are moved from the field to the warehouse. The following methods can be used to remove field heat: room cooling, package icing, forced-air cooling, hydrocooling, and vacuum cooling. The choice of cooling method for any of the cucurbitaceous fruits is dependent on the physiological and physical characteristics of the crop, economics, market requirements, available technology, and scale of operation. Examples for large-scale operations are given in Table 9.3.

Curing of the fruits follows field-heat removal to suberize and harden the rind. It heals superficial wounds and bruises, reduces fruit water content, and improves total soluble solids content. Curing also helps to prevent entry of pathogens that can potentially cause rotting, molds, and decay. Resultantly, shelf life is extended and eating quality of the cured fruit is improved. The cured fruits are then ready for storage under optimum climatic conditions using proper containers. Good ventilation around the fruits at all times, and particularly during storage, will minimize respiration rate and prevent the buildup of internal heat and ethylene.

Generally, cucurbitaceous fruits are cold sensitive and are susceptible to chilling injury (Table 9.3). Threshold temperatures for chilling injury vary among the different freshly harvested cucurbitaceous fruits. Threshold temperature ranges from −0.8°C to 5°C for lower limit depending on crop type and use. Exceeding the upper storage temperature limit of 15°C will hasten senescence, that is, ripening and deterioration processes of the fruits. Experiments with summer squash showed that 1-MCP (1-methylcyclopropene) can reduce chilling injury and extend to produce shelf life by inhibition of senescence-inducing ethylene. Modified atmosphere packaging for summer squash was tested with little success. However, controlled atmosphere storage technology works well with most of the cucurbitaceous plants.

9.2 SUMMARY

1. The *F* and *M* genes regulate phenotypic sex expression in the family Cucurbitaceae. Production of female flowers is aided by suppression of stamen primordia development by the *CS-ACS1G* gene.
2. Sex ratio, anthesis, fruit set, and potential yield in all the cucurbitaceous plants are influenced by environmental factors. These reproductive parameters are also affected by plant growth regulators, type of plant nutrient, photoperiod, and temperature.
3. At blooming, the first series of male flowers are formed, followed by a mixed phase of male–female flowers. This is often terminated by a phase of female flowers.
4. The optimum temperature for anthesis and pollen dehiscence is 13°C–18°C in pumpkin, cucumber, muskmelon, and watermelon. Snake gourd and pointed gourd set fruits in the morning, whereas bottle gourd and ridge gourd set fruits at midday when temperatures are high.
5. Cucurbitaceous plants are generally cross-pollinated due to the nature of their floral habit. Cross-pollination is about 60%–80%. Pollination is largely entomophilous.
6. The stage of development at which a cucurbitaceous fruit is said to be mature varies among the different crop types. Principally, maturity is determined by minimum acceptable quality as defined by the final consumer.
7. Yield and quality parameters are determined by inherent variability, environment, consumer demand, and market forces.
8. Cucurbitaceous fruits are highly perishable due to the high water content, high postharvest respiration rate, responsiveness to climatic conditions, and sensitivity to exogenous ethylene.
9. Field heat must be removed immediately after harvest before storage. The set temperature and relative humidity at storage must not cause chilling injury, dehydration, or decay.

REFERENCES

Abd El-Fattah, and M.E. Sorial. 2000. Sex expression and productivity responses of summer squash to bio-fertilizer application under different nitrogen levels. *Zagzig Journal of Agricultural Research* 27(2): 255–281.

Abduljabbar, M. and H.M. Ghurbat. 2010. Effect of foliar application of potassium and IAA on growth and yield of two cultivars of squash (*Cucurbita pepo* L.). *Journal of Tikrit University of Agricultural Science* 10(2): 222–232.

Agbagwa, I.O., B.C. Ndukwu, and S.I. Mensah. 2007. Floral biology, breeding system, and pollination ecology of *Cucurbita moschata* (Duch. Ex Lam) Duch. Ex Poir. varieties (Cucurbitaceae) from parts of the Niger Delta, Nigeria. *Turkish Journal of Botany* 31: 451–458.

Agbaje, G.O., F.M. Oloyede, and I.O. Obisesan. 2012. Effects of NPK fertilizer and season on the flowering and sex expression of pumpkin (*Cucurbita pepo* Linn.). *International Journal of Agricultural Sciences* 2(11): 291–295.

Arabsalmanik, K., A.H. Jalali, and J. Hasanpour. 2012. Control of sex expression in Cantaloupe (*Cucumis melo* L.) by ethephon application at different growth stages. *International Journal of Agricultural Sciences* 2(7): 605–612.

Biology Reference. 2014. Reproduction in plants. http://www.biologyreference.com/Re-Re/Reproduction-in-Plants.html. Accessed on December 13, 2014.

Dellaporta, S.L. and A. Calderon. 1994. The sex determination process in maize. *Science* 266: 1501–1504.

Devaraju, M.A.N., K.C. Reddy, M. Sivakumar, and D.D. Deepu. 2002. Effect of ethrel, cycocel and mepiquat chloride n flowering and sex expression in gherkin. *Current Research of University of Agricultural Sciences, Bangalore* 31(2): 16–17.

Dumas, C., M. Rougier, P. Zandonella, F. Ciampolini, M. Cresti, and E. Pacini. 1978. The secretory stigma in *Lycopersicum peruvianum* Mill.: Ontogenesis and glandular activity. *Protoplasma* 96: 173–187.

Edlund, A.F., R. Swanson, and D. Preuss. 2004. Pollen and stigma structure and function: The role of diversity in pollination. *The Plant Cell* 16: 84–97.

Gerrath, J.M., T.B. Guthrie, T.A. Zitnak, and U. Posluszny. 2008. Development of the axillary bud complex in *Echinocystis lobata* (*Cucurbitaceae*): Interpreting the cucurbitaceous tendril. *American Journal of Botany* 95(7): 773–781.

Ghani, M.A., M. Amjad, Q. Iqbal, A. Nawaz, T. Ahmad, O.B.A. Hafeez, and M. Abbas. 2013. Efficacy of plant growth regulators on sex expression, earliness and yield components in bitter gourd. *Pakistani Journal of Life and Social Sciences* 11(3): 218–224.

Grumet, R. and J. Taft. 2012. Sex expression in Cucurbits. In: Y. Hong-Wong, T.K. Behera, and C. Kole (eds.). *Genetics, Genomics and Breeding of Cucurbits*, pp. 353–375. Science Publishers, Enfield, NH.

Kamachi, S.S. 1997. Cucumber ACC synthase correlated with sex expression in flower. *Plant Physiology* 114: 809.

Lerner, B.R. and M.N. Dana. 2014. Growing cucumbers, melons, squash, pumpkins and gourds. Purdue University Cooperative Extension Service, West Lafayette, IN, Revised 4/01. http://www.hort.purdue.edu/ext/ho-8.pdf. Accessed on August 2, 2014.

Loy, J.B. 2004. Morpho-physiological aspects of productivity and quality in squash and pumpkins (*Cucurbita* spp.). *Critical Review in Plant Sciences* 23(4): 337–363.

Maynard, D.N. and G.J. Hochmuth. 2007. *Knott's Handbook for Vegetable Growers*. John Wiley & Sons Inc., Hoboken, NJ.

Mussen, E.C. and R.W. Thorp. 2014. Honey bee pollination of cantaloupe, cucumber, and watermelon. Division of Agriculture and Natural Resources, University of California, Berkeley, CA, Publication 7224. http://anrcatalog.ucdavis.edu/pdf/7224.pdf. Accessed on October 12, 2014.

Ntui, V.O., E.A. Ugoh, O. Udensi, and L.N. Enok. 2007. Response of pumpkin (*Cucurbita ficifolia* L.) to some growth regulators. *Journal of Food, Agriculture and Environment* 5(2): 211–214.

Okoli, B.E. 1984. Wild and cultivated cucurbits in Nigeria. *Economic Botany* 38: 350–357.

Peng, Y.B., S.N. Bai, Z.H. Xu, and Y.Q. Li. 2004. Glandular characteristics of the stigma during the development of *Cucumis sativus* female flowers. *Acta Bonica Sinica* 46(3): 319–327.

Putz, F.E. and N.M. Holbrook. 1991. Biomechanical studies of vines. In: F.E. Putz and H.A. Mooney (eds.). *The Biology of Vines*, pp. 73–97. Cambridge University Press, Cambridge, U.K.

Rai, N. and M. Rai. 2006. *Heterosis Breeding in Vegetable Crops*. New India Publishing Agency, New Delhi, India.

Rajan, S. and B.L. Markose. 2007. *Propagation of Horticultural Crops*. Vol. 6: Horticultural Science Series. New India Publishing Agency, New Delhi, India.

Reid, M.S. 2002. Maturation and maturity indices. In: A.A. Kader (ed.). *Postharvest Technology of Horticultural Crops*, pp. 55–65. University of California Agriculture and Natural Resources, Oakland, CA, Publication 3311.

Rubatzky, V.R. and M. Yamaguchi. 1997. *World Vegetables: Principles, Production, and Nutritive Value*. Chapman and Hall, NY.

Saito, S., N. Fujii, Y. Miyazawa, S. Yamasaki, S. Matsuura, H. Mizusawa, Y. Fujita, and H. Takahashi. 2007. Correlation between development of female flower buds and expression of the *CS-ACS2* gene in cucumber plants. *Journal of Experimental Botany* 58(11): 2897–2907.

Sedghi, M., A. Gholipouri, and R. SeyedSharifi. 2008. γ-Tocopherol accumulation and floral differentiation of medicinal pumpkin (*Cucurbita pepo* L.) in response to plant growth regulators. *Notulae Botanicae Horti Agrobotanici Cluj-Napoca* 36: 80–84.

Shiber, A., R.K. Gaur, R. Rimon-Knopf, A. Zelcer, and T. Trebitsh. 2008. The origin and mode of the female locus in cucumber. In: *Proceedings of the IXth EUCARPIA Meeting on Genetics and Breeding of Cucurbitaceae*, pp. 263–270. INRA, Avignon, France.

Sulochanamma, B.N. 2001. Effect of ethrel on sex expression in muskmelon (*Cucumis melo*) types. *Journal of Research NGRAU* 29: 91–93.

Susila, T., S.A. Reddy, M. Rajkumar, A. Padmaja, and P.V. Rao. 2010. Effect of plant growth regulators on flowering and yield of watermelon (*Citrullus lanatus* (Thunb.) Matsumura and Nakai). *Journal of Horticultural Science and Ornamental Plants* 2(1): 19–23.

Thompson, A.K. 2003. *Fruit and Vegetables: Harvesting, Handling and Storage*. Blackwell Publishing Ltd., Oxford, U.K.

Thompson, J.F., F.G. Mitchell, and R.F. Kasmire. 2002. Cooling horticultural commodities. In: A.A. Kader (ed.). *Postharvest Technology of Horticultural Crops*, pp. 97–112. University of California Agriculture and Natural Resources, Oakland, CA, Publication 3311.

Wang, Q.M. and G.W. Zeng. 1996. Effects of gibberellic acid and cycocel on sex expression of *Momordica charantia*. *Journal of Zhejiang Agriculture University* 22: 541–546.

Wang, Q.M. and G.W. Zeng. 1997. Morphological and histochemical study on sex differentiation on *Momordica charantia*. *Journal of Zhejiang Agriculture University* 23: 149–153.

Welbaum, G.E. 2014. Family Cucurbitaceae. In: G.E. Welbaum (ed.). *Vegetable Production and Practices*, pp. 10–49. CABI, Boston, MA. http://www.Cabi.org/Uploads/CABI/OpenResources/45346/Welbaum. Chapter-10.pdf. Accessed on September 28, 2014.

Westerfield, R.R. 2014. *Pollination of Vegetable Crops*. The University of Georgia Cooperative Extension, Athens, GA, Circular 934. www.extension.uga.edu/publications. Accessed on March 28, 2014.

Wien, H.C. 2007. The Cucurbits: Cucumber, melon, squash and pumpkin. In: H.C. Wien (ed.). *The Physiology of Vegetable Crops*, pp. 345–386. CAB International, New York.

Yang, L.L., M. Chen, F.Q. Liu, Y. Geng, C. Chen, Y.Q. Liu, Z.X. Cao, Z.H. Xu, and S.N. Bai. 2000. Carpel of cucumber (*Cucumis sativus* L.) male flowers maintains early primordial characteristics during organ development. *Chinese Science Bulletin* 45: 729–733.

10 Flowering and Its Modification in Cucurbits

Meenu Kumari, H.S. Singh, and T.K. Behera

CONTENTS

10.1 INTRODUCTION

Flowering in cucurbits is a major enigma since the eighteenth century, and it encompasses a large variation from sex expression to floral biology. Cucurbits are unique among other vegetable crops for sex expression and its regulation, hence amenable for manipulation, not only for economic production of hybrid seeds but also for development of new fruit types and maintenance of gynoecious lines in cucumber, muskmelon, and bitter gourd. During the last five decades, numerous efforts have been made to understand the sex mechanism of cucurbits at endogenous and exogenous levels by various physiological and molecular experiments. Flowering in cucurbits normally starts from 40 to 45 days after sowing of seeds, although it depends on soil and weather conditions for availability of mineral nutrition and photoperiodic regulations. Cucurbitaceous vegetable crops differ in their flowering time, and details on its major crops are given in Table 10.1.

Most cucurbits are monoecious (male and female in a single plant), many are dioecious (male and female on separate plants), and only a few of them are hermaphrodite (male and female organs in the same flower) as described by Roy and Saran (1990). In addition to these sex forms, cucurbits are diverse for a combination of sex types, for example, andromonoecy (separate male and bisexual flowers on the same plant), androdioecy (separate male and bisexual plants), gynomonoecy (separate female and bisexual flowers on the same plant), and gynodioecy (separate female and bisexual plants).

TABLE 10.1
Anthesis and Dehiscence Time of Important Cucurbits in Indian Conditions

Crops	Anthesis (h)	Dehiscence (h)
Cucumis melo (muskmelon)	5.30–6.30	5.00–6.00
Citrullus lanatus	6.00–7.30	5.00–6.30
Lagenaria siceraria	17.00–20.00	13.00–14.30
Momordica charantia	9.00–10.30	7.00–8.00
Trichosanthes anguina	18.00–21.00	Shortly before anthesis
Luffa acutangula (ridge gourd)	17.00–20.00	17.00–20.00

Source: Kalloo, G., Vegetable Breeding, Vol. I, CRC Press Inc., Boca Raton, FL, 1988, p. 239.

10.2 FLORAL ORGAN DEVELOPMENT AND SEX DIFFERENTIATION

Modification in flowering behavior of cucurbits is unique in nature; however, all flowering species follow the same evolutionary process. Hence, it is worthwhile to understand floral organ development and sex differentiation in ancestors. A typical flower of angiosperms consists of four whorls, that is, the sepals, petals, stamens, and carpels. Floral development studies have been extensively conducted in *Arabidopsis thaliana* and *Antirrhinum* spp. with their floral mutants and subsequently developed a widely accepted axillary bud complex model. The specific locations of floral organs in the flower are regulated by the overlapping expression of three types of floral organ genes:

1. *Floral organ identity genes*: Proteins encoded by these genes are transcription factors (e.g., MADS box) that are likely to control expression of other genes that are involved in floral organ development.
2. *Cadastral genes*: These regulate spatial identity of floral organs.
3. *Meristem identity genes*: Proteins encoded by these genes are important for initiation of organ development by cell division and differentiation. Therefore, these three genes are the key for floral organ development.

The floral organ identity genes were identified through homeotic mutations that altered floral organ identity, and eventually some of the floral organs appeared in the wrong place. Five different genes in *Arabidopsis*, *APETALA1* (*AP1*), *APETALA2* (*AP2*), *APETALA3* (*AP3*), *PISTILLATA* (*PI*), and *AGAMOUS* (*AG*), are known to specify floral organ identities that are grouped into three classes (A, B, and C) according to their functions. The A factors (AP1 and AP2) specify sepals, A and B factors (AP3 and PI) specify petals, B and C factors (AG) specify stamens, and the C factor alone specifies carpels. Thus, mutation in any factor leads to conversion of petals to sepals, sepals to carpels, or vice versa. A and C factors acting antagonistically and subsequently, the loss of one influences the other to allow work for all four whorls (Gustafson-Brown et al., 1994).

Sexual differentiation in cucurbits is driven by various factors, and there have been extensive investigations carried out to reveal this mystery. Here, we will classify all these factors into two major groups, that is, internal (genetical, cytological, and physiological) and external (environmental) factors. Manipulation is possible at both levels by breeding approach and controlling environmental factors, respectively. Therefore, before sex modification of flowering crops, it is essential to understand the genetics of sex mechanism in economically important cucurbitaceous vegetables.

10.3   SEX REGULATION BY GENETICAL FACTORS

Floral primordium differentiation in cucurbits depends on the combination of genetical, physiologi-cal, and environmental factors; however, transition of buds either as male or female is essentially affected by environmental and physiological factors. In this section, emphasis is given to the genetic mechanism of cucurbits.

10.3.1   CUCUMBER

It is considered to be a model crop for sex expression studies. Sex determination in cucumber has been reported to be controlled by three genes *F*, *A*, and *M*. The *F* gene is considered to be respon-sible for female phenotype and *M* gene for maintaining monoecy. The homozygous recessive alleles of the *A* and *F* genes (*aaff*) promote androecy, indicating that the *A* gene is responsible for maleness. Gynoecism is controlled by a single dominant gene and is influenced considerably by the modifying genes and designated as *Acr/acr*. However, intensifier for female sex expression (*In-F*) has also been reported by Kubicki (1996) from monoecious lines 18-1 and MSU 713-5.

10.3.2   MUSKMELON

In muskmelon, the predominant sex form is andromonoecious and monoecious up to some extent. Poole and Grimball (1939) reported two major loci (*g* and *a*) for the determination of sex type in case of hermaphrodite (*aagg*) sex form among some introductions from China. Plants carrying the domi-nant allele at both loci (*A-G-*) produce monoecious plants, and recessive homozygous at either the *g* or *a* locus (*A-gg* or *aaG-*) refers to gynoecy or andromonoecy, respectively. Later, the gynoecious line WI 998 of Wisconsin (the United States) was also studied by More and Seshadri (1988) in F_2 generation for the cross of the gynoecious and monoecious lines, which were found to be segregated in a 3:1 ratio.

10.3.3   WATERMELON

The flowering pattern of watermelon is either monoecious or andromonoecious and is fairly governed by a single pair of genes (Rudich and Zamski, 1985). However, Sugiyama et al. (1998) have developed hermaphroditic flower-bearing accession of andromonoecious watermelon that is expected to be useful in developing a high female bearing genetic stock. They also noticed the role of silver thiosulfate (STS) for sex modification.

10.3.4   PUMPKINS AND SQUASHES

Cucurbita species are monoecious and fairly stable with occasional perfect flowers as reported by Hayase (1956) in *Cucurbita moschata*. However, wild buffalo gourd could differ in flowering behavior due to gynoecious lines.

10.3.5   BOTTLE GOURD

Monoecy is the most prevalent sex form for this cucurbitaceous vegetable. "Andromon-6," the andromonoecious line, has been reported by Singh et al. (1996) in segregating lines of bottle gourd that is recessive in nature.

10.3.6   BITTER GOURD

This is a typical monoecious cucurbit and a considerable influence of the environment on its sex expression has been reported (Wang et al., 1997). Gynoecious sex form has been reported (Ram et al., 2002; Behera et al., 2006) in Indian germplasm that is governed by a single recessive gene "*gy-1*" (Ram et al., 2006; Behera et al., 2009), whereas Iwamoto and Ishida (2006) reported that gynoecious sex expression in bitter gourd is partially dominant.

10.3.7 SPONGE AND RIDGE GOURD

Both *Luffa acutangula* and *Luffa cylindrica* are monoecious and two gene loci are involved in sex determination (Choudhary et al., 1965). The genotypes in *Luffa* spp. with different combinations of these two genes reported were monoecious (*AG*), andromonoecious (*aG*), androecious (*aG*), gyno-monoecious (*Ag*), gynoecious (*Ag*), and hermaphrodite (*ag*).

Most of the cultivated cucurbits are monoecious, although some important cucurbit taxa are dioecious, that is, *Trichosanthes dioica*, *Coccinia indica*, *Momordica dioica*, and *Momordica cochinchinensis*. Besides these cultivated cucurbits, there are also a few feral species that contribute dioecious sex forms: *Luffa echinata*, *Melothria heterophylla*, *Edgeria darjeelingensis*, etc. (Seshadri, 1986).

10.4 SEX REGULATION BY PHYSIOLOGICAL FACTORS

Several studies have been conducted on exogenous application as well as endogenous concentration of hormones, which are involved in the physiology of flowering. Plant hormones including auxins (IAA), ethylenes, and gibberellins (GAs) have been shown to influence flower sex expression in cucurbits (Rudich, 1990).

Galun et al. (1963) demonstrated that application of IAA in cucumber at a very young stage (pre-sexual) can increase female flower formation and it has a direct effect on buds without any negative effect on leaves and other organs. However, endogenous IAA was reported to be lower in gynoecious than monoecious plants of cucumber and squash (Trebitsch et al., 1987). Also, treatment of gynoecious plant with antiauxin compound did not induce staminate flower, clearly indicating that IAA has played a secondary role in sex determination.

GAs are a class of plant hormones and a tetracyclic diterpenoid involved in plant growth and development, including seed germination, root growth, stem elongation, leaf expansion and floral induction, anther development, and seed and pericarp growth (Weiss et al., 2007). Exogenous application of GAs promotes maleness and prevents female flower development, while inhibitors of GA biosynthesis promote femaleness (Atsmon et al., 1968; Pike and Peterson, 1969).

Application of ethylene or ethylene-releasing compounds has been shown to increase pistillate flower development in cucumber, muskmelon, and squash (Robinson et al., 1969; Rudich et al., 1969; Augustine et al., 1973). Ethrel concentration up to 500 ppm delayed the male flowering in cucumber for 14 days and advanced the female flowering up to 9 days (Seshadri, 1986). In contrast to its feminizing effect in cucumber and melon, in watermelon, ethylene promotes male flower development as reported by Minkov et al. (2008). Along with exogenous ethylene application, endogenous concentration of ethylene has also been demonstrated to play a direct role in the determination of sex phenotype. Higher ethylene evolution was found in apices of gynoecious plants of cucumber as compared to those of monoecious ones and in female buds as compared to male ones (Rudich et al., 1972, 1976). There are ethylene perception inhibitors, for example, silver nitrate, STS, and inhibitors of ethylene synthesis, for example, aminoethoxyvinylglycine (AVG) increased maleness either through the production of both bisexual and male flowers in cucumbers, bitter gourds, and melons, demonstrating the possible role of ethylene in carpel development (Byers et al., 1972; Owens et al., 1980). Although a reverse effect by ethylene application was reported in watermelon that denies the universal role of ethylene for femaleness, ethylene levels also have been associated with fruit set via the production of female flower, but only the level of hormones is not considered a deciding factor for the bearing of a plant. Cucurbits also follow physiological balance and at a given time, during the flowering cycle, each shoot may bear a total of 3–45 male flowers and 1 or 2 female flowers (Seshadri, 1986).

Sex modification in cucurbits depends on the physiological growth stage of the plant and dosage of hormones. No effect on sex was observed by the treatment of ethrel on cucumber at the cotyledonary stage (Iwahori et al., 1970), and the critical stage for exogenous application of any growth hormone is reported to be the first true-leaf stage or the two- to four-true-leaf stage at which suppression or promotion of either sex primordium is possible (Robinson et al., 1969; Karchi, 1970). Hormones applied at later stages of plant development result in sex conversion on higher nodes, while higher doses lead to conversion at lower nodes (Robinson et al., 1969). Buds destined to be male could be converted to female *in vitro* only when they were removed prior to expansion of the stamen primordial (Galun et al., 1963). Therefore, all these studies and reports indicate that only certain stages in development of the floral primordium are receptive to sex determination.

Sex modification is not only the question for IAAs, ethylenes, or GAs but there are also various kinds of growth regulators that were reported from research. Maleic hydrazide (MH) at 450 µmol/L had a beneficial and feminine effect on sex expression in cucumber, whereas at higher concentration (100–200 ppm), it induces a number of male flowers in *cv.* "Straight Eight" (Ullah et al., 2011). B-9 and cycocel were reported for reducing the number of staminate flowers and consequently enhancing the number of female flowers and fruits in *cv.* "Long Green" of cucumber (Mishra and Pradhan, 1970). Gynoecious lines of cucumber, muskmelon, and bitter gourd could also be maintained by the application of silver ions in the form of silver nitrate or STS either by the production of staminate flowers or perfect flowers. Silver nitrate has been known to inhibit ethylene action (Beyer, 1976a), and silver ion is capable of specifically blocking the action of exogenously applied ethylene in responses such as abscission, senescence, and growth retardation (Beyer, 1976b). *In vivo* and *in vitro* silver ion-mediated responses seem to be involved in polyamines, ethylenes, and calcium-mediated pathways and play a crucial role in regulating physiological process including morphogenesis (Kumar et al., 2009). Possible mechanism for the action of silver ion has been reviewed and replacement of copper cofactor by silver ion serves to lock the receptor into a confirmation that continuously represses ethylene responses, perceived as most acceptable.

10.5 SEX REGULATION BY MINERALS AND NUTRIENTS

There are various other means of sex regulation besides chemicals or hormones. Some important effects have been noticed by photoperiod, temperature and light, minerals and nutrition, grafting, etc. (Heslop-Harison, 1956). The application of 100 kg nitrogen per hectare increased pistillate and staminate flowers and also their yield, but there was no change in the sex ratio of bottle gourd. The combined application of nitrogen and MH produced more female flowers and a greater yield (Pandey and Singh, 1973). A trial was conducted in a bitter gourd by Suresh and Pappiah (1991) for the combined effect of mineral nutrition and growth hormones, and the highest yield was obtained with 80 kg N, 45 kg P_2O_5, and 200 ppm MH/ha. Samdyan et al. (1994) conducted field trials where bitter gourd plants received N fertilizer at 25, 50, or 75 kg/ha, with cycocel at 100 or 250 ppm, ethrel at 50 or 100 ppm, GA_3 at 10 or 25 ppm, or MH at 25 or 50 ppm. Nitrogen at 75 kg/ha produced the thickest rind and highest fruit dry matter (DM) content, while 50 kg/ha gave the highest flesh weight, ascorbic acid, and total soluble solid (TSS) contents. Among the growth regulators, MH at 50 ppm gave the thickest and heaviest rind, while cycocel at 250 ppm gave the highest DM, ascorbic acid and TSS contents, and flesh thickness and weight. The combination of 75 kg N/ha + 50 ppm MH gave the thickest and heaviest rind and thickest flesh. We can conclude from various findings that the combined effects of mineral nutrients with plant growth regulators are superior to individual effects for sex regulation in cucurbits.

10.6 SEX REGULATION BY ENVIRONMENTAL FACTORS

It has been proved earlier that environmental factors affect fruit development and sex expression in a number of crop species. However, with a few exceptions (*L. cylindrica*, *Lagenaria siceraria*), cucurbits are day-neutral plants, that is, flowering is a photo-insensitive process in cucurbits, but sex expression of flowers is affected by day length and temperatures (Nitsch et al., 1952). In general, the number of pistillate flowers increased under short-day conditions and staminate flowers under long-day conditions for monoecious cultivars of cucumber (Matsuo, 1968), but *Cucumis hardwickii* (progenitor of *Cucumis sativus*) is an obligatory short-day plant. The effect of photoperiod can be evaluated by close examination of flowering shoots on which the first female flower appears, although different genotypes do not separate on the basis of node location. Usually at first, a male flush appears at the proximal region of flowering shoots, followed by a mix flush of male and female flowers at higher nodes, and only a female flush tends to appear at the distal portion of flowering shoots. Transition of phase in cucurbits depends on the photoperiod and the kind of species along with other external factors (Nitsch et al., 1952; Saito and Ito, 1964).

Temperature and photoperiod are often associated together, and they are supposed to influence sex determination in association. Separate studies were conducted by Staub et al. (1987) in cucumber and more number of staminate flowers were obtained on the plants that were grown at 30°C compared to those on plants kept at 16°C. In brief, low temperature can promote femaleness and high temperature facilitates maleness in cucumber (Fukushima et al., 1968; Cantliffe, 1981; Miao et al., 2011).

Low light intensity accompanied low temperature and short-day condition and high light intensity accompanied high temperature, so this matter of study was analyzed and found to produce more number of staminate flowers in a series of gynoecious processing cucumber at light intensity of 17,200 lx than 8,600, 12,900 and 25,800 lx, while a gynoecious inbred line showed no significant effect of light intensity (Cantliffe, 1981). The combined effects of light and air temperature (photothermal ratio) (Wang et al., 2014) clearly indicated that the total number of female nodes significantly increase by high photothermal ratio, and this is opposite for the number of male nodes in three cultivars of cucumber when compared with the plants grown under low photothermal ratio.

10.7   GRAFTING

Flowering induction can also be targeted by compatible grafting with either intergeneric or interspecific (*Cucumis* and *Luffa* spp.) grafts. The participation of floral stimulates in sex expression of flowers in cucurbits has been proved by Takahashi et al. (1982) for cucumber and a qualitative short-day plant, *Sicyos angulatus* L. In their study, *Sicyos* was induced to the flower by intergeneric grafting on the day-neutral plant *C. sativus* L. and the quantitative short-day plant *L. cylindrica* Roem was allowed to flower under noninductive long-day conditions. A substantial increase in the number of pistillate flowers was observed for those grafted onto *Sicyos* donors that had a sufficient number of leaves induced by short-day plants as compared to those grafted onto noninduced ones. Different species of cucumber and melon grafts had also given evidence that modifiers for sex expression can pass through a graft from stock to the scion and can modify the sex expression of the scion plant (Mockaitis and Kivilaan, 1964; Friedlander et al., 1977).

10.8   SUMMARY

The detailed study of sex regulation and its expression in cucurbits clearly demonstrated that the fundamental process of flowering is complex and is determined at internal and external levels. It is affected by a web of hormones, genes, environment (light, air, and temperature), minerals, and nutrition. These factors are not the ultimate ones because artificial modification is also possible by chemical spray in controlled conditions. There are several reports available to explain the role of ethylene and gibberellic acid at physiological and molecular aspects, but there are only a very few data for other hormones (jasmonic acid, brassinosteroids, etc.). Therefore, further studies need to continue on other potential hormones and their effects on different species of cucurbits.

REFERENCES

Atsmon, D., Lang, A., and Light, E. N. (1968). Contents and recovery of gibberellins in monoecious and gynoecious cucumber plants. *Plant Physiology*, **43**: 806–810.

Augustine, J. J., Bakier, L. R., and Sell, H. M. (1973). Female flower induction in androecious cucumber, *Cucumis sativus* L. *Journal of American Society of Horticultural Science*, **98**: 197–199.

Behera, T. K., Dey, S. S., Munshi, A. D., Gaikwad, A. B., Anand, P., and Singh, I. (2009). Sex inheritance and development of gynoecious hybrids in bitter gourd (*Momordica charantia* L.). *Scientia Horticulturae*, **120**: 130–133.

Behera, T. K., Dey, S. S., and Sirohi, P. S. (2006). DBGy-201and DBGy-202: Two gynoecious lines in bitter gourd (*Momordica charantia* L.) isolated from indigenous source. *Indian Journal of Genetics*, **66**: 61–62.

Beyer, E. M. (1976a). A potent inhibitor of ethylene action in plants. *Plant Physiology*, **58**: 268–271.

Beyer, E. M. (1976b). Silver ion: A potent antiethylene agent in cucumber and tomato. *HortScience*, **11**: 195–196.

Byers, R. E., Baker, L. R., Sell, H. M., Herner, R. C., and Dilley, D. (1972). Ethylene: A natural regulator of sex expression of *Cucumis melo* L. *Proceedings of the National Academy of Sciences of the United States of America*, **69**: 717–720.

Cantliffe, D. J. (1981). Alteration of sex expression in cucumber due to changes in temperature, light intensity and photoperiod. *Journal of the American Society for Horticultural Science*, **106**: 133–136.

Choudhary, B. and Thakur, M. R. (1965). Inheritance of sex forms in Luffa. *Indian Journal of Genetics and Plant Breeding*, **25**: 188.

Friedlander, M., Atsmon, D., and Galun, E. (1977). Sexual differentiation in cucumber. *Plant and Cell Physiology*, **18**: 261–269.

Fukushima, E., Matsuo, E., and Fujieda, K. (1968). Studies on the growth behaviour of cucumber, *Cucumis sativus* L. I. The types of sex expression and its sensitivity to various day length and temperature conditions. *Journal of the Faculty of Agriculture Kyushu University*, **14**: 349–366.

Galun, E., Yung, Y., and Lang, A. (1963). Morphogenesis of floral buds of cucumber cultured *in vitro*. *Developmental Biology*, **6**: 370–387.

Gustafson-Brown, C., Savidge, B., and Yanofsky, M. F. (1994). Regulation of the *Arabidopsis* floral homeotic gene *AP1*. *Cell*, **76**: 131–143.

Hayase, H. (1956). Cucurbita crosses. VII. The commencing time of pollen germination on stigma and anther dehiscence. *Japanese Journal of Breeding*, **5**: 261–267.

Heslop-Harrison, J. (1956). Auxin and sexuality in *Cannabis sativa*. *Physiologia Plantarum*, **9**: 588–597. doi: 10.1111/j.1399-3054.1956.tb07821.

Iwahori, S., Lyons, J. M., and Smith, O. E. (1970). Sex expression in cucumber plants as affected by 2-chloroethylphosphonic acid, ethylene, and growth regulators. *Plant Physiology*, **46**: 412–415.

Iwamoto, E. and Ishida, T. (2006). Development of gynoecious inbred line in balsam pear (*Momordica charantia* L.). *Horticultural Research (Japan)* **5**: 101–104.

Kalloo, G. (1988). *Vegetable Breeding*, Vol. I. CRC Press Inc., Boca Raton, FL, p. 239.

Karchi, Z. (1970). Effects of 2-chloroethanephosphonic acid on flower types and flowering sequences in muskmelon. *Journal of the American Society for Horticultural Science*, **95**: 575–578.

Kubicki, B. (1996). Investigations on sex determination in cucumber (*Cucumis sativus* L.). *Genetica Polonica*, **10**: 87–99.

Kumar, V., Parvatam, G., and Ravishankar, G. A. (2009). AgNO$_3$—A potential regulator of ethylene activity and plant growth modulator. *Electronic Journal of Biotechnology*, **12**: 1–15. doi: 10.2225/vol12-issue2-fulltext.

Matsuo, E. (1968). Studies on the photoperiodic sex differentiation in cucumber, (*Cucumis sativus* L.) photoperiodic and temperature conditions for sex differentiation. *Journal of the Faculty of Agriculture, Kyushu University*, **14**: 483–506.

Miao, M., Yang, X., Han, X., and Wang, K. (2011). Sugar signaling is involved in the sex expression response of monoecious cucumber to low temperature. *Journal of Experimental Botany*, **62**: 797–804.

Minkov, A. S., Levi, A., Wolf, S., and Trebtish, T. (2008). ACC Synthase genes are polymorphic in watermelon (*Citrullus* spp.) and differentially expressed in flowers and in response to auxin and gibberellins. *Plant and Cell Physiology*, **49**(5): 740–750.

Mishra, R. S. and Pradhan, B. (1970). The effect of (2-chloroethyl) trimethyl ammonium chloride on sex expression in cucumber. *The Journal of Horticultural Science and Biotechnology*, **45**: 29–32.

Mockaitis, J. M. and Kivilaan, A. (1964). Graft–induced sex changes in *Cucumis melo* L. *Nature*, **202**: 216.

More, T. A. and Seshadri, V. S. (1988). Development of tropical gynoecious lines in cucumber. *Cucurbit Genetics Cooperatives*, **11**: 17–18.

Nitsch, J., Kurtz, E. B., Livermann, J. L., and Went, F. W. (1952). The development of sex expression in cucurbit flowers. *American Journal of Botany*, **39**: 32–43.

Owens, K. W., Peterson, C. E., and Tolla, G. E. (1980). Production of hermaphrodite flowers on gynoecious muskmelon by silver nitrate and aminoethoxyvinylglycine. *HortScience*, **15**: 654–655.

Pandey, R. P. and Singh, K. (1973). Note on the effect of nitrogen and maleic hydrazide on sex expression, sex ratio and yield of bottle gourd. *Indian Journal of Agricultural Sciences*, **43**(3): 882–883.

Pike, L. M. and Peterson, C. E. (1969). Inheritance of parthenocarpy in the cucumber (*Cucumis sativus* L.). *Euphytica*, **18**: 106–109.

Poole, C. F. and Grimball, P. C. (1939). Inheritance of sex forms in *Cucumis melo* L. *Journal of Heredity*, **30**: 21–25.

Ram, D., Kumar, S., Banerjee, M. K., and Kalloo, G. (2002). Occurrence, identification and preliminary characterization of gynoecism in bitter gourd. *Indian Journal of Agricultural Science*, **72**(6): 348–349.

Ram, D., Kumar, S., Singh, M., Rai, M., and Kalloo, G. (2006). Inheritance of gynoecism in bitter gourd (*Momordica charantia* L.). *Journal of Heredity*, **97**: 294–295.

Robinson, R. W., Shannon, S., and La Guardia, M. D. (1969). Regulation of sex expression in the cucumber. *BioScience*, **19**: 141–142.

Roy, R. P. and Saran, S. (1990). Sex expression in Cucurbitaceae. In: Bates, D. M., Robinson, R. W., Jeffary, C. (eds.). *Biology and Utilization of Cucurbitaceae*. Cornell University Press, Ithaca, NY, pp. 2512–2568.

Rudich, J. (1990). Biochemical aspects of hormonal regulation of sex expression in cucurbits. In: Bates, D. M., Robinson, R. W., Jeffary, C. (eds.). *Biology and Utilization of the Cucurbitaceae*. Cornell University of Press, Ithaca, NY, pp. 269–280.

Rudich, J., Baker, L. R., Scott, J. W., and Sell, H. M. (1976). Phenotypic stability and ethylene evolution in androecious cucumber. *Journal of the American Society for Horticultural Science*, **101**: 48–51.

Rudich, J., Halevy, A., and Kedar, N. (1969). Increase in femaleness of three cucurbits by treatment with ethrel, an ethylene releasing compound. *Planta*, **86**: 69–76.

Rudich, J., Halevy, A., and Kedar, N. (1972). Ethylene evolution from cucumber plants related to sex expression. *Plant Physiology*, **49**: 998–999.

Rudich, J. and Zamski, E. (1985). *Citrullus lanatus*. In: Halevy, A. H. (ed.). *CRC Handbook of Flowering*, Vol. 2. CRC Press, Boca Raton, FL, pp. 272–274.

Saito, T. and Ito, H. (1964). Factors responsible for the sex expression of the cucumber plant. XIV. Auxin and gibberellin content in the stem apex and the sex pattern of flowers. *Tohoku Journal for Agricultural Research*, **14**: 227–239.

Samdyan, J. S., Srivastava, V. K., and Arora, S. K. (1994). Use of growth regulators in relation to nitrogen for enhancing quality indices of bitter gourd (*Momordica charantia*). *Haryana Agricultural University Journal of Research*, **24**: 102–106.

Seshadri, V. S. (1986). Cucurbits. In: Bose, T. K., Som, M. G. (eds.). *Vegetable Crops in India*. Naya Prakash, Calcutta, India, pp. 91–164.

Singh, S. P., Maurya, I. B., and Singh, N. K. (1996). Occurrence of andromonoecious form in Bottle gourd (*Lagenaria siceraria*) exhibiting monogenic recessive inheritance. *Current Science*, **70**: 458–459.

Staub, J. E. and Crubaugh, L. (1987). Imposed environmental stresses and their relationship to sex expression in cucumber (*Cucumis sativus* L.). *Cucurbit Genetics Cooperative Report* **10**: 11 (article 12).

Sugiyama, K., Kanno, T., and Morishita, M. (1998). Evaluation method of female flower bearing ability in watermelon using silver thiosulfate (STS). *Journal of the Japanese Society for Horticultural Science*, **67**: 185–189.

Suresh, J. and Pappiah, C. M. (1991). Growth and yield of bitter gourd as influenced by nitrogen, phosphorus and maleic hydrazide. *South Indian Horticulture*, **39**(5): 289–291.

Takahasi, H., Saito, T., and Suge, H. (1982). Intergeneric translocation of floral stimulus across a graft in monoecious cucurbitaceae with special reference to the sex expression of flowers. *Plant and Cell Physiology*, **23**(1): 1–9.

Trebitsch, T., Rudich, J., and Riov, J. (1987). Auxin, biosynthesis of ethylene and sex expression in cucumber (*Cucumis sativus* L.). *Plant Growth Regulation*, **5**: 105–113.

Ullah, H., Bano, A., Khokhar, K. M., and Mahmood, T. (2011). Effect of seed soaking treatment with growth regulators on phytohormone level and sex modification in cucumber (*Cucumis sativus* L.). *African Journal of Plant Science*, **50**: 599–608.

Wang, L. N., Yang, X. Y., Ren, Z. H., and Wang, X. F. (2014). The co-involvement of light and air temperature in regulation of sex expression in monoecious cucumber (*Cucumis sativus* L.). *Agricultural Sciences*, **5**: 858–863. doi:10.4236/as.2014.510092.

Wang, Q. M., Zhang, C. W., and Jiang, Y. T. (1997). Effects of temperature and photoperiod on sex expression of *Momordica charantia*. *China Vegetables*, **1**: 1–4.

Weiss, D. and Ori, N. (2007). Mechanisms of cross talk between gibberellin and other hormones. *Plant Physiology*, **144**: 1240–1246.

11 Pollination in Cucurbit Crops

*Isac Gabriel Abrahão Bomfim, Breno Magalhães Freitas,
Fernando Antonio Souza de Aragão, and Stuart Alan Walters*

CONTENTS

11.1 OVERVIEW

This chapter provides an overview of cucurbit pollination. It describes the importance of bees to cucurbits and provides aspects of cucurbit floral biology and pollination requirements targeting the most important wild and managed pollinators. Additionally, this chapter intends to give cucurbit growers, beekeepers, and scientists the best recommendations on how to choose the best pollinators and how to use them to optimize cucurbit production in greenhouse and open-field situations. Finally, we intend to make readers aware of the importance of minimizing pesticide use and to provide an update about the phenomenon regarding disappearance of bees and their impact on cucurbit production and agricultural systems.

11.2 IMPORTANCE OF POLLINATION TO CUCURBITS

Pollination is simply the transfer of pollen from the anthers to the stigma of a flower and is the first step in the sexual reproduction of plants (Kevan, 2007). Many crops require pollinator services to achieve fruit set, including most cucurbit crops (Robinson and Decker-Walters, 1997; Delaplane and

Mayer, 2000). Generally, pollination is a mutually beneficial relationship between the pollinators and the plant since pollinating insects receive some form of nutritional reward for visitation and pollen delivery.

Most cucurbits have imperfect flowers, meaning that the male and female reproductive parts are located in different flowers. Therefore, it is essential to transfer pollen grains from the anthers in male flowers to the stigmatic surface in female flowers, which is accomplished most often by several different insect pollinators (Free, 1993; Delaplane and Mayer, 2000). Thus, cucurbit crops are dependent on insect pollinators for fruit set. The absence of these vectors would result in more than 95% fruit production loss in these crops (Klein et al., 2007). However, pollination is oftentimes overlooked by cucurbit growers as many rely on natural feral insect pollinator populations to provide pollination services to their crops. In many cucurbit production fields, it is still common to see growers only depending on the wild pollinators present near production areas to provide pollination services, and this is oftentimes not sufficient to meet the pollination requirements of these crops (Siqueira et al., 2011).

In commercial cucurbit crops, the introduction of insect pollinators is critical to obtain population densities to maximize productivity as the feral insect pollinator populations are often at insufficient densities to provide optimum pollination services. It is important to note that pollination services result from the sum of the natural diversity of pollinator species that surround crop production fields as well as any pollinators that are added to maximize pollinator density. Pollinator-friendly practices such as the conservation and/or restoration of nearby areas that serve as habitats for feral insect pollinators, minimization or decrease of harmful pesticide applications to bees, and other management practices that aim to minimize negative impacts to pollinators are critical in maintaining high feral population densities (Garibaldi et al., 2013). Moreover, it is necessary to understand cucurbit plant floral biology as well as the pollinator(s) biology and behavior(s) to develop suitable management strategies to maximize fruit set of these important crops (Siqueira et al., 2011). This review will focus on providing an understanding of cucurbit floral biology and pollination requirements as well as the pollinators and pollination services they provide to these crops.

11.3　CUCURBIT FLORAL BIOLOGY

11.3.1　Sex Expression and Parthenocarpy

Cucurbit plants provide different types of sex expression (Table 11.1), and most commercial cultivars of these crops are monoecious, andromonoecious, or gynoecious. Monoecious cucurbits produce both staminate (male) or pistillate (female) flowers in separate flowers on the same plant. Andromonoecious cucurbits have both staminate and hermaphrodite (perfect) flowers on the same plant, while gynoecious plant types produce only pistillate flowers. It is known that specific environmental conditions affect sex expression and that gynoecious plants will often produce anywhere from a few to numerous staminate flowers depending on the environment. However, due to the low number of staminate flowers typically produced on gynoecious hybrids (such as cucumber, *Cucumis sativus*), they generally require supplemental sources of pollen to set fruit (Free, 1993; Delaplane and Mayer, 2000).

Although commercial cucurbit varieties are self-compatible, separation of the floral sexes requires that insect pollinators move pollen from staminate to pistillate flowers. This spatial separation in cucurbit flowers can occur by having male and female reproductive parts in the same flower (hermaphrodite) without contact or the male and female reproductive parts situated in different flowers, with staminate and pistillate flowers spatially separated on the same plant (Mussen and Thorp, 1997). Furthermore, in hermaphrodite flowers, insect pollinators are required since anthers often dehisce outward, causing pollen grains to fall and be deposited on the petals (Free, 1993). Thus, insects can transfer pollen from the staminate flowers of the

TABLE 11.1

Floral Characteristics of Many Different Cucurbit Crops

Crop	Corolla's Color	Flower Type	Sex Expression	Breeding System	Flower Ratio (S/P)	Floral Resource	Anthesis	References
Cucumber (*Cucumis sativus*)	Yellow	Staminate and pistillate (M) or only pistillate (G)	Monoecious or gynoecious	Self-pollination and cross-pollination (M), or only cross-pollination (G), or parthenocarpy (slicing type)	10:1 (M) or all or predominantly pistillate flowers (G)	Nectar and pollen (M) or only nectar (G)	Early morning until afternoon (during ≈ 7 h in open fields and ≈ 10 h in greenhouses)	Delaplane and Mayer (2000), Stanghellini et al. (2002), Nicodemo et al. (2012)
Calabash gourd (*Lagenaria siceraria*)	White or creamish	Staminate and pistillate (M)	Monoecious	Self-pollination and cross-pollination	5:1–15:1	Nectar and pollen	Early evening until afternoon of the next day (≈20 h)	Mc Gregor (1976), Burtenshaw (2003), Bhardwaj et al. (2012)
Sponge gourd (*Luffa cylindrica*)	Yellow	Staminate and pistillate (M)	Monoecious	Self-pollination and cross-pollination	20:1	Nectar and pollen	Early morning (before sunrise) until noon (during ≈ 8 h)	Mc Gregor (1976), Silva et al. (2012), Lima et al. (2014a)
Melon (*Cucumis melo*)	Yellow	Staminate and perfect (A) or staminate and pistillate (M)	Andromonoecious (American varieties) or monoecious (European varieties)	Self-pollination and cross-pollination	12:1	Nectar and pollen	Early morning until late afternoon (during ≈ 12 h)	McGregor (1976), Delaplane and Mayer (2000), Reyes-Carrillo and Cano-Ríos (2002)
Pumpkin (*Cucurbita* spp.)	Creamy white to deep orange yellow	Staminate and pistillate (M)	Monoecious	Self-pollination and cross-pollination	3.5–10:1	Nectar and pollen	Early morning until noon (during ≈ 6 h)	McGregor (1976), Delaplane and Mayer (2000)
Watermelon (*Citrullus lanatus*) Seeded	Greenish yellow	Staminate and pistillate (M) or staminate and perfect (A)	Monoecious or andromonoecious	Self-pollination and cross-pollination	5:1–13:1	Nectar and pollen nectar	Early morning until early afternoon (during ≈ 8 h)	Free (1993), Delaplane and Mayer (2000), Reyes-Carrillo and Cano-Ríos (2002), Bomfim et al. (2013)
Seedless		Staminate and pistillate (M)	Monoecious	Only cross-pollination with seeded varieties				

G, gynoecious; M, monoecious; A, andromonoecious; S, staminate flowers; P, pistillate flowers.

same plant (geitonogamous pollination) or from other neighboring plants, which results in cross-pollination (Delaplane and Mayer, 2000; Stanghellini et al., 2002).

Parthenocarpy is defined as fruit set without pollination. Although most cucumber cultivars grown commercially require pollination for fruit set, breeders have developed cultivars of cucumber (especially greenhouse types) that set fruit parthenocarpically (without pollination). This means that fruits develop without fertilization of ovules, and as a result, the fruits that develop are seedless. Pollination is undesirable for parthenocarpic cucumber as it will result in misshapen fruits that are not marketable. These cucumber types are grown and cultivated under protected environmental conditions to prevent pollination (Free, 1993; Delaplane and Mayer, 2000). Japanese salad-type greenhouse cucumbers are mostly monoecious and parthenocarpic, while Dutch salad-type greenhouse cultivars are parthenocarpic with gynoecious expression and have a high-yield potential. So, parthenocarpic cucumber hybrids with gynoecious expression do not need to be intermixed with monoecious varieties to achieve pollination and fruit set (Carvalho et al., 2013).

Environmental factors can also influence flower-type formation on cucurbit plants (Free, 1993). We have noticed that hermaphrodite flower formation will occur under shortening days and cool night-growing conditions in the autumn for certain squash (*Cucurbita pepo*) cultivars. Older cucurbit plants can also set fruit parthenocarpically.

11.3.2　FLOWER MORPHOLOGY

Most cucurbit plants have five yellow petals, although there are a few exceptions. The bottle gourd (*Lagenaria siceraria*) has white flowers, and pumpkins (*Cucurbita* spp.) can produce flowers colors from a creamy white to yellow to dark orange, depending on the species or cultivars that are grown (McGregor, 1976) (Table 11.1).

Generally, flowering begins in cucurbits with the production of staminate flowers, and depending on the species, pistillate or hermaphrodite flowers later appear on the plant, depending on the species or cultivars grown. The time required from seedling emergence to the onset and termination of flowering, as well as the time that elapses between the appearance of staminate and then either pistillate or hermaphrodite flowers vary with cucurbit species/cultivars and also depend on environmental conditions during the growth period. For example, in monoecious cucumber plants, staminate flowers are generally produced in the first three to six leaf axils with the first pistillate flowers produced a few days after the initiation of flowering in the fourth, fifth, sixth, or later leaf axils. After the first pistillate flowers appear, new ones are found at 4–5-day intervals throughout the growing season. Staminate flowers are generally produced every day or every other day and normally develop singly or in clusters in leaf axils, while pistillate flowers are only produced singly and at much less frequent intervals than males. The staminate to pistillate ratio on monoecious cucumber plants can vary from 4:1 to 20:1 or more. However, this sex ratio depends on plant growth, vigor, environmental conditions, and number of fruits that have already set (Delaplane and Mayer, 2000) (Table 11.1). Once plants have set fruits and the appearance of either hermaphrodite or pistillate flowers cease, staminate flower production continues for several days or weeks.

In cucurbits, fruit-producing flowers (hermaphrodite or pistillate) are easily distinguished from staminate flowers by the large ovary at the flower base, which when pollinated will develop into a fruit. Normally, the fruit-producing flower has one stigma divided into two to three lobes situated above a short broad style, which may (hermaphrodite) or may not (pistillate) contain the male structures. The staminate flower has three to five stamens with free or united anthers, whose pollen grains are large and sticky, meaning that wind does not play a role in the pollination process of these crops (McGregor, 1976; Free, 1993).

Most cucurbit flower stigmas are receptive during anthesis (from about 600 to 1400 each day) but are most receptive soon after the flower opens (Delaplane and Mayer, 2000). Pollen viability is also highest within a few hours of flower opening. However, in calabash gourd (*L. siceraria*), flowers begin opening at dusk and remain open until afternoon the following day, and pollen viability is still high even during the afternoon of that second day (Free, 1993; Morimoto et al., 2004). Environmental conditions, such as high temperatures and dry conditions during flowering, can reduce the amount of time that stigmas are receptive and pollen grains remain viable (Free, 1993).

Cucurbit flowers are only open for one day. The anthesis period of cucurbit flowers varies among species and even cultivars within a species (Table 11.1). Also, temperature, sunlight, and humidity have a great influence on the opening and closing of flowers. Once flowers close, they will never reopen and female flowers will not regain the ability to accept pollen after this time. The female flowers that are insufficiently pollinated and all male flowers will generally die and drop to the ground within a few days after opening (Delaplane and Mayer, 2000).

11.3.3 Attractiveness of Flowers to Pollinators

During the time that cucurbit flowers open, they are attractive to pollinators primarily as sources of either pollen or nectar. Both male and female flowers produce nectar, which provide a rich source of sugar to pollinators (Delaplane and Mayer, 2000). Also, cucurbit flowers can also offer pollen to insect pollinators via either staminate or hermaphrodite flowers (Table 11.1). Although pollen is required in the pollination and reproductive process, some pollinators will use pollen as a source of food, lipids, proteins, vitamins, and minerals. Although pollen grains adhere to anthers after dehiscence due their stickiness, they usually become scarce in flowers after midday. The speed at which pollen becomes scarce during the day is directly related to the intensity of flower visitation and pollen removal by insect visitors (Stanghellini et al., 2002).

Although floral visitors seek either nectar or pollen in cucurbit flowers, they are easily distracted to plants that provide more attractive flowers that are nearby, especially wildflowers. This is a problem that can oftentimes influence the effectiveness of cucurbit crop pollination. For this reason, growers are often interested in the use of chemical bee attractants; however, none has shown the ability to improve pollination in cucurbit crop pollination (Delaplane and Mayer, 2000). Therefore, most recommend increasing the number of bee colonies in a field to improve pollination services to cucurbit crops (Ambrose et al., 1995). Besides the factors previously discussed, the low attractiveness of cucurbit flowers to many insect pollinators is pronounced by the low number of flowers that open on plants each day, and also due to the green foliage that oftentimes covers and hides many flowers from insect pollinators.

11.4 CUCURBIT POLLINATION REQUIREMENTS

Pollination is defined as the process in which pollen from anthers is transferred to the stigma of a plant. The failure of cucurbit fruit set often results from inadequate pollination, which is directly related to low numbers of fertilized ovules. It is essential that fruit-producing cucurbit flowers (pistillate or hermaphrodite) receive adequate amounts of pollen for the pollination process to proceed, which will lead to successful fruit set and development. So, for marketable cucurbit fruits to develop, proper pollination activities in a flower must be completed, although parthenocarpic cultivars can set marketable fruit without pollination (Delaplane and Mayer, 2000).

Although cucurbits are plants with a xenogamous pollination system (outcrossing), they can also accept their own pollen for fruit set (self-compatible), which means that they have a mixed pollination system. Most cucurbits (except those that have parthenocarpy) require insect pollinators to carry pollen grains from staminate flowers to fruit-producing flowers so that fruit set occurs. A unique situation exists for pollination in seedless (triploid) watermelons. Triploid watermelons are seedless, but they are not parthenocarpic and require a pollination stimulus for fruit set. Since triploid watermelon produce mostly nonviable pollen, the pollination stimulus is provided by viable pollen grains from seeded diploid pollenizers that are planted in close proximity (Walters, 2005). Fertilization of triploid pistillate flowers only occurs when viable pollen grains from diploid pollenizers reach their stigmatic surface. So, a cross-pollination system is utilized with different watermelon genotypes to obtain triploid watermelon fruit set and development.

Pollination, pollen tube growth, and the eventual fertilization of ovules are responsible for the release of natural plant growth regulators (often referred to as plant hormones or phytohormones) that directly influence both cucurbit fruit set and production of fruit tissue and consequently fruit development. Most cucurbits require a minimum number of pollen grains be evenly spread across all stigma lobes in order for fruit to develop without deformities, and to achieve cucurbit fruit without deformities, the fruit-producing flowers must have a receptive stigma and receive several hundred viable pollen grains, which consequently results from multiple insect pollinator visits (Free, 1993; Delaplane and Mayer, 2000) (Table 11.2). However, the stigma of a cucurbit does not necessarily need to receive pollen grains on all three lobes to develop a well-formed fruit (Free, 1993). Fruit set is also affected by high humidity, which can possibly result from rainfall or excessive irrigation; this will not only decrease insect pollinator activity but can also result in coating or covering the stigma with water, resulting in poor pollen germination and tube development.

Earlier-setting cucurbit fruits have an inhibitory influence on those fruit that develop later from fruit-producing flowers on the same plant. This inhibition results from assimilates required for seed development in enlarging cucurbit fruit, which limits the ability of the plant to set additional fruits. Plants only have a limited pool of assimilates to allocate to tissues, so plants must restrict the allocation of assimilates to later developing fruits since seed development is still ongoing in those fruit that had set at an earlier date; once a fruit approaches maturity and has fully developed seed, the plant is then able to set additional fruits. However, parthenocarpic cucumber cultivars do not suffer from the first fruit inhibitory effect as they have no seeds that require high amounts of assimilates, and thus additional fruit set occurs all along the stem of a plant without interruption (Lower and Edwards, 1986; Free, 1993). Cucurbit pollination of a plant is also influenced by the time of day that flowers receive insect pollinator visits, the number of visits to a fruit-producing flower, the ovary size at the time of pollination, the vigor of the plant and branch where the flower is pollinated, and the number of fruits that have already set (Sanford and Ellis, 2010).

TABLE 11.2
Pollination Requirements for Optimum Fruit Set in Selected Cucurbits

Crop		Number of Visits		Number of Viable Pollen Grains over the Stigmatic Surface	References
		Honey Bee	Bumble Bee		
Cucumber	Pickling	>18	>18	—	Stanghellini et al. (1998)
(*Cucumis sativus*)	Slicing (salad type)	18	18	—	Stanghellini et al. (1998)
Melon (*Cucumis melo*)		≥12	—	≥400	McGregor et al. (1965)
Pumpkin	(*Cucurbita maxima*)	16	—	—	Nicodemo et al. (2009)
	(*Cucurbita pepo*)	12	4–8	1253	Vidal et al. (2010), Artz and Nault (2011)
Watermelon	Seeded	6–8	1	≥1000	Adlerz (1966), Stanghellini et al. (1997)
(*Citrullus lanatus*)	Seedless	16–24 (at 33% of pollenizer frequency)	—	—	Walters (2005)

11.5 POLLINATORS

11.5.1 GENERAL CONSIDERATIONS

The flowers of cucurbits are visited by a wide range of insect pollinators. Various species of bees, wasps, ants, butterflies, flies, and beetles have been reported to provide pollination services to cucurbit flowers (McGregor, 1976; Free, 1993; Delaplane and Mayer, 2000). However, an insect is not considered a pollinator if it visits a flower and does not touch the reproductive parts, does not carry pollen or carries nonviable pollen, or visits the flowers when the stigma is not receptive (Dafni et al., 2005). Thus, many insects can collect floral resources from cucurbits without touching the reproductive parts of flowers or only occasionally visit flowers and contribute nothing or very little to the pollination process.

Bees are the most studied and utilized pollinators throughout the world for cucurbit crops and provide the greatest contribution to the pollination of cucurbits (Delaplane and Mayer, 2000; Garibaldi et al., 2013). They are used to provide pollination services for either open-field or protected environment culture (McGregor, 1976; Free, 1993; Delaplane and Mayer, 2000) (Table 11.3). Their outstanding abilities as cucurbit pollinators are due to several adaptations: (1) they have an exclusive vegetarian diet (with rare exceptions), which allows them to be in constant close contact with the plants as they seek pollen and nectar in cucurbit flowers; (2) they have branched hairs that spread over their body surface, which increases the adherence of pollen grains to their bodies and transference to the stigmatic surfaces of fruit-producing flowers (Winston, 1987); and (3) they also have well-developed foraging activities and behaviors, which increases the chance of pollen grains being deposited on the stigma (Delaplane and Mayer, 2000; Slaa et al., 2006).

Cucurbit flowers encourage bee visitation for several reasons. All cucurbit flowers have the entire anthesis period (or at least a portion) during daylight hours, which favor visits by diurnal insects, and only a few cucurbits have flowers that open during nighttime. The calabash gourd is a cucurbit that receives visits from nocturnal insects. Although its flowers start to open during evening hours, this cucurbit also receives visits from diurnal insects since the flower anthesis

---

TABLE 11.3
Manageable Cucurbit Pollinators in Open-Field and Greenhouse Conditions

Crop	Manageable Pollinators		References
	Open Field	Greenhouse	
Cucumber (*Cucumis sativus*)	Honey bees (*A. mellifera*), bumble bees (*Bombus* spp.)	Honey bees, *Megachile rotundata*, bumble bees, stingless bees (*Scaptotrigona* aff. *Depilis* and *N. testaceicornis*)	Free (1993), Stanghellini et al. (1998), Santos et al. (2008), Nicodemo et al. (2013)
Calabash gourd (*Lagenaria siceraria*)	Honey bees (*A. mellifera*)	—	Free (1993)
Sponge gourd (*Luffa cylindrica*)	Honey bees (*A. mellifera*)	—	McGregor (1976)
Melon (*Cucumis melo*)	Honey bees (*A. mellifera*), bumble bees (*Bombus* spp.)	Honey bees, bumble bees, stingless bees (*Scaptotrigona* sp.), *Xylocopa pubescens*	Free (1993), Delaplane and Mayer (2000), Keasar et al. (2007), Sadeh et al. (2007), Bezerra (2014)
Pumpkin (*Cucurbita* spp.)	Honey bees (*A. mellifera*), bumble bees (*Bombus* spp.)	—	Free (1993), Walters and Taylor (2006), Woodcock (2012), Petersen et al. (2013)
Watermelon (*Citrullus lanatus*)	Honey bees (*A. mellifera*), bumble bees (*Bombus* spp.)	Honey bees, stingless bees (*Scaptotrigona* sp.), bumble bees	Free (1993), Stanghellini et al. (1998), Bomfim, 2013

period lasts until the afternoon of the following day. At night, calabash gourd flowers receive visits from *Cyrtopeltis tenuis*, which is a plant bug that is thought to provide pollination services, and during daytime, bees are the primary insects responsible for calabash gourd flower pollination (Free, 1993).

Although several solitary and social bee species have been reported to visit cucurbit flowers at high frequencies and provide pollination services (Free, 1993; Winfree et al., 2007) (Table 11.4), honey bees (*Apis mellifera*) and bumble bees (*Bombus* spp.) are the most widely utilized insect pollinators of cucurbit crops. However, the squash bee (*Peponapis pruinosa*) is a solitary bee that is well known as a cucurbit pollinator, especially squash and pumpkin (*Cucurbita* spp.). This species is a ground-nesting bee that forages during the early morning, when most cucurbit species have open flowers, which is the optimal time for pollination (Free, 1993; Delaplane and Mayer, 2000). Female squash bees are more effective pollinators than males, although male squash bees are also good pollinators as they feed on nectar and seek females inside cucurbit flowers (Cane et al., 2011). Although squash bees are rarely present at sufficient densities to provide adequate pollination services in large commercial cucurbit production fields, some cucurbit fields across North America have feral squash bee populations that can reach densities high enough to provide sufficient pollination services to

TABLE 11.4
Primary Bee Visitors to Flowers of Some Cucurbits

Crop	Bee Floral Visitors	References
Cucumber (*Cucumis sativus*)	Honey bees (*A. mellifera*), *A. dorsata*, *A. florea*, bumble bees (*Bombus* spp.), *Melipona* spp., *Scaptotrigona* aff. *depilis*, *N. testaceicornis*, *Melissodes* spp., *Pithitis smaragdula*, *Xylocopa fenestrata*; *Eucera hamata*, *Nomada* spp., *Agapostemon* spp., *Lasioglossum* spp., *Megachile* spp., *P. pruinosa*	Free (1993), Santos et al. (2008), Lowenstein et al. (2012), Woodcock (2012)
Calabash gourd (*Lagenaria siceraria*)	Honey bees (*A. mellifera*), *A. cerana*, bumble bees (*Bombus* spp.), *X. fenestrata*, *Xylocopa virginica*	McGregor (1976), Free (1993), Bhardwaj et al. (2012)
Sponge gourd (*Luffa cylindrica*)	Honey bees (*A. mellifera*), *X. fenestrata*, *X. virginica*	Bhardwaj et al. (2012), Lima et al. (2014b)
Melon (*Cucumis melo*)	Honey bees (*A. mellifera*), *A. florea*, bumble bees (*Bombus* spp.), *Hypotrigona* spp., *Melipona mandacaia*, *Plebeia mosquito*, *Scaptotrigona* sp., *Trigona carbonaria*, *Andrena wilkella*, *Augochlora* spp., *Callomegachile torida*, *Halictus* spp., *Hylaeus mesillae*, *Lasioglossum* spp., *Pseudomegachile lanata*, *Xylocopa grisescens*	Free (1993), Winfree et al. (2008), Kouonon et al. (2009), Coelho et al. (2012), and Bezerra (2014)
Pumpkin (*Cucurbita* spp.)	Honey bees (*A. mellifera*), *A. cerana*, *A. dorsata*, *A. florea*, bumble bees (*Bombus* spp.), *M. quadrifasciata*, *Trigona spinipes*, *Xylocopa* spp., *P. pruinosa*, *Xenoglossa* spp., *Agapostemon virescens*, *Augochlora pura*, *Dialictus* sp., *Halictus* sp., *Triepeolus remigatus*, *Melissodes bimaculatus*, *P. smaragdula*	Delaplane and Mayer (2000), Thapa (2006), Walters and Taylor (2006), Mélo et al. (2010), Cane et al. (2011), Petersen et al. (2013)
Watermelon (*Citrullus lanatus*)	Honey bees (*A. mellifera*), *A. cerana*, *A. florea*, bumble bees (*Bombus* spp.), *Melipona* spp., *Scaptotrigona* sp., *Trigona iridipennis*, *Agapostemon* spp., *Anthophora urbana*, *Augochlorella* spp., *Augochloropsis caerulea*, *Calliopsis andreniformis*, *Ceratina* spp., *Dialictus* spp., *Exomalopsis snowi*, *Florilegus condignus*, *Halictus* spp., *Hylaeus* spp., *Lasioglossum* spp., *Melissodes* spp., *Megachile* spp., *Nomada cruci*, *Triepeolus helianthi*	McGregor (1976), Free (1993), Kremen et al. (2002), Winfree et al. (2008), Henne et al. (2012), Bomfim (2013)

important cucurbit crops such as squash and pumpkin. It is important to note that special care must be taken to prevent damage to nesting sites in soil that can often occur by tillage (Woodcock, 2012).

The feral solitary and social stingless bees can also play an important role in pollination, depending on the area of the world to which they are adapted (Free, 1993; Kremen et al., 2002; Winfree et al., 2008). The presence of these unmanaged bees during flowering is most likely increased by suitable natural habitats that are nearby to cucurbits crop fields (Kremen et al., 2002; Winfree et al., 2008). The use of stingless bees (*meliponine*) as pollinators of agricultural crops has been quite promising in some tropical regions. Studies have indicated success in both rearing and use of certain stingless bee species (e.g., *Melipona subnitida*, *Melipona quadrifasciata*, *Nannotrigona testaceicornis*, *Scaptotrigona* spp., and *Tetragonisca angustula*) for pollination of agricultural crops, both in open-field (e.g., guava) and protected environment conditions (e.g., eggplant, melon, watermelon, strawberry, pepper, tomato, cucumber); however, the utilization and application of stingless bees for pollination purposes at a commercial level is still in the development phase (Heard, 1999; Malagodi-Braga et al., 2004; Cruz et al., 2005; Del Sarto et al., 2005; Slaa et al., 2006; Cruz, 2009; Roselino et al., 2010; Bomfim et al., 2013; Bezerra, 2014). These bees are unable to sting, have perennial colonies that can grow to high populations, and depending on the species, have the potential for rearing high populations in colonies; these characteristics suggest that stingless bees may have potential as pollinators in protected environment culture (Slaa et al., 2006; Venturieri et al., 2012).

11.5.2 Most Widely Used Pollinators

Although there are more than 20,000 described bee species around the world, only a few are commercially managed as pollinators of cultivated plants (Free, 1993; Cruz and Campos, 2009). Again, the primary managed bee species utilized in commercial cucurbit production are honey bees and bumble bees. Honey bees are utilized mostly in open-field cucurbit culture, while bumble bees are used in protected culture more than in open-field conditions (Guerra-Sanz, 2008). Honey bees are the most utilized manageable pollinators in commercial cucurbit crops around the world, since feral populations have spread throughout the world, including the Americas. They also have been widely studied and have well-defined colony management systems for utilization as pollinators for those crops that require pollination services. Moreover, they develop large populations, adapt well to various environmental conditions, forage and pollinate flowers from many different plant species across many plant families, and also have perennial colonies that are easily acquired, managed, and transported to the desired crop location (Stanghellini et al., 1998; Hogendoorn, 2004; Morais et al., 2012). Due to the large number of individuals in the same colony, high populations of foragers result that visit a wide variety of flowers to meet the food demands of the colony. Another important feature is their ability to produce high numbers of foragers that can visit flowers of a particular plant species, which is important for crop pollination and production (Winston, 1987). However, honey bees have problems adapting to enclosed environments, such as greenhouses, particularly those that are smaller in size; and this is due to the stress of confinement and temperature fluctuations that occur in these environments (Free, 1993; Guerra Sanz, 2008). Moreover, their defensive behavior can be accentuated in this type of environment, resulting in more stinging activity to workers in the facility. This oftentimes prevents workers from completing required cultural practice activities in greenhouses. This situation is even more pronounced for Africanized honey bees (Cruz and Campos, 2009).

Bumble bees are recognized worldwide as the largest pollinator of crops under protected cultivation. Unlike honey bees, bumble bees adapt easily and forage well in either small- or large-sized enclosed environments (Free, 1993). However, colonies purchased for pollination services cannot be reused since they are not perennial and perish after 8–12 weeks. Thus, new bumble bee colonies must be purchased each time they are to be used to provide pollination services (Slaa et al., 2006).

The rearing and marketing of these bees has reached very high levels due to implementation of technologies that allow mass production and importation of bumble bee colonies for crop pollinating purposes across all continents, which make this agribusiness enterprise worth several billion dollars (Velthuis and Van Doorn, 2006; Kevan, 2007; Guerra Sanz, 2008).

The importation of exotic bee species is prohibited by law in most countries, and anyway, locally abundant native bee species are usually preferred as pollinators instead of alien species (Hogendoorn, 2004; Cruz and Campos, 2009; Garófalo et al., 2012). The introduction of an alien species often provides competition and hybridization with native species, which can import new diseases and parasites, as well as potentially providing detrimental impacts to the local ecosystem (Saraiva et al., 2012). Furthermore, in some tropical areas, native bumble bees do not exist, while in other areas, they exhibit very defensive behaviors and become aggressive when disturbed; therefore, they are not typically used as managed pollinators in crop production. In Brazil, for example, the few existing bumble bee species are quite aggressive, which make them impossible to manage for pollination purposes (Imperatriz-Fonseca et al., 2006; Cruz and Campos, 2009).

11.5.3 POLLINATOR EFFICIENCY

Pollinator efficiency can be defined as the outcome provided by a single pollinator visit to a flower. This term is often used to determine or rank the importance of different species as floral pollinators. The measurement of pollinator efficiency in cucurbit crops can be determined by the number of pollen grains deposited on the stigma, pollen removed from anthers, and/or numbers of seeds or fruits produced (Stanghellini et al., 2002).

Pollinators differ in their pollination efficiency for a given crop and even within the same species (Dafni et al., 2005). Honey bee pollen collectors perform better pollination services to most crops, although nectar collectors are more efficient as pollinators for monoecious cucurbit crops. Most honey bee foragers leave their colonies early in the morning to collect pollen, while most are later attracted to flowers for the purposes of nectar collection during the rest of anthesis. This behavior influences the success of pollination in monoecious cucurbit plants. Since honey bees focus only on pollen collection early in the morning, this results in little to no pollen transfer from the anthers of staminate cucurbit flowers to the stigmas of pistillate fruit-producing flowers. Therefore, only honey bee nectar collectors are capable of pollen transfer from staminate to pistillate flowers, which results in pollination (Free, 1993).

A single pollinator cannot transfer sufficient pollen to the stigmatic surface of a flower in one visit to allow for the development of a perfectly formed fruit. Thus, each fruit-producing flower needs multiple pollinator visits to obtain sufficient amounts of pollen grains on the stigma to set fruit (Stanghellini et al., 1998). An increase in the number of visits per fruit-producing flower will result in cucurbit fruits having greater size, weight, sweetness, firmness, and more seeds (Free, 1993). Cucurbit fruit set, development, and quality are also influenced by the movement of the pollinator inside the flower as well as in the field. These behaviors define the manner in which pollen grains are distributed within the field (Walters and Schultheis, 2009) and on the stigmatic surface of a flower (Free, 1993). In addition, secondary pollen transfer can occur when pollen grains removed from anthers by pollinators drop onto the petals of the same flower or onto the petals of fruit-producing flowers after being transported some distance. Thus, sometimes pollen does not reach the stigmatic surface of fruit-producing flowers during the initial movement of pollen transfer, but in subsequent visits, these pollen grains can be transferred from petals to a stigmatic surface.

Bumble bees are more efficient pollinators (on an individual basis) compared to honey bees for many cucurbit crops (Stanghellini et al., 2002; Vidal et al., 2010; Artz and Nault, 2011) (Table 11.2). However, honey bees are generally more numerous due to high feral populations or colony placement in fields, which then overcompensate for their overall lower pollination efficiency. Squash bees have about the same pollination efficiency for cucurbit crops as honey bees, but they are solitary and their densities in fields are often very low. However, due to their ability to fly during low temperatures and at low light intensities, they can be more efficient pollinators than honey bees when cool or cloudy weather conditions hinder honey bee flight activity (Free, 1993).

11.6 POLLINATION SERVICES

11.6.1 BEHAVIOR AND MANAGEMENT OF POLLINATORS IN GREENHOUSE OR PROTECTED CULTURE

The cultivation of cucurbits in protected culture environments have been increasing during the last few decades. Although this system is primarily utilized to protect crops from adverse environmental conditions, it prevents the entrance of natural pollinators into the structure. Since cucurbits (except parthenocarpic cucumber cultivars) depend on pollinators to set fruits, some types of pollination system must be utilized for these crops when grown under protected culture conditions. Thus, growers have three options: pollinate flowers by hand or use of mechanical devices, use certain chemical products to induce fruit development, or introduce pollinators. In most cases, the most viable option is introduce pollinators to perform the required pollination services (Cruz and Campos, 2009).

Bumble bees are the most used pollinators in greenhouse or protected culture environments. Again, not all bees adapt well to this confined environment. Although honey bees are another option, they have many problems adapting to enclosed environments. Other promising pollinators for enclosed systems are stingless bees and a few solitary bees (Hogendoorn, 2004; Slaa et al., 2006).

Many bee species have specific behaviors that prevent them from adapting to enclosed environments (Free, 1993; Slaa et al., 2006). Once introduced into these environments, it is common for bees to not forage flowers, have behaviors of disorientation, and attempt to escape from the protected environment, especially during the first days after colony introduction (Fisher and Pomeroy, 1989; Free, 1993; Malagodi-Braga, 2002; Cauich et al., 2004; Cruz et al., 2004). Oftentimes, these behaviors result in death, especially in older bees, which are the most experienced members of the colony and have the task of foraging. However, after a few days of adaptation and the death of several foragers, they begin to develop less escape behaviors (Free, 1993; Cauich et al., 2004). Once bees adapt to the enclosed conditions, they will start to visit flowers occasionally and soon thereafter, will establish a regular floral visitation pattern that results in floral resource collection and pollination (Free, 1993; Cauich et al., 2004; Cruz et al., 2004). Visitation tends to begin on those flowers that are closer to the bee colonies, with the foraging area later expanding to the entire greenhouse (Cruz et al., 2004). The time required to adapt to enclosures and begin regular flower visits varies within and among bee species (Malagodi-Braga, 2002; Higo et al., 2004; Del Sarto et al., 2005), as well as with the type of covering used on greenhouses (Morandin et al., 2002). Bees belonging to the genus *Bombus* are more adaptive to confinement in enclosure structures compared to many other bee species (Fisher and Pomeroy, 1989; Guerra Sanz, 2008). Generally, bumble bees quickly adapt to the protected environment and begin pollination services soon after their introduction (Fisher and Pomeroy, 1989; Free, 1993).

Bumble bee colonies, used commercially to provide crop pollination services, suffer significant population reductions during the time inside greenhouses. This population decline primarily results from the limited poor quality food available in this environment, especially the protein portion. Some of the energy portion of the bumble bee diet is usually placed inside colonies before they are sold, and this will at least supply a portion of the energetic demands of the colony during the time it is providing pollination services inside a greenhouse. Other causes responsible for this population decline are temperature fluctuations that occur within some greenhouses and their own life cycle as they live only for 8–12 weeks, which makes it impossible to reuse the bumble bee colonies for subsequent or later pollination services (Velthuis and Van Doorn, 2006).

For honey bees, it takes at least 7 days to acclimatize under protected cultivation, and during this time, these bees learn to orient themselves and forage on the flowers (Higo et al., 2004). However, in this type of environment, honey bee foragers often have high mortality rates due to the lack of sufficient food resources to meet the needs of their colonies, or they are unable to pollinate the flowers as a result of disorientation, excitement, and/or stress caused by confinement (Free, 1993). It is also common to observe significant reductions within a colony that is placed into a protected culture environment. Sabara and Winston (2003) indicated that honey bee colonies placed for 3 weeks in a protected environment had reduced populations levels that was directly related to less amounts of eggs laid by the queen. Moreover, due to stress and weakened population by the protected cultivation, there is also a higher incidence and infestation of diseases and parasites, which would not normally cause problems to the colonies under natural environmental conditions (Cribb et al., 1993).

Honey bee colonies, used to provide pollination services in protected cultivation, should be replaced every 3 weeks and returned to outdoor field conditions to build populations for the next season or be sold in the spring as a new colony (Sabara and Winston, 2003). This is a different management strategy compared to bumble bee colonies that are discarded after 8 weeks or so in a greenhouse. A rotation system that replaces weak colonies is a more intensive management practice compared to the disposable colony system utilized for commercial bumble bee colonies (Sabara and Winston, 2003; Cauich et al., 2004). Furthermore, this rotation system can also be used for stingless bees (Cauich et al., 2004). It is important that the setup and replacement of colonies overlap in the greenhouse so that pollination services can be maintained for crops while newly placed bees are adapting to protected environments (Higo et al., 2004).

Smaller colonies with around 2000 bees are ideal for pollination in enclosures. Higo et al. (2004) suggest that the Langstroth bee box can be used to house honey bee colonies that are providing pollination services in greenhouses, and they should be prepared with at least five frames containing well-filled broods covered with adult bees and have an appropriate amount of honey (three or four frames). However, it is important to note that strong honey bee colonies are more difficult to adapt into a protected environment and the risk of stinging attacks to workers is increased due to their more aggressive defensive behaviors, especially for Africanized bees. Moreover, frames containing broods that have bees that are close to emerging can weaken the colony, and this factor should be considered when placing the bees into a protected environment. Also, sugar syrup and water should be provided as this is critical to regulate the temperature inside the hive (Free, 1993; Seeley, 1985). For honeydew melon pollination in enclosures, Keasar et al. (2007) compared the performance of honey bee mininucleus colonies (a small beehive that contained 4 small combs and a population of about 400 workers and a young queen) to a Langstroth beehive with about 25,000 workers. Although they found that colony population levels in Langstroth hives had deteriorated significantly while providing pollination services, the mininucleus colonies effectively pollinated melons in enclosed spaces, displayed nonaggressive behaviors, and maintained populations levels.

11.6.2 BASIC MANAGEMENT OF HONEY BEE COLONIES IN OPEN FIELDS

11.6.2.1 Colony Conditions

Generally, as the honey bee colony size increases, the proportion of the population that forage will also increase. So, smaller colonies have a smaller percentage of their bees as foragers. Larger colonies are able to provide not only more bees but also a higher proportion of the population as foragers, who are responsible to provide pollination (PNW, 2011). Moreover, it is important during the morning to track the movement of foragers in flowers to see if they are seeking floral resources in the target cucurbit crop (Bomfim et al., 2013).

Colonies used for pollination services should be healthy, having high populations, with a large brood area and a young queen (less than two years old) (Kevan, 2007; PNW, 2011). Thus, the hives should have at least six frames having combs that are well filled with the brood in various stages of development; this should be accompanied by adult bees that cover each comb, with a target population of about 25,000 adults. Furthermore, each colony should contain about two combs filled with honey (PNW, 2011; Bomfim et al., 2013). A colony will provide sufficient pollination services when it has at least 100 foragers per minute entering or leaving through its entrance (Delaplane and Mayer, 2000; PNW, 2011).

11.6.2.2 Introduction and Removal of Beehives

Honey bee colonies should be introduced in cucurbit fields about one week after the appearance of the first staminate flowers (Mussen and Thorp, 1997; Stanghellini et al., 2002; Reyes-Carrillo et al., 2009). If honey bees are introduced too early, they can become unproductive since they can be distracted by visiting nontarget plant species and establish flight patterns to more abundant and attractive food sources, such as wild flowers. Furthermore, each day that beehives are in the field, the rental costs increases for cucurbit growers. The timing of honey bee colony placement into fields is critical, and if their introduction is delayed by only a few days, yields can be reduced due to the lack of sufficient pollinators for the first fruit-producing flowers that develop on plants. For melon, Reyes-Carrillo et al. (2009) estimated a loss of 3.17 tons of fruit per hectare (or 7.16% of the total production) for each day that the introduction of honey bee hives into a field is delayed. This lack of pollination and fruit set causes a decrease in melon fruit weight, size, and number. Additionally, this delay can also cause the beginning of the melon harvest period to be delayed.

Sousa et al. (2014) concluded that growers can choose the appropriate time to introduce beehives into cucurbit fields according to the market they are targeting. For example, if the market demands large and heavy melons, beehives should be introduced into melon fields about 23 days after sowing; however, if the market demands small and lighter weight melons, growers can introduce beehives later (about 33 days and later after sowing). The introduction of beehives 28 days after sowing melons will cause an equal proportion of large and small melons in a field. This occurs because at the start of flowering (around 23 days after sowing) there are only a few hermaphrodite flowers at the crown area (or first few nodes) close to the base of the plant. Thus, melon plants will concentrate assimilates for the enlargement of first few fruits that set and provide a limited amount for later fruit set on laterals that causes a significant reduction in fruit size. In contrast, at 33 days after sowing, there will be few crown area flowers and abundant flowers on lateral branches, which results in great competition for assimilates, so most of these melons produced will be smaller in size and lighter in weight than when beehives are introduced 10 days earlier.

Honey bee hive removal from fields is based on whether a cucurbit crop has maximized its potential yield. For melons, honey bee colonies should be removed 28 days after the appearance of the first hermaphrodite flowers (Reyes-Carrillo et al., 2009), whereas in Calabash gourd, Bratsch (2009) suggested honey bee hive removal between 6 and 8 weeks after its introduction in a field.

11.6.2.3 Beehive Densities

Different honey bee hive densities per hectare have been suggested to provide optimal pollination services for cultivated cucurbits (McGregor, 1976; Free, 1993; Delaplane and Mayer, 2000). The number of beehives for a given area is variable since cucurbit pollination can be influenced by several factors, including the number of open flowers, the species and amounts of wild pollinators visiting flowers, the number of visits that a crop flower requires to be sufficiently pollinated, the number of bees within a hive that are actively foraging, the number of hours per day that bees

TABLE 11.5

Overall Average of Recommended Honey Bee Densities for Cucurbit Crop Pollination in Open-Field Conditions

Crop		Beehives (ha)	Bee/Flower or Plant	References
Cucumber (*Cucumis sativus*)	Monoecious (low plant population)	2.5	1 bee/100 flowers or 1 bee colony/50,000 plants	McGregor (1976), Delaplane and Mayer (2000)
	Gynoecious (high plant population)	7.5	1 bee/100 flowers or 1 colony/50,000 plants	McGregor (1976), Delaplane and Mayer (2000)
Calabash gourd (*Lagenaria siceraria*)		≈2.5	—	Bratsch (2009)
Sponge gourd (*Luffa cylindrica*)		≈4	—	Davis (2008)
Melon (*Cucumis melo*)		4.4	1 bee/10 perfect flowers	Delaplane and Mayer (2000)
Pumpkin (*Cucurbita* spp.)		3.8	—	Delaplane and Mayer (2000)
Watermelon (*Citrullus lanatus*)	Seeded	4.5	1 bee/100 flowers	Delaplane and Mayer (2000)
	Seedless	>Double of seeded watermelon	—	Walters (2005)

forage on the cucurbit crop, and the attractiveness of nontarget crops or wild flowers (Woodcock, 2012). Moreover, climatic conditions, the subsequent reutilization of colonies without a recovery period, and the types, amounts, application frequency, and timing of pesticides will all affect honey bee populations (Bomfim et al., 2013). The recommended honey bee densities for many important cucurbit crops grown in open-field conditions are shown in Table 11.5. Monoecious cucumbers grown at low plant populations and calabash gourd require 2.5 honey bee hives per hectare, while suggested densities for sponge gourd, melon, and seeded watermelon are between 4 and 4.5 hives per hectare. Gynoecious cucumber and seedless watermelon require high honey bee hive densities of 7.5 and >9 per hectare, respectively.

11.6.2.4   Distribution and Field Placement

The distribution of the honey bee hives in the field also needs to be considered. Honey bees can fly long distances to forage flowers for food, but they give preference to those that are closest to their colony. It is important to note that if flowers are closer to the colony, bees will recruit many other foragers from their colony to seek floral resources in these flowers, and this will result in more flower visitations per day (Jay, 1986). Therefore, a greater pollination service efficiency is achieved for a cucurbit crop when honey bee hives are placed in close proximity. Thus, it is recommended to distribute the colonies inside the field, following the appropriate density for each cucurbit crop. It is common in cucurbit fields to see honey bee hives distributed at only a few places inside a field. However, another possibility is to place them at the edges of the field, about 30 or 50 m from the first crop row. Mussen and Thorp (1997) indicated that groups of 10–20 hives spaced 160 m apart from each other along the edges of the field are more effective than placing all honey bee hives at one side of the field. An even better placement strategy for distribution (but more expensive and laborious) would be to place the recommended number of hives for each cucurbit crop inside each hectare so that honey bees can be evenly distributed throughout a field planting (Jay, 1986). Colonies should also be provided shade in some manner and have clean water near the hive area to prevent bees from interrupting their foraging activities to perform other activities for hive maintenance, which can have a negative influence on the overall pollination service that is provided (Seeley, 1985; Delaplane and Mayer, 2000).

11.6.2.5   Pesticides and Pollination Services

Honey bees collecting nectar and pollen in flowers contaminated by agrochemical application can simply die (lethal effect) or have changes in their physiology and behavior, which will affect the overall operation of the colony (sublethal effect) and in turn impair pollination services (Freitas and Pinheiro, 2012). The sublethal effects caused by pesticides can influence the division of labor within the hive, the life expectancy of individuals, and the foraging activity of workers and also reduce the immune systems of bees, resulting in lowering colony survival and consequently, less pollination services. For example, neonicotinoid insecticides, which are thought to be a class of pesticides responsible for colony losses around the world, have enzymatic activities that act physiologically in taste perception, olfactory learning, memory and motor activity of honey bees, which relate to problems in foraging activities, particularly navigation and orientation. Moreover, this insecticide can affect the bee immune system, making them more susceptible to pathogens, such as *Nosema microsporidian* (Freitas and Pinheiro, 2012; Fairbrother et al., 2014). This class of insecticide can also synergize with some fungicides and become more even more lethal to bees (Freitas and Pinheiro, 2012). Some countries have banned or tried to implement programs to reduce neonicotinoid use in an attempt to reduce bee mortality losses.

To minimize the risk of contamination with pesticides, the colonies should be introduced in the area only during the flowering and removed soon after obtaining the expected fruit set. Moreover, if the application of pesticides during flowering is unavoidable, products should be used that are less toxic to bees, with applications occuring in the late afternoon or evening, when there is low activity of bees in the field, and the entrance of the hives closed or pointed outward of the field in order to avoid contamination by drift. Another possibility is to cover the hives with large moistened cloths of some sort (e.g., burlap sacks), and this should be done in such a way to prevent pesticides from entering honey bee hives. This prevents those bees that are potentially contaminated from entering hives during and after pesticide application. If cloths covering the hives are watered down every few hours and consistently remain moist, the colonies can be confined for a period of up to two days with moistened sponges inside the hive as a water source for bees (Freitas and Pinheiro, 2012). However, pesticides having greater residual effect should always be avoided, although those that do not present risks to bees are most preferred (Mussen and Thorp, 1997; Reyes-Carrillo et al., 2009; Freitas and Pinheiro, 2012).

11.6.2.6 Impact of Colony Collapse Disorder on Pollination Services

Despite the worldwide increase in honey bee hive numbers during the last 50 years (~45%), the areas planted with crops that require honey bee pollination services have increased much more rapidly during the same period (>300%). This indicates that the demand for pollination services is higher than the number of beehives available worldwide. This situation is worse in Europe and North America, which have had a significant decline in honey bee hives in these 50 years (Aizen and Harder, 2009; Potts et al., 2010). This decline has become even more pronounced since 2006 and 2007, when some American and European beekeepers reported unexpected colony losses ranging from 30% to 90%, with the surviving colonies weakened and only able to provide minimal pollination services (Johnson, 2010). These high honey bee population losses were attributed to colony collapse disorder (CCD) events. This disorder is characterized by the sudden and rapid disappearance of adult bees from colonies and only contain a small cluster of young adult bees (which are insufficient to maintain the amount of brood inside the colony); other characteristics include a healthy and active queen, immature or capped brood and food stores, and lack of damaging levels of parasitic *Varroa destructor* mites or *N. microsporidian* (Ellis et al., 2010). It seems that healthy bees were simply abandoning their hives in high numbers and never returning. However, the fact that these bees never return to their hives is quite unusual because honey bees are a very social insect and colony-oriented, with a complex and organized colony system.

Many colonies that have died over the past 50–60 years display common symptoms of CCD. However, this phenomenon occurs fast, and there are no dead or dying workers bees in or around the hives; this disorder is not seasonal, it can manifest itself throughout the year (Johnson, 2010). Additionally, the remaining bees are reluctant to consume food stores within the colony, and also bees from different colonies or wax moths and small hive beetles ignore these hives full of food up to several weeks after their collapse (Ellis et al., 2010).

Currently, the cause of CCD remains under investigation by scientists of many parts of the world. The current consensus is that this disorder is the product of multiple factors and cannot be explained by a single cause. Most cited factors involved in CCD are (1) traditional bee pests and diseases (including American foulbrood, European foulbrood, chalkbrood, *N. microsporidian*, small hive beetles, and tracheal mites), (2) improper honey bee management practices (e.g., stress induced by increased traveling for pollination and overcrowded apiaries), (3) queen source (lack of genetic biodiversity promoted by high level of inbreeding queens that are commercialized, which makes bees more susceptible to any pest/disease), (4) unnecessary or excessive use of chemicals in bee colonies, (5) some pesticides used in agriculture, (6) *V. destructor mites* and associated pathogens, (7) poor diet (pollen and nectar scarcity; pollination of crops with low nutritional value), (8) undis-covered/new pests and diseases or increasing virulence of existing pathogens, (9) exploitation of food resources from genetically modified crops, (10) global warming (changing in temperatures and in timing of flower bloom), and (11) synergism between two or more of the aforementioned hypotheses. Other theories have been proposed to explain this phenomenon, including the effects of mobile phone signals and radiation on bee's navigational capabilities (Ellis et al., 2010; Johnson, 2010; Fairbrother et al., 2014).

Many beekeepers are no longer interested in the business of providing pollination services due to extra expenses associated with maintaining high population levels in colonies and the efforts needed to recover significant colony losses (Potts et al., 2010). Thus, fewer beekeep-ers translate into less beehives available to rent for crop pollination services, and in the last few years, the cost of renting hives for crop pollination has increased more than three times, although honey bee pollination rental fees vary from crop to crop. These fees are generally based on the distance from the beekeepers base of operation to the field in which crop pollina-tion services are desired, number of hives required, flowering time, number of days in the field, nutrition of the nectar and pollen provided by the crop, and volume and value of the honey that the crop could provide to beekeepers. Cucurbit crop pollination fees are in the average range among other pollination-dependent crops.

11.7 SUMMARY

Pollination services provided by insects are an important part of agriculture, having extensive economic consequences. The management of various pollinator species has allowed for enhanced productivity in many crops, including cucurbits. Although there are numerous pollinator species, only honey bees and bumble bees are actively managed to any degree in cucurbits, with honey bees the most important. However, managed honey bee populations have been declining in recent years, most likely due to multiple factors, which significantly reduce the availability of pollina-tors to crops that require pollination services, such as cucurbits. Although wild pollinators can in some instances provide pollination services to crops, such as cucurbits, they normally have low populations and cannot fully provide high enough field populations to adequately pollinate a crop. Therefore, the health of managed and wild insect pollinators is a concern, and anything that decreases their populations will definitely impact cucurbit production and other agricultural systems that depend of pollinators for fruit set.

REFERENCES

Adlerz, W.C. (1966). Honey bee visit numbers and watermelon pollination. *Journal of Economic Entomology*, 59: 28–30.

Aizen, M.A., Harder, L.D. (2009). The global stock of domesticated honey bees is growing slower than agri-cultural demand for pollination. *Current Biology*, 19: 1–4.

Ambrose, J.T., Schultheis, J.R., Bambara, S.B., Mangum, W. (1995). An evaluation of selected com-mercial bee attractants in the pollination of cucumbers and watermelons. *American Bee Journal*, 135: 267–272.

Artz, D.R., Nault, B.A. (2011). Performance of *Apis mellifera*, *Bombus impatiens*, and *Peponapis pruinosa* (Hymenoptera: Apidae) as pollinators of pumpkin. *Journal of Economic Entomology*, 104: 1153–1161.

Bezerra, A.D.M. (2014). Uso de abelha canudo (*Scaptotrigona* sp. nov.) na polinização do meloeiro (*Cucumis melo* L.) em ambiente protegido. Dissertação de mestrado (Mestrado em Zootecnia), Universidade Federal do Ceará, Fortaleza, Brazil, 91pp.

Bhardwaj, H., Thaker, P., Srivastava, M. (2012). Hymenopteran floral visitors as recorded from an agro-ecosystem near Bikaner, Rajasthan. *Global Journal of Science Frontier Research Agriculture & Biology*, 12: 19–34.

Bomfim, I.G.A. (2013). Uso de abelhas sem ferrão (Meliponinae: Apidae) em casa de vegetação para polinização e produção de frutos de minimelancia [*Citrullus lanatus* (Thunb.) Matsum. & Nakai] com e sem semente. Tese (Doutorado em Zootecnia), Universidade Federal do Ceará, Fortaleza, Brazil, 142pp.

Bomfim, I.G.A., Cruz, D.O., Freitas, B.M., Aragão, F.A.S. (2013). Polinização em melancia com e sem semente. Fortaleza, Brazil: Embrapa Agroindústria Tropical (Embrapa Agroindústria Tropical. Documentos, 168), 53pp.

Bratsch, A. (2009). Specialty crop profile: Ornamental gourds. Virginia Cooperative Extension, Publication 438-101, 6pp. Virginia Polytechnic Institute and State University, Blacksburg. http://pubs.ext.vt.edu/438/438-101/438-101.html. Accessed on March 2014.

Burtenshaw, M.K. (2003). The first horticultural plant propagated from seed in New Zealand: *Lagenaria Siceraria*. *New Zealand Garden Journal*, 6(1):10–16.

Cane, J.H., Sampson, B.J., Miller, S.A. (2011). Pollination value of male bees: The specialist bee *Peponapis pruinosa* (Apidae) at summer squash (*Cucurbita pepo*). *Environmental Entomology*, 40: 614–620.

Carvalho, A.D.F., Amaro, G.B., Lopes, J.F., Vilela, N.J., Filho, M.M., Andrade, R. (2013). A cultura do pepino. Circular técnica 113. Brasília, Brazil: Embrapa, 18pp.

Cauich, O., Quezada-Euán, J.J.G., Macias-Macias, J.O., Reyes-Oregel, V., Medina-Peralta, S., Parra-Tabla, V. (2004). Behavior and pollination efficiency of *Nannotrigona perilampoides* (Hymenoptera: Meliponini) on greenhouse tomatoes (*Lycopersicon esculentum*) in Subtropical México. *Horticultural Entomology*, 97: 475–481.

Coelho, M.S., Kiill, L.H.P., Costa, N.D., Pinto, J.M., Feitoza, E.A., Lima Junior, I.O. (2012). Diversidade de visitantes florais em cultivo orgânico de meloeiro. *Horticultura Brasileira*, 30: S1081–S1087.

Cribb, D.M., Hand, D.W., Edmondson, R.N. (1993). A comparative study of the effects of using the honeybee as a pollinating agent of glasshouse tomato. *Journal of Horticultural Science & Biotechnology*, 68: 79–88.

Cruz, D.O. (2009). Biologia floral e eficiência polinizadora das abelhas *Apis mellifera* L. (campo aberto) e *Melipona quadrifasciata* Lep. (ambiente protegido) na cultura da pimenta malagueta (*Capsicum frutescens* L.) em Minas Gerais, Brasil. Tese de doutorado (Doutorado em Entomologia), Universidade Federal de Viçosa, Viçosa, Brazil, 83pp.

Cruz, D.O., Campos, L.A.O. (2009). Polinização por abelhas em cultivos protegidos. *Revista Brasileira de Agrociência*, 15: 5–10.

Cruz, D.O., Freitas, B.M., Silva, L.A., Silva, E.M.S., Bomfim, I.G.A. (2004). Adaptação e comportamento de pastejo da abelha jandaíra (*Melipona subnitida* Ducke) em ambiente protegido. *Acta Scientiarum Animal Sciences*, 26: 293–298.

Cruz, D.O., Freitas, B.M., Silva, L.A., Silva, E.M.S., Bomfim, I.G.A. (2005). Pollination efficiency of the stingless bee *Melipona subnitida* on greenhouse sweet pepper. *Pesquisa Agropecuária Brasileira*, 40: 1197–1201.

Dafni, A., Kevan, P.G., Husband, B.C. (2005). *Practical Pollination Biology*. Cambridge, Ontário, Canada: Enviroquest, 590pp.

Davis, J.M. (2008). Commercial Luffa sponge gourd production. Horticulture Information Leaflet 120. Raleigh, NC: North Carolina State University, 3pp.

Delaplane, K.S., Mayer, D.F. (2000). *Crop Pollination by Bees*. Cambridge, U.K.: CABI, 344pp.

Del Sarto, M.C.L., Peruquetti, R.C., Campos, L.A.O. (2005). Evaluation of the neotropical stingless bee *Melipona quadrifasciata* (Hymenoptera: Apidae) as pollinator of greenhouse tomatoes. *Journal of Economic Entomology*, 98: 260–266.

Ellis, J.D., Evans, J.D., Pettis, J.S. (2010). Colony losses, managed colony population decline and colony collapse disorder in the United States. *Journal of Apicultural Research* 49(1): 134–136.

Fairbrother, A., Purdy, J., Anderson, T., Fell, R. (2014). Risks of neonicotinoid insecticides to honeybees. *Environmental Toxicology and Chemistry*, 33: 719–731.

Fisher, R.M., Pomeroy, N. (1989). Pollination of greenhouse muskmelons by bumble bees (Hymenoptera: Apidae). *Entomological Society of America*, 82: 1061–1066.

Free, J.B. (1993). *Insect Pollination of Crops*. London, U.K.: Academic Press, 684pp.

Freitas, B.M., Pinheiro, J.N. (2012). *Polinizadores e Pesticidas: Princípios de manejo para os ecossistemas brasileiros*, Vol. 1, 1st edn. Brasília, Brazil: Ministério do Meio Ambiente, 112pp.

Garibaldi, L.A., Steffan-Dewenter, I., Winfree, R., Aizen, M.A., Bommarco, R., Cunningham, S.A., Kremen, C. et al. (2013). Wild pollinators enhance fruit set of crops regardless of honey-bee abundance. *Science*, 339: 1608–1611.

Garófalo, C.A., Matins, C.F., Aguiar, C.M.L., Del Lama, M.A. Alves-Dos-Santos, I. (2012). As abelhas solitárias e perspectivas para seu uso na polinização no Brasil. In: *Polinizadores no Brasil: Contribuição e perspectivas para a biodiversidade, uso sustentável, conservação e serviços ambientais*. Imperatriz-Fonseca, V.L., Canhos, D.A.L., Alves, D.A., Saraiva, A.M. (Orgs.), São Paulo, Brazil: Edusp, pp. 183–202, Chapter 9.

Guerra Sanz, J.M. (2008). Crop pollination in greenhouses. In: *Bee Pollination in Agricultural Ecosystems*. James, R.R., Pitts-Singer, T.L. (eds.), New York: Oxford University Press, Inc., pp. 27–47, Chapter 3.

Heard, R.A. (1999). The role of stingless bees in crop pollination. *Annual Review of Entomology*, 44: 183–206.

Henne, C.S., Rodriguez, E., Adamczyk Jr., J.J. (2012). A survey of bee species found pollinating watermelons in the Lower Rio Grande Valley of Texas. *Psyche*, 2012: 1–5.

Higo, H.A., Rice, N.D., Winston, M.L., Lewis, B. (2004). Honey bee (Hymenoptera: Apidae) distribution and potential for supplementary pollination in commercial tomato greenhouses during winter. *Journal of Economic Entomology*, 97: 163–170.

Hogendoorn, K. (2004). On promoting solitary bee species for use as crop pollinators in greenhouses. In: *Solitary Bees: Conservation, Rearing and Management for Pollination*. Freitas, B.M., Pereira, J.O.P. (eds.), Fortaleza, Brazil: Imprensa Universitária, pp. 213–221.

Imperatriz-Fonseca, V.L., De Jong, D., Saraiva, A.M. (2006). *Bees as Pollinators in Brazil: Assessing the Status and Suggesting Best Practices*. Ribeirão Preto, Brazil: Holos, 112pp.

Jay, S.C. (1986). Spatial management of honey bees on crops. *Annual Review of Entomology* 31: 49–65.

Johnson, R. (2010). *Honey Bee Colony Collapse Disorder*. Diane Publishing: Darby, PA, 17pp.

Keasar, T., Shihadeh, S., Shmida, A., Majali, N., Weil, D., Reuven, N. (2007). An evaluation of mini-nucleus honey bee hives for the pollination of honeydew melons in enclosures. *Journal of Apicultural Research and Bee World*, 46: 264–268.

Kevan, P.G. (2007). *Bees: Biology & Management*. Cambridge, Ontario, Canada: Enviroquest Ltd., 345pp.

Klein, A.M., Vaissière, B., Cane, J.H., Steffan-Dewenter, I., Cunningham, S.A., Kremer C., Tscharntcke, T. (2007). Importance of pollinators in changing landscapes for world crops. *Proceedings of the Royal Society of London Biological Science*, 274: 303–313.

Kouonon, L.C., Jacquemart, A., Zoro Bi, A.I., Bertin, P., Baudoin, J., Dje, Y. (2009). Reproductive biology of the andromonoecious *Cucumis melo* subsp. *agrestis* (Cucurbitaceae). *Annals of Botany*, 104: 1129–1139.

Kremen, C., Williams, N.M., Thorp, R.W. (2002). Crop pollination from native bees at risk from agricultural intensification. *PNAS*, 99: 16812–16816.

Lima, C.J., Oliveira, F.L., Maracajá, P.B., Sousa, J.S. (2014b). Influência da concentração e o volume de néctar em flores de *Luffa cylindrica* (L.) M. Roem no comportamento de forrageio de *Apis mellifera*. *Agropecuária Científica no Semiárido*, 10: 39–50.

Lima, C.J., Oliveira, F.L., Maracajá, P.B., Sousa, J.S, Pereira, D.S. (2014a). Biologia floral e disponibilidade de néctar em cultivo convencional *Luffa cylindrica* (L.) M. Roem. *Revista Verde*, 9: 26–39.

Lowenstein, D.M., Huseth, A.S., Groves, R.L. (2012). Response of wild bees (Hymenoptera: Apoidea: Anthophila) to surrounding land cover in Wisconsin pickling cucumber. *Environmental Entomology*, 41: 532–540.

Lower, R.L., Edwards, M.D. (1986). Cucumber breeding. In: *Breeding Vegetable Crops*. Basset, M.J. (ed.), Westport, CT: AVI, pp. 173–207.

Malagodi-Braga, K.S. (2002). Estudo de agentes polinizadores em cultura de morango (*Fragaria × ananassa* Duch.—Rosaceae). Tese (Doutorado em Ecologia), Universidade de São Paulo, São Paulo, Brazil, 102pp.

Malagodi-Braga, K.S., Kleinert, A.M.P., Imperatriz-Fonseca, V.L. (2004). Abelhas sem ferrão e polinização. *Revista Tecnologia e Ambiente*, 10: 59–70.

Mcgregor, S.E. (1976). *Insect Pollination of Cultivated Crop Plants*. Washington, DC: United States Department of Agricultural Research Service, 496pp.

Mcgregor, S.E., Levin, M.D., Foster, R.E. (1965). Honey bee visitors and fruit setting of cantaloupes. *Journal of Economic Entomology*, 58: 968–970.

Mélo, D.B.M., Santos, A.L.A., Beelen, R.C., Lira, T.S., Almeida, D.A.S., Lima, L.P. (2010). Polinização da abóbora (*Curcubita moschata* D.): um estudo sobre a biologia floral e visitantes florais no município de Satuba-Al. *Revista Científica do IFAL*, 1: 47–57.

Morais, M.M., De Jong, D., Message, D., Gonçalves, L.S. (2012). Perspectivas e desafios para o uso das abelhas *Apis mellifera* como polinizadores no Brasil. In: *Polinizadores no Brasil: Contribuição e perspectivas para a biodiversidade, uso sustentável, conservação e serviços ambientais.* Imperatriz-Fonseca, V.L., Canhos, D.A.L., Alves, D.A., Saraiva, A.M. (Orgs.), São Paulo, Brazil: Edusp, pp. 203–212, Chapter 10.

Morandin, L.A., Laverty, T.M., Gegear, R.J., Kevan, P.G. (2002). Effect of greenhouse polyethelene covering on activity level and photo-response of bumble bees. *The Canadian Entomologist*, 134: 539–549.

Morimoto, Y., Gikungu, M., Maundu, P. (2004). Pollinators of the bottle gourd (*Lagenaria siceraria*) observed in Kenya. *International Journal of Tropical Insect Science*, 24: 79–86.

Mussen, E.C., Thorp, R.W. (1997). *Honey Bee Pollination of Cantaloupe, Cucumber and Watermelon.* University of California, Cooperative Extension, Division of Agriculture and Natural Resources, Publication 7224, Davis, CA. http://anrcatalog.ucdavis.edu/pdf/7224.pdf. Accessed on May 2014.

Nicodemo, D., Malheiros, E.B., De Jong, D., Nogueira Couto, R.H. (2012). Biologia floral de pepino (*Cucumis sativus* L.) tipo Aodai cultivado em estufa. *Científica*, 40: 41–46.

Nicomedo, D., Malheiros, E.B., De Jong, D., Nogueira-Couto, R.H. (2013). Enhanced production of parthenocarpic cucumbers pollinated with stingless bees and Africanized honey bees in greenhouses. *Semina: Ciências Agrárias*, 34: 3625–3634.

Nicodemo, D., Nogueira Couto, R.H., Malheiros, E.B., De Jong, D. (2009). Honey bee as an effective pollinating agent of pumpkin. *Scientia Agricola*, 66: 476–480.

Petersen, J.D., Reiners, S., Nault, B.A. (2013). Pollination services provided by bees in pumpkin fields supplemented with either *Apis mellifera* or *Bombus impatiens* or not supplemented. *PLoS One*, 8: 1–8.

PNW. 2011. Evaluating honey bee colonies for pollination: A guide for growers and beekeepers. Pacific Northwest Extension Publication No. 623. Corvallis, OR: Oregon State University, 8pp.

Potts, S.G., Biesmeijer, J.C., Kremen, C., Neumann, P., Schweiger, O., Kunin, W.E. (2010). Global pollinator declines: Trends, impacts and drivers. *Trends in Ecology & Evolution*, 25(6): 345–353.

Reyes-Carrillo, J.L., Cano-Ríos, P. (2002). *Manual de polinización apícola. Programa nacional para el control de la abeja africana.* Instituto Interamericano para la Cooperación Agrícola, San Jose, Costa Rica. 52pp.

Reyes-Carrillo, J.L., Cano-Ríos, P., Nava-Camberos, U. (2009). Período óptimo de polinización de melón con abejas melíferas (*Apis mellifera* L.). *Agricultura Técnica en México*, 35: 370–377.

Robinson, R.W., Decker-Walters, D.S. 1997. *Cucurbits.* Wallingford, U.K.: CABI, 226pp.

Roselino, A.C., Bispo Dos Santos, S.A., Bego, L.R. (2010). Qualidade dos frutos de pimentão (*Capsicum annuum* L.) a partir de flores polinizadas por abelhas sem ferrão (*Melipona quadrifasciata anthidioides* Lepeletier 1836 e *Melipona scutellaris* Latreille 1811) sob cultivo protegido. *Revista Brasileira de Biociências*, 8: 154–158.

Sabara, H.A., Winston, M.L. (2003). Managing honey bees (Hymenoptera: Apidae) for greenhouse tomato pollination. *Journal of Economic Entomology*, 96: 547–554.

Sadeh, A., Shmida, A., Keasar, T. (2007). The carpenter bee *Xylocopa pubescens* as an agricultural pollinator in greenhouses. *Apidologie*, 38: 508–517.

Sanford, M., Ellis, J.D. (2010). Beekeeping: Watermelon pollination. EDIS article ENY-154/AA-091. University of Florida, IFAS, Gainesville, FL, 5pp.

Santos, S.A.B., Roselino, A.C., Bego, L.R. (2008). Pollination of cucumber, *Cucumis sativus* L. (Cucurbitales: Cucurbitaceae), by the stingless Bees *Scaptotrigona* aff. *depilis* Moure and *Nannotrigona testaceicornis* Lepeletier (Hymenoptera: Meliponini) in greenhouses. *Neotropical Entomology*, 3: 506–512.

Saraiva, A.M., Acosta, A.L., Giannini, T.C., Imperatriz-Fonseca, V.L, Marco Junior, P. (2012). *Bombus terrestris* na América do Sul: Possíveis rotas de invasão deste polinizador exótico até o Brasil. In: *Polinizadores no Brasil: Contribuição e perspectivas para a biodiversidade, uso sustentável, conservação e serviços ambientais.* Imperatriz-Fonseca, V.L., Canhos, D.A.L., Alves, D.A., Saraiva, A.M. (Orgs.), São Paulo, Brazil: Edusp, pp. 203–212, Chapter 10.

Seeley, T.D. (1985). *Honeybee Ecology: A Study of Adaptation in Social Life.* Princeton, NJ: Princeton University Press, 216pp.

Silva, M.W.K.P., Ranil, R.H.G., Fonseka, R.M. (2012). *Luffa cylindrica* (L.) M. Roemer (Sponge Gourd-Niyan wetakolu): An emerging high potential underutilized curcurbit. *Tropical Agricultural Research*, 23(2):186–191.

Siqueira, K.M.M., Kiill, L.H.P., Gama, D.R.S., Araújo, D.C.S., Coelho, M.S. (2011). Comparação do padrão de floração e de visitação do meloeiro do tipo amarelo em Juazeiro-BA. *Revista Brasileira de Fruticultura*, 33: 473–478.

Slaa, E.J., Sánchez Chaves, L.A., Malagodi-Braga, K.S., Hofstede, F.E. (2006). Stingless bees in applied pollination: Practice and perspectives. *Apidologie*, 37: 293–315.

Sousa, R.M. Aguiar, O.S., Freitas, B.M., Maracaja, P.B., Azevedo, A.E.C. (2014). Period of introduction of Africanized honeybees (*Apis mellifera L.*) for pollination of yellow melon (*Cucumis melo L.*). *Revista Verde*, 9(4):1–4.

Stanghellini, M.S., Ambrose, J.T., Schultheis, J.R. (1997). The effects of honey bee and bumble bee pollination on fruit set and abortion of cucumber and watermelon. *American Bee Journal*, 137: 386–391.

Stanghellini, M.S., Ambrose, J.T., Schultheis, J.R. (1998). Using commercial bumble bee colonies as backup pollinators for honey bees to produce cucumbers and watermelons. *HortTechnology*, 8: 590–594.

Stanghellini, M.S., Schultheis, J.R., Ambrose, J.T. (2002). Pollen mobilization in selected cucurbitaceae and the putative effects on pollinator abundance on pollen depletion rates. *Journal of American Society for Horticultural Science*, 127: 729–736.

Thapa, R.B. (2006). Honeybees and other insect pollinators of cultivated plants: A review. *Journal of the Institute of Agriculture and Animal Science*, 27: 1–23.

Velthuis, H.H.W., Van Doorn, N.A. (2006). A century of advances in bumblebee domestication and the economic and environmental aspects of its commercialization for pollination. *Apidologie*, 37: 421–451.

Venturieri, G.C., Alves, D.A., Villas-Bôas, K., Carvalho, C.A.L, Menezes, C., Vollet-Neto, A. Contrera, F.A.L., Cortopassi-Laurino, M., Nogueira-Neto, P., Imperatriz-Fonseca, V.L. (2012). Meliponicultura no Brasil: Situação atual e perspectivas futuras para o uso na polinização agrícola. In: *Polinizadores no Brasil: Contribuição e perspectivas para a biodiversidade, uso sustentável, conservação e serviços ambientais*. Imperatriz-Fonseca, V.L., Canhos, D.A.L., Alves, D.A., Saraiva, A.M. (Orgs.), São Paulo, Brazil: Edusp, pp. 213–236, Chapter 11.

Vidal, M.G., De Jong, D., Wien, H.C., Morse, R.A. (2010). Pollination and fruit set in pumpkin (*Cucurbita pepo*) by honey bees. *Revista Brasileira de Botânica*, 33: 107–113.

Walters, S.A. (2005). Honeybee pollination requirements for triploid watermelon. *HortScience*, 40: 1268–1270.

Walters, S.A., Schultheis, J.R. (2009). Directionality of pollinator movements in watermelon plantings. *HortScience*, 44: 49–52.

Walters, S.A., Taylor, B.H. (2006). Effects of honey bee pollination on pumpkin fruit and seed yield. *HortScience*, 41: 370–373.

Winfree, R., Williams, N.M., Dushoff, J., Kremen, C. (2007). Native bees provide insurance against ongoing honey bee losses. *Ecology Letters*, 10: 1105–1113.

Winfree, R., Williams, N.M., Gaines, H., Ascher, J.S., Kremen, C. (2008). Wild bee pollinators provide the majority of crop visitation across land-use gradients in New Jersey and Pennsylvania, USA. *Journal of Applied Ecology*, 45: 793–802.

Winston M.L. (1987). *The Biology of the Honey Bee*. Cambridge, MA: Harvard University Press, 281pp.

Woodcock, T.S. (2012). *Pollination in the Agricultural Landscape: Best Management Practices for Crop Pollination*. Guelph, Ontario, Canada: University of Guelph, 113pp.

12 Sex Expression in Cucurbits
Special Reference to Cucumber and Melon

Puja Rattan and Sanjeev Kumar

CONTENTS

12.1 INTRODUCTION

The majority of flowering plants produce flowers that are "perfect." These flowers are both staminate (with stamens) and pistillate (with one or more carpels). In a small number of species, there is spatial separation of the sexual organs either as monoecy, where the male and female organs are found on separate flowers on the same plant, or dioecy, where male and female flowers are found on separate male (staminate) or female (pistillate) individuals. Sex determination systems in plants, leading to unisexuality as monoecy or dioecy, have undergone independent evolutions. In dioecious plant species, the point of divergence from the hermaphrodite pattern shows wide variations between species, implying that the genetic bases are very different. The Cucurbitaceae family is characterized by the presence of unisexual flowers. The sex expression in this family is highly variable, and a single genus can contain both monoecious and dioecious species. This family includes many important vegetables collectively referred to as cucurbits. The Cucurbitaceae is a distinct family without any close relatives. All the cultivated species are found in the subfamily Cucurbitoideae. The plants discussed in this chapter are dispersed among different tribes: Melothrieae (bur gherkins, melon, cucumber), Jaliffieae (bitter melon), Benincaseae (wax gourd, watermelon, angled luffa, smooth luffa), Cucurbitaceae (pumpkin, squash), and Sicyeae (chayote). Cucurbit plants have tailing or vining growth habit; they bear tendrils and are frost sensitive and appear annually.

12.2 EVOLUTION OF SEX FORMS IN CUCURBITS

In the course of time, numerous biochemical, morphological, and physiological phenomena have evolved, which promote the outcrossing. These phenomena include male sterility, self-incompatibility, and presence of unisexual flowers. The presence of unisexual flowers is the most extreme morphological modification that involves the development of either male (staminate) or female (pistillate) organs. Unisexuality has been estimated to arise more than 100 times within the plant kingdom (Renner and Ricklefs, 1995). In monoecy, separate male and female flowers are formed on the same plant, whereas in dioecy, male and female flowers are found in separate plants. Among the most important aspects of the biology of Cucurbitaceae, the flowering process is the integral element. Sex expression, flower formation, fruit set, and fruit growth and development are the necessary steps leading to good crop production in cucurbits. There are many factors that are responsible for the flowering process, and their occurrence in the proper combination is desired for the fulfillment of the goals. These characteristics would involve not only sexual differentiation but the time of appearance and proportion of pistillate or hermaphrodite flowers on which fruit production depends.

12.3 SEX EXPRESSION IN CUCUMBER

Cucumber plants usually produce separate male and female flowers in the same individual, but some cucumber lines produce bisexual flowers. Cucumber flower buds differentiate as lateral organs on the leaf axils. In general, the initiation of flower bud differentiation is observed on nodes situated three or four nodes below the apical meristem in the cucumber shoot (Fujieda, 1966). In the cucumber flower, buds, primordia of the sepals, petals, stamens, and pistils are formed centripetally from the outside (Atsmon and Galun, 1960; Malepszy and Niemirowicz-Szczytt, 1991). Flower buds contain primordia of both stamen and pistil at the early stages of development, and sex determination occurs due to the selective arrest of development of either the staminate or pistillate primordia just after the bisexual stage (Atsmon and Galun, 1960; Malepszy and Niemirowicz-Szczytt, 1991). Continued stamen development and the arrest of pistil development results in male flowers, whereas continued pistil development and the arrest of stamen development results in female flowers. Sex determination occurs in flower buds situated 10–12 nodes below the apical meristem irrespective of the age of the plant (Fujieda, 1966).

12.3.1 GENETIC CONTROL OF SEX EXPRESSION IN CUCUMBER

Depending on the ratio of male, female, and bisexual flowers produced by the plant, cucumbers are classified into five genotypes: monoecious, gynoecious, andromonoecious, hermaphrodite, and androecious (Galun, 1961; Shifriss, 1961; Kubicki, 1969; Malepszy and Niemirowicz-Szczytt, 1991). The monoecious line produces both male and female flowers, which are the commonest forms of cucumber. Gynoecious plants produce only female flowers. The andromonoecious line produces male and bisexual flowers, whereas the hermaphrodite line produces only bisexual flowers. Two major genes, *F* and *M*, control sex expression phenotypes in cucumber (Galun, 1961; Shifriss, 1961; Kubicki, 1969). That is, *M-F-*, *M-ff*, *mmff*, and *mmF-* confer gynoecious, monoecious, andromonoecious, and hermaphrodite phenotypes, respectively. The *F* gene is relatively dominant and promotes the production of female flowers. The *M* gene acts on the production of female flowers, and plants with recessive *m* alleles (*mm*) produce bisexual flowers. In addition to these genotypes, androecious cucumbers that produce only male flowers have been reported (Galun, 1961). The *A* gene acts downstream of the *F* gene, and plants become androecious when both *A* and *F* are recessive (*aaff*).

Galun et al. (1963) studied the inheritance of sex expression in cucumber. Through the interaction of major genes with modifying genetic and nongenetic factors, they investigated two major genes and a complex of polygenes affecting sex expression in *Cucumis sativus* L. The genetic factor *st* was found to affect sex by inducing a shift of the predetermined flowering pattern of this plant in the direction of its base. As this pattern is composed of a staminate stage followed by a mixed (staminate–pistillate) stage and a pistillate stage, a double dose of *st* will induce a change from the normal monoecious sex expression to absolute gynoecism. The second major gene studied, *m*, known previously to control sex in the individual flower (*m/m*, andromonoecious; *M*, monoecious), also interacts with factors affecting the flowering pattern by inducing male tendency.

Kamachi et al. (1997, 2000) reported that a high correlation existed between the evolution of ethylene from the apices and the development of female flowers. They isolated three distinct cDNA encoding 1-aminocyclopropane-1-carboxylate (ACC) synthases from cucumber. Among these, only CS-ACS2 mRNA was detected at the apices of gynoecious cucumber in which female flowers were developing. The expression of the *CS-ACS2* gene was examined in the apices of three cucumber cultivars (Rensei, Ougonmegami 2-gou, and Shimoshirazu) by RNA gel blot analysis. In these cultivars, both the timing and the levels of expression of the CS-ACS2 transcript were correlated with the development of female flowers at the nodes. Furthermore, the timing of the induction of expression of the *CS-ACS2* gene at the apex corresponded to that of the action of ethylene in the induction of the first female flower at the apex of gynoecious cucumber plants. These results suggest that the development of female flowers might be regulated by the level of CS-ACS2 mRNA at the apex.

Trebitsh et al. (1997) reported that sex determination in cucumber (*C. sativus* L.) is controlled largely by three genes: *F*, *m*, and *a*. The *F* and *m* loci interact to produce monoecious (*M_f_*) or gynoecious (*M_F_*) sex phenotypes. Ethylene and factors that induce ethylene biosynthesis, such as 1-aminocyclopropane-1-carboxylate (ACC) and auxin, also enhance female sex expression. They also suggested that CS-ACS1G is closely linked to the *F* locus and may play a pivotal role in the determination of sex in cucumber flowers.

Treves et al. (1998) cloned and sequenced three cDNA homologues of the homeotic gene *AGAMOUS* from early-stage floral buds of *C. sativus*. Their expression was studied by Northern analysis using two contrasting sex genotypes, an androecious line and a gynoecious one. It was observed that three genes were expressed at low levels at earlier bud stages, and the levels rise as the bud matures. Two of the clones, *CAG1* and *CAG3*, were expressed in the third and fourth whorl of mature flowers, while *CAG2* was restricted to the carpel; none was expressed in leaves. The transcript levels did not appear to be modulated by gibberellin or ethephon, two treatments that alter sex expression in cucumber. While MADS-box genes probably play an essential role in cucumber floral development, as they did in other plants, these findings may imply that the pathway leading to reproductive organ arrest in cucumber unisexual buds acts independently of MADS-box gene expression.

Kahana et al. (1999), in their study, noted that three full-length ACC oxidase cDNAs were isolated from cucumber floral buds. RFLP analysis of a population that segregates for the *F* (femaleness) locus indicated that CS-ACO2 is linked to *F* at a distance of 8.7 cM. The expression of two of the genes, CS-ACO2 and CS-ACO3, were monitored in flowers, shoot tips, and leaves of different sex genotypes. In situ mRNA hybridization indicated different patterns of tissue- and stage-specific expression of CS-ACO2 and CS-ACO3 in developing flowers. CS-ACO3 expression in midstage female flowers was localized to the nectaries and pistil and in the arrested staminoids, whereas CS-ACO2 transcript levels accumulated later and were found in placental tissues, ovaries, and staminoids. In male flowers, petals and nectaries expressed both genes, whereas ACO2 expression was strong in pollen of mature flowers. In young buds, a strong expression was observed along developing vascular bundles. Four sex genotypes were compared for CS-ACO2 and CS-ACO3 expression in the shoot apex and young leaf. FF genotypes had higher transcript levels in leaves but lower levels in the shoot apex and in young buds compared to ff genotypes; thus, it was inferred that the shoot-tip pattern is inversely correlated with femaleness.

Yamasaki et al. (2003b) observed that sex differentiation in cucumber plants (*C. sativus* L.) appears to be determined by the selective arrest of the stamen or pistil primordia. They investigated the influence of an ethylene-releasing agent (ethephon) or an inhibitor of ethylene biosynthesis (aminoethoxyvinylglycine) on sex differentiation in different developmental stages of flower buds. These treatments influence sex determination only at the stamen primordia differentiation stage in both monoecious and gynoecious cucumbers. To clarify the relationships between the ethylene-producing tissues and the ethylene-perceiving tissues in inducing female flowers in the cucumber, the localization of mRNA accumulation of both the ACC synthase gene (*CS-ACS2*) and the ethylene-receptor-related genes (*CS-ETR1, CS-ETR2,* and *CS-ERS*) were examined in flower buds by in situ hybridization analysis. *CS-ACS2* mRNA was detected in the pistil primordia of gynoecious cucumbers, whereas it was located in the tissues just below the pistil primordia and at the adaxial side of the petals in monoecious cucumbers. In flower buds of andromonoecious cucumbers, only *CS-ETR1* mRNA was detected and was located in the pistil primordia. The localization of the mRNAs of the three ethylene-receptor-related genes in the flower buds of monoecious and gynoecious cucumbers overlap but were not identical.

Mibus and Tatlioglu (2004) studied the molecular characterization and isolation of the *F/f* gene for femaleness in cucumber (*C. sativus* L.). They used nearly isogenic gynoecious (*MMFF*) and monoecious (*MMff*) lines (NIL) produced by backcross program. They confirmed the result of other groups that an additional genomic ACC synthase (key enzyme of ethylene biosynthesis) sequence (*CsACS1G*) should exist in gynoecious genotypes. A linkage was also verified between the *F/f* locus and the *CsACS1G* sequence. An exclusive amplification of the new isolated sequence (*CsACS1G*) in gynoecious (*MMFF*) and subgynoecious (*MMFf*) genotypes confirmed that the isolated gene is the dominant *F* allele.

Huiming et al. (2011) studied the inheritance of two novel subgynoecious genes in cucumber (*C. sativus* L.). Genetic analysis had showed the two subgynoecious inbred lines were controlled by one pair of recessive genes and one pair of incompletely dominant genes, which were designated as *mod-F2* and *Mod-F1*, respectively. Furthermore, the *mod-F2* and *Mod-F1* loci, which enhance the intensity of femaleness, also inherited independently with *F* and *M* genes.

Two ACC synthase genes (ACS), a key regulatory enzyme in the ethylene biosynthetic pathway were found to be involved in female flower production CsACS2 and CsACS1/G (Trebitsh et al., 1997; Saito et al., 2004). The expression of CsACS2 in individual flowers was suggested to be associated with differentiation and development of female flowers (Saito et al., 2004), and *CsACS1G* was mapped to the *F* locus (Trebitsh et al., 1997). Monoecious cucumbers producing mainly male flowers contain a single gene copy (CsACS1), while gynoecious plants have an additional female-specific copy (*CsACS1G*). A perfect correlation (100%) was found between female phenotypes possessing at least one dominant F allele and the presence of *CsACS1G* (Trebitsh et al., 1997).

12.3.2 Hormonal Control of Sex Expression in Cucumber

Cucumber flower buds, which differentiate in leaf axils of main shoots, are bisexual in their early developmental stages. Later, they develop into either staminate (male) or pistillate (female) flowers (Atsmon et al., 1960; Fujieda, 1966). In monoecious varieties, male flowers differentiate at the lower nodes, followed by female flowers at the higher nodes. The number of nodes to the first female flower and total number of female flowers are both reliable indices of sex expression. On monoecious wild-type cucumber plant, flowers are developed in a preset developmental sequence along the main stem: a phase of staminate flowers is followed by a mixed phase of staminate and pistillate flowers and terminated by a phase of pistillate flowers (Shifriss, 1961). The nature and length of each phase is genetically determined. Phenotypic expression of sex determining loci is strongly modified by environmental and hormonal factors. Byers et al. (1972) found that treating monoecious cucurbits with ethylene or compounds that promote ethylene synthesis causes earlier production of pistillate flowers, that is, promote feminization. They found that application of GA_3 promotes masculinization in cucumber gynoecious plants, but not in melon gynoecious plants. Silvernitrate ($AgNO_3$) and silver thiosulphate effectively induced male flowering in many nodes of 7 gynoecious cucumber genotypes Den Nijs and Visser (1980).

The genetically controlled sex expression in cucumbers varies in response to environmental and hormonal cues, and ethylene is a major regulator responsible for the sex of cucumber flowers (reviewed in Frankel and Galun, 1977; Yin and Quinn, 1995; Perl-Treves, 1999; Yamasaki et al., 2005).

In particular, plants or shoot apices of gynoecious lines produce more ethylene than monoecious lines (Rudich et al., 1972, 1976; Fujita and Fujieda, 1981; Trebitsh et al., 1987), and monoecious and andromonoecious lines treated with ethylene or an ethylene-releasing reagent produce an increased numbers of female and bisexual flowers, respectively (MacMurray and Miller, 1968; Iwahori et al.,

1969; Shannon and De La Guardia, 1969). In addition, treatment of cucumber plants with compounds that inhibit ethylene biosynthesis or that block ethylene signaling result in a decrease in the number of female or bisexual flowers (Beyer, 1976; Atsmon and Tabbak, 1979; Takahashi and Suge, 1980; Takahashi and Jaffe, 1984).

By treating monoecious and andromonoecious cucumber plants with various combinations of GA and ethrel or GA and ethylene inhibitors, Yin and Quinn (1995) demonstrated that ethylene is the main regulator of sex determination, with GA functioning upstream of ethylene, possibly as a negative regulator of endogenous ethylene production. These findings led them to propose a model for how sex determination might occur (Yin and Quinn, 1995), with ethylene serving both as a promoter of the female sex and an inhibitor of the male sex. The basic tenets of the model are that the *F* gene should encode a molecule that would determine the range and gradient of ethylene production along the shoot, thereby promoting femaleness, whereas the *M* gene should encode a molecule that perceives the ethylene signal and inhibits stamen development above threshold ethylene levels. This model is consistent with how unisexual flowers might arise very early and very late during shoot development; however, the model also predicts an entire range of intermediate types rarely or never seen in cucumber. As suggested by Perl-Treves (1999), variations in the model of Yin and Quinn and the incorporation of additional factors in the sex-determining process in cucumber could account for the observed lack of intermediate types.

Iwahori et al. (1969) studied the effects of 2-chloroethylphosphonic acid (Ethrel), ethylene, and some growth retardants on sex expression of cucumber plants (*C. sativus* L.) with the use of a monoecious cultivar (Improved Long Green), which has a strong tendency toward maleness. They reported that ethrel caused increased femaleness when applied at 50 mg/L at the first to the third leaf stage, but when applied at the cotyledon stage, it was ineffective. The later the time of application, the higher the node at which the first female flower appeared. The total number of female flowers was about the same regardless of application time. A mixture of gibberellins A_4 and A_7 caused maleness, and ethrel caused femaleness. However, when applied in combination at the first leaf stage, the interaction was not significant. It seems, therefore, that ethrel and gibberellins are not antagonistic but rather have different sites of action, although they have opposing effects on sex expression. Ethylene caused femaleness but was far less effective than Ethrel. Alar (*N*-dimethylaminosuccinamic acid), *CCC* ((2-chloroethyl)trimethylammonium chloride), phosphon D (2,4-dichlorobenzyl-tributylphosphonium chloride), and abscisic acid did not affect sex expression of cucumber.

Rudich and Halevy (1974) studied the interaction between abscisic acid (ABA), gibberellin (GA), and ethephon on sex determination in cucumber (*C. sativus* L.) flowers. ABA promoted the female tendency of gynoecious plants but did not change the sex expression of monoecious ones. When ABA was applied together with GA_{4+7}, the promoting activity of the GA on male flower formation in the gynoecious line was reduced. ABA also inhibited tendril appearance and internode length, characteristic of GA treatments. A combined ABA and ethephon treatment resulted in a synergistic activity, inhibiting growth and increasing the period of female flower appearance in the monoecious line. It is suggested that ABA participates in the sex regulation of the cucumber by inhibiting GA activity.

Trebitsh et al. (1997) observed that ethylene production, level of 1-aminocyclopropane-1-carboxylic acid (ACC), and activity of the ethylene forming enzyme (EFE) were higher in apices of gynoecious cucumber (*C. sativus* cv. Alma) compared to monoecious cucumber (*C. sativus* cv. Elem). Application of indole-3-acetic acid (IAA) enhanced ethylene and ACC production in both cultivars. The stimulatory effect of IAA was more pronounced in gynoecious apices. The induction of ethylene production and accumulation of ACC resulting from the treatment with IAA were effectively blocked by aminoethoxyvinylglycine (AVG). The content of endogenous IAA, measured by an enzyme immunoassay, was lower in gynoecious cucumber compared to monoecious one. They also observed that the treatment of gynoecious plants with the antiauxins α-(*p*-chlorophenoxy)isobutyric acid (PCIB) and β-naphthaleneacetic acid (β-NAA) did not inhibit female sex expression.

To investigate the action mechanism of ethylene in the induction of femaleness of cucumber flowers, Yamasaki et al. (2000) isolated three ethylene-receptor-related genes, *CS-ETR1*, *CS-ETR2*, and

CS-ERS, from cucumber (*C. sativus* L.) plants. Of these three genes, *CS-ETR2* and *CS-ERS* mRNA accumulated more substantially in the shoot apices of the gynoecious cucumber than those of the monoecious one. Their expression patterns correlated with the expression of the *CS-ACS2* gene and with ethylene evolution in the shoot apices of the two types of cucumber plants. The accumulation of *CS-ETR2* and *CS-ERS* mRNA was significantly elevated by the application of Ethrel, an ethylene-releasing agent, to the shoot apices of monoecious cucumber plants. In contrast, the accumulation of their transcripts was lowered when aminoethoxyvinylglycine (AVG), an inhibitor of ethylene biosynthesis, was applied to the shoot apices of gynoecious cucumber plants. Thus, the expression of *CS-ETR2* and *CS-ERS* is, at least in part, regulated by ethylene. The greater accumulation of *CS-ETR2* and *CS-ERS* mRNA in gynoecious cucumber plants may be due to the higher level of endogenous ethylene, which plays a role in the development of female flowers. Trebitsh et al. (1987) reported that although exogenous IAA enhances ACC and ethylene production, endogenous IAA might not have a major role in the control of sex expression in cucumber.

Galun (1962) inferred that two nongenetic factors, day length and gibberellic acid (GA), may mimic the genetic factors for sex expression. Furthermore, evidence was presented, indicating that this sex controlling ability of GA and the modifying genes may be based on physiological conditions common to those two genetic and nongenetic factors.

The concept that sex expression of cucumber plants may be regulated by a balance between native auxins and gibberellins (Atsmon et al., 1968) is supported by two types of evidence. Applied auxins, especially α-naphthaleneacetic acid, induce femaleness (Ito and Saito, 1956; Galun, 1959), whereas gibberellins induce maleness (Peterson and Anhder, 1960; Sarro and Ito, 1963). Secondly, the main shoots of andromonoecious plants bearing the male flowers contain more extractable auxins than hermaphrodite plants with bisexual flowers. Furthermore, monoecious plants contain more endogenous gibberellin-like substances than gynoecious plants (Atsmon et al., 1968). It was demonstrated that in some plant responses, auxin may exert its effect through ethylene evolution (Shannon and De La Guardia, 1969) and that gibberellins have generally an opposing effect to ethylene (Scott and Leopold, 1967). Since it has long been known that unsaturated hydrocarbons (Minina, 1938; Minina and Tylkina, 1947) induce femaleness in cucumber, it appears that ethylene, the most potent unsaturated hydrocarbon affecting plants (Burg and Burg, 1965), may be the active factor for inducing femaleness.

Stankovic and Prodanovic (2002) studied the effects of foliary-applied solutions of different silver nitrate concentrations and the effects of sowing seasons on sex expression of cucumber. It was observed that the number of male flowers increased with the increase in the silver nitrate (AgNO₃) concentration applied. In both sowing seasons, AgNO₃ reduced female flowers compared with the control.

Hallidri (2004) investigated the effect of AgNO₃ concentration and the number of sprays on sex expression of gynoecious, parthenocarpic cucumbers. The initial sprays were applied at the first true leaf stage of growth, and subsequent treatments were applied at weekly intervals. Treatment effects were evaluated through anthesis at the 10th node. The induction of staminate flowers is dependent upon AgNO₃ concentration and the number of sprays. All treatments using one spray were ineffective in producing staminate flowers. The treatments with 100 ppm failed to induce staminate flowers. The greatest number of staminate nodes was produced on plants sprayed 2× or 3× with 400–500 ppm AgNO₃. Plants showing injury a few days after spraying with 400–500 ppm recovered within 7–10 days. These results demonstrate that commercial seed production of gynoecious hybrids is feasible when the staminate parent is gynoecious, which permits maintaining gynoecious inbreeds.

12.3.3 Nongenetic Factors Controlling Sex Expression in Cucumber

Although sex expression in cucumber plants is determined genetically, it is modified by several environmental factors. High nitrogen, short days, low light intensity, and low night temperature are among the factors that favor femaleness. Reverse conditions tend to cause maleness (Harrison, 1957). Long days, high temperature, and GA promote the formation of male flowers, whereas short days, low temperature, ethylene, and auxin promote the formation of female flowers (Malepszy and Niermirowicz-Szczytt, 1991; Perl-Treves, 1999).

Takahashi and Suge (1980) studied the sex expression in cucumber plants, as affected by mechanical stress, using four cultivars with different genetic backgrounds for sex expression. Mechanical stress to the plant greatly reduced growth and increased the number of pistillate (female) flowers in the monoecious type, but it had no effect on the sex expression of the gynoecious type. The effect of mechanical stress on the growth and sex expression of the monoecious type was nullified by the foliar application of gibberellin A_{4+7}. Silver nitrate also was effective in nullifying the effect of mechanical stress on sex expression but not the effect of growth retardation. Pistillate flowers in a gynoecious strain were reduced by silver nitrate.

Miao et al. (2011) investigated how low temperature alters the sex expression of monoecious cucumbers (*C. sativus* L.). Plants were grown under different day/night temperature regimens, 28°C/18°C (12/12 h), 18°C/12°C, 28°C/12°C, and 28°C/(6 h 18°C + 6 h 12°C). It was found that plant femaleness is highest in the 28°C/(6 h 18°C + 6 h 12°C) treatment.

12.4   SEX EXPRESSION IN MUSKMELON

Muskmelon (*Cucumis melo* L.) is a valuable cash crop grown throughout the world, belonging to the Cucurbitaceae family. Gynoecious plants produce more ethylene than their monoecious counterparts in this.

In melon, Boualem et al. (2009) showed that the transition from monoecy to andromonoecy resulted from a mutation in 1-aminocyclopropane-1-carboxylic acid synthase (ACS) gene, *CmACS-7*.

It has been reported that andromonoecious plants (with male and hermaphrodite flowers) evolve low amounts of ethylene compared to monoecious plants (with male and female flowers) and hermaphrodite flowers (Jaiswal et al., 1985). Ethylene is a plant growth regulator known to alter sex expression in plants belonging to the Cucurbitaceae family, increasing the number of pistillate flowers when applied to monoecious plants (Papadopoulou et al., 2005). Ethylene is biologically active at very low concentrations measured in ppm and ppb ranges. Other molecules with specific configurations can mimic ethylene but are less effective. For example, C_2H_4 analogs propylene (C_3H_6) and acetylene (C_2H_2) and requires 100- and 2700-fold, respectively, of C_2H_4 concentration to elicit the same effect (Abeles et al., 1992). Most plants synthesize small amounts of ethylene that appear to coordinate growth and development. Ethephon, and similar C_2H_4 releasing chemicals, permit the commercial application of C_2H_4 in the field (Thappa et al., 2011). According to findings of the past three decades, researchers have emphasized that higher ethylene content is associated with female sex expression in plants. Exogenous application of plant growth regulators can alter a sex ratio and sequence if applied at a two or four leaf stage, the critical stage at which either suppression or promotion of either sex is possible (Hossain et al., 2006). It has been demonstrated that ethephon treatments with 100 mg/L concentration caused a significantly earlier production of female flowers in cucumber compared with untreated plants (Radwan, 1988). Conversely, ethephon at 300 mg/L decreased the yield of mature fruit and caused stunted growth and a decrease in fresh and dry weights of treated plants. Similarly, in another study, it was observed that increased applications of ethephon increased the number of days to first flowering, 50% flowering, first fruit set, and 50% fruit set. It was found that one ethephon application produced optimum percentage of culls and fruit quality rating since two ethephon applications resulted in a higher percentage of culls and lower rate fruit quality (Shetty and Whener, 2002).

Kenigsbuch and Cohen (1990), in a study, crossed the gynoecious *C. melo* (muskmelon) GY-4 (derived from WI-998 after four generations of selfing and selection) with the monoecious PI124111F or the andromonoecious "36." The F_1 progeny plants from both crosses were all monoecious. The F_2 progeny plants of the cross GY-4 × PI12411 IF (gynoecious × monoecious) segregated 12 monoecious, 3 gynomonoecious or trimonoecious, and 1 gynoecious, indicating a 2 recessive gene difference between gynoecy and monoecy. The backcross to the monoecious parent gave rise to monoecious plants, whereas the backcross to the gynoecious parent segregated two monoecious, one gynomonoecious, and one gynoecious. The F_2 progeny plants of the cross GY-4 × "36" (gynoecious × andromonoecious) segregated 36 monoecious, 12 andromonoecious, 9 gynomonoecious or trimonoecious, 4 hermaphrodite, and 3 gynoecious, indicating a 3 recessive gene difference between the parents. The backcross to the andromonoecious parent segregated one monoecious and one andromonoecious, whereas that to the gynoecious parent segregated one monoecious, one gynomonoecious or trimonoecious, and one gynoecious. The F_2 gynoecious plants from the cross GY-4 × "36" segregated in F_3 into 4 all progeny gynoecious and 10 mixed progeny of gynoecious (¾) and hermaphrodite (¼) plants. The results suggest that the monoecious, andromonoecious, and gynoecious parents carried the genes *AAGGMM*, *aaGGMM*, and *AAggmm*, respectively, for sex expression. The following phenotype–genotype relationships are proposed for sex expression in muskmelon: monoecious, *A-G*; andromonoecious, *aaG*; trimonoecious or gynomonoecious, *AA-ggM*; hemaphrodite, *aagg*; and gynoecious, *A-ggmm*.

REFERENCES

Abeles, F.B., Morgan, P.W., and Saltveit, Jr., M.E. 1992. *Ethylene in Plant Biology.* Academic Press, San Diego, CA.

Atsmon, D. and Galun, E. 1960. A morphogenetic study of staminate, pistillate and hermaphrodite flowers in *Cucumis sativus* L. *Phytomorphology* 10:110–115.

Atsmon, D., Lang, A., and Light, E.N. 1968. Contents and recovery of gibberellins in monoecious and gynoecious cucumber plants. *Plant Physiology* 43:806–810.

Atsmon, D. and Tabbak, C. 1979. Comparative effects of gibberellin, silver nitrate and aminoethoxyvinyl glycine on sexual tendency and ethylene evolution in the cucumber plant (*Cucumis sativus* L.). *Plant and Cell Physiology* 20:1547–1555.

Beyer, E. 1976. Silver ion: A potent antiethylene agent in cucumber and tomato. *HortScience* 11:195–196.

Boualem, A., Troadec, Ch., Kovalski, I., Sari, M.A., Perl-Treves, R., and Bendahmane, A. 2009. A conserved ethylene biosynthesis enzyme leads to andromonoecy in two *Cucumis* species. *PLoS ONE* 4:e6144.

Burg, S.P. and Burg, E.A. 1965. Ethylene action and the ripening of fruits. *Science* 148:1190–1196.

Byers, R.E., Baker, L.R., Dilley, D.R., and Sell, H.M. 1972a. Chemical induction of perfect flowers on a gynoecious line of muskmelon, *Cucumis melo* L. *HortScience* 913:321–331.

Byers, R.E., Baker, L.R., Sell, H.M., Herner, R.C., and Dilley, D.R. 1972b. Ethylene: A natural regulator of sex expression of *Cucumis melo* L. *Proceedings of the National Academy of Sciences of the United States of America* 69:717–720.

Den Nijs, A. and Visser, D. 1980. Induction of male flowering in gynoecious cucumbers (*Cucumis sativus* L.) by silver ions. *Euphytica* 29:237–280.

Frankel, R. and Galun, E. 1977. *Pollination Mechanisms, Reproduction, and Plant Breeding.* Heidelberg, Germany: Springer-Verlag.

Fujieda, K. 1966. A genecological study on the differentiation of sex expression in cucumber plants. *Bulletin of Horticultural Research Station, Japan, Series D* 4:43–86.

Fujita, Y. and Fujieda, K. 1981. Relation between sex expression types and cotyledon etiolation of cucumber in vitro. I. On the role of ethylene evolved from seedlings. *Plant and Cell Physiology* 22:667–674.

Galun, E. 1959. Effects of gibberellic acid and naphthaleneacetic acid on sex expression and some physiological characters in the cucumber plant. *Phyton* 13:1–8.

Galun, E. 1961. Study of the inheritance of sex expression in the cucumber: The interaction of major genes with modifying genetic and non-genetic factors. *Genetica* 32:134–163.

Galun, E., Yung, Y., and Lang, A. 1963. Morphogenesis of floral buds of cucumber cultured in vitro. *Developmental Biology* 6:370–387.

Hallidri, M. 2004. Effect of silver nitrate on induction of staminate flowers in gynoecious cucumber line (*Cucumis sativus* L.). *Acta Horticluture (ISHS)* 637:149–154.

Heslop-Harrison, J. 1957. The experimental modification of sex expression in flowering plants. *Biological Reviews* 32:38–89.

Hossain, D., Karim, M.A., Pramanik, M.H.R., Rahman, A.A.M. 2006. Effect of gibberellic acid (GA3) on flowering and fruit development of bittergourd (*Momordica charantia*). *International Journal of Botany* 2:329–332.

Humming, C., Yun, T., Xiangyang, L., and Xiaohong, L. 2011. Inheritance of two novel subgynoecious genes in cucumber (*Cucumis sativus* L.). *Scientia Horticulturae* 127:464–467.

Iwahori, S., Lyons, J.M., and William, L.S. 1969. Induced femaleness in cucumber by 2-chloroethanephosphonic acid. *Nature* 222:271–272.

Ito, H. and Saito, T. 1956. Factors responsible for the sex expression of Japanese cucumber. III. The role of auxin on the plant growth and sex expression (1). *Journal of the Horticulture Association of Japan* 25:101–110.

Jaiswal, V.S., Kumar, A., and Lal, M. 1985. Role of endogenous phytohormones and some macromolecules in regulation of sex differentiation in flowering plants. *Plant Sciences* 95(6):453–459.

Kahana, A., Silberstein, L., Kessler, N., Goldstein, R.S., and Perl-Treves, R. 1999. Expression of ACC oxidase genes differs among sex genotypes and sex phases in cucumber. *Plant Molecular Biology* 41:517–528.

Kamachi, S., Mizusawa, H., Matsura, S., and Sakai, S. 2000. Expression of two 1-aminocyclopropane-1-carboxylate synthase genes, *CS-ACS1* and *CS-ACS2*, correlated with sex phenotypes in cucumber plants (*Cucumis sativus* L.). *Plant Biotechnology* 17:69–74.

Kamachi, S., Sekimoto, H., Kondo, N., and Sakai, S. 1997. Cloning of a cDNA for a 1-aminocyclopropane-1-carboxylate synthase that is expressed during development of female flowers at the apices of *Cucumis sativus* L. *Plant and Cell Physiology* 38:1197–1206.

Kater, M.M., Franken, J., Carney, K.J., Colombo, L., and Angenent, G.C. 2001. Sex determination in the monoecious species cucumber is confined to specific floral whorls. *Plant Cell* 13:481–493.

Kenigsbuch, D. and Cohen, Y. 1989. The inheritance of gynoecy in muskmelon. *Genome* 33:317–320.

Kubicki, B. 1969. Investigation of sex determination in cucumber (*Cucumis sativus* L.). *Genetica Polonica* 10:69–143.

Malepszy, S. and Niemirowicz-Szczytt, K. 1991. Sex determination in cucumber (*Cucumis sativus*) as a model system for molecular biology. *Plant Science* 80:39–47.

McMurray, A.L. and Miller, C.H. 1968. Cucumber sex expression modified by 2-chlorocthanephosphonic acid. *Science* 162:1397–1398.

Miao, M., Yang, X., Han, X., and Wang, K. 2011. Sugar signalling is involved in the sex expression response of monoecious cucumber to low temperature. *Journal of Experimental Botany* 62:797–804. doi: 10.1093/jxb/erq315.

Mibus, H. and Tatlioglu, T. 2004. Molecular characterization and isolation of the *F/f* gene for femaleness in cucumber (*Cucumis sativus* L.). *Theoretical and Applied Genetics* 109:1669–1676.

Minina, E.G. 1938. On the phenotypical modification of sexual characters in higher plants under the influence of the conditions of nutrition and other external factors. *Proceedings of the USSR Academy of Sciences* (*Doklady*) 21:298–301.

Minina, E.G. and Tylkina, L.G. 1947. Physiological study of the effect of gases upon sex differentiation in plants. *Proceedings of the USSR Academy of Sciences* (*Doklady*) 55:165–168.

Papadopoulou, E. and Grumet, R. 1998. Exogenous brassinosteroids influence sex expression and vegetative development of cucumber (*Cucumis sativus* L.) plants. In: McCreight, J.D. (ed.), *Cucurbitaceae 98: Evaluation and Enhancement of Cucurbit Germplasm*. ASHS Press, Alexandria, VA, pp. 235–240.

Papadopoulou, E., Little, H.A., Hammar, S.A., and Grumet, R. 2005. Effect of modified endogenous ethylene production on sex expression, bisexual flower development and fruit production in melon (*Cucumis melo* L.). *Sexual Plant Reproduction* 18:131–142.

Perl-Treves, R. 1999. Male to female conversion along the cucumber shoot: Approaches to studying sex genes and floral development in *Cucumis sativus*. In: Ainsworth, C.C. (ed.), *Sex Determination in Plants*. Bios Scientific Publishers, Oxford, U.K., pp. 189–215.

Perl-Treves, R., Kahana, A., Rosenman, N., Xiang, Y., and Silberstein, L. 1998. Expression of multiple *AGAMOUS*-like genes in male and female flowers of cucumber (*Cucumis sativus* L.). *Plant and Cell Physiology* 39:701–710.

Peterson, C.E. and Anhder, L.D. 1960. Induction of staminate flowers on gynoecious cucumbers with gibberellic A3. *Science* 131:1673–1674.

Radwan, A.A. 1988. Effect of ethephon on growth, flowering and sex expression of monoecious cucumber. Cario University (Egypt), Faculty of Agriculture, Giza, Egypt, vol. 13, pp. 378–384.

Renner, S.S. and Ricklefs, R.E. 1995. Dioecy and its correlates in the flowering plants. *American Journal of Botany* 82:596–606.

Rudich, J., Baker, L.R., Scott, J.W., and Sell, H.M. 1976. Phenotypic stability and ethylene evolution in androecious cucumber. *Journal of the American Society for Horticultural Science* 101:48–51.

Rudich, J. and Halevy, A.H. 1974. Involvement of abscisic acid in the regulation of sex expression in the cucumber. *Plant and Cell Physiology* 15(4):635–644.

Rudich, J., Halevy, A.H., and Kedar, N. 1972. Ethylene evolution from cucumber plants related to sex expression. *Plant Physiology* 49:998–999.

Saito, Y., Yamasaki, S., Fujii, N., Hagen, G., Guilfoyle, T., and Takahashi, H. 2004. Isolation of cucumber CsARF cDNAs and expression of the corresponding mRNAs during gravity-regulated morphogenesis of cucumber seedlings. *Journal of Experimental Botany* 55:1315–1323.

Sarro, T. and Ito, H. 1963. Factors responsible for the sex expression of Japanese cucumber. XIL. Physiological factors associated with the sex expression of flowers (2). Role of gibberellin. *Journal of the Japanese Society for Horticultural Science* 32:278–290.

Scott, P.C. and Leopold, A.C. 1967. Opposing effects of gibberellin and ethylene. *Plant Physiology* 42:1021–1022.

Shannon, S. and De La Guardia, M. 1969. Sex expression and the production of ethylene induced by auxin in the cucumber (*Cucumis sativus* L.). *Nature* 223:186.

Shetty, N.V. and Wehner, T.C. 2002. Screening the cucumber germplasm collection for fruit yield and quality. *Crop Science* 42:2174–2183.

Shifriss, O. 1961. Sex control in cucumbers. *Journal of Heredity* 52:5–12.

Stankovic, L. and Prodanovic, S. 2002. Silver nitrate effects on sex expression in cucumber. *Acta Horticulture (ISHS)* 579:203–206.

Takahashi, H. and Jaffe, M.J. 1984. Further studies of auxin and ACC induced feminization in the cucumber plant using ethylene inhibitors. *Phyton* 44:81–86.

Takahashi, H. and Suge, H. 1980. Sex expression in cucumber plants as affected by mechanical stress. *Plant and Cell Physiology* 21(2):303–310.

Thappa, M., Kumar, S., and Rafiq, R. 2011. Influence of plant growth regulators on morphological, floral, and yield traits of cucumber (*Cucumis sativus* L.). *Kasetsart Journal—Natural Science* 45:177–188.

Trebitsh, T., Rudich, J., and Riov, J. 1987. Auxin, biosynthesis of ethylene and sex expression in cucumber (*Cucumis sativus*). *Plant Growth Regulation* 5(2):105–113.

Trebitsh, T., Staub, J.E., and O'Neill, S.D. 1997. Identification of a 1-aminocyclopropane-1-carboxylic acid synthase gene linked to the female (F) locus that enhances female sex expression in cucumber. *Plant Physiology* 113(3): 987–995.

Yamasaki, S., Fujii, N., and Takahashi, H. 2000. The ethylene-regulated expression of *CS-ETR2* and *CS-ERS* genes in cucumber plants and their possible involvement with sex expression in flowers. *Plant and Cell Physiology* 41:608–616.

Yamasaki, S., Fujii, N., and Takahashi, H. 2003a. Characterization of ethylene effects on sex determination in cucumber plants. *Sexual Plant Reproduction* 16:103–111.

Yamasaki, S., Fujii, N., and Takahashi, H. 2003b. Photoperiodic regulation of *CS-ACS2, CS-ACS4* and *CS-ERS* gene expression contributes to the femaleness of cucumber flowers through diurnal ethylene production under short day conditions. *Plant Cell and Environment* 26:537–546.

Yamasaki, S., Fujii, N., and Takahashi, H. 2005. Hormonal regulation of sex expression in plants. *Vitamins and Hormones* 72:79–110.

Yin, T. and Quinn, J.A. 1995. Tests of a mechanistic model of one hormone regulating both sexes in *Cucumis sativus* (Cucurbitaceae). *American Journal of Botany* 82:1537–1546.

Section V

Genetics, Genomics, and
Breeding of Cucurbits

13 Muskmelon Genetics, Breeding, and Cultural Practices

B.R. Choudhary and Sudhakar Pandey

CONTENTS

13.1 INTRODUCTION

Muskmelon (*Cucumis melo* L.) is an important horticultural crop across wide areas of the world. Within the genus *Cucumis*, it belongs to the subgenus *melo*. The fruits are used as a dessert that contains 0.6% protein, 0.2% fat, 3.5% carbohydrates, 32 mg calcium, 14 mg phosphorus, 1.4 mg iron, 16 mg carotene, and 26 mg vitamin C per 100 g fresh weight of fruit (Aykroyd, 1963). Seed kernels are also edible, tasty, and nutritious and are rich in oils and energy. Great morphological variation exits in the fruit characteristics such as size, shape, color and texture, taste, and composition; *C. melo* is, therefore, considered the most diverse species of the genus *Cucumis* (Bates and Robinson, 1995). The species comprises feral, wild, and cultivated varieties, the latter including sweet "dessert" melons, as well as nonsweet forms that are consumed raw, pickled, or cooked.

13.2 ORIGIN

Muskmelon (*C. melo* L.) probably originated in North Africa and Asia. Its center of development is believed to be Iran, with a secondary center that includes the northwest provinces of India, Kashmir, and Afghanistan. Based on cytogenetical studies of 43 wild species from south, east, and west Africa consisting of annual and perennials, diploids, tetraploids, and hexaploids, South Africa is considered the primary gene center of the genus. The oldest record of muskmelon is an Egyptian illustration of funerary offerings from 2400 BC in which it appears as "a fruit that some experts have identified as muskmelon, although others are not so sure." Muskmelons spread to the Mediterranean countries several hundred years BC were cultivated by the Romans and finally reached the Greeks by 300 BC Culture apparently spread next from Rome to Spain, and muskmelons were grown in the New World as early as 1535 AD.

The *Cucumis* genus comprises nearly 40 species (Whitaker and Davis, 1962). Muskmelon is a native of tropical Africa more specifically the eastern region, south of the Sahara Desert (Robinson and Whitaker, 1974). India, Persia, China, and southern Russia are considered secondary centers of diversity of muskmelon (Chadha and Lal, 1993).

13.3 BOTANY

Muskmelon (*C. melo* L.; 2n = 2x = 24) belongs to the Cucurbitaceae family and is commonly known as cantaloupe, melon, muskmelon, casaba, or winter melon. It is a dicotyledonous, annual, climbing, or trailing herb.

Habit	Annuals herb; climbing, creeping, and trailing with hairy stem.
Roots	Tap root; extensively branched; superficial.
Stem	Short, ribbed; with fine hairs; weak herbaceous.
Leaves	Alternate, orbicular; ovate to reniform; 3–7 lobed dentate; base cardate; hairy on under surface; petiole 4–8 cm in length.
Inflorescence	Solitary.
Flower	Many sex forms, that is, monoecious, andromonoecious, and gynomonoecious, are present. Andromonoecious is a dominant sex form in muskmelon. Flowers are actinomorphic and epigynous.
Calyx	Sepals five lobed, 5–9 mm in length.
Corolla	Five partite petals, round, 2–3 cm in length fused at the base.
Staminate flower	Staminate flowers born in cluster; 5 stamens; anther dehisce longitudinally.
Pistillate flower	Pistillate flower born solitary, 1.3–3.25 cm in diameter; yellow in color; pedicels with short pistil with 3–5 placentas and stigmata.
Fruit	Fleshy with many seeded pepo; very variable in size and shape; globular, oblong, flat round, smooth, or furrowed ring; glabrous and smooth to rough and reticulate; flesh is pale to deep yellow, yellowish brown or yellowish green, yellowish pink or green, or whitish.
Seeds	Flattened; white or cream in color; smooth; 5–15 cm in length.

13.4 SEX FORMS AND FLORAL BIOLOGY

Melons may be andromonoecious (perfect and staminate flowers separate on same plant), gynoecious (only pistillate flowers), gynomonoecious (perfect and pistillate flowers separate on same plant), hermaphrodite (only perfect flowers), or monoecious (pistillate and staminate flowers separate on same plant). Monoecious and andromonoecious are the most common sex forms; however, the andromonoecious form is predominant. Two major genes control sex expression in melon (A and G) and their genotypes are denoted as A-G- (monoecious), A-gg (gynomonoecious), aaG- (andromonoecious), and aagg (hermaphrodite). More et al. (1980) developed four monoecious lines (M$_1$, M$_2$, M$_3$, and M$_4$) for commercial utilization in heterosis breeding. Later, Choudhary et al. (2013) developed a stable monoecious line of muskmelon (AHMM/BR-8) with the objective to reduce cost of hybrid seed production that has national identity of IC 0599709 registered with Indian Council of Agricultural Research-National Bureau of Plant Genetic Resources (ICAR-NBPGR), New Delhi, as INGR 14043. The use of monoecious lines and genetic male sterility for hybrid seed production in muskmelon will reduce the cost of hybrid seed production because emasculation would not be required. Flowering starts 45 days after sowing. Anthesis takes place at around 5.30 a.m., and the stigma is receptive 12 h before the day of anthesis. The duration of pollen fertility is between 5:00 a.m. and 2:00 p.m., and dehiscence of the anther takes place between 5:00 and 6:00 a.m. Early morning pollination is desirable for maximum fruit setting. It is a strictly cross-pollinated crop, and honey bees are the chief insect pollinators.

13.5 BOTANICAL GROUPS OF MELON

The species *C. melo* is a large polymorphic taxon, encompassing a large number of botanical and horticultural varieties or groups. Naudin (1859) divided the cultivated melon into nine tribes and one wild form, and his work remained the basis of all subsequent studies. Munger and Robinson (1991) proposed a further simplified version of Naudin's taxonomy, dividing *C. melo* into a single wild variety, *C. melo* var. *agrestis*, and six cultivated varieties *cantalupensis*, *inodorus*, *conomon*, *dudaim*, *flexuosus*, and *momordica*. Kirkbride (1993) further divided *C. melo* on the basis of ovary pubescence into two subspecies: *C. melo* subsp. *agrestis* (with appressed hairs on the youngest fruits) and *C. melo* subsp. *melo* (with spreading hairs on the youngest fruits). Munger and Robinson (1991) proposed six subspecific cultivar groups (Spooner et al. 2003) encompassing cultivated and wild types that simplified the scheme described by Whitaker and Davis (1962), which was a condensation of previous classifications. The Munger and Robinson scheme came into use by many melonologists following the publication of the monograph by Robinson and Decker-Walters (1997). Pitrat et al. (2000) proposed a scheme consisting of two subspecies with 16 groups. Recently, Pitrat (2008) refined the 2000 proposal and reduced the number of groups to 15, 5 in subspecies *agrestis* and 10 in subspecies *melo*. In his classification, subsp. *agrestis* includes *acidulous, conomon, momordica* (nonsweet), *makuwa,* and *chinensis* (sweet), while subsp. *melo* includes *chate, flexuosus, tibish* (nonsweet), adana (Paris et al., 2012), *ameri, cantalupensis, chandalak, reticulatus, inodorus* (sweet), and *dudaim* (fragrance). Short descriptions of the botanical groups are as follows:

Botanical Groups	Characteristics
conomon	The fruit is elongated in shape with smooth, thin skin; its flesh nonsweet, nonaromatic, nonclimacteric, and white, firm. Fruits are eaten raw as salad or pickles. It has andromonoecious vines with dark-green foliage. It is cultivated in eastern Asia (China, Korea, and Japan).
makuwa	The plant's foliage is dark green in color; its flower, andromonoecious; and the fruit, flat, round, or oval with smooth thin skin that can be white, yellow, or light green with or without sutures. The fruit flesh is white and sweet with light aroma. Fruits are climacteric and the crop is grown in eastern Asia. Its cultivation is now decreasing.
chinensis	The plant's foliage is dark green in color; its flower is generally andromonoecious but few hermaphrodite accessions have been described. The fruit is pear shaped with a light-green or dark-green skin having spots. The fruit flesh is green or orange with medium sweetness and little or no aroma. Fruits are climacteric and nonclimacteric. This variety is found in China and Korea.
momordica	The plant's foliage is light green in color, with monoecious sex expression. The fruit is flat to round to elongated in form. The fruit rind is thin, smooth, or ribbed. Fruit bursting at maturity is its typical characteristic. The flesh is mealy and white in color at maturity and has mild aroma. Fruits are climacteric and this group is grown only in India.
acidulus	Its sex expression is monoecious. The fruit shape is oval or elliptic, with smooth and green or orange skin color, uniform or with spots. The fruit flesh is very firm and crisp with white color but without sugar and aroma.
tibish	This is one of the most primitive forms of melon, having andromonoecious sex form. The fruit shape is oval; size, small (300 g); and skin, dark green with light or yellow stripes. The fruit flesh is firm, white in color but without sugar and aroma. This is eaten raw in salad as cucumber. A very similar type (*seinat*) is cultivated for seed purposes. *Tibish* and *seinat* are cultivated only in Sudan.
chate	The plant is monoecious and sometimes andromonoecious. Its fruit shape is round to oval, with skin color of light to dark green with ribs. The fruit flesh is white to light orange without sugar and aroma. The fruit is climacteric. It is used as salad and harvested before maturity. It is cultivated in the Mediterranean basin and in western Asia.
flexuosus	Fruits are very elongated, nonsweet, and eaten even in its immature stage as cucumbers. They found in the Middle East and Asia, where similar, less elongated types, *adzhur* and *chate*, have also been reported as ancient vegetable crops (Hammer et al., 1986). Usually, they are monoecious in sex form.
cantalupensis	Its flowering behavior is andromonoecious. The fruit shape is flat to oval, strongly to moderately ribbed, with smooth skin, and sometimes with warts. The fruit flesh is orange in color but sometimes green, sweet, and aromatic. The fruit is climacteric and grown in Europe, western Asia, and North and South America. Dessert melon types such as Ananas, Charentais, Ogen, and Prescott belong to this variety group.
reticulatus	Its sex type is andromonoecious. The fruit shape is round to oval, with typically netted skin, with or without ribs. The fruit flesh is orange, sweet, and aromatic. It is climacteric and grown in Europe, Asia, and North and South America. Topmark, PMR 45, Galia, etc., belong to this variety group.
ameri	The flowering type is andromonoecious. The fruit shape is elongated and oval. The outer rind color is yellow to light green, usually without ribs. The skin can be slightly netted. The fruit flesh is white to light orange in color, juicy, less aromatic, and very sweet. The fruit is climacteric and grown in western and central Asia. Altajskaja, Khatoni, and Kzyl urum belong to this variety group.

(Continued)

Botanical Groups	Characteristics
inodorus	This group is very large and has an andromonoecious sex expression. Fruit shape is round to elliptical and sometimes pointed at the peduncle end. The fruit skin is white to yellow to dark green in color, uniform or with spots, and is often wrinkled, with or without ribs. The flesh is white, sweet, and less aromatic. The fruit is nonclimacteric. They are grown in central Asia, the Mediterranean basin, and North and South America. Piel de Sapo, Rochet, Amarillo (Canari), Casaba, Kirkagac, Yuva, Hasan bey, Tendral, Honeydew, Branco, and Baskavas belong to this variety group.
dudaim	The sex type is andromonoecious. The fruit is round, small (orange size), and yellow in color with ochre strips and a velvety skin. It has very thin white color flesh with strong typical aroma and is not sweet. The fruit is climacteric and this group is cultivated as ornamental or aromatic in central Asia, from Turkey to Afghanistan. Queen Anne's pocket melon, Dastanbou, and PI 177362 belong to this variety group.
adana	Fruits are round to oval and fairly large but with thin, rather dry, mealy flesh of low sugar content. Roman round melons, the melopepones, whose named is after a locality in a fertile plain in southern Turkey, are thought to be of the this variety group.

Horticulturally, the muskmelon and cantaloupe (of the United States) differ somewhat in physical characteristics and regional adaptation. Today, the "cantaloupe" simply refers to cultivars that are highly uniform in overall netting (corky tissue on the rind) with relatively indistinct ribs or vein tracts. Internally, its flesh is thick and salmon orange in color with characteristic flavor, and the seed cavity is small and dry. Muskmelon cultivars, on the other hand, have a stronger aroma, juicier flesh, and larger seed cavity.

Muskmelon (*C. melo* L.)

Kachri (*Cucumis callosus*)

C. melo var. *cantalupensis*

Snapmelon (*C. melo* var. *momordica*)

(*Continued*)

C. melo var. *chate*

C. melo var. *utilissimus*

13.6 BREEDING OBJECTIVES

High yield and uniform fruit shape and size are prerequisites for superior melon varieties and hybrids. An early and tough netted skin having small or negligible seed cavity, a mild, musky flavor, and an attractive outer color are the important traits. Being a dessert fruit, quality parameters, especially total soluble solids (TSSs), flesh thickness with good consistency, texture, and color, should be taken into account. The sugar content should vary from 11% to 13% but not less than 10%. The varieties and hybrid should be resistant to diseases (powdery mildew, downy mildew, mosaic virus) and insect pests (red pumpkin beetle, fruit fly).

13.7 GENETICS

The genetic architecture and the pattern of inheritance of traits are important considerations for determining the most appropriate breeding procedures applicable to any crop. The relevant information with respect to various qualitative and quantitative traits is as follows:

Traits	Mode of Inheritance
Qualitative	
Rind color	Yellow monogenically and partially dominant over cream (Bains and Kang, 1963)
	Green monogenically dominant over yellow (Chadha et al., 1972)
Flesh color	Orange monogenically dominant over light green (Bains and Kang, 1963)
	White monogenically dominant over green (Chadha et al., 1972)
Flesh thickness	Thinner partially dominant over thicker and monogenic (Bains and Kang, 1963)
Juiciness	Less juiciness monogenically dominant over juiciness (Chadha et al., 1972)
Fruit slipping	Slipping monogenically dominant over nonslipping (Chadha et al., 1972)
Sutures	Sutureless monogenically dominant over sutured (Bains and Kang, 1963)
	Absence of sutures digenically dominant over supplementary relation (Chadha et al., 1972)
Resistance to fruit fly	Susceptibility digenically dominant over complementary gene action (Sambandam and Chelliah, 1974)
Resistance to red pumpkin beetle	Resistance monogenically dominant over susceptibility (Vashistha and Choudhury, 1972)
Resistance to red powdery mildew	Resistance dominant over susceptibility (Nath et al., 1973)
	Resistance recessive to susceptibility (Choudhury and Sivakami, 1972)
Quantitative	
Earliness	Dominance variance more than the additive variance, overdominance also being presented (Chadha et al., 1972)
	Earliness partially dominant over lateness (Singh et al., 1976)
	Both additive and dominant genetic variances being significant, dominance variance exceeding additive variance (Singh et al., 1989)

(Continued)

Traits	Mode of Inheritance
Flesh thickness	Partial dominance along with adequate additive genetic variance (Chadha et al., 1972)
	Dominance gene effect (Kalloo and Dixit, 1983)
TSS	Partial dominance accompanied with adequate additive genetic variance (Chadha et al., 1972)
	Dominance gene effect (Kalloo and Dixit, 1983)
	More than 70 groups of genes with overdominance (Swamy and Dutta, 1985)
Number of fruits per vine	Partial dominance accompanied with adequate additive genetic variance (Chadha et al., 1972)
	Dominance gene effect (Kalloo and Dixit, 1983)
Fruit weight	39 genes with geometrically cumulative effects, small fruitedness partially dominant over large fruitedness (Sambandam and Chelliah, 1972)
Total yield	Dominance variance more than additive variance, overdominance also being presented (Chadha et al., 1972)
	Dominance gene effect (Kalloo and Dixit, 1983)
Vine length	Both additive and dominant genetic variances being significant, dominance variance higher than additive variance (Singh et al., 1989)

13.8 BIOTECHNOLOGICAL INTERVENTIONS

13.8.1 Genetic Diversity and Genetic Map

Most of the morphological characters are greatly influenced by environmental factors; therefore, the assessment of genetic diversity based on phenotype has limitations. In contrast, molecular markers based on DNA sequence polymorphism are independent of environmental conditions and show a higher level of polymorphism. This would allow a more efficient utilization of plant characters in developing suitable varieties for yield and quality and tolerance to biotic and abiotic stresses and stability. The genetic map based on morphological markers has been developed in *C. melo* (Garcia et al., 1998; Silberstein et al., 1999). Early melon mapping work employing classical markers was, however, limited to varietal surveys (Esquinas-Alcazar, 1977; Perl-Treves et al., 1985). The first melon genetic map was constructed using 23 markers with 8 linkage groups for disease resistance, reproductive biology, and vegetative characters by Pitrat (1991). Pitrat (2002) provided updated information on mapped gene for melon.

A phenotypic marker alone is not sufficient for the construction of a saturated map due to the low occurrence in genome and their unstable expression. Random Amplified Polymorphic DNA (RAPD) and Restricted Fragment Length Polymorphism (RFLP) molecular marker–based first genetic map of melon was constructed by Baudarcco-Arnas and Pitrat (1996), and the second map with 188 predominantly Amplified Fragment Length Polymorphism (AFLP) markers was developed by Wang et al. (1997). Oliver et al. (2001) mapped melon genome using 400 markers. Recombinant inbred line (RIL) population has been used to develop a high-density composite map with 12 linkage groups (Périn et al., 2002b), using 668 AFLP, inter-microsatellite amplification, and phenotypic markers. Another composite map of melon using Simple Sequence Repeats (SSR) markers was generated by Gonzalo et al. (2005). The mapping population used for such studies generally segregate for one or more than one disease resistance gene in addition to other horticultural traits.

13.8.2 Somaclonal Variations and Genetic Engineering

Genetic diversity in the melon allows the introduction of horticultural and resistant traits from wild to cultivated species/varieties by conventional breeding. Interspecific crosses between *C. melo* and *Cucumis metuliferus* were attempted by Norton and Granberry (1980) and Granberry and Norton (1980) without reproducible results. The viable reproductive embryo has been obtained from the interspecific cross of *Cucumis sativus* and *C. melo*, but complete regeneration of the plant has failed (Lebeda et al., 1996). Somatic hybridization through protoplast fusion may be an alternative for sexual hybridization of cross incompatible species of cucurbits. Somatic hybrid between *C. melo* and *Cucumis myriocarpus* was obtained through mesophyll protoplast fusion (Roig et al., 1986). Somatic hybrid between melon and pumpkin was developed through protoplast fusion, and hybridity was confirmed through isozyme pattern, chromosome number, and shape. The taste of the hybrid was toward melon, indicating that hybridity disappeared in the last stage of growth (Yamaguchi and Shiga, 1993). This technique is very useful to introduce the gene of interest from wild (incompatible to cross) to cultivated melon.

New varieties in melon are being developed using conventional breeding through hybridization and selection after evaluation. The genes of interest available in the other species cannot be introduced in melon due to cross incompatibility. Genetic engineering is a new tool that facilitates to introduce the specific gene from wild. This specific gene integrated as a transgenic plant has been regenerated and tested for its integrity. The economical yield and quality are complex traits and improvement of these traits requires more time. Quality traits are generally negatively correlated with high yield. Viral diseases of melon crop result in heavy yield losses and sometimes total failure of crops. The classical breeding for disease resistance in melon is hampered by the lack of natural sources of durable resistance to some of the most destructive viruses and other diseases. Genetic engineering has been successfully exploited to manipulate virus resistance. A full-length Zucchini yellow mosaic potyvirus (ZYMV) coat protein was introduced into melon plants, and coat proteins were found in transgenic plants expressing apparent immunity to ZYMV infection (Fang and Grumet, 1993). The homozygous transgenic plants of melon had more resistance compared to hemizygous transgenic plants under field conditions introduced with coat protein genes of cucumber mosaic virus (CMV), ZYMV, and WMV-2 (Fuchs et al., 1997). Among cucurbits more transgenic trials are undertaken in melon (54%) for yield and risk assessment (Gaba et al., 2004).

13.8.3 GENERATION OF MOLECULAR LINKAGE MAPS AND TAGGING OF USEFUL GENES

The number of genes for disease resistance and horticultural traits has been mapped in melon. The identification of marker(s) closely linked to the gene of interest is known as gene tagging. This process precedes gene mapping. Gene mapping is the localization of a marker on a chromosome. The identification of markers linked with monogenic traits is based on knowledge of the Mendelian segregation of the trait combined with screening a limited number of samples with a relatively large number of markers. Bulked segregant analysis method is a powerful approach to identify linked markers (Michelmore et al., 1991). In this approach, individuals from a segregating population are pooled on the basis of phenotypic expression; the pools are then fingerprinted to get a sufficient number of markers for both dominant and recessive monogenic traits. Research has been focused mainly on identifying markers linked to disease resistance and quality in melon. *Fusarium oxysporum* f. sp. *melonis* is a devastating disease all over the world. Wechter et al. (1995) tagged *Fom*-2 gene for *Fusarium* wilt disease resistance gene using RAPD molecular markers caused by *F. oxysporum* f. sp. *melonis* in a population MR-1 × AY, and a flanking marker 596 was identified at 2 cM. Later, this marker was converted into a sequence characterized amplified region (SCAR) marker (Wechter et al., 1998; Wang et al., 2000). Two flanked markers AAC/CAT_1 and ACC/CAT_1 were found at 1.7 and 3.3 cM for *Fom*-2 in the population MR-1 × AY (Wang et al., 2000). The same gene was also mapped using the RAPD marker and two flanking markers (E07-1.3 and G17-1.0) identified from the cross of Vedrantais × PI 161375 (Baudarcco-Arnas and Pitrat, 1996). Brotman et al. (2002) mapped the *Fusarium* wilt resistance gene *Fom-1* (races 0 and 2) and identified flanking markers NBS 47-3 and K-1180 at 2.6 and 7.0 cM apart, respectively. Molecular markers are also identified for several virus resistance genes in melon. Morales et al. (2002) mapped gene *nsv*, which is responsible for melon necrotic spot virus and identified flanking markers OPD08-0.80 and CTA/ACG115 and CTA/ACG120 at 4.4 and 1.5 cM. Two flanking markers X15L and M29 were identified at 0.25 and

2.0 cM for gene *nsv* from a population of PI 161375 × Pinyonet Piel de Sapo for melon necrotic spot virus resistance (Morales et al., 2003). Touyama et al. (2000) also mapped the same gene and identified the C2 marker at 2.0 cM. The gene *Zym-1*, conferring the resistance for ZYMV, was mapped in melon using the SSR marker CMAG36 (Danin-Poleg et al., 2002). A gene (*Vat*), conferring the resistance against melon aphid (*Aphis gossypii*), has been mapped, and RFLP flanked markers NSB 2 and AC 39 are identified at 31 and 6.4 cM, respectively (Klingler et al., 2001).

Besides disease resistance, monogenic inherited agronomic traits have been mapped in melon. Danin-Poleg et al. (2002) mapped pH locus that confers flesh acidity utilizing SSR marker and identified flanked marker CMAT141 at 1.7 cM and another marker 269–0.9 at 8.7 cM on the other side. Perin et al. (1998) identified marker EIU5 at 6 cM for *p* gene that confers five carpels in flower. Silberstein et al. (2003) mapped the gene *a* that confers a monoecious phenotype with stamenless flowers, as opposed to hermaphrodite flowers in homozygous recessive state and a flanking RFLP marker CS-DH 21 identified at 7 cM. Other important horticulture traits, that is, nonyellowing and having long shelf life (Touyama et al., 2000) and *ms-3* (GMS gene, Park and Crosby, 2004), were mapped in melon.

13.8.4 Mapping of Quantitative Trait Loci

Several agronomic traits are governed by quantitative trait loci (QTLs). The analysis of flanking molecular markers for QTLs was done to find out the complex inheritance of polygenic traits into Mendelian-like factors that help in selection. RILs, or near isogenic lines (NILs), or any segregating population of wide cross are ideal population for QTL mapping. RILs and NILs are near to homozygous and little effects of environment help in increasing the accuracy to detect the QTL. The gene underlying the QTL may be cloned and identified markers can be used in molecular breeding program. The QTL-controlled horticultural traits can be introduced into desirable parents through molecular marker-assisted selection, just like a Mendelian inherited trait. Perin et al. (2002a) mapped the QTL for fruit length (*fl*) and fruit shape (*fs*) and eight QTLs were reported for fruit shape (*fs1.1, fs2.1, fs8.2*). It was concluded that fruit shape is generally determined during early ovary development stage as QTLs for fruit and ovary shape are cosegregated. QTLs for fruit ethylene production were mapped and four QTLs (eth1.1, eth2.1, eth3.1, and eth111.1) were detected, which were tightly linked to ethylene receptor gene ERS1 (Perin et al., 2002b). Monforte et al. (2004) detected the QTL for earliness, TSS, fruit weight, and fruit shape from the population (PI 161375 × Pinyonet Piel de Sapo). Nine QTLs for earliness (closely linked ea1.1, ea9.2), eight for fruit shape (closely linked *fs1.1, fs9.1, fs11.1*), six for fruit weight (closely linked fw4.1, fw5.2), and five for TSS (closely linked ssc4.1, ssc8.1) were reported. Singh et al. (2015) reported five QTLs for fruit characters, namely, fruit length, fruit weight, number of fruits per plant, and ascorbic acid content in infraspecific mapping population of melon. Three NILs were developed from PI 161375 × Pinyonet Piel de Sapo (Eduardo et al., 2004) to verify the effect of alleles at *fs1.1, fs9.1*, and *fs11.1* on fruit shape. The NILs carrying alleles at *fs1.1* and *fs9.1* (PI 161375) produced more elongated fruits than NILs in a Pinyonet Piel de Sapo background. The alleles at *fs1.1* produce growth only in the longitudinal direction compared to *fs9.1* that cause growth in the transversal direction. QTLs conferring disease resistance in melon have also been identified. More recently, Perchepied et al. (2005) identified the QTLs for *Fusarium* wilt resistance gene. They used Isabelle as the resistant and Védrantais as susceptible parents and mapped the *Fom-1.2* gene (*Fusarium* wilt race 1.2) that confers polygenic partial resistance and detected nine QTLs in five linkage groups.

13.8.5 Map-Based Gene Cloning and Molecular-Assisted Selection

Several disease resistance genes are cloned in muskmelon. The downy mildew resistance in melon was governed by partially dominant complementary genes (*At1* and *At2*) that were cloned (Balaas et al., 1992; Taler et al., 2004). Joobeur et al. (2004) cloned the *Fom-2* gene conferring résistance against *Fusarium* wilt in melon. The colonization of melon/cotton aphid was inhibited by *Vat*, and this process resists the transmission of unrelated CMV and potyviruses. Pauquet et al. (2004) reported that the cloned *Vat* gene (6 kb in size with 3 introns and encodes a protein of 1473 amino acids) belongs to the coiled-coil NBS/LRR class of plant disease resistance genes. They also validated and confirmed the function of the *Vat* gene using susceptible melon varieties. The cloned gene conferring the resistance can be introduced to susceptible lines. The introduction of resistance genes should be through transformation or traditional breeding aided by MAS. The efficiency of MAS can be increased by converting them to SCARs or sequence-tagged sites. Wang et al. (2000) used the *Fom-2* marker for screening 45 melon genotypes for *Fusarium* wilt resistance. These markers correctly predicted disease phenotypes in the population. Molecular-assisted selection is successfully demonstrated for phenotypic screening for *Fom-2* gene (*Fusarium* wilt disease resistance in melon) linked markers (Wang et al., 2000; Burger et al., 2003).

The melon genome was compared with *Arabidopsis thaliana*, and the microsynteny between both has been reported; therefore, *A. thaliana* ESTs can be readily adapted for genomic research. A number of known genes responsible for disease resistance and horticultural traits are mapped and are being utilized for MAS. The focus should be on new genes for fruit quality and disease resistance.

13.9 BREEDING METHODS

Usually, inbreeding depression has not been reported in melons; therefore, melons can be handled as self-pollinated crop for breeding purposes through judicious application of selfing and selection for desirable traits.

13.9.1 PEDIGREE METHOD

This method is used to develop genotypes by crossing parental lines having complementary traits followed by single plant selection of desirable types in segregating generations F_2, F_3, etc., up to 5–6 generations till homozygosity is attained. For characters with a simple inheritance, selection is efficient in the first generations (F_2 or F_3). For characters with more complex inheritance, such as fruit quality, earliness, or yield, selection is more efficient if applied in more advanced generations (F_4 or F_5).

13.9.2 BACKCROSS METHOD

Backcross breeding is used to transfer one qualitative (highly heritable) trait, namely, powdery mildew resistance, downy mildew resistance, and nematode resistance into an otherwise superior inbred, which is referred to as the recurrent parent. Backcrossing is simple when the trait to be transferred is governed by a single dominant gene but becomes complex in case of one or more recessive genes. Often six generations of selection and backcrossing to the recurrent parent are required to recover the desired genotype (recurrent parent with the additional trait) and eliminate the undesirable traits inherited from the nonrecurrent (donor) parent. Using this breeding approach, powdery mildew resistant lines have been evolved (Whitaker, 1979).

13.9.3 HETEROSIS BREEDING

Muskmelon has the advantage of producing large number of seeds in a single cross-coupled with lesser requirement of seed for planting per unit area. Munger (1942) was the first to report heterosis in muskmelon and observed 30% higher yield and uniform fruits. Despite the lack of inbreeding depression in muskmelon, heterosis has been observed for earliness, fruit size, fruit weight, flesh thickness, TSSs, fruit flavor, transportability, and total yield by various researchers. The hybrids generally showed heterosis for earliness (Gaurav et al., 2000; Choudhary et al., 2003). Significant and desirable heterosis for flesh thickness has been observed by Chadha and Nandpuri (1980), Dhaliwal and Lal (1996), and Choudhary et al. (2003).

Resistance against certain specific diseases can also be transmitted to the hybrids. Foster (1960) transferred crown blight resistance to F_1 hybrids. Resistance had been reported to be dominant (Dhillon and Wehner, 1991), which might be utilized in the breeding program. Kesavan and More (1987, 1991) utilized three monoecious lines (M_1, M_2, and M_5) in a hybrid program and identified four promising hybrids resistant to powdery mildew along with desirable horticultural characters.

13.10 MECHANISMS OF HYBRID PRODUCTION

Heterosis breeding has been extensively explored and utilized in muskmelon; being a cross-pollinated crop and having its ability to produce plenty of seeds per fruit facilitate heterosis breeding and make it economically viable. The presence of different pollination mechanisms like monoecy, gynoecy, and genetic male sterility has made possible phenomenal progress in the exploitation of F_1 hybrids (Choudhary and Pandey, 2010).

13.10.1 HAND EMASCULATION AND HAND POLLINATION

If the female parent is andromonoecious, then hand emasculation and hand pollination are essential. The seed parent and pollen parent are grown in separate plots. Emasculation should be done just before the anthesis of selected buds on the seed parent 1 day before in the evening, and it should be bagged by a cotton layer or butter paper.

13.10.2 USE OF MONOECIOUS LINES

Generally, muskmelon is andromonoecious but some genotypes have monoecious sex forms. Monoecism has been assessed for its use in heterosis breeding and excludes emasculation (More et al., 1980; Kesavan and More, 1991). They also advocated the use of monoecious lines in hybrid seed production that reduces the time required by 50% and enhances fruit set by 40%–70% compared to 5%–10% in andromonoecious parent. Monoecious character is controlled by a single pair of recessive genes. Considering the problems associated with the use of male sterile gene(s) for production of muskmelon hybrids (maintenance of single recessive gene due to heterozygous condition and problem of identifying and rouging of 50% male fertile plants from the female row at the time of flowering) and also to overcome the problems of andromonoecious sex form, More et al. (1980) developed four monoecious lines (M_1, M_2, M_3, and M_4) for commercial utilization in heterosis breeding. The main problem in using monoecious lines in heterosis breeding is the undesirable linkage between genes controlling sex expression and fruit shape (Risser, 1984). Due to this limitation, F_1 hybrids with round fruits cannot be obtained. This problem is also found in Pusa Rasraj; hence, this hybrid is not accepted commercially due to its undesirable fruit shape and poor external appearance (Sandha and Lal, 1999).

In this method, male buds of monoecious parental line are removed before they open and only female flowers are left on the plant for pollination by insects. The production of staminate flowers on monoecious plants can also be temporarily suppressed by repeated spray of ethephon (2-CEPA) 250 ppm at the two leaf stage, four leaf stage, and again once before flowering. Male and female parents should be planted (1 male/4 female) side by side in adjacent rows in isolation and allowed for free pollination by honey bees. F_1 seeds are extracted from female parents only.

13.10.3 USE OF GYNOECIOUS LINES

Gynoecious lines produce only female flowers and ensure 100% pure hybrid seed production. Peterson et al. (1983) was the first to develop a gynoecious line in muskmelon from the progeny of gynomonoecious lines, that is, monoecious × hermaphrodite cross. Such lines can be maintained through the stimulation of male flowers by chemicals. Silver compounds like silver thiosulfate at 300–400 ppm (More and Seshadri, 1987) have been used successfully to induce perfect flowers in gynoecious lines for their maintenance. The chief advantage of using gynoecious lines over *ms* lines in heterosis breeding is that the tedious job of identification and rouging of 50% male fertile plants from the mixed population is avoided. Therefore, Loy et al. (1979) advocated the use of gynoecious lines in hybrid seed production. Dhaliwal and Lal (1996) evaluated a large number of crosses involving monoecism, gynoecism, and male sterility in the female parent. The cross W-321 × N-233 was out-yielded in comparison to Punjab Hybrid (check) by more than 50% and was also significantly earlier. Thus, the hybrid W-321 × N-233 has good potential for commercial exploitation. It has been observed that gynoecious lines are poor in TSS, so it is essential to improve the TSS level of available gynoecious lines through selection for the production of better quality hybrids. The instability of gynoecious lines at high temperature also poses a problem in their utilization.

13.10.4 USE OF GENETIC MALE STERILITY

The use of male sterile line as female parent is an economical method of hybrid seed production. Seeds of male sterile line are sown at double seed rate, and 50% male fertile plants are identified immediately after flowering and removed from the field. So far, seedling marker linked with any of the male sterile gene has not been reported that could facilitate the early identification and removal of male fertile plants. Five male sterile genes (ms-1, ms-2, ms-3, ms-4, and ms-5) have been identified in muskmelon, and all of them are recessive and nonallelic. It has been observed that male sterile plants in ms-1 and ms-2 progenies are difficult to identify as the aberrant flowers are observed on genetically fertile siblings, and thus the expression of these genes is unstable (McCreight, 1984), which could lead to genetic impurity in F_1 hybrid seed. In India, the male sterile gene ms-1 was introduced in 1978 and used to release two commercial cultivars Punjab Hybrid (Nandpuri et al., 1982) and Punjab Anmol (Lal et al., 2007). Due to the instability of this ms-1 gene in subtropical conditions of India, the seed production of these hybrids has posed numerous problems consistently (Dhatt and Gill, 2000). Kumar and Dhillon (2008) reported that the lines ms-1 and ms-2 are phenotypically unstable, and this will reduce the genetic purity of hybrid seed. It is safer to exploit ms-3, ms-4, and ms-5 genes for developing genetically pure hybrids.

13.11 SALIENT BREEDING ACHIEVEMENTS

In India, many high-yielding varieties and hybrids of muskmelon have been evolved:

Pusa Madhuras: It has been developed at Indian Council of Agricultural Research-Indian Agricultural Research Institute (ICAR-IARI), New Delhi. Fruits are round in shape, salmon flesh in color, and juicy and sweet (12%–14% TSS). It is moderately resistant to *Fusarium* wilt.

Pusa Sharbati: This has been developed at ICAR-IARI, New Delhi, from a cross of Kutana × PMR-6 of the United States. This fruit is round with netted skin, and its flesh is thick and orange in color. The TSS content is 11%–12% and the yield potential is 15 t/ha.

Pusa Rasraj: A monoecious line M_3 developed at ICAR-IARI, New Delhi, and crossed with Durgapura Madhu gave rise to the F_1 hybrid, Pusa Rasraj. Fruits are oblong in shape, with 11%–12% TSS content.

Arka Rajhans: It is a selection from local collection (IIHR-107) of Rajasthan at ICAR–Indian Institute of Horticultural Research (IIHR), Bengaluru. Fruits are round and slightly oval in shape and medium to large in size with white and firm flesh color and 11%–14% TSS content. The average fruit weight is 1.25–2 kg. It is moderately resistant to powdery mildew.

Arka Jeet: It is a selection from "Bati" strain of Lucknow (UP), which has been developed at ICAR-IIHR, Bengaluru. Fruits are small, flat, and round, with an attractive orange rind, weighing 300–500 g. Flesh is white and sweet with medium soft texture.

Kashi Madhu: It has been developed at ICAR-Indian Institute of Vegetable Research, Varanasi. This fruit shape is round, with open prominent green sutures and weight of 650–725 g. It has half slip in nature, with thin rind. The fruit is smooth and pale yellow at maturity. The flesh is salmon orange (mango color), thick, and very juicy. Its TSS content is 13%–14% and the seeds are loosely packed in the seed cavity. It is tolerant to powdery and downy mildew and of medium size at maturity and yields 200–270 q/ha.

Hara Madhu: This variety has been developed at Punjab Agricultural University (PAU), Ludhiana, from the local material of Kutana type (a local collection of UP). Vines are 3–4 m long and vigorous. The fruits are large, round, and slightly tapering toward the stalk end. There are 10 prominent green sutures and the average fruit weight is 1 kg. The flesh is green with small seed cavity and 12%–15% TSS content. The yield potential is 12 t/ha.

Punjab Sunehri: It is a selection from the cross between Hara Madhu and Edisto developed at PAU, Ludhiana. Fruits are round to elliptical, devoid of sutures, netted, and thick skinned. The flesh is salmon orange, thick, and sweet (11% TSS). Its shelf life and transport quality are excellent. The average yield is 16 t/ha. It is moderately resistant to *Fusarium* wilt.

Punjab Hybrid: This is an F_1 hybrid developed at PAU, Ludhiana, having its parentage as male sterile line (*ms-1*) × Hara Madhu. It has 2–2.5 m long vines of vigorous luxuriant growth and bears globular fruits (weighing about 800 g) with distinct sutures. The flesh is creamy yellow. Its rind is netted. The TSS content is about 12%. It has early maturity and good postharvest life and transportability. It is moderately resistant to powdery mildew.

Durgapura Madhu: This is a very early cultivar with a pale-green rind and an oblong shape. It has a light-green flesh with dry texture and is developed at Agricultural Research Station (ARS), Durgapura (Rajasthan). The fruits are very sweet (13%–14% TSS) and resistant to *Fusarium* wilt.

RM 43: The plant is developed at ARS, Durgapura (Rajasthan). Its fruits are oblong, with a small seed cavity and TSS content of 12%–14%. It is of good keeping quality and transportability. These are field resistant to powdery mildew and root rot.

RM-50: The plant is developed through hybridization (Durgapura Madhu × Sel-1), followed by pedigree selection at ARS, Durgapura (Rajasthan). Its fruits are oval in shape, with a pale-green rind having 10 green sutures. The fruit flesh is green with 14%–16% TSS content. The average fruit weight is 500 g. Its yield potential is 17–20 t/ha.

MHY-3: It is a selection from a cross between Durgapura Madhu and Pusa Madhuras developed at ARS, Durgapura (Rajasthan). Fruits are flattened and round with a light-green flesh and 13%–15% TSS content. It is moderately resistant against powdery mildew, downy mildew, root rot, and virus under field conditions.

MHY-5: It is developed through hybridization (Durgapura Madhu × Hara Madhu), followed by pedigree selection at ARS, Durgapura (Rajasthan). The fruits have a smooth rind and medium seed cavity and are roundish flat and tapering at the end. The rind is greenish yellow in color. The flesh is light green; texture, soft; and TSS content, 13%–16%. The average fruit weight is 700–800 g. Its yield potential is 18–20 t/ha.

13.12 CLIMATE AND SOIL

Muskmelon flourishes well under warm climate and cannot tolerate frost. Optimum germination temperature is 25°C–28°C during day and night temperature not lower than 18°C. Optimum growth temperatures at night are 18°C–20°C, during the day 24°C–30°C, and for ripening 15°C–25°C.

The muskmelon thrives best and develops the highest flavor in a hot dry climate. Dry weather with clear sunshine during ripening ensures a high sugar content, better flavor, and high percentage of marketable fruits. High humidity increases the incidence of diseases, particularly those affecting foliage. Continuous rain or cloudy weather will not only stunt the plant growth but also reduce flowering and fruit setting. Short day length and low-temperature conditions favor the expression of perfect/female flowers.

A well-drained, loamy soil is ideal. When sandy soils are used, they should be supplemented with humus or compost. Poorly drained soils should be avoided. Lighter soils that warm up quickly in spring are usually utilized for early crop. In heavier soils, vine growth is more and fruit maturity is delayed. On sandy riverbeds, alluvial substrata and subterranean moisture of river streams support its growth. In riverbeds, its long tap root system is adapted to growth. Muskmelon is sensitive to acid soils. The crop cannot be grown successfully below a pH of 5.5. A soil pH between 6.0 and 7.0 is ideal. Muskmelon should not be grown on the same soil year after year because of *Fusarium* and nematode problem.

13.13 CULTIVATION

13.13.1 SOWING

Muskmelon is propagated through seeds. The optimum seed rate is 2.0–2.5 kg/ha; however, for sowing hybrids, about a one-third seed rate is sufficient. Sowing of muskmelon is done in the summer season (mid-February to mid-March). It can also be grown during the winter season (December–January) using riverbed cultivation system to fetch a high price from off-season crop. Muskmelon can be directly seeded to the field or grown as transplants in polybags/pro-trays and then transplanted to the field. Prior to sowing, the seeds must be treated with Captan or Thiram at 2 g/kg of seed. Seeds should be soaked in water overnight and kept in moist gunny bags for 2–3 days in warm place to initiate germination. The soil surface should be smooth, loose, friable, and free from clods at the time of sowing. Generally, 2–3 pregerminated seeds/hill are sown at 2–2.5 cm depth. After sowing, the seeds should be covered with a thin layer of well-rotten fine Farm Yard Manure (FYM). At the time of sowing, the soil must have sufficient moisture for better germination, so the channels should be irrigated 2 days before seed sowing. The important methods for sowing muskmelon are as follows:

1. *Shallow pit or flat bed method*: Shallow pits of 45 × 45 × 45 cm size are dug and left open for 3 weeks before sowing for partial soil solarization. Each pit is filled with a mixture of soil and 4 to 5 kg of FYM or compost, 30–40 g urea, 40–50 g single super phosphate, and 80–100 g muriate of potash. After filling, the pit circular basins are made and 2–3 seeds are sown per basin at 2–2.5 cm deep and covered with fine soil, FYM, or compost.

2. *Deep pit or trench method*: The deep pit method is commonly practiced for raising muskmelon in riverbeds. In this method, circular pits of 60–75 cm diameter and 1.5 m depth are dug at a recommended distance or about 60 cm wide trenches are made at a distance of about 2.5 m across the slope up to the depth of the clay layer. These pits or trenches are filled with a mixture of top soil, well-rotten FYM, and recommended NPK. The pregerminated seeds are sown in trenches 60–90 cm apart.

3. *Sowing of seeds on ridges*: In this method, 50–60 cm wide channels are prepared manually or mechanically, maintaining 2.0–2.5 m distance between two channels, depending upon cultivars. Seed is sown on both ridges of the channel at a spacing of about 60–90 cm. Generally, two to three sprouted seeds are sown per hill in spring–summer and adequate moisture is maintained at the time of emergence. The vines are allowed to spread in between space of channels.

13.13.2 Nutrition

Muskmelon is grown best in soils that have high organic matter content. The doses of fertilizers depend on nutritional status of the soil, climate, and variety to be grown. Pre plant soil analysis should be done at least 6 months prior to planting, as this forms the basis for planning the fertilizer program. The crop is also sensitive to low availability of Mg and the micronutrients Fe, Mo, and B. The soils should be analyzed for electrical conductivity (EC), pH, and Na, P, K, Ca, Mg, Zn, S, B, and Mo micronutrient content. Melons are reasonably sensitive to salt, displaying a 50% yield reduction in the range of EC 4–6 (mMhos/cm at 25°C). Apply well-decomposed FYM at 20–25 t/ha in the furrows or pits made for sowing and mix thoroughly. This is supplemented by half dose of N (75–100 kg/ha) and full dose each of P_2O_5 (60–80 kg/ha) and K_2O (40–50 kg/ha). The remaining nitrogen is divided in to two equal parts and applied at the time of vine initiation and 10–15 days later. Foliar sprays of boron 25 ppm along with urea 1% at 2–3 times beginning from 2–4 true leaf stage to flower initiation increase the number of female flowers and yield. The complete doses of fertilizers should be applied before fruit set. Excessive application of nitrogen should be avoided as it results in more number of male flowers, which affect the fruit yield adversely. While applying fertilizers, care must be taken not to cause damage to roots, and there should be sufficient moisture in the soil. Irrigation should be supplied shortly after applying fertilizers to move them toward the roots.

13.13.3 Aftercare

Thinning of plants should be done 10–15 days after sowing, retaining only two healthy seedlings per hill. Two or three hoeing may be done during the early stage of growth to keep down the weeds and to conserve soil moisture. First weeding should be done at 30 days after sowing and subsequent weeding is done at an interval of one month followed by hoeing and earthing up. Deep hoeing should be avoided as it may destroy many of the fine roots near the soil surface. The application of herbicides like Fluchloralin or Trifluralin (0.75–1.5 kg/ha) as preplant soil at two weeks before sowing and incorporation of Butachlor (1 kg/ha) as postemergence after first weeding proved better in controlling the weeds. Mulching is also found effective for controlling weeds. Mulching the bed surface with rice straw or sugarcane leaves not only retains the soil moisture, prevents nutrient leaching, and improves soil aeration but also controls the weeds and provides support for tendrils. The weeds can also be controlled effectively by the use of biodegradable plastic mulch. Mulching with silver-colored ultraviolet reflective plastic mulch reflects thrips and aphids, hence reducing the incidence of viral diseases.

13.13.4 Pruning and Training

Removal of all secondary growth up to the seventh node in Hara Madhu, third node in Punjab Sunehri, fourth node in Punjab Hybrid and Pusa Madhuras, and sixth node in Pusa Sharbati has been reported to enhance fruit yield in all varieties (Mangal and Pandita, 1979) when compared with unpruned controls. Under greenhouse conditions, muskmelon plants are trained upward so that the main stem of plants is allowed to climb to the overhead wires along with a polythene twine. The side branches are pruned up to 45–60 cm above the bed surface, and then they are pruned only after leaving one or two fruit buds (female flowers) that will bear the fruits.

13.13.5 Irrigation

Irrigation should be done at critical stages, namely, vine development, preflowering, flowering, and fruit development periods to get a high yield. When the female flower appears, it is advocated to withhold water supply in order to improve the fruit setting. When fruits start to develop, irrigation should be applied to produce good size of fruits. During ripening, excess soil moisture reduces the sugar content and adversely affects the development of flavor. Therefore, irrigation should be completely stopped during fruit ripening. Irrigation in furrows needs more quantity of irrigation water that also frequently wets the vines or vegetative parts and promotes the incidence of diseases. Among the irrigation systems, drip irrigation is the most efficient method for muskmelon cultivation. The single lateral lines (12–16 mm size) with in-line or on-line drippers of 2–4 L/h discharge capacity can successfully be used. Crops sown in riverbeds do not need any irrigation unless the soil is too dry. The crops should be irrigated at the following three stages:

1. *Sowing to emergence*: Irrigate with plain water to field capacity, to a depth of at least 1 m before sowing/transplanting. Keep the soil profile at field capacity until seedlings have emerged or roots are growing strongly from the seedling module.
2. *Emergence to first fruit set*: Plants should be watered more heavily at a lower frequency prior to fruit set. Allow plants to get a little stressed in order to induce deep root growth. When areas of stressed plants develop in the field at midday, apply water.
3. *First fruit set to harvest*: During fruit enlargement, irrigation should be frequent and light. Irrigation should be reduced or stopped 7–10 days prior to harvest.

13.13.6 CROP REGULATION

Excessive vine growth due to high nitrogen nutrition coupled with high temperature and high soil moisture condition promotes staminate flowers in the vine, resulting in low fruit set and low yield. The best way to control vine growth within reasonable limits is by adjusting nitrogen fertilizer doses and frequency of irrigation. Exogenous application of 300–400 ppm silver thiosulfate twice, first at two true leaves of the plants, induces the staminate flower in the gynoecious line that enables maintenance of the gynoecious line in pure form (More and Seshadri, 1987).

13.13.7 OFF-SEASON CULTIVATION

It is a kind of vegetable forcing where muskmelon is sown in December–January under plastic low tunnels to get early crop. Low tunnels are installed over the rows of directly sown muskmelons; transplanted muskmelon enhances plant growth by warming the temperature around the plants during the winter season. These tunnels are quite cost-effective for the growers in the northern parts of the country, where the night temperature during the winter season falls below 8°C for a period of 30–40 days. Low tunnels offer the advantage of crop advancement from 30 to 40 days over their normal season of cultivation and ultimately high prices of the produce.

Seedlings are raised in the greenhouse in plastic protrays having 1.5 in. cell size in soil-less media during December–January. A nursery can be raised even in polythene bags under very simple and low-cost protected structures. About 30–32-day-old seedlings at four true leaf stages are transplanted with a ball of earth in the open field in January and covered with plastic low tunnels. When the seedlings are transplanted at the end of February, there is no need to cover with plastic low tunnels. Transplanting should be done during evening hours followed by irrigation. Direct sowing of seed under plastic low tunnels can also be done beginning from mid-December to mid-January.

Transplanting or sowing is done in rows at a recommended distance using drip system of irrigation. The flexible galvanized iron hoops are fixed manually at a distance of 2.0–2.5 m. The width of two ends of hoop is kept 1 m, with a height of 60–70 cm above the level of ground for covering the plastic on the rows or beds for making low tunnels. Transparent, 30 μm biodegradable plastic is generally used for making low tunnels that reflects infrared radiation to keep the temperature of the low tunnels higher than outside. The 3–4 cm size vents are made on the eastern side of the tunnels just below the top at a distance of 2.5–3.0 m when temperature increases within the tunnels during the peak day time. These vents also facilitate pollination of crop by visiting bees. Plastic is completely removed from the plants in February–March, depending upon the prevailing night temperature in the area.

13.13.8 HARVESTING AND YIELD

Muskmelon is a climacteric fruit that ripens during transit and storage. Hence, it is harvested before it is fully ripe so that it will reach the consumer at fully ripe stage. Muskmelon fruits will be ready for picking in about 90–110 days depending upon the variety and agroclimate. The stage at which the fruits are picked also affects the quality of fruits; hence, they should be picked in time. The stage of maturity is generally judged by the change in the external color of fruit, softening of the rind, development of the abscission layer (the stem begins to separate or slip from the fruit), and strong odor. If left long enough, the stem will naturally separate from the fruit, which is called "full slip." Fruit at this stage should be used within 36–48 h as it will spoil soon. For better quality, harvest fruits at the "half slip" stage when the stem is partially separated from the fruit. Do not harvest melons too early because the sugar content does not increase after harvest. Muskmelon can improve in flavor after harvest, but this is caused by the mellowing of the flesh. Yield varies with the varieties used; however, on an average 15–20 t/ha fruit yield can be harvested under good management practices.

13.14 PHYSIOLOGICAL AND NUTRITIONAL DISORDERS

13.14.1 DROUGHT STRESS

Cucurbits are particularly sensitive to drought. Fruits typically contain 85%–90% water and can suffer under drought conditions. Muskmelon often produces many large-sized leaves and can transpire large quantities of water during hot summer days. Severe drought stress affects fruit development, resulting in unmarketable produce. The fruits become soft and wrinkled and fail to gain appropriate size. A loss of foliage during drought will also result in sunburn of the fruit.

Management: Avoid moisture stress during critical stages of crop growth, namely, vine development, preflowering, flowering, and fruit development stages. Protect the crop from hot winds during the summer season.

13.14.2 MAGNESIUM DEFICIENCY

It is more likely to occur on sandy soils with a low pH, especially in dry years. Sandy soils often have a low cation exchange capacity and may not contain adequate levels of magnesium. Deficiency symptoms are more commonly observed in muskmelon than in other cucurbits. Symptoms first appear as a yellowing between the leaf veins (interveinal chlorosis), beginning with the oldest leaves and slowly spreading to newer growth. Yellowed tissues may turn brown, die, and drop out, giving the leaf a shot-hole pattern. Magnesium deficiency usually appears during periods of rapid growth, when the fruit is enlarging.

Management: Maintain the soil pH near 6.5. Potential sources of preplant magnesium include magnesium oxide and dolomitic lime. If necessary, fertigate Epsom salts (magnesium sulfate) and magnesium oxide through a drip irrigation system. Avoid heavy applications of fertilizers containing competing cations (K^+, Ca^{2+}, NH_4^{2+}). Foliar sprays are generally ineffective in correcting significant deficiencies.

13.14.3 MANGANESE TOXICITY

Symptoms include water-soaked areas on the underside of leaves and yellow or bronzed spots on the upper leaf surface. Although manganese is an essential plant micronutrient, high levels of it can lead to toxicity symptoms in cucurbits. Manganese toxicity is generally the result of a low soil pH that allows manganese to become available to plants in toxic levels.

Management: Check the fall in the soil pH prior to planting; if it is below 6.0, apply lime in the fall and disk in.

13.14.4 MOLYBDENUM DEFICIENCY

Muskmelon is affected when grown on dark heavy soils with a pH below 6.0. Heavy application of ammonium nitrate through drip irrigation may lower the pH in the plant root zone and contribute to either manganese toxicity or molybdenum deficiency. Molybdenum deficiency usually is seen in the crown leaves when the plants begin to vine. Leaves become pale green to slightly chlorotic between the veins. As symptoms progress, the leaf margins become necrotic and plant growth ceases.

Management: Maintain the soil pH between 6.0 and 6.5; foliar treatments with sodium molybdate will help alleviate symptoms and permit normal growth.

13.14.5 NITROGEN DEFICIENCY

Nitrogen deficiency generally appears as a yellowing of older foliage on plants. Nitrogen is the most abundant nutrient in the plant and often the most limiting nutrient for plant growth. Although cucurbits are not particularly heavy nitrogen feeders, they can experience nitrogen deficiencies during periods of rapid growth or fruit set.

Management: Broadcast the 1/3–1/2 of the total nitrogen requirement prior to forming beds and fertigate the remainder throughout the season if the crop is grown on drip irrigation and plastic mulch. When drip irrigation and black plastic is not used, the remaining nitrogen can be banded in one or two side-dressings prior to fruit formation.

13.14.6 BLOSSOM END ROT

It typically appears as a general rot at the blossom end of the developing fruit. Blossom end rot is usually the result of inadequate or uneven irrigation, high humidity, or other factors that slow the movement of water through the plant. Since calcium is taken into the plant with the transpiration stream (water), slow water movement can often lead to temporary calcium deficiencies, resulting in blossom end rot.

Management: Provide adequate calcium fertility and proper irrigation. Do not use high levels of ammonia fertilizer that can aggravate this problem. Avoid root injury.

13.14.7 POOR POLLINATION

Being a cross-pollinated crop, muskmelon requires pollination to produce fruit. Several visits from pollinators on the day a flower is open are often required to ensure appropriate fruit development. Many fruits will appear misshapen and small when pollination is poor. Very high and low temperatures can also affect pollen viability, resulting in poor pollination. If too much nitrogen is used (resulting in excessive vegetative growth) or plants were improperly spaced, bees may have difficulty in locating the flowers.

Management: Provide pollinators to ensure good fruit set and high yields. Place one or two beehives per hectare under normal conditions in the field when 5%–10% plants have open flowers. Beehives should be placed in clusters around the periphery of the field, with additional hives inside the larger fields. Do not spray insecticides during morning hours when flowers are open and insects are actively pollinating plants.

13.14.8 WIND DAMAGE/SANDBLASTING

High winds often cause stem damage and drying of transplants, particularly on the area of the stem facing prevailing winds. Excessive winds will desiccate leaves, causing them to die from the margins toward the center. Entire fields can be affected, leading to significant losses.

Management: Employ windbreaks along fields and avoid transplanting in high winds whenever possible.

13.15 DISEASES

13.15.1 POWDERY MILDEW (*ERYSIPHE CICHORACEARUM* AND *SPHAEROTHECA FULIGINEA*)

Tiny white to dirty grey spots on the foliage, leaves, and green stem appear. Later, these spots become powdery and enlarge into patches. Under severe infection, the fruits may also be covered with powdery mass. The humid weather is favorable for the spread of this disease.

Control: Grow resistant varieties and spray the crop with Karathane at 0.1% 2–3 times at an interval of 10–15 days just after the appearance of the disease.

13.15.2 DOWNY MILDEW (*PSEUDOPERONOSPORA CUBENSIS*)

Angular yellow-colored spots often restricted by the veins appear on the upper surface of leaves giving purplish growth on the lower surface. The affected leaves die quickly.

Control: Give hot water treatment to seeds before sowing (55°C for 15 min). Follow prophylactic spray of Dithane M-45 at 0.2% at a 15-day interval.

13.15.3 ANTHRACNOSE (*COLLETOTRICHUM LAGENARIUM*)

Reddish brown spots are formed on the affected leaves and become angular or round when many spots collapse. It results in shriveling of leaves that later die.

Control: Treat the seeds with Agrosan GN or Bavistin at 2 g/kg seed. Spray the crop with Dithane M-45 at 0.2% or Bavistin at 0.1% or Copper Oxychloride at 0.2%, and repeat at 7–10-day interval if necessary.

13.15.4 FRUIT ROT (*PYTHIUM APHANIDERMATUM* AND *PYTHIUM BUTLERI*)

Water-soaked lesions girdle the stem, later extending upward and downward. The rotting of the affected tissues occurs and even grown-up plants collapse. The fungus causes rotting of fruits that have direct contact with soil.

Control: Avoid flood irrigation and raise the crop on drip system. Treat the seeds with Thiram or Bavistin at 2 g/kg seed. Apply *Trichoderma* at 5 kg/ha in soil before sowing. Spray the crop with Bavistin at 0.1% or Copper Oxychloride at 0.2%, and repeat the spray if necessary.

13.15.5 FUSARIUM WILT (*FUSARIUM OXYSPORUM* F. SP. *NIVEUM*)

The fungus is seed-borne as well as a persistent soil inhabitant. Seedling injury is high at 20°C–30°C temperature. Wilt development is also favored by a temperature of about 27°C. No infection occurs at temperatures below 15°C and above 35°C. The plants are prone to attack at all stages of growth. Germinating seeds may also rot in the soil. The affected plants turn yellow and show wilting, and later, the whole plant dies. It causes damping-off disease of seedlings. Small leaves lose their green color and wilt.

Control: Dip the seeds in hot water (55°C) for 15 min to kill the seed-borne infection. Treat the seeds with Captan or Thiram or Bavistin at 2 g/kg seed before sowing. Drench the soil around the roots with Captan or Bavistin at 0.2% solution.

13.15.6 MOSAIC VIRUS

The young leaves develop small greenish-yellow areas, and they become more translucent than those in the remaining parts of the leaf. Yellow mottling is seen on leaves and fruits. Leaf distortion and stunting of infected plants occur. The virus is transmitted through saps, seeds, and aphids.

Control: Use the seed collected from virus-free plants. Rough out the infected plants from the field as soon as they are noticed. Eliminate the weed hosts from the field. Spray Imidacloprid 17.8 SL mixed with 0.5–0.6 mL/L water to control the vectors.

13.16 INSECT PESTS

13.16.1 RED PUMPKIN BEETLE (*AULACOPHORA FOVEICOLLIS*)

Both grub and adult attack the crop at seedling stage and make holes in the cotyledonary leaves. When the attack is severe, the crop is totally destroyed.

Control: Spray the crop with Spinosad 45 SC (0.5–0.7 mL/L water) or Dimethoate 30 EC (1.5–2 mL/L water) or Malathion 50 EC (1.5–2 mL/L water).

13.16.2 APHID (*APHIS GOSSYPII* OR *MYZUS PERSICAE*)

Both nymphs and adults of this tiny insect suck the cell sap from the tender leaves, reducing plant vigor. As a result, the leaves curl up and ultimately wilt. The aphids excrete honeydew on which black sooty mold develops, which hampers photosynthetic activity. Besides, these aphids act as vectors for the transmission of many viruses. The attack is severe during March–April.

Control: Spray *Neem* seed kernel extract (5%), *Neem* oil (2%), or tobacco decoction (0.05%). Spray the crop with Imidacloprid 17.8 SL at 0.5–0.6 mL/L water and repeat after 15 days, if necessary.

13.16.3 FRUIT FLY (*BACTROCERA CUCURBITAE*)

The damage is caused by maggots of the fly. The adult fly punctures tender fruits and lays eggs below the fruit skin (epidermal layer). The maggots feed inside the fruits and make them unfit for consumption. The hot and humid weather is most suitable for its attack.

Control: Install pheromone traps in the field (8 trap/ha). Apply bait spray containing Malathion 50 EC (0.05%) and jaggery (10%) mixed in 20 L water. Spray Spinosad 45 SC (0.5–0.7 mL/L water) or Dimethoate 30 EC (1.5–2 mL/L water) or Malathion 50 EC (1.5–2 mL/L water).

13.16.4 MITE (*POLYPHAGOTARSONEMUS LATUS*)

Both nymphs and adults suck the sap from young foliage and growing tips. Downward curling and crinkling of leaves giving an inverted boat-shaped appearance, stunted growth, and elongation of petiole are the characteristic symptoms.

Control: Spray the crop with Propargite mixed in 2–3 mL/L water.

REFERENCES

Aykroyd, W.R. 1963. ICMR special report. Series no. 42.
Bains, M.S. and U.S. Kang. 1963. Inheritance of some flower and fruit characters in muskmelon. *Indian J. Genet. Plant Breed.* 23: 101–106.
Balass, M., Y. Cohen, and M. Bar-Joseph. 1992. Identification of a constitutive 45 kD soluble protein associated with resistance to downy mildew in muskmelon (*Cucumis melo* L.) line PI 124111F. *Physiol. Mol. Plant Pathol.* 41: 387–396.
Bates, D.M. and R.W. Robinson. 1995. Cucumbers, melons and watermelons. In: Smartt, J. and Simmonds, N.W. (eds.). *Evolution of Crop Plants*, 2nd edn. Longman Scientific, Essex, U.K., pp. 89–96.
Baudracco-Arnas, S. and M. Pitrat. 1996. A genetic map of melon (*Cucumis melo* L.) with RFLP, RAPD, isozyme, disease resistance and morphological markers. *Theor. Appl. Genet.* 93: 57–64.

Brotman, Y., L. Silberstein, I. Kovalski, C. Perin, C. Dogimont, M. Pitrat, J. Klingler, G.A. Thompson, and R. Perl-Treves. 2002. Resistance gene homologs in melon are linked to genetic loci conferring disease and pest resistance. *Theor. Appl. Genet.* 104: 1055–1063.

Burger, Y., N. Katzire, G. Tzuri, V. Portnoy, U. Saar, S. Shriber, R. Perl-Treves, and R. Cohen. 2003. Variation in the response of melon genotypes to *Fusarium oxysporum* f. sp. *melonis* race 1 determined by inoculation test and molecular markers. *Plant Pathol.* 52: 204–211.

Chadha, M.L. and T. Lal. 1993. Improvement of cucurbits. In: *Advances in Horticulture.* Vol. 5. Vegetable Crops. K.L. Chadha and G. Kalloo, (eds.). Malhotra Publishing House, New Delhi, India.

Chadha, M.L. and K.S. Nandpuri. 1980. Hybrid vigour studies in muskmelon. *Indian J. Hort.* 37(3): 276–282.

Chadha, M.L., K.S. Nandpuri, and S. Singh. 1972. Inheritance of quantitative characters in muskmelon. *Indian J. Hort.* 29: 174–178.

Choudhary, B.R., R.S. Dhaka, and M.S. Fageria. 2003. Heterosis for yield and yield related attributed in muskmelon (*Cucumis melo* L.). *Indian J. Genet. Plant Breed.* 63(1): 91–92.

Choudhary, B.R., S.M. Haldhar, R. Bhargava, S.K. Maheshwari, and S.K. Sharma. 2013. Monoecious line of muskmelon developed. *ICAR News*, July–September, Vol. 19(3), pp. 9–10.

Choudhary, B.R. and S. Pandey. 2010. Breeding of F₁ hybrids in muskmelon: Accomplishment and prospects. *Indian J. Arid Hort.* 5(1–2): 1–5.

Choudhury, B. and N. Sivakami. 1972. Screening muskmelon (*Cucumis melo* L.) for breeding resistant to powdery mildew. In: *Third International Symposium on Subtropical and Tropical Horticulture*, Vol. 2, p. 10.

Danin-Poleg, Y., Y. Tadmor, G. Tzuri, N. Reis, J. Hirschberg, and N. Katzir. 2002. Construction of a genetic map of melon with molecular markers and horticultural traits, and localization of genes associated with ZYMV resistance. *Euphytica* 125: 373–384.

Dhaliwal, M.S. and T. Lal. 1996. Genetics of some important characters using line × tester analysis in muskmelon. *Indian J. Genet. Plant Breed.* 56(2): 207–213.

Dhatt, A.S. and S.S. Gill. 2000. Effect of genetic male sterility on flowering behavior of muskmelon. *Veg. Sci.* 27: 31–34.

Dhillon, N.P.S. and T.C. Wehner. 1991. Host plant resistance in insects in cucurbits-germplasm resources, genetics and breeding. *Trop. Pest Manage.* 37: 421–428.

Eduardo, I., P. Arús, and A.J. Monforte. 2004. Genetics of fruit quality in melon, verification of QTLs involved in fruit shape with near-isogenic lines (NILs). In: Lebeda, A. and Paris, H.S. (eds.). *Progress in Cucurbit Genetics and Breeding Research. Proceedings of Cucurbitaceae 2004, Eighth Eucarpia Meeting on Cucurbit Genetics and Breeding*, Olomouc, Czech Republic, July 12–17, 2004, pp. 499–502.

Esquinas-Alcazar, J.T. 1977. Alloenzyme variation and relationships in the genus *Cucumis*. Dissertation, University of California, Davis, CA.

Fang, G.W. and R. Grumet. 1993. Genetic engineering of potyvirus resistance using constructs derived from the zucchini yellow mosaic virus coat protein gene. *Mol. Plant Microbe Int.* 6: 358–367.

Foster, R.E. 1960. Breeding for disease resistance and variety. Cantaloupe Research in Arizona. Arizona University Report, Vol. 195, pp. 1–4.

Fuchs, M., J.R. McFerson, D.M. Tricoli, J.R. McMaster, R.Z. Deng, and M.L. Boeshore. 1997. Cantaloupe line CZW-30 containing coat protein genes of cucumber mosaic virus, zucchini yellow mosaic virus, and watermelon mosaic virus-2 is resistant to these three viruses in the field. *Mol. Breed.* 3: 279–290.

Gaba, V., A. Zelcer, and A. Gal-On. 2004. Cucurbit biotechnology: The importance of virus resistance. *In Vitro Cell. Develop. Biol. Plant* 40: 346–358

Garcia, L., M. Jamilena, J.I. Alvarez, T. Arnedo, J.L. Oliver, and R. Lozano. 1998. Genetic relationship among melon breeding lines revealed by RAPD markers and agronomic traits. *Theor. Appl. Genet.* 96: 878–885.

Gonzalo, M.J., M. Oliver, J. Garcia-Mas, A. Monfort, R. Dolcet-Sanjuan, N. Katzir, P. Arús, and A.J. Monforte. 2005. Simple-sequence repeat markers used in merging linkage maps of melon (*Cucumis melo* L.). *Theor. Appl. Genet.* 110: 802–811.

Granberry, D.M. and J.D. Norton. 1980. Response of progeny for interspecific cross of *Cucumis melo* × *Cucumis metuliferus* to *Meloidogyneincognita acrita*. *J. Am. Soc. Hort. Sci.* 119: 345–355.

Gurav, S.B., K.N. Wavhal, and P.A. Navale. 2000. Heterosis and combining ability in muskmelon (*Cucumis melo* L.). *Agric. Univ. J. Maharashtra* 25(2): 149–152.

Hammer, K., R. Hanelt, and E. Perrino. 1986. *Carosello* and the taxonomy of *Cucumis melo* L. especially of its vegetable races. *Kulturpflanzen* 34: 249–259.

Joobeur, T., J.J. King, S.J. Nolin, C.E. Thomas, and R.A. Dean. 2004. The *Fusarium* wilt resistance locus *Fom-2* of melon contains a single resistance gene with complex features. *Plant J.* 39: 283–297.

Kalloo, G. and J. Dixit. 1983. Genetic components of yield and its contributing traits in muskmelon (*Cucumis melo* L.). *Haryana J. Hort. Sci.* 12: 218–220.

Kesavan, P.K. and T.A. More. 1987. Powdery mildew resistance in F₁ hybrid in muskmelon. In: *National Symposium on Heterosis Exploitation, Accomplishment and Prospects*, Parbhani, Maharashtra, October 15–17, 1987, p. 54.

Kesavan, P.K. and T.A. More. 1991. Use of monoecious lines in heterosis breeding muskmelon (*Cucumis melo* L.). *Veg. Sci.* 18: 59–64.

Kirkbride, J.H. 1993. Biosystematic monograph of the genus Cucumis (*Cucurbitaceae*). Parkway Publishers, Boone, NC.

Klingler, J., I. Kovalski, L. Silberstein, G.A. Thompson, and R. Perl-Treves. 2001. Mapping of cotton-melon aphid resistance in melon. *J. Am. Soc. Hort. Sci.* 126: 56–63.

Kumar, J. and N.P.S. Dhillon. 2008. Assessment of stability of expression of various male sterile genes in melon in subtropical field conditions. In: *Proceedings of the IXth EUCARPIA Meeting on Genetics and Breeding of Cucurbitaceae*, INRA, Avignon, France, pp. 535–538.

Lal, T., V. Vashisht, and N.P.S. Dhillon. 2007. Punjab Anmol—A new hybrid of muskmelon (*Cucumis melo* L.). *J. Res. Punjab Agric. Univ.* 44: 83.

Lebeda, A., E. Kristkova, and M. Kubalakova. 1996. Interspecific hybridization of *Cucumis sativus* × *Cucumis melo* as a potential way to transfer resistance to *Pseudoperonospora cubensis*. In: Gomes-Guillamon, M.L., Soria, C., Caurtero, J., Tores, J.A., and Fernandez-Munoz, R. (eds.). *Cucurbit towards 2000, Proceeding of the VIth Eucarpia Meeting on Cucurbit Genetics and Breeding*, Malaga, Spain, May 28–30, 1996, pp. 31–37.

Loy, J.B., T.A. Natti, C.D. Zack, and S.K. Fritts. 1979. Chemical regulation of sex expression in a gynomonoecious line of muskmelon. *J. Am. Soc. Hort. Sci.* 104(1): 100–101.

Mangal, J.L. and M.L. Pandita. 1979. Effect of pruning and fruit position on growth, flowering, fruit yield and quality of muskmelon variety Hara Madhu. *Haryana J. Hort. Sci.* 8 (3/4): 194–197.

McCreight, J.D. 1984. Phenotypic variation of male fertile and male sterile segregates of *ms*-1 and *ms*-2 muskmelon hybrids. *J. Hered.* 75: 51–54.

Michelmore, R.W., I. Paran, and R.V. Kesseli. 1991. Identification of markers linked to disease resistance genes by bulked segregant analysis: A rapid method to detect markers in specific genomic regions by using segregating populations. *Proc. Natl. Acad. Sci. USA* 88: 9828–9832

Monforte, A.J., M. Oliver, M.J. Gonzalo, J.M. Alvarez, R. Dolcet-Sanjuan, and P. Arus. 2004. Identification of quantitative trait loci involved in fruit quality traits in melon (*Cucumis melo* L.). *Theor. Appl. Genet.* 108: 750–758.

Morales, M., M. Luis-Arteaga, J.M. Alvarez, R. Colcet-Sanjuan, A. Monfort, P. Arus, and J. Garcia-Mas. 2002. Marker saturation of the region flanking the gene NSV conferring resistance to the melon necrotic spot Carmovirus (MNSV) in melon. *J. Am. Soc. Hort. Sci.* 127: 540–544.

Morales, M., H. van Leeuwen, P. Puigdomenech, P. Arus, A. Monfort, and J. Garcia-Mas. 2003. Towards the positional cloning of the nsv resistance gene against MNSV in melon. In: *Plant and Animal Genome XI*, San Diego, CA, January 11–15, 2003.

More, T.A. and V.S. Seshadri. 1987. Maintenance of gynoecious muskmelon with silver thiosulphate. *Veg. Sci.* 14: 138–142.

More, T.A., V.S. Seshadri, and J.C. Sharma. 1980. Monoecious sex forms in muskmelon. *Cucurbit Genet. Coop. Rpt.* 3: 32–33.

Munger, H.M. 1942. The possible utilization of first generation muskmelon hybrids and an improved method of hybridization. *Proc. Am. Soc. Hort. Sci.* 40: 405–410.

Munger, H.M. and R.W. Robinson. 1991. Nomenclature of *Cucumis melo* L. *Cucurbit Genet. Coop. Rpt.* 14: 43–44.

Naudin, C. 1859. Essais d'une monographie des espèces et des varieties du genre *Cucumis*. *Ann. Sci. Nat.* 4(11): 5–87.

Nandpuri, K.S., S. Singh, and T. Lal. 1982. Punjab hybrid—A new variety of muskmelon. *Prog. Fmg.* 18: 3–4.

Nath, P., S. Subramanyam, and O.P. Dutta. 1973. Improvement of cucurbitaceous crops—A review. *SABRAO J.* 8: 117–119.

Norton, J.D. and D.M. Granberry. 1980. Characterization of progeny from an interspecific cross of *Cucumis melo* with *Cucumis metuliferus*. *J. Am. Soc. Hort. Sci.* 105(2): 174–180.

Oliver, M., J. Garcia-Mas, M. Cardus, N. Pueyo, A.I. Lopez-Sese, M. Arroyo, H. Gomez-Paniagua, P. Arus, and M.C. de Vicente. 2001. Construction of a reference linkage map for melon. *Genome* 44: 836–845.

Paris, H.S., Z. Amar, and E. Lev. 2012. Medieval emergence of sweet melons, *Cucumis melo* (Cucurbitaceae). *Ann. Bot.* 110(1): 23–33. doi:10.1093/aob/mcs098, available online at www.aob.oxfordjournals.org.

Park, S.O. and K.M. Crosby. 2004. Development of RAPD markers linked to the male-sterile ms-3 gene in melon. *Acta Hort.* 637: 243–249.

Pauquet, J., E. Burget, L. Hagen, V. Chovelon, A. Menn, N. Valot, S. Desloire et al. 2004. Map-based cloning of the *Vat* gene from melon conferring resistance to both aphid colonization and aphid transmission of several viruses. In: Lebeda, A. and Paris, H.S. (eds.). *Progress in Cucurbit Genetics and Breeding Research. Proceedings of Cucurbitaceae 2004, Eighth Eucarpia Meeting on Cucurbit Genetics and Breeding*, Olomouc, Czech Republic, July 12–17, 2004, pp. 325–329.

Perchepied, L., C. Dogimont, and M. Pitrat. 2005. Strain-specific and recessive QTLs involved in the control of partial resistance to *Fusarium oxysporum* f. sp. *melonis* race 1.2 in a recombinant inbred line population of melon. *Theor. Appl. Genet.* 111: 65–74.

Perin, C., M.C. Gomez-Jimenez, L. Hagen, C. Dogimont, J.C. Pech, A. Latche, M. Pitrat, and J.M. Lelievre. 2002b. Molecular and genetic characterization of a non-climacteric phenotype in melon reveals two loci conferring altered ethylene response in fruit. *Plant Physiol.* 129: 300–309.

Perin, C., L. Hagen, C. Dogimont, V. De Conto, and M. Pitrat. 1998. Construction of genetic map of melon with molecular markers and horticultural traits. In: MacCreight, J.D. (ed.). *Cucurbitaceae 98*. ASHS Press, Alexandria, VA, pp. 370–376.

Perin, C., L.S. Hagen, N. Giovinazzo, D. Besombes, C. Dogimont, and M. Pitrat. 2002a. Genetic control of fruit shape acts prior to anthesis in melon (*Cucumis melo* L.). *Mol. Gen. Genet.* 266: 933–941.

Perl-Treves, T., D. Zamir, N. Navot, and E. Galun. 1985. Phylogeny of *Cucumis* based on isozyme variability and its comparison with plastome phylogeny. *Theor. Appl. Genet.* 71: 430–436.

Peterson, C.E., K.E. Owens, K.E. Rowe, and P.R. Rowe. 1983. Wisconsin-998 muskmelon germplasms. *HortScience* 18: 116.

Pitrat, M. 1991. Linkage groups in *Cucumis melo* L. *J. Hered.* 82: 406–411.

Pitrat, M. 2002. 2002 gene list for melon. *Cucurbit Genet. Coop. Rpt.* 2: 76–93.

Pitrat, M. 2008. Melon. In: Prohens, J. and Nuez, F. (eds.). *Handbook of Plant Breeding*. Springer, New York, pp. 283–315.

Pitrat, M., P. Hanelt, and K. Hammer. 2000. Some comments on intraspecific classification of cultivars of melon. *Acta Hort.* 510: 29–36.

Risser, G. 1984. Correlation between sex expression and fruit shape in muskmelon (*Cucumis melo* L.). Plovdiv, Bulgaria, July 1984, pp. 100–103.

Robinson, R.W. and Decker-Walters, D.S. 1997. *Cucurbits*. CAB International, Oxon, GB.

Robinson, R.W. and Whitaker, T.W. 1974. Cucumis. In: *Handbook of Genetics*, R.C. King (ed.). vol. 2. Plenum, New York, pp. 145–150.

Roig, L.A., M.V. Roche, M.C. Orts, L. Jubeldia and V. Moreno. 1986. Plant regeneration from cotyledons protoplast of *Cucumis melo* L. cultivar Cantaloup charentais. *Cucurbit Genet. Coop. Rpt.* 9: 74–77.

Sambandam, C.N. and S. Chelliah. 1972. *Cucumis callosus* (Rottl) Logn. a valuable material for resistance breeding in muskmelon. In: *Third International Symposium on Horticulture*, Bangalore, India, p. 7.

Sandha, M.S. and T. Lal. 1999. Heterosis breeding in muskmelon-a review. *Veg. Sci.* 26(1): 1–5.

Silberstein, L., I. Kovalski, Y. Brotman, C. Perin, C. Dogimont, M. Pitrat, J. Klingler et al. 2003. Linkage map of *Cucumis* melo including phenotypic traits and sequence-characterized genes. *Genome* 46: 761–773.

Silberstein, L., I. Kovalski, R. Huang, K. Anagnostou, M.M.K. Jahn, and R. Perl-Treves. 1999. Molecular variation in melon (*Cucumis melo* L.) as revealed by RFLP and RAPD markers. *HortScience* 79: 101–111.

Singh, D., K.S. Nandpuri, and B.R. Sharma. 1976. Inheritance of some economic quantitative characters in an intervarietal cross of muskmelon (*Cucumis melo* L.). *Punjab Agric. Univ. J. Res.* 13: 172–176.

Singh, M.J., K.S. Randhawa, and T. Lal. 1989. Genetic analysis for maturity and plant characteristics in muskmelon. *Veg. Sci.* 16: 181–184.

Singh, S., S. Pandey, R. Raghuwanshi, V. Pandey, and M. Singh. 2015. SSR analysis for fruit and quality characters in infra-specific mapping population of melon. *Indian J. Agric. Sci.* 85(1): 32–37.

Spooner, D.M., R.G. Vanden Berg, W.L.A. Hetterscheid, and W.A. Brandenburg. 2003. Plant nomenclature and taxonomy: A horticultural and agronomic perspective. *Hort. Rev.* 28: 1–60.

Swamy, K.R.M. and O.P. Dutta. 1985. Diallel analysis of total soluble solids in muskmelon. *Madras Agric. J.* 72: 399–403.

Taler, D., M. Galperin, I. Benjamin, Y. Cohen, and D. Kenigsbuch. 2004. Plant *eR* genes that encode photorespiratory enzymes confer resistance against disease. *Plant Cell* 16: 172–184.

Touyama, T., I. Asami, T. Oyabu, K. Yabe, S. Sugahara, and M. Kanbe. 2000. Development of PCR-based markers linked to a long-shelf-life-character "nonyellowing" in muskmelon. *Res. Bull. Aichi Ken Agric. Res. Center* 32: 47–52.

Vashistha, R.N. and B. Choudhury. 1974. Studies on growth and yield potentials of muskmelon cultivars to red pumpkin beetle. *Haryana J. Hort. Sci.* 1: 55–61.

Wang, Y.H., C.E. Thomas, and R.A. Dean. 1997. A genetic map of melon (*Cucumis melo* L.) based on amplified fragment length polymorphism (AFLP) markers. *Theor. Appl. Genet.* 95: 791–798.

Wang, Y.H., C.E. Thomas, and R.A. Dean. 2000. Genetic mapping of a *Fusarium* wilt resistance gene (*Fom-2*) in melon (*Cucumis melo* L.). *Mol. Breed.* 6: 379–389.

Wechter, W.P., R.A. Dean, and C.E. Thomas. 1998. Development of sequence-specific primers that amplify a 1.5-kb DNA marker for race 1 *Fusarium* wilt resistance in *Cucumis melo* L. *HortScience* 33: 291–292.

Wechter, W.P., M.P. Whitehead, C.E. Thomas, and R.A. Dean. 1995. Identification of a randomly amplified polymorphic DNA marker linked to the *Fom-2 Fusarium* wilt resistance gene in muskmelon MR-1. *Phytopathology* 85: 1245–1249.

Whitaker, T.W. 1979. The breeding of vegetable crops: Highlights of the past seventy-five years. *HortScience* 14: 359–363.

Whitaker, T.W. and G.N. Davis. 1962. *Cucurbits*. Interscience, New York, p. 250.

Yamaguchi, J. and T. Shiga. 1993. Characterization of regenerated plants via protoplast electrofusion between melon (*Cucumis melo*) × pumpkin (interspecific hybrid, *Cucurbita maxima* and *C. moschata*). *Jpn. J. Breed.* 43: 173–182.

14 Ash Gourd (*Benincasa hispida*) Breeding and Cultivation

Sudhakar Pandey and Suhas G. Karkute

CONTENTS

14.1 INTRODUCTION

Ash gourd belongs to the genus *Benincasa* of the Cucurbitaceae family with a single cultivated species *hispida*, synonym *cerifera*. *Hispida* refers to the hirsute pubescence on the foliage and immature fruit, whereas *cerifera* means wax bearing. Its primary basic chromosome number is x = 6. It has eight pairs of medium chromosome and four pairs of subterminal ones. The total length of chromosome complement is 57.36 μm with an average chromosome length of 2.39 μm. It has been reported that *Benincasa hispida* may be a stable tetraploid derived from an ancestor having a basic chromosome number x = 6. Ash gourd is known by different names in different parts of the world such as white gourd, winter melon, fallow gourd, white pumpkin, hairy melon, Chinese preserving melon, and wax gourd (Robinson and Deckers-Walters, 1997). Besides, it is also known by several other names as koosmanda (Sanskrit), pazadaba (Persian), fakwa, funggwa, mokwa, doonqua, chamkwa (Chinese), tougan (Japanese), tankoy, kundol (Filipino), and petha kaddu (Indian). The gourd was named as ash or wax gourd because of the presence of waxy cuticles that develop on mature fruits. This helps in increasing the shelf life of the fruit. High temperature favors the formation of waxy bloom. Ash gourd is an important vegetable in China, India, the Philippines, and other parts of Asia. It is also cultivated in Latin America and the Caribbean, usually by immigrants of Chinese descent. It is believed to have originated in Java and Japan. It has been cultivated in China for more than 2300 years. The diversity of cultivars in China suggested that this crop might be the indigenous to southern China (Yang and Walters, 1992). Simmond (1976) reported that ash gourd is indigenous to Asian tropics and distributed from Japan to India by foreign navigators and missionaries. It is grown throughout the plains of India, China, Malaysia, Taiwan, Bangladesh, and Sri Lanka up to an altitude of 1500 m. It is widely cultivated in various parts of India, but the cultivation area is concentrated in and around Uttar Pradesh.

14.2 COMPOSITION AND USES

14.2.1 COMPOSITION

The fruits of ash gourd are a rich source of vitamins, fibers, and minerals. The edible portion of ash gourd constitutes around 98%. Nutritional composition of the fruit is given in Table 14.1.

TABLE 14.1

Nutritive Composition of Ash Gourd

Constituents	Contents	Constituents	Contents
Edible portion (%)	98	Sodium (mg)	2.0
Water (%)	93.8	Iron (mg)	0.3
Dietary fiber (g)	2.1	Zinc (mg)	0.2
Protein (mg)	467[a]	Vitamin A (mg)	0.02
Fat (mg)	102[a]	Thiamin (mg)	0.07
Carbohydrates (g)	2.0	Riboflavin (mg)	0.05
Ash (g)	0.7	Niacin (mg)	0.20
Calcium (mg)	36.16[a]	Vitamin C (mg)	10.76[a]
Magnesium (mg)	15	Organic acid (g)	0.04
Potassium (mg)	250	Energy (kcal)	44
Crud fiber (g)	0.80[a]	Dry matter (g)	3.71[a]
Total sugar (%)	2.61[a]	Pectin (g)	0.97[a]

Source: Wills, R.B.H. et al., *J. Agric. Food Chem.*, 32(2), 413, 1984.

Note: Values are in 100 g fresh edible portion.

[a] Pandey et al. (2009).

14.2.2 Uses

Both mature and immature fruits of ash gourd are consumed usually by cooking or pickling. Mature fruits of ash gourd are used in making confectioneries such as candies, preserves, sweets, and pickles and immature fruits (young) as culinary vegetable in various parts of India. In China, it is specially used for preparing a variety of winter melon soups or Dong-gua soups and Chinese stewed winter melon. It is also sold as cut slices. Occasionally, the entire fruit is steamed and stuffed with lotus seeds, vegetables, meat, or other ingredients. Chinese also use it as elegant serving bowls for the soup. For ceremonial occasions and formal banquets, the interior of the fruit is scooped out and filled with soup. Generally, specific cultivars or land races are grown for vegetable preparation at immature stage and for preparation of *sweets* at mature fruit stage. The ash gourd enzyme may have potential as an alternative for calf rennet in cheese manufacture (Eskin and Landman, 1975).

Young leaves, vine tips, and flower buds are also boiled and eaten as greens. Seeds can be consumed fried but usually as a medication instead of a food. The fruit wax that develops even after the fruit is harvested is used for making candles. Malaysians also use the wax as a vehicle for carrying poison for homicide.

The ash gourd is mentioned in ancient Ayurvedic texts like *Charaka Samhita* and *Ashtanga Hridaya Samhita* for its many nutritional and medicinal properties. Ayurveda physicians call it "kushmanda," and a preparation with the fruit is called "kushmanda avaleham." The raw fruit is useful for alleviating "vatha" and "pitta," while the ripe fruit alleviates all three humors. The paste of the leaf or fruit is applied on burns and wounds to get relief. For relief from headache caused by vatha or pitta, seed oil is applied on head. It is also useful in treating respiratory disorders like asthma, blood-related diseases, and urinary diseases like kidney stones. In the gastrointestinal system, it acts as a laxative. The seeds are anthelmintic and have special effect on tapeworm. It is the best fruit for use as a brain food to treat mental illnesses and nervous disorders such as epilepsy and insanity. Its high potassium content makes this a good vegetable for maintaining a healthy blood pressure. It strengthens a weak heart and lungs if consumed as a curry. Eating curry also helps in dysuria (difficulty in passing urine) and reduces stones in the bladder. It is an effective medicine for women suffering from excessive bleeding and excessive vaginal discharge. Its usual intake in the diet increases sperms and treats disorders or defects in sperms. Extract of ash gourd has acid-neutralizing property and is recommended for the management of peptic ulcer. Ash gourd juice is used to treat diabetes, and the cucurbitacin from its juice prevents kidney damage caused by poisoning of mercuric chloride. Besides these, it has a cooling effect on the body and helps to induce healthy and sound sleep. The paste of the fruit prepared with ghee is the excellent food for gaining weight.

14.3   BOTANY AND FLORAL BIOLOGY

The ash gourd is a trailing or climbing vine with a tap root system. The stem is thick, furrowed (angular), and hispid with coarse hairs. The tendrils of vine are cylindrical and hairy with two or three branches. Leaves of ash gourd are 22–25 cm long and are irregularly five to seven lobed with long petiole, reniform orbicular, deeply cordate, and scabrous. The lower surface of the leaf is rigidly hispid and margins are serrated. Leaves emit an unpleasant odor when bruised. Ash gourd is a monoecious plant having solitary yellow flowers in leaf axils, but pistillate and staminate flowers are present on different nodes. Staminate flowers are with long pedicles and pistillate flowers are almost sessile. Pistillate flowers are generally larger than staminate flowers. The ratio of staminate to pistillate flowers is 34:1. Flowers are calyx teethed at early stage and are often narrow and serrated.

The pistillate flowers born on 1.8–4 cm long hairy stalk are 5–15 cm in length. The 2–4 cm long ovary is densely hairy with three curved and bilobed stigmas. Stalks of staminate flowers are 5–15 cm long with a flower diameter of 8–10 cm. Ash gourd fruits are of two types: purple green and green. They are also classified based on shape as round and oblong. Fruit weight ranges from 0.5 to 40 kg. The fruits are fleshy, succulent, and densely hairy at immature stage but thickly deposited with white, easily removable waxy layer, which is triterpenol acetates and triterpenol (Meusel et al. 1994) at maturity. The waxy layer makes the fruit resistant to insects or pathogens and also preserves the moisture. The fruit flesh is white, juicy, and generally spongy. Seeds are $1–1.5 \times 5.7$ cm in size and are flat and numerous. They are smooth with a narrow base and white, yellow, or pale brown in color. Seeds contain pale-yellow oil.

The anthesis of flowers takes place in the early morning at 4:30 a.m. and continues up to 7:30 a.m.; dehiscence of pollen takes place between 3:00 and 4:30 a.m. The pollen is viable for 8 h before and 18 h after anthesis. Similarly, the stigma is receptive for 18 h before anthesis and continues to be so till 24 h after anthesis.

14.4   GENETIC RESOURCES AND DIVERSITY

Germplasms of ash gourd have been collected and maintained at the Institute of Plant Breeding, the Philippines; Vavilov Institute of Plant Industry, Petersburg; Zentralinstitut fur Genetik und Kulturpflanzenforschung, Gatersleben, Germany; South Regional Plant Introduction Station, Georgia; Cornell University, New York; Imperial Valley Conservation Research Center, California. Germplasms are also being maintained in India at the Indian Institute of Vegetable Research, Varanasi; National Bureau of Plant Genetic Resources, New Delhi; Kerala Agricultural University, Vellanikkara, Kerala; and Tamil Nadu Agricultural University, Coimbatore, Tamil Nadu.

Based on vegetative and fruit diversity, four major groups are recognized in ash gourd by Rifai and Reyes (1993). These groups are as follows:

1. *Unridged*: The plant seed is with unridged margins; the fruit shape is cylindrical, up to 2 m long; its maturing period is 3 months after pollination; the plant rind is dark green, almost waxless.
2. *Ridged*: The plant seed is with ridged margins; other attributes are similar to those in group one.
3. *Fussy gourd*: The seeds are with ridged margins; the fruit shape is cylindrical, 20–25 cm long; its maturing period is within two months after pollination; the rind is green, almost waxless, covered with white soft hairs.

4. *Wax gourd*: The plant seed is with ridged margins; the fruit shape is globose to oblong, the fruit set is two months after pollination; its rind is light green, covered with a white waxy bloom, and glabrous or finely hairy. The wax gourd group is categorized into three prominent cultivars as (1) fruits nearly round and essentially hairless, (2) fruits nearly round and hairy, and (3) fruits oblong and hairy. The ash gourd germplasm of India has good variability and has been characterized on morphological and molecular basis (Pandey et al., 2008; Singh 2002).

14.5 GENETICS AND GENE ACTION

Ash gourd crop has very limited reports on genetical studies. Additive and dominance type of gene action controls the expression of days to first male and female flowering, number of fruits per plant, average fruit weight, polar and equatorial circumference of fruit, and yield per plant; therefore, population improvement procedure through selection would be useful. The ratio of recessive and dominant alleles was distributed more frequently than the recessive ones for vine length, number of fruit, fruit weight, equatorial and polar circumference of fruit. High heritability coupled with high genetic advance is present for yield per plant, average fruit weight, number of the node at which the first male flower appears and number of the node at which the first female flower appears. The highest positive direct effect on yield was exerted by fruit weight, fruit number, and polar and equatorial circumference of fruit.

14.6 BREEDING METHODS

Breeding program for ash gourd aims at various traits ultimately for increasing yield, resistance, or tolerance to biotic and abiotic stresses and enhancing quality for processing, particularly in the areas where petha sweets are prepared. The diseases targeted during breeding are downy mildew, powdery mildew, fruit rot and viruses, and insect pests like red pumpkin beetle and fruit fly. Ash gourd is a cross-pollinated crop, and hence traditional breeding practices like selection, pedigree method, and bulk population methods can be employed in ash gourd. The following methods are employed for improvement of this crop.

14.6.1 SELECTION

Ash gourd being a highly cross-pollinated crop has a high degree of heterozygosity. Therefore, selfing followed by selection improves the population and helps to develop a new variety with distinct features from the original. Different selection methods can be applied for population improvement.

14.6.1.1 Mass Selection

The selection of superior plants from the base population and mixing of their seeds for raising the next generation are followed in this method. This selection procedure is repeated in uniform growing conditions up to the selection as a new variety/type. Mass selection is effective in improving the sugar content of muskmelon and watermelon. This method is effective in improving simply inherited and highly heritable qualitative characters.

14.6.1.2 Single Plant Selection

Single plant selection is a very common method of selection. The selfed individual plants are selected from the heterozygous and nonuniform material. The seeds of selected plants are raised individually in the next generation for evaluation and maintained by selfing. The advanced progenies do not show loss of vigor due to inbreeding. Therefore, homozygosity for the concerned characters can be attained in the individuals of the progeny by selfing. After necessary evaluation, the best selection can be treated as a new type. Several varieties have been developed by single plant selection in ash gourd.

14.6.2 Inbreeding and Selection

Inbreeding is the mating of closely related individuals, that is, either selfing or sib mating. Individual selection of inbred is practiced after attaining maximum uniformity and claimed as a variety.

14.6.3 Hybridization

Hybridization creates new genetic variability in the F_2 and subsequent generations that helps efficient selection of desirable types. The hybrid progenies are advanced in the subsequent generations by selfing to select desired plant type. Single plant selection by maintaining the pedigree of segregating generation is applied.

14.6.4 Backcross Breeding

Backcross breeding consists of crossing of F_1 with one of the parents (recurrent parent) followed by selection of genotypes for specific characters. One to three backcrosses may be made till homozygosity is attained. This method is generally applied to transfer simple inherited characters like resistance and some morphological traits to the unimproved variety. Two types of parents are involved in this breeding method: one as a recurrent parent (high yielding) and the other as a donor parent (low yielding but possessing specific desirable trait). This method is commonly used for the development of resistant varieties.

14.6.5 Heterosis Breeding

Heterosis breeding is also employed in the improvement of ash gourd. Heterosis for earliness, fruit size, number of fruits, and fruit weight has been reported. The heteroses for the number of fruits and total yield were found to be 46% and 26%, respectively. High yield of F_1 hybrid was contributed by various traits such as early maturity, longer vine, more fruits per plant, bigger fruit size, and higher fruit weight. In most of the crosses, all the characters were under the control of duplicated epistasis indicating high promise for heterosis breeding in ash gourd. It is a highly cross-pollinated crop; therefore, it is necessary to produce inbred lines. If varieties/lines are maintained as pure lines by inbreeding, these lines can be used as inbred lines. These inbred lines will be crossed in different mating designs (diallel, line × tester) to test the specific and general combining abilities of lines. Once a good heterotic parental combination has been identified, the hybrid seed can be easily produced by adopting any one of the following methods:

1. Pinching of staminate flowers before anthesis and hand pollination where isolation distance is not available
2. Pinching of staminate flowers before anthesis on pistillate parent and insect pollination where isolation distance is available
3. Chemical suppression of staminate flowers and insect pollination (at proper isolation)

14.6.6 Mutation Breeding

Generally, mutations are heterozygous and recessive. Therefore, the mutant phenotypes are not expressed in the M_1 generation. Several chemical mutagens (ethyl methane sulfonate, diethyl sulfate ethyl amine, etc.) may be used for inducing the mutation. Besides, physical mutagens such as x-rays, gamma rays, and neutrons also may be used for the induction of mutation. In ash gourd, the development of seedless fruits may be one of the areas of mutation breeding.

14.6.7 BREEDING FOR RESISTANCE

Breeding for resistance to biotic and abiotic stress is an integral part of any breeding program. It is a continuous process; therefore, due attention must be paid to develop new varieties with acceptable quality. Management of diseases and insects involves development of resistant varieties. The first requirement of such breeding is to find out the resistance sources against diseases and insects. *Fusarium* wilt (*Fusarium oxysporum*) is the most serious disease of ash gourd. Kaur et al. (1985) identified fruit rot-resistant varieties caused by different species of *Fusarium*. Resistance to watermelon virus type 1 (WMV-1) and tolerance to WMV-2 are reported in accession PI 391545 of ash gourd.

14.6.8 BREEDING FOR QUALITY

Fully ripened or mature fruits are used for the preparation of candy and sweet (petha) or bari (fruit flesh mixed in pulse) in Uttar Pradesh. For petha preparation, big-sized (10–15 kg), oval to cylindrical fruits are required, while for household consumption, a small cylindrical cultivar (1–2 kg) without ash is in demand. The quality requirement of ash gourd variety for petha processing is high pulp recovery, high content of dry matter, less crude fiber, high flesh diameter, and less seed with linear arrangement in the fruit. The breeding for quality improvement should be concentrated on these parameters.

14.7 BIOTECHNOLOGICAL INTERVENTIONS

Assessment of genetic diversity based on the phenotype has limitations, since most of the morphological characters are greatly influenced by environmental factors and the developmental stage of the plant. In contrast, molecular markers based on DNA sequence polymorphism are independent of environmental conditions and show a higher level of polymorphism. Genetic diversity was assessed among 34 accessions of *B. hispida* using quantitative traits and Random Amplified Polymorphic DNA (RAPD) data (Pandey et al. 2008). There are few other reports on assessment of genetic diversity using RAPD markers for assessing genetic diversity (Meng et al., 1996, Sureja et al., 2006). Jiang et al., (2013) examined transcriptome in ash gourd, and more than 44 million of high-quality reads were generated from five different tissues of ash gourd using Illumina paired-end sequencing technology. About 6242 microsatellites (simple sequence repeats) were detected as potential molecular markers in ash gourd. The result showed that 170 of the 200 primer pairs were successfully amplified and 49 (28.8%) of them exhibited polymorphisms.

14.8 CULTIVARS

Cultivars of ash gourd have evolved locally and vary among geographic regions. A majority of cultivars are available from seed companies. Small-fruited cultivars in China include fuzzy squash or Chinese fuzzy guard and Beijing Yi-chuan-lin. Mature fruits of these varieties are small (20–25 cm long), cylindrical, hairy, and with little or no waxy bloom. Fruits are usually eaten while immature. Large-fruited cultivars in China include Guang-dung-Quig-pi, Hui-pi Dong-gua, and Beijing Di-dong-gua. These fruits are consumed when mature. Two hybrids were also developed in China through heterosis breeding, namely, Qingza 1 and Qingza, and are successfully commercialized.

In India, cultivars with large, round to oblong fruits and a thick waxy coating are cultivated. Public sector institutes are engaged in breeding programs and have developed the following cultivars for different parts of India.

Kashi Ujawal: Fruits are globular in shape, each weighing 10–12 kg and less seeded. This variety is suitable for the preparation of petha sweets. It has a yield potential of 40–50 t/ha in 130–140 days of crop duration.

Kashi Dhawal: This variety has been developed through selection from a local germplasm collection. Its vine length is 7.5–8 meter. It bears oblong fruits, each of 11–12 kg. The fruit flesh is white with 8.5–8.7 cm thickness and the seed arrangement is linear; its crop duration is 120 days. Due to high flesh recovery, this variety is suitable for the preparation of petha sweets. The variety has a yield potential of 58–60 t/ha.

Kashi Surbhi: The fruit shape is oblong, medium in size (9.5–10 kg); its flesh is white with 8.5–8.7 cm thickness; the fruit yield is 2.5–3.0 fruit/plant, with high flesh recovery, suitable for the preparation of petha sweets. This variety has yield potential of 70–75 t/ha.

CO-2: Fruits are small and long and spherical with an average weight of 3 kg. It has a yield potential of 23–25 t/ha in 120 days of crop duration.

Pusa Ujjwal: The fruit shape is oblong or ellipsoid; its rind is greenish white; the fruit flesh is white with an average fruit weight of 7 kg. Its fruits are ideal for long-distance transportation. It has a yield potential of 48–50 t/ha (kharif season) and 41–42 t/ha (summer season).

CO-1: The fruits are globular (35 cm long and 22 cm in girth) and large with an average weight of about 6.8 kg. First harvesting is done about 100 days after sowing, with an average yield of 20–25 t/ha in 140–150 days of crop duration.

Indu: The fruits are round, 24.3 cm long, and 23.78 cm in breadth, with an average weight of 4.8 kg. It gives an average yield of 24.5 t/ha.

Kau Local: Its fruits are oval to oblong with green color at tender immature stage and white at full maturity. The fruits are 45–55 cm long and the length/breadth ratio is 2.05 with an average weight of 6–8 kg. The flesh thickness is 5.1–6.2 cm. It gives an average yield of 28 t/ha in crop duration of 105–120 days for mature fruit production; however, harvesting at tender stage is advisable for better yield and culinary purpose.

PAG-3: The fruits are globular in shape and medium sized with an average weight of 8–10 kg. The fruit color is green at immature stage and covered with white shining coat at maturity. It yields 70–75 t/ha in a crop duration of about 145 days.

Other varieties, that is, KAG-1, PAG-72, and Mudliar, and hybrids, that is, DAGH-14 and DAGH-16, have been recommended for cultivation.

14.9 SUGGESTED CULTURAL PRACTICES

14.9.1 CLIMATE AND SOILS

Ash gourd requires relatively stable high temperature, long days, and moderate, humid climate for good growth. The most optimum temperature range for its growth is from 24°C to 30°C. The development of more pistillate flowers is stimulated by low night temperature, short days, and humid climate, while male flower production is encouraged by high temperature, long days, and dry climate. High temperature and long day length are the important environmental factors that delay flowering in ash gourd but relative humidity and rainfall have little influence. Plants are sensitive to cold but can tolerate drought.

The crop can be grown on a variety of soils, ranging from sandy loam to clay loam rich in organic matter, with good drainage. The soils with lighter texture that warm quickly are good for raising early crop of ash gourd. It can be grown in soils with a pH range from 5.0 to 7.5 but the most optimum pH range is 6–7. It can be grown in the soils that maintain temperature in the range of 15.5°C–33.5°C, with an optimum of 21.1°C. The soil moisture should be at least 10%–15% above the permanent wilting point for successful cultivation. Soil must be prepared well by repeated plowing.

14.9.2 SEED RATE AND SEED TREATMENT

The seed rate depends upon the variety, growing season, soil type, agroclimatic condition of the growing region, seed size, and sowing distance between rows and plants. The 1000 seed weight of ash gourd varies between 62 and 75 g. If four seeds per pit with 60% germination are sown, the seed rate would be about 2.5–3.0 kg/ha at recommended plant spacing, keeping population density of 5000–6000 plants per hectare and leaving 10% land area used in irrigation channels and paths. The seed rate can be reduced to 1.5 kg/ha by raising seedlings in potting plug and polythene and transplanting. Seed treatment is done with carbendazim in 2.5 g/kg of seed.

14.9.3 Sowing Time

Ash gourd is cultivated mainly in the summer season after the danger of frost has passed in most of the world with temperate climate. In subtropical regions, the best time for sowing is the second fortnight of February and June–July. In the countries having frost-free tropical climate, sowing is done mainly in June–August but the crop can be grown round the year.

14.9.4 Sowing/Planting System and Distance

The seeds are generally sown either on raised bed or in pits. Besides, following systems may be followed for the sowing.

14.9.4.1 Shallow Pit Method or Flat Bed Method

In this method, shallow pits of 30 × 30 × 30 cm size are dug at recommended distance. The pits after digging are left open for 3 weeks before sowing for partial solarization. Then each pit is filled with a mixture of soil and 4–5 kg compost and a part of fertilizers like urea 50–60 g, SSP 100–120 g, and MOP 80 g.

14.9.4.2 Raised Bed Method

In this method, ridge and furrow is prepared manually or mechanically keeping 2.5 m distance between two rows with a channel width of 40–50 cm. The seeds are sown on the edges of raised beds after mixing the aforementioned quantity of compost, fertilizers, and insecticide thoroughly.

14.9.4.3 Mound Method

In this method, 15–20 cm raised mounds are generally prepared keeping 2.5 m distance in both directions between two mounds. Similar mixture of soil, compost, and fertilizers should be mixed as in shallow pit method. Seeds (three to four) are sown on each mound at proper depth.

14.9.5 Direct Seeding

For direct sown crop, generally, three to four seeds are sown in each pit or per hill in trenches. The seed should be sown 2 cm deep in a vertical orientation (Krishnaswamy, 1992). The seed takes 7–8 days for normal germination at a temperature range of 25°C–30°C.

14.9.6 Raising Indoor Nursery and Transplanting

A small potting plug polythene bag (size 7 × 5 in.) should be filled with potting mixture that has good water-holding capacity. The potting plug and polythene bag should have good drainage. Two seeds per pot or bag should be sown and thinned to a single seedling when they have four to six true leaves. Moisture should be maintained by light watering every morning. Seedlings are ready for transplanting 20–22 days after sowing or when they have two to three true leaves. Seedlings should be pulled along with a rooting medium for transplanting. If a polythene bag is used for raising the nursery, remove the bag before transplanting. Bare root seedling transplanting has very less survival. Transplant seedlings into the field at distances similar to those used for the direct seeding method.

14.9.7 Training and Pruning

Training and pruning at proper stages is required for maintaining the balance between vegetative and reproductive growths. In ash gourd, prune all secondary shoots and leaves up to 3 ft of vine from the base. Vines are generally not trained; however, staking of plants particularly in April sown crop is helpful in checking the rotting of fruits as it will be ready in the rainy season.

14.9.8 Weeding

At initial stages of crop growth, the weeds compete with crops for nutrients, moisture, light, and space and cause huge yield reduction. Firsthand weeding may be done about 15–20 days after sowing, followed by shallow hoeing. A total of two to three weedings are needed in cropping duration. Herbicides can also be used for weed control in ash gourd. Preemergence application of pendimethalin or alachlor (1 kg/ha) may solve the problem of weeds up to 30–35 days after sowing. If weeds appear later in the season, they may be controlled by hand weeding.

14.9.9 Mulching

Organic mulches, which protect the root of the plant from heat, conserve soil moisture, reduce weed infestation, and modify the soil and air microclimate, are successfully used in ash gourd. Mulch also improves the quality of fruits by avoiding their direct contact with soil.

14.9.10 Irrigation

Ash gourd has extensive root system and responds well to irrigation. In rainy season crop, no irrigation is required unless there is a long break in rain; however, in summer season crop, light irrigation is given just after seed sowing/planting to facilitate germination and establishment of plant, and subsequent irrigations are given at 8–10-day interval. There should be sufficient soil moisture during flowering and fruit development. Overirrigation causes excessive vegetative growth, while water logging causes severe damage to crop. Irrigation is provided through channels in ridge sown crop and flood irrigation is given during summer to create the microclimate that will favor the growth, flowering, and fruiting. In ash gourd, initial vegetative growth, flowering, and fruit enlarging are the critical stages for irrigation. Any moisture stress at these stages will reduce the crop yield and fruit quality.

14.9.11 Nutrition

In general, 100–120 kg nitrogen, 60–80 kg phosphorus, and 60–80 kg potash are recommended. The total amount of phosphorus and potash and ⅓ amount of nitrogen are mixed and applied as basal dressing near the expected root zone at the time of furrow or pit preparations, while the remaining ⅔ amount of nitrogen is given as top dressing in two equal splits for about 20 days (start of vine growth) and 40 days (flower initiation) after seed sowing followed by hoeing and earthing up. The spray of water-soluble fertilizer at the rate of 5 g at 15-day intervals up to fruit set gives good response.

14.10 MANAGEMENT OF DISEASES AND PESTS

14.10.1 Diseases

14.10.1.1 Anthracnose

This disease is caused by *Colletotrichum orbiculare* or *Colletotrichum lagenarium*. Symptoms are found on all aboveground plant parts. Light-brown circular spots appear on the leaf that later turn to deep brown. Elongated lesions observed on the stem and circular to oval sunken lesions appear on the fruit, and subsequently fruits shrivel, darken, and finally dry up.

14.10.1.1.1 Control

All infected leaves should be collected and burned. Seeds should be treated with carbendazim or captan at 2.5 g/kg seed before sowing. Two foliar sprays of carbendazim (1 g/L of water) are effective to control this disease.

14.10.1.2 Fruit Rots

The decaying of fruits is a major problem during storage and caused by *Fusarium solani*, *Fusarium moniliforme*, *Verticillium dahliae*, *Sclerotium rolfsii*, and *Phomopsis cucurbitacearum*. Rotting starts from the lower fruit portion that had been in contact with soil. White fungal growth, which later turned into brown circular scelerotia, is observed on fruit surface. This is more common in fields having high moisture. The infected fruits contain completely rotten seeds that turn white, become hollow, and fail to germinate.

14.10.1.2.1 Control

Soil should be exposed to the sun by repeated deep summer plowing. Cereal-based crop rotation should be adopted. Seeds should be treated with *Trichoderma viride* along with *neem* cake application.

14.10.1.3 *Fusarium* Wilt

This disease is more common in sandy soil and caused by *F. oxysporum* f. sp. *benincasae*. The symptom is characterized by yellowing of lower leaves, which gradually progresses to upper leaves. The plant, soon after infection, starts drooping followed by wilting, and later the wilting becomes permanent. Vascular browning was observed as a characteristic symptom after splitting of roots and lower portion of the stem. The severity of disease is more if the soil is infected with root knot nematode.

14.10.1.3.1 Control

Apply *Trichoderma* at 5–8 kg/ha, depending upon the soil structure, to minimize the disease and treat the seed with carbendazim at 2.5 g/kg of seed.

14.10.1.4 Downy Mildew

This disease is caused by *Pseudoperonospora cubensis*. Generally, at initial stage, small, irregular, yellowish lesions appear on leaves. Old lesions become necrotic and are clearly demarcated with light-yellow areas. In high humid weather, faint, white downy fungus growth is observed on the lower surface of the leaves.

14.10.1.4.1 Control

Infected leaves should be removed before spraying and the crop debris should be burned. Protective spray of zineb mixed with 2 g/L of water at 7-day intervals gives good control. In severe cases, one spray of cymoxanil 8% + mancozeb 64% mixed with 2 g/L of water may be used.

14.10.1.5 Powdery Mildew

This disease is caused by *Erysiphe cichoracearum*. White, long patches or coatings appear first on the under surface of the leaves, which later spread to both surfaces and stem, petiole, and other succulent plant parts.

14.10.1.5.1 Control

Infected leaves should be removed before spraying and burned. Regular spray of fungicides like thiophanate methyl or carbendazim mixed with 1 g/L of water at 10–15-day intervals gives better control.

14.10.1.6   Watermelon Virus

Most of the cucurbits and cowpea plants act as hosts for this virus. The leaves of diseased plants develop mosaic or mottling accompanied by green vein banding and reduction in leaf size. The petiole and internodes also get shortened. The mosaic in ash gourd was reported from Gorakhpur and the virus was identified as WMV.

14.10.1.6.1   Control

Avoid the overlapping of vines by maintaining proper sowing distance and spray the insecticide imidacloprid mixed with 3 mL/10 L of water for controlling vectors.

14.10.2   INSECT PESTS

14.10.2.1   Red Pumpkin Beetle

The red pumpkin beetle (*Aulacophora foveicollis*, Syn. *Raphidopalpa foveicollis*) is very destructive, particularly in the summer season when the plants have two to four leaves. Adults feed on the cotyledonary leaves, damaging the seedling and foliage by biting and making holes.

14.10.2.1.1   Control

Two to three sprays of carbaryl (50 WP) 2–2.5 g/L or dichlorvos 76 EC mixed with 0.75 mL/L of water at 10-day intervals is recommended at early growing stage. Drenching of soil near the root zone with chloropyrophos (20EC) mixed with 2.5 mL/L of water at 15-day intervals is helpful to manage the larval infestation in roots.

14.10.2.2   Leaf Miner

Leaf miner (*Liriomyza trifolii*) attack is observed more in the early crop. Leaf miner mines the leaves, especially mature leaves. The larvae scrap the chlorophyll and leaf tissues.

14.10.2.2.1   Control

The infected old leaves containing pupae of leaf miner should be collected and destroyed. Spraying of imidacloprid 17.8 SL mixed with 0.35 mL/L of water during the early stage of the crop before flowering or the application of dichlorvos 76EC mixed with 0.75 mL/L of water in severe infestation during reproductive phase of crop is recommended to manage the leaf miner. Before spraying, mature fruits must be harvested.

14.10.2.3   Fruit Fly

Fruit fly (*Bactrocera cucurbitae*) infects ash gourd throughout India. The adult female lays eggs in/on the fruit surface using a conical ovipositor. After hatching, the maggots feed inside the fruits, causing rotting and premature dropping of fruits, making them unsuitable for consumption. Mature fruits of ash gourd are damaged more than immature ones. The flies are most active after summer rain (June) and the rainy season (July–August) in northern India.

14.10.2.3.1   Control

At the initial stage, the infected fruits should be collected and destroyed. Application of bait containing 10% molasses along with carbaryl 50 WP at 2 g/L or malathion 50 EC at 2 mL/L of water was found more effective. Such bait in 1 ha should be sprayed at 250 spots. Male annihilation technique of adult flies through plastic bottle trap with ethanol, any insecticide (carbaryl/malathion), or cuelure (6:1:2) coated in wooden block is also very effective to manage fruit flies. Such traps should be installed at 25–30 places in a hectare.

14.10.2.4   Root Knot Nematode

The root knot nematode (*Meloidogyne incognita*) is very common in sandy soil. The symptoms are characterized as stunted and unthrifty condition of plant. Infected plants become pale to yellow. The plant canopy and fruiting capacity of infected plants is reduced drastically. The characteristic symptoms are clearly observed after uprooting the plants as swelling and gall formation on roots and rootlets. Small, round, oblong to irregular galls are formed as a result of hypertrophy and hyperplasia of the cell.

14.10.2.4.1   Control

Deep summer plowing and crop rotation with cereals like puddled paddy, bajara, sorghum, and marigold are helpful to minimize the nematode infestation. *Neem* cake should be applied at 20–25 q/ha in the soil.

14.11   HARVESTING

The crop matures in about 90–140 days after seed sowing. Small, solid, green, hairy, immature fruits, which are usually harvested after 10–12 days of anthesis, are best for culinary purposes. The immature fruit should have uniform color and size, with partially developed seeds. Development of a thick layer of white wax is the index for judging the right stage for the harvesting of fruits at full maturity. Using a sharp knife, harvesting is done, leaving a long peduncle attached to the fruit. Fruits harvested at fully mature stage have a better keeping quality than fruits harvested at immature stage and can be marketed to distant places.

The yield depends on several factors such as variety, growing season, soil type, and climatic conditions of the growing region. In general, the average yield of open-pollinated varieties of ash gourd varies from 30 to 70 t/ha and of hybrid from 35 to 70 t/ha.

14.12   POSTHARVEST MANAGEMENT

14.12.1   Storage

Green, immature fruits can be stored for 10–14 days at 10°C–12.5°C temperature and 85%–90% relative humidity. However, mature fruits having a thick layer of wax may be stored up to 3 months at room temperature in a cool, dry place. For prolonged shelf life, the waxy surface should not be washed off unless it needs cleaning. During storage, dehydration may induce sponginess, an unpleasant acidic flavor may result, and physical damage may occur. Other common defects during storage include small brown pits, seed germination, and change in flavor toward sourness. The green, immature fruits can be packed into Styrofoam-ventilated trays or in wooden containers of 15 kg or more for long-distance transportation.

14.12.2   Value Addition

14.12.2.1   Processing Technique and Value Addition

Several value-added products are prepared from mature fruits, and the most common preparations are preserves (sweets), candies, and bari. Kushmanda avaleham (like chyawanprash), an Ayurvedic preparation, is also made from the fruit paste along with ghee and other ingredients.

14.12.2.2   Type of Petha and Processing Protocol

A wide range of petha products (a type of sweet) is available in the market in several styles and flavors. The name of petha sweet is generally associated with the materials used, that is, kesar elaichi petha (use of kesar and elaichi) and dry fruit petha (use of dry fruits). In general, based on preparation techniques, petha is broadly classified as crystallized (plain petha) and preserved (Punchi or Kashi petha).

TABLE 14.2
Ingredients Used to Get 100 kg of a Final Product of Petha

Ingredient	Quantity	Ingredient	Quantity
Crystallized petha			
Ash gourd fruits	220 kg	Alum	150 g
Sugar	75 kg	Sodium hydrogen sulfite	150 g
Water	50–55 L	Milk powder	250 g
Lime	5 kg	Rose water	40 mL
Kashi petha (petha with sugar syrup)			
Ash gourd fruit	400 kg	Alum	150 g
Sugar	60–65 kg	Sodium hydrogen sulfite	150 g
Water	60–65 L	Milk powder	250 g
Lime	5 kg	Rose water	50 mL
Citric acid	20 g		

The processing method of crystallized petha has been standardized with modification in lime water treatment time and blanching time compared to commercial method of preparation of petha (candy). The quantity of sugar was also standardized and recommended to be comparatively lesser compared to existing commercial processing methods. The processing method of Punchi petha or Kashi petha (preserve with sugar syrup) has also been standardized. The material required for the preparation of a 100 kg final product of crystallized petha and Kashi petha is given in Table 14.2.

REFERENCES

Eskin, N. A. M. and Landman, A. D. 1975. Study of milk clotting by an enzyme from ash gourd (*Benincasa cerifera*). *Journal of Food Science*. 40(2): 413–414.

Jiang, B., Xie, D., Liu, W., Peng, Q., and He, X. 2013. De Novo assembly and characterization of the transcriptome, and development of SSR markers in wax gourd (*Benincasa hispida*). *PLoS ONE*. 8(8): e71054.

Kaur, S., Randhawa, K. S., Singh, M., and Arora, S. K. 1985. Comparative pathogenicity of three species of *Fusarium* causing fruit rot of ash gourd. *Indian Journal of Mycology and Plant Pathology*. 15: 308–310.

Krishnasamy, V. 1992. Effect of orientation of seed placement in soil on seedling emergence in some cucurbitaceous vegetables. *Seed Research*. 20(2): 70–73.

Meng, X. D., Wei, Y. Y., Ma, H., Zhang, W. H., and Li, J. R. 1996. Identification of Chinese wax gourd and chieh-qua cultivars using RAPD markers. *Acta Agriculturae Shanghai*. 12: 45–49.

Meusel, I., Leistner, E., and Barthlott, W. 1994. Chemistry and micromorphology of compound. *Plant Systematics and Evolution*. 193(1–4): 115–123.

Pandey, S., Jha, A., and Rai, M. 2009. Screening of advance breeding lines/cultivars for shelf-life and biochemical changes during storage of ash gourd (*Benincasa hispida*). In: *Proceedings of the International Symposium on Underutilized Plants for Food Security, Nutrition, Income and Sustainable Development* (eds. H. Jaenicke et al.), Arusha, Tanzania. *Acta Horticultureae*. 806(1): 249–255.

Pandey, S., Kumar, S., Mishra, U., Rai, A., Singh, M., and Rai, M. 2008. Genetic diversity in Indian ash gourd (*Benincasa hispida*) accessions as revealed by quantitative traits and RAPD markers. *Scientia Horticulturae*. 118: 80–86 (doi:10.1016/j.scienta.2008.05.031).

Rifai, M. A. and Reyes, M. E. C. 1993. *Benincasa hispida* (Thunberg ex Murray) Cogniaux. In: *Plant Resources of South-East Asia*, eds. J. S. Siemonsma and K. Piluek, pp. 95–97. Wageningen, the Netherlands: Pudoc Scientific Publishers.

Robinson, R. W. and Decker-Walters, D. S. 1997. *Cucurbits*. New York: CAB International.

Simonds, N. W. 1976. *Evaluation of Crop Plants*, 1st edn. New York: Longman.

Singh, D. K. 2002. Genetic analysis of yield and its components in ash gourd [*Benincasa hispida* (Thunb.) Cogn.]. PhD thesis, UP College, Varanasi, India.

Sureja, A. K., Sirohi, P. S., Behera, T. K., and Mohapatra, T. 2006. Molecular diversity and its relationship with hybrid performance and heterosis in ash gourd [*Benincasa hispida* (Thunb.) Cogn.]. *Journal of Horticultural Science and Biotechnology*. 81: 33–38.

Wills, R. B. H., Wang, A., Scriven, F. M., and Green Field, H. 1984. Nutrient composition of chinese vegetables. *Journal of Agriculture and Food Chemistry*. 32(2): 413–416.

Yang, S. L. and Walter, T. W. 1992. Ethnobotany and the economic role of the Cucurbitaceae of China. *Economic Botany*. 46: 349–367.

Section VI

Cucurbit Grafting

15 Cucurbit Grafting
Methods, Physiology, and Responses to Stress

*Maryam Haghighi, Atena Sheibanirad,
and Mohammad Pessarakli*

CONTENTS

15.1 INTRODUCTION

Grafting has a long history in fruit trees, but commercial grafting in vegetable crops is relatively new (Ashtia et al., 1977; Sakata et al., 2007). The first reports of vegetable grafting were published in China, Japan, and Korea, and recently, it has been commercially performed. For extending grafting methods, plastic covers were used following cultures under greenhouse conditions. For the first time, cucurbit grafting was done in Japan and Korea, and 40 years later, around 1990, it was introduced to the Western countries. The first grafting practice was reported in Japan in 1930; *Citrullus lanatus* was grafted on *Lagenaria siceraria* L. rootstocks. Grafting is a successful, professional, and positive method in vegetable production, so some producers try to introduce suitable seeds of scion and rootstock for each specific purpose, although it has some negative points, such as more facilities are required (Sakata et al., 2007; Cushman and Huan, 2008; Davis et al., 2008).

Korea and Japan were the most popular countries for cucurbit production; both countries used the same number of seedlings per hectare, but the cultivation area was reduced in 2005 compared to that in 2000. Mostly, watermelon (69 × 1000 seedling/ha), melon (710 × 1000 seedling/ha), and cucumber (2030 × 1000 seedling/ha) grafting transplants were cultivated in these countries (Lee et al., 2010).

For the first time in Iran at Tehran University in 1971, the Charleston gray watermelon was grafted on the Mahbubi cultivar to make blossom and rot resistance in Charleston gray cultivar. It has been proven that grafting has some good effects in decreasing the dangers of cultural practices and abiotic stresses (Lee et al., 1993; Lee, 1994; Lee and Oda, 2003). Nowadays, people are more conscious about using healthy and organic products with a lower amount of fertilizers and pesticides, and the increasing world population has forced producers to use grafting methods more than before. Asia, Japan, Korea, China, and Taiwan were the countries that used grafting methods more than other Asian countries for vegetable production. In Japan, more than 13,000 ha is used for watermelon production and 92% of these fields used grafting methods like hole insertion grafting and splice grafting (Lee, 1994; Lee and Oda, 2003). *L. siceraria* and *C. lanatus* were used as watermelon rootstock in Japan. In Korea, nearly all fields of watermelon used grafting methods such as hole insertion grafting and tongue approach grafting, and mostly *Cucurbita maxima* × *Cucurbita moschata* and *L. siceraria* were used as rootstock (Lee, 1994; Lee and Oda, 2003; Lee, 2007, 2008). In China, only 20% of the watermelon fields used grafting methods (Lee, 1994; Lee and Oda, 2003). Also, hole insertion grafting, tongue approach grafting, and splice grafting methods on *L. siceraria* and *C. lanatus* rootstocks have been used. In Taiwan, similar methods as China were used, but *C. maxima* × *C. moschata* and *L. siceraria* were the most common rootstocks (Lee et al., 2010). In cucumber production, 75% of cultivation area in Japan and Korea used two methods: tongue approach grafting and splice grafting. Also, *L. siceraria*, *C. maxima* × *C. moschata*, and *C. lanatus* are used there as rootstocks. In China, 30% of 1,702,777 ha used grafting methods for cucumber production. Hole insertion grafting and tongue approach grafting methods were used on *C. moschata*, *Sicyos angulatus*, and *Cucurbita ficifolia* rootstocks for cucumber scions (Lee et al., 2010). In melon production, Japan has the highest cultivation area, and 30% of this area used tongue approach grafting and splice grafting methods on *C. moschata* and *C. maxima* × *C. moschata* rootstocks. In Korea, splice grafting and tongue approach grafting techniques are used, but only *C. maxima* × *C. moschata* is a common rootstock used for cucumber. In China, *C. moschata* and *C. maxima* × *C. moschata* *and C. lanatus* rootstocks with hole insertion grafting are common for cucumber production (Lee et al., 2010). In bitter melon production, *Luffa cylindrica* was used as rootstock and hole insertion grafting with tongue approach grafting methods were used (Lee et al., 2010).

15.2 GRAFTING PURPOSES

15.2.1 RESISTANCE TO SOILBORNE DISEASES

Reports have shown that the most suitable rootstocks had a substantial resistance to soilborne diseases such as *Fusarium, Verticillium, Pseudomonas, Didymella bryoniae, Monosporascus cannonballus*, and nematodes (Edelstein et al., 1999; Cohen et al., 2000, 2005, 2007; Ioannou, 2001). Naturally, the resistances shown in various scions and rootstocks were different and also were enhanced by their resistance to virus diseases (Sakata et al., 2007). Furthermore, using resistant rootstocks prevent spreading diseases in hydroponic systems. The resistance mechanisms are not clearly known yet; however, scientists believe that some compounds are made in rootstocks and transferred by the xylem to the scion and these compounds would not be produced in scions.

15.2.2 ENHANCING PLANT GROWTH VIGOR

By selecting right rootstocks with strong and extended roots, plants are able to absorb more water and nutrient elements under stress conditions. For example, in grafted watermelons, chemical fertilizer applications decreased to 1/2 to 2/3 of standard recommendation rate for engrafted plants (Lee et al., 2003; Salehi-Mohammadi et al., 2009). Especially, providence in nitrogen fertilizer usage in early growth stage and flower set happened in lower nodes of the grafted rather than the ungrafted plants. In the aforementioned reports, most of the grafted plants had more effective water and nutrient absorption, and early fruit set also happened on grafted plants and caused an increased in crop yield. In addition, xylem sap cytokinin contents were higher in the grafted plants compared with the ungrafted ones, and this higher amount of cytokinin promoted plant growth. Under less frequent irrigation condition in cucumber plants, the grafted ones absorbed more water, and root systems were more vigorous. The cytokinin compositions of xylem sap in grafted and ungrafted plants have been shown some differences: 86%, 19%, and 17% increased in *Cucumis sativus* zeatin, zeatin riboside, and dehydozeatin riboside, respectively, in grafted compared to ungrafted plants (Cushman and Huan, 2008). In *C. moschata* ungrafted plant, there was not much zeatin observed but in grafted plant (1.56 ng/mg sap) have been seen (Cushman and Huan, 2008; Lee et al., 2010). In *C. maxima*, the zeatin content did not change, but dihydrozeatin riboside and isopentyl adenine compounds were reduced in grafted plants. The total cytokinin content of *C. ficifolia* reduced 50% in grafted plant compared to ungrafted ones (Lee et al., 2010).

15.2.3 INCREASE YIELD

During the past years, the primary objective of horticulture has been to increase yield and productivity in order to provide the vegetables needed by a growing world population (Lee, 1994). In grafted vegetable fruit crops regardless the effects of soilborne disease on yield reduction, grafting increases crop yield. In grafted oriental melon compared to ungrafted one, fruit yield increased by 25%–55%, and it was related to the maintenance of good plant vigor until late in the growing season (Chung and Lee, 2007). Clearly, the scion variety affects final size, yield, and quality of fruit in grafted plants (Flores et al., 2010). Also, plants yield increased was related to the higher growth vigor under suboptimal conditions, particularly when *Fusarium* fungi attacked plants wherein no marketable plant yield was obtained (Chung and Lee, 2007). Furthermore, under soilborne disease condition, similar results in cucumbers and melons were found (Lee and Oda, 2003).

15.2.4 TOLERANCES TO ABIOTIC STRESSES

Some abiotic stresses like high and low temperatures can be prevented by grafting methods (Rivero et al., 2003; Venema et al., 2008). It has been reported that grafting improved water use efficiency (Rouphael et al., 2008a), increased endogenous hormones (Dong et al., 2008), improved nutrient uptake (Colla et al., 2010a), caused less organic pollutant absorption from soil (Otani and Seike, 2007), increased alkalinity resistance (Colla et al., 2010b), increased salt tolerance (Romero et al., 1997; Colla et al., 2006a,b; Yetisir et al., 2006), and prevented negative effects of heavy metals (Edelstein et al., 2005, 2007; Rouphael et al., 2008b; Savvas et al., 2009). In greenhouse cucurbit production, resistance to the high temperature during summer and low temperature during winter is very important. In order to have early crop production, seedlings were transferred to the greenhouse in early winter and then harvested in spring and early summer (Tachibana, 1982). In some greenhouses, the only heating source is sunlight and in some others, heating systems are used. In the traditional greenhouses, supplying adequate soil temperature in the early growth stage is difficult. Seed germination and seedlings at early growth stage need a high temperature in cucurbits; therefore, grafting cucumbers, watermelons, and oriental melons on tolerance rootstocks such as *C. maxima* Duch. × *C. moschata* Duch. or fig leaf gourd can greatly reduce the risk of severe growth inhibition caused by the low soil temperatures in winter in the greenhouses (Tachibana, 1982). Also, many physiological disorders can be minimized effectively by using grafted plants.

15.2.5 Improve/Enhance Fruit Quality

There are some contradictory reports about the advantages and disadvantages of grafting on fruit quality (Proietti et al., 2008; Flores et al., 2010). These differences are related to the various compounds of the rootstocks and scions, various climatic factors, different cultural conditions, and different growth stages and harvesting times. The grafted watermelons had a bigger size and heavier fruits than ungrafted ones, and it made producers choose the grafted type and have a higher yield. Furthermore, some other quality factors such as fruit shape, skin color, rind thickness, and total soluble solid concentrations were enhanced by the rootstock type. For example, in the exported cucumbers, important quality factors, including fruit skin color and flower development that are although genetically controlled, are greatly enhanced by the rootstocks. In addition, reports have shown, in cucumber, rootstocks caused reduction in soluble solids but increased rind thickness (Cushman and Huan, 2008; Ko, 2008).

15.2.6 Other Advantages of Cucurbit Grafting

In order to physiologically investigate the early blooming and flower initiation, scientists have used grafting that is also reported useful for bioassays of some diseases (Lee and Oda, 2003; Davis et al., 2008). Grafting has many benefits for cucurbit production, including yield increase; shoot growth initiation; resistance to nematodes, viruses, and disease; tolerance to high and low temperatures; improved water and nutrient absorption; resistance to high salt concentration and waterlogging; prolonged harvesting period and frequency; enhanced crop quality; convenient production of organic wastes; and ornamental values for exhibition and education (Masuda and Gomi, 1982). On the other hand, some disadvantages, including extra seed requirements for root stock production, required specialties, correct rootstock and scion selection, different combinations for cropping seasons and cropping methods, high seedlings cost, high seed-borne disease risk, excessive vegetative growth, delayed fruit harvesting, low fruit quality (taste, color, and TSS), increased physiological disorders, symptoms of incompatibility at later growth stages, different cultural methods used for different rootstocks and scions, have been reported (Lee et al., 2010; Lee, 2007, 2008).

15.3 GRAFTING METHODS

Different grafting methods for herbaceous plants such as vegetables were introduced. For different species, different methods were used and even among similar species, grafting methods were varied (Lee et al., 2010). In Korea, watermelons were grafted 60% with tongue, 35% with hole, and 1% with split grafting method. In Japan, mostly hole grafting is used, and only 7% cleft grafting is used. In cantaloupe production, the Japanese have used only hole insertion and tongue grafting, and the Koreans mostly have used these two methods (hole insertion and tongue grafting) and 15% split grafting. Also, in Korea, cucumber is usually grafted with split grafting, but in Japan, 86% have used tongue grafting method (Lee et al., 2010).

Grafting methods have some variations in details, such as grafting consolidating, necessary time for a successful grafting, and the amount of live plantlet, have been mentioned (Lee et al., 2010).

The most important and common grafting methods in vegetable crops include hole insertion grafting, tongue approach grafting, cleft grafting, splice grafting, pin grafting, tube grafting, and mechanical grafting, including semiautomatic and automatic grafting. These grafting methods are briefly described in the following sections.

15.3.1 HOLE INSERTION GRAFTING

Japanese farmers use very small watermelon hypocotyl scion on summer squash rootstocks. Seven to eight days before grafting, watermelon seeds are planted; then 3–4 days later, rootstock seeds are planted; scion and rootstock should be ready at the same time. Rootstock real leaves are clearly cut and a 1–1.5 cm hole is made inside them; stem is then prepared; and scion is inserted into the hole. Since this method does not need any excessive costs for eclipsing and removing them, it has been acceptable and popular among producers. The highest productive layer connection is one of the most important advantages of this method.

15.3.2 TONGUE APPROACH GRAFTING

One of the oldest grafting methods is tongue approach grafting. This method is even used in unsuitable and harsh conditions. Also, this method is more expensive and needs more space, but the number of the survived seedlings is higher in this method and no specific facilities are needed. In order to have more successful grafting, it is better that the scion and the rootstock have the same diameters. Scion seeds (cucumbers, melons, and watermelons) are usually planted earlier than the rootstock seeds. Rootstock growth point must be clearly cut before grafting. A notch with 30–40 degrees angle is made on the rootstock from top to bottom, and the same notch with the same size from bottom to top is made on the scion. To make a connection, both cuts should be put in the same point and fixed with grafting pins. Grafted seedlings should then be transferred to the pot with 9–12 cm diameter in a net house or in a cool greenhouse (Kashi et al., 2008).

15.3.3 SPLICE GRAFTING

Splice grafting is an acceptable method among Korean and Japanese producers. In this method, vascular systems of both parts are completely connected to each other and have a strong grafting point in the rootstock of one of the cotyledon leaves and the growth point of the cut. The grafting part is fixed with special pins (Kashi et al., 2008).

15.3.4 CLEFT GRAFTING

This kind of grafting in herbaceous plants is more difficult than in woody plants. The head of the rootstock must be cut and a notch of 1–1.5 cm height must be made inside it. The scion end must be shaped like a wedge and inserted into the rootstock notch. Finally, the grafted point should be fixed with grafting pins (Kashi et al., 2008; Lee, 2007).

15.3.5 PIN GRAFTING

This method is similar to splice grafting, but instead of using clips, special, natural ceramic pins are designed that can be kept inside the plant without any changes. Some pins that are made from bamboo can be an alternative for the ceramic pins.

15.3.6 TUBE GRAFTING

Using this method, many young and small plants can be used and grafting speed is increased. Japanese scientists are interested in using this method. Plants are grafted at the early growth stage; first, the rootstock part is cut in italic form, then, the scion is cut in the same way and put into the grafting point, and fixed with special pins (Kashi et al., 2008; Liu et al., 2004).

15.4 MECHANICAL GRAFTING

Based on rootstocks and scions structure, the best grafting methods are selected. Success in grafting is related to its method and speed. Big producers are interested in automatical grafting methods and grafting robots. These increase plant production above 10 million and reduce labor cost. The first grafting robots were made in Japan in 1993 and are still used. Presently, complete automatic and semiautomatic grafting robots are produced. A semiautomatic robot performs 600–800 grafts/h but needs an operator head and a simple worker; its maximum capacity equals 5–6 exerted workers (Lee, 1994). Recently, completely automatic robots produce 750 grafts/h and 90% of them are successful (Lee et al., 2010); they can feed rootstocks and seedlings and only need a skillful worker. The two kinds of grafting robots are described as follows.

15.4.1 SEMIAUTOMATIC

The semiautomatic grafting machine is the first machine used widely in Asia and North America. It can do 650–900 grafts/h with 90% success (Lee et al., 2010).

15.4.2 AUTOMATIC

The automatic grafting machine was made in Japan and can make 800 grafts/h with more than 95% success; it only needs an expert to lead it (Lee et al., 2010).

15.5 ROOTSTOCKS

Different rootstocks are used for various cultivars. Also, seed producers and breeders have introduced different rootstocks with specific characteristics for different situations. Some cucurbit rootstock characteristics that are taken from several reports are described here (Kato and Lou, 1989; Ko, 1999; Lee et al., 2008):
Watermelon grafted on some rootstocks

- *Lagenaria siceraria* L.: Cultivars Dongjanggoon, Bulrojangsaeng, and Sinhwachangjo in Korea and FR Dantos, Renshi, Friend, and Super FR Power in Japan because of their vigorous root systems. They have *Fusarium* and low-temperature tolerance.
- *C. moschata*: In Korea on Chinkyo, No. 8, Keumkang cultivars. These cultivars have some advantages such as vigorous root systems and *Fusarium* and low-temperature tolerance.
- *C. maxima* × *C. moschata*: Common cultivars of this hybrid in Japan, China, Taiwan, and Korea are Shintos wa #1, Shintos wa #2, and Chulgap. These rootstocks have some benefits, including vigorous root systems, strong vigor, and high- and low-temperature and *Fusarium* tolerance.
- *Cucurbita pepo*: This rootstock has three common cultivars, including Keumssakwa, Unyoung, and Super unyoung. This rootstock is mainly used because of its advantages of having vigorous root systems and *Fusarium* and low-temperature tolerance.
- *Benincasa hispida*: Lion, Best, and Donga cultivars are used because of their good disease resistance but may show incompatibility signs.

- *Citrullus lanatus*: The common rootstock in Japan is Tuffines and in Korea Ojakkyo. Both are used because of their *Fusarium* tolerance (Heo, 2000).
- *Cucumis metuliferus*: The only common cultivar of this rootstock is NHRI-1 and is used because of its *Fusarium* and nematode tolerance.

Cucumber rootstocks

- *C. ficifolia*: Black seeded fig leaf (Heukjong) cultivar, because of its low temperature and good disease resistance, is used for cucumber scions.
- *C. moschata*: Butternut, Unyoung #1, and Super unyoung cultivars have fruit quality modification and *Fusarium* tolerance properties; therefore, they are used as rootstocks for cucumber plants.
- *C. maxima* × *C. moschata*: Some common cultivars like Shintos wa, Keumtozwa, Fero RZ, 64-05 RZ, and Gangryuk Shinwha, because of their *Fusarium* and low-temperature tolerance, are used for common cucumber rootstocks.
- *S. angulatus*: Only the Andong cultivar, because of some of its advantages such as *Fusarium*, high soil moisture, nematode, and low-temperature tolerance, is used in cucumber production. However, reduction in yield may result from using this cultivar.
- *C. metuliferus*: Like watermelon, NHRI-1 cultivar is used as rootstock because of improved *Fusarium* and nematode tolerance, but plants may show weak temperature resistance.

Melons rootstocks:

- *C. moschata*: Because of *Fusarium* and low-temperature tolerance, Beakkukzwa, No. 8, Keumkang, Hongtoz wa cultivars *C. moschata* are used as rootstocks for melons.
- *C. maxima* × *C. moschata*: *Fusarium*, soil moisture, high- and low-temperature resistance result with grafting on Shintos wa, Shintos wa #1, and Shintos wa #2 cultivars. However, this rootstock may result in lower fruit quality.
- *C. pepo*: Some cultivars, including Keumsakva, Unyoung, and Super unyoung, like the previous rootstock, have *Fusarium*, soil moisture, and high- and low-temperature resistance.
- *Cucumis melo*: *Fusarium* tolerance and fruit quality improvement result by using Rootstock #1, Kangyoung, Keonkak, and Keumgang cultivars.
- AH cucumber (*E. May. ex Naud*): This is an African cucumber that has *Fusarium* and nematode tolerance, low temperature, and high soil moisture resistance in grafted scion, but plants may show low-temperature resistance.

Compared to classical breeding methods, grafting is one of the fastest ways of producing resistance rootstocks. Every year, many rootstocks are introduced that are stress resistant (Lee and Oda, 2003; Flores et al., 2010). Six rootstocks have been introduced using *L. siceraria* L. for watermelon and cucumber grafting. Also, several cultivars of watermelon, cucumber, melons, bitter melon, and wax gourd have been introduced using *C. moschata*. *C. maxima* × *C. moschata* cultivars are mostly used for watermelon, cucumber, and melon rootstocks. *Luffa* spp. is not a common rootstock for cucurbit plants; it is used only for bitter melon scions. Except for water melon, *C. ficifolia* is used as a strong rootstock for many cucurbits, especially cucumber plants (Lee et al., 2010).

15.6 EFFECTS OF STRESSES ON GRAFTING

15.6.1 Low-Temperature Stress

Temperature is one of the most important climatic factors that affect plant growth and crop yield. For crop production in early spring or during fall and autumn, the plant is faced with low-temperature stress. Melons and cucumbers can tolerate maximum temperature of 12°C–18°C when they are growing (Lee et al., 2010). In higher temperatures, more than 25°C–35°C, metabolic function rate is progressively decreased. Temperatures lower than 25°C–35°C in temperate zones causes increase in plant physiological disorders from slight structural changes to plant death. High temperature during vegetative growth reduces leaf growth and prevents leaf initiation; thus, the rate of early crop growth is decreased (Venema et al., 2008). Hormonal signals are released from roots, and nutrient absorption and cell wall flexibility are reduced. According to Lee et al. (2008), during tomato production, if the temperature is decreased, the following result:

1. Pollen quality reduces; thus, fruit set is decreased.
2. The time period of pollination and fruit set is increased.
3. Undesirable enlargement occurs in fruit size.
4. Fruit hardening rate is reduced.

Low fruit set and hardening rate reduce crop yield. Fruit quality is greatly related to the temperature (Dorais et al., 2001). Nowadays, breeders are interested in producing cultivars with higher-energy usage potential in order to increase crop yield and also save energy source (Van de et al., 2005, 2007). Also, production of some cultivars that are adapted to the special greenhouses with no heating system is suitable, and these cultivars are reported to be more adaptive to low temperatures (den Nijs, 1980; Zijlstra et al., 1994; Rivero et al., 2003; Venema et al,. 2008). Producers use rootstocks to increase crop yield, increase scion growth, and transfer disease tolerance. Recently, grafting has been used due to its positive effect on fruit quality (Davis et al., 2008; Flores et al., 2010; Rouphael et al., 2010).

The advantages of increasing vegetable crop resistance to suboptimal temperature include the following:

- Increasing growth period that increases harvesting period
- Crop adoption to short growth season
- Reduced fruit disorders under low temperature (Rick, 1983)
- Reduced irrigation needs
- Earlier seedling settlement (Foolad and Lin, 2001)
- Efficient use of greenhouse facilities
- Reduced extra CO_2 usage (Venema et al., 2008)

15.6.1.1 Rootstock Selection for Increasing Low-Temperature Stress Tolerance

Over 60 years ago, producers have used grafting to increase crop yield in plastic greenhouses or outside with no specific heating system, but plant growth was limited during winter by low temperature and even resulted in seedlings' death (Ahn et al., 1999; Lee et al., 2005). In 1970, the first cucumber rootstocks were introduced (Okimura et al., 1986; Bloon et al., 1998). For cucumber grafting, *C. ficifolia* Bouch. and *S. angulatus* L. were the first rootstocks used. The only cucurbit species that grows in 15°C is fig leaf gourd (Tachibana, 1982; Lee et al., 1998; Ahn et al, 1999; Rivero et al., 2003). Different researches indicated that these two rootstocks have improved vegetative growth and early fertilization under suboptimal temperatures (den Nijs, 1980; Tachibana, 1982; Bulder et al., 1991; Zhou et al., 2007). Also, the same result was found when the roots were exposed to 8°C temperature (Ahn et al., 1999). Recently, cucumber grafting on *C. moschata* Douch., which is grown under warm nutrient solution, 30°C, showed better growth under stress condition (Shibuya et al., 2007).

Watermelon grafting on *C. maxima* × *C. moschata* hybrid (Shin-tosa) improved planting time in cool season (Davis et al., 2008); another rootstock that can change the planting time for watermelon is the species *Torvum vigor* (Okimura et al., 1986). New breeding methods for rootstocks were related on a trial-and-error approach to construct interspecific hybrids of selected well-rooted wild species and vigorous cultivated species.

15.6.1.1.1   Rootstock Metabolisms under Low-Temperature Stress

Low root zone temperature improves plant appearance and crop yield. Root yield under low temperature changes water viscosity, root pressure and hydraulic conductance, metabolic activity, production, and upward transport of phytohormones (Nagel et al., 2009) as well as the ability of the root to absorb nutrients. Savvas et al. (2010) reported that the interaction of rootstock and scion, their physiological age, and stress period indicate the rootstock's ability to alleviate the negative effects of low root zone temperature.

15.6.1.2   Root Growth and Structure

The root is the hidden half of the plant and less research is done on it compared to the foliage part. New methods and facilities make it easier to study root changes under different temperatures (Nagel et al., 2009). In resistance cultivars, root development increased under low root zone temperatures (Tachibana, 1982, 1987; Vnema et al., 2008). Physiological mechanisms that are related to the energy division from leaves (source) to root (sink) were not clear till now (Perez-Alfocea et al., 2010). Compatibility reactions caused an even root and shoot function so that roots could absorb water and nutrient elements efficiently (Nakano et al., 2002). In another research, Lee et al. (2004) indicated no significant differences between sensitive cucumbers and fig leaf gourd root structure, but the root elongation rate decreased in some other studies (Zamir and Gadish, 1987; Vnema et al., 2008). Based on acid growth theory in pH 5, H-ATPase is activated and H^+ is released; thus, the cell wall would be extended (Cosgrove, 2000). H^+-ATPase activity was enhanced by hormonal signals like auxins (Rayle and Cleland, 1992) and environmental factors like temperature (Sze et al., 1999). Cell elongation was also affected by other internal factors like ethylene (Le et al., 2001; De Cnodder et al., 2006), abscisic acid (Sharp and LeNoble, 2002), cytokinins (Werner and Schmülling, 2009), gibberellic acid (Tanimoto, 2005), and exogenous root condition like calcium (Kiegle et al., 2000), phosphate, and iron concentration (Ward et al., 2008).

15.6.1.3   Nutrient Absorption

There is a great deal of information published on nutrient absorption in plants, especially on cucurbits, including fig leaf gourd. Cucumber rootstocks were sensitive to low temperature, although fig leaf gourd was resistant to this condition, and high use of nutrient elements, that is, nitrogen and phosphorus absorption, increased with this rootstock (Tachibana, 1982, 1987; Masuda and Gomi, 1984; Choi et al., 1995). Some micronutrients, Mn, Cu, and Zn content, increased by decreasing temperature (Li and Yu, 2007). Iron content was not significantly affected by low temperature; thus, no special changes in iron content among grafted and ungrafted plants were observed. Since under low temperature fewer roots were produced, nutrient absorption per root unit decreased (Starck et al., 2000).

15.6.1.4   Water Uptake and Osmolyte Transportation

Water uptake is one of the most important elements that is affected by low temperature, and a lot of research has been done about the chilling sensitive plant species that suffer from water stress (Choi et al., 1995; Ahn et al., 1999). The fastest and common visible symptom of water stress is leaf wilting (Bloom et al., 1998; Lee et al., 2005b). In chilling resistance rootstocks, by increasing root hydraulic conductance, cell wall suberin layers, lipid peroxidation, and closure of the stomata were reported to be reduced; thus, they could overcome the chilling temperature (Bloom et al., 1998; Lee et al., 2005a). Root hydraulic conductivity in fig leaf gourd roots was two times less than that in cucumbers under 8°C, which indicates less radial water transport (Lee et al., 2004b, 2005b, 2008). The cucumber aquaporins were more sensitive than the fig leaf gourd under low root zone temperature (Lee et al., 2005c). Furthermore, in chilling resistance species, root plasma membrane H^+-ATPase activity was involved in osmotic potential reduction for subsequent water uptake. In some species, by activating proton pump water absorption, the ability of the plants increased so that plants could tolerate the chilling condition (Ahn et al., 1999, 2000; Lee et al., 2004b, 2005b). Research indicated biochemical differences between the fig leaf gourd and cucumber's aquaporin and H^+-ATPase (Rhee et al., 2007). By using a suitable chilling reissuance rootstock, plant stomata stayed open longer and transpiration process continued; therefore, the plant lived longer (Yu et al., 1997, 1999). Under low temperature, in cucumber grafted plants, the water potential of leaves reduced; then some osmoregulation compounds such as amino acids, quaternary ammonium compounds, polyols, and sugars were accumulated to increase osmotic pressure in leaves, and there is no disrupting activity in plant metabolisms (Tachibana, 1982).

15.6.1.5 Antioxidants and Lipid Peroxidation

Reactive oxygen species (ROS) compounds such as hydrogen peroxide, superoxide, and hydroxide radicals that increase unsaturated membrane lipids under low temperature were higher in sensitive species (Tachibana, 1982; Guy et al., 2008). Electrolyte leakage and lipid peroxidation in watermelons and cucumbers that were grafted on a chilling tolerance rootstock decreased (Gao et al., 2008). Electrolyte leakage was measured by malondialdehyde concentration and increased under chilling stress with ROS compounds in sensitive species (Zhou et al., 2004, 2006, 2007; Rhee et al., 2007). Antioxidant activity was not the same as ROS scavenger enzymes, and for detoxifying these, it increased as reported by several investigators (Feng et al., 2002; Rivero et al., 2003; Li and Yu, 2007; Li et al., 2008; Gao et al., 2009; Zhou et al., 2009).

15.6.1.6 Sink–Source Relations

Root growth and metabolic activity were prevented by low root zone temperature, so the absorption ability of photosynthetic compounds reduced, and the consequence was carbohydrate accumulation in leaves and reduced leaf area (Venema et al., 1999, 2008). In this situation, thicker leaves were produced; thus, less light was fixed and less photosynthetic compounds were produced, so biomass production decreased (Paul and Foyer, 2001). Root sink ability was affected by suboptimal temperatures; these temperatures also affect leaf morphology, photosynthetic ability, and growth (Tachibana, 1987; Venema et al., 2008). Different researches elucidated low-temperature tolerance rootstocks increased photosynthetic activity under stress conditions (Ahn et al., 1999; Zhou et al., 2007; Gao et al., 2008; Li et al., 2008; Miao et al., 2009). Besides scion sink ability increase, roots of chilling resistance rootstocks should supply more water and nutrient elements for plant growth (Zhou et al., 2009).

15.6.1.7 Phytohormones

Phytohormone production is affected by low temperature; thus, root growth, photosynthetic ability, and shoot morphology are changed (Albacete et al., 2008, 2009; Perez-Alfocea et al., 2010). Researches on grafted cucumbers indicated that chilling-sensitive species in temperatures less than 7°C had higher ABA increase in xylem sap, but cytokinins content decreased (Zhou et al., 2007). In contrast, cucumbers' cytokinins concentration in fig leaf gourd xylem sap increased (Tachibana, 1987). Results of increasing cytokinins concentrations were more at mRNA levels of the large and small subunits of RuBisCO and the activities of RuBisCO and FBPase (Zhou et al., 2007).

15.6.2 HIGH-TEMPERATURE STRESS

Fruit production in vegetable plants was limited under supraoptimal temperature (Abadelhafeez et al., 1975; Abdelmageed and Gruda, 2009). According to Wang et al. (2007), with soilless culture and controlled conditions in greenhouses, vegetable's fruit production in hot regions is possible. High temperature, like other environmental stresses, causes a lot of changes in morphological, biological, biochemical, and molecular characteristics of plants (Wang et al., 2007).

15.6.2.1 Rootstocks as a Tool to Increase Supraoptimal Temperature Tolerance

Grafting in the *Cucurbitaceae* family did not have a significant effect on the high-temperature tolerance, but in the *Solanaceae* family, by grafting eggplant rootstock, the scion was able to tolerate temperatures higher than 28°C (Wang et al., 2007).

15.6.2.2 Rootstock Metabolism under Supraoptimal Temperature

High root zone temperature caused root elongation reduction and roots thicker in diameter (Qin et al., 2007). Ethylene biosynthesis increased; thus, root elongation was limited and resulted in an increase in leaf water content and stomata opening (Abeles et al., 1992; Qin et al., 2007). Some researchers showed phosphorus and iron limitation under high temperature in shoots and roots, and this caused an increase in ethylene production (Tan et al., 2002; Ward et al., 2008). High-temperature damages include changes in nutrient uptake, lipid phase transitions, and metabolism inhibition (Hansen et al., 1994).

15.6.3 WATER STRESS

15.6.3.1 Drought Stress

In arid and semiarid zones of the world, water sources are limited and water resources security is one of the most important issues. This is a great concern in supplying water for commercial vegetable production. An effective way to conserve water consumption is to use genetic engineering and transfer drought-resistant genes from a transporter like *Arabidopsis thaliana* to the main crop. In a study on cucumber, after expressing dehydration stress resistance, genes like *CBF1* and *CBF3* leaves produced more proline and compatible solute compounds, and photochemical efficiency also improved, so water stress resistance was enhanced (Beck et al., 2007). A way to decrease water losses in production and enhanced water use efficiency under drought conditions is to graft high-yielding genotypes onto drought-resistant rootstocks that are capable of reducing the effect of water stress on the shoot (García-Sánchez et al., 2007; Satisha et al., 2007). One of the benefits of drought-tolerant rootstocks is to improve nitrogen fixation (Serraj and Sinclair, 1996). Miniwatermelons that are grafted on a commercial rootstock (PS 1313 *C. maxima* Duchesne × *C. moschata* Duchesne) indicated 60% higher marketable yield when grown under drought conditions compared with ungrafted melons (Rouphael et al., 2008). This higher yield was related to more nutrient absorption, especially N, K, and Mg elements, and higher CO_2 assimilation (Holbrook et al., 2002).

15.6.3.2 Flooding Stress

When the bitter melon (*Momordica charantia* L. cv. New Known You #3) was grafted on luffa (*L. cylindrica* Roem cv. Cylinder #2), flooding resistance increased (Liao and Lin, 1996). A moderate reduction in photosynthetic rate, stomatal conductance, transpiration, soluble proteins, and/or activity of RuBisCO was possibly related to this difference in flooding tolerance. On the other hand, the depression of chlorophyll content in cucumber leaves induced by waterlogging was enhanced by grafting onto squash rootstocks (Kato et al., 2001). A biochemical signal in xylem exudate stimulated ethylene biosynthesis in foliage part, so it resulted in less commercial yield in watermelons (*C. lanatus* (Thunb.) Matsum and Nakai cv. 'Crimson Tide'), and also less chlorophyll content and aerenchyma formation in ungrafted plants were observed (Liao and Lin, 1996; Yetisir et al., 2006).

15.6.4 SALT STRESS

One of the biggest worldwide abiotic stresses is salinity in water sources and soils. Salt stress causes a worldwide crop productivity reduction (Arzani, 2008). Rhizosphere salt concentration increased due to high evaporation, low rainfall, bad water management, and indiscriminate use of chemical fertilizers (Mahjan and Tuteja, 2005). Plant response to the salinity differs by salt concentration, time of exposure, and growth stage and type of the cultural medium (Munns, 2002).

Breeders seek several methods to increase plant salinity tolerance, but the nature of the genetically complex mechanisms of abiotic stress tolerance and potential detrimental side effects makes this task extremely difficult (Wang et al., 2003; Flowers, 2004). In the *Cucurbitaceae* family, grafting salt-sensitive scion onto salt-resistant rootstocks ameliorated salt-induced damages to the shoot (Fernández-García et al., 2002, 2004; Santa-Cruz et al., 2002; Colla et al., 2005, 2006a,b; Estañ et al., 2005; Wei et al., 2007; Goreta et al., 2008; Martinez-Rodriguez et al., 2008; Zhu et al., 2008a,b; He et al., 2009; Huang et al., 2009a,b,c, 2010; Uygur and Yetisir, 2009; Yetisir and Uygur, 2010; Zhen et al., 2010). This environment-friendly technique also makes plant breeders mix favorite shoot characteristic with good root characteristics (Zijlstra et al., 1994; Pardo et al., 1998). Some mechanisms by grafting induced salt tolerance: leaves produce higher amounts of proline and sugar (Ruiz et al., 2005), leaves' antioxidant capacity increased (López-Gómez et al., 2007), and lower accumulation of Na^+ and/or Cl^- in the leaves is observed (Fernández-García et al., 2004; Estañ et al., 2005; Goreta et al., 2008; Zhu et al., 2008a,b). Scion and rootstock both impress the salt tolerance level of grafted plants (Etehadnia et al., 2008). In melons, determining the salt tolerance with root characteristics was possible (Romero et al., 1997). Zhu et al. (2008a,b) indicated that salt tolerance of grafted cucumber seedlings is related to the shoot genotype. Also, better crop performance in watermelon, melon, and cucumber has been seen. When watermelon ('Fantasy') was grafted onto 'Strongtosa' rootstock (*C. maxima* Duch. × *C. moschata* Duch.), the reductions in shoot weight and leaf area due to exposure to salinity were less than ungrafted plants. Furthermore, another research demonstrated that grafted 'Crimson Tide' watermelon (*C. lanatus* (Thunb.) Matsum et Nakai) onto *C. maxima* and two *L. siceraria* rootstocks showed higher growth performance than ungrafted plants under saline conditions (8.0 dSm^{-1}, Yetisir and Uygur, 2010). The reduction in shoot dry weight was 41% in ungrafted plants, while it varied from 22% to 0.8% in grafted plants under the same saline conditions (Goreta at al., 2008). A comparison of two melon varieties that were grafted onto three hybrids of squash under 4.6 dSm^{-1} salinity condition indicated that grafted melons were more resistant to salinity and showed higher yields than ungrafted ones (Romero et al., 1997). Although in some researches sensitivity to salinity was similar between grafted and ungrafted melon plants (Edelstein et al., 2005; Colla et al, 2006b). Bottle gourd rootstock 'Chaofeng 8848' enhanced cucumber 'Jinchun No. 2' shoot dry weight under salinity condition. Moreover, higher fruit number and marketable fruit yield were observed (Huang et al., 2009a,b).

15.6.5 ORGANIC POLLUTANTS AS STRESS

Organic pollution, known as "drind," includes

- Aldrin
- Dieldrin
- Endrin

These three groups have high toxicity levels, high bioaccumulation, and high persistency rate in the environment and are known as organic pollutants. In a research from Otani and Seike (2007), dieldrin and endrin absorptions from soil by cucumber plants were investigated in four different cultivars of *Cucurbita* spp., and 16 grafting combinations were made. Their results indicated that dieldrin and endrin contents in the foliage tissues of the grafted plants were mainly influenced by the rootstock variety. The highest dieldrin concentration was found in 'Schintosa-1gou' (*C. maxima* Duchesne × *C. moschata* Duchesne), which was 1.6 times less than the recorded value of 'Yuyuikki-black' rootstock (*C. moschata* Duchesne). Similar results were obtained for the endrin content, but the differences between grafted plants on *Cucurbita* rootstocks were smaller than those for dieldrin concentration.

REFERENCES

Abadelhafeez, A.T., Harssema, H., Verkerk, K., 1975. Effects of air temperature, soil temperature and soil moisture on growth and development of tomato itself and grafted on its own and eggplant rootstock. *Sci. Hortic.* 3, 65–73.

Abdelmageed, A.H.A., Gruda, N., 2009. Influence of grafting on growth, development and some physiological parameters of tomatoes under controlled heat stress conditions. *Eur. J. Hortic. Sci.* 74(1), 16–20.

Abeles, F.B., Morgan, P.W., Saltveit, M.E., 1992. *Ethylene in Plant Biology*, 2nd edn. Academic Press, San Diego, CA.

Ahn, S.J., Im, Y.J., Chung, G.C., Cho, B.H., Suh, S.R., 1999. Physiological responses of grafted-cucumber leaves and rootstock roots affected by low root temperature. *Sci. Hortic.* 81, 397–408.

Ahn, S.J., Im, Y.J., Chung, G.C., Seong, K.Y., Cho, B.H., 2000. Sensitivity of plasma membrane H^+-ATPase of cucumber root system in response to low root temperature. *Plant Cell Rep.* 19, 831–835.

Albacete, A., Ghanem, M.E., Martinez-Andujar, C., Acosta, M., Sanchez-Bravo, J., Martinez, V., Lutts, S., Dodd, I.C., Perez-Alfocea, F., 2008. Hormonal changes in relation to biomass partitioning and shoot growth impairment in salinized tomato (*Solanum lycopersicum* L.) plants. *J. Exp. Bot.* 59, 4119–4131.

Albacete, A., Martinez-Andujar, C., Ghanem, M.E., Acosta, M., Sanchez-Bravo, J., Asins, M.J., Cuartero, J., Lutts, S., Dodd, I.A., Perez-Alfocea, F., 2009. Rootstock-mediated changes in xylem ionic and hormonal status are correlated with delayed leaf senescence, and increased leaf area and crop productivity in salinized tomato. *Plant Cell Environ.* 32, 928–938.

Arzani, A., 2008. Improving salinity tolerance in crop plants: A biotechnological review. *In Vitro Cell. Dev. Biol. Plant* 44, 373–383.

Beck, E.H., Heim, R, Hansen, J. 2004. Plant resistance to cold stress: Mechanisms and environmental signals triggering frost hardening and dehardening. *J. Biosci.* 29, 449–459

Bloom, A.J., Randall, L.B., Meyerhof, P.A., St. Clair, D.A., 1998. The chilling sensitivity of root ammonium influx in a cultivated and wild tomato. *Plant Cell Environ.* 21, 191–199.

Bulder, H.A.M., den Nijs, A.P.M., Speek, E.J., van Hasselt, P.R., Kuiper, P.J.C., 1991. The effect of low root temperature on growth and lipid composition of low temperature tolerant rootstock genotypes for cucumber. *J. Plant Physiol.* 138, 661–666.

Choi, K.J., Chung, G.C., Ahn, S.J., 1995. Effect of root-zone temperature on the mineral composition of xylem sap and plasma membrane K^+-Mg^{2+}-ATPase activity of grafted-cucumber and figleaf gourd root systems. *Plant Cell Physiol.* 36, 639–643.

Chung, H.D., Lee, J.M., 2007. Rootstocks for grafting. In: *Horticulture in Korea*. Korean Society for Horticultural Science Technical Bulletin, Horticultural Science, pp. 162–167.

Cohen, R., Burger, Y., Horev, C., Koren, A., Edelstein, M., 2007. Introducing grafted cucurbits to modern agriculture: The Israeli experience. *Plant Dis.* 91, 916–923.

Cohen, R., Burger, Y., Horev, C., Porat, A., Edelstein, M., 2005. Performance of Galia type melons grafted onto Cucurbita rootstock in *Monosporascus cannonballus* infested and non-infested soils. *Ann. Appl. Biol.* 146, 381–387.

Cohen, R., Pivonia, S., Burger, Y., Edelstein, M., Gamliel, A., Katan, J., 2000. Toward integrated management of Monosporascus wilt of melons in Israel. *Plant Dis.* 84, 496–505.

Colla, G., Fanasca, S., Cardarelli, M., Rouphael, Y., Saccardo, F., Graifenberg, A., Curadi, M., 2005. Evaluation of salt tolerance in rootstocks of Cucurbitaceae. *Acta Hortic.* 697, 469–474.

Colla, G., Rouphael, Y., Cardarelli, M., Massa, D., Salerno, A., Rea, E., 2006b. Yield, fruit quality and mineral composition of grafted melon plants grown under saline conditions. *J. Hortic. Sci. Biotechnol.* 81, 146–152.

Colla, G., Rouphael, Y., Cardarelli, M., Rea, E., 2006a. Effect of salinity on yield, fruit quality, leaf gas exchange, and mineral composition of grafted watermelon plants. *HortScience* 41, 622–627.

Colla, G., Rouphael, Y., Cardarelli, M., Salerno, A., Rea, E., 2010b. The effectiveness of grafting to improve alkalinity tolerance in watermelon. *Environ. Exp. Bot.* 68, 283–291.

Colla, G., Suárez, C.M.C., Cardarelli, M., Rouphael, Y., 2010a. Improving nitrogen use efficiency in melon by grafting. *HortScience* 45, 559–565.

Cosgrove, D.J., 2000. Loosening of plant cell walls by expansins. *Nature* 407, 321–326.

Cushman, K.E., Huan, J., 2008. Performance of four triploid watermelon cultivars grafted onto five rootstock genotypes: Yield and fruit quality under commercial growing conditions. *Acta Hortic.* 782, 335–342.

Davis, A.R., Perkins-Veazie, P., Sakata, Y., López-Galarza, S., Maroto J.V., Lee, S.G., Huh, Y.C. et al., 2008. Cucurbit grafting. *Crit. Rev. Plant Sci.* 27, 50–74.

De Cnodder, T., Verbelen, J.P., Vissenberg, K., 2006. The control of cell size and rate of elongation in the *Arabidopsis* root. *Plant Cell Monogr.* 5, 249–269.

den Nijs, A.P.M., 1980. The effect of grafting on growth and early production of cucumbers at low temperature. *Acta Hortic.* 118, 57–63.

Dong, H.H., Niu, Y.H., Li, W.J., Zhang, D.M., 2008. Effects of cotton rootstock on endogenous cytokinins and abscisic acid in xylem sap and leaves in relation to leaf senescence. *J. Exp. Bot.* 59, 1295–1304.

Dorais, M., Papadopoulos, A.P., Gosselin, A., 2001. Greenhouse tomato fruit quality. *Hortic. Rev.* 26, 239–319.

Edelstein, M., Ben-Hur, M., Cohen, R., Burger, Y., Ravina, I., 2005. Boron and salinity effects on grafted and non-grafted melon plants. *Plant Soil* 269, 273–284.

Edelstein, M., Ben-Hur, M., Plaut, Z., 2007. Grafted melons irrigated with fresh or effluent water tolerate excess boron. *J. Am. Soc. Hortic. Sci.* 132, 484–491.

Edelstein, M., Cohen, R., Burger, Y., Shriber, S., Pivonia, S., Shtienberg, D., 1999. Integrated management of sudden wilt of melons, caused by *Monosporascus cannonballus*, using grafting and reduced rate of methyl bromide. *Plant Dis.* 83, 1142–1145.

Estañ, M.T., Martinez-Rodriguez, M.M., Perez-Alfocea, F., Flowers, T.J., Bolarin, M.C., 2005. Grafting raises the salt tolerance of tomato through limiting the transport of sodium and chloride to the shoot. *J. Exp. Bot.* 56, 703–712.

Etehadnia, M., Waterer, D., Jong, H.D., Tanino, K.K., 2008. Scion and rootstock effects on ABA-mediated plant growth regulation and salt tolerance of acclimated and unacclimated potato genotypes. *J. Plant Growth Regul.* 27, 125–140.

Feng, X., Yu, X., Guo, H., Ma, H., Wei, M., 2002. Effect of low temperature stress on the protective-enzyme activity of grafted cucumber seedlings and own-rooted cucumber seedlings. *J. Shandong Agric. Univ.* 33(3), 302–304.

Fernández-García, N., Martínez, V., Carvajal, M., 2004. Effect of salinity on growth, mineral composition, and water relations of grafted tomato plants. *J. Plant Nutr. Soil Sci.* 167, 616–622.

Fernández-García, N., Martínez, V., Cerdá, A., Carvajal, M., 2002. Water and nutrient uptake of grafted tomato plants grown under saline conditions. *J. Plant Physiol.* 159, 899–905.

Flores, F.B., Sanchez-Bel, P., Estañ, M.T., Martinez-Rodriguez, M.M., Moyano, E., Morales, B., Campos, J.F. et al., 2010. The effectiveness of grafting to improve tomato fruit quality. *Sci. Hortic.* 125, 211–217.

Flowers, T.J., 2004. Improving crop salt tolerance. *J. Exp. Bot.* 55, 307–319.

Foolad, M.R., Lin, G.Y., 2001. Genetic analysis of cold tolerance during vegetative growth in tomato, *Lycopersicon esculentum* Mill. *Euphytica* 122, 105–111.

Gahoonia, T.S., Nielsen, N.E., 1997. Variation in root hairs of barley cultivars doubled soil phosphorus uptake. *Euphytica* 98, 177–182.

Gao, J.J., Qin, A.G., Yu, X.C., 2009. Effect of grafting on cucumber leaf SOD and CAT gene expression and activities under low temperature stress. *Chin. J. Appl. Ecol.* 20(1), 213–217.

Gao, Q.H., Xu, K., Wang, X.F., Wu, Y., 2008. Effect of grafting on cold tolerance in eggplant seedlings. *Acta Hortic.* 771, 167–174.

García-Sánchez, F., Syvertsen, J.P., Gimeno, V., Botia, P., Perez-Perez, J.G., 2007. Responses to flooding and drought stress by two citrus rootstock seedlings with different water-use efficiency. *Biol. Plant* 130, 532–542.

Goreta, S., Bucevic-Popovic, V., Selak, G.V., Pavela-Vrancic, M., Perica, S., 2008. Vegetative growth, superoxide dismutase activity and ion concentration of salt stressed watermelon as influenced by rootstock. *J. Agric. Sci.* 146, 695–704.

Guy, C., Kaplan, F., Kopka, J., Selbig, J., Hincha, D.K., 2008. Metabolomics of temperature stress. *Physiol. Plant* 132, 220–235.

Hansen, L.D., Afzal, M., Breidenbach, R.W., Criddle, R.S., 1994. High- and low temperature limits to growth of tomato cells. *Planta* 195, 1–9.

He, Y., Zhu, Z.J., Yang, J., Ni, X.L., Zhu, B., 2009. Grafting increases the salt tolerance of tomato by improvement of photosynthesis and enhancement of antioxidant enzymes activity. *Environ. Exp. Bot.* 66, 270–278.

Heo, Y.C., 2000. Disease resistance of Citrullus germplasm and utilization as watermelon rootstocks. PhD Dissertation, Kyung Hee University, Seoul, South Korea (in Korean with English summary).

Holbrook, N.M., Shashidhar, V.R., James, R.A., Munns, R., 2002. Stomata control in tomato with ABA deficient roots: Response of grafted plants to soil drying. *J. Exp. Bot.* 53(373), 1503–1514.

Huang, Y., Tang, R., Cao, Q.L., Bie, Z.L., 2009. Improving the fruit yield and quality of cucumber by grafting onto the salt tolerant rootstock under NaCl stress. *Sci. Hortic.* 122, 26–31.

Huang, Y., Tang, R., Cao, Q.L., Bie, Z.L., 2009b. Improving the fruit yield and quality of cucumber by grafting onto the salt tolerant rootstock under NaCl stress. *Sci. Hortic.* 122, 26–31.

Huang, Y., Bie, Z.L., Liu, Z.X., Zhen, A., Wang, W.J., 2009c. Exogenous proline increases the salt tolerance of cucumber by enhancing water status and peroxidase enzyme activity. *Soil Sci. Plant Nutr.* 55, 698–704.

Huang, Y., Zhu, J., Zhen, A., Chen, L., Bie, Z.L., 2009a. Organic and inorganic solutes accumulation in the leaves and roots of grafted and ungrafted cucumber plants in response to NaCl stress. *J. Food Agric. Environ.* 7, 703–708.

Huang, Y., Bie, Z., He, S., Hua, B., Zhen, A., Liu, Z., 2010. Improving cucumber tolerance to major nutrients induced salinity by grafting onto Cucurbita ficifolia. *Environ. Exp. Bot.* 69, 32–38.

Ioannou, N., 2001. Integrating soil solarization with grafting on resistant rootstocks for management of soil-borne pathogens of eggplant. *J. Hortic. Sci. Biotechnol.* 76, 396–401.

Kashi, A., Salehi, R., Javanpoor, R., 2008. Grafting technology in vegetable production. Agriculture Training Publisher, Tehran, Iran (in Farsi).

Kato, C., Ohshima, N., Kamada, H., Satoh, S., 2001. Enhancement of the inhibitory activity for greening in xylem sap of squash root with waterlogging. *Plant Physiol. Biochem.* 39, 513–519.

Kato, T., Lou, H., 1989. Effect of rootstock on the yield, mineral nutrition and hormone level in xylem sap in eggplant. *J. Jpn. Soc. Hortic. Sci.* 58, 345–352.

Kiegle, E., Gilliham, M., Haseloff, J., Tester, M., 2000. Hyperpolarisation-activated calcium currents found only in cells from the elongation zone of *Arabidopsis thaliana* roots. *Plant J.* 21, 225–229.

Ko, K.D., 1999. Response of cucurbitaceous rootstock species to biological and environmental stresses. PhD dissertation, Seoul National University, Seoul, South Korea.

Ko, K.D., 2008. Current status of vegetable seedling production in Korea and its prospects. In: *Inauguration Seminar of Korean Plug Growers Association*, June. Seoul, South Korea.

Le, J., Vandenbussche, F., Van Der Straaten, D., Verbelen, J.P., 2001. In the early response of arabidopsis roots to ethylene cell elongation is up- and downregulated and uncoupled from differentiation. *Plant Physiol.* 125, 519–522.

Lee, J.M., 1994. Cultivation of grafted vegetables. I. Current status, grafting methods, and benefits. *HortScience* 29, 235–239.

Lee, J.M., 2008. Vegetable grafting: A powerful aid for cultivation of environmentally friendly produce. *KAST Rev. Modern Sci. Technol.* 4, 68–85 (The Korean Academy of Science & Technology).

Lee, J.M., Bang, H.J., Ham, H.S., 1998. Grafting of vegetables. *J. Jpn. Soc. Hortic. Sci.* 67, 1098–1114.

Lee, S.H., Chung, G.C., 2005. Sensitivity of root system to low temperature appers to be associated with the hydraulic properties trough aquaporin activity. *Sci. Hortic.* 105, 1–11.

Lee, J.M., Oda, M., 2003. Grafting of herbaceous vegetable and ornamental crops. *Hortic. Rev.* 28, 61–124.

Lee, S.H., Singh, A.P., Chung, G.C., 2004a. Rapid accumulation of hydrogen peroxide in cucumber roots due to exposure to low temperature appears to mediate decreases in water transport. *J. Exp. Bot.* 55, 1733–1741.

Lee, S.H., Singh, A.P., Chung, G.C., Ahn, S.J., Noh, E.K., Steudle, E., 2004b. Exposure of roots of cucumber (*Cucumis sativus*) to low temperature severely reduces root pressure, hydraulic conductivity and active transport of nutrients. *Physiol. Plant.* 120, 413–420.

Lee, S.H., Ahn, S.J., Im, Y.J., Cho, K., Chung, G.C., Cho, B.-H., Han, O., 2005a. Differential impact of low temperature on fatty acid unsaturation and lipoxygenase activity in figleaf hourd and cucumber roots. *Biochem. Biophys. Res. Commun.* 330, 1194–1198.

Lee, S.H., Chung, G.C., Steudle, E., 2005b. Gating of aquaporins by low temperature in roots of chilling-sensitive cucumber and chilling-tolerant figleaf gourd. *J. Exp. Bot.* 56, 985–995.

Lee, S.H., Chung, G.C., Steudle, E., 2005c. Low temperature and mechanical stresses differently gate aquaporins of root cortical cells of chilling-sensitive cucumber and -resistant figleaf gourd. *Plant Cell Environ.* 28, 1191–1202.

Lee, J.M., Kubota, C., Tsao, S.J., Bie, Z., Hoyos Echevarria, P., Morra, L., Oda, M., 2010. Current status of vegetable grafting: Diffusion, grafting techniques, automation. *Sci. Hortic.* 127, 93–105.

Lee, J.M., Kubota, C., Tsao, S.J., Vinh, N.Q., Huang, Y., Oda, M., 2008. Recent progress in vegetable grafting. In: *International Workshop on Development and Adaptation of Green Technology for Sustainable Agriculture and Enhancement of Rural Entrepreneurship*, IRRI, Los Banós, Laguna, Philippines, September 28–October 2, 2009, 21pp.

Lee, J.M., Oda, M., 2003. Grafting of herbaceous vegetable and ornamental crops. *Hortic. Rev.* 28, 61–124.

Lee, S.G., 2007. Production of high quality vegetable seedling grafts. *Acta Hortic.* 759, 169–174.

Li, T., Yu, X., 2007. Effect of Cu^{2+}, Zn^{2+}, and Mn^{2+} on SOD activity of cucumber leaves extraction after low temperature stress. *Acta Hortic. Sin.* 34(4), 895–900.

Li, Y.T., Tian, H.X., Li, X.G., Meng, J.J., He, Q.W., 2008. Higher chilling-tolerance of grafted-cucumber seedling leaves upon exposure to chilling stress. *Agric. Sci. China* 7(5), 570–576.

Liao, C.T., Lin, C.H., 1996. Photosynthetic response of grafted bitter melon seedling to flood stress. *Environ. Exp. Bot.* 36, 167–172.

Liu, H.Y., Zhu, Z.J., Lü, G., 2004. Effect of low temperature stress on chilling tolerance and protective system against active oxygen of grafted watermelon. *Chin. J. Appl. Ecol.* 15, 659–662 (in Chinese).

López-Gómez, E., San Juan, M.A., Diaz-Vivancos, P., Mataix Beneyto, J., García-Legaz, M.F., Hernández, J.A., 2007. Effect of rootstocks grafting and boron on the antioxidant systems and salinity tolerance on loquat plants (*Eriobotrya japonica* Lindl.). *Environ. Exp. Bot.* 60, 151–158.

Mahjan, S., Tuteja, N., 2005. Cold, salinity and drought stresses: An overview. *Arch. Biochem. Biophys.* 444, 139–158.

Martinez-Rodriguez, M.M., Estañ, M.T., Moyano, E., Garcia-Abellan, J.O., Flores, F.B., Campos, J.F., Al-Azzawi, M.J., Flowers, T.J., Bolarín, M.C., 2008. The effectiveness of grafting to improve salt tolerance in tomato when an 'excluder' genotype is used as scion. *Environ. Exp. Bot.* 63, 392–401.

Masuda, M., Gomi, K., 1982. Diurnal change of the exudation rate and the mineral concentration in xylem sap after decapitation of grafted and non-grafted cucumbers. *J. Jpn. Soc. Hortic. Sci.* 51, 293–298.

Masuda, M., Gomi, K., 1984. Mineral absorption and oxygen consumption in grafted and non-grafted cucumbers. *J. Jpn. Soc. Hortic. Sci.* 54, 414–419.

Miao, M., Zhang, Z., Xu, X., Wang, K., Cheng, H., Cao, B., 2009. Different mechanisms to obtain higher fruit growth rate in two cold-tolerant cucumber (*Cucumis sativus* L.) lines under low night temperature. *Sci. Hortic.* 119, 357–361.

Munns, R., 2002. Comparative physiology of salt and water stress. *Plant Cell Environ.* 25, 239–250.

Nagel, K.A., Kastenholz, B., Jahnke, S., van Dusschoten, D., Aach, T., Muhlich, M., Truhn, D. et al., 2009. Temperature responses of roots: Impact on growth, root system architecture and implications for phenotyping. *Funct. Plant Biol.* 36, 947–959.

Nakano, Y., Watanabe, S.I., Okano, K., Tatsumi, J., 2002. The influence of growing temperatures on activity and structure of tomato roots hydroponically grown in wet atmosphere or in solution. *J. Jpn. Soc. Hortic. Sci.* 71, 683–690.

Okimura, M., Matso, S., Arai, K., Okitsu, S., 1986. Influence of soil temperature on the growth of fruit vegetable grafted on different stocks. *Bull. Veg. Ornam. Crops Res. Stn. Jpn.* C9, 43–58.

Otani, T., Seike, N.M., 2007. Rootstock control of fruit dieldrin concentration in grafted cucumber (*Cucumis sativus*). *J. Pest. Sci.* 32, 235–242.

Pardo, J.M., Reddy, M.P., Yang, S., 1998. Stress signaling through Ca^{2+}/calmodulin-dependent protein phosphatase calcineurin mediates salt adaptation in plants. *Proc. Natl. Acad. Sci. USA* 95, 9681–9686.

Paul, M.J., Foyer, C.H., 2001. Sink regulation of photosynthesis. *J. Exp. Bot.* 52, 1383–1400.

Perez-Alfocea, F., Albacete, A., Ghanem, M.E., Dodd, I.A., 2010. Hormonal regulation of source-sink relations to maintain crop productivity under salinity: A case study of root-to-shoot signalling in tomato. *Funct. Plant Biol.* 37, 592–603.

Proietti, S., Rouphael, Y., Colla, G., Cardarelli, M., De Agazio, M., Zacchini, M., Moscatello, S., Battistelli, A., 2008. Fruit quality of mini-watermelon as affected by grafting and irrigation regimes. *J. Sci. Food Agric.* 88, 1107–1114.

Qin, L., He, J., Lee, S.K., Dodd, I.C., 2007. An assessment of the role of ethylene in mediating lettuce (*Lactuca sativa*) root growth at high temperatures. *J. Exp. Bot.* 58, 3017–3024.

Rayle, D., Cleland, R.E., 1992. The acid-growth theory of auxin-induced cell elongation is alive and well. *Plant Physiol.* 99, 1271–1274.

Rhee, J.Y., Lee, S.H., Singh, A.P., Chung, G.C., Ahn, S.J., 2007. Detoxification of hydrogen peroxide maintains the water transport activity in figleaf gourd (*Cucurbita ficifolia*) root system exposed to low temperature. *Physiol. Plant* 130, 177–184.

Rick, C.M., 1983. Genetic variability in tomato species. *Plant Mol. Biol. Rep.* 1, 81–87.

Rivard, C.L., Louws, F.J., 2008. Grafting to manage soilborne diseases in heirloom tomato production. *HortScience* 43, 2104–2111.

Rivero, R.M., Ruiz, J.M., Sanchez, E., Romero, L., 2003. Does grafting provide tomato plants an advantage against H$_2$O$_2$ production under conditions of thermal shock? *Physiol. Plant* 117, 44–50.

Romero, L., Belakbir, A., Ragala, L., Ruiz, J.M., 1997. Response of plant yield and leaf pigments to saline conditions: Effectiveness of different rootstocks in melon plants (*Cucumis melo* L.). *Soil Sci. Plant Nutr.* 43, 855–862.

Rouphael, Y., Cardarelli, M., Colla, G., Rea, E., 2008. Yield, mineral composition, water relations, and water use efficiency of grafted mini-watermelon plants under deficit irrigation. *Hortscience* 43(3), 730–736.

Rouphael, Y., Cardarelli, M., Colla, G., Rea, E., 2008a. Yield, mineral composition, water relations, and water use efficiency of grafted mini-watermelon plants under deficit irrigation. *HortScience* 43, 730–736.

Rouphael, Y., Cardarelli, M., Rea, E., Colla, G., 2008b. Grafting of cucumber as a means to minimize copper toxicity. *Environ. Exp. Bot.* 63, 49–58.

Rouphael, Y., Schwarz, D., Krumbein, A., Colla, G., 2010. Impact of grafting on product quality of fruit vegetable crops. *Sci. Hortic.* 127, 172–179.

Ruiz, J.M., Blasco, B., Rivero, R.M., Romero, L., 2005. Nicotine-free and salt tolerant tobacco plants obtained by grafting to salinità-resistant rootstocks of tomato. *Physiol. Plant* 124, 465–475.

Sakata, Y., Ohara, T., Sugiyama, M., 2007. The history and present state of the grafting of cucurbitaceous vegetables in Japan. *Acta Hortic.* 731, 159–170.

Salehi-Mohammadi, R., Khasi, A., Lee, S.G., Huh, Y.C., Lee, J.M., Delshad, M., 2009. Assessing survival and growth performance of Iranian melon to grafting onto Cucurbita rootstocks. *Korean J. Hortic. Sci. Technol.* 27(1), 1–6.

Santa-Cruz, M.M., Martinez-Rodriguez, F., Perez-Alfocea, R., Romero-Aranda, R., Bolarin, M.C., 2002. The rootstock effect on the tomato salinity response depends on the shoot genotype. *Plant Sci.* 162, 825–831.

Satisha, J., Prakash, G.S., Bhatt, R.M., Sampath Kumar, P., 2007. Physiological mechanisms of water use efficiency in grape rootstocks under drought conditions. *Int. J. Agric. Res.* 2, 159–164.

Savvas, D., Papastavrou, D., Ntatsi, G., Ropokis, A., Olympios, C., Hartmann, H., Schwarz, D., 2009. Interactive effects of grafting and Mn-supply level on growth, yield and nutrient uptake by tomato. *HortScience* 44, 1978–1982.

Savvas, D., Colla, G., Rouphael, Y., Schwarz, D., 2010. Amelioration of heavy metal and nutrient stress in fruit vegetables by grafting. *Sci. Hortic.* 127, 156–161.

Serraj, R., Sinclair, T.R., 1996. Processes contributing to N2-fixation insensitivity to drought in the soybean cultivar Jackson. *Crop Sci.* 36, 961–968.

Sharp, R.E., LeNoble, M.E., 2002. ABA, ethylene and the control of shoot and root growth under water stress. *J. Exp. Bot.* 53, 33–37.

Shibuya, T., Tokuda, A., Terakura, R., Shimizu-Maruo, K., Sugiwaki, H., Kitaya, Y., Kiyota, M., 2007. Short-term bottom-heat treatment during low-air-temperature storage improves rooting in squash (*Cucurbita moschata* Duch.) cuttings used for rootstock of cucumber (*Cucumis sativus* L.). *J. Jpn. Soc. Hortic. Sci.* 76(2), 139–143.

Starck, Z., Niemyska, B., Bogdan, J., Akour-Tawalbeh, R.N., 2000. Response of tomato plants to chilling in association with nutrient or phosphorus starvation. *Plant Soil* 226, 99–106.

Sze, H., Li, X., Palmgren, M.G., 1999. Energization of plant cell membranes by H$^+$-pumping ATPases: Regulation and biosynthesis. *Plant Cell* 11, 677–689.

Tachibana, S., 1982. Comparison of root temperature on the growth and mineral nutrition of cucumber cultivars and figleaf gourd. *J. Jpn. Soc. Hortic. Sci.* 51, 299–308.

Tachibana, S., 1987. Effect of root temperature on the rate of water and nutrient absorption in cucumber and figleaf gourd. *J. Jpn. Soc. Hortic. Sci.* 55, 461–467.

Tan, L.P., He, J., Lee, S.K., 2002. Effects of root-zone temperature on the root development and nutrient uptake of *Lactuca sativa* L. cv. Panama grown in an aeroponic system in the tropics. *J. Plant Nutr.* 25, 297–314.

Tanimoto, E., 2005. Regulation of root growth by plant hormones—Roles for auxin and gibberellin. *Crit. Rev. Plant Sci.* 24, 249–265.

Uygur, V., Yetisir, H., 2009. Effects of rootstocks on some growth parameters, phosphorous and nitrogen uptake by watermelon under salt stress. *J. Plant Nutr.* 32, 629–643.

Van der Ploeg, A., Heuvelink, E., 2005. Influence of sub-optimal temperature on tomato growth and yield: A review. *J. Hortic. Sci. Biotechnol.* 80, 652–659.

Van der Ploeg, A., Heuvelink, E., Venema, J.H., 2007. Wild relatives as a source for sub-optimal temperature tolerance in tomato. *Acta Hortic.* 761, 127–133.

Venema, J.H., Posthumus, F., Van Hasselt, P.R., 1999. Impact of suboptimal temperature on growth, photosynthesis, leaf pigments, and carbohydrates of domestic and high aötitude, wild Lycopersicon species. *J. Plant Pysiol.* 155, 711–718.

Venema, J.H., Dijk, B.E., Bax, J.M., van Hasselt, P.R., Elzenga, J.T.M., 2008. Grafting tomato (*Solanum lycopersicum*) onto the rootstock of a high-altitude accession of *Solanum habrochaites* improves suboptimal-temperature tolerance. *Environ. Exp. Bot.* 63, 359–367.

Wang, S., Yang, R., Cheng, J., Zhao, J., 2007. Effect of rootstocks on the tolerance to high temperature of eggplants under solar greenhouse during summer season. *Acta Hortic.* 761, 357–360.

Wang, W., Vinocur, B., Altman, A., 2003. Plant responses to drought, salinity and extreme temperatures: Towards genetic engineering for stress tolerance. *Planta* 218, 1–14.

Ward, J.T., Lahner, B., Yakubova, E., Salt, D.E., Raghothama, G.K.G., 2008. The effect of iron on the primary root elongation of arabidopsis during phosphate deficiency. *Plant Physiol.* 147, 1181–1191.

Wei, G.P., Zhu, Y.L., Liu, Z.L., Yang, L.F., Zhang, G.W., 2007. Growth and ionic distribution of grafted eggplant seedlings with NaCl stress. *Acta Bot. Boreal-Occident. Sin.* 27, 1172–1178.

Werner, T., Schmülling, T., 2009. Cytokinin action in plant development. *Curr. Opin. Plant Biol.* 12, 527–538.

Yetisir, H., Caliskan, M.E., Soylu, S., Sakar, M., 2006. Some physiological and growth responses of watermelon [*Citrullus lanatus* (Thunb.) Matsum. and Nakai] grafted onto *Lagenaria siceraria* to flooding. *Environ. Exp. Bot.* 58, 1–8.

Yetisir, H., Uygur, V., 2010. Responses of grafted watermelon onto different gourd species to salinity stress. *J. Plant Nutr.* 33, 315–327.

Yu, X., Xing, Y., Ma, H., Wei, M., Feng, X., 1997. Study on the low temperature tolerance in grafted cucumber seedlings. *Acta Hortic. Sin.* 24(4), 348–352.

Yu, X., Xing, Y., Ma, H., Wei, M., Li, B., 1999. Changes of hormone in grafted and nongrafted cucumber seedlings under low temperature stress. *Acta Hortic. Sin.* 26(6), 406–407.

Zamir, D., Gadish, I., 1987. Pollen selection for low temperature adaptation in tomato. *Theor. Appl. Genet.* 74(5), 545–548.

Zhen, A., Bie, Z.L., Huang, Y., Liu, Z.X., Li, Q., 2010. Effects of scion and rootstock genotypes on the antioxidant defense systems of grafted cucumber seedlings under NaCl stress. *Soil Sci. Plant Nutr.* 56, 263–271.

Zhou, Y.H., Yu, J.Q., Huang, L.F., Nogues, S., 2004. The relationship between CO_2 assimilation, photosynthetic electron transport and water–water cycle in chill exposed cucumber leaves under low light and subsequent recovery. *Plant Cell Environ.* 27, 1503–1514.

Zhou, Y.H., Huang, L.F., Zhang, Y., Shi, K., Yu, J.Q., Nogues, S., 2007. Chill-induced decrease in capacity of RuBP carboxylation and associated H_2O_2 accumulation in cucumber leaves are alleviated by grafting onto Figleaf Gourd. *Ann. Bot.* 100, 839–848.

Zhou, Y.H., Wu, J.X., Zhu, L.J., Shi, K., Yu, J.Q., 2009. Effects of phosphorus and chilling under low irradiance on photosynthesis and growth of tomato plants. *Biol. Plant* 53, 378–382.

Zhou, Y.H., Yu, J.Q., Mao, W.H., Huang, L.F., Song, X.S., Nogues, S., 2006. Genotypic variation of rubisco expression, photosynthetic electron flow and antioxidant metabolism in the chloroplasts of chill-exposed cucumber plants. *Plant Cell Physiol.* 47, 192–199.

Zhu, J., Bie, Z.L., Huang, Y., Han, X.Y., 2008a. Effect of grafting on the growth and ion contents of cucumber seedlings under NaCl stress. *Soil Sci. Plant Nutr.* 54, 895–902.

Zhu, S.N., Guo, S.R., Zhang, G.H., Li, J., 2008b. Activities of antioxidant enzymes and photosynthetic characteristics in grafted watermelon seedlings under NaCl stress. *Acta Bot. Boreal-Occident. Sin.* 28, 2285–2291.

Zijlstra, S., Groot, S.P.C., Jansen, J., 1994. Genotypic variation of rootstocks for growth and production in cucumber; possibilities for improving the root system by plant breeding. *Sci. Hortic.* 56, 185–196.

16 Effects of Grafting on Nutrient Uptake by Cucurbits Irrigated with Water of Different Qualities

Menahem Edelstein and Meni Ben-Hur

CONTENTS

16.1 INTRODUCTION

Vegetable plant grafting is an old practice, with grafting of cucurbitaceous plants dating back to the seventeenth century. The primary motivation for the grafting of vegetable plants is to prevent damage caused by soil-borne pests and pathogens. However, grafted plants also exhibit changes in nutrient absorption and translocation. Cucurbits are grown mainly under irrigation, and the quality of the irrigation water (e.g., freshwater [FW], saline water, or treated wastewater [TWW]) can differ with the region. This chapter, therefore, focuses on the effects of cucurbit plant grafting on nutrient uptake under irrigation with various water qualities. Many studies have shown that under FW irrigation, the grafting of vegetable plants onto suitable rootstocks can increase the absorption and translocation of nitrogen (N), phosphorus (P), and potassium (K) to the plants, resulting in a higher yield. However, the mechanisms responsible for this increase in nutrient content in grafted plant shoots are still not completely known. It has been suggested that the main factor controlling the tolerance of grafted plants to low K concentration is the rootstock's high efficiency at absorbing K and translocating it to the shoot, rather than higher utilization of K. In contrast, the effects of plant grafting on nutrient uptake under irrigation with saline water and TWW have only been investigated in a few studies; under these conditions, the effects of grafting on nutrient absorption are more complex. In a field experiment under irrigation with a drip system, an interactive effect was found between plant grafting and water qualities (FW and TWW) on the content of some nutrients in the plants shoots. In other field experiments, it was found that under irrigation with saline water, the contents of magnesium (Mg), zinc (Zn), and manganese (Mn) are significantly lower in the leaves of grafted versus nongrafted melon plants.

16.2　GRAFTING

Plant grafting is the process of uniting two living plant parts, rootstock and scion, so that they grow as a single plant (Figure 16.1). The main procedures involved in this process are (1) choosing the proper rootstock and scion species; (2) creating a graft union by physical manipulation; (3) healing the union; and (4) acclimating the grafted plant. Grafting of vegetable plants is an old practice, with grafting of Cucurbitaceae being mentioned in a Korean book in the seventeenth century. The first commercial use of grafting of vegetable plants was in Asia in the twentieth century. Grafting of eggplants (*Solanum melongena* L., Solanaceae) was initiated in the 1950s, followed by grafting of cucumber (*Cucumis sativus* L., Cucurbitaceae) and tomato (*Lycopersicon esculentum* L., Solanaceae) around 1960 and 1970, respectively. More recently, a number of tools and machines have been developed by various companies to perform grafting and to hold the graft unions together (Lee and Oda 2003), enabling the production of large numbers of grafted plants in a relatively short time (Kurata 1994; Lee 2003; Lee et al. 2010). Consequently, grafting of vegetable plants has become common practice in many countries, such as Japan, Korea, the United States, Spain, Italy, Turkey, and Greece (Cohen et al. 2007). For example, in the year 2008, there were 3000 million grafted vegetable plants in Japan and Korea (Lee et al. 2010).

The primary motivation for grafting vegetable plants is to prevent damage caused by soil-borne pests and pathogens (Oda 2002). However, grafting plants has also been found to (1) increase the efficiency of water and nutrient use (Colla et al. 2010b; Ruiz and Romero 1999; Shimada and Moritani 1977; Yamasaki et al. 1994); (2) improve fruit yield amount and quality (Lee and Oda 2003; Moncada et al. 2013; Nisini et al. 2002; Oda 2002); and (3) enhance the plants' tolerance to environmental stresses (Rivero et al. 2003; Schwarz et al. 2010) such as low soil temperatures (Ahn et al. 1999), high salinity (Edelstein et al. 2011b; Fernandez-Garcia et al. 2003; Romero et al. 1997), high boron concentration (Edelstein et al. 2005, 2011a), and other toxic elements (Edelstein and Ben-Hur 2012). Despite the advantages of plant grafting, there are some limitations to applying this technique; for example, graft incompatibility can cause physiological disorders in the plants and fruit (Lee 1994; Pina and Errea 2005; Rouphael et al. 2010). Rootstocks can affect the growth and yield of scions as a result of their influence on the plant's nutrient uptake. Cucurbits are grown mainly under irrigation, and the quality of the irrigation water can differ with the region. This chapter focuses on the effects of grafting of cucurbit plants on nutrient uptake under irrigation with water of different qualities.

FIGURE 16.1　(**See color insert.**) Melon plant grafted onto pumpkin rootstock in a nursery.

16.3 IRRIGATION WITH DIFFERENT WATER QUALITIES

Water quality is determined mainly by the concentration and composition of its solutes and the concentration of inorganic and organic suspended solids. In regions that do not suffer from limited water, mainly FW (drinking water quality, permitted for human consumption) is used for irrigation, pumped mostly from surface water sources and ground water. In contrast, arid and semiarid regions suffer from a shortage of FW resources, constituting one of their main environmental problems. Due to the increasing population in these regions and global climate change, the pressure on water resources is expected to increase in the near future. Therefore, one of the challenges facing agriculture in these regions is to find new water sources for irrigation. The most common options involve the use of marginal water (saline water and TWW) for irrigation.

The main differences between saline and FW are in

1. The total concentration of electrolytes (C) in the water, which can be determined by measuring the electrical conductivity (EC). In general, the EC increases linearly with increasing C concentration in the water, and the approximate relationship between EC (dS/m) and C (mmol$_c$/L) in water is defined as C = 10·EC.
2. The sodicity of the water, which is determined by the sodium adsorption ratio (SAR) as

$$SAR = \frac{(Na^+)}{(Ca^{2+} + Mg^{2+})^{0.5}}$$

where the concentrations of the ions Na^+, Ca^{2+}, and Mg^{2+} are measured in mmol/L.
3. The anion composition in the water: whereas in FW, Cl^- is practically the only ion, the concentration of SO_4^{2-} can be relatively high in saline water.

Inland saline water usually has an EC of 0.7–42 dS/m and SAR > 5. In Israel, for example, the EC and SAR values of FW are commonly ~0.8 dS/m and ~2.5 (mmol$_c$/L)$^{-0.5}$ (Ben-Hur 2004), respectively, and those of saline water are commonly ~5 dS/m and ~20 (mmol$_c$/L)$^{-0.5}$, respectively (Shainberg and Letey 1984).

TWW is defined by its higher salinity, sodicity, and concentrations of macro- and microelements, nutrients, suspended solids, and dissolved organic matter (DOM) relative to FW (Ben-Hur 2004). Until recently, most of the TWW used for irrigation had undergone secondary (biological) treatment, and the quality of this TWW from different treatment plants in Israel is presented in Table 16.1. Today, new treatments for TWW are being developed (Tal 2006), which include a tertiary treatment consisting of filtration of the secondary effluent through various membranes to remove suspended solids, organic molecules, pathogens, parasitic worms, and salts from the water (Shannon et al. 2008). These treatments aim to produce TWW that is suitable for unrestricted irrigation and aquifer recharge. The quality of the TWW produced from domestic sewage from the city of Arad in the south of Israel after consecutive treatments in oxidation ponds (OP, secondary treatment), polishing ponds (PP, secondary treatment), ultrafiltration (UF, membrane treatment), and reverse osmosis (RO, membrane treatment) is presented in Table 16.2. The UF treatment removed most of the suspended solids and part of the DOM but did not change the EC or SAR of the TWW (Table 16.2). In contrast, the RO treatment significantly reduced all of the quality parameters of the TWW, including EC and SAR (Table 16.2).

TABLE 16.1

Values of pH, Electrical Conductivity (EC) and Sodium Adsorption Ratio (SAR), and Concentrations of HCO_3^-, Cl^-, Dissolved Organic Matter (DOM), and Macro- and Micronutrients in Secondary Treated Wastewater (TWW) and Freshwater (FW) Used in Different Locations in Israel

Site Location	Water Type	pH	EC (dS/m)	SAR	HCO_3^- (mmol/L)	Cl^-	DOM	Macronutrients (mg/L) NH_4–N	NO_3–N	PO_4–P	K	Micronutrients (µg/L) Fe	Mn	Zn	Cu	Mo
Coastal plain	TWW	7.5 (0.3)	1.8 (0.4)	3.8 (0.4)	7.9 (0.3)	6.1 (0.5)	138.2 (67)	32.4 (10.1)	0.05 (0.09)	7.5 (3.0)	31.2 (5.0)	320 (160)	ND[a]	144 (88)	88.0	5.0
	FW	7.2 (0.2)	0.9 (0.1)	1.5 (0.3)	3.6 (0.2)	3.0 (0.2)	4.3 (1.7)	0.16 (0.04)	4.9 (1.9)	0.04 (0.03)	0.2 (0.08)	20 (8)	8.0	ND	1.0 (0.4)	ND
Yizre'el valley	TWW	7.8 (0.2)	2.6 (0.36)	4.8 (0.04)	9.2 (0.07)	9.6 (2.0)	35.3 (0.22)	42.6 (0.07)	0.3 (0.74)	13.7 (0.86)	74.1 (3.0)	418 (0.17)	ND	145 (45)	31.1 (0.58)	ND
	FW	7.2 (0.2)	1.0 (0.1)	2.5 (0.2)	4.9 (0.3)	7.1 (0.9)	~0	~0	1.5 (0.2)	0.04 (0.02)	6.2 (0.7)	9.0 (0.8)	ND	54 (7.8)	BDV[b]	ND
Afula	TWW	8.5 (0.1)	2.0 (0.3)	5.3 (0.2)	ND[a]	9.9 (0.3)	39.0 (6)	13.7 (0.2)	3.8 (0.3)	7.9 (0.4)	26.6 (4)	4.0	8.0	32 (3.5)	12 (4)	4.0
	FW	7.2	1.0	2.5	ND	7.0	~0	~0	1.5	0.04	6.3	9	7	54	BDV	0
Akko	TWW	7.5 (0.2)	1.3 (0.1)	2.5 (0.2)	11.4 (0.3)	3.7 (0.3)	130 (45)	34.4 (8.2)	6.5 (0.5)	6.4 (0.4)	0.6 (0.1)	106 (4.2)	72 (2.1)	119 (8.5)	5.2 (0.2)	0.8 (0.1)
	FW	7.7	0.8	0.3	7.5	1.6	~0	~0	0.1	~0	0.1	9.3	3.7	19.5	~0	~0

Notes: Data for Coastal plain and Yizre'el valley follow Lado et al. (2012), for Afula, Edelstein et al. (2007), and for Akko, Edelstein and Ben-Hur (unpublished data). Numbers in parentheses are standard deviations.

[a] ND, not determined.

[b] BDV, below detection value.

TABLE 16.2

Average Values of Electrical Conductivity (EC), pH, Sodium Adsorption Ratio (SAR), and Concentrations of Cl⁻, HCO₃⁻, Dissolved Organic Matter (DOM), and Total Suspended Solids in Effluents after Oxidation Pond (OP), Ultrafiltration (UF), and Reverse Osmosis (RO) Treatments

Treated Sewage Water Type	EC (dS/m)	pH	SAR $(mmol_c/L)^{0.5}$	Cl^-	HCO_3^-	DOM	Total Suspended Solids >1 μm	<1 and >0.45 μm
				(mmol/L)		(mg/L)		
OP	2.0 (0.2)	7.7 (0.2)	6.0 (1.2)	7.4 (1.4)	12.3 (3.0)	266.9 (121.4)	69 (3.0)	2.3 (1.5)
UF	1.8 (0.3)	7.8 (0.3)	5.5 (0.9)	6.8 (1.4)	11.1 (1.9)	51.6 (32.5)	5.3 (0.6)	1.7 (0.6)
RO	0.2 (0.2)	7.1 (0.5)	2.5 (1.1)	0.7 (0.4)	2.2 (0.6)	8.4 (3.0)	0	0

Source: Lado, M. and Ben-Hur, M., *Soil Sci. Soc. Am. J.*, 74, 23, 2010.
Note: Numbers in parentheses are standard deviations.

16.4 NUTRIENT-USE EFFICIENCY

The nutrient-use efficiency of plants is controlled by three main components: (1) absorption—the nutrient-uptake capacity of the plant roots; (2) translocation—the plant's ability to deliver the nutrients from the root system to the shoot; and (3) utilization—the plant's efficiency at using nutrients for dry matter production. Nutrient-use efficiency (E_n) can be defined as

$$E_n = \frac{Y_p}{N}$$

where
Y_p is the total dry biomass of the plant
N is the available amount of the nutrient in the root zoon

Nutrient-use efficiency varies among species (Ahmad et al. 2001; Baligar et al. 2001), and grafting can increase E_n values in some plants. Many studies have investigated the effects of plant grafting on nutrient-use efficiency under irrigation with FW, but much less is known about these effects under irrigation with marginal water.

16.4.1 IRRIGATION WITH FRESHWATER

16.4.1.1 Nitrogen

N is needed for high yields. However, N fertilizers are relatively expensive, and excess amounts may contribute to ground water and surface water pollution. Using plants with high N-use efficiency could mitigate these problems. Ruiz and Romero (1999) showed that the concentration of organic N in leaves of melon (*Cucumis melo* L.) cv. Yuma plants grafted onto 'Shintoza' (*Cucurbita maxima* Duchesne × *Cucurbita moschata* Duchesne) rootstock was double that in nongrafted melons. In this case, the fruit yield also doubled in the grafted plants. Colla et al. (2010b) reported that grafting melon 'Proteo' plants on 'P360' or on 'PS1313' (*C. maxima* Duchesne × *C. moschata* Duchesne) increased the amount of N in the plant shoot by 17% and 25%, respectively, compared to the non-grafted melon; the yields of the grafted plants were also significantly increased. Similar results were reported by Ruiz et al. (1997), where the N concentration in the leaves of 'Yuma' melons grafted onto 'Shintoza' rootstock was 23% higher than in nongrafted melons.

Pulgar et al. (2000) conducted an experiment with watermelon (*Citrullus lanatus* [Thunb.] Matsum. & Nakai) 'Early Gray', which was either nongrafted or grafted onto 'Brava', 'Shintoza', or 'Kamel' rootstocks. The grafted watermelon plants exhibited significantly lower NO_3^- concentration in the leaves accompanied by higher nitrate reductase activity and higher concentrations of total N, free amino acids, and soluble proteins than the nongrafted plants. An increase in leaf N concentration was also found in mini-watermelon 'Ingrid' plants grafted onto 'PS1313' rootstock compared to non-grafted plants growing under field conditions (Rouphael et al. 2008). Results showing that fruit yield and N concentration in the xylem sap are higher for melon plants grafted onto *Cucurbita* than for non-grafted melons (Table 16.3, Salehi et al. 2010) led to the conclusion that *Cucurbita* rootstocks enhance not only the absorption of N and its translocation to the shoot but also its utilization, depending on the type of rootstock. Yamasaki et al. (1994) also reported that N concentration in xylem exudate is higher in watermelon grafted onto various rootstocks than in nongrafted plants (Table 16.4).

16.4.1.2 Phosphorus

Like N, P absorption is usually higher in grafted versus nongrafted plants. Ruiz et al. (1996) showed that in melons grafted on *Cucurbita*, rootstocks have a positive effect on total P level in the leaves, causing greater shoot vigor in the grafted plants as well as higher carbohydrate (glucose, sucrose, fructose, and starch) content in the plant tissues. Those authors concluded that higher P uptake

---

TABLE 16.3

Nutrient Concentration in the Xylem Sap and Marketable Yield of Melon, 85 Days after Transplanting as Influenced by Rootstocks

Scion/Rootstock	Mineral Ion Concentration in Sap (mg/L)				Marketable Yield (t/ha)
	NO_3^-	NH_4^+	PO_4	K^+	
Khatooni (control)	960 c	107 c	279 b	300 c	24.3 c
Khatooni/ShintoHongto	1677 a	266 a	469 a	649 a	39.2 b
Khatooni/Shintoza	1581 a	179 b	450 a	450 b	61.9 a
Khatooni/Ace	1257 b	207 b	457 a	470 b	59.7 a

Source: Salehi, R. et al., *HortScience*, 45, 766, 2010.

Note: Different letters in each column indicate statistically significant differences at the 0.05 level between scion/rootstock combinations.

TABLE 16.4

Nutrient Concentrations in Xylem-Sap Exudate from Grafted Watermelons as Affected by Rootstocks

Scion/Rootstock	Mineral Ion Concentration (mg/L)			
	NO_3–N	P	Ca	Mg
Fujihikari (control)	503.0 b	74.4 c	225.6 c	42.1 c
Fujihikari/bottle gourd	621.0 a	96.5 b	426.3 a	55.1 b
Fujihikari/squash	611.2 a	123.0 a	360.5 b	86.1 a

Source: Yamasaki, A. et al., *J. Jpn. Soc. Hortic. Sci.*, 62, 817, 1994.

Note: Different letters in each column indicate statistically significant differences at the 0.05 level between scion/rootstock combinations.

by the root system of the grafted plants leads to higher P translocation from the root to the shoot, which consequently increases shoot biomass in those plants. The scion can also be involved in P uptake; Ruiz et al. (1997) reported that in melons grafted on *Cucurbita* rootstocks, P concentration is affected by both the scion and its interaction with the rootstock. Colla et al. (2010a) reported that grafting cucumber 'Ingrid' onto 'P360' or 'P313' rootstocks significantly increases P concentration in the leaves by 12% and 13%, respectively, compared to nongrafted plants. The same pattern of P absorption was reported in eggplant grafted onto tomato 'Beaufort' (Leonardi and Giuffrida 2006), cucumber grafted onto *Cucurbita* 'Shintoza' (Rouphael et al. 2008), and melon grafted onto three *Cucurbita* rootstocks, 'ShintoHongto', Shintowza, and Ace (Table 16.3). In fact, there is general agreement among researchers that the bigger and more vigorous the *Cucurbita* rootstock, the higher the P absorption by the plant. To understand the effect of root functioning on P absorption, the concentration of this element in xylem-sap exudate of 'Fujihikari' watermelon grafted onto bottle gourd or *Cucurbita* rootstocks was determined during anthesis (Yamasaki et al. 1994). P concentration in the exudate of the watermelons grafted on bottle gourd or *Cucurbita* rootstocks was 30% and 65% higher, respectively, than in nongrafted plants (Table 16.4). This indicated that the *Cucurbita* rootstock is more efficient at absorbing P than the bottle gourd rootstock.

16.4.1.3 Potassium

Many studies have found that grafting improves K absorption by the plant roots, for example, melon grafted on *Cucurbita* '1 Shengzhen' (Qi et al. 2006), mini-watermelon grafted on pumpkin (Rouphael et al. 2008), and cucumber grafted on *C. moschata* (Zhu et al. 2008). The concentration of K in the xylem-sap exudate of 'Fujihikari' watermelon grafted on *Cucurbita* rootstocks at the mature fruit stage was 33% higher than in nongrafted watermelon (Yamasaki et al. 1994). However, when the same watermelon was grafted on bottle gourd, K concentration in the xylem-sap exudate was lower than in the nongrafted watermelon. The mechanisms responsible for the increased K concentration in the grafted plants are still unknown. However, since K movement in the soil is based mainly on diffusion, it is expected that a plant with a more developed root system will absorb more K. In a recent research conducted by Huang et al. (2013), watermelon plants were exposed to a low K concentration (0.6 mM); the shoot dry weight of nongrafted 'Zaochunhongyu' was 23% lower than normal K concentration (6.0 mM), whereas in the same watermelon grafted onto 'Jingxinzhen' (*C. moschata*) rootstock, this reduction amounted to only 3% (Table 16.5). A similar pattern was observed for K-accumulation and K-absorption efficiency in the shoot. However, when the same watermelon was grafted onto 'Nabizhen' (Lagenaria) rootstock, the difference between grafted and nongrafted plants was smaller (Table 16.5). Huang et al. (2013) suggested that (1) grafting can increase the tolerance of watermelon to low K concentration, depending on the rootstock genotype, and (2) the main factor controlling this tolerance to low K concentration is the rootstock's high absorption of K and its translocation to the shoot, rather than its higher utilization.

16.4.1.4 Other Nutrients

Grafting also influences the absorption and translocation of nutrients other than N, P, and K. Yamasaki et al. (1994) reported that Ca concentration in the xylem exudate of watermelon grafted onto bottle gourd and squash increased by 89% and 160%, respectively, compared to nongrafted plants. In the same experiment, Mg in the exudate increased by 31% when watermelon was grafted onto bottle gourd and by 109% when grafted onto squash compared to nongrafted plants (Table 16.4). Other studies (Nie and Chen 2000) also found that Ca and Mg increase more in the exudates of grafted watermelons versus nongrafted plants. In contrast, Edelstein et al. (2005) found insignificant differences in leaf Ca and Mg concentrations between nongrafted 'Arava' melons and melons grafted onto *Cucurbita* 'TZ-148' rootstock. Concentrations of the micronutrients Mn, Cu, and Zn increased in the leaves of cucumbers grafted on *Cucurbita* compared to nongrafted plants (Li and Yu 2008). On the other hand, Ikeda et al. (1986) found a decrease in Zn concentration in grafted cucumber plants compared to nongrafted ones.

TABLE 16.5
Effect of Graft Combinations and K Fertilization Treatments on Shoot Dry Weight and K Accumulation and Uptake Efficiency in Grafted and Nongrafted Watermelons

Scion/Rootstock	K Treatment (mM)	Shoot Dry Weight		Shoot K Accumulation		Plant K-Uptake Efficiency (mg/Plant)
		(g/Plant)	Reduction (%)	(mg/Plant)	Reduction (%)	
Zaochunhongyu (Z) (control)	6	2.55		139.6		162.8
	0.6	1.96	23	91.2	37	105.1
	Mean	2.25 d		115.4 d		133.9 c
Z/Hongdum	6	4.16		238.4		287.5
	0.6	3.95	5	194.6	18	226.8
	Mean	4.05 b		216.5 b		257.2 a
Z/Jingxinzhen	6	4.96		246.1		288.2
	0.6	4.79	3	221.8	10	255.3
	Mean	4.88 a		233.9 a		271.8 a
Z/Nabizhen	6	3.81		210.6		269.5
	0.6	3.12	18	158.7	25	201.7
	Mean	3.46 c		184.6 c		235.6 b

Source: Huang, Y. et al., *Sci. Hortic.*, 14, 80, 2013.
Note: For each column, different letters indicate statistically significant differences at the 0.05 level between the means values.

16.4.2 IRRIGATION WITH SALINE AND TREATED WASTEWATER

The uptake of macro- and microelements by nongrafted melon and melon grafted onto *Cucurbita* rootstock irrigated with FW or secondary TWW (Table 16.1) was examined in a field experiment near the city of Akko in northern Israel under a randomized block design with four replicates for each treatment. The experimental field consisted of clay soil, and the treatment plots were irrigated via a drip irrigation system in which the same amount and type of fertilizers were applied with the irrigation water (fertigation) to all plots during three growing seasons. The contents of the different nutrients in the shoots of the grafted and nongrafted plants (plant treatments) irrigated with FW or TWW (water treatments) for the third growing season are presented in Figure 16.2. The effects of the plant and water treatments on nutrient content in the plant shoot varied among the different nutrients. The contents of Fe, Cu, and Zn in the plant shoot did not differ significantly between the nongrafted and grafted plants or between irrigation water qualities (Figure 16.2). Mn content was significantly higher in the shoots of the nongrafted versus grafted plants for each water quality. Mo content was significantly lower in the nongrafted versus grafted plants for each water quality, but the effect of water quality was insignificant for each plant treatment. The N–NO$_3$ contents were similar in the shoots of the grafted and nongrafted plants under the two water qualities but were significantly higher in the grafted plants under irrigation with TWW versus FW. The contents of Ca, Mg, and K in the grafted and nongrafted plants were similar under irrigation with FW and TWW; the Ca content in the nongrafted plants was significantly higher than in the grafted plants under irrigation with FW; the Mg content in the grafted plants was significantly higher than in nongrafted plants under irrigation with TWW; and the K content in the grafted plants was significantly higher than in nongrafted plants under irrigation with FW. The results in Figure 16.2 indicate an interaction between the effects of plant grafting and water quality (FW or TWW) on the contents of some nutrients in the plant shoots.

The effects of plant grafting on nutrient uptake under irrigation with saline water have been investigated in a few studies (e.g., Colla et al. 2006; Edelstein et al. 2005). Edelstein et al. (2005)

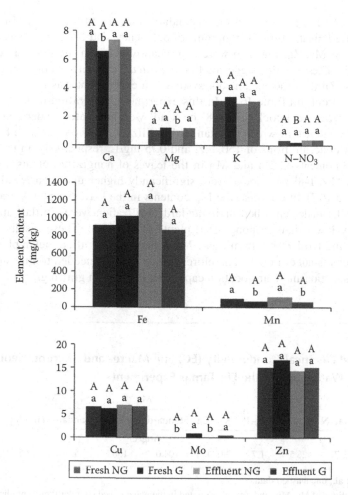

FIGURE 16.2 **(See color insert.)** Total elemental content in plant foliage of grafted and nongrafted plants irrigated with fresh and treated sewage water. Different lowercase or uppercase letters indicate statistically significant differences at the 0.05 level between plant types for each irrigation water type or between irrigation water types for each plant type, respectively (Edelstein et al., unpublished data).

studied the effects of grafting melon cv. Arava onto *Cucurbita* rootstock 'TZ 148' under irrigation with FW (EC = 1.8 dS/m) or saline water (EC = 4.6 dS/m) on nutrient contents in the plant shoot in a greenhouse experiment. That study showed that for each water quality and nutrient, the Ca, Mg, and K contents in the shoots of the nongrafted and grafted plants were similar. In contrast, in a greenhouse experiment, Colla et al. (2006) found a significant interaction between grafting of 'Tex' watermelon and irrigation water salinity in their effect on K content in the leaves; increasing the EC of the irrigation water from 2.0 to 5.2 dS/m significantly decreased the K content from 14.8 to 12.1 mg/kg, but insignificant differences were found in the K contents of the grafted plants under irrigation with the two water qualities.

The effects of grafting plants on nutrient uptake under irrigation with saline water were also studied in the years 2013 and 2014 in field experiments at Ein-Tamar, in the northern part of the Arava valley, Israel (Edelstein et al. unpublished data). In these experiments, grafted and non-grafted melon plants were grown in soil in walk-in tunnels, irrigated with saline water (Table 16.6) via a drip system, and fertilizers were applied with the irrigation water (fertigation). In the 2013 experiment, nongrafted melon and melon grafted onto *Cucurbita* rootstock 'TZ 148' were fertigated (Table 16.6) as per the common (standard) practice in the region. In this experiment, leaves of the grafted plants showed symptoms of deficiency, and chemical analysis indicated that the leaf contents of Mg, Zn, and Mn were significantly lower in the grafted versus nongrafted plants (Table 16.7). These results suggested that the *Cucurbita* root system absorbs or translocates less Mg, Mn, and Zn than the melon root system. An experiment was run in 2014 to find ways of mitigating this problem (Table 16.7). In this experiment, nongrafted melon '6023' and '6023' grafted onto *Cucurbita* rootstock 'TZ 148' or 'Gad' were fertigated under standard practice or with an enriched treatment in which the standard fertilization was enriched with Mg, Zn, and Mn to obtain nutrient concentrations of 150, 7.5, and 0.75 mg/L, respectively, in the irrigation water (Table 16.6). The contents of Zn and Mn in the leaves of nongrafted plants and plants grafted on 'Gad' and on 'TZ 148' rootstocks were significantly higher under enriched versus standard fertilization (Table 16.7). In contrast, the Mg contents in the leaves of plants grafted on 'Gad' or 'TZ 148' rootstocks under enriched and standard fertilization were similar, and these contents were significantly lower than in nongrafted plants (Table 16.7). These results indicate that the practice of enriching fertilization to mitigate Mg deficiency in grafted melon plants onto 'Gad' or 'TZ 148' rootstocks is not effective. Therefore, further study is needed to find suitable rootstocks with high Mg absorption and translocation capabilities for use in grafting.

TABLE 16.6
Values of pH and Electrical Conductivity (EC) and Macro- and Micronutrient Concentrations in the Irrigation Water Used in the Ein-Tamar Experiment

								Mg		Mn		Zn	
pH	EC	Cl⁻	NH₄–N	NO₃–N	PO₄–P	K	Ca	Standard	Enriched	Standard	Enriched	Standard	Enriched
	(dS/m)							(mg/L)					
6.6	3.5	776	2.72	26.5	1.7	22.0	179	104.0	150	5.0	7.5	0.05	0.75

Source: Edelstein et al., unpublished data.

Note: The concentrations of Mg, Mn, and Zn are presented in the nonenriched (standard) and enriched waters.

TABLE 16.7
Concentration of Elements in Leaves of Grafted (G) Melon, Nongrafted (NG) melon, Melon Grafted on 'Gad' Rootstock (G/GAD), and Melon Grafted on 'TZ 148' Rootstock (G/TZ 148) in 2013 and 2014 Field Experiments in Ein-Tamar

	2013		2014					
			NG		G/GAD		G/TZ 148	
	G	NG	Standard	Enriched	Standard	Enriched	Standard	Enriched
Element				(mg/kg)				
Mg	0.81 b	1.60 a	1.27 a	1.27 a	0.81 b	0.83 b	0.83 b	0.78 b
Zn	27.92 b	52.46 a	25.63 b	32.68 a	18.94 c	27.13 ab	23.25 bc	25.63 b
Mn	28.25 b	49.63 a	46.94 c	107.56 a	29.38 d	76.00 b	31.00 d	65.19 b

Source: Edelstein et al., unpublished data.

Note: Different letters indicate statistically significant differences at the 0.05 level between plant types for each element in the 2013 experiment and between plant types and fertilization treatments for each element in the 2014 experiment.

16.5 CONCLUSIONS

Under irrigation with FW, plant grafting onto cucurbits generally increases the uptake of N, P, K, Ca, Mg, Cu, and Zn. However, some reports have shown a decrease in the uptake of micronutrients such as Zn and Mg in grafted plants. The effects of grafting on the absorption and distribution of minerals in the plant depend mainly on the rootstock. Thus, the selection of rootstock should be based on characteristics related to nutrient uptake. It is concluded that the use of grafted plants on selected rootstocks presents a potential strategy for increasing mineral-use efficiency and yield, and coping with soil fertility problems under "low-input" conditions.

REFERENCES

Ahmad, Z., M.A. Gill, and R.H. Qureshi. 2001. Genotypic variation of phosphorous utilization efficiency of crops. *J. Plant Nutr.* 24:1149–1171.

Ahn, S.J., Y.J. Im, G.C. Chung, B.H. Cho, and S.R. Suh. 1999. Physiological responses of grafted-cucumber leaves and rootstock root affected by low root temperature. *Sci. Hortic.* 81:397–408.

Baligar, V.C., N.K. Fageria, and Z.L. He. 2001. Nutrient use efficiency in plants. *Commun. Soil Sci. Plant Anal.* 32:921–950.

Ben-Hur, M. 2004. Sewage water treatments and reuse in Israel. In: *Water in the Middle East and in North Africa: Resources, Protection, and Management* (F. Zereini and W. Jaeschke, eds.), pp. 167–180. Springer-Verlag, New York.

Cohen, R., Y. Burger, C. Horev, A. Koren, and M. Edelstein. 2007. Introducing grafted cucurbits to modern agriculture, the Israeli experience. *Plant Dis.* 91:916–923.

Colla, G., Y. Rouphael, and A. Cardarelli. 2006. Effect of salinity on yield, fruit quality, leaf gas exchange, and mineral composition of grafted watermelon plants. *HortScience* 41:622–627.

Colla, G., Y. Rouphael, M. Cardarelli, A. Salerno, and A. Rea. 2010a. The effectiveness of grafting to improve alkalinity tolerance in watermelon. *Environ. Exp. Bot.* 68:283–291.

Colla, G., C.M.C. Suarez, and M. Cardarelli. 2010b. Improving nitrogen use efficiency in melon by grafting. *HortScience* 45:559–565.

Edelstein, M. and M. Ben-Hur. 2012. Use of grafting to mitigate chemical stresses in vegetables under arid and semiarid conditions. In: *Advances in Environmental Research*, Vol. 20 (J.A. Daniels, ed.), pp. 163–179. Nova Science Publishers, Inc., New York.

Edelstein, M., M. Ben-Hur, R. Cohen, Y. Burger, and I. Ravina. 2005. Boron and salinity effects on grafted and non-grafted melon plants. *Plant Soil* 269: 273–284.

Edelstein, M., M. Ben-Hur, L. Leib, and Z. Plaut. 2011a. Mechanism responsible for restricted boron concentration in plant shoots grafted on pumpkin rootstocks. *Isr. J. Plant Sci.* 59:207–215.

Edelstein, M., M. Ben-Hur, and Z. Plaut. 2007. Grafted melons irrigated with fresh or effluent water tolerate excess boron. *J. Am. Soc. Hortic. Sci.* 132:484–491.

Edelstein, M., M. Ben-Hur, and Z. Plaut. 2011b. Sodium and chloride exclusion and retention by non-grafted and grafted melon and *Cucurbita* plants. *J. Exp. Bot.* 62:177–184.

Fernandez-Garcia, N., V. Martinez, A. Creda, and M. Carvajal. 2003. Fruit quality of grafted tomato plants grown under saline conditions. *J. Hortic. Sci. Biotechnol.* 79:995–1001.

Huang, Y., J. Li, H. Bin, L. Zhixiong, F. Moling, and B. Zhilong. 2013. Grafting onto different rootstocks as a mean to improve watermelon tolerance to low potassium stress. *Sci. Hortic.* 14:80–85.

Ikeda, H., S. Okitsu, and K. Arai. 1986. Comparison of magnesium deficiency of grafted and non-grafted cucumbers in eater culture and soil culture and the effect of increased magnesium application on the prevention of magnesium deficiency disorder. *Bull. Veg. Ornam. Crops Res. Stn.* C9:31–34.

Kurata, K. 1994. Cultivation of grafted vegetables. II. Development of grafting robots in Japan. *HortScience* 29:240–244.

Lado, M., A. Bar-Tal, A. Azenkot et al. 2012. Changes in chemical properties of semiarid soils under long-term secondary treated wastewater irrigation. *Soil Sci. Soc. Am. J.* 76:1358–1369.

Lado, M. and M. Ben-Hur. 2010. Effects of irrigation with different effluents on saturated hydraulic conductivity of arid and semiarid soils. *Soil Sci. Soc. Am. J.* 74:23–32.

Lee, J.M. 1994. Cultivation of grafted vegetables. I. Current status, grafting methods, and benefits. *HortScience* 29:235–239.

Lee, J.M. 2003. Advances in vegetable grafting. *Chron. Hortic.* 43:13–19.

Lee, J.M., C. Kubota, S.J. Taso, P.H. Echevarria, L. Morra, and M. Oda. 2010. Current status of vegetable grafting: Diffusion, grafting techniques, automation. *Sci. Hortic.* 127:93–105.

Lee, J.M. and M. Oda. 2003. Grafting of herbaceous vegetable and ornamental crops. *Hortic. Rev.* 28:61–124.

Leonardi, C. and F. Giuffrida. 2006. Variation of plant growth and macronutrient uptake in grafted tomatoes and eggplants on three different rootstocks. *Eur. J. Hortic. Sci.* 71:97–101.

Li, T. and X. Yu. 2008. Effect of Cu^{2+}, Zn^{2+}, and Mn^{2+} on SOD activity in cucumber leaves extraction after low temperature stress. *Acta Hortic. Sin.* 34:895–900.

Moncada, A., A. Miceli, F. Vetrano, V. Mineo, D. Planeta, and F. D'Anna. 2013. Effect of grafting on yield and quality of eggplant (*Solanum mellongena* L.). *Sci. Hortic.* 149:108–114.

Nie, L.C. and L.G. Chen. 2000. Study on growth trends and physiological characteristics of grafted watermelon seedlings. *Acta Agric. Boreali-Occidentalis Sin.* 9:100–103.

Nisini, P.T., G. Colla, E. Granati, O. Temperini, P. Crino, and F. Saccardo. 2002. Rootstock resistance to Fusarium wilt and effect on fruit yield and quality of two muskmelon cultivars. *Sci. Hortic.* 93:281–288.

Oda, M. 2002. Grafting vegetable crops. *Sci. Rep. Agric. Biol. Sci., Osaka Pref. Univ.* 54:49–72.

Pina, A. and P. Errea. 2005. A review of new advances in mechanism of graft compatibility-incompatibility. *Sci. Hortic.* 106:1–11.

Pulgar, C., G. Villora, D.A. Moreno, and L. Romero. 2000. Improving the mineral nutrition in grafted watermelon plants: Nitrogen metabolism. *Biol. Plant.* 43:607–609.

Qi, H.Y., Y.F. Liu, D. Li, and T.L. Li. 2006. Effects of grafting on nutrient absorption, hormone content in xylem exudation and yield of melon (*Cucumis melo* L.). *Plant Physiol. Commun.* 42:199–202.

Rivero, R.M., J.M. Ruiz, and L. Romero. 2003. Role of grafting in horticultural plants under stress conditions. *Sci. Technol.* 1:70–74.

Romero, L., A. Belakbir, L. Ragala, and M.J. Ruiz. 1997. Response of plant yield and leaf pigments to saline conditions: Effectiveness of different rootstocks in melon plant (*Cucumis melo* L.). *Soil Sci. Plant Nutr.* 43:855–862.

Rouphael, Y., M. Cardarelli, G. Colla, and E. Rea. 2008. Yield, mineral composition, water relations, and water use efficiency of grafted mini-watermelon plants under deficit irrigation. *HortScience* 43:730–736.

Rouphael, Y., D. Schwartz, A. Krumbein, and G. Colla. 2010. Impact of grafting on product quality of fruit vegetables. *Sci. Hortic.* 127:172–179.

Ruiz, J.M., A. Belakbir, I. Lopez-Cantarero, and L. Romero. 1997. Leaf-macronutrient content and yield in grafted melon plants. A model to evaluate the influence of rootstock genotype. *Sci. Hortic.* 71:227–234.

Ruiz, J.M., A. Belakbir, and L. Romero. 1996. Foliar level of phosphorus and its bioindicators in *Cucumis melo* grafted plants. A possible effect of rootstock. *J. Plant Physiol.* 149:400–404.

Ruiz, J.M. and L. Romero. 1999. Nitrogen efficiency and metabolism in grafted melon plants. *Sci. Hortic.* 81:113–123.

Salehi, R., A. Kashi, J.M. Lee et al. 2010. Leaf gas exchange and mineral ion composition in xylem sap of Iranian melon affected by rootstocks and training methods. *HortScience* 45:766–770.

Schwarz, D., Y. Rouphael, G. Colla, and J.H. Venema. 2010. Grafting as a tool to improve tolerance of vegetables to abiotic stresses: Thermal stress, water stress and organic pollutants. *Sci. Hortic.* 127:162–171.

Shainberg, I. and J. Letey. 1984. Response of soils to sodic and saline conditions. *Hilgardia* 52:1–57.

Shannon, M.A., P.W. Bohn, M. Elimelech, J.G. Georgiadis, B.J. Mariñas, and A.M. Mayes. 2008. Science and technology for water purification in the coming decades. *Nature* 452:301–310.

Shimada, N. and M. Moritani. 1977. Nutritional studies on grafting of horticultural crops. (2) Absorption of minerals from various nutrient solutions by grafted cucumber and pumpkin plants. *J. Jpn. Soc. Soil Sci. Plant Nutr.* 48:396–401.

Tal, A. 2006. Seeking sustainability: Israel's evolving water management strategy. *Science* 313:1081–1084.

Yamasaki, A., M. Yamashita, and S. Furuya. 1994. Mineral concentrations and cytokinin activity in the xylem exudate of grafted watermelons as affected by rootstocks and crop load. *J. Jpn. Soc. Hortic. Sci.* 62:817–826.

Zhu, J., Z. Bie, Y. Huang, and X. Han. 2008. Effect of grafting on the growth and ion concentration of cucumber seedlings under NaCl stress. *Soil Sci. Plant Nutr.* 54:895–902.

17 Melon Grafting

Xin Zhao, Wenjing Guan, and Donald J. Huber

CONTENTS

17.1 INTRODUCTION

Vegetable grafting technology, originally developed in east Asia over half a century ago, has been practiced in many countries today around the world, particularly in the production of high-value solanaceous and cucurbitaceous crops, including tomato (*Solanum lycopersicum*), eggplant (*Solanum melongena*), pepper (*Capsicum annuum*), melon (*Cucumis melo*), watermelon (*Citrullus lanatus*), and cucumber (*Cucumis sativus*) (Lee and Oda, 2003; Lee et al., 2010). By connecting vascular bundles at the graft union between the scion and rootstock plants, a new plant can be created that combines desirable aboveground traits provided by the scion and superior belowground traits provided by the rootstock. As an effective tool for managing soilborne diseases, grafting plays an important role in maintaining and improving crop productivity under intensive cultivation and protected culture especially where crop rotation is rather limited and disease pressure may easily build up (Guan et al., 2012; Louws et al., 2010). With the phase-out of methyl bromide and the challenge in seeking alternative soil fumigants, along with the increasing concerns and regulations regarding chemical inputs in agricultural systems, grafting as an environment-friendly cultural practice is becoming more widely recognized for improving sustainability of vegetable production.

In addition to disease management using resistant/tolerant rootstocks, other major benefits of vegetable grafting include promoting vigorous plant growth and development; improving crop tolerance to abiotic stresses, for example, salinity, sub-optimal temperature, and drought; and enhancing nutrient and water uptake and absorption. While rootstock selection and grafted vegetable production systems are still being optimized toward increased production efficiency and crop yield, the use of vegetable grafting is also limited by the relatively high cost of grafted vegetable seedlings, possible rootstock–scion incompatibility, rootstock-induced imbalance between vegetative and reproductive phases, and the potential adverse impact of rootstock on fruit quality.

Focusing on the application of grafting in melons, this chapter explores the history of melon grafting, describes major methods used in melon grafting, and discusses grafting for disease management and yield improvement in melon production as well as fruit quality of grafted melons and rootstock–scion interactions.

17.2 HISTORY OF MELON GRAFTING

Tracing the history of melon grafting helps reveal how rootstocks have been developed temporally and geographically and how the grafting technology and production systems have evolved. This information can provide an insightful view for future use of grafting in commercial melon production. The primary events that occurred in the history of melon grafting are summarized in Table 17.1.

17.2.1 ASIA

Research in cucurbit grafting was initiated in Japan in the 1920s, when great interest emerged in grafting watermelons onto *Cucurbita moschata* for Fusarium wilt control (Sakata et al., 2007). Large-scale melon grafting did not begin until the 1960s. Although there was less interest in commercial melon grafting compared with watermelon grafting in the early days, rootstock evaluation among cucurbit species from 1930 to 1940 identified that *Cucurbita* species (*C. moschata*, *C. maxima*, *C. pepo*, and *C. ficifolia*), *C. sativus*, *Lagenaria siceraria*, *Benincasa hispida*, *C. lanatus*, and *Luffa cylindrica* were compatible with melons (Imazu, 1949; Matsumoto, 1931). Cleft grafting method, the same method as tree grafting, was the early method used in vegetable grafting (Sakata et al., 2008).

During the early 1960s, oriental melons (makuwa group) grafted with *C. moschata* and *C. maxima* × *C. moschata* rootstocks were widespread in Japan and Korea, mainly for improved Fusarium wilt resistance and cold tolerance (Lee et al., 2010). Anatomy studies on graft union were conducted to assist with identification of compatible rootstocks in Korea (Pyo and Sung, 1970).

In Japan, a hybrid melon cultivar between *makuwa* and 'Charentais' (var. *cantalupensis*) melons became popular in the 1970s. Grafting the hybrid melon cultivars with squash rootstocks showed more vigorous plant growth; however, fruit quality was compromised (Sakata et al., 2008). Cultural practices such as pruning and thinning exhibited few effects on improving fruit quality of the grafted plants (Matsuda and Honda, 1981). During the 1970s, netted melons with green flesh were released for greenhouse production in Japan. Some of these melons contained Fusarium wilt resistance gene *Fom*-1, which were almost never grafted. For melon cultivars that were susceptible to Fusarium wilt, resistant *C. melo* rootstocks instead of squash rootstocks were used to reduce the negative impacts of grafting on melon fruit quality (Sakata et al., 2008). In the 1980s, the growing interest in cucumber grafting led to the development of automated or robotic grafting (Suzuki, 1990). Grafting robots largely increased grafting efficiency, but due to the high cost and less flexibility compared with manual grafting, its application was still limited. According to Lee et al. (2010), grafted melons accounted for about 30% of melon production in Japan, with *C. moschata* and *C. maxima* × *C. moschata* as the most commonly used rootstocks. Melons produced in Korea are primarily oriental melons, which are sensitive to low soil temperature. Up to 90% of the melon production in Korea used grafted transplants with *C. maxima* × *C. moschata* rootstocks. In China, grafted plants contributed to approximately 5% of the melon cultivation area, using mainly *C. moschata* and *C. maxima* × *C. moschata* rootstocks.

17.2.2 EUROPE

Cucurbit grafting was introduced to Europe soon after World War II to control Fusarium wilt on cucumbers. This technique first appeared in studies conducted in the Netherlands for greenhouse production (Groenewegen, 1953). Besides combating Fusarium wilt, cold tolerance, higher yield, and early harvest were reported in grafted cucumbers (Groenewegen, 1953). *C. ficifolia* was the recommended cucumber rootstock. Tongue-inarching method (tongue approach grafting) was developed in the Netherlands (Davis et al., 2008).

TABLE 17.1

Major Events That Occurred in the History of Melon Grafting in Asia, Europe, and Russia

Time	Asia	Europe	Russia
1920s	Research in cucurbit grafting was initiated because of the growing interest in grafting watermelons for Fusarium wilt control; cleft grafting method was the early used method.		
1930s	Evaluation of cucurbit rootstocks compatible with melon scion.		Watermelon and melon grafting was introduced to Russia for enhanced cold tolerance.
1940s		Cucumber grafting with *Cucurbita ficifolia* rootstock was introduced to Europe for Fusarium wilt control; tongue approach grafting method was invented in the Netherlands.	Assessment of local melon cultivars grafted onto *C. maxima* rootstock.
1950s		Melon was found incompatible with *C. ficifolia* rootstock.	Vegetative hybridization through grafting became a focus of melon grafting studies.
1960s	Melon (oriental melon) grafting was widespread in Japan and Korea for improved Fusarium wilt resistance and cold tolerance.	Grafting melons onto *Benincasa hispida* rootstock became popular and widespread in European countries.	
1970s	Grafting newly introduced hybrid melons with *Cucurbita* rootstocks resulted in reduced melon fruit quality.	Interest in melon grafting declined because of emerging diseases, improved melon cultivars, and use of methyl bromide for controlling soilborne diseases.	
1980s	Newly introduced melon cultivars were grafted onto *Cucumis melo* rootstocks to reduce negative impacts of grafting on fruit quality; grafting robot was developed in Japan.		
1990s to present	Majority of oriental melons were grafted.	Phase-out of methyl bromide stimulated worldwide interest in cucurbit grafting. Emerging diseases drive the research interest in developing new melon rootstocks.	

In contrast to cucumber grafting, melon grafting received less attention before the 1950s (Groenewegen, 1953). One of the concerns was the graft incompatibility between melon and *C. ficifolia* (de Stigter, 1956). After grafting muskmelon onto *C. ficifolia*, plants wilted and died during production. Interestingly, it was noticed that this incompatibility issue could be prevented by keeping at least four to five leaves of the rootstock in the grafted plants. It was speculated that some substances essential for *C. ficifolia* could not be provided by the muskmelon scion and thus caused the incompatibility (de Stigter, 1956; Wellensiek, 1949). *C. pepo* var. *ovifera* was suggested as a better rootstock than *C. ficifolia* as it produced stronger grafted plants. However, melons grafted onto *C. pepo* delayed fruit development and harvest (Groenewegen, 1953; Naaldwijk, 1952).

Completely removing rootstock foliage after grafting was possible with *B. hispida* (*Benincasa cerifera*) rootstock (Louvet and Peyriere, 1962). When grafting Charentais melons onto 30 cucurbitaceous species and varieties, *B. hispida* showed the most promising results in controlling Fusarium wilt (Louvet and Lemaitre, 1961). In addition, grafted melon plants exhibited longer fruiting period and higher yield (Louvet and Peyriere, 1962). Because of the desirable characteristics of grafted melon plants, melon grafting was widespread in European countries, including the Netherlands, France, and Italy. With increasing interest in greenhouse production with the newly established artificial light system at that time, growth of grafted melons under different light conditions was evaluated (Chavagnat, 1971).

Interest in melon grafting for Fusarium wilt management declined after 1970, partly due to the emergence of new *Fusarium* races that broke rootstock resistance, which resulted in reoccurrence of the disease (Benoit, 1974). Furthermore, melons grafted onto *Benincasa* spp. did not control diseases caused by *Phomopsis sclerotioides* and *Verticillium dahliae*. *Cucurbita* spp. rootstock exhibited partial resistance to these pathogens, but performance was inconsistent (Alabouvette et al., 1974). Broad-spectrum soil fumigant methyl bromide became a more reliable approach to controlling soilborne pests in greenhouse melon production since 1970, which also decreased the popularity of melon grafting.

Detrimental effects of using methyl bromide were widely recognized at the end of the 1990s. (Ristaino and Thomas, 1997). Because of environmental and human health concerns, 168 countries agreed to gradually decrease the application of methyl bromide in agricultural production and eventually phase it out in 2005 except for critical use (Crinò et al., 2007). The loss of methyl bromide greatly simulated the employment of grafting in vegetable production for managing soilborne diseases. Similarly, interest in vegetable grafting for open field production emerged in the United States since 2000 and has been fueled toward building sustainable vegetable production systems.

17.2.3 RUSSIA

The grafting technique was introduced to Russia around 1930. Grafting experiments with solanaceous and cucurbitaceous vegetables were conducted in Voronezh and then in Moscow areas. Using cold-tolerant *C. maxima* rootstocks, grafting made it possible to grow warm-season vegetables in the northern areas (Lebedeva, 1937). The grafted plants not only showed improved cold tolerance, but also increased flowering, better rooting, extended plant life, and earlier ripening and better flavor of fruit. Interestingly, fruits were harvested from both rootstock and scion of the grafted plants. Total yield including melons from the scion and pumpkins from the rootstock were greatly increased compared with melon fruit alone from the nongrafted plants (Lebedeva, 1937). It was also found that the melon progeny from grafted plants produced better seed quality, earlier maturity, higher yield, and hardier plants (Lebedeva, 1937).

The melon grafting research taking place in Russia had a different focus from that in Asian and European countries. Instead of exploring the practical application of grafting in melon production, grafting hybrid was an active research topic in Russia. Researchers believed that intermediate characteristics derived from two species very different in their genetic makeup could be achieved by so-called vegetative hybridization through grafting, which might provide a solution for incompatibility of interspecific crosses (Weiss, 1940). It was speculated that grafting could heritably affect traits of both rootstock and scion, while continuously grafting the progenies of scion onto the same rootstock (similar to backcrossing) would eventually generate new plants with combined favorable characteristics of both rootstock and scion (Ludilov, 1964, 1969; Parhomenko, 1941). Many of these studies in cucurbit grafting reported that various plant characteristics, either in the first generation or several following generations, were changed by grafting (Galun, 1958; Hohlaceva, 1955; Mustafin, 1961; Sanaev, 1966). The observed morphological alterations were attributed to the exchange of substances across the graft union (Gaskova, 1944). Radiographic techniques were used to study translocation of substances between rootstock and scion, which confirmed the transport of C^{14} labeled photoassimilates from scion to rootstock (de Stigter, 1961). Despite the improved understanding of rootstock and scion interactions, none of these studies provided direct evidence to support the hypothesis that grafting could genetically modify plant characteristics of either rootstock or scion plants.

17.3   GRAFTING METHODS

In general, three types of melon grafting methods are currently adopted by propagators and growers at various production scales. These methods, that is, hole insertion, one cotyledon (splice), and tongue approach (Figure 17.1), differ considerably in the grafting procedure and have their own advantages and disadvantages (Davis et al., 2008; Guan and Zhao, 2014; Lee et al., 2010). Although grafting machines and robots have been used by some large operations, manual grafting remains popular in the industry largely due to management and economic considerations. The one cotyledon method can be easily adapted to mechanized grafting, whereas hole insertion and tongue approach are mainly limited to hand grafting. Overall, hole insertion and one cotyledon grafting are more likely to be adopted by commercial propagators and experienced growers, while tongue approach grafting is often preferred by inexperienced grafters due to the ease of healing and relatively high graft survival rate.

(a)   (b)   (c)

FIGURE 17.1   Well-healed muskmelon plants grafted onto interspecific hybrid squash rootstock using different grafting methods. (a) Hole insertion grafting. (b) One cotyledon grafting. (c) Tongue approach grafting.

Tongue approach grafting was initially developed in the Netherlands. It requires rootstock and scion plants with similar stem diameters to ensure the best result. Usually, both the root system of scion and growing tip of rootstock remain intact during grafting. Grafting takes place below the rootstock cotyledons, which eliminates the rootstock sucker issue at planting. The rootstock is often cut downward halfway through the hypocotyl, while the scion is cut upward, with the cut surfaces aligned and supported by the grafting clip. A highly controlled humid environment is not necessary for graft healing. Toward the end of healing, the rootstock shoots and scion roots (with lower part of the hypocotyl) are completely removed from the grafted plant. Despite the high success rate, tongue approach grafting typically needs more space and labor than hole insertion and one cotyledon methods.

Hole insertion is the most popular method used in China for cucurbit grafting. This method demands a higher level of grafting skill in comparison with tongue approach and one cotyledon methods. Rootstock seedlings with thicker hypocotyls are preferred as a slit (hole) needs to be made between the two cotyledons of the rootstock plant while removing the apical meristem tissue. The hypocotyl of the young scion seedling is cut to make a wedge or a tapered end in order to get inserted into the hole made in the rootstock. The rootstock age needs to be carefully controlled since the central pith cavity may start to develop in the hypocotyl following the emergence of cotyledons (Figure 17.2). The grafting window may also be constrained by the age of scion as small seedlings are always preferred. Grafting clips are not required with the hole insertion method given the strong contact between rootstock and scion cut surfaces at the graft union and the support provided by the rootstock cotyledons, which is advantageous in improving grafting efficiency. The main issue is that as incomplete elimination of rootstock growing tips often occurs during grafting, rootstock suckers need to be monitored and removed prior to and following transplanting, thus increasing the cost of labor.

To date, one cotyledon grafting is becoming more widely accepted by the industry owing to its simple process and flexible adaptability to diverse systems for grafted transplant production. It is also suggested that one cotyledon method may produce stronger graft union than tongue approach grafting because of more complete connection of vascular bundles between the scion and the rootstock. Having similar stem diameters may also allow for better fusion of the rootstock and scion cut surfaces. The rootstock plant is usually cut at an angle between the two cotyledons by removing

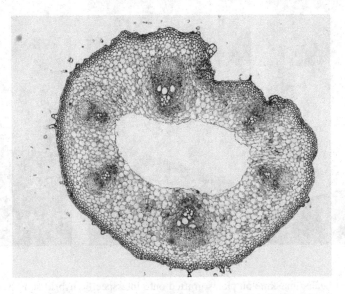

FIGURE 17.2 A cross-section of the hypocotyl of an interspecific hybrid squash rootstock (*C. maxima* × *C. moschata*) seedling.

one cotyledon and the growing point. The scion plant is cut at a similar angle below the cotyledons with the length of the cut matching that of the rootstock. A grafting clip is used to fix the graft union by attaching the scion to the rootstock. Sometimes, roots are excised from the rootstock plants before grafting to accommodate automated grafting and increase production efficiency, since new root systems can be conveniently regenerated from the grafted plants. Similar to hole insertion grafting, rootstock suckering issues may be encountered if complete removal of the rootstock apical meristem tissue cannot be achieved at grafting.

Preference for specific grafting techniques may differ based on geographic locations and operations, primarily determined by the facility design, space, labor, and the ultimate production cost. Regardless of the grafting method used, seeding of the rootstock and the scion needs to be carefully scheduled to ensure the suitable timeframe for grafting, depending upon the type of rootstock and scion cultivars. Moreover, post-graft healing is essential for producing high quality grafts with high survival. It is critical to maintain high humidity during the first 48–72 h after grafting, particularly with one cotyledon and hole insertion methods. Appropriate temperature and light intensity need to be provided as well. During the final phase of healing, grafted seedlings should be gradually exposed to normal greenhouse conditions and acclimated prior to field planting. At transplanting, the graft union needs to be placed well above the soil line to prevent rooting from the scion.

17.4 USING GRAFTING FOR DISEASE MANAGEMENT AND YIELD IMPROVEMENT

Currently, the most commonly used rootstocks in grafted melon production are developed from *C. melo*, *Cucurbita* spp., and *C. maxima* × *C. moschata*. The effectiveness of grafting in managing a variety of soilborne diseases in melons has been extensively reported and described in recent reviews (Guan et al., 2012; King et al., 2008; Louws et al., 2010). The major diseases that can be successfully controlled by using resistant rootstocks include Fusarium wilt (caused by *Fusarium oxysporum* f. sp. *melonis*), Verticillium wilt (caused by *V. dahliae*), Monosporascus root rot and vine decline (caused by *Monosporascus cannonballus*), and gummy stem blight (caused by *Didymella bryoniae*). Commercially available rootstocks lack resistance to root-knot nematodes (*Meloidogyne* spp.), another pathogen threatening melons in many production regions. Recent studies demonstrated the potential of *Cucumis metulifer* as a promising melon rootstock for reducing root galling and suppressing nematode damage (Guan et al., 2014; Sigüenza et al., 2005). Host and nonhost resistances have been widely used in breeding of resistant rootstocks. In addition, changes in composition of the rhizosphere microorganisms and microbial diversity, vigorous root system with enhanced nutrient and water uptake, as well as systemic resistance induced by grafting may contribute to the improved defense response of grafted plants to pathogen attack (Guan et al., 2012). Identifying the prevalent devastating pathogens in the cropping systems and selecting the appropriate rootstocks to cope with disease problems is the key to successful use of grafting for disease management. Rotation of rootstock cultivars may be considered to reduce the loss of disease resistance. With the emerging pathogens and appearance of new races of an existing pathogen, it is a continuous effort to develop new rootstocks. Rootstocks with a more complete disease resistance package will be more effective when multiple diseases occur in the field during the production season. It is also noteworthy that as an integrated pest management practice, grafting needs to be well incorporated into the management system that involves a suite of control measures such as crop rotation, cover cropping, and organic amendment.

Fruit yield improvement is more pronounced under higher disease pressure in grafted melon production. However, grafting with selected rootstocks may also lead to growth and yield enhancement in the absence of biotic stresses. Uptake and translocation of N, P, Mg, Fe, and Ca can be influenced by rootstocks in cucurbits (Davis et al., 2008) and a reduction of fertilizer inputs by up to one-half of the recommended rate is suggested for growing certain grafted cucurbits (Lee et al., 2010). Potential for yield increase in grafted melons varied with the rootstock genotypes

(Ruiz et al., 1997). Melon plants exhibited enhanced N assimilation and higher yields (Ruiz and Romero, 1999) and improved N use efficiency and N uptake efficiency (Colla et al., 2010) when grafted onto certain *C. maxima* × *C. moschata* rootstocks. Yield improvement of grafted cucurbits has also been attributed to improved endogenous plant hormone (e.g., cytokinins) status as a result of grafting with vigorous rootstocks (Davis et al., 2008). Furthermore, enhancement of tolerance to environmental stresses including drought, flooding, low temperature, and salinity has been widely observed in grafted vegetables including melons (Davis et al., 2008; Lee et al., 2010; Schwarz et al., 2010). *C. maxima* × *C. moschata* rootstocks with salt tolerance resulted in lower concentrations of Na in the leaf tissue and improved photosynthetic capacity in grafted melons under saline (NaCl) conditions (Rouphael et al., 2012). To fully utilize the yield improvement potential of *C. maxima* × *C. moschata* rootstocks and solve the rootstock–scion incompatibility issue with certain melon cultivars, double grafting onto a vigorous rootstock with an intermediate melon scion has been tested with positive results in increasing nutrient uptake and fruit yield (San Bautista et al., 2011).

17.5 FRUIT QUALITY AND ROOTSTOCK–SCION INTERACTIONS

While the multifaceted benefits of grafted melon production are being elucidated, certain grafting combinations have revealed compromised fruit quality (Guan et al., 2015; King et al., 2010), which may be rather challenging to study owing to the diverse genotypes of melon scions and complexity of fruit quality attributes. Melons are categorized into 16 groups within 2 subspecies (Burger et al., 2010). The sweet melons are in the groups of makuwa, cantalupensis, reticulatus, and inodorus (Guan et al., 2013). Melons in the makuwa group belong to the subspecies *agrestis* that is believed to originate from southern and eastern Asia. Besides the makuwa group, the other sweet melon groups are in the subspecies *melo*. They are associated with forms derived from Africa and western and central Asia (Burger et al., 2010), such as muskmelons (cantaloupe) in the reticulatus group, Charentais melons in the cantalupensis group, and honeydew melons in the inodorus group. Melons in the reticulatus and cantalupensis groups are aromatic. They develop abscission zones during ripening that are commonly used as harvest index (Shellie and Lester, 2004). Inodorus melons have a less aromatic flavor, and they do not slip from vines when ripening (Lester and Shellie, 2004).

The majority of cultivated melons have high sugar content (Burger et al., 2010). In general, organic acids do not play a significant role in determining melon fruit quality. Melon taste and flavor, therefore, is primarily determined by sugar content and aroma for melons in the groups of cantalupensis and reticulatus. In orange flesh muskmelons, fruits are also rich in mineral nutrients, particularly potassium, vitamin C, and *beta*-carotene (Lester, 1997).

Undesirable graft compatibility can possibly lead to reduced fruit quality. It is most noticeable when a landrace or a genetically distant species is used as a rootstock. For example, 'Thraki' grafted onto winter squash landrace 'Kolokitha' compromised fruit taste. The grafting combination showed high variability in survival rate, which could be an indicator of poor graft compatibility (Koutsika-Sotiriou and Traka-Mavrona, 2002). *C. metulifer* was a promising rootstock material as it shows resistance to root-knot nematodes and Fusarium wilt. Grafting galia melon 'Arava' onto *C. metulier* reduced fruit flesh firmness and total soluble solids contents (SSC). The smaller stem diameter of *C. metulifer* compared with that of scion might imply a potentially incomplete graft connection between the two species (Guan et al., 2015).

Intraspecific grafting, that is, melon scion grafted onto melon rootstock, largely eliminates graft incompatibility issues, but changes in fruit quality could also occur in some cases. Melon 'Piñonet Torpedo' grafted onto *C. melo* ssp. *agrestis* 'Pat 81', a rootstock with high resistance to *M. cannonballus* root rot, exhibited reduced SSC (Fita et al., 2007). 'Energia' and 'Sting' rootstocks (*C. melo*) had resistance to *F. oxysporum* f. sp. *melonis*, while they decreased the *beta*-carotene content of melon scion 'Proteo' (*C. melo* var. *reticulatus*) (Condurso et al., 2012).

Commercial hybrid squash rootstocks (*C. maxima* × *C. moschata*) are generally compatible with melon scions. However, fruit quality decline in some grafting combinations has been reported. Grafting 'Piel de Sapo' melon 'Piñonet Torpedo' and galia melon 'London' onto the 'RS841' rootstock resulted in reduced fruit SSC (Balázs, 2010; Fita et al., 2007). Reduced SSC and consumer evaluated sensory properties were also observed in galia melon 'Arava' and 'London' grafted onto the *C. maxima* × *C. moschata* rootstock (Balázs, 2010; Guan et al., 2015). The early Japanese literature reported that some oriental melons grafted onto interspecific hybrid squash had preharvest internal decay, internal breakdown, alcoholic fermentation, fibrous flesh, and poor netting (Rouphael et al., 2010). Because of fruit quality concerns, 'Shin-tosa', a popular interspecific hybrid squash rootstock widely used in Japan for watermelon grafting, was almost avoided in grafting Japanese melons (Davis et al., 2008).

Known for disease resistance, abiotic stress tolerance, and vigorous root systems, interspecific hybrid squash is one of the most valuable cucurbit rootstocks worldwide. Unfortunately, considering the potential negative impacts on melon fruit quality, several commercial hybrid squash rootstocks are not recommended for melon grafting. To facilitate the development of melon rootstocks with desirable traits and minimal influence on fruit quality, it is important to understand factors associated with quality decline.

Harvest maturity is one of the factors to consider. Since grafted and nongrafted plants were harvested simultaneously in many cases, the difference in fruit quality can be a reflection of harvest maturity if grafted plants delayed fruit ripening as observed in watermelons grafted with hybrid squash rootstock (Soteriou et al., 2014). On the other hand, this may not be the case for cantaloupe melons and other melons that are harvested constantly based on the development of abscission zones (Zhao et al., 2011).

Adjusting crop management practices, particularly water management, may help potentially reduce negative quality issues for grafted melons (Balázs, 2010). Hybrid squash rootstocks are well known for their vigorous root systems and large vessel elements. Grafted melon plants often show enhanced transpiration rate and increased leaf water potential, stomata conductance, and amount of xylem saps (Agele and Cohen, 2009; Jifon et al., 2010). Given the improved overall water status of grafted melon plants compared with nongrafted controls, reduced fruit SSC in some grafting combinations might be a consequence of water dilution effect.

Plant hormones are important endogenous molecules that regulate the formation of flowers, stems, and leaves and the development and ripening of fruit (Vicente and Plascncia, 2011). Grafting with squash rootstocks changed hormone levels of cucumbers (Aloni et al., 2010) that might affect fruit quality either directly or indirectly. Galia melon 'Arava' grafted onto 'Strong Tosa' rootstock delayed blooming of female flowers for 8–9 days as opposed to that of nongrafted plants. Surprisingly, the first harvest date and the early and total yields of grafted plants were not affected (unpublished data). As fruit sugar content tends to correlate with the duration of fruit development (Burger and Schaffer, 2007), accelerated fruit development of grafted melons might partially explain the reduced SSC. Ethylene is known to promote the natural process of fruit development, ripening, and senescence (Saltveit, 1999). Whether ethylene is involved in the differential regulation of fruit ripening in grafted versus nongrafted plants and how it may contribute to the rootstock effects on fruit quality remain unknown. Grafted melon plants with hybrid squash rootstocks began to collapse during the late fruit development stage (unpublished data). Exogenous application of auxins increased the number of collapsed plants (Minuto et al., 2010). It was speculated that auxins transported from the scion to auxin-sensitive rootstock triggered production of reactive oxygen species, which led to the degeneration of roots in the grafted plants (Aloni et al., 2008). The declined root systems might provide an explanation to the previous observation that plant photosynthesis was greatly decreased at the late fruit development stage (Xu et al., 2005). The reduced photosynthesis rate, therefore, might cause the decline of fruit quality especially during late harvest.

There are also hypotheses that fruit quality constituents that affect fruit quality could translocate from rootstock to scion (Fita et al., 2007). Although the chemical and biological nature of the fruit quality constituents is largely unknown, studies did show phloem proteins and small RNAs can translocate across the graft union between rootstock and scion (Golecki et al., 1998; Tiedemann and Carstens-Behrens, 1994).

Not all melon scions respond in the same manner to the grafting practice, even with the same rootstocks, indicating the role of rootstock–scion interactions in fruit quality modification following grafting. Galia melon 'Arava' (reticulatus group) and honeydew melon 'Honey Yellow' (inodorus group) were grafted onto both hybrid squash and *C. metulifer* rootstocks. Although SSC, flesh firmness, and sensory properties of 'Arava' were negatively affected by grafting, quality attributes of grafted 'Honey Yellow' melons were similar to fruit from nongrafted plants (Guan et al., 2015). When examining the effects of *Cucurbita* spp. rootstocks on fruit quality of melon cultivars in the inodorus and the cantalupensis groups, the quality of honeydew melon 'Lefko Amynteou' (inodorus group) was unaffected by all the tested *Cucurbita* spp. rootstocks (Traka-Mavrona et al., 2000). It was interesting to notice that in the studies involving scion cultivars from different melon groups, quality attributes of melons in the reticulatus and cantalupensis groups are more likely to be affected by grafting in contrast to the honeydew melons. Melons in the former groups are climacteric fruit that are more aromatic than nonclimacteric honeydew melons (Obando-Ulloa et al., 2008). It is possible that sensory properties for melons that are rich in aroma compounds are more sensitive to grafting practices (Guan et al., 2015). Nevertheless, the alteration of fruit ripening patterns as a result of grafting and rootstock–scion interactions as well as the consequent impacts on fruit quality attributes deserves more in-depth studies.

17.6 PROSPECTS FOR FUTURE RESEARCH

Integrated use of grafting in melon production will vary substantially by country, region, and farming operation. Essentially, economic feasibility of using grafted plants is at the center of long-term sustainability assessment of grafting technology and production system. At present, research interest in vegetable grafting in the United States is on the rise; however, the high cost of using grafted plants remains a major concern for wider adoption of grafting among vegetable growers with large production acreage. With the continuous effort of advancing the grafting technology, economic analysis needs to be taken into consideration in the first place. In the case of melon grafting, further improvement of grafting techniques and automation may help increase grafting efficiency and reduce transplant cost. Meanwhile, more research is warranted to optimize management practices (e.g., fertilization and irrigation) for grafted melon production in order to maximize the benefits of rootstock vigor and capacity in dealing with environmental stresses. Rootstock development for disease management will still be an important focus of the future rootstock breeding programs. As pointed out by King et al. (2010), new melon rootstocks need to offer resistances to wilt diseases and root and stem rot diseases. The new rootstocks are expected to have not only a wide range of grafting compatibility but also nondetrimental impacts on fruit quality attributes. Rootstock–scion interaction, especially with regard to its influence on plant growth and development, is an intriguing area in melon grafting research. Understanding the communication between scion and rootstock and signaling at physiological, biochemical, and molecular levels will build a solid foundation for more effective and targeted selections of rootstocks that contribute to long-term sustainability of grafted melon production.

REFERENCES

Agele, S. and S. Cohen. 2009. Effect of genotype and graft type on the hydraulic characteristics and water relations of grafted melon. *J. Plant Interact.* 4:59–66.

Alabouvette, C., F. Rouxel, J. Louvet, P. Bremeersch, and M. Mention. 1974. The search for a rootstock resistant to *Phomopsis sclerotioides* and *Verticillium dahliae* for greenhouse melon and cucumber growing. *Pepinieristes Horticulteurs Maraichers* 152:19–24 (in French).

Aloni, B., R. Cohen, L. Karni, H. Aktas, and M. Edelstein. 2010. Hormonal signaling in rootstock-scion interactions. *Sci. Hortic.* 127:119–126.

Aloni, B., L. Karni, G. Deventurero, Z. Levin, R. Cohen, N. Katzir, M. Lotan-Pompan et al. 2008. Possible mechanisms for graft incompatibility between melon scions and pumpkin rootstocks. *Acta Hortic.* 782:313–324.

Balázs, G. 2010. Rootstock selection in grafted muskmelon production in Hungary. In S. Marić and Z. Lončarić (eds.), *Vegetable Growing*, pp. 536–539. *45th Croatian and Fifth International Symposium on Agriculture*, Josip Juraj Strossmayer University, Opatijia, Croatia.

Benoit, F. 1974. The Fusarium problem in melon growing in Belgium and the relative value of certain rootstocks. *Tuinbouwberichten, Belgium* 38:16–20 (in Dutch).

Burger, Y., H.S. Paris, R. Cohen, N. Katzir, Y. Tadmor, E. Lewinsohn, and A.A. Schaffer. 2010. Genetic diversity of Cucumis melo. *Hortic. Rev.* 36:165–198.

Burger, Y. and A.A. Schaffer. 2007. The contribution of sucrose metabolism enzymes to sucrose accumulation in *Cucumis melo. J. Am. Soc. Hortic. Sci.* 132:704–712.

Chavagnat, A. 1971. Effects of artificial light on melon and grafted melon plants. *Pepinieristes, Horticulteurs, Maraichers, Revue Horticole* 117:45–50 (in French).

Colla, G., C.M. Cardona Suárez, M. Cardarelli, and Y. Rouphael. 2010. Improving nitrogen use efficiency in melon by grafting. *HortScience* 45:559–565.

Condurso, C., A. Verzera, G. Dima, G. Tripodi, P. Crinò, A. Paratore, and D. Romano. 2012. Effects of different rootstocks on aroma volatile compounds and carotenoid content of melon fruits. *Sci. Hortic.* 148:9–16.

Crinò, P., C.L. Bianco, Y. Rouphael, G. Colla, F. Saccardo, and A. Paratore. 2007. Evaluation of rootstock resistance to Fusarium wilt and gummy stem blight and effect on yield and quality of a grafted 'Inodorus' melons. *HortScience* 42:521–525.

Davis, A.R., P. Perkins-Veazie, Y. Sakata, S. López-Galarza, J.V. Maroto, S.-G. Lee, Y.-C. Huh, Z. Sun, A. Miguel, and S.R. King. 2008. Cucurbit grafting. *Crit. Rev. Plant Sci.* 27:50–74.

de Stigter, H.C.M. 1956. Studies on the nature of the incompatibility in a cucurbitaceous graft. *Mededelingen van de Landbouwhogeschool* 56:1–51.

de Stigter, H.C.M. 1961. Translocation of C14-Photosvnthates in the graft muskmelon/*Cucurbita ficifolia. Acta Bot. Neerl.* 10: 466–473.

Fita, A., B. Pico, C. Roig, and F. Nuez. 2007. Performance of *Cucumis melo* ssp. *agrestis* as a rootstock for melon. *J. Hortic. Sci. Biotechnol.* 82:184–190.

Galun, E. 1958. On the physiology of sex expression in the cucumber. In *XVth International Horticultural Congress*, Nice, France, pp. 11–18.

Gaskova, O. 1944. Grafting as a method of changing plants. *Doklady Vsesoyuznoi Akademii sel'sko-khozyaistvennykh Nauk im V. I. Lenina* 8:12–18 (in Russian).

Golecki, B., A. Schulz, U. Carstens-Behrens, and R. Kollmann. 1998. Evidence for graft transmission of structural phloem proteins or their precursors in heterografts of Cucurbitaceae. *Planta* 206:630–640.

Groenewegen, J.H. 1953. Grafting of cucumbers and melons. *Mededel Directeur Tuinbouw* 16:169–182.

Guan, W. and X. Zhao. 2014. Techniques for melon grafting. Oct. 6, 2015. http://edis.ifas.ufl.edu/hs1257 University of Florida/Institute of Food and Agricultural Sciences Extension EDIS Publication HS1257, Gainesville, FL.

Guan, W., X. Zhao, W.D. Dickson, M.L. Mendes, and J. Thies. 2014. Root-knot nematode resistance, yield and fruit quality of specialty melons grafted onto *Cucumis metulifer. HortScience* 49:1046–1051.

Guan, W., X. Zhao, R. Hassell, and J. Thies. 2012. Defense mechanisms involved in disease resistance of grafted vegetables. *HortScience* 47:164–170.

Guan, W., X. Zhao, D.J. Huber, and C.A. Sims. 2015. Instrumental and sensory analyses of quality attributes of grafted specialty melons. *J. Sci. Food Agric.* DOI: 10.1002/jsfa.7050.

Guan, W., X. Zhao, D.D. Treadwell, M.R. Alligood, D.J. Huber, and N.S. Dufault. 2013. Specialty melon cultivar evaluation under organic and conventional production in Florida. *HortTechnology* 23:905–912.

Hohlaceva, N.A. 1955. Using pumpkin as a stock for the cucumber. *Sad i Ogorod* 11:29 (in Russian).

Imazu, T. 1949. On the symbiotic affinity caused by grafting among Cucurbitaceous species. *J. Jpn. Soc. Hortic. Sci.* 18:36–42 (in Japanese).

Jifon, J.L., D.I. Leskovar, and K.M. Crosby. 2010. Rootstock effects on the water relations of Grafted watermelons. *HortScience* 45: S213.

King, S.R., A.R. Davis, W. Liu, and A. Levi. 2008. Grafting for disease resistance. *HortScience* 43:1673–1676.

King, S.R., A.R. Davis, X. Zhang, and K. Crosby. 2010. Genetics, breeding and selection of rootstocks for Solanaceae and Cucurbitaceae. *Sci. Hortic.* 127:106–111.

Koutsika-Sotiriou, M. and Traka-Mavrona, E. 2002. The cultivation of grafted melons in Greece, current status and prospects. *Acta Hortic.* 579:325–330.

Lebedeva, S.P. 1937. Changing the nature of plants (melon) by grafting. *Novoe v sel'skom khozyaistve* 16:42 (in Russian).

Lee, J.M., C. Kubota, S.J. Tsao, Z. Bie, P. Hoyos Echevarria, L. Morra, and M. Oda. 2010. Current status of vegetable grafting: Diffusion, grafting techniques, automation. *Sci. Hortic.* 127:93–105.

Lee, J.M. and M. Oda. 2003. Grafting of herbaceous vegetable and ornamental crops. *Hortic. Rev.* 28:61–124.

Lester, G. 1997. Melon (*Cucumis melo* L.) fruit nutritional quality and health functionality. *HortTechnology* 7:222–227.

Lester, G. and K. Shellie. 2004. Honeydew melon. In: K.C. Gross, C.Y. Wang, and M. Saltveit (eds.), *The Commercial Storage of Fruits, Vegetables, and Florist and Nursery Stocks*. U.S. Department Agriculture, Washington, DC, December 28, 2014. http://ww.ba.ars.usda.gov/hb66/honeydewMelon.pdf.

Louvet, J. and C. Lemaitre. 1961. The use of melon grafts for the control of Fusarium. *Rev. Hortic.* 133:8–10 (in French).

Louvet, J. and J. Peyriere. 1962. The advantage of grafting melons on *Benincasa cerifera* Savi. In *Proceedings of the 16th International Horticultural Congress*, Brussels, Belgium, Vol. 1, p. 70 (in French).

Louws, F.J., C.L. Rivard, and C. Kubota. 2010. Grafting fruiting vegetables to manage soilborne pathogens, foliar pathogens, arthropods and weeds. *Sci. Hortic.* 127:127–146.

Ludilov, V.A. 1964. Variations in some characters of seedling progeny of watermelon grafted on squash. *Agrobiologija* 4:616–617 (in Russian).

Ludilov, V.A. 1969. Distant grafting as a mutagenic factor. *Vestnik sel'sko-khozyaistvennoi Nauki* 5:66–68 (in Russian).

Matsuda, T. and F. Honda. 1981. Studies on physiological disorders of melon fruits. I. Influence of grafting and plant- and fruit-pruning in 'Pince' melon. *Bull. Veg. Ornam. Crops Res. Station Ser. C* 5:31–50 (in Japanese).

Matsumoto, S. 1931. Grafting of cucurbitaceous vegetables. *Jissaiengei, Jissaiengei-sya, Tokyo, Japan* 11:288–291 (in Japanese).

Minuto, A., C. Bruzzone, G. Minuto, G. Causarano, G. La Lota, and S. Longombardo. 2010. The physiological sudden collapse of grafted melon as a result of a not appropriate growing procedure. *Acta Hortic.* 883:229–234.

Mustafin, A.M. 1961. Hermaphrodite forms of cucumber. *Priroda* 50:113–114 (in Russian).

Naaldwijk, 1952. Trials on the grafting of cucumbers and melons. Jaarverslag. *Proefstation voor de Groente-en Fruitteelt onder Glas te Naaldwijk*, pp. 28–29 (in Dutch).

Obando-Ulloa, J.M., E. Moreno, J. García-Mas, B. Nicolai, J. Lammertyn, A.J. Monforte, and J.P. Fernández-Trujillo. 2008. Climacteric or non-climacteric behavior in melon fruit: 1. Aroma volatiles. *Postharvest Biol. Technol.* 49:27–37.

Parhomenko, M.P. 1941. New developments in the grafting of herbaceous plants. *Vernalisation* 3:121–124 (in Russian).

Pyo, H.K. and I.J. Sung. 1970. The anatomy of the graft union in cucurbits. *Res. Rep. Off. Rur. Dev., Suwon (Hortic.)* 13:71–76 (in Korean).

Ristaino, J.B. and W. Thomas. 1997. Agriculture, methyl bromide, and the ozone hole: Can we fill the gaps? *Plant Dis.* 81:964–977.

Rouphael, Y., M. Cardarelli, E. Rea, and G. Colla. 2012. Improving melon and cucumber photosynthetic activity, mineral composition, and growth performance under salinity stress by grafting onto Cucurbita hybrid rootstocks. *Photosynthetica* 50:180–188.

Rouphael, Y., D. Schwarz, A. Krumbein, and G. Colla. 2010. Impact of grafting on product quality of fruit vegetables. *Sci. Hortic.* 127:172–179.

Ruiz, J.M., A. Belakbir, I. López-Cantarero, and L. Romero. 1997. Leaf-macronutrient content and yield in grafted melon plants. A model to evaluate the influence of rootstock genotype. *Sci. Hortic.* 71:227–234.

Ruiz, J.M. and L. Romero. 1999. Nitrogen efficiency and metabolism in grafted melon plants. *Sci. Hortic.* 81:113–123.

Sakata, Y., T. Ohara, and M. Sugiyama. 2007. The history and present state of the grafting of Cucurbitaceous vegetables. *Acta Hortic.* 731:159–170.

Sakata, Y., T. Ohara, and M. Sugiyama. 2008. The history of melon and cucumber grafting in Japan. *Acta Hortic.* 767:217–228.

Saltveit, M.E. 1999. Effect of ethylene on quality of fresh fruits and vegetables. *Postharvest Biol. Technol.* 15:279–292.

San Bautista, A., A. Calatayud, S.G. Nebauer, B. Pascual, J. Vicente Maroto, and S. López-Galarza. 2011. Effects of simple and double grafting melon plants on mineral absorption, photosynthesis, biomass and yield. *Sci. Hortic.* 130:575–580.

Sanaev, N.F. 1966. The inheritance of characteristics acquired by the cucumber Nezin and the melon Gruntovaja gribovskaja as a result of their vegetative interaction. *Uc. Zap. mordovsk. Univ.* 55:78–88 (in Russian).

Schwarz, D., Y. Rouphael, G. Colla, and J.H. Venema. 2010. Grafting as a tool to improve tolerance of vegetables to abiotic stresses: Thermal stress, water stress and organic pollutants. *Sci. Hortic.* 127:162–171.

Shellie, K.C. and G. Lester. 2004. Netted melons. In: K.C. Gross, C.Y. Wang, and M. Saltveit (eds.), *The Commercial Storage of Fruits, Vegetables, and Florist and Nursery Stocks*. U.S. Department Agriculture, Washington, DC, December 28, 2014. http://www.ba.ars.usda.gov/hb66/nettedMelon.pdf.

Sigüenza, C., M. Schochow, T. Turini, and A. Ploeg. 2005. Use of *Cucumis metuliferus* as a rootstock for melon to manage *Meloidogyne incognita*. *J. Nematol.* 37:276–280.

Soteriou, G.A., M.C. Kyriacou, A.S. Siomos, and D. Gerasopoulos. 2014. Evolution of watermelon fruit physicochemical and phytochemical composition during ripening as affected by grafting. *Food Chem.* 165:282–289.

Suzuki, M. 1990. Automated grafting. *Agric. Hortic.* 65:123–130 (in Japanese).

Tiedemann, R. and U. Carstens-Behrens. 1994. Influence of grafting on the phloem protein patterns in cucurbitaceae. I: Additional phloem exudate proteins in *Cucumis sativus* grafted on two *Cucurbita* species. *J. Plant Physiol.* 143:189–194.

Traka-Mavrona, E., M. Koutsika-Sotiriou, and T. Pritsa. 2000. Response of squash (*Cucurbita* spp.) as rootstock for melon (*Cucumis melo* L.). *Sci. Hortic.* 83:353–362.

Vicente, M.R. and J. Plasencia. 2011. Salicylic acid beyond defense: Its role in plant growth and development. *J. Exp. Bot.* 62:3321–3338.

Weiss, F.E. 1940. Graft hybrids and chimaeras. *J. R. Hortic. Soc.* 65:237–243.

Wellensiek, S.J. 1949. The prevention of graft-incompatibility by own foliage on the stock. *Mededel. Landbouwhogeschool Wageningen* 49:257–272.

Xu, C.Q., T.L. Li, and H.Y. Qi. 2005. Effects of grafting on the photosynthetic characteristics, growth situation, and yield of netted muskmelon. *China Watermelon Melon* 2:1–3 (in Chinese).

Zhao, X., Y. Guo, D.J. Huber, and J. Lee. 2011. Grafting effects on postharvest ripening and quality of 1-methylcyclopropene-treated muskmelon fruit. *Sci. Hortic.* 130:581–587.

Section VII

Cucurbit Pathology and Diseases

18 Important Diseases of Cucurbitaceous Crops and Their Management

Akhilesh Sharma, Viveka Katoch, and Chanchal Rana

CONTENTS

18.1 INTRODUCTION

Cucurbitaceous crops comprise a large and diverse group of crops. Cucurbits are warm-season crops that are cultivated and harvested over spring, summer, and autumn seasons. They constitute an important part of a diverse and nutritious diet throughout the world, which are used as salad and pickled (cucumber), cooked (all gourds and squashes), candied, or preserved (ash gourd) vegetables or as dessert fruits (muskmelon and watermelon). Also, cucurbits are used as fiber source and for utensil preparations, decorations, and ceremonial and medicinal purposes (McGrath 2004).

A wide range of pathogens affect the productivity of cucurbits, which constitute over 200 diseases (Zitter et al. 1996). The diseases may be caused by fungi, bacteria, viruses, or mycoplasma-like organisms. The disease may be soilborne, seedborne (carried by seed), spread by wind, or transmitted by insect vectors. The moisture content of the seed, storage period, prevailing temperature, and degree of invasion influence the development of seedborne fungi (Anjorin and Mohammed 2009). The pathogen may cause seed abortion and rot, necrosis, reduction, or elimination of germination capacity as well as seedling damage at later stages of plant growth, resulting in the development of the disease as systemic or local infection (Khanzada et al. 2001). The diseases cause heavy loss in terms of yield and quality. The diseases caused by pathogen are controlled by fungicides against fungal diseases, bactericide for bacterial diseases, or insecticide in case of viruses transmitted by insect vectors. The diseases can be managed by adopting cultural, biological, and chemical methods of disease management. These days, consumers are quite cautious about pesticide residue effects and organic cultivation of crops is in vogue. Moreover, "prevention is better than cure." The following management strategies are required to be taken into consideration for raising healthy and disease-free cucurbitaceous crops:

1. Visit the field daily and carefully observe the crop. Disease management in the initial stages would help to reduce chemical load and shall be very economical.
2. Always use certified seed from authenticated sources.
3. Grow resistant varieties/hybrids.
4. Treat the seed before sowing.
5. Crop rotation with crops other than cucurbits should be done. Follow a minimum of 3-year crop rotation.
6. Maintain field sanitation, that is, discard plant debris and keep the cropping area free from weeds as they may serve as alternate host to disease-causing organisms.
7. Ensure proper drainage especially during rainy season.
8. Avoid overcrowding of crop plants. Maintain desired plant population.
9. Avoid injury during intercultural operations, harvesting, and handling.
10. Discard the diseased portion carefully in a way that inoculum does not spread.
11. Use clean irrigation water and well-sterilized tools.

12. Control insect vector to check viral diseases.
13. Do not spray in the daytime as pollinators are quite active in the day time.
14. Do not use a single chemical to combat disease repeatedly.
15. Adjust harvesting and spray schedule in a way that the waiting period required for a particular pesticide is followed.

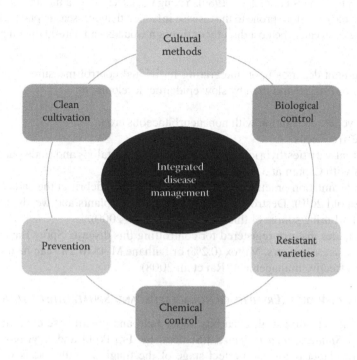

The major diseases of cucurbits are described as follows:

18.2 FUNGAL DISEASES

18.2.1 *Alternaria* Leaf Spot or *Alternaria* Blight
(*Alternaria cucumerina* and *Alternaria alternate*)

The fungus, *Alternaria* leaf spot, occurs throughout the world and can infect most cucurbits. It is also known as target leaf spot or *Alternaria* blight, caused by *Alternaria cucumerina* [(Ellis & Everh.) Elliot] and *Alternaria alternate* [(Fr.) Keisslar]. The disease is seed- and soilborne. It appears on bottle gourd, watermelon, muskmelon, cucumber, snake gourd, and vegetable marrow. High humidity and temperature in the range of 15°C–32°C is quite favorable for its development. The symptoms appear as small circular tan spots 1–2 mm in diameter on the upper surface of the older or crown leaves that may be surrounded by a yellow halo. These spots later spread to the younger leaves toward the tips of the vines. Lesions formed on the lower leaf surface tend to be more diffused. These spots later coalesce to form large lesions, which may be more than 10 mm in diameter. These lesions show a target-like pattern of rings that is typical of most *Alternaria* spp. These lesions bear concentric rings that may cause severe leaf drop. Dark sunken brown spots initiate the disease on fruits, which later develop dark powdery lesions. It causes damage by defoliating the vines and leads to reduction in fruit yield, size, and quality. Partial defoliation may even cause the fruit to sunscald and ripen prematurely. The first symptoms usually were observed at the time of initial fruit development as fruits turn brown and reduce in size, which later turn black and mummified.

The fungus overwinters as dormant mycelium (saprophyte) in diseased and partly decayed crop debris, in weeds of the cucurbit family, and possibly in the soil. Conidia are produced in the spring and act as the primary inoculum. The inoculum is carried by wind for long distances and splashing water from diseased plants to susceptible tissues (Seebold 2010; Watt 2013). The conidia can survive under warm and dry conditions for several months. The period between infection and the appearance of symptoms varies from 3 to 12 days (Babadoost 1989). Young plants less than a month old and fruit-bearing plants (70–75 days old) are more prone to the disease infection than midseason plants (45–60 days old). Germinating spores can enter the host directly or through wounds and natural openings.

Management
Effective management depends upon integrating individual control measures that help to reduce leaf spot in the initial stages and thereby slow epidemic development:

1. Follow a 3-year crop rotation with noncucurbitaceous crop.
2. Use disease-free seeds.
3. Grow resistant varieties/hybrids. Consult current seed catalogs and trade publications.
4. Treat seeds with Captan at 3 g/kg of seed.
5. Follow good sanitation practices such as cleaning up crop debris at the end of the growing season (Seebold 2010). Destroy volunteer cucurbit crop plants and weeds, which may be capable of harboring spores of this disease (Kucharek 2000).
6. Apply fungicides that are registered for controlling this disease. Spray Bavistin (0.1%) as soon as the disease appears. Miltox (0.2%) or Dithane M-45 (0.2%) can be used at 15 days interval for effective management (Rai et al. 2008).

18.2.2 POWDERY MILDEW (*ERYSIPHE CICHORACEARUM* AND *SPHAEROTHECA FULIGINEA*)

Powdery mildew affects almost all cucurbits under field and greenhouse conditions. It is caused by airborne fungi *Sphaerotheca fuliginea* [(Schlect. ex. Fr.) Poll.] and *Erysiphe cichoracearum* (D. ex. Merfat), the latter being the perfect stage of the fungi. The disease is widely distributed and destructive among cucurbits in most areas of the world and can be a major production problem causing yield losses of 30%–50% (El-Naggar et al. 2012). The disease first appears on older leaves. Conidia are produced profusely in the white powdery mycelium, and these spores spread quickly through wind to the adjacent leaves or plants as well as travel over long distances.

Powdery mildew appears as talc-like (white fluffy) colonies or circular patches on the under surface of leaves. As the disease progresses, the entire leaf surface is colonized by the fungus. In case of severe infection, the patches on the leaves coalesce and become yellow and necrotic. Such leaves die within a short span of time. The disease is most severe after fruit set and in densely planted fields. Symptoms and signs can also develop on stems and fruit. Powdery mass on the leaves decreases the photosynthetic rate (Queiroga et al. 2008), causing reduction in plant growth, premature foliage loss, and consequently reduction in yield. The infected plant parts remain stunted and distorted and may drop prematurely. The infected fruits do not develop fully; as a result, yield and quality of fruits is affected. The yield loss is proportional to the severity of the disease and the length of time that plants have been infected (Mossler and Nesheim 2005). If this disease is not controlled in time, symptoms can be severe enough to cause extensive premature defoliation of older leaves and wipe out the crop (Nunez-Palenius et al. 2012).

Disease development is favored by vigorous plant growth and moderate temperature. The most favorable conditions for disease development are 35°C temperature and high relative humidity of more than 70% (Ali et al. 2013). Temperature and humidity must be examined together since it is the water vapor pressure deficit that has the greatest effect on host–parasite interactions (Jarvis et al. 2002). During periods of intensive dew on leaf surfaces, the severity of this disease is enhanced. However, excessive water on the leaf surface is often detrimental to the development of powdery mildew disease.

Management

- Monitor crops closely to identify the disease early in its cycle. Monitor even powdery mildew-resistant varieties because there are many races of the pathogen and some of them may break resistance. Summer squash (spring planted) is affected earlier and can be used as an indicator for examination in other crops.
- Grow resistant/tolerant varieties. Genetic resistance is used extensively in cucumber and melon and has been incorporated into other cucurbit crops (McGrath 2011). However, there are several fungal races of powdery mildew, and hence some resistant cultivars might be susceptible to a specific fungal race (Zitter et al. 1996). According to Konstantinidou-Doltsinis and Schmitt (1998), the need to control powdery mildew disease is one of the reasons for the increased use of fungicides in cucurbits.
- Biological control involves the use of fungal spores of *Ampelomyces quisqualis* Ces., which parasitize and destroy the powdery mildew. Similarly, bacteria *Bacillus subtilis* and fungus *Sporothrix flocculosa* (syn. *Pseudozyma flocculosa*) gave promising results (Nunez-Palenius et al. 2012).
- Follow field sanitation.
- Petroleum spray oils, milk, and bicarbonates might offer some control. Neem oil helps in effective management, but the harmful effects on beneficial insects render it impractical. However, cinnamon oil has been found effective (Nunez-Palenius et al. 2012). Most of these natural oils have to be applied repeatedly during the active growing season in order to achieve a reliable level of control. Thus, they cost much higher than the chemicals. Cow's milk sprayed onto the leaves of greenhouse-grown zucchini was found to be effective (Bettiol 1999). White mustard oil at 1% was also found effective (Stoyka et al. 2014).
- Fungicides should be applied at an interval of 7–10 days with the appearance of early disease symptoms. Most of the fungicides to control powdery mildew are primarily preventive, that is, they are effective when applied before the disease appears. In addition, powdery mildew fungi can develop resistance to specific fungicides (Brown 2002). Spray sulfur-based fungicides like Karathane (0.05%) or Hexconazole (0.1%) for effective control. Addition of 2.3 mM silicon to the nutrient solution (fertigation) can significantly delay and reduce the incidence of disease (Menzies et al. 1991). Moreover, foliar sprays of chlorite mica clay that contains silicon have also shown suppression of powdery mildew in cucumber (Ehret et al. 2001). Follow rotation of protective and systemic fungicides to reduce the chance of fungicide resistance development.

18.2.3 Downy Mildew (*Pseudoperonospora cubensis*)

Downy mildew caused by *Pseudoperonospora cubensis* [(Berk. Curt.) Rostow zew] is airborne and the major foliar disease that infects most of the cucurbits. It is particularly serious in warm weather conditions or tropical environment. Symptoms first appear as pale green areas on the upper surface of leaves that changes to bright yellow angular or rectangular spots. Leaf spots are irregular or blocky in appearance and are delimited by leaf veins. In case of severe infection, the lesions expand and coalesce, tissues turn necrotic and brown, and results in shriveling and death of large areas of leaf surface giving a scorched appearance. Slight yellowing may be seen around the edges of the spots or on other parts of the already infected leaf. The affected foliage in the early plant development stage causes a reduction in photosynthetic activity that results in stunted plant growth and yield reduction, especially in cucumber (Colucci and Holmes 2010). When humidity is high, the lower surface of infected leaves shows water-soaked lesions, slightly sunken with profuse growth of light to dark gray or purple sporulation that is also delimited by leaf veins, which is evident as a fuzzy or downy growth. Premature defoliation may also result in sunscalding of fruits due to direct exposure to sunlight. The disease progresses aggressively and kills the plant quickly through rapid defoliation.

High humidity, fog, and heavy dew favor the disease development (Keinath 2014). After infection, downy mildew continues to get worse even under dry weather conditions. In contrast to powdery mildew, spores of downy mildew are dark purplish gray and appear only on the underside of leaves (McGrath 2006).

Symptoms on watermelon and cantaloupe are not as distinctive as on cucumber and squash and are mostly mistaken for other diseases such as anthracnose, target spot, *Alternaria* leaf spot, or gummy stem blight (Colucci and Holmes 2010). The symptoms of downy mildew infection are variable in different cucurbit crops, for example, the lesions are angular and are limited by the leaf veins on cucumber and squash, whereas these are typically irrregularly shaped on the foliage of watermelon and cantaloupe and turn a brown rapidly. As the disease progresses, the lesions expand and multiply, causing the field to have a brown and brittle look (Celetti and Roddy 2010).

It is an obligate parasite that requires living host tissue to grow and reproduce. The organism overwinters on the infected cucurbit plants in areas without hard frost. Sporangia are the primary source of inoculum that spread from the infected plants through wind currents or rain splashes to local or distant places. Once the sporangia reach a susceptible host, they germinate and directly infect the leaf within an hour. The optimum temperature for sporulation is 15°C–20°C with at least 6 h of high humidity. The plants with yellow lesions have high sporulation ability. The symptoms appear after 3–12 days of infection, depending upon the inoculum load, temperature, and relative humidity. High temperatures of more than 35°C are not favorable though cooler nights favor disease development.

Management

- Grow resistant/tolerant varieties. With the development of new races, different varieties, especially cucumber, do not exhibit a high level of resistance as they did to previous races. However, resistant varieties are still a valuable component of downy mildew management.
- Monitor crops to identify the disease early in its cycle.
- Prophylactic measures should be followed on the basis of the weather forecast as the disease typically coincides with the onset of rains.
- Grow cucurbits in environments where humidity can be manipulated to manage downy mildew. For example, vining cultivars should be tied to the trellises so that the dew on leaves dries quickly. Also, increasing plant spacing or growing in greenhouses can help in decreasing relative humidity and leaf wetness.
- Early planting can also avoid the disease (Keinath 2014).
- Avoid excess overhead irrigation and apply irrigation during the late morning hours to facilitate rapid leaf drying.
- Wash hands before moving from one field to another, and if possible, wear clean clothes.
- Neem oil derived from *Azadirachta indica* is a botanical control for both downy and powdery mildews on cucurbits, but negative impacts have been reported as it has been found to be toxic to ladybeetles (Banken and Stark 1998) and harmful to bees (so, apply when they are not active in the field). Therefore, it should not be used without clear need and caution. Biocontrol agent *B. subtilis* is available in a wettable powder formulation that can be used for downy mildew control (Fravel 1999). However, Keinath (2014) reported that no organic-approved fungicides or biofungicides prevent or control cucurbit downy mildew.
- Broad-spectrum contact protectant fungicides provide some degree of control. Systemic fungicides are recommended in the initial stages when there is forecast of occurrence or symptoms have just appeared. Apply them at 5–7-day intervals depending on disease severity. Spray Ridomil (0.3%), Blitox (0.2%), or Mancozeb (0.2%). The application of Dithane M-45 (0.3%) at a 15-day interval is good to control the disease (Rai et al. 2008). Rotate fungicide products to reduce the risk of resistance to fungicides.

18.2.4 ANTHRACNOSE (*COLLETOTRICHUM ORBICULARE*)

Anthracnose caused by *Colletotrichum orbiculare* [(Berk & Mont.) Arx.] is both air- and seedborne fungal disease that occur on most cucurbits during warm and moist seasons. The symptoms occur on all aboveground parts of cucurbit plants. The initial symptoms appear as water-soaked lesions on leaves, which then turn as yellowish circular spots and later as brown to black. The symptoms of infection are variable in different cucurbits. The spots are irregular and turn dark brown or black on watermelon foliage, while the spots turn brown on cucumber and muskmelon and can enlarge considerably. The petioles, fruit pedicel, and stem can also become infected, resulting in vine defoliation, fruit decline, and death of plant. Long dark spots develop on stems. The lesions on muskmelon stem can girdle it and cause wilting of vines. The most noticeable symptoms appear on the fruits. The young, growing fruits may turn black, shrivel, and drop if fruit pedicels are infected. The symptoms appear as roughly circular, sunken, and water-soaked spots with dark borders on mature fruits. The size of the spots may vary from 6 to 13 mm in diameter and 6 mm deep on watermelon. Within the lesion, small, black fruiting structures (acervuli) are formed. Under humid conditions, black stromata with a gelatinous mass of salmon pink spores may be observed on muskmelon, watermelon, and cucumber. Cankers lined with this characteristic color can never be mistaken for any other disease (Zitter 1987). The disease is primarily promoted under humid and moist weather conditions. Seedborne anthracnose infection results in drooping and wilting of cotyledons, and lesions may appear on the stem near the soil line. Huge losses in storage or shipment can occur if infected fruits are packed.

The fungus overwinters on infected residue of the previously grown cucurbit crop and may also be passed on to cucurbit seeds. In the spring, under humid conditions, the fungus releases conidia (airborne spores) that infect foliage of cucurbit vines. High humidity and temperature of 24°C is optimum for disease multiplication though conidia do not germinate below 4.4°C or above 30°C or under nonhumid conditions. The pathogen needs water to liberate the conidia from their sticky covering. Anthracnose usually establish midseason after the plant canopy has developed. The fungus can penetrate leaves directly and does not require natural openings (e.g., stomates) or wounds.

Management

- Grow resistant/tolerant varieties, which should be the first step in managing any plant disease and can greatly reduce yield losses due to anthracnose. Several public institutes and seed companies offer cultivars with varying levels of resistance to this disease that can be grown. Two distinct races (and perhaps as many as seven), which vary in their ability to infect a range of cucurbit genera, species, and cultivars, have been reported (Goldberg 2004). Race specialization poses limitations for breeding resistant cultivars as it is difficult to obtain resistance for all pathogenic races. However, resistance is still an important way to manage this disease.
- It is desirable to grow a clean seed by obtaining it from production areas that are not having anthracnose problems.
- Treat seeds with Thiram or Captan at 3 g/kg seed.
- Monitor crops closely to identify the disease early in its cycle, especially on the young plants. The management of early infections and protecting young plants from infection prevent inoculum buildup in the field and reduce losses (Thompson and Jenkins 1985).
- Follow good sanitation practices such as cleaning up crop debris and wild hosts at the end of the growing season. Also, cucurbit volunteers and alternative hosts in and around the field should be destroyed.
- Carry out deep plowing immediately after the final harvest to destroy all infected cucurbit plants and debris in the field (Palenchar et al. 2012).
- Follow long crop rotation with noncucurbit crops.
- Ensure proper drainage.

- Spraying of Difolatan (0.2%), Dithane M-45 (0.2%), or Bavistin (0.2%) has been found effective (Rai et al. 2008). Fungicide applications generally are not required within 2–3 weeks of the final harvest.
- Copper products and biological control agent, *B. subtilis* strain, QST713, have also been recommended (Li 2014). Shimizu et al. (2009) reported that endophytic *Streptomyces* sp. strain, MBCu-56, has strong potential for controlling cucumber anthracnose.

18.2.5 SCAB OR GUMMOSIS (*CLADOSPORIUM CUCUMERINUM*)

It is caused by seedborne fungi *Cladosporium cucumerinum*. The disease appears on leaves, petioles, stems, and fruits (Ogorek et al. 2012). Numerous water-soaked spots occur on leaves and runners, which eventually turn gray to white and become angular, often with yellow margins. The center of the spots could then drop out to give irregular-shaped holes in the leaves. The lesions on the stem and petiole are comparatively oblong in shape and may be covered with dark-green velvety mold under moist conditions. Lesions on fruit are often confused with *Anthracnose*. The spots might ooze a gummy substance, which could then be invaded by secondary rotting bacteria, and these spots cause foul smell (Yuan 1989; Watson and Napier 2009).

The disease overwinters on crop debris and seed. The spores are blown long distances in moist air and are also spread by insects, implements, and clothes of workers. The spores penetrate the host within 9 h of germination and infection can be seen in 3 days (Zitter 1986). The optimum temperature for disease development is 17°C–27°C along with high humidity. The spread of pathogen is favored by low night temperature, fog, and low light intensity.

Management

- Grow resistant cultivars.
- Seed should be harvested from disease-free plants.
- Treat seeds with Thiram or Captan at 3 g/kg seed.
- After the crop is harvested, plant debris and especially the fruits should be removed.
- Prophylactic measures should be taken particularly if cool, wet weather is expected.
- Crop rotation with noncucurbitaceous crops should be followed.
- Avoid low-lying, shaded areas prone to heavy fogs and dews.
- Avoid overhead sprinkler irrigation (Koike et al. 2006).
- Photodynamic dyes (bengal rose, toluidine blue, and methylene blue) systemically protect cucumber plants from cucurbit scab. It was found that all dyes at 0.5–200 μM significantly suppressed symptoms of the disease (AverYanov et al. 2011).

18.2.6 CERCOSPORA LEAF SPOT (*CERCOSPORA CITRULLINA*)

The disease is caused by soil- and airborne fungi, *Cercospora citrullina* (Cooke). The symptoms can be typically found on the foliage but may appear on petioles and stems in case the disease is severe and conditions are highly favorable for development. Dark spots are seen first on the older leaves. The spots are circular to irregular in shape with pale or light centers and dark margins. As lesions expand, they coalesce and give a blighted appearance. If the disease is severe, then defoliation occurs and the yield is affected. It can reduce fruit size and quality, but economic losses are rarely severe (Schwartz and Gent 2007).

The pathogen survives on infested crop debris and on weeds between cucurbit crops. The disease infection through its spores (conidia) is readily spread by wind currents, splashing rain, and irrigation water. The infection begins after the deposition of spores onto the leaves and petioles. The disease development is favored by temperatures in the range of 26°C–32°C and high humidity. The new cycles of infection and sporulation occur every 7–10 days during warm and humid weather conditions (Schwartz and Gent 2007).

Management

- Crop rotation with noncucurbitaceous crops should be followed.
- Avoid overhead irrigation and irrigation in the morning hours.
- Follow good sanitation practices by removing old diseased crop residues and cucurbitaceous weeds. Deep plowing may also be practiced immediately after final harvest.
- Spray the crop with Dithane M-45 at 0.2%.

18.2.7 GUMMY STEM BLIGHT (*DIDYMELLA BRYONIAE*)

Gummy stem blight/leaf blight/black rot is a major disease of cucumber, cantaloupe, pumpkin, and watermelon. The disease is transmitted through seed, soil, and air. The leaf symptoms appear as small, circular tan spots, sometimes surrounded by a yellow halo. The disease may be confused with downy mildew, but the leaf spots are larger than downy mildew. Under favorable conditions, the leaf lesions coalesce and the entire leaf may become blighted. The spots have a ringed appearance and can be observed on the midvein of leaves and petioles as water-soaked reddish-brown spots. The infection often begins at leaf margins (Koike 1997). When the spots dry, they become cracked and may tear off, causing the leaves to have tattered appearance. Stem infections consist of oblong, water-soaked brown lesions. Main stem lesions enlarge and slowly girdle the main stem. The stems may split open to form open wounds or cankers. Characteristic red or brown gummy fluid oozes out of the cankers. The fungus also form tiny black pimple-like fruiting bodies on the stem or nodes. The cankers on the infected stems/vines girdle the entire stem and cause wilting.

The most important form of the disease is crown rot, which may kill plants. On fruits, the disease is known as black rot. Initially pale brown, then bleached patches develop on the fruit and reddish gum oozes out from cracks. The affected areas are studded with pycnidia of the fungus. The infected fruits develop irregular, water-soaked, small oval-to-circular spots that are greasy green in color and turn dark brown as the spots enlarge. Gummy exudates and black fruiting bodies may develop on the spots. Fruits can rot within 2–3 days after infection. The fungus survives on crop residue and in the soil for 2 years without a host. The disease-infected conidia migrate through splashing rain, irrigation water, and winds from the source to the new crops. The optimum temperature for disease development is 20°C–24°C along with free moisture. The fungus is most devastating in warm and humid weather.

Management

- Procure disease-free seeds from reliable sources.
- Treat seeds with Thiram or Captan at 3 g/kg seed.
- Monitor crops and ensure proper disease diagnosis as charcoal rot and *Fusarium* wilt show similar symptoms.
- Crop rotation with nonhost crops should be followed.
- Overhead irrigation should be avoided.
- Remove and destroy infected fruits and vines at the end of the season.
- Spray Bavistin (0.2%) as soon as the disease is noticed. In case the disease is not controlled, spray Dithane M-45 (0.25%) or Propiconazole (0.1%).
- Combined use of green manure with Melcast-scheduled fungicide application, chlorothalonil (Bravo Ultrex 82.5 WDG at 3.0 kg/ha), *B. subtilis* (Serenade 10WP at 4.5 kg/ha), and pyraclostrobin plus boscalid (Pristine 38 WG at 1.0 kg/ha) could effectively manage the disease and reduce fungicide use. Biofungicides alternated with chlorothalonil also minimized use of synthetic fungicides (Zhou and Everts 2008).
- Avoid injury to fruit before or during harvest as wounds enable the pathogen to enter the fruit during storage.

18.2.8 Charcoal Rot (*Macrophomina phaseolina*)

The disease is caused by the soilborne fungus *Macrophomina phaseolina* [(Tassi) Goid]. Infected leaves and stems near the crown of the plants have a bleached appearance and later turn brown to black. Symptoms of charcoal rot are very similar to gummy stem blight as the gum exudes from the infected plant tissues. But the symptoms of charcoal rot appear late in the season. The black sclerotia are visible on removal of the epidermis of the stem, while black streaks can be observed within the pith on cutting open the collar region. This disease has a wide range of hosts. The affected plants die under hot and dry weather conditions. The plants carrying latent infection survive under wet and cool weather conditions but they die as soon as favorable weather conditions for disease development prevail. The fungus persists in soil and crop residue as microsclerotia for 3–12 years and can infect 500 plant species (Davis et al. 2012a). Roots become infected first, and the fungus later invades the plant crown. It occurs mainly in hot climate with prevailing temperature of at least 28°C. In infected tissues, profuse microsclerotia having irregular shape and black color are produced.

Management

- Grafting techniques that graft the susceptible scions onto resistant cucurbit rootstock can be utilized, which could be an effective management strategy for the control of soilborne root-infecting pathogens where the use of chemicals is not feasible or economical.
- Follow long-term crop rotation with nonhost crops.
- Destroy infected plant debris at the end of growing season.
- Biological control through seed treatment of melon with *Streptomyces* strain effectively regulates the mycelia growth of *M. phaseolina* (Etebarian 2006a). Different strains of *Trichoderma harzianum* (Bi), *T. harzianum* (T39), *Trichoderma virens* (DAR74290), *Trichoderma viride* (MO), *T. harzianum* (M), and *Trichodermin* (B) as a commercial formulation have been reported to be potential biological agents for the control of charcoal stem rot in melon (Etebarian 2006b).
- Management of saline soils can effectively decrease charcoal rot severity (Roustaee et al. 2012).

18.2.9 Damping-Off of Seedlings and Fungal Root Rots (*Pythium* spp., *Rhizoctonia* spp., and *Fusarium* spp.)

The seed and seedlings of cucurbits are affected by a number of soilborne pathogens. Of these, damping-off and root rots caused by a complex of fungi, namely, *Pythium*, *Rhizoctonia*, and *Fusarium*, are most common. Damping-off affects the crop before (preemergence damping-off) and after the germination of the seed (postemergence damping-off). Preemergence infection causes the rotting of seed inside the seed coat. The seed may germinate but the radical and cotyledon turn brown and soft and fail to grow further. The initial symptoms of postemergence damping-off appear as yellow to dark brown, water-soaked lesions on the root and hypocotyl tissues. With time, the hypocotyl tissues shrivel, roots decay further, and the seedlings topple down or may wilt and eventually collapse. Plants that survive may show symptoms of root rot. Roots can have a watery gray appearance, particularly the fine feeder roots. Cool temperatures, high soil moisture, and poor aeration favor the disease development. When plastic mulches are used under moist, hot conditions, the roots of even older plants rot. The fungal complex can invade many plants species and can survive even on decaying plant material. Usually, sporadic outbreaks are difficult to control.

Management

- Treat seeds with Ceresan or Thiram at 3 g/kg seed. Phosphonate seed treatment is a cost-effective way of protecting cucumber plants from *Pythium* damping-off (Abbasi and Lazarovits 2006).
- Ensure proper drainage and avoid overwatering.

- Drench the soil with Bavistin (0.1%).
- Field sanitation should be ensured by carefully removing the plant debris and weeds after the final harvest.
- Follow crop rotation with noncucurbitaceous crops.
- Highly effective biological control of soilborne pathogens can be attained only with the combined application of organic amendments and microbial biocontrol agents (Chang et al. 2007; Singh et al. 2007).

18.2.10 FUSARIUM WILT (FUSARIUM SPP.)

Fusarium wilt of cucurbits is caused by seed- and soilborne fungi of the genus *Fusarium*, that is, *Fusarium oxysporum* f. sp. *niveum* (EF Smith Snyder & Hansen) affects watermelon (Owen 1955), *Fusarium oxysporum* f. sp. *melonis* (Snyder & Hansen) affects muskmelon, and *Fusarium oxysporum* f. sp. *cucumerinum* (Owen) affects cucumber. They cause wilt and root rot. The fungi invade the roots of the plant and progress into the stems. Plants in early stages of their growth can develop damping-off due to lower stem infections; as a result, the seedlings often show hypocotyl rot or may topple down at the soil line. It is characterized by loss of turgor pressure of the vines. The wilting generally starts on the older leaves and advances to the young foliage. Older plants may first exhibit temporary wilting during peak heat periods of the day and revive during cool nights but eventually die within a few days. Wilt symptoms develop on one or few lateral vines in the beginning, while other branches remain apparently unaffected. However, the whole plant may wilt and die within a short time under high inoculum conditions or in highly susceptible host species. Vascular browning, gummosis, and tylosis in xylem vessels occur in mature plants. The disease may invade the fruits through the stem end.

In wet weather, white-to-pink fungal growth may be visible on the surface of the dead tissues. High nitrogen, especially ammoniacal form, less than 25% soil moisture, and light, sandy, and slightly acidic soils (pH 5–5.5) favor disease development (Zitter 1998). The disease occurs when the soil temperature is between 20°C and 30°C and the weather is dry. The causal fungi survive in old infected plant debris, other host plants, seed, or soil. It is generally presumed to be monocyclic, that is, it does not spread from plant to plant during the season. The spread within a field can occur by the movement of infested soil, while the spread across the fields can occur by using infected equipments and plants. Many *Fusarium* wilt pathogens including *F. oxysporum* f. sp. *niveum* are capable of being seedborne, although the extent of the contamination varies widely (Egel and Martyn 2013). Long-term survival of the pathogen in the soil and the evolution of new races make the management of *Fusarium* wilt difficult.

On the other hand, *Fusarium* crown and foot rot [*Fusarium solani* f. sp. *cucurbitae* (Snyder & Hansen)] causes root rot in squashes. The early symptoms appear as wilting of leaves, while the entire plant may wilt and die with the passage of time. The rot develops initially as water-soaked, light-colored areas that progressively turn darker. The fungus is generally confined to the crown area of the plant and the infection starts in the cortex of the root. It causes the cortex tissue to slough off and ultimately wipe out all the tissues except the fibrous vascular strands. The affected plants split easily about 2–4 cm below the soil line.

Management

- The exclusion of the pathogen is the best means to manage disease and, accordingly, procure disease-free seeds and seedlings from reliable sources.
- Grow resistant varieties/hybrids, which is the best and most economical method of control. However, there is no complete resistance to all races available in any commercial watermelon lines (Egel and Martyn 2013).
- Treat seeds with Benlate or Bavistin (2 g/kg seed) or with hot water at 52°C for 30 min.

- Drench the soil with Captan (0.2%–0.3%) or Bavistin (0.1%).
- Rotate with noncucurbitaceous crops such as garlic, radish, onion, and beetroot.
- Soil solarization (Martyn 1986) can be used to lower infection in soil sufficiently to delay the onset of wilt symptoms as well as to reduce the disease incidence; however, the disease cannot be eliminated.
- Use other cucurbit species resistant to *Fusarium* wilt as rootstocks for grafting. The most common rootstocks for cucurbit crops include *Lagenaria siceraria* and *Cucurbita moschata × Cucurbita maxima* hybrids, both of which are highly resistant to the common pathovars of *F. oxysporum* affecting cucurbits (King et al. 2008).
- Destroy plant debris and weeds, especially cucurbitaceous weeds, after crop harvest.
- Soil amendment with *B. subtilis* (strain-B006 powder) as nursery substrate followed by drenching with B006 suspension during transplanting and 1 week after transplanting and addition of 10 g organic fertilizer at the time of transplanting significantly suppressed the disease development in cucumber (Yang et al. 2014). Use of antibiotic-producing soil fungi and bacteria, that is, *Gliocladium* spp., *Trichoderma* spp., and *Pseudomonas* spp. were found effective (Srinon et al. 2006). Use nonpathogenic strains of *F. oxysporum* that compete with pathogenic forms for root colonization (Freeman et al. 2002). Though none of these methods give adequate control in the field, integration of these management strategies may be useful. Cover crops such as hairy vetch as soil amendments can also be useful (Zhou and Everts 2004).

18.2.11 VINE DECLINE OR CROWN BLIGHT (*MONOSPORASCUS CANNONBALLUS*)

It affects a number of cucurbit crops but is most severe on melons, squash, and pumpkin. The symptoms appear approximately 2 weeks prior to fruit maturity. The initial symptoms of the disease appear as stunting of young plants or severe yellowing of foliage followed by the chlorosis of lower older leaves that later wilt and collapse. The collapse of the vine reduces fruit size, fruit number, and fruit quality or fruit may exhibit sunburn symptoms due to lack of foliage. The symptoms on roots appear as tan to red-brown lesions that further expand and result in the death of small feeder roots. In severe cases, the root system may rot and the plant collapse with no other symptoms.

The characteristic feature of disease development is minute black, spherical, and erumpent perithecia at the perithecial openings on the entire root along with brown rot lesions that are easily visible with naked eyes or a hand lens. On maturity, the perithecia rupture and the ascospores fall into the soil, and thus the disease is soilborne. The optimum temperature range for the germination and colonization of ascospores on the roots is 24°C–27°C though they can germinate at temperature as low as 20°C. In a spring-planted crop, root infection occurs about 57 days after planting, whereas in a late-spring- or fall-planted crop, root infection can occur within 9 days after planting (Davis et al. 2012b).

Management

- Preplant soil fumigation with methyl bromide, chloropicrin, 1,3-dichloropropene, or metam sodium significantly reduces the buildup of inoculum (ascospores) in soil field (Martyn 2002).
- Sanitize the field by destroying infected roots immediately after the final harvest.
- Apply metam sodium in the sick field immediately after harvest to prevent disease development.
- Use grafted melons on resistant squash (*Cucurbita* spp.) or bottle gourd (*Lagenaria* spp.) rootstocks.
- Follow long crop rotation with noncucurbitaceous crops.
- Avoid overirrigation and ensure proper drainage.

- Cover crops such as canola, winter rape, and mustards may help in reducing the disease severity (Egel and Martyn 2010).
- Treatment of melon seeds with methyl jasmonate (an inducer of pathogen defense mechanisms in plants) reduced the severity of disease under experimental conditions and offers an additional potential control strategy for the future (Davis et al. 2012b).

18.2.12 PHYTOPHTHORA BLIGHT OR PHYTOPHTHORA CROWN AND ROOT ROT (PHYTOPHTHORA CAPSICI, PHYTOPHTHORA DRECHSLERI, AND PHYTOPHTHORA PARASITICA)

The saprophytic fungi *Phytophthora capsici* affect most cucurbit crops but squash and pumpkin are the most affected. The disease is also common on tomato, eggplant, capsicum, beet, Swiss chard, lima bean, turnip, spinach, and many common weeds. It has also been reported to be caused by *Phytophthora drechsleri* and *Phytophthora parasitica*. The disease infects more than 50 plant species in more than 15 families (Babadoost 2005). The pathogen may afflict the crop right from preemergence stage to maturity and cause a wide variety of symptoms. It may cause pre- and postemergence damping-off, stem and vine blight, and wilting of young shoots and leaves, followed by wilting and collapse of whole foliage or fruit rot. Mature plants may wilt suddenly even without appearance of any symptoms like stem or vine lesions. Upon uprooting the plants, tan to brown rotted root system can be observed. The fine feeder roots break off along with the withering of outer tissues of the tap root. Stem and leaf petiole lesions are light to dark brown, water soaked, soft, and irregular, which gradually rot, dry out, and become papery (Miller et al. 1996). Early symptoms on fruits appear as large, water-soaked or slightly sunken, circular lesions that enlarge and cover the fruit with white mold. The mold is the source of further infection as it consists of sporangia (spores) that are deciduous and can be dispersed by wind and rain. Fruit rot can develop even after harvesting the fruits. The disease development is favored by high soil moisture (frequent and heavy rains/irrigation), warm temperature (optimum being 24°C–33°C), and poor drainage. In general, disease initiation is from the area where the plant is exposed to high moisture conditions for long periods as zoospores infect the crown and root tissues of the host plant. The initial inoculum consists of oospores that survive as dormant propagules in soil for prolonged period. *Phytophthora* foliar blight and fruit rot may result in total loss of the crop (Babadoost 2005).

Management

- Do not select low-lying damp area or poor draining and heavy textured soils for raising the crop.
- Follow long crop rotation with nonsusceptible host crops.
- Avoid overirrigation and ensure proper drainage. Schedule irrigations so that excess water is not applied and fields drain properly (Koike et al. 2006). Evade irrigation from water sources (i.e., ponds or reservoirs) that receive runoff water from an infested field (Babadoost 2005).
- Carefully discard the crop debris after the host crops are harvested.
- Ensure proper trellising of the vines so that the fruits on the vines do not come in contact with the soil.
- Preferably raise the bush type of cucurbits on dome-shaped ridges or raised beds and do not allow planting depressions that collect water near plants (Louws et al. 2008).
- Do not harvest seeds from the infested fruits.
- Seed treatment with fungicides mefenoxam (Apron XL LS at 0.42 mL/kg seed) or metalaxyl (Allegiance FL at 0.98 mL/kg seed) can protect the seedlings up to 5 weeks of transplanting. Spray dimethomorph (Acrobat 50WP at 448 g/ha) plus copper sulfate (Cuprofix Disperss 36.9F at 2.25 kg/ha) at weekly intervals for foliar blight and fruit rot. Combined application of Apron XL LS as seed treatment along with spray applications of Acrobat plus copper can reduce the damage (Babadoost 2005).

18.3 BACTERIAL DISEASES

18.3.1 ANGULAR LEAF SPOT (*PSEUDOMONAS SYRINGAE* PV. *LACHRYMANS*)

Angular leaf spot is the most widespread bacterial disease of cucurbits that causes reduction in fruit number, fruit yield, and quality (Pohronezny et al. 1977). This disease is caused by the bacteria *Pseudomonas syringae* pv. *lachrymans*. The disease primarily affects cucumber, but it may occur on melons, squashes, and pumpkins. The bacterium infection occurs on all aboveground parts of cucurbit plants. Initially, the symptoms appear on leaves as small, water-soaked lesions that later enlarge. The shape of older lesions tends to be angular as they enlarge and encounter veins. As the infection progresses, the affected tissues often dry and fall, and thus, the leaves are left with torn irregular-shaped holes. The shredded leaves have lower photosynthetic efficiency, causing indirect yield losses. Under severe conditions, the leaves turn yellow, and occasionally, the growing tips of vines become water-soaked and yellow and growth ceases. On fruits, the infection appears as small, circular spots with a yellow halo. Later, the spots turn dull white. Dry cracks may occur, which are a little deeper, rendering the fruits unmarketable. The cracks expose fruits to other secondary infections like soft rot, which usually follows the bacterial infection. Sometimes, the infection progresses up to the seeds.

Angular leaf spot is most active between 24°C and 28°C and is favored by high humidity. Under very humid conditions and warm temperatures, white bacterial ooze may be found on the underside of lesions, which dries to form a thin, white crust. The disease is seed- and soilborne. The bacterium does not have spores adapted to carry them over long unfavorable periods and survive as vegetative cells in seed, soil, diseased plant debris, or weed plants (Leben 1981). The pathogen is disseminated by splashing rain, irrigation water, soil, insects, farm tools, and field workers. Infection occurs through natural openings and wounds. The bacterium is usually transmitted by water.

Resistance to *P. lachrymans* is controlled by a large number of recessive genetic factors (Klossowska 1976). However, it has also been reported that the disease is controlled by a single recessive gene "pl" (Dessert et al. 1982). The use of resistant cucumber cultivars is effective in reducing damage and losses due to this disease.

Management

- Use only disease-free seeds.
- Treat seeds with mercuric chloride solution (1:1000) for 5–10 min. Also, hot water treatment of seeds at 50°C ± 2°C can reduce the incidence but cannot eliminate.
- Practice a 3-year or longer crop rotation with noncucurbitaceous crops.
- Maintain field sanitation/hygiene during the crop-growing period and remove crop debris immediately after final harvest.
- Do not work in the crop when it is wet. Cultivation in dry soil is most effective in reducing bacterial survival (Kritzman and Zutra 1983).
- Also, follow deep plowing and soil solarization in summer months.
- Avoid overhead irrigation as the spread of bacteria takes place by water and splashes.
- Avoid injury to the plants during intercultural operations.
- Umekawa and Watanabe (1982) reported suppression to an appreciable extent by managing nighttime humidity (to 80%–90%) with dehumidifiers under protected structures.
- Chemical controls are most effective when integrated with sound cultural control practices. Streptocycline is an effective antibiotic against bacterial plant pathogens at a rate of 400 ppm. Copper-based bactericides are often necessary at an interval of 4–7 days to reduce the severity of the disease (Schwartz and Gent 2007).
- The biological control agent pentaphage (lysate of the virulent strain of *P. syringae*) was most effective when applied at high relative humidity (90%) in the morning and evening at intervals of 12–14 days (Korol and Bylinskii 1994). Avagimov and Panteleev (1984)

reported reduced incidence and increased yield by using an isolate of the L33 strain of *Pseudomonas geniculata*. Systemic induction of resistance by infection with tobacco necrosis virus on first leaves was reported by Jenns and Kuc (1979).

18.3.2 BACTERIAL LEAF SPOT (*XANTHOMONAS CAMPESTRIS* PV. *CUCURBITAE*)

The disease is not common but can occur during persistent warm and humid conditions on different cucurbits. This disease is caused by the seed- and soilborne bacterium *Xanthomonas campestris* pv. *cucurbitae* (Bryan). The initial symptoms appear on seedlings as small brown spots on cotyledons. The disease appears as necrotic spots on leaves and characteristic lesions on fruits. The symptoms appear as small water-soaked or greasy patches on the underside of leaves and as indefinite yellow patches on the upper surface of leaves. As the disease cycle progresses, the patches turn brown, round to angular with translucent centers and a typical yellow halo. The necrotic areas on the leaves do not fall, unlike the angular leaf spot. The disease occasionally occurs on young stems, but young fruits may be infested. The appearance of disease on fruits is erratic and can be correlated with moisture content at rind maturity (Goldberg 2012). The affected fruits bear water-soaked, sunken and cracked areas with light-brown exudates. Later, they become distinctly sunken and the rind may crack, leading to fruit rotting and other secondary infections. Viscous liquid drops of a yellowish-brown color are formed on the surface of these spots in damp weather. The infection can progress into the seed cavity of fruits and result in the rotting of flesh and contamination of seed. The disease can be transmitted by seed and infected crop residue, rain splash, and infected tools. The disease development is favored by high temperature in the range of 25°C–30°C and high relative humidity in the range of 90%. The disease incidence increases during the vegetative growth period. Severe disease incidence takes place due to precipitation/overhead irrigation. Yield losses may exceed 50% under humid/moist conditions (Babadoost 2012).

Management

- The management practices are similar to the angular leaf spot except for the biological control as there are no reports of biological control for bacterial spot of cucurbits.

18.3.3 BACTERIAL FRUIT BLOTCH (*ACIDOVORAX AVENAE* SUBSP. *CITRULLI*)

Bacterial fruit blotch is caused by the bacterium *Acidovorax avenae* subsp. *citrulli*. It is a serious problem in watermelon, though all cucurbits are susceptible. Under favorable conditions, the bacterium spreads rapidly in the field, leading to seedling blight or fruit rot at later stages. The initial symptoms appear on seedlings after 5–8 days of planting as greasy, water-soaked lesions on the undersurface of cotyledons and persist under dry conditions. The lesions coalesce and extend along the veins of cotyledons and along stems to tissues of true leaves. In severe cases, seedlings may collapse and die in a damping-off fashion (Walcott 2005). At later stages, elongated and dark to reddish-brown lesions develop on cotyledon veins. On leaves, the water congestion symptoms look similar to the disease symptoms but do not persist till daytime.

Fruit symptoms begin as small, irregular, olive-colored, water-soaked spots, which appear just prior to harvest maturity (2–3 weeks old) and enlarge rapidly to cover upper fruit surface. Later, the lesions turn reddish brown and may develop cracks. White bacterial ooze may be observed on the lesions under high humid conditions. Being a bacterial disease, it spreads by mechanical means or by rain or sprinklers. The disease severity in watermelon is affected by differences in rind color, for example, dark-colored fruits are least susceptible, whereas striped and light-green fruits are moderately and highly susceptible, respectively (Webb and Goth 1965). Infected seeds/seedlings are likely to be the primary source of inoculum for field outbreaks.

Mild temperatures, high humidity, and use of overhead irrigation are ideal for the spread, proliferation, and establishment of the bacterium on young seedlings (Latin and Hopkins 1995). Disease outbreaks are favored by heavy rainfall and windy conditions (Schaad et al. 2003). Foliar symptoms are not always clear and can be confused with abiotic stresses, and these apparently healthy plants might serve as inoculum for maturing fruits. Similarly, fruits sometimes also do not look infected but may carry infection and make it fairly difficult to produce disease-free seeds (Bahar and Burdman 2010).

Management

- Procure seeds from disease-free mother fruits.
- Treat seeds with HCl (1%) or $CaOCl_2$ (1%) for 15 min (Hopkins et al. 1996). Seed treatments including thermotherapy, NaOCl, fermentation, HCl, and peroxyacetic acid significantly reduce transmission but can adversely affect seed physiology (Walcott 2005).
- Disease can be controlled in protected structures by maintaining low temperature and low humidity.
- Follow long crop rotations.
- Avoid overhead irrigation and do not work in fields while the foliage is wet.
- Observe for infection regularly, destroy the affected seedlings immediately, or burn the crop under severe infection.

18.3.4 BACTERIAL RIND NECROSIS (*ERWINIA CARNEGIEANA*)

The disease occurs sporadically and is thought to be caused by bacteria that are naturally present in watermelon fruits. The external symptoms are usually absent, with only misshapen fruits in few severely affected melons. The symptoms include brown, dry, and hard necrosis of the rind that rarely extends into the flesh. In cross section, corky layers appear between the outer rind and the edible fruit tissues. There are no external symptoms that reduce the losses since the fruits generally look good and marketable. Severely affected melons are unmarketable only when sliced.

Management

- There is no chemical treatment for this disease and varietal resistance is controversial (Anonymous 2009). The only effective measure is to avoid planting melons in fields where rind necrosis has been encountered.

18.4 VIRAL DISEASES

Cucurbits are very sensitive to viral infection. A complex of viruses is able to infect cucurbits, and over thirty viruses have been reported (Zitter et al. 1996). The majority of these viruses cause huge losses in cucurbit production. The important viruses are cucumber mosaic virus (CMV), squash mosaic virus (SqMV), watermelon mosaic virus (WMV), zucchini yellow mosaic virus (ZYMV), and papaya ring spot virus (PRSV) (Tobias and Tulipan 2002). These viruses are transmitted by aphids except SqMV, which is spread by beetles and is seedborne. The symptoms caused by different cucurbit viruses are very similar, and it is very difficult or even impossible to identify the causal virus. Enzyme-linked immunosorbent assay is widely used for the detection of plant viruses (Clark and Adams 1977). The other viruses that are not very common include cucumber green mottle mosaic virus and melon necrotic spot virus.

18.4.1 CUCUMBER MOSAIC VIRUS

CMV is probably the most widely distributed and important virus disease of cucurbits (Ferreira and Boley 1992). It infects all cucurbit crop plants and has a very wide range of natural hosts, including many other noncucurbit crop plants and weeds belonging to different crop families.

On cucurbits, symptoms may occur on about 6-week-old plants at vigorous growth stage. The first symptoms appear on young leaves that exhibit mottled mosaic leaf pattern of alternate light-green and dark-green patches with edges that curl downward. The characteristic symptoms are yellow-colored mottling, leaf distortion, and stunted plant growth due to shortening of stem internodes. Older leaves develop chlorotic areas, which turn necrotic along the margins and later cover the entire leaf. Dead leaves either fall off or droop; wilting of the petioles occurs leaving the older vine mostly bare. The new leaves in case of muskmelon and cucumber may wilt and die, while older crown leaves may turn yellow and later dry up, resulting in a slow decline of plant health (MacNab et al. 1983).

The stem end portion of the infected fruits becomes mottled with yellowish-green and dark-green spots, and this mottling pattern gradually covers the whole fruit. The wartlike raised dark spots are usually formed on the fruit, and thus the fruit appears distorted. Fruits produced by the plants in the later stages of the virus infection are somewhat misshapen but have a smooth gray-white color with some irregular green areas, often called white pickle. The infected fruits of cucumber may have a bitter taste and make soggy pickles (Agrios 1978). In squash, there is blotchy appearance of fruits in nonyellow varieties. The infected plants usually produce few runners, flowers, and fruits.

CMV is transmitted through aphid vectors, namely, *Aphis gossypii* and *Myzus persicae*, and also through seeds and several weed hosts. The virus overwinters in many perennial weed sources especially attractive to aphids in spring. It can be transmitted through sap adhered on the hands and clothes of workers harvesting fruit. Agrios (1978) has reported that the entire field of cucurbits sometimes turns yellow due to CMV immediately after the first harvest.

18.4.2 SQUASH MOSAIC VIRUS

SqMV infects cultivated as well as wild or native cucurbits. However, it does not infect noncucurbitaceous crops or weeds. The virus causes stunting, chlorotic leaf mottling, green vein banding, and leaf distortion of infected seedlings. There are two strains of this virus, namely, strain 1 (greater effect on melons than squash) and strain 2 (reverse action of strain 1) (Haudenshield and Palukaitis 1998). Older infected plants develop distorted margins, blistering, hardening, and mild to severe dark-green mosaic pattern on leaves. This pattern may be confused with hormonal herbicide effect. Infected fruits may have mottling on the skin and become malformed or distorted (Dukic et al. 2002). The virus is seedborne and spread by beetles (*Acalymma* and *Diabrotica*). Upon feeding, the beetles inject the virus through saliva and infect plants. The virus can also spread readily through a mechanical injury. The virus is carried within the seed and cannot be eliminated by hot water or chemical treatment (Zitter and Banik 1984).

18.4.3 POTYVIRUSES

These include WMV, PRSV, and ZYMV. All three viruses are transmitted by different aphid species, namely, Aphis spp., *Brachycaudus* spp., *Cavariella* spp., *Myzus* spp., and *Liriomyza* spp. In general, high activity of aphids causes potyviruses in the field, which often spread very fast, and symptoms are localized.

18.4.3.1 Zucchini Yellow Mosaic Virus

All cucurbitaceous crops are susceptible to ZYMV, and it also infects certain noncucurbitaceous weeds and wild cucurbits. The infection causes blistered, deformed leaves with severe mosaic symptoms besides reduction in size and stunted plant growth. Melon fruits get discolored with hardening of flesh, seed deformation, and external cracks (Desbiez and Lecoq 1997). Pumpkin and zucchini fruits become discolored and deformed due to knobbiness, while rockmelon fruits have poor surface netting (Blua and Perring 1989). This virus is transmitted by a wide range of aphid species of which the green peach (*M. persicae*) and melon (*A. gossypii*) aphids are the most important. The cowpea aphid has been reported to transmit the virus more efficiently than the cotton or melon aphid (Yuan and Ullman 1996).

18.4.3.2 Papaya Ring Spot Virus

PRSV biotype "W" infects many cucurbit crops but does not infect noncucurbitaceous crops (Bateson et al. 2002). Some wild and native cucurbit species act as infection reservoirs. Characteristic mosaic symptoms appear as profuse mottling, puckering, and deformation of crown leaves, while the symptoms do not occur on lower mature leaves. Leaf distortion and blistering have also been reported. Zucchini squash was the most susceptible to PRSV, followed by watermelon and cucumber (Mansilla et al. 2013). Infected zucchini and pumpkin develop lumpy, distorted fruits, while rockmelon fruits may have poor-quality surface netting. Watermelon fruits may develop uneven surface or typical ring spot patterns on the skin. This virus is spread by a number of aphid species including the green peach and melon aphids (Jensen 1949).

18.4.3.3 Watermelon Mosaic Virus

The virus infects almost all cucurbit crops, mainly *Cucurbita pepo*, *C. maxima*, and *C. moschata* (Greber et al. 1987), and also wild cucurbits. The diseased vines bear a typical petunia-like appearance, that is, the tips of vines and a proliferation of shoots around the crown extend beyond the general level of the vines. The internodal length of the shoots gets shortened, resulting in crowding of leaves that become rolled, blistered, and small in size with mild mosaic symptoms and have little effect on fruits. It produces less severe symptoms on cucurbit leaves and fruits than PRSV, ZYMV, and SqMV (Coutts 2006). Flowers of the severely infected plants are abnormal in size, shape, and color, resulting in mottled, misshapen small fruits.

18.4.4 OTHER MINOR VIRAL DISEASES

18.4.4.1 Tobacco Ring Spot Virus

It mainly affects melons and cucumbers and is transmitted by nematodes (*Xiphinema americanum*). The newly infected leaves show a very bright mosaic with stunted plant growth.

18.4.4.2 Tomato Ring Spot Virus

It is most severe in squashes and is also transmitted by nematodes. It can overwinter on many weed species without expressing symptoms.

18.4.4.3 Clover Yellow Vein Virus

It infects summer squash and shows a conspicuous symptom of yellow specking on the foliage. It is transmitted by aphids.

18.4.4.4 Aster Yellow Mycoplasma

The disease symptoms appear as yellowing of plants with stunted growth which is usually confused as viral disease. It is transmitted by leafhoppers.

Management of CMV, PRSV, SqMV, WMV, and ZYMV

Identification is very important for the management of viral diseases. Viral diseases can be confused with many other diseases or sometimes nutrient deficiency symptoms or physiological disorders. Early identification provides an effective management; otherwise, huge losses may be encountered. The management of the virus transmitting vector is more important than the pathogen itself. Prevention of viral disease is the first strategy that may include using resistant cultivars and pathogen-free planting material, preventing spread and overwintering of virus, and reducing pathogen/vector population. The general control measures are summarized as follows:

- *Grow resistant varieties*: Planting resistant cultivars can serve as an effective control for viral diseases. The most effective and simplest method of controlling viral diseases is to utilize genetic resistance. Even though resistance has been found in some cucurbit species, genetic barrier due to incompatibility among species poses a problem for transferring this resistance to other cucurbits (Zitter and Murphy 2009).
- Use virus-indexed seeds and seedlings.
- Seed treatment of *C. pepo* with hot air (70°C for 2 days) or hot water (50°C for 60 min) prevents the inoculum, but the seed coat prevents systemic inoculum.
- In the early stages of crop growth, monitor the crop carefully and remove the plants with symptoms of virus.
- Grow taller, nonsusceptible barrier crops such as corn that may delay initial infections.
- Remove all weeds and volunteer cucurbit crop plants within and around cucurbit crops as these can harbor aphids and viruses. Eradication of weed hosts is often a difficult task because of the extensive host range of different viruses, but clean cultivation can reduce the incidence and intensity.
- Avoid planting cucurbit crops consecutively in close proximity to each other.
- Destroy the plant debris promptly once the crop is harvested.
- Follow long crop rotations with noncucurbitaceous crops.
- Use of superreflective plastic mulch deters vector aphids from landing and thus limits the spread of virus. Floating row covers or reflective mulches may help exclude or repel aphids (Stapleton and Summers 2002; Barbercheck 2014).
- The severely infected crop should be burnt.
- Parasitoids can be used effectively to manage the aphid vectors such as *Aphidius colemani*, *Aphidius matricariae*, *Lysiphlebus fabarum*, and *Binodoxys angelicae* (Kos et al. 2008).
- Vector management is not very simple since aphids need only probe the plant to transmit the virus that infects instantly. To control the insect effectively, insecticides would need to be applied at regular intervals, leading to huge monetary inputs and may not be economically feasible. For CMV, spray dimethoate (0.05%), monocrotophos (0.05%), Confidor (0.03%), or Metasystox (0.02%) at weekly intervals. Though managing viruses through insecticides is not a good approach because insecticides do not act fast enough to prevent the rapid spread of the viruses by aphids, once they have well-established colonies, they may increase the disease spread. However, the aphid control can reduce the infection intensity and the losses.

18.5 CONCLUSIONS

Vegetables in the cucurbit family include cucumber, gourd, muskmelon (cantaloupe), watermelon, summer squash, winter squash, and pumpkin. The production of these crops is being challenged by many new diseases as well as earlier described diseases that are appearing in new areas. The goal of plant disease management is to reduce the economic damage and to maintain the quality of produce. The enormous use of chemicals has led to the development of resistance that poses a problem in cucurbit disease management. Therefore, use of multifaceted or integrated disease management (IDM) approach is essential to cope with the problem. IDM consists of scouting and management strategies, which may include, site selection, field preparation, use of resistant cultivars, altering planting time, optimum plant density, optimum irrigation, proper drainage, mulching, optimum use of nutrients and judicious use of pesticides. In addition, monitoring environmental factors (temperature, moisture, soil pH, nutrients, etc.), forecasting diseases, and establishing economic thresholds are important to the management scheme (Maloy 2005). These measures should be harmonized properly to maximize the benefits of each component.

REFERENCES

Abbasi, P.A. and G. Lazarovits. 2006. Seed treatment with phosphonate (AG3) suppresses *Pythium* damping-off of cucumber seedlings. *Plant Disease.* 90: 459–464.

Agrios, G.N. 1978. *Plant Pathology*, 2nd ed. Academic Press Inc., San Diego, CA, pp. 466–470.

Ali, M., A. Hannan, J. Shafi, W. Ahmad, C.M. Ayyub, S. Asad, H.T. Abbas, and M.A. Sarwar. 2013. Nutrient supplement efficacy against powdery mildew of pumpkin (*Sphaerotheca fuliginea*) and its correlation with environmental factors. *International Journal of Advanced Research.* 10(1): 17–20.

Anjorin, S.T. and M. Mohammed. 2009. Effect of seed borne fungi on germination and seedling growth of water melon (*Citrullus lanatus*). *Journal of Agriculture and Social Sciences.* 5: 77–80.

Anonymous. 2009. *Bacterial Rind Necrosis.* Aggie Horticulture, TAMU. Texas A&M University, TX. http://aggie-horticulture.tamu.edu/vegetable/problem-solvers/cucurbit-problem-solver/cucurbit-fruit-disorders/bacterial-rind-necrosis.

Avagimov, V.A. and A.A. Panteleev. 1984. Control of angular spot of cucumber. *Zashchita Rastenii.* 7: 15.

AverYanov, A., V. Lapikova, T. Pasechnik, and C.J. Baker. 2011. Phytodynamic dyes may systematically reduce cucumber scab severity. In: *PR Proteins and Induced Resistance against Pathogens and Insects*, Neuchatel, Switzerland, September, 4–8, 2011, p. 60.

Babadoost, M. 1989. *Alternaria* leaf spot or blight of cucurbits. http://web.aces.uiuc.edu/vista/pdf_pubs/918.pdf. University of Illinois. © The American Phytopathological Society. pp. 1-8.

Babadoost, M. 2005. Phytophthora blight of cucurbits. *The Plant Health Instructor.* doi:10.1094/PHI-I-2005-0429-01. University of Illinois Extension. Department of Crop Sciences, University of Illinois.

Babadoost, M. 2012. Bacterial spot of cucurbits. Report on plant disease, RPD no. 949. extension.cropsci.illinois.edu/fruitveg/pdfs/949_bacterial_spot.pdf.

Bahar, O. and S. Burdman. 2010. Bacterial fruit blotch: A threat to the cucurbit industry. *Israel Journal of Plant Sciences.* 58: 19–31.

Banken, J. and J. Stark. 1998. Multiple routes of pesticide exposure and the risk of pesticides to biological controls: A study of neem and the seven-spotted lady beetle (Coleoptera: Coccinellidae). *Journal of Economic Entomology.* 91(1): 1–6.

Barbercheck, M.E. 2014. Biology and management of aphids in organic cucurbit production systems. www.extension.org/pages/60000/biology-and-management-of-aphids-in-organic-cucurbit-production-systems.

Bateson, M.F., R.E. Lines, P. Revill, W. Chaleeprom, C.V. Ha, A.J. Gibbs, and J.L. Dale. 2002. On the evolution and molecular epidemiology of the poty papaya ringspot virus. *Journal of General Virology.* 83(10): 2575–2585.

Bettiol, W. 1999. Effectiveness of cow's milk against zucchini squash powdery mildew (*Sphaerotheca fuliginea*) in greenhouse conditions. *Crop Protection.* 18:489–492.

Blua, M.J. and T.M. Perring. 1989. Effect of Zucchini Yellow Mosaic Virus on development and yield of cantaloupe (*Cucumis melo*). *Plant Disease.* 73: 317–320.

Brown, J.K.M. 2002. Comparative genetics of avirulence and fungicide resistance in the powdery mildew fungi. In: Belanger, R.R., W.R. Bushnell, A.J. Dik, and T.L.W. Carver (eds.), *The Powdery Mildews: A Comprehensive Treatise.* APS Press, St. Paul, MN, pp. 56–65.

Celetti, M. and E. Roddy. 2010. Downy mildew in cucurbits. http://www.omafra.gov.on.ca/english/crops/facts/10-065.htm.

Chang, W.T., Y.C. Chen, and C.L. Jao. 2007. Antifungal activity and enhancement of plant growth by *Bacillus cereus* grown. *Journal of Plant Pathology.* 93(1): 43–50.

Clark, M.F. and A.N. Adams. 1977. Characteristics of the microplate method of enzyme linked immunosorbent assay for the detection of plant viruses. *Journal of General Virology.* 34: 475–483.

Colucci, S.J. and G.J. Holmes. 2010. Downy mildew of cucurbits. *The Plant Health Instructor.* © The American Phytopathological Society. pp. 1–8.

Coutts, B. 2006. *Virus Diseases of Cucurbit Crops.* Department of Agriculture Australia, Farmnote 166. Sydney, Australia. ISSN: 0726-934X. http://archive.agric.wa.gov.au/objtwr/imported_assets/content/hort/veg/pw/fn2006_viruscucurbits_bcoutts.pdf?noicon.

Davis, R.M., T.A. Turini, B.J. Aegerter, and J.J. Stapleton. 2012a. Cucurbits: Charcoal rot-*Macrophomina phaseoli*. http://www.ipm.ucdavis.edu/PMG/r116101311.html.

Davis, R.M., T.A. Turini, B.J. Aegerter, and J.J. Stapleton. 2012b. Cucurbits: Vine decline (crown blight). http://www.ipm.ucdavis.edu/PMG/r116102411.html.

Desbiez, C. and H. Lecoq. 1997. Zucchini yellow mosaic virus. *Plant Pathology.* 46: 809–829.

Dessert, J.M., L.R. Baker, and J.F. Fobes. 1982. Inheritance of reaction to *Pseudomonas lachrymans* in pickling cucumber. *Euphytica.* 31: 847–55.

Dukic, N., B. Krstic, I. Vico, N.I. Katis, C. Papavssilious, and J. Berenji. 2002. Biological and serological characterization of viruses on summer squash crops in Yugoslavia. *Journal of Agricultural Science.* 47: 149–160.

Egel, D.S. and R.D. Martyn. 2010. *Vegetable Diseases. Mature Watermelon Vine Decline and Similar Vine Decline Diseases of Cucurbits.* Purdue Extension BP-65-W, Purdue Extension Education Store, Purdue University, USA.

Egel, D.S. and R.D. Martyn. 2013. *Fusarium* wilt of watermelon and other cucurbits. *The Plant Health Instructor.* © The American Phytopathological Society. pp. 1–11.

Ehret, D.L., C. Koch, J. Menzies, P. Sholberg, and T. Garland. 2001. Foliar sprays of clay reduce the severity of powdery mildew on long English cucumber and wine grapes. *HortScience.* 36: 934–936.

El-Naggar, M., H. El-Deeh, and S. Ragab. 2012. Applied approach for controlling powdery mildew disease of cucumber under plastic houses. *Pakistan Journal of Agriculture: Agricultural Engineering Veterinary Sciences.* 28: 52–61.

Etebarian, H.R. 2006a. Evaluation of *Streptomyces* strains for biological control of charcoal stem rot of melon caused by *Macrophomina phaseolina. Plant Pathology.* 5(1): 83–87.

Etebarian, H.R. 2006b. Evaluation of *Trichoderma* isolates for biological control of charcoal stem rot in melon caused by *Macrophomina phaseolina. Journal of Agricultural Science and Technology.* Hawaii. 8: 243–250.

Ferreira, S.A. and R.A. Boley. 1992. *Crop Knowledge Master.* Department of Plant Pathology, CTAHR. http://www.extento.hawaii.edu/Kbase/Crop/Type/cucvir.htm.

Fravel, D. 1999. Commercial biocontrol products for use against soilborne crop diseases. http://www.barc.usda.gov/psi/bpdl/bpdlprod/bioprod.html.

Freeman, S., A. Zveibil, and M. Maymon. 2002. Isolation of non-pathogenic mutants of *Fusarium oxysporum* f. sp. *melonis* for biological control of *Fusarium* wilt in cucurbits. *Phytopathology.* New Mexico. 92: 164–168.

Goldberg, N. 2012. Bacterial leaf spot. *Extension Plant Pathology: News You Can Use.* NMSU. aces.nmsu.edu/ces/plant_sciences/.../bacterial-leaf-spot-of-cucurbits.pdf.

Goldberg, N.P. 2004. Anthracnose of cucurbits. http://aces.nmsu.edu/pubs/_h/H247.pdf.

Greber, R.S., G.D. McLean, and M.S. Grice. 1987. Zucchini yellow mosaic virus in three States of Australia. *Australasian Plant Pathology.* 16(1): 19–21.

Haudenshield, J.S. and P. Palukaitis. 1998. Diversity among isolates of Squash Mosaic Virus. *Journal of General Virology.* 79: 2331–2341.

Hopkins, D.L., J.D. Cucuzza, and J.C. Watterson. 1996. Wet seed treatments for the control of bacterial fruit blotch of watermelon. *Plant Disease.* 80: 529–532.

Jarvis, W., W.G. Gubler, and G.G. Grove. 2002. Epidemiology of powdery mildews in agricultural ecosystems. In: Belanger, R., W.R. Bushnell, A.J. Dik, and T.L.W. Carver (eds.), *The Powdery Mildews: A Comprehensive Treatise.* The American Phytopathological Society, St. Paul, MN, pp. 169–199.

Jenns, A. and J. Kuc. 1979. Graft transmission of systemic resistance of cucumber to anthracnose induced by *Colletotrichum lagenarium* and tobacco necrosis virus. *Phytopathology.* 69: 753–7563.

Jensen, D.D. 1949. Papaya virus diseases with special reference to papaya ring spot. *Phytopathology.* 39: 191–211.

Keinath, A.P. 2014. Cucurbit downy mildew management for 2014. *Clemson University Extension Information Leaflet,* p. 90.

Khanzada, K.A., M.A. Rajput, G.S. Shah, A.M. Lodhi, and F. Mehboob. 2001. Effect of seed dressing fungicides for the control of seed borne mycoflora of wheat. *Asian Journal of Plant Sciences.* 1: 441–444.

King, S.R., A.R. Davis, W. Liu, and A. Levi. 2008. Grafting for disease resistance. *HortScience.* 43(6): 1673–1676.

Klossowska, E. 1976. Studies on the inheritance and selection advance regarding resistance to angular leaf spot of cucumber (*Pseudomonas lachrymans*). *Genetica Polonica.* 17(2): 181–190.

Koike, S.T. 1997. First report of gummy stem blight, caused by *Didymella bryoniae,* on watermelon transplants in California. *Plant Disease.* 81: 1331.

Koike, S.T., P. Gladders, and A.O. Paulus. 2006. *Vegetable Diseases, A Colour Handbook.* CRC Press/Taylor & Francis Group, Boca Raton, FL. pp. 228–229.

Konstantinidou-Doltsinis, S. and A. Schmitt. 1998. Impact of treatment with plant extracts from *Reynoutria sachalinensis* (F Schmidt) Nakai on intensity of powdery mildew severity and yield in cucumber under high disease pressure. *Crop Protection.* 17: 649–656.

Korol, A.L. and A.F. Bylinskii. 1994. Pentaphage against angular leaf spot. *Zashchita Rasteniĭ (Moskva).* 4: 16–17.

Kos, K., Z. Tomanovic, O. Petrovic-Obradovic, Z. Laznik, M. Vidrih, and S. Trdan. 2008. Aphids (Aphididae) and their parasitoids in selected vegetable ecosystems in Slovenia. *Acta Agriculturae Slovenica*. 91(1): 15–22.

Kritzman, G. and D. Zutra, 1983. Systemic movement of *Pseudomonas syringae* pv. *lachrymans* in the stem, leaves, fruits and seed of cucumber. *Canadian Journal of Plant Pathology*. 273–278.

Kucharek, T. 2000. *Alternaria* leaf spot of cucurbits. http://plantpath.ifas.ufl.edu/extension/fact-sheets/pdfs/pp0032.pdf.

Latin, R.X. and D.L. Hopkins. 1995. Bacterial fruit blotch of watermelon: The hypothetical exam question becomes reality. *Plant Disease*. 79: 761–765.

Leben, C. 1981. Survival of *Pseudomonas syringae* pv. *lachrymans* with cucumber seed. *Canadian Journal of Plant Pathology*. 3: 247–249.

Li, H.N. 2014. *Anthracnose* of cucumber. www.ct.gov/caes.

Louws, F.J., G.J. Holmes, and K.L. Ivors. 2008. *Cucurbits—Phytophthora Blight*. North Carolina State University, North Carolina. http://www.cals.ncsu.edu/plantpath/extension/fact_sheets/Cucurbits_-_Phytophthora_blight.html.

MacNab, A.A., A.F. Sherf, and J.K. Springer. 1983. *Identifying Diseases of Vegetables*. The Pennsylvania State University, Pennsylvania.

Maloy, O.C. 2005. Plant disease management. *The Plant Health Instructor*. © The American Phytopathological Society. pp. 1–10. http://www.apsnet.org/edcenter/intropp/topics/Pages/PlantDiseaseManagement.aspx.

Mansilla, P.J., A.G. Moreira, A.P.O.A. Mello, J.A.M. Rezende, J.A. Ventura, V.A. Yuki, and F.J. Levatti. 2013. Importance of cucurbits in the epidemiology of Papaya ringspot virus type p. *Plant Pathology*. 62(3): 571–577.

Martyn, R.D. 1986. Use of soil solarization to control *Fusarium* wilt of watermelon. *Plant Disease*. 79: 762–766.

Martyn, R.D. 2002. *Monosporascus* root rot and vine decline of melons. *The Plant Health Instructor*. © The American Phytopathological Society. pp. 1–11.

McGrath, M.T. 2004. Diseases of cucurbits and their management. In: Naqvi, S.A.M.H. (ed.). *Disease of Fruits and Vegetables*, Vol. 1. Kluwer Academic Publishers, Dordrecht, Netherlands. pp. 455–510.

McGrath, M.T. 2006. *Update on Managing Downy Mildew in Cucurbits*. http://vegetablemdonline.ppath.cornell.edu/NewsArticles/Cuc_Downy.htm.

McGrath, M.T. 2011. Powdery mildew of cucurbits. *Vegetable Crops*. Fact Sheet Page: 732.30. http://vegeta-blemdonline.ppath.cornell.edu/factsheets/Cucurbits_PM.html.

Menzies, J.G., D.L. Ehret, A.D.M. Glass, T. Helmer, C. Koch, and F. Seywerd. 1991. Effects of soluble silicon on the parasitic fitness of *Sphaerotheca fuliginea* on *Cucumis sativus*. *Phytopathology*. 81: 84–88.

Miller, S.A., R.C. Rowe, and R.M. Riedel. 1996. *Phytophthora* blight of pepper and cucurbits. http://ohioline.osu.edu/hyg-fact/3000/pdf/3116.pdf.

Mossler, M.A. and O.N. Nesheim. 2005. *Florida Crop/Pest Management Profile: Squash*. Electronic Data Information Source of UF/IFAS Extension (EDIS). CIR 1265. February 3, 2005. http://edis.ifas.ufl.edu/.

Nunez-Palenius H.G., D. Hopkins, and D.J. Cantliffe. 2012. Powdery mildew of cucurbits in Florida. http://edis.ifas.ufl.edu/hs321.

Ogorek, R., A. Lejman, W. Pusz, A. Miłuch, and P. Miodynska. 2012. Characteristics and taxonomy of *Cladosporium* fungi. *Mikologia Lekarska*. 19(2): 80–85.

Owen, J.H. 1955. *Fusarium* wilt of cucumber. *Phytopathology*. 45: 435–439.

Palenchar, J., D.D. Treadwell, L.E. Datnoff, A.J. Gevens, and G.E. Vallad. 2012. Cucumber anthracnose in Florida. Publication #PP266. http://edis.ifas.ufl.edu/pp266.

Pohronezny, K., P.O. Larsen, D.A. Ematty, and J.D. Farley. 1977. Field studies of yield losses in pickling cucumber due to angular leafspot. *Plant Disease Reporter*. 61: 386–390.

Queiroga, R., M. Puiatti, P.C.R. Fontes, and P.R. Cecon. 2008. Produtividade e qualidade de frtos de meloeir-ovariandonumero de frutos e de folhasporplanta. *Horticultura Brasileira*. 26: 209–215.

Rai, M., S. Pandey, and S. Kumar. 2008. Cucurbit research in India: A retrospect. In: Pitrat, M. (ed.), *Proceedings of the IXth EUCARPIA Meeting on Genetics and Breeding of Cucurbitaceae*. INRA, Avignon, France.

Roustaee, A., M.K. Reyhanb, and M. Jafaric. 2012. Study of interaction between salinity and charcoal rot diseases of melon (*Macrophomina phaseolina*) in Semnan and Garmsar. *DESERT*. 16(2): 111–180, Article 11.

Schaad, N., W.G. Sowell, R.W. Goth, R.R. Colwell and R.E. Webb. 1978. *Pseudomonas pseudoalcaligenes* subsp. *citrulli* subsp. nov. *International Journal of Systematic Bacteriology*. 28: 117–125.

Schaad, N.W., E. Postnikova and P.S. Randhawa. 2003. Emergence of Acidovorax avenae subsp. citrulli as a crop threatening disease of watermelon and melon. In: Iacobellis, N.S., A. Collmer, S.W. Hutcheson, J.W. Mansfield, C.E. Morris, J. Murillo, N.W. Schaad, D.E. Stead, G. Surico and M.S. Ullrich. eds. Pseudomonas syringae and related pathogens. Kluwer Academic Publishers, Dordrecht, The Netherlands, pp. 573–581.

Schwartz, H.F. and D.H. Gent. 2007. *Cercospora* leaf spot (Cucumber, Melon, Pumpkin, Squash, and Zucchini). http://wiki.bugwood.org/uploads/CercosporaLeafSpot-Cucurbits.pdf.

Seebold, K. 2010. Foliar diseases of cucurbits. http://www2.ca.uky.edu/agcollege/plantpathology/ext_files/PPFShtml/PPFS-VG-10.pdf.

Shimizu, M., S. Yazawa, and Y. Ushijima. 2009. A promising strain of endophytic *Streptomyces* sp. for biological control of cucumber anthracnose. *Journal of General Plant Pathology.* 75(1): 27–36.

Singh, A., S. Srivastava, and H.B. Singh. 2007. Effect of substrates on growth and shelf life of *Trichoderma harzianum* and its use in biocontrol of diseases. *Bioresource Technology.* 98: 470–473.

Srinon, W., K. Chuncheen, K. Jirattiwarutkul, K. Soytong, and S. Kanokmedhakul. 2006. Efficacies of antagonistic fungi against *Fusarium* wilt disease of cucumber and tomato and the assay of its enzyme activity. *Journal of Agricultural Technology.* 2(2): 191–201.

Stapleton, J.J. and C.G. Summers. 2002. Reflective mulches for management of aphids and aphid-borne virus diseases in late-season cantaloupe (*Cucumis melo* L. var. *cantalupensis*). *Crop Protection.* 21: 891–898.

Stoyka, M., L. Tsvetana, V. Nikolay, and V. Georgi. 2014. Botanical products against powdery mildew on cucumber in greenhouses. *Turkish Journal of Agricultural and Natural Sciences.* 2: 1707–1712.

Thompson, D.C. and S.F. Jenkins. 1985. Influence of cultivar resistance, initial disease, environment, and fungicide concentration and timing on *anthracnose* development and yield loss in pickling cucumbers. *Phytopathology.* 75: 1422–1427.

Tobias, I. and Tulipan M. 2002. Results of virological assay on cucurbits in 2001. *Novenyvedelem.* 38: 23–27.

Umekawa, M. and Y. Watanabe. 1982. Relation of temperature and humidity to the occurrence of angular leaf spot of cucumber grown under plastic house. *Annals of the Phytopathological Society of Japan.* 48: 301–307.

Walcott, R.R. 2005. Bacterial fruit blotch of cucurbits. *The Plant Health Instructor.* © The American Phytopathological Society. pp. 1–8. http://www.apsnet.org/edcenter/intropp/lesson/sprokaryotes/Pages/BacterialBlotch.aspx.

Watson, A. and Y.T. Napier. 2009. Diseases of cucurbit vegetables. Primefact 832. http://www.dpi.nsw.gov.au/_data/assets/pdf_file/0003/290244/diseases-of-cucurbit-vegetables.pdf.

Watt, B.A. 2013. Alternaria leaf blight of cucurbits. Pest Management Fact Sheet #5086. http://extension.umaine.edu/ipm/ipddl/publications/5086e.

Webb, R.E. and R.W. Goth. 1965. A seed borne bacterium isolated from watermelon. *Plant Disease Report.* 49: 818–821.

Yang, Q.Y., K. Jia, W.N. Gena, R.J. Guo, and S.D. Li. 2014. Management of cucumber wilt disease by *Bacillus subtilis* f. sp. *cucumerinum* in rhizosphere. *Plant Pathology Journal.* 13: 160–166.

Yuan, C. and D.E. Ullman. 1996. Comparison of efficiency and propensity as measures of vector importance in Zucchini yellow mosaic potyvirus transmission by *Aphis gossypii* and *A. craccivora. Phytopathology.* 86: 698–703.

Yuan, M. 1989. The occurrence of cucumber scab and its control. *Acta Agriculture University Jilinensis.* 11: 7–13.

Zhou, X.G. and K.L. Everts. 2004. Suppression of *Fusarium* wilt of watermelon by soil amendment with hairy vetch. *Plant Disease.* 88: 1357–1365.

Zhou, X.G. and K.L. Everts. 2008. Integrated management of gummy stem blight of watermelon by green manure and Melcast-scheduled fungicides. *Plant Health Progress.* © 2008 Plant Management Network. pp. 1–8. doi: 10.1094/PHP-2008-1120-01-RS (Online).

Zitter, T.A. 1986. Scab of cucurbits. *Vegetable Crops.* Fact Sheet Page 732.50. http://vegetablemdonline.ppath.cornell.edu/factsheets/Cucurbit_Scab.html.

Zitter, T.A. 1987. Anthracnose of cucurbits. *Vegetable Crops.* Fact Sheet Page 732.60. http://vegetablemdonline.ppath.cornell.edu/factsheets/Cucurbit_Anthracnose.html.

Zitter, T.A. 1998. Fusarium diseases of cucurbits. http://vegetablemdonline. ppath.cornell.edu/factsheets/Cucurbits_Fusarium.html.

Zitter, T.A. and M.T. Banik. 1984. Vegetable crops: Virus diseases of cucurbits. Fact Sheet Page: 732.40. Vegetable MD online. Cornell University. Cornell, New York. http://vegetablemdonline.ppath.cornell.edu/factsheets/Viruses_Cucurbits.htm.

Zitter, T.A. and J.F. Murphy. 2009. Cucumber mosaic. *The Plant Health Instructor.* © The American Phytopathological Society. pp. 1–10. http://www.apsnet.org/edcenter/intropp/lessons/viruses/Pages/Cucumbermosaic.aspx.

Zitter, T.A., D.L. Hopkins, and C.E. Thomas. 1996. *Compendium of Cucurbit Diseases.* APS Press, St. Paul, MN.

Section VIII

Weed Control, Pest Control, and Insects of Cucurbits

19 Important Insect Pests of Cucurbits and Their Management

Akhilesh Sharma, Chanchal Rana, and Kumari Shiwani

CONTENTS

19.1 INTRODUCTION

Cucurbits belong to the family Cucurbitaceae, which includes about 118 genera and 825 species. Cucumbers, muskmelons, watermelons, squashes, gourds, and pumpkins are commonly grown cucurbits in most parts of the world. These crops are attacked by a variety of insect pests from seeding until harvest. A lot of time, money, and natural resources are invested to grow these vegetables. Good pest management practices can save this investment by avoiding losses. Successful cultivation of cucurbits requires an effective and economical control of insect pests. Commercial vegetable growers must produce a quality product that is attractive and safe to the consumer at a minimum cost. Insect pest infestations in cucurbits bring about heavy losses through reduction in yield, lowered quality of produce, and increased cost of production and harvesting besides expenditure incurred on materials and equipments to apply control measures. These losses individually and collectively reduce the income of growers and are unacceptable. Effective and economic pest control/management requires the use of cultural, mechanical, biological, and chemical methods. The combination of these different methods is necessary for achieving good management of pests. Insecticides are highly effective in controlling most insect pests. However, a limited number of generally effective pesticides may be used that are safe to apply, handle, and store. The different agencies at their respective countries at the world level regulate the registration and use of pesticides on vegetables and set the tolerance labels for miniscule amounts of residues that are allowed on a crop at the time of harvest. The continual tightening of pesticide regulations has resulted in the present tendency for growers to use a minimum of pesticides and those that disappear rapidly and are readily biodegradable. Consequently, renewed interest is being devoted to research on biological and cultural methods of control besides breeding insect pest-resistant cultivars. Pest management can be achieved only by a long-term assurance to integrated pest management practices (IPM). IPM involves the strategic use of resistant varieties, cultural measures, crop rotations, biological control, and selective pesticides. IPM requires an understanding of the interaction between pests, plants, and the environment. IPM must ensure optimal use of chemical pesticides and minimum environmental contamination to maintain crop production.

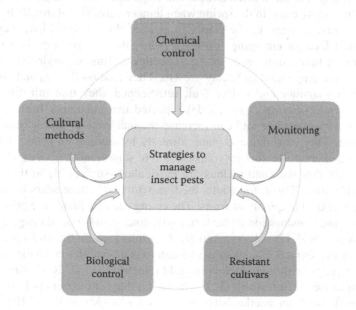

Cucurbits are attacked by a number of insect pests at different growth stages that cause defoliation of leaves, damage roots or flowers, contribute to poor crop stand, transmit bacterial and viral diseases, and generate wounds that help the invasion of fungal pathogens. Major insect pests include cucumber beetles, red pumpkin beetles, fruit flies, hadda beetles, squash bugs, aphids, white flies, squash vine borers, two-spotted spider mites, and nematodes.

19.2 CUCUMBER BEETLES (*ACALYMMA VITTATUM* AND *DIABROTICA UNDECIMPUNCTATA*)

The striped cucumber beetle (*Acalymma vittatum*) and the spotted cucumber beetle (*Diabrotica undecimpunctata*) are the most common insect pests on all cucurbit crops. They cause significant damage right from young seedling growth till fruit ripening. The second generation of these insects may appear close to harvesting and feed on the rind of developing fruits. This reflects that they are present throughout the growing season and feed on all parts of the plant, including the flowers and fruits.

19.2.1 IDENTIFICATION AND SYMPTOMS OF DAMAGE

The larva of spotted cucumber beetle (*D. undecimpunctata*) is yellowish white with a brown head and is 0.6–1.2 cm long (1.9 cm at full growth) with a brown spot at the tail end. The adult has a black head with black antennae and 12 black spots on its yellow-green body from where its common name is derived and is about 6 mm long. On the other hand, the larva of *A. vittatum* is white and 8.5 mm in length, while adults are about 5 mm long with three longitudinal black stripes on its top wings. Another type of beetle is the banded cucumber beetle (*Diabrotica balteata*); the adult is yellowish green with three bright green stripes or bands running across the wing covers. The adult beetles enter reproductive diapauses in early autumn and hibernate in trash, soil, or woodland litter. The beetles emerge in spring with the appearance of host plants, and mating beetles can be seen on seedlings of young cucurbits. Eggs are oval and orange yellow in color that are laid in the soil and hatch in 7–10 days. In spring, the larva often attacks the roots and stems within 2–6 weeks (Capinera 2008). They become active early in the spring when temperatures rise above 50°F.

These beetles cause damage to the plants in at least three ways. First, they enter the soil through cracks and feed on emerging seedlings below the soil surface; as a result, the plant is either killed or may have stunted growth. Adult beetles feeding on cotyledons and young leaves can cause reduction in crop stand and delay growth. They feed on flowers and pollen in the later growth stage during summer and reduce fruit set. Second, they transmit diseases like bacterial wilt (*Erwinia tracheiphila*). Diver (2008) reported that cucumber beetles transmit bacterial wilt and squash mosaic virus and can increase the incidence of powdery mildew, black rot, and *Fusarium* wilt. They also damage plants directly by feeding on roots, stems, leaves, and fruits. They are the alternate vectors for transmission of squash mosaic virus while feeding from one plant to another. Squashes and melons are particularly susceptible to this disease because of heavy seed infestation. Third, the adults feed on fruits and cause scars on the fruit surface, thereby lowering marketable quality of fruits. The young cucurbit plants are primarily vulnerable to stunted growth and transmission of bacterial wilt disease, whereas damage to older plants is due to scars on fruits. With rise in temperature, the adults feed on the underside of young fruits and cause scars on the surface. The older plants can tolerate defoliation to the extent of 25% due to the feeding of beetles, with no reduction in yield (Hoffmann et al. 2002, 2003).

The bacterium causes bacterial wilt (*E. tracheiphila*) that overwinters only in the intestines of some of the striped cucumber beetles between 1% and 10% (Mitchell and Hanks 2009). When beetles become active in the spring and start feeding, they spread the bacterium through their feces. Wounds caused due to feeding on young leaves or cotyledons make entry points for the pathogen from the feces via moisture, that is, rain and irrigation. With its entry inside the plant, the bacterium multiplies in the vascular system, producing blockages that cause the wilting of plants within 2–5 weeks after infection. Beetles are attracted to infected plants and transmit the bacterium to healthy plants. The first symptom of bacterial wilt on cucumber and muskmelon is a distinct flagging of lateral and individual leaves. Adjacent leaves wilt thereafter, and finally, the whole plant wilts and dies.

19.2.2 Management

19.2.2.1 Monitoring of Pest and Control Strategies

Cucumber beetles are difficult to control. Early treatment is crucial for its management in large-scale commercial production of cucurbits. Pesticide sprays must be directed at adult beetles. Larvae of spotted cucumber beetle develop outside cucurbit fields, while those of striped cucumber beetle are to be found on roots where their control is difficult (Anonymous 2012). Population monitoring methods like crop scouting and sticky traps are commonly used. Monitoring should be initiated after planting or with the emergence of seedlings up to the fruiting stage. Special attention must be paid to the undersides of leaves and plants at the corners of the field. Population counts are used to calculate the average number of beetles per plant (Petzoldt 2008). Economic threshold for cucumber beetle control depends on the type of cucurbit, age of plants, and susceptibility to bacterial wilt. Well-established plants can tolerate 25%–50% loss of foliage without a reduction in yield (Diver 2008). However, the infestation is usually more damaging in the initial stages of plant growth as seedlings may be seriously affected or killed by heavy feeding. It is imperative to keep beetle population low, especially before the five-leaf stage. Application of pesticides for controlling adults may be essential if there is an average of one beetle per plant during seeding till plants attain a height of 10 cm. The beetles should not be allowed to exceed one beetle for every two plants to prevent bacterial wilt in highly susceptible crops like melons and cucumbers, while less wilt-susceptible crops (mostly squashes) can tolerate one or two beetles per plant without yield losses (Hazzard and Cavanagh 2010). Spray the crop within 24 h after the threshold level is reached. Apply insecticides before the introduction of beehives or bee activity in the field to facilitate pollination in cucurbits. Different management strategies to control beetles are explained further.

19.2.2.2 Cultural Methods

It is imperative to destroy crop residues after harvest, especially those from the roots and fruits, which helps in reducing the overwintering population of cucumber beetles. Follow thorough deep cultivation and shredding of crop residue accelerates the decomposition of top and bottom crop residues. Early plowing/disking that removes unwanted plants and discourages egg laying, delayed planting, and high seed rates helps to minimize the impact of these beetles (Sorensen 1999). Avoid planting of cucurbits close to preferred host plants of beetle larvae, for example, beans, corn, asparagus, brinjal, potato, tomato, and other grasses and weeds. The most significant damage can be avoided by simply delaying the planting of summer cucurbits by a few weeks until the dispersion of beetles and laying of their eggs at some other sites (Keszey 2012). This tactic can also provide extra time for seedlings to grow into vigorous and mature plants capable of withstanding beetle pressure. This strategy can help to reduce the need for insecticides.

Crop rotation can also help in reducing the incidence of cucumber beetles. Grow cucurbits far away from fields used the previous year to reduce pest problems since these beetles often overwinter in such fields. However, the beetles are highly mobile, so crop rotation alone is unlikely to entirely control cucumber beetles (Snyder 2012). Bach (1980) observed that intercropping cucumber with corn and broccoli reduced the incidence of striped cucumber beetles substantially compared to monoculture of cucumber. In another study, Cline et al. (2008) reported that intercropping watermelons or muskmelons with radish (*Raphanus sativus* L.), tansy (*Tanacetum vulgare* L.), nasturtium (*Tropaeolum* spp. L.), buckwheat (*Fagopyrum esculentum* Moench), cowpea (*Vigna unguiculata* L.), and sweet clover (*Melilotus officinalis* L.) had similar benefits of reducing beetle densities on cucurbits. For small-scale cultivation of cucurbits, young plants can be protected by mechanical means using row covers, screens, or cones to keep beetles away (Bessin 2010).

Trap crops may give some degree of control. The goal of trapping tactics is to lure beetles away from the main crop with attractive odors and colors. Treat the trap crop with insecticides before the adults lay eggs. An early planting of cucurbits can be made in order to attract the overwintering cucumber beetles where they can be destroyed by insecticides (Day 2008). This reduces the numbers of cucumber beetles that could feed on the late-planted main cucurbit crop. It is important to pull out and burn the remaining vines of the trap crop after destroying the cucumber beetles. Trap crops should probably not be relied solely for control. Luna and Xue (2009) reported that cucumber beetles are generally gathered at field edges and raising attractive trap crops may further accentuate this tendency. So, plantings of highly attractive cucurbit variety other than the main crop in the perimeter can help in reducing the incidence. Recent research indicates that *Cucurbita maxima* and *Cucurbita pepo* are particularly attractive to cucumber beetles (Adler and Hazzard 2009), and accordingly, application of recommended insecticides can be done on the trap crop only, which help in reducing insecticide use (Cavanagh et al. 2009). Use of trap baits containing insect pheromones (species-specific communication chemicals) and/or kairomones (host plant chemicals) is effective in controlling pests, which can be sprayed separately or in combination with synthetic or botanical pesticides (Alston and Worwood 2008). Cidetrak® CRW (a feeding stimulant) and eugenol (a pheromone) attract cucumber beetles to feed on baits and be killed by the insecticide.

Plastic or organic mulches can prevent cucumber beetles from laying eggs in the field near plant stems and may prevent movement of larvae feeding from roots to fruits. It was reported that aluminum plastic mulches were effective in repelling beetles and aphids from plants. Using mulch and drip irrigation helped in reducing soil moisture beneath the fruits and also lowered beetle feeding as well (Alston and Worwood 2008). Lower cucumber beetle density was found on cucumber plants grown in richly mulched soil compared to lesser ones (Yardim et al. 2006) because organic matter promotes diverse community of beneficial soil microorganism that activates the internal defense system of plants (Zehnder et al. 1997). Straw mulch also reduce cucumber beetle problem directly by their slow movement from one plant to another (Cranshaw 1998; Olkowski 2000) and help in conservation of predators like wolf spiders and others by providing shelter for them from hot and dry conditions (Snyder and Wise 2001; Williams and Wise 2003). Further, straw mulch acts as a food for springtails and other insects that eat decaying plant material, which help to increase spider numbers as they are important nonpest prey for spiders (Halaj and Wise 2002). Caldwell and Clarke (1998) suggested the use of metallic-colored plastic mulches that repel cucumber beetles and reduce their damage through feeding and also the transmission of bacterial wilt disease. Cucumber beetle attack was less on zucchini plants in plots where sunn hemp (*Crotalaria juncea* L.) was interplanted as living mulch compared with no sunn hemp plots (Hinds and Hooks 2013).

19.2.2.3 Resistant Cultivars

Cucumber beetles are attracted to host plants by the chemical cucurbitacin (gives a bitter flavor to cucurbits) that is used as a defense against less-specialized herbivores (Deheer and Tallamy 1991). These beetles ingest cucurbitacin that makes them unpleasant and thereby protects them from predators and parasitoids (Gould and Massey 1984; Tallamy et al. 1998). Therefore, it is important to select those cucurbit varieties having low cucurbitacin contents to decrease their attractiveness for cucumber beetles.

19.2.2.4 Biological Control

Beneficial insects can attack adults, eggs, and larvae on plants or on the soil surface. Some of the important natural enemies that attack cucumber beetles include parasitoids (tachinid flies, *Celatoria diabrotica* Shimer), fungi (*Beauveria*), and nematodes (*Howardula benigna* Cobb) (Capinera 2008). Predators like wolf spiders (*Hogna helluo* and *Rabidosa rabida*) have been shown to feed heavily on these beetles in cucurbit crops (Snyder and Wise 2001). Also, when spiders are around, beetles avoid feeding on the crop even when the spiders do not actually kill them (Snyder and Wise 2001; Williams and Wise 2003). Another predator namely, ground beetles also feed on adult cucumber beetles (Snyder and Wise 2001) like other big predators such as bats (Whitaker 1995). So, a biodiverse community of predators (harvestmen or "daddy long legs," ground and rove beetles, several kinds of spiders, predatory mites and bats) may be important for biological control of cucumber beetles rather than relying on any single predator species (Snyder 2012). Entomopathogenic nematodes have been found to suppress larvae and pupae of beetles in the soil (Alston and Worwood 2008). Fungal pathogens and entomopathogenic nematodes are available commercially to control larvae. These biopesticides and their soil-drenching formulations have shown some action against cucumber beetle larvae in the soil (Choo et al. 1996; Ellers-Kirk et al. 2000). A tachinid fly and a braconid parasitoid wasp parasitize striped cucumber beetle, and sometimes both of them have large impacts on striped cucumber beetles (Smyth and Hoffmann 2010). Entomopathogenic fungi, namely, *Beauveria bassiana* and *Metarhizium anisopliae*, are most widely used as biopesticides and have been evaluated for suppression of larvae with varying levels of biocontrol (Diver 2008). Commercially available parasitic nematodes, namely, *Steinernema* spp. and *Heterorhabditis* spp., are effective in controlling spotted beetle larvae. Ellers-Kirk et al. (2000) reported 50% reduction in striped cucumber beetle larvae using *Steinernema riobravis* in cucumbers, which resulted in better root growth. Natural populations of these beneficial agents can be conserved by avoiding the use of broad spectrum toxic insecticides and enhancing crop and soil health through cultural practices.

Botanical and biorational insecticides like azadirachtin (extract from the *neem* tree) combined with karanja (oil derived from the tree *Pongamia glabra*) reduced cucumber beetle populations by 50%–70% overnight (Diver 2008). Soil drenching with *neem* oil was effective against larval damage as it acts as an ovicide. The botanical pesticides sabadilla, rotenone, or pyrethrum showed moderate effectiveness in controlling cucumber beetles.

19.2.2.5 Organic Chemicals

Treatment of plants before transplanting in the field could help to get past the early vulnerable stages (Yao et al. 1996). Kaolin clay, pyrethrum, and spinosad (all formulations of spinosad are not organic-based) are some of the organic chemicals that can be used to manage cucumber beetles to some extent (Snyder 2012). Kaolin clay makes the cucurbit crops unattractive to cucumber beetles as it gums up their antennae and otherwise irritates them. Pyrethrum (extracted from the dried flower heads of African chrysanthemum) should be used carefully as it kills both harmful pests and beneficial pests. Apply pyrethrum only on the trap crop or specific hotspots within the main crop.

19.2.2.6 Chemical Control

Insecticides can be effective in controlling cucumber beetles but should not be used as the sole management tool. It is best to combine insecticides with other options such as cultural and biological control for long-term and sustainable management. It is also important to rotate the chemicals with different modes of action to avoid development of pesticide resistance in beetles (Alston and Worwood 2008). Early treatment is essential both for beetle control and management of bacterial wilt. The use of chemicals should be initiated at the emergence of seedlings or immediately after transplanting. Beetles are most active in the spring, so application of foliar insecticides is required twice a week during that period when the plants are small (Bessin 2010). A foliar insecticide application at the cotyledon stage is important to hinder beetle feeding. Additional foliar applications may be needed to prevent cucumber mosaic and bacterial wilt diseases depending upon beetle intensity (Sorensen 1999). A single posttransplant soil drenching with neonicotinoids gives protection for 3–4 weeks. Systemic neonicotinoid products, namely, imidacloprid and thiamethoxam, may be used as an in-furrow, banded, drench, or drip irrigation application to the seed/seedling root zone during or after planting/transplanting operations and are not applied as a foliar spray (Hazzard and Cavanagh 2010). The important insecticides that are effective in suppressing cucumber beetles are synthetic pyrethroid (bifenthrin, cypermethrin, lambda-cyhalothrin), carbamate (Sevin), organophosphate (malathion), and chloronicotinyl (imidacloprid, a systemic insecticide; apply as a sidedress or drench at planting or to young plants).

19.3   RED PUMPKIN BEETLE (*RAPHIDOPALPA FOVEICOLLIS/AULACOPHORA FOVEICOLLIS*)

It is a common and major pest of a wide range of cucurbits, especially sweet gourd, bottle gourd, bitter gourd, watermelon, and muskmelon. It is polyphagous in nature (Doharey 1983). Both larval and adult stages are injurious to the crop and cause severe damage to seedlings and young and tender leaves and flowers (Rahaman and Prodhan 2007; Rahaman et al. 2008). The pest is widely distributed in different parts of the world, especially in Asia, Africa, Australia, and south Europe. In India, it occurs throughout the country but is more common in the northwestern parts.

19.3.1   IDENTIFICATION AND SYMPTOMS OF DAMAGE

The dorsal part of the body of the adult beetle is deep orange, while the ventral side is black (plain beetle). The beetle appears to be oblong about 5–8 mm in length and 3.5–3.75 mm in width, with six legs. The posterior part of the abdomen bears soft white hairs. The female lays eggs in clusters of 8–9 in the moist soil near the base of the host plant or dead leaves that hatch into larvae in 6–15 days. The eggs are elongated and brown in color. Larvae are creamy white in color with brown heads and are about 10–12 mm long. Beetles start laying eggs after about 7 days of emergence and complete five generations from March to October. Another beetle known as the banded pumpkin beetle is also orange colored but has four very distinct large black spots on its back (one on each corner of the wing cover).

Creamy, yellow-colored larvae feed on the roots, stems, and fruits touching the soil. The damaged roots and underground stems may rot due to infection by the saprophytic fungi. Adult beetles feed voraciously on the leaf lamina by scrapping off the chlorophyll and making irregular holes or leaves with netlike appearance. The maximum damage is done during the cotyledon stage, which revealed that the first generation is more injurious than the subsequent generations. The attacked plants may shrivel, and resowing/planting may become essential under severe infestation. The young and smaller fruits of the infested plants may dry up, whereas the bigger and mature fruits become unfit for human consumption. Sweet gourd was the most suitable and bitter gourd was the least suitable host for red pumpkin beetles (Kamal et al. 2014).

They are strong fliers, very active in hot weather, and take flight quickly when disturbed. Sometimes, damage becomes very severe if it is not controlled timely. Losses due to infestation are quite evident, which may reach up to 35%–75% at the seedling stage (Yamaguchi 1983). In some cases, the losses due to this pest have been reported up to 30%–100% in the field (Khan et al. 2012). At the advent of spring, the beetles defoliate the cucurbit seedlings to such an extent that the crop has to be sown repeatedly three to four times (Parsad and Kumar 2002; Mahmood et al. 2005). It delays marketing of the produce of different cucurbits and hence reduces the income of the growers.

19.3.2   MANAGEMENT

Seedlings should be monitored twice a week to check their infestation. In the initial stages, it is good to collect the beetles and destroy them; otherwise, alternative methods as suggested for cucumber beetle can be employed to control it. The older plants should also be monitored and would be treated if defoliation is severe. Preventive measures like burning of old plants, plowing, and harrowing of field after harvest of the crops are followed for the destruction of adult, larvae, and pupae. Khan (2012) reported the preferential host plants for red pumpkin beetles and categorized bitter gourd, sponge gourd, ribbed gourd, and snake gourd as nonpreferred hosts (resistant), while muskmelon, cucumber, and sweet gourd were the most preferred hosts (susceptible) along with bottle gourd and ash gourd as moderately preferred hosts (moderately susceptible). Raising cucurbits earlier than the normal planting time could also be effective as the plants pass the cotyledonary stage by the time the beetles become active.

Use of neem oil cake in the soil is effective in killing the pest larvae. Vishwakarma et al. (2011) observed that treatment with entomopathogenic fungi *B. bassiana* resulted in maximum reduction of beetle infestation along with the highest fruit yield in bottle gourd. Khan and Wasim (2001) observed maximum repellency against beetles in treatment comprising of neem extracts mixed with benzene. The application of plant extract of *Parthenium* spp. was found to be highly effective in controlling the red pumpkin beetle (Ali et al. 2011). These beetles especially attack the crop at the cotyledonary stage when adults skeletonize the young leaves. During initial infestation, applications of carbaryl (0.1%) or malathion (0.5%) suppress the damage successfully (Hasan et al. 2011). Mahmood et al. (2010) reported that permethrin dust (0.5%) alone and ash + permethrin dust (2000:1 a.i. w/w) effectively control beetles on the cucumber crop, with no mortality of plants. Synthetic pyrethroids (deltamethrin 0.004%, cypermethrin 0.012%, and fenvalerate 0.01%) were effective in controlling the beetle for about a week (Thapa and Neupane 1992).

19.4 FRUIT FLY (*BACTROCERA CUCURBITAE*)

Melon fruit fly is distributed widely in temperate, tropical, and subtropical regions of the world (Sapkota et al. 2010). It has been reported to damage 81 host plants and is a major pest of cucurbitaceous vegetables. Doharey (1983) reported that it infests over 70 host plants of which fruits of bitter gourd (*Momordica charantia*), muskmelon (*Cucumis melo*), snap melon (*Cucumis melo* var. *momordica*), and snake gourd (*Trichosanthes anguina* and *T. cucumerina*) are the most preferred hosts. The extents of loss vary between 30% and 100%, depending upon the cucurbit species and the season (Shooker et al. 2006). Singh et al. (2000) reported around 30% damage on bitter gourd and watermelon in India. The infestation increases at temperatures below 32°C with a relative humidity range between 60% and 70% (Dhillon et al. 2005). It prefers to infest young, green, and soft-skinned fruits.

19.4.1 IDENTIFICATION AND SYMPTOMS OF DAMAGE

Generally, the females prefer to lay eggs in soft, tender fruit tissues by piercing them with the ovipositor. After egg hatching, the larvae bore into the pulp tissue and make the feeding galleries. The fruit subsequently rots or becomes distorted. Young larvae leave the necrotic region and move to healthy tissue, where they often introduce various pathogens and hasten fruit decomposition (Dhillon et al. 2005). The full-grown larvae come out of the fruit by making one or two exit holes for pupation in the soil. The larvae pupate in the soil at a depth of 0.5–15 cm. It also lays eggs inside the corolla of the flower, tap root, leaf stalk, and stem. The larvae successfully develop in these plant parts and feed inside (Weems et al. 2001). The fruits attacked in the early stages fail to develop properly and drop or rot on the plant. Since the maggots damage the fruits internally, it is difficult to control this pest with insecticides. Therefore, there is a need to explore alternative methods of control or formulating an integrated control strategy for effective management of this pest.

19.4.2 MANAGEMENT

The fruits of cucurbits are harvested at short intervals for marketing and self-consumption. Therefore, it is not fair to rely on insecticides to control this pest. Under severe infestation, it is important to use soft insecticides with low residual toxicity and short waiting periods. Different strategies should be used for the management of fruit fly as follows.

19.4.2.1 Local Area Management

Local area management means the minimum scale of pest management over a restricted area to suppress the pest population below economic threshold rather than its eradication (Dhillon et al. 2005). The options include bagging of fruits, field sanitation, protein baits, and cue-lure traps; host plant resistance, biological control, and soft insecticides can be employed without health and environmental hazards.

19.4.2.2 Monitoring and Control with Pheromone Lures/Cue-Lure Traps

The sex attractant cue-lure traps are more effective for monitoring the *fruit fly*. Methyl eugenol and cue-lure traps have been used to attract males for monitoring and mass trapping (Liu and Lin 1993; Seewooruthun et al. 1998). They can also be controlled by using *Ocimum sanctum* as the border crop sprayed with protein bait containing spinosad as a toxicant (Roomi et al. 1993). A number of commercially produced attractants are available in the market and can be used effectively in controlling this pest.

Installation of used water bottle baited with cue-lure (as MAT) saturated wood blocks (ethanol/cue-lure/carbaryl in a ratio 8:1:2) at 25 traps/ha prior to flower initiation is quite effective in trapping male flies. Use of a repellent (NSKE 4%) could enhance trapping and luring in bait spots. Application of neem as a repellent increased the catch in *para*-pheromone traps and enhanced the luring ability of *para*-pheromone by 52%. However, along with bait spray and repellents, other practices like removal and destruction of maggots in early infested fruits and field sanitation must be adopted.

19.4.2.3 Cultural Methods

Field sanitation should be ensured by removal and destruction of infested fruits daily to minimize pest intensity. Bury the infested fruits 0.46 m deep into the soil to break the reproduction cycle and population increase (Klungness et al. 2005). Akhtaruzzaman et al. (1999) suggested to bag cucumber fruits after 3 days of anthesis and retain for 5 days for effective control. Bagging of 3–4 cm long fruits on the plant with two layers of paper bags minimizes fruit fly infestation and increases the net returns by 40%–58% (Jaiswal et al. 1997) and is also environmentally safe. Grow 2–3 rows of maize as a trap crop between the cucurbits, which act as a resting site for the adult fruit fly. Any contact insecticides can be sprayed on maize during the evening hours to kill adult fruit flies.

19.4.2.4 Biological Control

Opius fletcheri Silv. is a dominant parasitoid of *Bactrocera cucurbitae*, and the efficacy varies from 0.2% to 1.9% in *M. charantia* (Wong et al. 1989). Another parasitoid, *Fopius arisanus*, has also been included in the IPM program of *B. cucurbitae* at Hawaii (Wood 2001). The cultural filtrate of the fungus *Rhizoctonia solani* was found to be effective against larvae (Sinha 1997), whereas the fungus *Gliocladium virens* has been reported to be effective against the adult fruit fly (Sinha and Saxena 1998). Culture filtrates of the fungi *R. solani*, *Trichoderma viride*, and *G. virens* affected the oviposition and development of fruit fly adversely (Sinha and Saxena 1999).

19.4.2.5 Host Plant Resistance

It is an important component in IPM programs. It does not cause any adverse effects to the environment. The success in developing high yielding and fruit fly-resistant varieties is limited. The resistance genes from the wild relatives of cucurbits can be transferred in the cultivated genotypes to develop resistant varieties to fruit fly using wide hybridization (Dhillon et al. 2005).

19.4.2.6 Botanicals

The continuous feeding with extract of *Acorus calamus* (0.15%) mixed with sugar (at 1 mL/g sugar) reduced the adult longevity from 119.2 to 26.6 days (Nair and Thomas 1999). Ranganath et al. (1997) reported neem oil (1.2%) and neem cake (4%) for its effective control.

19.4.2.7 Chemical Control

The use of chemicals to manage fruit fly is relatively ineffective. Apply malathion (0.05%) as cover spray to kill the insects on contact or a bait spray by adding 50 g gur + 10 mL malathion in 10 L water that attract and kill the adults. The application of malathion + molasses + water in the ratio of 1:0.1:100 provides good control of fruit fly (Akhtaruzzaman et al. 2000). This technique is economical and there is very low contamination of fruits from insecticides. Gupta and Verma (1982) reported that fenitrothion (0.025%) in combination with protein hydrolysate (0.25%) reduced fruit fly damage to 8.7% compared to 43.3% damage in nontreated control. Reddy (1997) reported triazophos as the most effective insecticide to manage it on bitter gourd. On the other hand, Borah (1998) obtained the highest yield and minimum damage in pumpkins when treated with carbofuran at 1.5 kg a.i./ha at 15 days after germination. Oke (2008) reported that lambda-cyhalothrin reduced more number of fruit fly pupae and increased the quality of harvested cucumber fruits in relation to the infestation of fruits with ovipositor marks.

19.4.2.8 Wide Area Management

The aim of wide area management is to coordinate and combine different characteristics of an insect eradication program over an entire area within a defensible perimeter. The wide area management program includes a three-tier model, that is, initial reduction in population using bait spray, suppression of reproduction using *para*-pheromone lure blocks to eradicate males to prevent oviposition by females, and intensive survey using traps and fruit inspection until it could be ascertain that the pest is entirely eradicated (Mumford 2004). Male-sterile technique, also covered under this program, comprises of releasing sterile males in the fields for mating with the wild females. The transmission of dominant lethal mutations kills the progeny. The females either do not lay eggs or lay sterile eggs. Ultimately, the pest population can be eradicated by maintaining a barrier of sterile flies (Dhillon et al. 2005).

19.5 HADDA BEETLE (*EPILACHNA IMPLICATE, E. VIGINTIOCTOPUNCTATA,* AND *E. BOREALIS*)

The squash beetle is an occasional pest on cucurbits and does not feed on other insects. It is a serious pest of bitter gourd, squash, and pumpkin. Larvae only feed on the underside of the leaf, while adults may feed on both surfaces or even on the rind of fruit, leaving spiral-shaped scars and deteriorating fruit quality and market appeal (Boucher 2014).

19.5.1 Identification and Symptoms of Damage

Both adults and larvae feed voraciously by scrapping the chlorophyll of the leaves causing characteristic skeletonization of leaf lamina, leaving a fine net of veins. The affected leaves gradually dry and drop down. A severe infestation kills the young plants overnight. The adult beetles are medium-sized (6–8 mm long), yellowish-brown, and globular, bearing 12–28 black spots on the elytra. The eggs are laid in clusters on the undersurface of the leaves that hatch into yellowish larvae. The female may lay 300–400 eggs. The larva (7–9 mm long) is found on the underside of leaves and is yellow with branched black spines covering the body. Full-grown larvae pupate below the leaf or at the base of the stems. The pupa hangs from the leaf, is yellow in color, and lacks spines. The development stages are completed in 4–6 weeks under optimal conditions. Scrapping of the epidermis indicates the feeding manner of the grubs, while adults make semicircular cuts in rows. Young plants can be entirely destroyed, while older plants can tolerate considerable leaf damage. Adults overwinter under loose tree bark or under leaf litter near the edge of fields.

19.5.2 Management

19.5.2.1 Cultural Methods

The beetles are not strong fliers, so crop rotation to distant fields tends to limit colonization and populations (Boucher 2014). Both the larvae and adults are not very aggressive defoliators and therefore, handpicking is recommended for the management of the pest on a small scale like home gardening. The best time of day to inspect for the beetles on cucurbits is around noon. Harrowing and destroying vines and larvae after harvesting early cucurbits can help to minimize pest population. Row covers can protect cucurbit crops from the beetles.

19.5.2.2 Botanicals

The weekly foliar sprays of aqueous *neem* kernel extracts at concentrations of 25, 50, and 100 g/L and neem oil applied with an ultralow-volume sprayer at 10 and 20 L/ha significantly reduced feeding by *Epilachna* beetles in squash and cucumber (Ostermann and Dreyer 1995). Mondal and Ghatak (2009) reported that the seed extracts of *Annona squamosa* (3 mL/L of water) as botanical pesticide help in reducing population buildup to the extent of 76%, followed by 64% and 57% through *NeemAzal* (5 mL/L of water) and petroleum ether extracts of rhizome of *A. calamus* (2 mL/L of water), respectively. *Tephrosia* leaf extract (20 g/100 mL water) effectively controls *Epilachna* beetle by killing adults and inhibiting pupae formation along with the highest yield and is an environment friendly pest control method (Rahaman et al. 2008). Swaminathan et al. (2010) have demonstrated the antifeedant and lethal effects of *Azadirachta indica*, *P. glabra*, and *Madhuca latifolia* on this beetle. Larvicidal bioassays with crude aqueous leaf extracts of three plants, namely, *Ricinus communis*, *Calotropis procera*, and *Datura metel* showed significant toxicity against the experimental *Epilachna* beetles by adversely affecting both oviposition and egg hatching besides prolonged larval duration, pupae formation, and adult emergence (Islam et al. 2011).

19.5.2.3 Chemical Control

This beetle may also be controlled through foliar applications of some synthetic pesticides like malathion, parathion, lambda-cyhalothrin, pyrethrin, spinosad methoxychlor, and rotenone.

19.6 APHIDS (*MYZUS PERSICAE* AND *APHIS GOSSYPII*)

Many aphid species including green peach aphid (*Myzus persicae*) and melon aphid (*Aphis gossypii*) feed on cucurbits and cause similar damage. They suck plant fluids from stems, leaves, and other tender plant parts by piercing with their slender mouthpart.

19.6.1 IDENTIFICATION AND SYMPTOMS OF DAMAGE

Aphids are small about 3 mm long, soft-bodied, pear-shaped insects with long legs and antennae. They are pale green, yellow, brown, red, or black in color. Some secrete a waxy white or gray material that covers their body, giving them a waxy or wooly appearance. The adults are usually wingless but also appear in winged forms especially when populations are high during spring and autumn season. Most species have a pair of tubelike structures called cornicles projecting rearwards from their abdomen, and these cornicles distinguish aphids from all other insects. Aphids can disperse long distances through the wind. The majority of aphid species reproduce asexually. Adult females give birth to nymphs (immature aphids), which are always wingless and become adults after molting and shedding the skin multiple times within a week. Aphid populations can increase rapidly as each adult reproduces numerous nymphs in a short period of time. The green peach aphid is slender, dark green to yellow, and has no waxy bloom. They tend to cluster on succulent plant parts and take just 10–12 days to complete one generation and reproduce over 20 generations annually under mild climates (Capinera 2005). It is considered to be the most important vector for the transmission of viruses throughout the world with equal capability of both nymphs and adults (Barbercheck 2014).

The first sign of aphid damage is a downward curling and crinkling of the leaves. Aphids are often found on lower leaves, soft-growing tips, flower buds, and flowers. Aphids pierce the plant tissue and extract sap, which causes a variety of symptoms, including reduced plant growth and vigor, mottling, yellowing, browning, curling, or wilting of leaves, which result in low yields and sometimes death of plant. Saliva injected into plant tissues by aphids can cause puckering and curling of leaves. Such curled and distorted leaves protect them from natural enemies or treatment applied for their control. Aphids feeding on flower buds and fruits can cause malformed flowers or fruits.

They also excrete honeydew (sticky and sugary liquid waste) as a result of feeding on the plant sap. Ants feed on the honeydew and aggressively defend aphids from predators and parasites, that is, interfere with their control by natural enemies. Accumulation of honeydew also creates a growth substrate for sooty molds (fungi) on leaves and other plant parts, thereby hindering photosynthesis by blocking light. They also involve in the spread of several viruses that affect all cucurbits causing a high rate of crop failure and huge economic losses. These viruses include cucumber mosaic virus, watermelon mosaic virus, zucchini yellow mosaic virus, and papaya ring spot virus. The symptoms of a virus infection are mottling, yellowing, or curling of leaves and stunted plant growth.

19.6.2 MANAGEMENT

Aphids have the ability to multiply at a rapid rate, which must be taken into account while monitoring for this pest. Plants should be checked at least twice a week with special attention to the undersurface of the leaves. Most problems occur toward the end of the growing season. Yellow sticky traps would be helpful in detecting aphids 2–3 weeks prior to planting.

19.6.2.1 Cultural Methods

Floating row covers or reflective mulches may help exclude or repel aphids (Barbercheck 2014). Aluminum foil mulches can repel invading aphid populations on young plants and check transmission of viruses. In very hot or arid areas, reflective mulches can cause conditions that are too warm for good crop growth and are not recommended. Reflective mulches were found effective in repelling aphids from plants by consistently reducing aphid population than those growing over bare soil and also help in delaying symptoms of cucumber mosaic cucumovirus and watermelon mosaic and zucchini yellow mosaic potyviruses by 3–6 weeks (Stapleton and Summers 2002). Living mulches reduce the contrast between the bare ground and the plant foliage, and as a result, the aphids do not detect their host. These mulches provide additional feeding sites for viruliferous aphids (aphids carrying virus) around the crop and hence reduce the incidence and spread of aphid-borne nonpersistently transmitted viruses (Toba et al. 1977). Reflective polyethylene and biodegradable synthetic latex spray mulches were found effective for the management of aphids and aphid-borne virus diseases on late-season melons (Stapleton and Summers 2002).

19.6.2.2 Biological Control

Biological control by natural enemies through predators, parasitoids, and pathogens has a significant impact on aphid populations. Beneficial insects are attracted to plants with moderate to heavy aphid infestations. These natural enemies may eat large numbers of aphids. However, the reproductive capability of aphids may be so high that the impact of the natural enemies may not be enough to keep the aphids at or below acceptable levels. Some common aphid predators include species of lady beetles and their larvae (*Harmonia axyridis, Hippodamia convergens, Coleomegilla maculata*) (Koch 2003), minute pirate bug (*Orius insidiosus* and *O. tristicolor*), larvae of the syrphid fly (Laska et al. 2006), and green (*Chrysoperla carnea, Chrysopa rufilabris, Chrysopa* spp.) and brown (*Hemerobius* spp.) lacewing larvae and larvae of the aphid midge (*Aphidoletes aphidimyza*) (Markkula et al. 1979).

Among natural enemies of aphids, parasitoids have an important place (Tomanovic and Brajkovic 2001). Some common aphid parasitoids (parasitic wasps) include *Aphidius colemani, Aphidius matricariae, Lysiphlebus fabarum,* and *Binodoxys angelicae* (Kos et al. 2008). Aphids are very susceptible to fungal diseases in humid weather. Some fungi that infect and provide biological control of aphids are *B. bassiana, M. anisopliae, Verticillium lecanii* (Hall 1982), and *Neozygites fresenii*. *B. bassiana* must be applied three times at an interval of 5–7 days for effective control. Fungal diseases can be more effective than insecticides for controlling large populations of aphids.

19.6.2.3 Chemical Control

Organic chemical controls include potassium soap and petroleum oil or primicarb. Endosulfan, Dimethoate, Lannate, Fulfill, and Actara are recommended for aphid control. Kill aphids using a detergent and vegetable oil solution before destroying old crops to avoid winged virus-infected aphids from getting to nearby crops. Cypermethrin (0.01%) or acetamiprid (0.01%) or bifenthrin (0.01%) or malathion (0.05%) can be used to control aphids.

19.7 SQUASH BUG (*ANASA TRISTIS*)

19.7.1 IDENTIFICATION AND SYMPTOMS OF DAMAGE

The squash bug has been reported to attack nearly all cucurbits but squash and pumpkin are the most preferred for oviposition and high rates of reproduction and survival. It causes severe damage to cucurbits by secreting highly toxic saliva into the plant. The primary site of infestation is foliage that wilts, becomes blackened, and dies upon feeding. The fruits are also infested. The wilting of plants is sometimes called "anasa wilt." More importantly, this bug is the vector to cause yellow vine decline disease that affects melons, watermelon, and pumpkins (Bruton et al. 2003; Pair et al. 2004). The bronze eggs are round and laid in clusters of 12 or more, which hatch in 1–2 weeks. In the early stage, the nymphs are dark with light-green abdomen and turn to light gray with black legs in the later stage. Young nymphs feed together in groups and grow into adults in 5–6 weeks (Palumbo et al. 1991). The control is very difficult with population buildup. The intensity of damage is directly proportional to the bug density. The adults emit a strong odor when crushed.

19.7.2 MANAGEMENT

19.7.2.1 Cultural Methods

Apply row covers (plastic and spun-bonded materials) at planting and gradually remove them at first bloom or earlier if needed. Straw mulch helps in controlling squash bugs by providing cover and alternative prey for ground beetles, which are key pests of squash bug eggs (Snyder and Wise 2001). However, mulches also provide cover for squash bugs themselves (Cranshaw et al. 2001), so the indirect benefits of mulching due to greater biological control may not always offset direct benefits to the squash bugs themselves. Pumpkin, watermelon, and squash are the most preferred crop and hence are seriously damaged, while zucchinis are less susceptible (Bonjour and Fargo 1989; Bonjour et al. 1993). On account of its preference for squash, squash planting can be used as a trap crop near other cucurbits and the trap crop can be treated with an insecticide to control the infestation.

19.7.2.2 Biological Control

Several natural enemies of squash bug are known, principally wasp egg parasitoids (Hymenoptera, Encyrtidae, and Scelionidae). The best-known natural enemy is a common parasitoid *Trichopoda pennipes* (Snyder and Wise 2001; Rondon et al. 2003; Decker and Yeargan 2008). A tachinid fly parasitoid *T. pennipes* attacks nymphs and adults of the squash bug (Worthley 1923), and parasitism rates can approach 100% in certain fields in particular years (Pickett et al. 1996). The female fly lays egg on the underside of the bug nymphs or adults. The larva feeds and develops inside the bug and eventually kills it. Several parasitoid wasps (in the families Encyrtidae and Scelionidae) attack squash bug eggs (Olson et al. 1996).

19.7.2.3 Organic Products

PyGanic, insecticidal soaps, and certain oils can be used for its control.

19.7.2.4 Chemical Control

Timing is the key to successful squash bug control and eliminating squash bugs is the key to management of yellow vine decline. Systemic insecticides suppress the bugs up to 3 weeks. Foliar sprays targeting newly hatched nymphs are more effective than sprays used against larger stages. Multiple foliar sprays are often needed for extended periods of control. Some recommended insecticides for squash bugs are dinotefuran in soil application and pyrethrin in foliar application (Palumbo et al. 1993).

19.8 WHITEFLY (*BEMISIA TABACI, BEMISIA ARGENTIFOLII,* AND *TRIALEURODES VAPORARIORUM*)

Several species of whiteflies infest cucurbits. Of these, tobacco whitefly (*Bemisia tabaci*), silverleaf whitefly (*Bemisia argentifolii*), and greenhouse whitefly (*Trialeurodes vaporariorum*) are the most serious. They have a wide host range that includes many weeds and crops.

19.8.1 IDENTIFICATION AND SYMPTOMS OF DAMAGE

These are small insects about 1–1.5 mm long. The body and wings of adult flies are covered with fine whitish powdery wax. The silverleaf whitefly often holds its wings vertically tilted with a visible space between them, while the greenhouse whitefly usually holds its wings flatter over the back, touching the abdomen or slightly overlapping. Whiteflies colonize the underside of leaves. Adults and eggs are commonly present on younger leaves and nymphs are found on older leaves. A female can lay around 300 eggs (Nyoike 2007). The eggs are oval and are laid into a slit in the leaf surface. The eggs are initially white, changing to brown, and are hatched within 8–10 days. The first instar (called crawler) is the only mobile instar having legs and antennae that moves to look for feeding sites, while the other instars are sessile and complete their life cycle on the same leaf (McAuslane and Smith 2000). Whiteflies can complete one generation in about 3–4 weeks.

The silverleaf whitefly gets its name because it injects a toxin into the plant that causes whitening of the undersurface of newly emerging leaves. Damage may be more severe on younger plants compared to plants near harvest. It can affect the crop directly by its feeding and by acting as a vector of viruses. They suck sap from the phloem and excrete honeydew (a sugar-rich substrate) that promotes the growth of sooty mold (*Capnodium* spp.) on leaves and economic plant parts. They also damage the plant by transmitting viruses and induce physiological disorders. Criniviruses, namely, cucurbit chlorotic yellow virus, cucurbit yellow stunting disorder virus, beet pseudo yellow virus, and lettuce infectious yellows virus, are exclusively transmitted by whiteflies in field- or greenhouse-grown cucurbits (Abrahamian and Abou-Jawdah 2014).

19.8.2 MANAGEMENT

19.8.2.1 Monitoring

Yellow sticky traps can be used to detect and monitor the activity of whiteflies in the field. This is an important step for their management. Yellow sticky traps provide a good correlation between catches and actual whitefly numbers in the field in some systems, and hence their use is very popular in whitefly IPM programs (Basu 1995). Traps should be placed strategically with one trap every 100 m² and should be placed just above the canopy of the crop as whiteflies are most attracted toward young foliage (Jelinek 2010). In young crops, the warning threshold is 10 adults per trap per week. *In situ* count conducted in the morning is quite effective because of least mobility of whiteflies.

19.8.2.2 Cultural Methods

The cultural control methods include mulching, crop rotation, floating row covers, cover crops, non-infested transplants, and good field sanitation to prevent the buildup of whiteflies. Delayed planting or host-free periods may decrease severity of attack as temperature and rainfall influence whitefly population dynamics (Horowitz et al. 1984; Horowitz 1986). Soil ground covers (synthetic and living mulches) have been found effective in reducing whitefly infestation (Frank and Liburd 2005; Nyoike 2007). UV-reflective plastic mulch was found effective for the management of silverleaf whitefly by repelling the adults, which reduce their colonization and nymph population on pumpkin and zucchini squash (Summers and Stapleton 2001).

Field hygiene is an integral part of the overall strategy to manage whiteflies, whitefly-transmitted virus incidence, and insecticide resistance. Destroy crop residues that shelter whiteflies immediately after final harvest to decrease their population and sources of plant viruses (Webb et al. 2014). Use a contact desiccant/herbicide in conjunction with a heavy application of oil (not less than 3% emulsion) and a nonionic adjuvant to destroy crop plants and whiteflies. Destroy crops block by block as per completion of harvest rather than waiting and destroying the entire field at one time. Eradicate weeds regularly from the crop as they can support large populations of whiteflies.

19.8.2.3   Biological Control

Natural or introduced biological controls provide the best long-term solution to keep most of the whitefly species at low levels. Biological control agents such as *Encarsia formosa*, *Encarsia luteola*, and *Eretmocerus californicus* have been found fairly effective in the greenhouse (Cranshaw 2014). The products based on insect pathogenic fungi/mycoinsecticide, namely, *V. lecanii*, *Paecilomyces fumosoroseus*, and *B. bassiana*, have the capacity to suppress whiteflies in both greenhouse and field crops (Faria and Wraight 2001). The important predators affecting whiteflies are beetles (Coccinellidae), true bugs (Miridae, Anthocoridae), lacewings (Chrysopidae, Coniopterygidae), mites (Phytoseiidae), and spiders (Araneae) (Gerling et al. 2001). The dusty lacewing *Conwentzia africana* has been considered one of the most important predators of *B. tabaci* (Legg et al. 2003). Several fungi, namely, *V. lecanii*, *B. bassiana*, and *P. fumosoroseus*, can be useful control agents under high humid conditions. Fungal pathogen *V. lecanii* is reported to be capable of infecting whitefly only for a short period of time after its application. In general, losses due to feeding by whiteflies can be managed through natural enemies to some extent, whereas they are generally insufficient to prevent virus spread and transmission.

19.8.2.4   Organic Products

Neem-based pesticide formulations containing azadirachtin (NeemAzal and Azatin) have been reported to control young nymphs, inhibit growth and development of older nymphs, and reduce egg laying by adult whiteflies. Soap and certain oil sprays are acceptable to use in an organically certified crop. The efficacy of neem-based pesticides can be enhanced by adding 0.1%–0.5% soft soap.

19.8.2.5   Chemical Control

Whiteflies are usually difficult to control by using chemicals as the adults at immature stage reside on the underside of leaves, particularly older leaves which makes the coverage of chemical spray important for good control. A soil application of neonicotinoid insecticide (imidacloprid or thiamethoxam) at planting effectively controls whiteflies, while spraying of neonicotinoid insecticide such as acetamiprid should be done in the early stages of growth before flower initiation. Spiromesifen is effective against immature stages of the whitefly as is buprofezin and pyriproxyfen (insect growth regulators). Although spiromesifen, pyriproxyfen, and buprofezin affect mostly the reproduction and survival of immature nymphs, they help in reducing secondary spread by slowing the multiplication of the whitefly population (Webb et al. 2014). Adult whiteflies are usually best controlled with pyrethrins or various pyrethroid insecticides. Avoid consecutive applications of the same insecticide, that is, rotate insecticidal applications among chemicals of different classes to delay resistance buildup. Never follow soil application of any neonicotinoid with a foliar application of another neonicotinoid. The adult and crawler stages are the most susceptible to contact insecticides but the egg, scale, and pupa vary in their resistance to these chemicals. A single spray of any chemical will affect only the susceptible stages at the time of chemical application or during the time it remains active, while all other stages will survive and continue their life cycle. Thus, combinations of two to three chemical applications are usually required during the cropping season to control whiteflies.

19.9 SQUASH VINE BORER OR CLEAR WINGED MOTH (*MELITTIA CUCURBITAE* AND *MELITTIA EURYTION*)

It is an important native pest of cucurbits, especially summer squash (*C. pepo* L.) and some winter squashes, pumpkins, and gourds (*C. pepo* and *C. maxima* Duchesne) that have large and hollow stems (Howe and Rhodes 1973). Cucumbers and melons are less affected. Cultivars of *Cucurbita moschata* are the least satisfactory hosts (Howe and Rhodes 1973).

19.9.1 IDENTIFICATION AND SYMPTOMS OF DAMAGE

Adults are stout, dark gray moths (1.25 cm long) with orange abdomen, having black dots, hairy red hind legs, opaque front wings, and clear hind wings with dark veins. Unlike most moths, they attack the crop during day time and appear more like a paper wasp than a moth (Canhilal et al. 2006). The small, brown, flat eggs about 1 mm in size are laid individually on leaf stalks and vines. They hatch in 7–10 days and the newly born larva immediately bores into the stem. The cream-colored larvae, about 2.5 cm in size, enter and feed in the stems of cucurbit vines blocking the flow of water to the rest of the plant, which cause wilting of plants and finally mortality (Britton 1919). A larva feeds for 14–30 days before exiting the stem to pupate in the soil (Delahut 2005). Symptoms appear in midsummer when an entire plant wilts suddenly. Infested vines usually die beyond the point of attack. Sawdust-like frass (insect excrement) near the base of the plant is the best evidence of squash vine borer activity (McKinlay and Roderick 1992). The primary feature for distinguishing squash vine borer from Bacterial wilt and *Fusarium* wilt is frass accumulating at the entrance to the larval tunnel.

19.9.2 MANAGEMENT

19.9.2.1 Monitoring

Prevention of squash vine borer in the early stages is most crucial to save the plant as it may be too late once the eggs are laid and hatched. Therefore, monitoring for the presence of adult borers is very important, which can be observed by close attention or yellow-colored pots filled with water may be helpful. Moth flights (and egg laying) may be predicted by using pheromone traps. Small wire cone traps, nylon mesh *Heliothis* traps, and unitrap bucket were found most successful for monitoring squash vine borer (Jackson et al. 2005).

19.9.2.2 Cultural Methods

In a borer-infested field, destroy infected crops and avoid planting cucurbits the following year. Disking the soil in early fall should expose cocoons that are buried 2.5–15 cm deep. Early planting helps to escape losses as crop matures before egg laying. Floating row covers placed over the crops prevent the moths from laying eggs in the initial stages. If early infestations are observed on the vine, larvae may be removed by slitting the stem longitudinally near the entrance hole with a fine blade and cover the wounded stem with soil to promote new root growth (Welty 2009).

The incidence of squash vine borer was reduced to the extent of 88% in zucchini or summer squash by raising Blue Hubbard squash in the perimeter as a trap crop (Boucher and Durgy 2004). Apply an effective insecticide on the trap crop to prevent borer infestation or destroy the trap crop after the peak egg-laying period is over to kill borers. *C. maxima* was the most susceptible, followed by *C. pepo*, while *C. moschata* and *Cucurbita mixta* were the least preferred (Grupp 2015).

19.9.2.3 Biological Control

Parasitic wasps can attack at the most susceptible stage (egg) of the squash vine borer, but they are not able to destroy significant numbers of their eggs. Ground beetles can attack the larvae but do not cause significant mortality (Welty 2009). Thus, natural control is marginal. Several species and strains of entomopathogenic nematodes have been tested against the squash vine borer (Canhilal and Carner 2006) of which *Steinernema carpocapsae* or *Steinernema feltiae* applied to the stem and soil provided control similar to a conventional insecticide. Stem injections and spray application of *Bacillus thuringiensis* (Bt) provided similar control to that of the conventional insecticide (Canhilal and Carner 2007).

19.9.2.4 Botanicals

Several insecticidal active ingredients approved for organic production are azadirachtin (neem), neem oil, kaolin clay, geraniol, thyme oil, pyrethrin, and spinosad.

19.9.2.5 Chemical Control

The key to management of squash vine borer is controlling the borers before the larvae develop/ enter the stem, otherwise insecticides are not effective. The time of insecticide application is very crucial once the early signs of larval feeding are detected. Hence, monitoring of plants from mid-June to August at weekly intervals is required for initial signs of borer frass. Apply insecticides twice at weekly intervals to control newly hatched larvae. Acetamiprid, bifenthrin, and carbaryl are effective in controlling squash vine borer.

19.10 MELONWORMS OR RINDWORMS (*DIAPHANIA HYALINATA*)

It is an uncommon late-season pest of cucurbits (Zehnder 2011). Melonworms feed primarily on foliage especially summer or winter squash. Indirect yield loss up to 23% due to foliage damage and 9%–10% direct yield reduction due to fruit damage has been reported (McSorley and Waddill 1982).

19.10.1 IDENTIFICATION AND SYMPTOMS OF DAMAGE

Eggs are oval and flattened in shape, about 0.7 mm in length and 0.6 mm in breadth. They are laid at night in a cluster of 2–6 eggs on different plant parts like buds, stems, and underside of leaves and hatch within 3–4 days (Capinera 2005). The larva is yellow green, about 2.5 cm long, and has fine yellow stripes running down its back in the last instar. High populations defoliate plants giving a lacelike appearance as only leaf veins remain intact. The larva may feed on the surface of the fruit or even burrow into the fruit in case of nonavailability of foliage or nonpreference crops like cantaloupe, leading to the name rindworm. Snake cucumber was found to harbor larvae the most, while pumpkin was the least preferred (Mohaned et al. 2013).

19.10.2 MANAGEMENT

19.10.2.1 Monitoring

Monitoring is difficult due to the nocturnal behavior of the adults. The moths are not attracted to light traps, while pheromone traps are not commercially available though identified (Raina et al. 1986). Therefore, the most effective ways to monitor crops are to check leaf damage in the early growth stage and also for the presence of larvae.

19.10.2.2 Cultural Methods

Early plantings often escape serious damage except in tropical areas where melonworms over-winters (Capinera 2005). Row covers can be used effectively to exclude melonworm adults (Webb and Linda 1992). Intercropping of squash with corn and beans helps to reduce its damage

(Letourneau 1986). Squash is the most preferred host among cucurbits and can be used as a trap crop. In addition, removal and destruction of crop residue containing melonworm pupae following harvest is a good cultural practice to reduce populations (Smith 1911). Removing host plants like creeping cucumber (*Melothria pendula*) and wild balsam apple (*Momordica charantia*) prevents overwintering (Elsey 1985).

19.10.2.3 Biological Control

B. thuringiensis is commonly recommended for suppression. The Bt variety 'kurstaki' is most effective on caterpillars of pickleworm and melonworm (Zhender 2011). The entomopathogenic nematode *S. carpocapsae* provides only moderate suppression where larvae are found resting and feeding (Shannag and Capinera 1995). Important parasites like *Apanteles* sp., *Hypomicrogaster diaphaniae*, *Pristomerus spinator*, *Casinaria infesta*, *Temelucha* sp., and trichogrammatids (all hymenopterans) also suppress the pest (Pena et al. 1987; Capinera 1994). Other reported parasitizing agents include *Gambrus ultimus*, *Agathis texana*, *Nemorilla pyste*, and *Stomatodexia cothurnata*. Medina-Gaud et al. (1989) reported that predators such as *Calosoma* spp., *Harpalus* spp., *Chauliognathus pennsylvanicus* (soldier beetle), and *Solenopsis invicta* (red fire ant) caused mortality of melonworms up to 24%.

19.10.2.4 Chemical Control

Important chemical active ingredients recommended are lambda-cyhalothrin 5% EC or WG and malathion 50% EC.

19.11 MITES (*TETRANYCHUS URTICAE*)

There are a number of mite species that attack cucurbits including two-spotted spider mite (*Tetranychus urticae*), bean spider mite (*Tetranychus ludeni*), red-legged earth mite (*Halotydeus destructor*), broad mite (*Polyphagotarsonemus latus*), blue oat mite (*Penthaleus major*), and clover mite (*Bryobia cristata*). Of these, two-spotted mites (TSMs, *T. urticae*) are the most serious, while the other species are of less importance. Frequently, infestations include a mixture of spider mite species (Natwick et al. 2012).

19.11.1 IDENTIFICATION AND SYMPTOMS OF DAMAGE

TSMs are rarely detectable with the naked eyes. The eggs are small, round, translucent, straw-colored, shiny, and pearl-like and are laid one at a time near veins on the undersurface of leaves. Each female can lay around 200 eggs. These eggs hatch into a nymph (first-stage larvae, having only three pairs of legs). The larva feeds for a few days and molts into the first nymphal stage (protonymph), having four pairs of legs. It then molts into the second nymphal stage (deutonymph), turning into an adult. The young adults are typically pale green, orange, or yellow in color with two dark spots on the body and about 0.3–0.5 mm long. When colonies of two spider mites are large in numbers, they spin fine silky webs for protection from predators that are often filled with cast skins, dust, and debris. They also move from leaf to leaf or from one plant to another. During unfavorable or cold weather conditions, they may change to an orange/red color and are commonly known as red spider mites. Depending upon the ambient temperature and other biotic and abiotic factors, the life cycle may vary from 6 to 40 days. Under optimum temperature of more than 27°C and relative humidity less than 50%, the mites can complete their life cycle within 5–7 days. The life cycle tends to stop below 10°C and above 40°C. The life cycle is faster at higher temperatures, and hence, the multiplication of spider mites is alarmingly rapid in greenhouses and can complete 5–6 generations per year (Naved 2012).

Dusty conditions also favor early infestations. Hot and dry weather conditions favor rapid development of eggs, increase feeding of nymphs and adults, and decrease the abundance of pathogenic fungi. Females find their way into fields by climbing to the top of their feeding site and releasing a long string of silk from their abdomen that catches the breeze and becomes airborne. They have such a wide host range that they usually start feeding wherever they land. Mite infestations usually start on the field edge and move toward the center over time. The mites feed on the underside of the leaves by piercing and sucking the leaf surface. Both immature mites and adults have rasping and sucking mouthparts. The initial symptoms appear as pale or bronzed areas along the midrib and veins of the leaves. Leaves turn speckled, yellowish, or grayish in appearance. The initial symptoms appear as tiny, light spots on the leaves (stippling), which later turn brown. The severe infestation causes mottling of leaves with silvery-yellow appearance and leads to premature leaf fall (Gupta 1985). Loss of leaves can lead to sunburning and have a significant impact on yield. There are approximately 14% reductions in the total leaf area that results in significant yield loss (Park and Lee 2005).

Red-legged earth mite (*H. destructor*) is another important mite of cucurbits. Adults and nymphs have a black, somewhat flattened body and red legs. It spends most of its time on the soil surface rather than on plant foliage. Mite feeding causes damage to plant cell and cuticle, which promotes desiccation, retards photosynthesis, and produces the characteristic silvering that is often mistaken as frost damage. The seedlings are seriously damaged and some times crop may need resowing. The adults of bean spider mite (*T. ludeni*) are uniformly red and cause bronzing or whitening of the upper leaf surface. Leaves may turn red or yellow and drop due to heavy infestations.

19.11.2 MANAGEMENT

19.11.2.1 Monitoring

The crop should be checked at least twice a week using a magnifying glass especially on the under-surface of the leaf as mite damage or presence is not noticeable on the top surface of the leaf. Inspect the field for mites initially on the borders of crop fields. Carefully examine yellow and speckled crown leaves for mites by using a 10× hand lens as these leaves are a prime site for mite development. If mite infestation appears along the field edges, it is essential to examine the whole field. The application of miticide is justified on account of heavy mite infestation, otherwise spot spraying may be effective. The first miticide application should be followed by another application within 5 days as the first spray kills all mite stages except the eggs, whereas the second application kills the newly hatched nymphs.

19.11.2.2 Cultural Methods

Cultural control methods for mites include weed control, crop rotation, clean fallowing, mixed cropping, trap, or border crops, minimizing dust and changes in tillage practices. Destroy weeds in and around fields during the fall or early spring and crop growth season. Proper cleanup at the end of the crop is to be done to reduce initial infestations in the next crop, which would reduce or eliminate mite populations just before the overwintering or diapausing phase (Murphy et al. 2014). Heavy rains increase relative humidity favorable for the development of fungal diseases in the mites and may also wash the pests off of the leaves. Also, irrigation with an overhead sprinkler may provide some short-term relief of mite infestations. Good water management increases plant tolerance to these pests.

19.11.2.3 Biological Control

Biological control is an important component of mite management. Take measures to ensure the survival of predators and parasites. TSMs are commonly attacked by predator mites such as *Phytoseiulus persimilis*, *Amblyseius californicus*, *Typhlodromus occidentalis*, *Mesoseiulus longipes*, *Neoseiulus californicus*, *Galendromus occidentalis*, and *Amblyseius fallicus*. Predatory mites can be distinguished from spider mites because of their longer legs, more active and faster movement. *P. persimilis* is the most common predator and preys on all stages of mites and can be used effectively to control TSM on cucurbits. Ten thousand predators over 200 m² of crop should be released at the first sign of TSM. Besides, the predatory bug *Macrolophus caliginosus* is also an efficient biocontrol agent (Naved 2012). Minute pirate bugs (*O. tristicolor*), big-eyed bugs (*Geocoris* spp.), six-spotted thrips (*Scolothrips sexmaculatus*), western flower thrips (*Frankliniella occiden-talis*), lady beetles/spider mite destroyers (*Slethorus* spp.), and lacewing larvae (*C. carnea*) also predate upon mites.

B. bassiana effectively controls mites and should be applied three times at a 5–7-day interval. Biorationals at higher concentrations like *B. bassiana* (10¹⁰ spores/mL), *Withania somnifera* (7.5%), and *Glycyrrhiza glabra* (7.5%) have the potential to be employed in pest management programs designed for *T. urticae* control in cucumber (Tehri and Gulati 2014) though maximum population reduction was recorded with omite (0.05%), followed by Nimbecidine (5 mL/L). Dimetry et al. (2013) reported that good reduction in aphids and TSMs population was achieved through the four successive alternate sprays of botanical insecticide Nimbecidine and entomopathogenic fungus (Bio-Catch).

19.11.2.4 Organic Controls

Organic measures include soaps and oils (moderate to poor results). Petroleum-based horticultural oils and plant-based oils such as neem, canola, or cottonseed oils can be used. Recommended plant extract-based formulations include garlic extract, clove oil, mint oils, rosemary oil, and cinnamon oil (Godfrey 2011).

19.11.2.5 Chemical Control

Mites can be difficult to control by chemical means due to their short life cycle and quick buildup of resistance to chemicals (Aguiar et al. 1993). Many aspects of the biology of TSM including rapid development, high fecundity, and haplodiploid sex determination seem to facilitate rapid evolu-tion of pesticide resistance (Van et al. 2010). Chemicals do not kill mite eggs, so it is important to spray when most mites have emerged (Anonymous 2014). TSM can be controlled with potassium soap or imidacloprid or the miticide dicofol or a mix of dicofol and tetradifon or abamectin or spi-romesifan, etc. Tetradifon controls eggs as well as the adults and nymphs, whereas dicofol controls only the adults and nymphs (Brown 2003). Bifenazate was also found to be effective against TSM (Al-Antary et al. 2012). Spider mites exposed to carbaryl (Sevin) in the laboratory have been shown to reproduce faster than untreated populations (Godfrey 2011). Pesticides combined with preda-tory mites were more effective in controlling TSSM than pesticides or predatory mites alone, for example, a combination of bifenthrin and *P. persimilis* restricted mite populations to a low level (Alzoubi and Cobanoglu 2010).

19.12 NEMATODES (*MELOIDOGYNE* SPP., *PRATYLENCHUS* SPP., *TRICHODORUS* SPP., *PARATRICHODORUS* SPP., AND *LONGIDORUS AFRICANUS*)

Nematodes are typically microscopic, elongated roundworms. Plant parasitic nematodes obtain their food only from living plant tissues. They feed by puncturing cells and withdrawing the contents with a needlelike mouthpart called a stylet. Several nematodes are pests of cucurbita-ceous crops like root-knot nematodes (*Meloidogyne incognita*, *M. arenaria*, *M. javanica*, and *M. hapla*), lesion nematode (*Pratylenchus* spp.), stubby root nematode (*Trichodorus* sp. and *Paratrichodorus* sp.), and needle nematode (*Longidorus africanus*) (Westerdahl and Becker 2011). Of these, root-knot nematodes are the most serious and alarming as they cause tremen-dous yield losses (Hussain et al. 2011a). Most cucurbits, especially muskmelon, cucumber, pumpkin, bottle gourd, and bitter gourd, are extremely susceptible to plant parasitic nematode infestation. Both *Cucumis* species (cucumber and melon) are the more preferred hosts for nematode invasion and reproduction than zucchini squash, followed by watermelon (Lopez-Gomez and Verdejo-Lucas 2014).

19.12.1 LIFE CYCLE AND SYMPTOMS OF DAMAGE

Nematodes usually live for 3–4 weeks only. Each mature female can produce more than 1000 eggs during its relatively short life. The eggs hatch immediately followed by three or more moltings before they become adults. The juvenile root-knot nematodes enter the roots and continue feeding and complete most of their life cycle within the roots of their host plant although they can survive in the soil as eggs or as second stage juveniles. These are encased in a gelatinous sac that protects them from dehydration. The optimum soil temperature for root-knot nematodes development is 25°C–28°C, and they complete their life cycle in 3–4 weeks at this range. However, development will take more time at lower temperatures (Westerdahl and Becker 2011).

Root-knot nematode is a problem mainly in areas with lightly textured or sandy soils. They establish a permanent feeding site of "giant cells" within the root and become immobile. This lifestyle is called sedentary endoparasitism (Westerdahl and Becker 2011). These hypertrophic giant cells adjacent to the head of sedentary females serve as food factories for nematodes. The "knots" (hyperplastic galls) formed on infected roots are due to an increased number of plant cells around the feeding site. Each gall can grow more than 2.5 cm in diameter and inhibits the uptake of water and nutrients to the plant. The symptoms include stunting, flagging, and chlorosis between veins of the leaves, which may be confused for nutrient deficiency symptoms. Plants show premature wilting and watering cannot relieve either the problem or symptoms. Plants have poor root system and can be uprooted easily. Symptoms usually occur in patches of nonuniform growth rather than as an overall decline of plants within an entire field. Quantitative studies on cantaloupe determined that a root-knot population density of 40 second stage juveniles per 100 cm of soil before planting can cause a minimum of 30% yield loss (Westerdahl and Becker 2011).

Lesion nematodes are migratory endoparasites that invade roots. They move and feed within the root cortex. In contrast to root-knot nematodes, they are able to leave the host if conditions become unfavorable. Infestation may cause reddish brown to dark brown lesions on roots. *Paratylenchus* species occur in production fields but are not known to cause significant damage to cucurbits (Westerdahl and Becker 2011). Sting nematode can be very injurious, causing infected plants to form a tight mat of short roots due to cessation of root elongation and root necrosis. The roots show a swollen appearance because of damage to new root initials under heavy infestations. These symptoms seem like fertilizer salt burn. All other potentially crop-damaging nematodes are ectoparasites. Stubby root nematodes prefer to feed on root tips. Symptoms include short feeder roots, stunting, and yellowing of plants. Nematode-infected plants may have small leaves and bear fewer flowers than nematode-free plants that result in low yield and poor quality fruits. Davis (2007) reported 24%–30% reduction in fruit weight of watermelon. Nematode damage may cause some cucurbits to produce large amounts of ethylene gas, which causes premature fruit ripening (Noling 2009). Plants stressed by inadequate nutrition or moisture may be more susceptible to damage by root-knot nematodes and are also vulnerable to infection by other pathogens (Noling 2009).

19.12.2 MANAGEMENT

The wide host range of these pathogens makes them almost impossible to eradicate once they infest a soil. It is best to prevent an infestation as complete eradication is almost impossible (Dorman and Nelson 2012). Nematicides and soil fumigation effectively suppress many nematode species but are most effective if combined with cultural practices (Schwartz and Gent 2007). Soil treatment with chemicals using biocontrol agents (Rahoo et al. 2011; Vagelas and Gowen 2012) combined with cultural practices such as crop rotation are common methods of nematode control.

Once a nematode problem is confirmed, the affected areas and plants should be isolated as transplants, machinery, and irrigation water can spread nematode infection. Sampling is necessary to confirm and to quantify the nematode population. Sampling and management are to be taken before planting or after harvest because the corrective measures are not efficient to rectify the problem completely afterward. Collect soil and root samples from 10 to 20 different locations at soil depth of 15–25 cm as most of the nematode species are concentrated in the crop root zone. Alternatively, live plant root samples can be collected for further examination. Also, roots may be checked for root gall severity.

19.12.2.1 Cultural Methods

Various cultural practices like crop rotation, fallowing, sanitation, water management, organic manures, and mulching help in the management of nematodes. Crop rotation with nonhost crops is very effective, for example, grasses, cowpea, and winter cereals are generally poor hosts and little nematode reproduction occurs during the cold winter months. Follow a 3-year or longer crop rotation with nonhost crops. Marigold serves as a good rotation crop as its roots exude chemicals (allelopathic effects) that kill root-knot nematodes (Wang et al. 2007). Control weeds that may serve as alternate hosts of nematodes, for example, hairy and black nightshade, pigweed, and purple and yellow nut sedge. Higher soil organic matter content protects plants against nematodes by increasing soil water-holding capacity and enhancing the activity of naturally occurring biological organisms that compete with nematodes in the soil (Noling 2009). Crab meal compost potentially suppresses nematode and must be applied before planting to maximum root zone depth for more effectiveness. Thoroughly cleaned equipment must be used between fields to prevent spread. Root-knot nematodes are not seed-borne, and it is imperative to grow nematode-free seedlings (Noling 2009). Many cowpea cultivars are poor hosts to root-knot nematodes (Wang and McSorely 2004) and should be used as cover crops to protect the main crop. Allowing a period of dry fallowing, deep plowing during the warmer months, and solarization of soil by covering with a transparent polythene mulch help in reducing nematode population (Noling 2009; Westerdahl and Becker 2011), but it is neither as effective nor as reliable as chemical fumigation.

19.12.2.2 Resistant Cultivars

The cultivars resistant to root-knot nematodes have comparatively better crop yield than susceptible varieties and can be employed as a component of integrated nematode management (Mukhtar et al. 2013). According to Oostenbrink (1966), the cultivation of a resistant variety may suppress nematode population to 10%–50% of its harmful density. Norton and Granberry (1980) reported that *Cucumis metuliferus* was highly resistant to root-knot nematodes. Resistance to root-knot nematodes was reported in cucumber cv. 'Capris' (Khelu et al. 1989) and several watermelon lines, including 'Crimson Sweet' (Zhang et al. 1989). Other *Cucumis* species, including *C. anguria*, *C. ficifolius*, *C. longipes*, and *C. heptadactylus*, are resistant to *Meloidogyne* spp. (Fassuliotis 1967). Cucurbit varieties can be grafted onto rootstocks resistant to root-knot nematodes (Thies et al. 2010). Siguenza et al. (2005) reported that *C. metuliferus* can be used as a rootstock for melon to prevent both growth reduction and a strong nematode buildup in *M. incognita*-infested soil.

19.12.2.3 Biological Control

Zhang et al. (2008) preinoculated the cucumber plants with three arbuscular mycorrhizal fungi, namely, *Glomus intraradices*, *G. mosseae*, and *G. versiforme*, which significantly reduced root galling. The use of antagonistic plants (*A. indica*, *C. procera*, *Datura stramonium*, and *Tagetes erecta*) as soil amendments has also been found effective (Ahmad et al. 2004; Hussain et al. 2011b; Kayani et al. 2012). Marigold produces a substance called alpha-terthienyl, which can aid in the reduction of root-knot nematodes (Soule 1993), and can be used as a cover or rotation crop. Several microbial pathogen formulations like *Pasteuria penetrans* (*Bacillus penetrans*), *B. thuringiensis*, *Burkholderia cepacia*, *Trichoderma harzianum*, *Hirsutella rhossiliensis*, *Hirsutella minnesotensis*, *Verticillium chlamydosporium*, *Arthrobotrys dactyloides*, *Myrothecium verrucaria*, and *Paecilomyces lilacinus* were found to be highly effective in the control of nematodes (Messenger and Braun 2000). A strain of *B. thuringiensis* (Bt) that reduced damage by the root-knot nematode species *Rotylenchulus reniformis*. *B. penetrans* has been shown to attack root-knot nematodes, although not yet commercially available (Anonymous 2015). Xalxo et al. (2013) reported maximum nematode mortality due to *T. viride* and *Aspergillus flavus*. The results suggest that the various fungi associated with the rhizosphere soil of different vegetable crops could be used as a biocontrol agent of nematodes and the culture filtrates can serve as a source of novel nematicidal compound of fungal origin, which is more environmental friendly. The fungus *P. lilacinus* parasitizes the eggs of some nematodes, including *M. incognita* and is relatively effective (Anonymous 2015).

19.12.2.4 Chemical Control

To mitigate nematode damage, apply nematicides before or at planting. Postplant applications can be considered only for supplemental suppression of plant parasitic nematodes. Nematicides (Mocap 15% Granular and Mocap EC) and soil fumigation (Telone C-17 and Vapam) effectively suppress many nematode species but are most effective in combination with cultural practices. A significant reduction of nematode juveniles is recorded with the application of metham sodium, dazomet, and 1,3-dichloropropene (Giannakou et al. 2002). Nematode injury often occurs in localized areas in the fields and can be effectively managed by spot treatments with nematicides like chloropicrin, 1,3-dichloropropene + chloropicrin, and oxamyl (Sharma and Trivedi 1985). Fumigant application requires uniform diffusion with the soil, and thus fumigants need irrigation management, proper drainage, and mulching. In general, the use of soil fumigants has been more consistently effective than nonfumigants (Noling 2009). Methyl bromide is very effective against nematodes. All of the fumigants are phytotoxic to plants and as a precautionary measure, should be applied at least 3 weeks before crops are planted.

Fumigants nematicides: Telone (1,3-dichloropropene), Pic-Clor, Vapam (metam sodium), KPam HL, dimethyl disulfide, methyl bromide, metam potassium, chloropicrin, ethylene dibromide, Dazomet, and Di-Trapex (methyl isothiocyanate).

Nonfumigant nematicides: Nemafos (thionazin), Vydate (oxamyl), Mocap (ethoprophos), and Furadan (carbofuran). They have an advantage as the application is relatively simple (Wright 1981).

19.13 CONCLUSIONS

Cucurbits are an important part of the fresh market vegetable crops. The attacks made by the insect pests in cucurbits result in severe yield and quality losses. The average economic loss caused by insect pests in vegetable crops is around 40%, whereas fruit fly alone causes damage to the extent of 70%–80% in cucurbits. The current pest management still relies heavily on chemical pesticides. An important risk of using pesticides alone leads to the development of resistance to pesticides that can greatly affect production economics and pesticide use patterns. There are also the real and perceived risks to human health and environment. Therefore, an integrated approach involving monitoring of pests; cultural methods, such as planting dates, field rotation, and soil cultivation; resistant cultivars; biological control; botanicals; and judicious use of chemicals can reduce these risks. An effective integrated strategy for pest management is, therefore, affected by economics of production, government regulations, effective management strategies, and education programs that transfer current research results to the stakeholders.

REFERENCES

Abrahamian, P.E. and Y. Abou-Jawdah. 2014. Whitefly-transmitted criniviruses of cucurbits: Current status and future prospects. *Virusdisease*. 25(1): 26–38.

Adler, L.S. and R.V. Hazzard. 2009. Comparison of perimeter trap crop varieties: Effects on herbivory, pollination, and yield in butternut squash. *Environmental Entomology*. 38: 207–215. doi:10.1603/ 022.038.0126.

Aguiar, E.L., G.A. Carvalh, E.B. Menezes, and C.A. Machado. 1993. Efficacy of the acaricide/insecticide diafenthiuron in the control of the two-spotted spider mite *Tetranychus urticae* (Koch) on roses. *Anais da Sociedade Entomologica do Brasil*. 22: 577–582.

Ahmad, M.S., T. Mukhtar, and R. Ahmad. 2004. Some studies on the control of Citrus nematode (*Tylenchulus semipenetrans*) by leaf extracts of three plants and their effects on plant growth variables. *Asian Journal of Plant Sciences*. 3: 544–548.

Akhtaruzzaman, M., M.Z. Alam, and M.M. Ali-Sardar. 1999. Suppressing fruit fly infestation by bagging cucumber at different days after anthesis. *Bangladesh Journal of Entomology*. 9: 103–112.

Akhtaruzzaman, M., M.Z. Alam, and M.M. Ali-Sardar. 2000. Efficiency of different bait sprays for suppressing fruit fly on cucumber. *Bulletin of the Institute of Tropical Agriculture, Kyushu University*. 23: 15–26.

Al-Antary, T.M., M.R. Al Lala, and M.I. Abdel-Wali. 2012. Response of seven populations of the two-spotted spider mite (*Tetranychus urticae* Koch) for bifenazate acaricide on cucumber (*Cucumis sativus* L.) under plastic houses in Jordan. *Advances in Environmental Biology*. 6(7): 2203–2207.

Ali, H., S. Ahmad, G. Hassan, A. Amin, and M. Naeem. 2011. Efficacy of different botanicals against red pumpkin beetle (*Aulacophora foveicollis*) in bitter gourd (*Momordica charantia* L.). *Journal of Weed Science Research*. 17(1): 65–71.

Alston, D.G. and D.R. Worwood. 2008. Western striped cucumber beetle western spotted cucumber beetle. http://extension.usu.edu/files/publications/factsheet/cucumber-beetle08.pdf.

Alzoubi, S. and S. Cobanoglu. 2010. Integrated control possibilities for two-spotted spider mite *Tetranychus urticae* Koch. (Acarina: Tetranychidae) on greenhouse cucumber. *International Journal of Acarology*. 36(3): 259–266.

Anonymous. 2012. UC IPM Pest Management Guidelines 2012: Cucurbits. UC ANR Publication 3445. http://www.ipm.ucdavis.edu/PMG/r116300511.html.

Anonymous. 2014. Mites—An overview. http://ausveg.com.au/intranet/technical-insights/cropprotection/mites.html.

Anonymous. 2015. Nematode: Management. TNAU Agritech Portal, Crop Protection. http://agritech.tnau.ac.in/crop_protection/crop_prot_nematode_management.html.

Bach, C.E. 1980. Effects of plant-density and diversity on the population-dynamics of a specialist herbivore, the striped cucumber beetle *Acalymma vittata*. *Ecology*. 61: 1515–1530. http://www.jstor.org/stable/1939058.

Barbercheck, M.E. 2014. Biology and management of aphids in organic cucurbit production systems. www.extension.org/pages/60000/biology-and-management-of-aphids-in-organic-cucurbit-production-systems.

Basu, A.N. 1995. *Bemisia Tabaci (Gennadius) Crop Pest and Principal Whitefly Vector of Plant Viruses*. Boulder, CO: Westview Press.

Bessin, R. 2010. Cucumber beetles. ENTFACT-311. College of Agriculture Food and Environment, University of Kentucky, Lexington, KY (September 12, 2013).

Bonjour, E.L. and W.S. Fargo. 1989. Host effects on the survival and development of *Anasa tristis* (Heteroptera: Coreidae). *Environmental Entomology*. 18: 1083–1085. http://www.ingentaconnect.com/content/esa/envent/1989/00000018/00000006/art00028.

Bonjour, E.L., W.S. Fargo, A.A. Al-Obaidi, and M.E. Payton. 1993. Host effects on reproduction and adult longevity of squash bugs (Heteroptera: Coreidae). *Environmental Entomology*. 22: 1344–1348.

Borah, R.K. 1998. Evaluation of an insecticide schedule for the control of red pumpkin beetle and melon fruit fly on red pumpkin in the hills zone of Assam. *Indian Journal of Entomology*. 60: 417–419.

Boucher, J. 2014. Squash beetle: *Epilachna borealis*. U. Conn. Extension. http://ipm.uconn.edu/documents/raw2/672/Squash%20beetle%20article.pdf.

Boucher, T.J. and R. Durgy. 2004. Moving towards ecologically based pest management: A case study using perimeter trap cropping. *Journal of Extension* 42 (6)/Feature Articles/A2/ www.joc.org/joc/2004 December/a2.php.

Britton, W.E. 1919. Insects attacking squash, cucumber, and allied plants in Connecticut. *Connecticut Agricultural Experiment Station Bulletin*. 216: 33–51.

Brown, H. 2003. Common insect pests of cucurbits. Agnote 805. No. 159. Primaryindustries, Australia. www.
    nt.gov.au/d/Primary_Industry/Content/File/horticulture/805.pdf.
Bruton, B.D., F. Mitchell, J. Fletcher, S.D. Pair, A. Wayadande, U. Melcher, J. Brady, B. Bextine, and T.W.
    Popham. 2003. *Serratia marcescens*, a phloem-colonizing, squash bug-transmitted bacterium: Causal
    agent of cucurbit yellow vine disease. *Plant Disease*. 87: 937–944. doi:10.1094/PDIS.2003.87.8.937.
Caldwell, J.S. and P. Clarke. January–February 1998. Aluminum-coated plastic for repulsion of cucumber
    beetles. *Commercial Horticulture Newsletter*. Virginia Cooperative Extension, Virginia Tech. www.ext.
    vt.edu/news/periodicals/commhort/1998-02/1998-02-01.html.
Canhilal, R. and G.R. Carner. 2006. Efficacy of entomopathogenic nematodes (Rhabditida: Steinernematidae
    and Heterorhabditidae) against the squash vine borer, *Melittia cucurbitae* (Lepidoptera: Sesiidae)
    in South Carolina. *Journal of Agricultural and Urban Entomology*. 23: 27–39. http://scentsoc.org/
    Volumes/JAUE/v23/27.pdf.
Canhilal, R. and G.R. Carner. 2007. *Bacillus thuringiensis* as a pest management tool for control of the squash
    vine borer, *Melittia cucurbitae* (Lepidoptera: Sesiidae) in South Carolina. *Journal of Plant Diseases
    and Protection*. 114: 26–29.
Canhilal, R.G., G.R. Carner, and R.P. Griffin 2006. Life history of the squash vine borer, *Melittia cucurbitae*
    (Harris) (Lepidoptera: Sesiidae) in South Carolina. *Journal of Agricultural and Urban Entomology*
    23: 1–16.
Capinera, J.L. 1994. Pickleworm and melonworm. In: *Pest Management in the Subtropics: Biological
    Control—A Florida Perspective*. Rosen, D., F.D. Bennett, and J.L. Capinera (eds.). Andover, U.K.:
    Intercept, pp. 140–145.
Capinera, J.L. 2005. *Melon Worm. Diaphania hyalinata. Featured Creatures*. Division of Plant Industry,
    Department of Entomology and Nematology Florida Cooperative Extension Service, Institute of Food
    and Agricultural Sciences, University of Florida, Gainesville, FL. Publication # EENY-163. http://
    edis.ifas.ufl.edu/in320.
Capinera, J.L. 2008. Spotted cucumber beetle or southern corn rootworm, *Diabrotica undecimpunc-
    tata* Mannerheim (Coleoptera: Chrysomelidae). *Encyclopedia of Entomology*. Capinera, J.L. (ed.)
    2nd Edition. Springer, the Netherlands. pp. 3519–3522.
Cavanagh, A., R. Hazzard, L.S. Adler, and J. Boucher. 2009. Using trap crops for control of *Acalymma vit-
    tatum* (Coleoptera: Chrysomelidae) reduces insecticide use in butternut squash. *Journal of Economic
    Entomology*. 102: 1101–1107.
Choo, H.Y., A.M. Koppenhofer, and H.K. Kaya. 1996. Combination of two entomopathogenic nematode spe-
    cies for suppression of an insect pest. *Journal of Economic Entology*. 89: 97–103.
Cline, G.R., J.D. Sedlacek, S.L. Hillman, S.K. Parker, and A.F. Silvernail. 2008. Organic management of
    cucumber beetles in watermelon and muskmelon production. *HortTechnology*. 18: 436–444.
Cranshaw, E., M. Bartolo, and F. Schweissing. 2001. Control of squash bug (Hemiptera: Coreidae) injury:
    Management manipulations at the base of pumpkin. *Southwestern Entomologist*. 26: 147–150.
Cranshaw, W. 1998. *Pests of the West*. Revised: *Prevention and Control for Today's Garden and Small Farm*.
    Golden, CO: Fulcrum Publishing.
Cranshaw, W.S. 2014. Greenhouse whitefly. http://www.ext.colostate.edu/pubs/insect/05587.html.
Davis, R.F. 2007. Effect of *Meloidogyne incognita* on watermelon yield. *Nematropica*. 37(2): 287–293.
Day, E. 2008. Cucumber beetles. Department of Entomology, Virginia Cooperative Extension, Virginia State
    University. https://pubs.ext.vt.edu/2808/2808-1009/2808-1009_pdf.pdf.
Decker, K.B. and K.V. Yeargan. 2008. Seasonal phenology and natural enemies of the squash bug (Hemiptera:
    Coreidae) in Kentucky. *Environmental Entomology*. 37: 670–678.
Deheer, C.J. and D.W. Tallamy. 1991. Affinity of spotted cucumber beetle (Coleoptera: Chrysomelidae) larvae
    to cucurbitacins. *Environmental Entomology*. 20: 1173–1175.
Delahut, K. 2005. Squash vine borer. University of Wisconsin Garden Facts X1024. http://hort.uwex.edu/
    articles/squash-vine-borer; Grupp, S.M., The bug review: Squash vine borer. University of Illinois
    Extension. http://www.urbanext.uiuc.edu/bugreview/squashvineborer.html
Dhillon, M.K., R. Singh, J.S. Naresh, and H.C. Sharma. 2005. The melon fruit fly, *Bactrocera cucurbitae*:
    A review of its biology and management. *Journal of Insect Science*. 5: 40.
Dimetry, N.Z., A.Y. El-Laithy, A.M.E. AbdEl-Salam, and A.E. El-Saiedy. 2013. Management of the major
    piercing sucking pests infesting cucumber under plastic house conditions. *Archives of Phytopathology
    and Plant Protection*. 46(2): 158–171.
Diver, S. 2008 (Updated by Tammy Hinman). Cucumber beetles: Organic and biorational integrated pest
    management. ATTRA Publication #IP212. http://attra.ncat.org/attra-pub/PDF/cucumberbeetle.pdf.

Doharey, K.L. 1983. Bionomics of red pumpkin beetle, *Aulacophora foveicollis* (Lucas) on some fruits. *Indian Journal of Entomology*. 45: 406–413.

Dorman, M. and S. Nelson. 2012. Root-knot nematodes on cucurbits in Hawaii. *Plant Disease*. UH–CTAHR, PD-84. Published by College of Tropical Agriculture and Human Resources, University of Hawaii at Manoa. http://www.ctahr.hawaii.edu/oc/freepubs/pdf/pd-84.pdf.

Ellers-Kirk, C.D., S.J. Fleischer, R.H. Snyder, and J.P. Lynch. 2000. Potential of entomopathogenic nematodes for biological control of *Acalymma vittatum* (Coleoptera: Chrysomelidae) in cucumbers grown in conventional and organic soil management systems. *Journal of Economic Entomology*. 93(3): 605–612.

Elsey, K.D. 1985. Resistance mechanisms in *Cucurbita moschata* to pickleworm and melonworm. *Journal of Economic Entomology*. 78: 1048–1051.

Faria, M. and S.P. Wraight. 2001. Biological control of *Bemisia tabaci* with fungi. *Crop Protection*. 20: 767–778.

Fassuliotis, G. 1967. Species of Cucumis resistant to the root-knot nematode, *Meloidogyne incognita* Acrita. *Plant Disease Reporter*. 51: 720–723.

Frank, D.L. and O.E. Liburd. 2005. Effects of living and synthetic mulch on the population dynamics of whiteflies and aphids, their associated natural enemies and insect-transmitted plant diseases in zucchini. *Environmental Entomology*. 34: 857–865.

Gerling, D., O. Alomar, and J. Arno. 2001. Biological control of *Bemisia tabaci* using predators and parasitoids. *Crop Protection*. 20: 779–799.

Giannakou, I.O., A. Sidiropoulos, and D.P. Athanasiadou. 2002. Chemical alternatives to methyl bromide for the control of root-knot nematodes in greenhouses. *Pest Management Science*. 58: 290–296.

Godfrey, L.D. 2011. Pest notes: Spider mites. UC ANR Publication 7405. UC Statewide Integrated Pest Management Program, University of California, Davis, CA. http://www.ipm.ucdavis.edu/PMG/PESTNOTES/pn7405.html.

Gould, F. and A. Massey. 1984. Cucurbitacins and predation of the spotted cucumber beetle, *Diabrotica undecimpunctata* howardi. *Entomologia Experimentalis et Applicata*. 36: 273–278.

Grupp, S.M. 2015. The bug review: Squash vine borer. Cooperative Extension, University of Illinois, Urbana, IL. www.urbanext.uiuc.edu/bugreview/squashvineborer.html.

Gupta, J.N. and Verma, A.N. 1982. Effectiveness of fenitrothion bait sprays against melon fruit fly, *Dacus cucurbitae* Coquillett in bitter gourd. *Indian Journal of Agricultural Research*. 16: 41–46.

Gupta, S.K. 1985. *Handbook: Plant Mites of India*. Calcutta, India: Zoological Survey of India, pp. 89, 429–430, 520.

Halaj, J. and D.H. Wise. 2002. Impact of a detrital subsidy on trophic cascades in a terrestrial grazing food web. *Ecology*. 83: 3141–3151. http://www.jstor.org/stable/3071849.

Hall, R.A. 1982. Control of whitefly, *Trialeurodes vaporariorum* and cotton aphid, *Aphis gossypii* in glasshouses by two isolates of the fungus, *Verticillium lecanii*. *Annals of Applied Biology*. 101(1): 1–11.

Hasan, M.K., M.M. Uddin, and M.M. Haque. 2011. Efficacy of malathion for controlling red pumpkin beetle, *Aulacophora foveicollis* in cucurbitaceous vegetables. *Progress Agriculture*. 22(1 & 2): 11–18.

Hazzard, R. and A. Cavanagh. 2010. *Managing Striped Cucumber Beetle in Vine Crops*. The Centre for Agriculture, Food and the Environment, the College of Natural Sciences, University of Massachusetts Amherst, MA. https://extension.umass.edu/vegetable/articles/managing-striped-cucumber-beetle-vine-crops.

Hinds, J. and C.R.R. Hooks. 2013. Population dynamics of arthropods in sunn-hemp zucchini interplanting system. *Crop Protection*. 53: 6–12.

Hoffmann, M.P., R. Ayyappath, and J. Gardner. 2002. Effect of striped cucumber beetle (Coleoptera: Chrysomelidae) foliar feeding on winter squash injury and yield. *Journal of Entomological Science*. 37: 236–243.

Hoffmann, M.P., R. Ayyappath, and J. Gardner. 2003. Effect of striped cucumber beetle on pumpkin yield. *Journal of Entomological Science*. 38: 439–448.

Horowitz, A.R. 1986. Population dynamics of *Bemisia tabaci* (Gennadius): With special emphasis on cotton fields. *Agriculture, Ecosystems and Environment*. 17: 37–47.

Horowitz, A.R., H. Podoler, and D. Gerling. 1984. Life table analysis of the tobacco whitefly *Bemisia tabaci* (Gennadius) in cotton fields in Israel. *Acta Ecological/Ecologia Applicata*. 5: 221–233.

Howe, W.L. and A.M. Rhodes. 1973. Host relationships of the squash vine borer, *Melittia cucurbitae* with species of Cucurbita. *Annals of the Entomological Society of America*. 66: 266–269. doi:10.1603/029.102.0331, http://entnemdept.ufl.edu/creatures/veg/leaf/silverleaf_whitefly.htm, http://www.ingentaconnect.com/content/esa/envent/1993/00000022/00000006/art00017, http://www.ingentaconnect.com/content/esa/jee/1996/00000089/00000001/art00016.

Hussain, M.A., T. Mukhtar, and M.Z. Kayani. 2011a. Assessment of the damage caused by *Meloidogyne incognita* on okra (*Abelmoschus esculentus*). *Journal of Animal and Plant Sciences*. 2: 857–861.

Hussain, M.A., T. Mukhtar, and M.Z. Kayani. 2011b. Efficacy evaluation of *Azadirachta indica*, *Calotropis procera*, *Datura stramonium* and *Tagetes erecta* against root-knot nematodes *Meloidogyne incognita*. *Pakistan Journal of Botany*. 43: 197–204.

Islam, K., M.S. Islam, and Z. Ferdousi. 2011. Control of *Epilachna vigintioctopunctata* Fab. (Coleoptera: Coccinellidae) using some indigenous plant extracts. *Journal of Life and Earth Science*. 6: 75–80.

Jackson, D.M., R. Canhilal, and G.R. Carner. 2005. Trap monitoring squash vine borers in cucurbits. *Journal of Agricultural and Urban Entomology*. 22: 27–39. http://scentsoc.org/Volumes/JAUE/v22/27.pdf.

Jaiswal, J.P., T.B. Gurung, and R.R. Pandey. 1997. Findings of melon fruit fly control survey and its integrated management 1996/97, Kashi, Nepal. Lumle Agriculture Research Centre Working Paper No. 97(53), pp. 1–12.

Jelinek, S. 2010. Whitefly management in greenhouse vegetable crops. Primefact 1007, pp. 1–6. http://www.dpi.nsw.gov.au/_data/assets/pdf_file/0009/339588/Whitefly-management-in-greenhouse-vegetable-crops.pdf.

Kamal, M.M., M.M. Uddin, M. Shajahan, M.M. Rahman, M.J. Alam, M.S. Islam, M.Y. Rafii, and M.A. Latif. 2014. Incidence and host preference of red pumpkin beetle, *Aulacophora foveicollis* (Lucas) on Cucurbitaceous vegetables. *Life Science Journal*. 11(7): 459–466.

Kayani, M.Z., T. Mukhtar, and M.A. Hussain. 2012. Evaluation of nematicidal effects of *Cannabis sativa* L. and *Zanthoxylum alatum* Roxb. against root-knot nematodes, *Meloidogyne incognita*. *Crop Protection*. 39: 52–56.

Keszey, J. 2012. Organic control measures for striped cucumber beetles. http://www.highmowingseeds.com/blog/organic-control-measures-for-striped-cucumber-beetles.html.

Khan, M.M.H. 2012. Host preference of pumpkin beetle to cucurbits under field conditions. *Journal of the Asiatic Society of Bangladesh, Science*. 38(1): 75–82.

Khan, M.M.H., M.Z. Alam, M.M. Rahman, M.I. Miah, and M.M. Hossain. 2012. Influence of weather factors on the incidence and distribution of red pumpkin beetle infesting cucurbits. *Bangladesh Journal of Agricultural Research*. 37(2): 361–367.

Khan, S.M. and M. Wasim. 2001. Assessment of different plant extracts for their repellency against red pumpkin beetle (*Aulacophora foveicollis* Lucas.) on muskmelon (*Cucumis melo* L.) crop. *Journal of Biological Sciences*. 1(4): 198–200.

Khelu, A.Z., V.G. Zaets, and A.A. Shesteperov. 1989. Tests on resistance to root-knot nematodes in tomato, cucumber and pepper varieties grown under cover in central Lebanon. *Byulleten' Vsesoyuznogo Instituta Gel'mintologii im. K. I. Skryabina*. 50: 85–89.

Klungness, L.M., E.B. Jang, R.F.L. Mau, R.I. Vargas, J.S. Sugano, and E. Fujitani. 2005. New approaches to sanitation in a cropping system susceptible to tephritid fruit flies (Diptera: Tephritidae) in Hawaii. *Journal of Applied Science and Environmental Management*. 9: 5–15.

Koch, R.L. 2003. The multicolored Asian lady beetle, *Harmonia axyridis*: A review of its biology, uses in biological control, and non-target impacts. *Journal of Insect Science*. 3: 16–32.

Kos, K., Z. Tomanovic, O. Petrovic-Obradovic, Z. Laznik, M. Vidrih, and S. Trdan. 2008. Aphids (Aphididae) and their parasitoids in selected vegetable ecosystems in Slovenia. *Acta Agriculturae Slovenica*. 91(1): 15–22.

Laska, P., C. Perez-Banon, L. Mazanek, S. Rojo, G. Stahls, M.A. Marcos-Garcia, V. Bicik, and J. Dusek. 2006. Taxonomy of the genera *Scaeva*, *Simosyrphus* and *Ischiodon* (Diptera: Syrphidae): Descriptions of immature stages and status of taxa. *European Journal of Entomology*. 103: 637–655.

Legg, J., D. Gerling, and P. Neuenschwander. 2003. Biological control of whiteflies in sub-Saharan Africa. In: *Biological Control in IPM System in Africa*. Neuenschwander, P. C. Borgemeister and J. Langewald, (eds.) International Institute of Tropical Agriculture, Benin Station, Cotonou, Benin.

Letourneau, D.K. 1986. Associational resistance in squash monocultures and polycultures in tropical Mexico. *Environmental Entomology*. 15: 285–292.

Liu, Y.C. and J.S. Lin. 1993. The response of melon fly, *Dacus cucurbitae* Coq. to the attraction of 10% MC. *Plant Protection Bulletin*. 35: 79–88.

Lopez-Gomez, S. and M. Verdejo-Lucas. 2014 Penetration and reproduction of root-knot nematodes on cucurbit species. *European Journal of Plant Pathology*. 138: 863–871.

Luna, J.M. and L. Xue. 2009. Aggregation behavior of Western spotted cucumber beetle (Coleoptera: Chrysomelidae) in vegetable cropping systems. *Environmental Entomology*. 38: 809–814. doi:10.1603/022.038.0334.

Mahmood, T., K.M. Khokhar, S.I. Hussain, and M.H. Laghari. 2005. Host preference of red pumpkin beetle, *Aulacophora* (*Raphidopalpa*) *foveicollis* (Lucas) among cucurbit crops. *Sarhad Journal of Agriculture*. 21(3): 473–475.

Mahmood, T., M.S. Tariq, K.M. Khokar, Hidayatullah, and S.I. Hussain. 2010. Comparative effect of different plant extracts and insecticide application as dust to control the attack of red pumpkin beetle on cucumber. *Pakistan Journal of Agricultural Research*. 23(3–4): 196–199.

Markkula, M., K. Tiitanen, M. Hamalainen, and A. Forsberg. 1979. The aphid midge *Aphidoletes aphidimyza* (Diptera: Cecidomyiidae) and its use in biological control of aphids. *Annales Entomologici Fennici*. 45(4): 89–98.

McAuslane, H.J. and H.A. Smith. 2000. Sweetpotato Whitefly B Biotype, Bemisia tabaci (Gennadius) (Insecta: Hemiptera: Aleyrodidae). EENY-129, University of Florida/IFAS Extension. University of Florida, Gainesville, FL. https://edis.ifas.ufl.edu/in286.

McKinlay, R.G. 1992. *Vegetable Crop Pests*. Boston, MA: CRC Press, pp. 98–101.

McSorley, R. and V.H. Waddill. 1982. Partitioning yield loss on yellow squash into nematode and insect components. *Journal of Nematology*. 14: 110–118.

Medina-Gaud, S., E. Abreu, F. Gallardo, and R.A. Franqui. 1989. Natural enemies of the melonworm *Diaphania hyalinata* L. (Lepidoptera: Pyralidae) in Puerto Rico. *Journal of Agriculture of the University of Puerto Rico*. 73: 313–320.

Messenger, B. and A. Braun. 2000. Alternatives to Methyl Bromide for the Control of SoilBorne Diseases and Pests in California. Pest Management Analysis and Planning Program. http://cdpr.ca.gov/docs/emon/methbrom/alt-anal/sept2000.pdf

Mitchell, R.F. and L.M. Hanks. 2009. Insect frass as a pathway for transmission of bacterial wilt of cucurbits. *Environmental Entomology*. 38(2): 395–403.

Mohaned, M.A., M. Mohamed, H.Z. Elabdeen, and S.A. Ali. 2013. Host Preference of the Melon Worm, *Diaphania hyalinata* L. (Lepidoptera: Pyralidae), on Cucurbits in Gezira State, Sudan. *Persian Gulf Crop Protection*. 2(3): 55–63.

Mondal, S., and S.S. Ghatak. 2009. Bioefficacy of some indigenous plant extracts against epilachna beetle (*Henosepilachna vigintioctopunctata* Fabr.) infesting cucumber. *Journal of Plant Protection Sciences*. 1(1): 71–75.

Mukhtar, T., M.Z. Kayani, and M.A. Hussain. 2013. Response of selected cucumber cultivars to *Meloidogyne incognita*. *Crop Protection*. 44: 13–17.

Mumford, J.D. 2004. Economic analysis of area-wide fruit fly management. In: *Proceedings of the Sixth International Symposium on Fruit Flies of Economic Importance*, May 6–10, 2002. Barnes, B. and M. Addison (eds.). Stellenbosch, South Africa: Infruitec Press.

Murphy, G., G. Ferguson, and L. Shipp. 2014. Mite pests in greenhouse crops: Description, biology and management. www.omafra.gov.on.ca/english/crops/facts/14-013.html.

Nair, S. and J. Thomas. 1999. Effect of *Acorus calamus* L. extracts on the longevity of *Bactrocera cucurbitae* Coq. *Insect Environment*. 5: 27.

Natwick, E.T., J.J. Stapleton, and C.S. Stoddard. 2012. Cucurbits. UC ANR Publication 3445. http://www.ipm.ucdavis.edu/PMG/r116400111.html.

Naved, S. 2012. Management of greenhouse mites. In: *Proceedings of the 26th Training Diseases and Management of Crops under Protected Cultivation*. Dubey, K.S. and R.P. Singh (eds.). Pantnagar, India. pp. 65–71, Chapter 7. www.gbpuat.ac.in/26%20CAFT%20Proceeding.pdf.

Noling, J.W. 2009. Nematode management in cucurbits (cucumbers, melons, squash). http://edis.ifas.ufl.edu/ng025.

Norton, J.D. and D.M. Granberry. 1980. Characteristics of progeny from an interspecific cross of *Cucumis melo* with *Cucumis metuliferus*. *Journal of the American Society for Horticultural Science*. 105: 174–180.

Nyoike, T.W. 2007. Evaluation of living and synthetic mulches with and without a reduced-risk insecticide for suppression of whiteflies and aphids, and insect transmitted viral diseases in zucchini squash. A thesis submitted to the Graduate school in partial fulfillment for MS in integrated pest management. University of Florida, Gainesville, FL, p. 90.

Oke, O.A. 2008. Effectiveness of two insecticides to control melon fruit fly (*Bactrocera Cucurbitae* Coq.) in cucumber (*Cucumis sativu*s L.). crop at Anse Boileau Seychelles. *European Journal of Scientific Research*. 22: 84–86.

Olkowski, W. 2000. *Mass Trapping Western Spotted Cucumber Beetles*. OFRF Information Bulletin No. 8 (Summer). Santa Cruz, CA: Organic Farming Research Foundation, pp. 17–22.

Olson, D.L., J.R. Nechols, and B.W. Schurle. 1996. Comparative evaluation of population effect and economic potential of biological suppression tactics versus chemical control for squash bug (Heteroptera: Coreidae) management on pumpkins. *Journal of Economic Entomology*. 89: 631–639.

Oostenbrink, M. 1966. Major characteristics of the relation between nematodes and plants. *Mededeelingen van de Landbouwhogeschool Wageningen*. 66(4): 46.

Ostermann, H. and M. Dreyer. 1995. The Neem tree *Azadirachta indica*, A. Juss. and other meliaceous plants sources of unique natural products for integrated pest management, industry and other purposes. In: *Vegetables and Grain Legumes*. Schmutterer, H. in collaboration with K.R.S. Ascher, M.B. Isman, M. Jacobson, C.M. Ketkar, W. Kraus, H. Rembolt, and R.C. Saxena (eds.). VCH Weinheim, Germany. pp. 392–403. ISBN: 3-527-30054-6.

Pair, S.D., B.D. Bruton, F. Mitchell, J. Fletcher, A. Wayadande, and U. Melcher. 2004. Overwintering squash bugs harbor and transmit the causal agent of cucurbit yellow vine disease. *Journal of Economic Entomology*. 97: 74–78. doi:10.1603/0022-0493-97.1.74.

Palumbo, J.C., W.S. Fargo, and E.L. Bonjour. 1991. Colonization and seasonal abundance of squash bugs (Heteroptera, Coreidae) on summer squash with varied planting dates in Oklahoma. *Journal of Economic Entomology*. 84: 224–229. http://www.ingentaconnect.com/content/esa/jee/1991/00000084/00000001/art00039.

Palumbo, J.C., W.S. Fargo, R.C. Berberet, E.L. Bonjour, and G.W. Cuperus. 1993. Timing insecticide applications for squash bug management: Impact on squash bug abundance and summer squash yields. *Southwest Entomology*. 18: 101–111.

Park, Y.L. and J.H. Lee. 2005. Impact of twospotted spider mite (Acari: Tetranychidae) on growth and productivity of glasshouse cucumbers. *Journal of Economic Entomology*. 98(2): 457–463.

Parsad, K. and P. Kumar. 2002. Insect pest status on summer vegetables in hill tracts. *Karnataka Journal of Agricultural Sciences*. 15(1): 156–157.

Pena, J.E., V.H. Waddill, and K.D. Elsey. 1987. Survey of native parasites of the pickleworm, *Diaphania nitidalis* Stoll, and melonworm, *Diaphania hyalinata* (L.) (Lepidoptera: Pyralidae), in southern and central Florida. *Environmental Entomology*. 16: 1062–1066.

Petzoldt, C. 2008. Cucurbits: Insects and weeds (Chapter 18, Part 2). In: *Integrated Crop and Pest Management Guidelines for Commercial Vegetable Production*. Cornell University.

Pickett, C.H., S.E. Schoenig, and M.P. Hoffmann. 1996. Establishment of the squash bug parasitoid, *Trichopoda pennipes* Fabr. (Diptera: Tachinidae), in northern California. *Pan-Pacific Entomologist*. 72: 220–226.

Rahaman, M.A. and M.D.H. Prodhan. 2007. Effects of net barrier and synthetic pesticides on red pumpkin beetle and yield of cucumber. *International Journal of Sustainable Crop Production*. 2(3): 30–34.

Rahaman, M.A., M.D.H. Prodhan, and A.K.M. Maula. 2008. Effect of botanical and synthetic pesticides in controlling Epilachna beetle and the yield of bitter gourd. *International Journal of Sustainable Crop Production*. 3(5): 23–26.

Rahoo, A.M., T. Mukhtar, S.R. Gowen, and B. Pembroke. 2011. Virulence of entomopathogenic bacteria *Xenorhabdus bovienii* and *Photorhabdus luminescens* against *Galleria mellonella* larvae. *Pakistan Journal of Zoology*. 43: 543–548.

Raina, A.K., J.A. Klun, M. Schwarz, A. Day, B.A. Leonhardt, and L.W. Douglass. 1986. Female sex pheromone of the melonworm, *Diaphania hyalinata* (Lepidoptera: Pyralidae), and analysis of male responses to pheromone in a flight tunnel. *Journal of Chemical Ecology*. 12: 229–237.

Ranganath, H.R., M.A. Suryanarayana, and K. Veenakumari. 1997. Management of melon fly *Bactrocera cucurbitae* (Zeugodacus) in cucurbits in South Andaman. *Insect Environment*. 3: 32–33.

Reddy, A.V. 1997. Evaluation of certain new insecticides against cucurbit fruit fly (*Dacus cucurbitae* Coq.) on bitter gourd. *Annals of Agricultural Research*. 18: 252–254.

Rondon, S.I., D.J. Cantliffe, and J.F. Price. 2003. *Anasa tristis* (Heteroptera: Coreidae) development, survival and egg distribution on beit alpha cucumber and as prey for *Coleomegilla maculata* (Coleoptera: Coccinellidae) and *Geocoris punctipes* (Heteroptera: Lygaeidae). *Florida Entomologist*. 86: 488–490.

Roomi, M.W., T. Abbas, A.H. Shah, S. Robina, A.A. Qureshi, S.S. Hussain, and K.A. Nasir. 1993. Control of fruit flies (*Dacus* spp.) by attractants of plant origin. *Anzeiger fur Schadlingskunde, Aflanzenschutz, Umwdtschutz*. 66: 155–157.

Sapkota, R., K.C. Dahal, and R.B. Thapa. 2010. Damage assessment and management of cucurbit fruit flies in spring-summer squash Use of pheromone traps. *Journal of Entomology and Nematology*. 2(1): 7–12.

Schwartz, H.F. and D.H. Gent. 2007. Nematodes (Cucumber, Melon, Pumpkin, Squash, and Zucchini). *Cucurbits*. High Plains IPM Guide, A Cooperative Effort of the University of Wyoming, University of Nebraska, Colorado State University and Montana State University. p. 22.

Seewooruthun, S.I., P. Sookar, S. Permalloo, A. Joomaye, M. Alleck, B. Gungah, and A.R. Soonnoo. 1998. An attempt to the eradication of the Oriental fruit fly, *Bactrocera dorsalis* (Hendel) from Mauritius. In: *Proceedings of the Second Annual Meeting of Agricultural Scientists*, August 12–13, 1997. Lalouette, J.A., D.Y. Bachraz, N. Sukurdeep, and B.D. Seebaluck (eds.). Reduit, Mauritius: Food and Research Council, pp. 181–187.

Shannag, H.K. and J.L. Capinera. 1995. Evaluation of entomopathogenic nematode species for the control of melonworm (Lepidoptera: Pyralidae). *Environmental Entomology*. 24: 143–148.

Sharma, P.K. and P.C. Trivedi. 1985. Nematotoxicity of some organophosphates and carbamates against *Meloidogyne incognita* and their effect on plant growth and rhizobial nodulation in pea. *International Nematology Network Newsletter*. 2: 10–12.

Shooker, P., F. Khayrattee, and S. Permalloo. 2006. Use of maize as a trap crops for the control of melon fly, *B. cucurbitae* (Diptera: Tephritidae) with GF-120 Bio-control and other control methods. http://www.fcla.edu/FlaEnt/fe87p354.pdf.

Siguenza, C., M. Schochow, T. Turini, and A. Ploeg. 2005. Use of *Cucumis metuliferus* as a rootstock for melon to manage *Meloidogyne incognita*. *Journal of Nematology*. 37(3): 276–280.

Singh, S.V., A. Mishra, R.S. Bisan, Y.P. Malik, and A. Mishra. 2000. Host preference of red pumpkin beetle, *Aulacophora foveicollis* and melon fruit fly, *Dacus cucurbitae*. *Indian Journal of Entomology*. 62: 242–246.

Sinha, P. 1997. Effects of culture filtrates of fungi on mortality of larvae of *Dacus cucurbitae*. *Journal of Environmental Biology*. 18: 245–248.

Sinha, P. and S.K. Saxena. 1998. Effects of culture filtrates of three fungi in different combinations on the development of *Dacus cucurbitae in vitro*. *Indian Phytopathology*. 51: 361–362.

Sinha, P. and S.K. Saxena. 1999. Effect of culture filtrates of three fungi in different combinations on the development of the fruit fly, *Dacus cucurbitae* Coq. *Annals of Plant Protection Service*. 7: 96–9.

Smith, R.I. 1911. Two important cantaloupe pests. *North Carolina Agricultural Experiment Station Bulletin*. 214: 101–146.

Smyth, R.R. and M.P. Hoffmann. 2010. Seasonal incidence of two co-occurring adult parasitoids of *Acalymma vittatum* in New York State: *Centistes diabroticae* (Syrrhizus) and *Celatoria setosa*. *Biocontrol*. 55: 219–228. doi:10.1007/s10526-009-9232-y.

Snyder, W.E. 2012. *Managing Cucumber Beetles in Organic Farming Systems*. Cornell University Cooperative Extension, Ithaca, New York. http://www.extension.org/pages/64274/managing-cucumber-beetles-in-organic-farming-systems#.VMs4jNKUdEA.

Snyder, W.E. and D.H. Wise. 2001. Contrasting trophic cascades generated by a community of generalist predators. *Ecology*. 82: 1571–1583. http://www.jstor.org/stable/2679801.

Sorensen, K.A. 1999. Cucumber beetles, Coleoptera: Chrysomelidae. Greenshare Fact Sheets. University of Rhode Island Landscape Horticulture Program (September 12, 2013). http://www.uri.edu/ce/factsheets/sheets/cucbeetles.html.

Soule, J. 1993. Tagetes minuta: A potential new herb from South America. In: *New Crops*. Janick, J. and J.E. Simon (eds.). New York: Wiley Library, pp. 649–654. http://www.hort.purdue.edu/newcrop/proceedings1993/v2-649.html.

Stapleton, J.J. and C.G. Summers. 2002. Reflective mulches for management of aphids and aphid-borne virus diseases in late-season cantaloupe (*Cucumis melo* L. var. *cantalupensis*). *Crop Protection*. 21: 891–898.

Summers, C.G. and J.J. Stapleton. 2001. Use of UV reflective mulch to delay the colonization and reduce the severity of *Bemisia argentifolii* (Homoptera: Aleyrodidae) infestations in cucurbits. *Crop Protection*. 21: 921–928.

Swaminathan, R., S. Manjoo, and T. Hussain. 2010. Anti-feedant activity of some biopesticides on *Henosepilachna vigintioctopunctata* (Fab.) (Coleoptera: Coccinellidae). *Journal of Biopesticides*. 3(1): 77–80.

Tallamy, D.W., D.P. Whittington, F. Defurio, D.A. Fontaine, P.M Gorski, and P. Gothro. 1998. The effect of sequestered cucurbitacins on the pathogenicity of *Metarhizium anisopliae* (Moniliales: Moniliaceae) on spotted cucumber beetle eggs and larvae (Coleoptera: Chrysomelidae). *Environmental Entomology*. 27: 366–372.

Tehri, K. and R. Gulati. 2014. Field efficacy of some biorationals against the two spotted spider mite *Tetranychus urticae* Koch (Acari: Tetranychidae). *Journal of Applied and Natural Science*. 6(1): 62–67.

Thapa, R.B. and F.P. Neupane. 1992. Incidence, host preference and control of the red pumpkin beetle, *Aulacophora foveicollis* (Lucas) (Coleoptera: Chrysomelidae) on Cucurbits. *Journal of the Institute of Agriculture and Animal Science*. 13: 71–77.

Thies, J.A., J.J. Ariss, R.L. Hassell, S. Olson, C.S. Kousik, and A. Levi. 2010. Grafting for management of southern root-knot nematode, *Meloidogyne incognita*, in watermelon. *Plant Disease.* 94(10): 1195–1199.

Toba, H.H., A.N. Kishaba, G.W. Bohn, and H. Hield. 1977. Protecting muskmelon against aphid-borne viruses. *Phytopathology.* 67: 1418–1423.

Tomanovic, Z. and M. Brajkovic. 2001. Aphid parasitoids (Hymenoptera: Aphidiidae) of agroecosystems of the south part of the Pannonian area. *Archives of Biological Sciences.* 53: 57–64.

Vagelas, I. and S.R. Gowen. 2012. Control of *Fusarium oxysporum* and root-knot nematodes (*Meloidogyne* spp.) with *Pseudomonas oryzihabitans. Pakistan Journal of Phytopathology.* 24: 32–38.

Van, L.T., J. Vontas, A. Tsagkarakou, W. Dermauwa, and L. Tirry. 2010. Acaricide resistance mechanisms in the two-spotted spider mite *Tetranychus urticae* and other important Acari: A review. *Insect Biochemistry and Molecular Biology.* 40: 563–572.

Vishwakarma, R., P. Chand, and S.S. Ghatak. 2011. Potential plant extracts and entomopathogenic Fungi against Red pumpkin beetle, *Raphidopalpa foveicollis* (Lucas). *Annals of Plant Protection Science.* 19(1): 84–87.

Wang, K.H. and R. McSorely. 2004. *Management of Nematodes with Cowpea Cover Crops.* UF IFAS Extension, ENY-712. Electronic. http://edis.ifas.ufl.edu/pdffiles/IN/IN51600.pdf.

Wang, K.-H., C.R.R. Hooks, and A. Ploeg. 2007. *Protecting Crops from Nematode Pests: Using Marigold as an Alternative to Chemical Nematicides.* Cooperative Extension Service, College of Tropical Agriculture and Human Resources, University of Hawai'i at Mänoa, Honolulu, HI, PD-35. http://www.ctahr.hawaii.edu/oc/freepubs/pdf/PD-35.pdf.

Webb, S. and S. Linda. 1992. Evaluation of spunbonded polyethylene row covers as a method of excluding insects and viruses affecting fall-grown squash in Florida. *Journal of Economic Entomology.* 85: 2344–2352.

Webb, S.E., F. Akad, T.W. Nyoike, O.E. Liburd, and J.E. Polston. 2014. *Whitefly-Transmitted Cucurbit Leaf Crumple Virus in Florida.* IFAS Extension. http://edis.ifas.ufl.edu/pdffiles/IN/IN71600.pdf.

Weems, H.V., J.B. Heppner and T.R. Fasulo. 2001. Melon fly, *Bactrocera cucurbitae* Coquillett (Insecta: Diptera: Tephritidae). Florida Department of Agriculture and Consumer Services, Division of Plant Industry, and T.R. Fasulo, University of Florida. University of Florida Publication EENY-199. https://edis.ifas.ufl.edu/in356.

Welty, C. 2009. Squash vine borer. Agriculture and Natural Resources Fact Sheet HYG-2153-09. The Ohio State University Extension. http://ohioline.osu.edu/hyg-fact/2000/pdf/2153.pdf.

Westerdahl, B.B. and J.O. Becker. 2011. UC IPM Pest Management Guidelines: Cucurbits. http://www.ipm.ucdavis.edu/PMG/r116200111.html.

Whitaker, J.O. 1995. Food of the big brown bat *Eptesicus fuscus* from maternity colonies in Indiana and Illinois. *American Midland Naturalist* 134: 346–360. http://www.jstor.org/stable/2426304.

Williams, J.L. and D.H. Wise. 2003. Avoidance of wolf spiders (Araneae: Lycosidae) by striped cucumber beetles (Coleoptera: Chrysomelidae): Laboratory and field studies. *Environmental Entomology.* 32: 633–640. http://dx.doi.org/10.1603/0046-225X-32.3.633.

Wong, T.T.Y., R.T. Cunningham, D.O. Mcinnis, and J.E. Gilmore. 1989. Seasonal distribution and abundance of *Dacus cucurbitae* (Diptera: Tephritidae) in Rota, Commonwealth of the Mariana Islands. *Environmental Entomology.* 18: 1079–1082.

Wood, M. 2001. Forcing exotic, invasive insects into retreat: New IPM program targets Hawaii's fruit flies. *Agricultural Research Washington.* 49: 11–13.

Worthley, H.N. 1923. The squash bug in Massachusetts. *Journal of Economic Entomology.* 16: 73–79. http://www.ingentaconnect.com/content/esa/jee/1923/00000016/00000001/art00010.

Wright, D.J. 1981. Nematicides: Mode of action and new approaches to chemical control. In: *Plant-Parasitic Nematodes.* Zuckerman, B.M. and R.A. Rohde (eds.). New York: Academic Press, pp. 421–449.

Xalxo, P.C., D. Karkun, and A.N. Poddar. 2013. Rhizospheric fungal associations of root knot nematode infested cucurbits: In vitro assessment of their nematicidal potential. *Research Journal of Microbiology.* 8: 81–91.

Yamaguchi, M. 1983. World Vegetables: Principles, Production and Nutritive Values, Springer, Netherlands, p. 415.

Yao, C.B., G. Zehnder, E. Bauske, and J. Klopper. 1996. Relationship between cucumber beetle (Coleoptera: Chrysomelidae) density and incidence of bacterial wilt of cucurbits. *Journal of Economic Entomology.* 89: 510–514.

Yardim, E.N., N.Q. Arancon, C.A. Edwards, T.J. Oliver, and R.J. Byrne. 2006. Suppression of tomato hornworm (*Manduca quinquemaculata*) and cucumber beetles (*Acalymma vittatum* and *Diabotrica undecimpunctata*) populations and damage by vermicomposts. *Pedobiologia.* 50: 23–29. http://dx.doi.org/10.1016/j.pedobi.2005.09.001.

Zehnder, G. 2011. Biology and management of pickleworm and melonworm in organic curcurbit production systems. Electronic. http://www.extension.org/pages/60954/biology-and-management-of-pickleworm-and-melonworm-in-organic-curcurbit-production-systems#.VNmbv3uH6yc.

Zehnder, G., J. Kloepper, C.B. Yao, and G. Wei. 1997. Induction of systemic resistance in cucumber against cucumber beetles (Coleoptera: Chrysomelidae) by plant growth-promoting rhizobacteria. *Journal of Economic Entomology.* 90: 391–396. http://www.ingentaconnect.com/content/esa/jee/1997/00000090/00000002/art00022.

Zhang, L., J. Zhang, P. Christie, and X. Li. 2008. Pre-inoculation with arbuscular mycorrhizal fungi suppresses root knot nematode (*Meloidogyne incognita*) on cucumber (*Cucumis sativus*). *Biology and Fertility of Soils.* 45: 205–211.

Zhang, X.W., X.L. Qian, and J.W. Liu. 1989. Evaluation of the resistance to root-knot nematode of watermelon germplasm and its control. *Journal of Fruit Science.* 6: 33–38.

20 Insect Pest Management in Cucurbits

Research Development and Perspective

Ruparao T. Gahukar

CONTENTS

20.1 INTRODUCTION

Cucurbits are regularly attacked by an array of insect pests in all crop growing areas. Considering the current economic importance, fruit flies, leaf miners, squash vine borers, leaf beetles, and squash bugs are comparatively important. For cultivating pest-free crops, synthetic pesticides have been used from the seedling to harvest stage. However, some of the insects could be controlled by adopting good agricultural practices and efficient monitoring of pest populations and plant damages. Also, trapping with baits, planting of pest-resistant/pest-tolerant crop varieties, and spraying of botanicals have been widely accepted by farmers. In this review, integrated pest management (IPM) is discussed while considering economic and ecological aspects along with crop productivity.

Further studies on insect pheromones, bioefficacy of new pesticide formulations, and regulatory measures are needed for validating and refining current IPM strategies.

Cucurbit crops (family: Cucurbitaceae) include several species of melons (watermelon, bitter melon, musk melon, snap melon), gourds (sponge gourd, ridge/ribbed gourd, wax gourd, bottle gourd, bitter gourd, ash gourd/sweet gourd, snake gourd), cantaloupe, cucumber, gherkin, pumpkin, squash, and chayote. Cucurbits form a part of important vegetables worldwide although the culinary uses differ from one geographic region to another and within regions and countries. Among pests that cause a substantial yield loss and deterioration in quality of harvestable fruits, insects are the most destructive crop enemies (Coveillo et al., 2005; Dhillon et al., 2005a; Webb, 2013). Common pests include cucumber beetles, fruit flies, epilachna beetles, pumpkin beetles, aphids, thrips, and squash bugs. Others are specific, such as squash vine borer that attacks pumpkins and squash but rarely watermelon, cucumber, or cantaloupe (Brust, 2009). Studies on pest ecology and biology in different agrosystems demonstrated that the economic importance of each insect species differs as per crop and variety, cultivation practices, market price, and climatic conditions (Capinera, 2001; Rai, 2008). Some pests are cosmopolitan in distribution and polyphagous and attack several field crops other than cucurbits. Therefore, cucurbit pests are inadvertently controlled to some extent when other crops are protected from these pests. Generally, the growers resort to chemical pesticides, but excessive sprays lead to secondary pest outbreaks and the development of resistance. Of course, new insecticides have been introduced in cucurbits to minimize toxic residues and the risk to environment and consumers. But they act against a narrow range of pest species rather than the old broad-spectrum molecules. Therefore, monitoring of pest populations and establishment of economic threshold levels (ETLs) can play an important role while deciding the control strategy. Most of the recommendations referred in this review are taken from extension/information bulletins published by agricultural universities. The current economic status and geographic distribution of common pests are shown in Table 20.1. In this review, I have discussed the current control practices of major pests of worldwide importance with an objective to implement the IPM approach in order to reduce treatment cost and disturbance to environment.

20.2   INSECT PESTS AND THEIR MANAGEMENT

20.2.1   COLEOPTERA

20.2.1.1   Cucumber Beetles (Chrysomelidae)

Two species of cucumber beetles are major pests in the United States. The striped cucumber beetle (SCB) *Acalymma vittatum* (Fb.) feeds on a majority of cucurbits (e.g., squash, cucumber, cantaloupe, pumpkin), mostly on newly emerged cotyledons, tender shoots, seedlings, and occasionally on stems near or below the soil surface. Larvae can develop only on cucurbits, whereas adult beetles feed on other plants until cucurbit plants appear. Larvae feed exclusively on cucurbit roots by chewing holes and tunneling into the roots in the soil and underground stems. Feeding injury to roots results in stunted growth of aged plants and killing of seedlings or young plants. Adults overwinter in residue from the previous year's cucurbit crop or nearby and later move into a new crop and feed on young plants, whereas second generation adults damage leaves and flowers. They also scar the fruit by damaging the rind and flesh, thereby reducing the marketability and storage life of harvested fruits. Overwintering beetles transmit bacterial wilt (*Erwinia tracheiphila* Smith) (Capinera, 2001).

The spotted cucumber beetle *Diabrotica undecimpunctata* Barber, being polyphagous, feeds on a variety of plants including cucurbits and is less severe than SCB. The larvae are commonly known as rootworms as they feed on roots of corn, peanuts, legumes, and small grains and pupate in the soil. Adult beetles generally appear 2–3 weeks later than SCB and feed on cucurbit leaves and sometimes on soft fruits. Beetles can spread squash mosaic virus and can increase the incidence of powdery mildew, black rot, and Fusarium wilt (Capinera, 2001).

TABLE 20.1

Insect Species, Host Plants, Geographic Distribution, and Economic Importance of Common Cucurbit Pests

Pest Species	Common Name	Order Family	Geographical Distribution	Host Plants Attacked (Cucurbits)
Acalymma (Diabrotica) vittatum (Fb.)	Striped cucumber beetle	Coleoptera Chrysomelidae	USA	All cucurbits[a]
Anasa tristis (De Geer)	Squash bug	Heteroptera Coreidae	USA	Cucumber, squash,[a] pumpkin[a]
Aphis gossypii (Glover)	Melon aphid	Homoptera Aphididae	COS	Cantaloupe, squash, watermelon,[a] cucumber[a]
Apomecyna spp.	Vine borer	Coleoptera Cerambycidae	AS	Pointed gourd, snake gourd, bottle gourd, pumpkin
Aulacophora cinta (Fb.)	Red pumpkin beetle	Coleoptera Chrysomelidae	AS, AF	All cucurbits
Aulacophora foveicollis (Lucas)	Red pumpkin beetle	Coleoptera Chrysomelidae	AS, AF, OC, EU	All cucurbits, cucumber,[a] water melon,[a] round gourd,[a] musk melon,[a] bottle gourd,[a] sweet gourd[a]
Aulacophora hilaris (Boisd.)	Pumpkin beetle	Coleoptera Chrysomelidae	AS, OC	Cucumber
Aulacophora intermedia (Jacoby)	Pumpkin beetle	Coleoptera Chrysomelidae	AS, AF	All cucurbits
Batrocera cucurbitae (Coquillett)	Melon fruit fly	Diptera Tephritidae	COS	All gourds,[a] luffa, squash, pumpkin, snap melon, cucumber,[a] water melon, musk melon
Batrocera ciliatus (Loew)	Fruit fly	Diptera Tephritidae	AS, AF	Cucumber, melon,[a] gourds, chayote
Batrocera dorsalis (Hendel)	Fruit fly	Diptera Tephritidae	USA, SEA	Cucumber,[a] all gourds[a]
Batrocera tau (Walker)	Fruit fly	Diptera Tephritidae	SEA	Cucumber,[a] all gourds[a]
Bemisia tabaci (Genn.)	Whitefly	Homoptera Aleyrodidae	COS	All cucurbits[a]
Delia platura (Meigen)	Seed corn maggot	Diptera Anthomyiidae	USA	Cucumber
Diabrotica balteata (Le Conte)	Banded cucumber beetle	Coleoptera Chrysomelidae	USA	All cucurbits
Diaphania nitidalis (Stoll)	Pickle worm	Lepidoptera Pyralidae	USA	Cucumber,[a] cantaloupe,[a] squash[a]
Diaphania hyalinata L.	Melon worm	Lepidoptera Pyralidae	LA, ME, USA	Pumpkin,[a] all cucurbits
Diaphania indica (Saunders)	Cucumber moth	Lepidoptera Pyralidae	AS	Cucumber[a]
Epilchna chrysomelina (Fb.)	Epilachna beetle	Coleoptera Coccinellidae	AS, ME, AF, EU	All cucurbits, cucumber[a]
Diabrotica undecimpunctata (Barber)	Spotted cucumber beetle	Coleoptera Chrysomelidae	USA	All cucurbits[a]

(Continued)

TABLE 20.1 (*Continued*)
Insect Species, Host Plants, Geographic Distribution, and Economic Importance of Common Cucurbit Pests

Pest Species	Common Name	Order Family	Geographical Distribution	Host Plants Attacked (Cucurbits)
Helicoverpa armigera (Hbn.)	Corn earworm	Lepidoptera Noctuidae	OC	Cucumber[a]
Henosepilachna vigintioctopunctata (Fb.)	Spotted lady-bird beetle	Coleoptera Coccinellidae	OC, AS	All cucurbits, cucumber,[a] bitter gourd[a]
Liriomyza trifolii (Burgess)	Serpentine leaf miner	Diptera Agromyzidae	COS	Cucumber[a]
Liriomyza sativa (Blanchard)	Leaf miner	Diptera Agromyzidae	USA	Cucumber[a]
Margaronia indica (Saunders)	Pumpkin caterpillar	Lepidoptera Pyralidae	AS	Pumpkin,[a] gourds, melon, cucumber
Melittia cucurbitae (Harris)	Squash vine borer	Lepidoptera Sesiidae	USA, AS	Musk melon, squashes,[a] pumpkin, gourds, cucumber
Myopardalis pardalina (Bigot)	Cucumber fly	Diptera Tephritidae	AS	Cucumber,[a] musk melon, water melon
Myzus persicae (Sulzer)	Green peach aphid	Homoptera Aphididae	USA, AS	Cucumber[a]
Plusia peponis L.	Snake gourd semilooper	Lepidoptera Noctuidae	AS	Snake gourd,[a] cucumber
Podagrica uniforma (Jacoby)	Flea beetle	Coleoptera Chrysomelidae	AF, EU	Cucurbits
Sphenarches caffer (Zeller)	Snake gourd plume moth	Lepidoptera Pteromalidae	AS	Snake gourd,[a] cucumber
Thrips tabaci (Lindeman)	Onion thrip	Thysanoptera Thripidae	AS, OC	Water melon[a]
Thrips palmi (Karny)	Onion thrip	Thysanoptera Thripidae	OC	Water melon[a]

Sources: Capinera, J.L., *Handbook of Vegetable Pests*, Academic Press, San Diego, CA, 2001, 729pp.; Rai, N., *Handbook of Vegetable Science*, NIPA Publications, New Delhi, India, 2008, 765pp.; Cline, G.R. et al., *HortTechnology*, 18, 436, 2008; Brust, G., *Cucurbit Pest Management*, Extension Bulletin, University of Maryland, College Park, MD, 2009, 8pp.; Groves, R.L., *Insect Pest Management in Cucurbit Vegetable Crops*, Department of Entomology, University of Wisconsin, Madison, WI, 2011; University of Delaware, *IPM—Cucurbit Scouting Guidelines*, College of Agriculture and Natural Resources, Newark, DE, 2012; Webb, S.E., *Insect Management for Cucurbits*, Florida Cooperative Extension Service, University of Florida, Gainesville, FL, Doc. No. ENY-460, 2013; Anonymous, *AgriTech Portal*, Tamil Nadu Agricultural University, Coimbatore, Tamil Nadu, India, 2014.

Geographical distribution: COS, Cosmopolitan; AF, Africa; AS, Asia; LA, Latin America; OC, Oceania; USA, United States of America.

[a] Major pest on a specific crop.

20.2.1.1.1 Cultural and Mechanical Methods

As far as possible, rotation or at least planting cucurbits far away (at least >700 m) from last year's crop or other host plants (corn, beans, small grains, grasses, etc.) can minimize the size of the beetle population (Brust, 2009). Field sanitation including destruction of weeds, weedy edges, and crop residue, especially roots and fruits after harvest, and pulling out and burning the remaining vines after destroying beetles helps to reduce the population of overwintering beetles. Also, residue shredding is effective to accelerate the decomposition of above- and belowground crop leftover. Preplanting deep plowing destroys pest harboring sites particularly weeds and grasses within and around field borders. Limiting irrigation close to crop harvest restricts contact of maturing fruits with soil. Drip irrigation may, therefore, be advised to avoid water spread. Otherwise, drip irrigation can be combined with mulching to reduce soil moisture under fruits and to avoid pest damage.

Soil under organic mulch harbors a fewer number of beetles, and mulches deter them from laying eggs in the soil near plant stems (Yardim et al., 2006). Metallic-colored or aluminum-coated reflexive plastic mulches repel beetles, thereby reducing their number and disease transmission. Larval movement from roots is also hindered, resulting in less insect feeding. For this purpose, straw mulch is advantageous because this mulch directly slows beetle movement from one plant to another. It provides refuge to wolf spiders and other predators from hot and dry conditions, and thus helps their conservation in the field (Snyder and Wise, 2001). The straw serves as food for spring tails and other insects that are the prey of spiders, ultimately increasing the spider number (Halaj and Wise, 2002). Because the straw mulch should not contain herbicide residue or weed seeds, weed control by soil incorporation of pre-emergent herbicides or manual weeding is important.

Transplanting rather than direct sowing helps to avoid crop exposure to beetles during the susceptible plant stages, especially seedlings. Transplanting also reduces the total time that cucurbit plants are in the field each season, providing less time for the beetles to build their population and for disease symptoms to develop. Thus, delayed planting or transplanting after beetles have laid their eggs can reduce crop injury (Groves, 2011). This practice may, however, be difficult for crops with short growing seasons. Also, delayed planting would eliminate early harvest of cucumber and summer squash for marketing. Intercropping watermelon or musk melon with radish, nasturtium, tansy, buckwheat, cowpea, or sweet clover reduces pest population densities (Cline et al., 2008).

When trap crops are planted as border strips or adjacent plots about 2 weeks before planting the main crop, insecticide spraying on trap crops kills beetles that aggregate at field edges. Even preferred or susceptible varieties of the main crop can be planted as a trap crop (Pair, 1997; Groves, 2011); for example, zucchini (cvs. President, Black Jack, Green Eclipse, Senator, Super Select, Dark Green, Buttercup), squash (cvs. Cocozelle, Blue Hubbard, Lemondrop, Caserta), buttercup squash (cv. Amber Cup), melon (cv. Classic), and pumpkin (cvs. Big Max, Baby Poo) (Bellows and Diver, 2002; Adler and Hazzard, 2009). Planting squash as a trap crop reduced insecticide use by up to 94% compared with conventional methods because overwintering beetles were attracted to the trap crop before the main crop was attractive (Cavanagh et al., 2009).

Handpicking to remove the beetles is time-consuming and laborious but effective. Otherwise, beetles can be sucked up and collected with a large-scale vacuum pump such as D-vac suction sample or reverse leaf blower. This practice is valid particularly for areas with trap crops where a limited area needs to be treated. However, the practicality, including manpower requirement and economics, of this mechanical method have not been studied.

Covering plants with cloth, plastic, or spun-bonded polyester is a practical and less costly means to prevent beetles from landing on plants (Groves, 2011), but the covers can block access to the crop for weeding and other operations. On the contrary, floating row covers are quite effective. They should be removed before flowering, otherwise crop pollination by bees may be affected (Brust, 2009).

20.2.1.1.2 Traps

Yellow sticky cards or cups are useful tools for population monitoring since beetles are attracted to yellow color (NYS, 2014), and beetle populations can be reduced if the number of traps is increased in the area. Traps containing insect pheromones and/or kairomones can be combined with spraying of synthetic or botanical pesticides. Traps containing a volatile lure and poison bait and provided with a cup to collect the dead beetles are available in the market. Beetles enticed to feed on baits are killed by insecticides. The addition of spinosad (a commercial product containing the bacterium

Saccharopolyspora spinosa) or carbaryl to the bait, although exerted lethal effects on beetles, did not reduce beetle population and fruit damage (Pedersen and Godfrey, 2011). These results need further experimentation for confirmation.

20.2.1.1.3   Pest-Resistant Cultivars

Plant susceptibility is associated with the release of cucurbitacin (a feeding stimulant) and several floral volatiles. Plants containing low cucurbitacin being less attractive to beetles, crop varieties with low preference/tolerance should be planted, for example, summer squash (cvs. Slender Gold, Sunbar, Goldbar, Seneca Prolific, Peter Pan), winter squash (cvs. Carnival, Table Ace, Zenith), pumpkin (cvs. Baby Pam, Jackpot, Munchkin, Tom Fox, Seneca Harvest Moon) (Bellows and Diver, 2002).

20.2.1.1.4   Natural Enemies

Predators such as ground beetles, long-horned soldier beetles (particularly *Chauliognathus penn-sylvanicus* De Geer), and wolf spiders sometimes feed heavily on beetles (Snyder and Wise, 2001). Two tachinid flies, *Celatoria setusa* (Coquillett) and *Celatoria diabroticae* (Shimer), and a braconid wasp, *Centistes (Syrrhizus) diabroticae* (Gahan), parasitize larvae and often have significant impact on SCB populations (Smyth and Hoffman, 2010). The other entomopathogenic nematode, *Steinernema riobrave* Cabanillas (Poinar & Raulston), reduced larval population by up to 50% (Ellers-Kirk et al., 2000). Likewise, soil drenching of commercial products containing microorganisms has been recommended by Ellers-Kirk et al. (2000). The entomopathogens living in the soil can also provide effective control of larvae, but data on pest mortality are not available.

20.2.1.1.5   Chemical Pesticides

In the spring, beetles migrate into cucurbits for mating and egg laying. During that period, before feeding starts, pesticide applications are significantly effective. In the late spring and early summer, the crop should be treated when eggs hatch and before larvae move to the roots for feeding. During mid and late summer, application of pesticide can prevent feeding damage by adults and larvae to leaves, stems, flowers, and fruits. In any case, sprays must penetrate the whole crop canopy so that spray droplets get deposited on the entire plant (particularly on the top and underside of leaves). When larvae are active, drenching the soil with an insecticide solution can kill them effectively and can help to prevent damage during the seedling stage. Utmost care is, however, needed while treating as resistance development in the larvae and adults is possible and should be strictly monitored. For this, the best way is to rotate the chemical class of insecticide or mode of its action. Also, as far as possible, botanicals and microbial pesticides may be preferred between chemical applications. To control *A. vittatum*, a precision band application to inject a solid stream of imidacloprid solution in furrows directly over the seed during planting (Jasinski et al., 2009) or post-transplant soil drenching with imidacloprid or thiamethoxam (Brust, 2009) provided 3–4 weeks of control. Enhanced pest mortality was observed in shorter band lengths combined with high volume sprays of insecticide solution. This system reduced leaf feeding by 70%–100% and saved US $215/ha (Jasinski et al., 2009).

20.2.1.1.6   Integrated Control

Early treatment is essential for SCB management in commercial fields (Brust, 2009) particularly when beetles are active. In watermelon, protection is necessary only when plants are small and pest population is high. On the contrary, seedlings of cantaloupe and cucumber need protection only when the plants reach the five-leaf stage or beyond (Brust, 2009). Since chemicals should be avoided to protect bees, application of reduced-risk pesticides (azadirachtin, pyrethrin, kaolin clay, suspension of *Beauveria bassiana* Vuill.) may be preferred (Groves, 2011), and foliar spraying with pyrethroids may be necessary only when ETL has surpassed (Brust, 2009).

20.2.1.2 Pumpkin Beetles (Chrysomelidae)

Red pumpkin beetles *Aulacophora* (*Raphidopalpa*) *foveicollis* Lucas in Asia and *Aulacophora hilaris* (Boisd.) in Oceania are considered as the most important pests. Adult beetles feed voraciously on leaves, flowers, and fruits. They make holes in the plant tissues causing death or growth retardation. In the case of heavy infestation, replanting is required. Larvae live in the soil and feed on the roots and underground stem of the plant.

20.2.1.2.1 Cultural and Mechanical Methods

Keeping the fields clean especially by burning old creepers can prevent pest entry into fields. Plowing and harrowing of field just after harvesting kill larvae and hibernating adults. When the crop is planted early, the plant passes the cotyledon stage by the time the beetles become active. In early stages of infestation, collection and destruction of beetles in the morning hours when adults remain sluggish is effective on small farms (Anonymous, 2014). Chaudhary (1995) used polythene cages/bags containing a mixture of soil + sand + farm yard manure in cucumber for collecting beetles. Few scattered plants grown early in the season can be treated with insecticide so that beetles attracted to these plants get killed. Recently, the olfactory bioassay revealed that from 13 fatty acids isolated from *Momordica cochinchinensis* Spreng, palmitic acid at 5.42 µg as bait is effective (Mukherjee et al., 2014). This crop can probably be used to attract and trap the beetles.

20.2.1.2.2 Pest-Resistant Cultivars

Among the 12 genotypes of cucumber in India, cvs. Nepal local and Sikkim cucumber received the lowest infestation at the fruiting stage (1.5%–5.0%) compared to 15.0% infestation on susceptible cv. Khira Paprola (Khurseed et al., 2013c). On the basis of plant damage, Rai et al. (2004) evaluated 68 genotypes and reported 8 entries as resistant (PCVC nos. 7, 36, 47, 66, 99, 102, 108, and 110). In another attempt, among 27 genotypes of bitter gourd, the lowest damage 15 days after sowing was noticed on cv. VRBG-50 (Shivalingaswamy et al., 2008). In Pakistan, pest-resistant sources have been identified for sponge gourd (cvs. RKS-6, RKS-7), bitter gourd (cvs. Jaunpuri, Jhalri), and bottle gourd (cvs. DIK round green, sweet yellow, bottle gourd long) (Saljoqi and Khan, 2007). Further testing in different regions would only reveal more resistant sources and mechanisms of plant resistance.

20.2.1.2.3 Plant-Derived Products

In India, the pest was effectively controlled with traditional products as antifeedants, that is, cow dung (1:5, w/v) and tobacco extract (100 g leaves in 10 L water) as repellents, and capsicum pepper powder (50%) (1:10, w/v), black soil powder (25%) (1:10, w/v), fly ash powder (250 g in 15 L water), and a mixture of tobacco leaves (100 g) + turmeric powder (250 g) in 15 L water (Nath and Ray, 2012). Likewise, neem cake (NC) incorporation into soil can kill larvae (Anonymous, 2014).

Among eight commercial neem products, Gronim® (0.5%) caused the highest beetle mortality of 49.89% followed by Neemazal-F® (0.1%) compared to traditional neem leaf extract (NLE, 10%) with 20.2% mortality (Rathod et al., 2009). In another comparison, water extract (10%) of *Parthenium hysterophorus* L. leaves was less effective than neem seed kernel extract (NSKE, 5%) and leaf extract (5%) of *Eucalyptus* sp. The best treatments caused significant reduction in beetle population (1.41 beetles/m² versus 3.89 beetles/m² in control) and plant damage (31.11% versus 41.64% in control) (Ali et al., 2011). Methanolic extract (10%) of *Momordica charantia* L. acts as a strong antifeedant due to the presence of momordicin I and II in the fruit (Abe and Matsuda, 2000). Neem leaf extract in benzene showed a maximum repellency of 60% in musk melon (Khan and Wasim, 2001). Aqueous extract (3%) and ethanolic extract (1%) of *Melia azedarach* L. were more effective than neem-based Econeem® and Nimbecidine®, both sprayed at 4 mL/L in fields of cucumber, ribbed gourd, and bottle gourd (Luna et al., 2008). In the laboratory, neem seed extract (NSE, 5%) in water showed greater repellency (60%–80%) than water extract of *Annona squamosa* L. or

M. azedarach seeds (Tandon and Sirohi, 2009). Antifeeding effect of *Coleus amboinicus* Lour., *Mentha piperata* L., and *Pogostemon heyneaus* Benth has been reported by Chandel et al. (2009). In Pakistan, NSE or NLE (5%), water extract (5%) of tobacco leaves or permethrin (0.5%), mixed with cow dung ash (0.05%) controlled the pest in a cucumber field (Mahmood et al., 2010). When three sprays of the vegetable-based pesticide oxymetrine 0.5EC @ 750 mL/ha were assayed in the field, it proved significantly better than the neem product containing 1500 ppm of azadirachtin (AZ) @ 1.5 L/ha or endosulfan 35EC @ 1 L/ha, for reducing the beetle population (80.8% versus 0% in control) and fruit damage (16.8% versus 51.8% in control) and for increasing the yield of watermelon (151 kg/ha versus 44 kg/ha in control) (Kumar et al., 2013).

20.2.1.2.4 Chemical Pesticides

Weekly spraying of 0.2% carbaryl 50WP, 0.05% endosulfan 35EC, 0.05% malathion 50EC, or 0.05% methyl parathion 50EC @ 500 L/ha or dusting with 5% malathion or 4% endosulfan @10 kg/ha during the seedling stage is the current recommendation in India (Anonymous, 2014). In a 2-year field study on cucumber, Verma (2012) reported minimum pest population (1.64 beetles/plant versus 5.6 beetles/plant in control) with soil application of carbofuran @ 500 g a.i./ha at the time of sowing followed by seed treatment with thiamethoxam 70WS @ 3 g/kg seed and dusting with rice husk ash @ 30 kg/ha. This treatment resulted in the highest yield (106 q/ha versus 69.7 q/ha in control) with a cost–benefit ratio of 1:1.75. In a comparison of chemicals with botanicals, 0.004% lambda-cyhalothrin 5EC was at par with plant infestation reduction of 74% followed by 0.2% carbaryl 50WP (71% reduction) and 0.0045% AZ 1EC (59% reduction) (Khurseed and Raj, 2013b). In the Middle East, Mahmood et al. (2006) recommended spraying of carbaryl (0.2%) or lambda-cyhalothrin (0.004%) or dusting of carbaryl (10%). On the contrary, deltamethrin (0.0025%) in India (Rajak and Singh, 2002) and carbaryl dust (2%) in Pakistan (Said and Muhammad, 2000) were the best treatments. Among nine chemical treatments evaluated by Rathod et al. (2009), carbaryl (0.2%) caused the highest larval mortality of 63.4% followed by fenvalerate (0.01%) with 59.9% mortality.

Spraying on the soil around the root zone with carbaryl (0.2%), lindane (0.1%), methyl parathion (0.02%), or malathion (0.05%) or dusting carbaryl (5% dust) repelled adults and killed developing larvae and pupae (Anonymous, 2014). Conclusively, carbamates, being relatively cheap and effective, can be recommended for cucurbits in all crop growing areas.

20.2.1.2.5 Integrated Control

In IPM of pumpkin beetles, various methods were combined and evaluated in field trials (Mahmood et al., 2006). A treatment consisting of mosquito net barriers and soil application of carbofuran was the best in a cucumber field (Rahaman and Prodhan, 2007). In the glasshouse, sweet gourd seedlings were covered with mosquito net and sprayed with NSKE (5%) at weekly intervals. This treatment was more effective than carbofuran granules (1 kg a.i./ha) applied into the soil 3 days before planting @ 5 g/plant or spraying NO (10 mL/L) after reaching economic threshold (Khorsheduzzaman et al., 2010). In bottle gourd crop, spraying of *B. bassiana* wettable powder (3 g/L) resulted in 72.2% mortality, followed by aqueous extract of *Strychnos nux-vomica* L. (4 mL/L) with 65.4% mortality, and *Metarhizium anisopliae* (Metch.) Sorokin wettable powder (3 g/L) spraying resulted in 64.7% mortality (Vishwakarma et al., 2011b). In Bangladesh, significant reduction in beetle population and leaf infestation were achieved by a dose of 2 g/L of thiodicarb 75WP (larvin®), 1.5 mL/L of diazinon 60EC, or 3 mL/L of neem product 7.5EC (Osman et al., 2013). In another field trial, methomyl 40SP (1 mL/L) and water extract (5%) of *P. hysterophorus* L. leaves showed a significantly low pest number (1.28 beetles/plant versus 3.89 beetles in control) and low plant infestation (27.8% versus 41.65% in control) (Ali et al., 2011). From a trial with three synthetic pesticides, one microbial pesticide, and one neem product, Lakshmi et al. (2005) recommended spraying of 0.2% carbaryl 50WP as this treatment reduced the pest population by up to 46.5% followed by 0.054% monocrotophos 36EC with 39.9% reduction. Even carbaryl (2% dust) gave significant control up to 7 days (Khan and Jehangir, 2000).

Repetitive applications are, however, needed for satisfactory pest control. Conclusively, carbaryl dusting or spraying as soon as a pest attack is noticed may be an ideal remedy.

20.2.1.3 Epilachna Beetles (Coccinellidae)

Among beetles found on curcurbits, the spotted ladybird beetle or epilachna beetle *Henosepilachna vigintioctopunctata* Fb. is a polyphagous insect attacking cucurbits and other vegetables, and currently, it is considered a major pest in the Indian subcontinent. The other two species, *Epilachna chysomelina* Fb. and *Podagrica uniforma* (Jacoby), are occasional pests in the Middle East. Adult beetles and larvae feed on leaves and snip a circular trench that arcs from one leaf edge to another and then feed on tissues isolated by the trench. Since leaf tissues of lower surface are damaged leaving more or less intact, the leaf gives a characteristic lacelike skeletonized appearance on the upper surface. Consequently, plant growth is affected.

20.2.1.3.1 Cultural and Mechanical Methods

Amaranthus (*Amaranthus cruentus* L.) planted as an intercrop in cucumber lowered the population of *Epilchna chrysomelina* and *P. uniforma* by up to 50%–75% (Pitan and Esan, 2014). If the crop area is small, regular picking of egg masses, larvae, and beetles can be practiced. Also, shaking plants early in the morning to dislodge larvae and adults in to a container filled with water and kerosene is effective.

20.2.1.3.2 Plant-Derived Products

In the laboratory, NSKE (5%) was effective against second and fourth instar larvae of *E. chysomelina* (Abdul-Moniem et al., 2004). Similarly, an extract of *A. squamosa*, *Croton tiglium* L., or *S. nux-vomica* in ethyl acetate, crude methanol, or petroleum ether at 0.1%–0.5% acted as oviposition deterrent in a dose-dependent manner (Sankaraiyah, 2010). Crude aqueous extract of *Ricinus communis* L., *Calotropis procera* (Ait.) W.T. Aiton or *Datura metel* L. reduced oviposition and egg hatching, prolonged larval duration, and inhibited pupation and adult emergence with LC_{50} values of 18.4%, 23.75%, and 29.6%, respectively (Islam et al., 2011).

In a comparison of botanicals and synthetic pesticides in the cucumber field, Mondal and Chatak (2009) reported 75% reduction in pest population by spraying endosulfan (2 mL/L), whereas methanol extract (6 mL/L) of *A. squamosa* seeds, Neemazal® (6 mL/L), and petroleum ether extract (3 mL/L) of *Acorus calamus* L. rhizomes gave a mortality of 76.4%, 64%, and 57%, respectively. On the contrary, water extract (10%, w/v) of *Tephrosia vogelii* Hook. leaves is a better treatment than diazinon 60EC (2 mL/L) for reducing pest population in bitter gourd (1.67 larvae/m² versus 4.9 larvae/plant in control, 0.53 beetles/m² versus 2.1 beetles/plant in control, 2.67 eggs/plant versus 3.27 eggs/plant in control) with a higher crop yield (9.8 t/ha with *T. vogelii*) than diazinon and control with 9.4 and 8.4 t/ha, respectively (Rahaman et al., 2008).

20.2.1.3.3 Chemical Pesticides

In the cucumber field, 0.2% carbaryl, 0.0045% AZ, and 0.004% lambda-cyhalothrin showed persistence up to 2 weeks with mean reduction of 87%–89% in fruit damage and 83%–85% in pest populations (Khurseed and Raj, 2013a). It appears from field studies that epilachna beetles have developed resistance to malathion and endosulfan. Therefore, Kumar and Kumar (1998) recommended cypermethrin, fenvalerate, or carbaryl for effective pest control. Experimentation on reduced-risk new pesticides is needed to replace or at least minimize the use of broad-spectrum pesticides.

20.2.1.3.4 Integrated Control

In a comparison of petroleum ether extract of seeds of *S. nux-vomica* or *Pachyrhizus erosus* (L.) at 4 mL/L with the powder of entomopathogenic fungi *B. bassiana* or *M. anisopliae* at 3 g/L, maximum reduction in population (74.9%) was achieved with *B. bassiana* and its residual persistence lasted up to 10 days after spraying (Vishwakarma et al., 2011a). It may, therefore, be possible to recommend a combination of biopesticides and plant-derived products against epilachna beetles.

20.2.2 DIPTERA

20.2.2.1 Leaf Miners (Agromyzidae)

The serpentine leaf miner *Liriomyza trifolii* (Burgess) attacks all cucurbits worldwide and is a major pest, whereas an attack of *Liriomyza sativa* (Blanchard) is occasional. Maggots feed between the upper and lower leaf surfaces creating meandering mines that enlarge as the larvae grow. The injury affects leaf development and results in yield reduction.

20.2.2.1.1　Cultural and Mechanical Methods

Vigorous plant growth can be maintained by the destruction of crop residue, and by proper soil fertilization and irrigation frequency. In small fields, clipping of heavily infested leaves can be executed if cheap labor is available at the time of operation. Yellow sticky traps have been experimented on limited areas in India. Large-scale application may be recommended after verifying application norms, field efficacy, and practicality.

20.2.2.1.2　Natural Enemies

A generalist predator, *Cyrtopeltis modestus* (Distant) (Homoptera: Miridae); three larval parasitoids, *Diglyphus intermedius* (Gir.), *Diglyphus begini* (Ashm.), and *Diglyphus isaea* (Walker) (Hymenoptera: Eulophidae); two larval–pupal parasitoids, *Opius dissitus* Muesebeck (Braconidae) and *Chrysocharis parksi* (Craw.) (Eulophidae); and a nematode, *Neoplectana carpocapsiae* (Weiser), have been reported from different agroecosystems. Therefore, field experimentation in local conditions is needed to quantify natural parasitism and measures to augment the number of these natural enemies. Direct application of abamectin showed a negative effect on the survival of *D. isaea* adults. Survival was also affected when abamectin-contaminated leaf miner larvae were consumed by a predator. However, there was no effect on emergence and longevity of adults; abamectin was therefore considered safe and compatible (Kaspi and Parrella, 2006). Similarly, augmentative releases of *D. isaea* combined with sterile male technique showed a synergistic effect in IPM in the United States (Kaspi and Parrella, 2005). In Australia, cyromazine and abamectin provided effective leaf miner control. Cyromazine being safe and compatible with two local major parasitoids, *Hemiptarsenus varicornis* (Girault) and *D. isaea*, its integration has been suggested by Bjorksten and Robinson (2005).

20.2.2.1.3　Chemical Pesticides

In a 2-year field study on cucumber in India, Verma (2012) reported minimum plant damage (14.6% versus 31.7% in control) and lowest pest population (1.64 beetles/plant) with soil application of carbofuran @ 500 g a.i./ha at the time of planting followed by seed treatment with thiamethoxam 70WS at 3 g/kg seed and dusting with rice husk ash @ 30 kg/ha. This treatment resulted in highest yield (106 q/ha versus 69.7 q/ha in control) with a CB ratio of 1:1.75. In chemical pesticides, resistance is unstable and there is no cross-resistance among cyromazine or abamectin and a biopesticide (spinosad) currently used in developed countries (Ferguson, 2004). Therefore, these insecticides can be recommended in cucurbits or else, in order to avoid very toxic chemicals. NSKE (5%) also can be considered as a preventive treatment at the initial period of pest attack (Anonymous, 2014).

20.2.2.2　Fruit Flies (Tephritidae)

Two fruit fly species, *Batrocera cucurbitae* (Coquillett) and *Batrocera dorsalis* (Hendel) (=*Batrocera invadens* Drew, Tsura & White), are cosmopolitan in geographic distribution, and the economic loss they cause made them major pests of most of the cucurbits. From Thailand, two more species, *Batrocera tau* (Walker) and *Batrocera diversa* (Coquillett), have been reported as major pests of luffa and bitter gourd by Chinajaviyawong et al. (2003). Larvae, after hatching, feed on the pulp of the fruit. Resinous fluid oozing from the fruit renders it unfit for human consumption. In a severe attack, the fruit is distorted and malformed and drops prematurely.

20.2.2.2.1   Cultural and Mechanical Methods

Rotation with noncucurbitaceous crops is an effective preventive measure. Deep plowing and turning over the soil after harvest can expose hibernating pupae. Before sowing, digging pits in the soil and dusting them with carbaryl (10%) helps to destroy fly pupae. In pest endemic areas, sowing dates should be changed after studying the seasonal population dynamics of the fly species. Intercropping of cucumber with amaranth (*A. cruentus* L.) lowered populations of fruit flies (*Dacus ciliatus* (Loew), *B. dorsalis* (Hemdel)) by up to 50%–75% when amaranth was established 2 weeks before or on the day of cucumber planting (Pitan and Esan, 2014).

Corn and ribbed gourd act as trap crops (Shooker et al., 2006; Anonymous, 2014) and can be sprayed with 0.15% carbaryl 50WP or 0.1% malathion 50EC to kill congregating flies settled on the underside of leaves (Anonymous, 2014). Daily collection of fallen and infested fruits and burning them in deep pits reduce the fly population. Also, bagging and harvesting of the fruits before they start ripening are practical measures.

20.2.2.2.2   Pest-Resistant Cultivars

In bitter gourd, wild genotypes showed significantly lower fruit infestation and larval densities and were positively correlated (r = 0.96) with the depth of ribs, flesh thickness, fruit diameter and length, and negatively correlated with fruit toughness. Overall, content of moisture, potassium, reducing sugars, fruit diameter, and flesh thickness are major factors for fruit damage, while content of moisture, phosphorus, protein, reducing and total sugars, and fruit length and flesh thickness are governing factors for larval density (Dhillon et al., 2005b). Gogi et al. (2010) also studied various physical characters of varietal resistance in bitter gourd and reported fruit toughness followed by fruit diameter and number of longitudinal ribs as major factors. Minor factors include fruit length, height of small ridges and longitudinal ribs, and pericarp thickness.

In round gourd, greater content of total phenols and silica are negatively correlated with fruit infestation of *B. cucurbitae* (Verma et al., 2013), whereas greater content of total sugars, total free amino acids, and low content of total phenols are positively correlated with varietal susceptibility (Ingoley et al., 2005). Other characteristics of varietal resistance include tenderness of fruit rind, fruit shape and color, and hair density. These studies revealed that flies preferred young, green, and tender fruits with a soft rind (Sapkota et al., 2010). In the screening tests in India, only 2 varieties among 20 cucumber entries were found moderately resistant to the attack of *B. cucurbitae* (Ingoley et al., 2005).

20.2.2.2.3   Lure and Kill Traps

These traps can be manufactured from wood, plastic, or metal. In Thailand, major fruit flies (*B. cucurbitae*, *B. tau*, *B. diversa*) in luffa and bitter gourd were controlled by using cardboard funnel traps with Australian Pinnacle protein bait + trichlorfon (95SP) @ 6 g a.i./ha or Thai yeast bait alone (Chinajariyawong et al., 2003). In lure and kill traps, more females were trapped than males in Thailand (Chinajariyawong et al., 2003). The use of improved wick-type trap and the addition of imidacloprid, acetamiprid, or naled to the bait resulted in male mortality in *B. dorsalis* to the extent of 40%, 64%, and 80%–98%, respectively. Therefore, neonicotinoid insecticides can be an alternative for broad-spectrum insecticides (Chuang and Hou, 2007). The specialized pheromone and lure application technology (SPLAT) with spraying attractant (methyl eugenol, Cue-lure®) and killing product (spinosad) used with dispenser is a promising substitute for liquid synthetic pesticides (Vargas et al., 2008). This is a valid and practical measure to recommend to even marginal farmers. Similarly, for detecting and monitoring purposes, Farmatech® traps containing methyl eugenol and cue-lure wafers with dichlorvos are more convenient to handle than Jackson traps with naled or dichlorvos (Vargas et al., 2009).

20.2.2.2.4 Bait Application Technique (BAT)

Baiting is successful in fruit fly management, particularly if baits are sprayed as coarse droplets on the underside of leaves. Spot spraying is also significantly better than broadcasting. Alternatively, baits can be used for bottle trapping. The best time is when the first flowers appear and later, baiting is repeated every week and after heavy rain.

In Hawaii, a bait with spinosad was sprayed on border rows of melon crop (Prokopy et al., 2003). Among protein-based baits, two commercial products containing autolyzed yeast extract (Provesta-621® and Mazoferm E820®) were found more effective than other products (GF-120® and GF-120 NF/naturalyte®) (Barry et al., 2006; Shooker et al., 2006). Since commercial baits are costly, locally prepared baits with organic and inorganic materials in different proportions have been extensively used by poor growers, for example (1) jaggery (crude sugar) 100 g + carbaryl 2 g + water 1 L (Patel and Mondal, 2011), (2) molasses + malathion + water (1:0.1:100) (Akhtaruzzaman et al., 2000), (3) pulp of over-ripe banana + furadan 10 g + citric acid 1 g (Satpathy and Rai, 2002), (4) pulp of ripe banana + water extract of *Ocimum sanctum* L. leaves (Satpathy and Rai, 2002), (5) citronella oil, eucalyptus oil, vinegar (acetic acid) or lactic acid (Anonymous, 2014), (6) fermented rice 200 g + molasses 5 mL + borax 4 g + malathion 1 mL, (7) pulp of over-ripe banana 500 g + molasses 10 mL + borax 10 g + malathion 2.5 mL (Sapkota et al., 2010). Local baits are effective, cheap, and simple to prepare, and indigenous plant material is easily available. However, frequent applications are needed for satisfactory pest mortality and the cost of treatment increases. Kumar and Agarwal (2005) recorded greater numbers of male flies in Steiner traps lured with a mixture of soybean powder and Cue-lure (containing 4-*p*-acetoxyphenyl-2 butanone) than with fresh fermented palm (*Borassus flabellifer* L.) or its juice.

In sponge gourd in India, 0.03% dichlorvos 76EC mixed with fermented jaggery (500 g in 10 L water) recorded minimum fruit damage of *B. cucurbitae* (6.8% versus 36.7% in control) and highest fruit yield (86.7 q/ha) followed by 0.05% malathion 50EC + jaggery (coarse sugar) (Desai et al., 2014). In summer squash fields in Nepal, fly populations were significantly reduced from 21.8% to 10.6% and from 54.3% to 29.2% with local baits (Sapkota et al., 2010). Likewise, a mixture of fermented palm juice + sugar (1:1) caught a significantly greater number of flies in the Philippines (e.g., 30 flies/trap) than a mixture of coconut + molasses (1.3 flies/trap) or coconut wine alone (7 flies/trap) (Barba and Tablizo, 2014).

20.2.2.2.5 Male Annihilation Technique (MAT)

In this technique, a lure is prepared by soaking plywood blocks for 48 h in a mixture of ethyl alcohol, methyl eugenol, and malathion in a ratio of 6:4:1 (Stonehouse et al., 2007). These blocks are placed in the field to attract flies. Square and oblong blocks were more effective than round and hexagonal blocks (Stonehouse et al., 2002). Alternatively, a lure can be prepared by mixing methyl eugenol and malathion 50EC (1:1). This mixture (10 mL) or wet fish meal (5 g) is kept in polythene bags (20 cm × 15 cm) with six holes (3 mm diameter) and 0.1 mL of dichlorvos is added. Normally, 25 polythene bags or 5 fish meal traps are needed for 1 ha area. Dichlorvos is added every week and fish meal is renewed once in 20 days (Anonymous, 2014). In India, when same bottles were used for MAT + BAT, the reduction in insecticide use by 15 times and application cost by 1.7 times was possible (Patel and Mondal, 2011). In Sudan, the plywood blocks impregnated with methyl eugenol and malathion were found suitable and cheaper by 50% than conventional sponge or cotton wicks (Sidahmed et al., 2014). Generally, MAT kills four times more flies than traps, is cheaper and less vulnerable to weather, and requires no replacement (Stonehouse et al., 2002). Moreover, the BAT and MAT techniques employed together reduced the number of insecticide sprays and thereby the treatment cost by 1.7 times and increased the CB ratio 3 times higher than chemical treatment alone (Patel and Mondal, 2011).

Recently, Vargas et al. (2014) suggested MAT formulation consisting of amorphous polymer matrix in combination with methyl eugenol and spinosad because spinosad has low contact toxicity compared to naled. When mixed with an attractant, the matrix offers a reduced risk to humans and nontarget organisms and assures slow release of chemical. Earlier, Vargas et al. (2012) compared a solid mallet containing three lures (trimedlure, methyl eugenol, raspberry ketone formate [RTF]) with dichlorvos and recommended RTF as it is safe, convenient to handle, and can be used in place of several individual lures and trap systems. Similarly, from both laboratory and field trials in Hawaii, these workers recommended the SPLAT with spinosad cue-lure (5%) because of its effectiveness (pest mortality, longer residual persistence) and low toxicity to human and nontarget organisms compared to RTF and SPLAT melo-cure® (Vargas et al., 2010). In India, a vial filled with methyl eugenol nanogel was found effective for 30 weeks under ambient field conditions (Herlekar, 2014). Another advantage with nanogels is that they are not water soluble like pheromone hydrogels (Herlekar, 2014).

20.2.2.2.6 Sticky Traps

Chartreuse green sticky traps could be a useful device for monitoring and managing flies in China (Xue and Wu, 2013). The ideal height of a trap should be 30 cm above ground level in cucumber crops (Jiji et al., 2009). This is a new technique being addressed in fruit fly management and needs intensive studies.

20.2.2.2.7 Plant-Derived Products

In the laboratory, a dose of 625 ppm of extract (in acetone or water) of *Acacia auriculiformis* A. Cunn. bark prolonged the larval period and affected pupation, adult emergence, oviposition, and egg hatching in *B. cucurbitae* (Kaur et al., 2010). On the contrary, commercial neem-based products (Achook®, Econeem, Neemjeevan®) were less effective in protecting summer squash from the pest probably because they were less persistent (3 days) than synthetics (7 days) (Sood and Sharma, 2004).

20.2.2.2.8 Chemical Pesticides

When oxymetrine 0.5EC was assayed in the field at 300, 500, and 750 mL/ha, it was significantly better at 750 mL/ha than neem product (1500 ppm of AZ) sprayed at 1.5 L/ha or endosulfan (35EC) at 1 L/ha in lowering fruit fly populations (80.8% mortality) and fruit damage (16.8% infestation versus 51.8% in control) and increasing the yield of watermelon (151 kg/ha versus 44 kg/ha in control) (Kumar et al., 2013). In another trial on synthetic insecticides, deltamethrin 37.5 g a.i./ha, cypermethin 75 g a.i./ha, or fenvalerate 75 g a.i./ha controlled the pest effectively (Sood and Sharma, 2004). Spraying 0.2% carbaryl, 0.05% malathion, or 0.05% endosulfan during the flowering stage at 10-day intervals, starting from flower initiation, or spraying of 0.025% cypermethrin, 0.05% profenophos, or 0.15% carbaryl at 15-day intervals partially checked fly incidence (Sood and Sharma, 2004). In round gourd, sequential sprays of 0.07% endosulfan 35EC followed by 0.05% malathion 50EC or 0.03% acephate 75SP were effective in reducing plant damage (11.3% infestation versus 41.65% in control) and increasing crop yield (102.5 q/ha versus 49.7 q/ha in control) with a CB ratio of 1:37 (Verma et al., 2010). Recently, Bhowmik et al. (2014) reported that chlorfenapyr 10SC @ 50 g a.i./ha, when sprayed thrice on pointed gourd, reduced fruit infestation (6.3%–9.8% versus 51.6%–60.5% in control) and increased the yield (9.0–9.5 q/ha compared to control with 3.5–3.7 q/ha). The best treatment was followed by deltamethrin 2.8EC (10 g a.i./ha) and spinosad 45SC (60 g a.i./ha). In bitter gourd, bioefficacy of water extract (10%) of ailanthus (*Ailanthus triphysa* Dennst) or cashew (*Anacardium occidentale* L.) leaves was equal to that of synthetic pesticides (Jacob et al., 2007).

After studying nine field-collected ecotypes of *B. dorsalis* in China, Zhang et al. (2014) reported that field populations displayed narrow variations in tolerance to cyantraniliprole compared to laboratory-reared insects that developed 19.44-fold resistance after 14 generations of selection (LC_{50} = 3.29–15.83 µg/g). This resistance level in due course decreased due to synergistic effect when piperonyl butoxide or maleate was added to insecticide (Zhang et al., 2014).

20.2.2.2.9　Natural Enemies

Several natural enemies attacking larvae have been reported from Hawaii by Dhillon et al. (2005a). The major parasitoids are two braconid wasps, *Opius fletcheri* Silv. and *Fopius arisanus* (Sonan); a nematode, *Steinernema carpocapsae* (Weiser); and three fungi, *Rhizoctonia solani* Kuhn, *Gliocladium virens* Origen, and *Trichoderma viride* Pers. In French Polynesia, the release of *F. arisanus* lowered fruit fly numbers by up to 97.9% with 51.9% parasitism (Vargas et al., 2007). For other natural enemies, data on parasitism and impact on pest mortality are lacking.

20.2.2.2.10　Integrated Control

Since fruit flies neither suck nor chew the foliage, pesticide sprays become ineffective. Larvae inside fruit are difficult to control by nonsystemic chemicals. Therefore, Dhillon et al. (2005a) recommended bagging of fruits, field sanitation, and cue-lure traps. Application of NSKE (10%), NO (1%) or Neemgold® (2.5 L/ha) has been suggested for IPM in round gourd in India (Verma et al., 2010). For IPM in squash crop in Nepal, a new local product Jholmal® was sprayed 40 days after transplanting. It was prepared from fresh cow dung, cow urine, garlic, and capsicum pepper (100 g each) + extract of leaves (500 g each) of neem, *Adhatoda vasica* Nees, *O. sanctum* L., *Lycopersicon esculentum* Mill., *Artremisia vulgaris* L., *A. calamus* L., *Tagetes* sp., *Sapium insigne* Trimen, *Chysanthemum* sp., and *Vitex negundo* L. This product was compared with dichlorvos @ 2 mL/L, cue-lure and two indigenous baits. The local baits contained a mixture of (a) fermented rice 200 g + molasses 5 mL + borax 4 g + malathion 1 mL, or (b) pulp of ripe banana 500 g + molasses 10 mL + borax 10 g + malathion 2.5 mL. Spraying Jholmal has effectively controlled fruit flies and has been recommended by Sapkota et al. (2010). In India, a combination of BAT and MAT reduced fruit fly losses by up to 59%. Furthermore, when these techniques were incorporated in a village-level area-wide management adoption, no yield loss was reported (Stonehouse et al., 2007). Of course, intensive efforts would be needed to validate and refine the existing package of practices to be cost-effective and appropriate to the local, cheap resources (Mehta et al., 2002).

20.2.3　Homoptera

20.2.3.1　Aphids (Aphididae)

The most destructive aphid species are the melon/cotton aphid *Aphis gossypii* Glover and the green peach aphid *Myzus persicae* (Sulzer), which cause considerable damage to cucurbits, especially melons and cucumbers. Both species are polyphagous, but *A. gossypii* has a very wide host range although some strains or biotypes exhibit preferences for certain plants. Cucurbits are not attacked by aphids until the vines form runners. Nymphs and adults suck sap from the underside of young leaves and growing tips of vines. Attack on cotyledon leaves makes them crinkled and in severe cases, seedlings wither. Leaves of fully grown vines turn yellow and chlorotic and the plant loses its vigor. Plants may die prematurely. Feeding also causes considerable distortion and leaf curling, blossom shedding, and hinders photosynthesis. Honey dew secreted by aphids gives a glossy appearance and attracts sooty mold that impairs fruit quality, making them unmarketable. Aphids are vectors for the transmission of cucurbit potyviruses (cucumber mosaic virus, water melon mosaic virus-2, zucchini yellow mosaic virus).

20.2.3.1.1　Control

Since aphids attack several crops, intensive research studies have been carried out on control tactics and are well documented. In cucurbits, silver reflexive mulches, when plants are small, can repel aphids and prevent pest entry into the crop, thus reducing or delaying virus transmission by 2–4 weeks (Brust, 2009). The wide host range makes crop rotation rather difficult but may be tried on small areas. Collection and destruction of residue of infested crop immediately after harvest can restrict insect dispersal. Likewise, removal of weeds and uncultivated plants from the field does not favor retention of the pest populations, particularly where continuous cropping is practiced. Row covers and planting in aluminum foil-covered beds inhibit pest development. The time of planting may influence the potential of aphid population. However, the data on this method are not available. Sticky cards are not very effective but can indicate a possible source of infestation. On the contrary, yellow pans filled with water to trap aphids can be installed in infested fields at the start of the crop season (NYS, 2014).

Many naturally occurring generalist predators and parasitoids are known to be effective against melon aphid but their efficiency is reduced in hot, dry summers. Lady bird beetles, syrphid flies, and chrysopas are potential predators (Hoffmann and Frodsham, 1993). Eventually, ants found associated with aphids may hinder these predators. The convergent lady beetle, *Hippodamia convergens* Guerin-Meneville, a common species in North America, may provide effective control throughout the summer (Brust, 2009). Releases of the predatory mite *Neosieulus cucumeris* (Oudemans) reduced aphid population by up to 68%–79% in plastic houses (Hassan et al., 2008). A braconid, *Lysiphlebus testaceips* (Cresson), sometimes causes significant parasitism, resulting in swollen aphid mummies and tanned and hardened dead aphids (Hoffmann and Frodsham, 1993).

Methanol or ethanol extract (5%) of *Chenopodium ficifolium* Sm. leaves controlled >80% aphids on cucumber plants (Quang et al., 2010). Likewise, less toxic materials (mineral oil, insecticidal soap, soap products), pyrethroids, or neonicotinoids can be included in IPM (Dreistadt et al., 2001). Oil sprays interfere with virus transmission and may kill aphids. Otherwise, spraying 0.1% malathion, methyl demeton, or dimethoate is effective when the crop is treated sufficiently early before the attack becomes severe but does not stop virus transmission (Anonymous, 2014). Distorted leaves provide a safe hiding place for aphids. Therefore, systemic insecticides are best suited for treating plants. But due attention to the development of insecticide resistance should be given because resistance to organophosphate and pyrethroid insecticides has been reported by Dreistadt et al. (2001). Alternatively, reduced-risk insecticides including imidacloprid, thiamethoxam, endosulfan, insecticidal soaps, horticultural oil, and biopesticides such as *B. bassiana* should be preferred (Brust, 2009). In glasshouse experiments, melons grown on soil containing 30% vermicompost showed low pest infestation and may be included in IPM (Razmjou et al., 2011).

20.2.3.2 White Flies (Aleyrodidae)

Among several species, the polyphagous silverleaf whitefly, *Bemisia tabaci* (Genn.), is a major pest of cucurbits. Nymphs and adults suck sap from the underside of leaves. Heavy infestation can deplete plants of sap, causing malformation of leaves, leaf wilting, or the infested leaves dry and fall off the plant. Chlorotic yellow spots sometimes appear at feeding sites on leaves. The sticky honeydew they produce supports the growth of sooty mold on leaves. The resulting dark splotches on the leaves may reduce photosynthesis and other physiological functions of the plant. Whiteflies are also vector of viruses.

20.2.3.2.1 Control

From studies on cultural measures carried out in Egypt, Mohamed (2012) confirmed that early planting, wide plant spacing, and selecting improved pest tolerant varieties of cucumber receive low rate of infestation of nymphs. Placing aluminum foil on the ground under the plant can deter flies from settling. Washing plants with clean water to remove insects can be tried on a limited scale. Yellow sticky traps, destruction of crop residue soon after harvest, and reflexive mulches are quite effective. In India, planting row crops as a physical barrier for pest movement and early planting of the main crop to escape pest attack have been advocated (Anonymous, 2014). In plastic houses, release of the predatory bug *Neosieulus zaheri* (Yousuf & El-Borolossy) reduced the population of *B. tabaci* by up to 68%–90% (Hassan et al., 2008).

Plastic nets (0.9 mm pore diameter and 40–80 mesh size) treated with alpha-cypermethrin (as repellent to flies) allowed movement of only 19% of flies compared to 35%–46% in nonprotected crop (Matin et al., 2014). Thiacloprid at 30 kg/ha reduced white flies Biotype B populations by about

37.2%–95.3% within 21 days of spraying, improved plant vigor, and increased the yield of marketable cucumber fruits. Also, when compared with avermectin (7.5 L/ha), thiacloprid (15 kg/ha) proved superior and therefore, has been recommended by Dong et al. (2014). This whitefly species is cosmopolitan in distribution and has a wide range of host plants. The recommendation against this pest for other crops may be evaluated for field bioefficacy, and later an IPM can be planned as per crop and agroecosystem.

20.2.4 HETEROPTERA

20.2.4.1 Squash Bug (Coreidae)

The squash bug *Anasa tristis* (De Geer) is one of the most common and severe pests of cucurbits, especially squash and pumpkins in the United States. Adults and nymphs may be found clustered about the crown of the plant, beneath damaged leaves, under clods or any other protective ground cover. They suck plant juices from the leaves. Young plants are more prone to pest attack than aged plants.

Young nymphs are gregarious and feed together in groups. They inject a toxin that causes plant wilting (sometimes called "anasa wilt"), leaf blackening, collapse, and dieback. Feeding on fruits causes scarring. Squash bugs are vectors of the virus (cucurbit yellow vine decline) in watermelon and pumpkin (Brust, 2009).

20.2.4.1.1 Cultural Methods

Field sanitation and timely cultural operations can reduce pest damage (Hoffmann and Frodsham, 1993). Postharvest deep tillage and destruction by burning crop residues and debris that provide shelter for overwintering beetles should be practiced after crop harvest (Groves, 2011; NYS, 2014). Delayed planting and annual crop rotation with noncucurbit crops could reduce bug populations in the United States by up to 80% (Bauernfeind and Nechols, 2006) because rotation can delay buildup of populations in the season. However, the scope is limited in long crop seasons when adults move easily between early and late planted fields. Mulching with plastic sheets, newspapers, and hay combined with tightly secured row covers prevent bugs from reaching the plants in large numbers to lay eggs at the time of planting and flowering (Bauernfeind and Nechols, 2006). Planting of repellent crops (viz. tansy, marigold, mint, radish, nasturtiums, catnip) or trap crops/varieties that are attractive to bugs (cvs. Lemon drop and Blue Hubbard squash) have proved effective in cantaloupe, squash, and watermelon crops (Pair, 1997). Hand-picking eggs and nymphs provide the best prevention and restrict the pest outbreaks. Few crop genotypes have been identified as tolerant to pest attack. For example, squash (cvs. Butternut, Royal acorn, Sweet cheese, Pink banana, Black zucchini) and pumpkin (cv. Green striped cushaw). Winter varieties of squash such as Hubbard and Marrows are much more severely damaged than other varieties (Groves, 2011).

20.2.4.1.2 Traps

Cardboards of 30 cm × 30 cm can be placed between plants to trap moving bug (NYS, 2014). Further evaluation of the boards can confirm effectiveness and subsequent integration in pest management strategy.

20.2.4.1.3 Natural Enemies

The generalist predators including ground beetles, spiders, robber flies, and mites have been recorded, but data are not yet available on predation rate or pest mortality. The tachinid fly *Trichopoda pennipes* Fb. is a potential parasitoid in the United States. The parasitoid is present throughout the production period of squash and pumpkins. The maggot feeds on the body fluids of the bug, and the bug dies after emergence of the fly. The fly is most effective when it parasitizes nymphs because

50% of the larvae die before becoming adults, and 65% of the remaining population that become adults will die before laying eggs, resulting in high rate parasitism (even up to 95%) (Brust, 2009). Buckwheat (*Fagopyrum esculentum* Moench), being attractive to this parasitoid, can be planted as an intercrop. This system encourages conservation and augmentation of parasitoids. Otherwise, the release methodology will have to be studied for large-scale application in a particular crop and region. Another important parasitoid is the wasp *Gryon pennsylvanicus* (Ashmead) that parasitizes bugs to the extent of 31% in the United States (Decker and Yeargan, 2006). The major hurdle in biological control is that bugs are prolific in breeding and continue feeding and laying eggs even after a parasite attack (Brust, 2009). Consequently, even high level of parasitism may not prevent heavy crop damage. Thus, pest control cannot be relied upon naturally occurring predators and parasitoids and alternative measures need to be studied.

20.2.4.1.4 Chemical Pesticides

Bugs are not equally susceptible to insecticide treatments at all developmental stages. Eggs are impervious, whereas nymphs and adults have thick cuticles that inhibit insecticide penetration. Among synthetics, pyrethroids are effective on early instar nymphs and before populations build and when sprays against adults are directed at the base of the plant (Brust, 2009). Molecules with low toxicity and biopesticides may be combined with other cost-effective methods for integrated pest control.

20.2.5 LEPIDOPTERA

20.2.5.1 Squash Vine Borer (Sesiidae)

The squash vine borer *Melittia cucurbitae* (Harris) is becoming an important pest in the United States, Canada, and Latin America on winter squash, summer squash, pumpkin, gourds, and rarely on cucumber and melons. Damage is caused by larvae tunneling into stems. The tunneling often kills plants, especially when the larvae feed in the basal portions of vines. Sometimes fruits are also attacked. Sudden wilting of a vine, leaves turning yellow and eventually brown around the leaf margins, and insect excreta coming from holes in the stem are characteristic symptoms of pest attack.

20.2.5.1.1 Control

Soil tillage and destruction of vines expose overwintering larvae or pupae in the soil (Williamson, 2014). Since the borers do not emerge until favorable weather sets in, early planting is a valid measure. Overall, pest management should be oriented to newly hatched larvae because once the larvae attack the stem, control becomes difficult. Planting of pumpkin and other cucurbits year after year should invariably be avoided (Williamson, 2014). Floating row covers over the plants and firmly anchoring them to the ground can prevent pest entry into the main crop. However, removal of the row covers during flowering is needed to allow pollination by bees. Pheromone traps and lures attract male moths and may be used to reduce moth populations in the crop area. Butternut and green-striped cushaw varieties of squash exhibit a higher level of resistance than currently cultivated cucurbits and susceptible cv. Hubbard (Williamson, 2014). These varieties may be recommended for large-scale planting. For chemical control, whenever the need arises, sprays should be directed at the base of the plant.

20.2.6 THYSANOPTERA

20.2.6.1 Thrips (Thripidae)

Common thrip species *Frankliniella occidentalis* (Pergande), *Frankliniella schultzei* (Trybon), *Thrips palmi* (Karny), and *Thrips tabaci* (Lindeman) are generally a problem on flowers and shoot tips of cucurbits early in the season, particularly when plants are drought stressed. They are found on the underside of leaves producing silver flecking near the large leaf veins and sometimes deforming leaves (edges tend to curve downward). Nymphs and adult thrips scrape the leaf surface of the plant disrupting cells and feeding on the cell contents. In Taiwan, Peng et al. (2011) reported the incidence of watermelon silver mottle virus and melon yellow spot virus transmitted by thrips.

20.2.6.1.1 Control

Removal of weeds, before they flower, and unwanted vegetation reduces the hiding places. Prophylactic introduction of naturally occurring predators can be experimented such as mites *N. cucumeris* (Oudemans) and *Amblyseius swirskii* (Athias-henriot) (against first instar larvae) and *Hypoaspis aculifer* (Canestrini) (against pupae); rove beetle *Atheta coriaria* Kraatz; and minute pirate bug *Macrotracheliella nigra* Parshley (Dreistadt et al., 2001). Repeated spraying on the underside of leaves with a strong stream of water reduces population density (Dreistadt et al., 2001). Chemical treatments are needed when pest population is heavy or leaf feeding is severe. To control *T. tabaci* in the plastic houses, releases of *N. cucumeris* and spraying of pesticides (abamectin/avermectin 1.8EC [1 mL/L] or fenpyroximate 5SC [2 mL/ha]) were effective as pest mortality up to 85%–89% and 64%–72%, respectively, was achieved (Hassan et al., 2008). Unnecessary applications of synthetic pesticides can cause spider mite buildup. Spraying reduced-risk and compatible insecticides (neem oil, azadirachtin, insecticidal soaps, pyrethrins, spinosad) can prove effective at the start of infestation. In IPM in all cucurbits, spraying *B. bassiana* early in the cropping season and before plants flower can be experimented.

20.3 PERSPECTIVE

Currently, the ETLs of insect populations or the severity of plant damage caused by insect feeding on different plant parts are available for some of the important pests (Table 20.2). The threshold depends upon cultivation system, crop variety, plant organ attacked, insect life stage, etc. The threshold should however be revised from time to time for each region due to changes in climatic conditions, introduction of new crop varieties, and adoption of improved cultural practices. The timing of spraying operation can also be judged and advocated after frequent and careful field scouting. Scouting with proper sampling is important for new pests such as soybean aphid (*Aphis glycines* Matsumura) as the pest status may change rapidly (Groves, 2012).

Pest management can be planned for a particular pest or a complex of pests in a given crop. Likewise, it may work in a particular crop growing region or elsewhere with reasonable adaptation.

TABLE 20.2
Examples of Economic Thresholds of Some Major Pests of Cucurbits

Insect Pest	Plant Infestation or Insect Number
Melon aphid	5–19 aphids/20–50 leaves, 5 or more aphids/leaf (University of Delaware, 2012)
	20% squash or cucumber runners with live aphids (NYS, 2014)
Cucumber beetles	1 beetle/plant (Brust, 2009; Webb, 2013)
	1 beetle/plant for melons, cucumbers, and young pumpkins (Groves, 2012)
	2 beetles/plant (University of Delaware, 2012)
	5 beetles/plant for water melon, squash, and old pumpkins (Webb, 2013)
Squash vine borer	2 moths/pheromone trap (Brust, 2009)
Squash bug	>1 egg mass/plant during early flowering stage (Groves, 2011; NYS, 2014)
	2 overwintering adults feeding on small plants (<5 leaves) or 2 egg masses/aged plant (Brust, 2009)
Thrips	20% vine tips infested or 8 thrips/leaf (Dreistadt et al., 2001)
Whiteflies	3 whiteflies on third youngest leaf on cantaloupe (Anonymous, 2014)

As a prophylactic measure, insects (particularly beetles) can be collected manually when pest population is quite low and then destroyed. Spot spraying for localized infestation can be economical on small farms. In watermelon, among six synthetic pesticides, spraying abamectin alone (20 g a.i./ha) significantly reduced the incidence of aphids by 96.2%, thrips by 81.1%, and leaf miners by 30.9% whereas beta-cyfluthrin (18.75 g a.i./ha) was most effective against red pumpkin beetles (6.9% plant infestation) (Babu et al., 2002). Similarly, spraying abamectin (20 g a.i./ha) or fenpyroximate (30 g a.i./ha) reduced the population of *A. gossypii*, *B. tabaci*, and *T. tabaci* by 75%–85%, 59%–90%, and 64%–72%, respectively, in the plastic houses (Hassan et al., 2008). A pest complex of bitter gourd that comprised of *Batrocera* sp., *Helicoverpa armigera* (Hbn.), *Diaphania indica* (Saunders), *Aulacophora* sp., and *L. trifolii* was controlled with three sprays of chlorantraniprole @ 30 g a.i./ha. It reduced the population of *D. indica* (0.72 larvae/plant versus 7.30 larvae in control), *Aulacophora* sp. (0.70 beetles/plant versus 5.04 beetles/plant in control), and *L. trifolii* (7.52% versus 36.2% in control). The reduction in plant damage of all pests (9.7% versus 55.7% in control) resulted in a higher yield (11.98 t/ha versus 6.45 t/ha in control) (Bharathi et al., 2011). For melon, weekly applications of pesticides (WAP) worked well but there is less need for WAP when prophylactic treatments are applied (Lima et al., 2014). In the case of chemical sprays, application should be done in the evening hours to avoid spray drift and repellent action of insecticides to pollinating bees. Thus, the integrated approach reduces cost of treatment due to single application against several pests.

Among botanicals, neem products (5 mL/L) were moderately effective against leaf miners and fruit flies (Babu et al., 2002). In cucumber, spraying 10% water extract of *Swertia chirata* L. or *Swietenia mahogoni* Jacq. significantly reduced leaf perforation (3.44% infestation versus 14.22% in control) and the number of defoliators (1.33 insects/leaf versus 4.66 insects/leaf in control) and increased crop yield by 1.5 times (Azad et al., 2013). In Indonesia, water extract of local plant, *Elsholtzia pubescens* Benth, containing camphor showed promising results as attractant to fruit fly. Whenever indigenous plants are available, locals exploit it for traditional preparations and commercial products (Hasyium et al., 2007). Generally, plant-derived products give effective control of various pests of vegetable crops (Gahukar, 2007). Further research is now needed to isolate, identify, and formulate phytochemicals, and traditional preparations including crude extracts may be included in IPM in cucurbits.

20.4 CONCLUSIONS

Since cucurbits are important vegetables, organic farming should be promoted with the use of spinosad, botanical pesticides, kaolin clay, etc. Otherwise, less toxic pesticides such as 0.2% carbaryl compared to recommended endosulfan (0.05%) or monocrotophos (0.05%) seem to be proper for all cucurbits (Anonymous, 2014). Against coleopteran insects, carbaryl dusting (5%) or spraying (0.2%) should be undertaken by the growers as soon as symptoms of damage appear. Abamectin is recommended against several pests, but it is highly toxic to honey bees. Other insecticides may be sprayed at 2-week intervals and residual toxicity should be monitored *a priori* on fresh fruits. It may be more economical and ecofriendly to use botanicals or biopesticides by replacing or minimizing the use of synthetic chemicals. The combined strategy of BAT and MAT for fruit flies dramatically reduces the cost of treatment and avoids direct contact of insecticides (drift or runoff) with produce. Obviously, further research on pest management tactics are needed for each crop and agroregion to make plant protection as efficient and cost-effective as possible even for the use of small and marginal farmers.

ACKNOWLEDGMENT

I thank Dr. S.S. Nilakhe, Entomologist, Texas Agriculture Department, Austin, TX, USA, for his critical review and valuable suggestions on the draft.

REFERENCES

Abdul-Moniem, A.S.H., Gomma, A.A., Dimetry, N.Z., Wetzel, T., and Volkmar, C. 2004. Laboratory evaluation of certain natural compounds against the melon lady bird beetle, *Epilachna chrysomelina* F. attacking cucurbit plants. *Archives of Phytopathology and Plant Protection*, 37, 71–81.

Abe, M. and Matsuda, K. 2000. Feeding deterrent from *Momordica charantia* leaves to cucurbitaceous feeding beetle species. *Journal of Applied Entomology and Zoology*, 35(1), 143–149.

Adler, L.S. and Hazzard, R.V. 2009. Comparison of perimeter trap crops varieties: Effects on herbivory, pollination and yield in butternut squash. *Environmental Entomology*, 38, 207–215.

Akhtaruzzaman, M., Alam, A.Z., and Ali Sardar, M.M. 2000. Efficacy of bait sprays for suppressing fruit fly on cucumber. *Bulletin of the Institute of Tropical Agriculture, Kyushu University (Japan)*, 23, 15–26.

Ali, H., Ahmad, S., Hassan, G., Amin, A., and Naeem, M. 2011. Efficacy of different botanicals against red pumpkin beetle (*Aulacophora foveicollis*) in bitter gourd (*Mamordica charantia* L.). *Pakistan Journal of Weed Science Research*, 17(1), 65–71.

Anonymous, 2014. *AgriTech Portal*. Tamil Nadu Agricultural University, Coimbatore, Tamil Nadu, India.

Azad, A., Sardar, A., Yesmin, N., Rahman, M., and Islam, S. 2013. Ecofriendly pest control in cucumber (*Cucumis sativa* L.) field with botanical pesticides. *Natural Resources*, 4(5), 404–409.

Babu, P.G., Reddy, D.G., Jadhav, D.R., Chiranjeevi, C., and Khan, M.A.M. 2002. Comparative efficacy of selected insecticides against pests of water melon. *Pesticide Research Journal*, 14, 57–62.

Barba, R.B. and Tablizo, R.P. 2014. Organic-band attractant for the control of fruit flies (Diptera: Tephritidae) infesting ampalaya. *International Journal of Scientific & Technology Research*, 3(9), 348–355.

Barry, J.D., Miller, N.W., Pinero, J.C., Tuttle, A., Mau, R.F.L., and Vargas, R.I. 2006. Effectiveness of protein baits on melon fly and oriental fruit fly (Diptera: Tephritidae): Attraction and feeding. *Journal of Economic Entomology*, 99(4), 1161–1167.

Bauernfeind, R.J and Nechols, J.R. 2006. *Squash Bugs and Squash Vine Borers*. Bulletin MF-2508. Kansas State University Agriculture Experiment Station, Garden city, KS, 8pp. http://www.oznet.ksu.edu/library/entomol2/MF2508.pdf.

Bellows, B.C. and Diver, S. 2002. *Cucumber Beetles: Organic and Biorational IPM*. ATTRA Publication No. IP 212. National Sustainable Agriculture Information Service, Garden city, KS, ATTRA. http://attra.ncat.org/attra.pub.cucumberbeetle.html.

Bharathi, K., Mohanraj, A., and Rajvel, D.S. 2011. Field efficacy of chlorantraniliprole 20% EC against pests of bitter gourd. *Pestology*, 35(7), 32–35.

Bhowmik, P., Mandal, D., and Chatterjee, M.L. 2014. Biorational management of melon fruit fly, *Batrocera cucurbitae* on pointed gourd. *Indian Journal of Plant Protection*, 42(1), 34–37.

Bjorksten, A.T. and Robinson, M. 2005. Juvenile and sublethal effects of selected pesticides on the leaf miner parasitoids, *Hemiptarsenus varicornis* and *Diglyphus isaea* (Hymenoptera: Eulophidae) from Australia. *Journal of Economic Entomology*, 98(6), 1831–1838.

Brust, G. 2009. *Cucurbit Pest Management*. Extension Bulletin, University of Maryland, College Park, MD, 8pp.

Capinera, J.L. 2001. *Handbook of Vegetable Pests*. Academic Press, San Diego, CA, 729pp.

Cavanagh, A., Hazzard, R., Adler, L.S., and Boucher, J. 2009. Using trap crops for control of *Acalymma vittatum* (Coleoptera: Chrysomelidae) reduces insecticide use in butternut squash. *Journal of Economic Entomology*, 102(3), 1101–1107.

Chandel, B.S., Singh, V., Trivedi, S.S., and Katiyar, A. 2009. Anifeedant bioefficacy of *Coleus amboinicus, Mentha piperata, Pogostemon heyneaus* and *Mentha longifolia* against red pumpkin beetle, *Raphidopalpa foveicollis* Lucas (Coleoptera: Chrysomelidae). *Journal of Environmental and Biological Sciences*, 23(2), 147–151.

Chaudhary, R.N. 1995. Management of red pumpkin beetle, *Aulacophora foveicollis* (Lucas) using polythene cages on cucumber. *Pest Management in Horticultural Ecosystems*, 1(1), 55–57.

Chinajariyawong, A., Kritsaneepaiboon, S., and Drew, R.A.I. 2003. Efficacy of protein bait sprays in controlling fruit flies (Diptera: Tephritidae) infesting angled luffa and bitter gourd in Thailand. *The Raffles Bulletin of Zoology*, 51(1), 7–15.

Chuang, Y.Y. and Hou, R.F. 2007. Effectiveness of attract and kill systems using methyl eugenol incorporated with insecticides against the oriental fruit fly (Diptera: Tephritidae). *Journal of Economic Entomology*, 100(3), 352–359.

Cline, G.R., Sedlacek, J.D., Hillman, S.L., Parker, S.K., and Silvernail, A.F. 2008. Organic management of cucumber beetles in water melon and muck melon production. *HortTechnology*, 18, 436–444.

Coveillo, R.L., Natwick, E.T., Godfrey, L.D., Fouche, C.B., Summers, C.G., and Stapleton, J.J. 2005. Cucumber beetles. In: *UC IPM Pest Management Guidelines: Cucurbits*. UC ANR Publication No. 3445. University of California, Davis, CA.

Decker, K. and Yeargan, K. 2006. Seasonal phenology, natural enemies and control of squash bug, *Anasa tristis* (DeGeer) on summer squash in Kentucky. Paper presented at the *2006 ESA Annual Meeting*, Indianapolis, IN, December 10–13, 2006.

Desai, S., Jakhar, B.L., and Patel, R.K. 2014. Bioefficacy of different insecticides against melon fly, *Batrocera cucurbitae* (Coquillett) on sponge gourd. *Pestology*, 38(2), 49–53.

Dhillon, M.K., Naresh, J.S., Singh, R., and Sharma, N.K. 2005b. Influence of physico-chemical traits of bitter gourd, *Momordica charantia* L. on larval density and resistance to melon fruit fly, *Batrocera cucurbitae* (Coquillett). *Journal of Applied Entomology*, 129(7), 393–399.

Dhillon, M.K., Singh, R., Naresh, J.S., and Sharma, H.C. 2005a. The melon fruit fly, *Batrocera cucurbitae*: A review of its biology and management. *Journal of Insect Science*, 5, 1–16.

Dong, S., Qiao, K., Zhu, Y., Wang, H., Xia, X., and Wang, K. 2014. Managing *Meloidogyne incognita* and *Bemisia tabaci* with thiacloprid in cucumber crops in China. *Crop Protection*, 58, 1–5.

Dreistadt, S.H., Clark, J.K., and Flint, R.L. 2001. *Integrated Pest Management for Floriculture and Nurseries*. Publication No. 3402. University of California Agriculture and Natural Resources, Oakland, CA.

Ellers-Kirk, C.D., Fleischer, S.T., Snyder, R.H., and Lynch, J.P. 2000. Potential of entomopathogenic nematodes for biological control of *Acalymma vittatum* (Coleoptera: Chrysomelidae) in cucumbers grown in conventional and organic soil management systems. *Journal of Economic Entomology*, 93, 605–612.

Ferguson, S.J. 2004. Development and stability of insecticide resistance in the leaf miner, *Liriomyza trifolii* (Diptera: Agromyzidae) to cyromazine, abamectin and spinosad. *Journal of Economic Entomology*, 97(1), 112–119.

Gahukar, R.T. 2007. Botanicals for use against vegetable pests and diseases: A review. *International Journal of Vegetable Science*, 13(1): 41–60.

Gogi, M.D., Ashfaq, M., Arif, M.J., and Khan, M.A. 2010. Biophysical basis of antixenotic mechanism of resistance in bitter gourd (*Momordica charantia* L., Cucurbitaceae) against melon fruit fly, *Batrocera cucurbitae* (Coquillett) (Diptera: Tephritidae). *Pakistan Journal of Botany*, 42(2), 1251–1266.

Groves, R.L. 2011. *Insect Pest Management in Cucurbit Vegetable Crops*. Department of Entomology, University of Wisconsin, Madison, WI.

Halaj, J. and Wise, D.H. 2002. Impact of detrital subsidy on tropic cascades in a terrestrial grazing food web. *Ecology*, 83, 3141–3151.

Hassan, M.F., Ali, F.S., Hussein, A.M., and Mahgoub, M.H. 2008. Biological and chemical control of the plant piercing-sucking insect pests on cucumber in plastic houses. *Egyptian Journal of Biological Pest Control*, 18(1), 167–170.

Hasyium, A., Muryati, M.I., and de Kogel, W.J. 2007. Male fruit fly, *Batrocera tau* (Diptera: Tephritidae) attractants from *Elsholtzia pubescens* Bth. *Asian Journal of Plant Sciences*, 6(1), 181–183.

Herlekar, I. 2014. Using nano technology to control pests: Trapping fruit fly using pheromone gels. *Current Science*, 106(1), 14–15.

Hoffmann, M.P. and Frodsham, A.C. 1993. *Natural Enemies of Vegetable Insect Pests*. Cooperative Extension, Cornell University, Ithaca, NY, 63pp.

Ingoley, P., Mehta, P.K., Chauvan, Y.S., Singh, N., and Awasthi, C.P. 2005. Evaluation of cucumber genotypes for resistance to melon fruit fly, *Batrocera cucurbitae* Coq. under mid-hill conditions of Himachal Pradesh. *Journal of Entomological Research*, 29(1), 57–60.

Islam, K., Islam, M.S., and Ferdousi, Z. 2011. Control of *Epilachna vigintioctopunctata* (Fab.) (Coleoptera: Coccinellidae) using some indigenous plant extracts. *Journal of Life and Earth Sciences*, 6, 75–80.

Jacob, J., Leela, N.K., Sreekumar, K.M., Anesh, R.Y., and Hema, M. 2007. Phytotoxicity of leaf extracts of multipurpose trees against insect pests of bitter gourd (*Momordica charantia*) and brinjal (*Solanum melongena*). *Allelopathy Journal*, 20(2), 411–417.

Jasinski, J., Darr, M., Ozkan, E., and Precheur, R. 2009. Applying imidacloprid via a precision banding system to control striped cucumber beetle (Coleoptera: Chrysomelidae) in cucurbits. *Journal of Economic Entomology*, 102(6), 2255–2264.

Jiji, T., Simna, T., and Verghese, A. 2009. Effect of heights of food bait trap on attraction of melon flies (Diptera: Tephritidae) in cucumber. *Pest Management in Horticultural Ecosystems*, 15(1), 48–50.

Kaspi, R. and Parrella, M.P. 2005. Improving the biological control of leaf miners (Diptera: Agromyzidae) using the sterile male technique. *Journal of Economic Entomology*, 99(4), 1161–1167.

Kaspi, R. and Parrella, M.P. 2006. Abamectin compatibility with the leaf miner parasitoid, *Diglyphus isaea*. *Biological Control*, 35(2), 172–179.

Kaur, A., Sohal, S.K., Singh, R., and Arora, S. 2010. Development inhibitory effect of *Acacia auriculiformis* extracts on *Batrocera cucurbitae* (Coquillett) (Diptera: Tephritidae). *Journal of Biopesticides*, 3(2), 499–504.

Khan, S.M. and Jehangir, M. 2000. Efficacy of different concentrations of sevin dust against red pumpkin beetle, *Aulacophora foveicollis* (Lucas) causing damage to muskmelon (*Cucumis melo* L.) crop. *Pakistan Journal of Biological Sciences*, 3(1), 183–185.

Khan, S.M. and Wasim, M. 2001. Assessment of different plant extracts for their repellency against red pumpkin beetle (*Aulacophora foveicollis* Lucas) on musk melon (*Cucumis melo* L.). *Journal of Biological Sciences*, 1(4), 198–200.

Khorsheduzzaman, A.K.M., Nessa, Z., and Rahman, M.A. 2010. Evaluation of mosquito net barrier on cucurbit seedling with other chemical, mechanical and botanical approaches for suppression of red pumpkin beetle damage in cucurbit. *Bangladesh Journal of Agricultural Research*, 35(3), 395–401.

Khurseed, S. and Raj, D. 2013a. Efficacy of insecticides and biopesticides against hadda beetle, *Henosepilachna vigintioctopunctata* (Fabricius) (Coleoptera: Coccinellidae) on bitter gourd. *Indian Journal of Entomology*, 75(2), 163–166.

Khurseed, S. and Raj, D. 2013b. Efficacy of some insecticides and biopesticides against red pumpkin beetle, *Aulacophora foveicollis* on cucumber. *Indian Journal of Plant Protection*, 41(2), 184–186.

Khurseed, S., Raj, D., and Ganie, N.A. 2013c. Resistance to red pumpkin beetle, *Aulacophora foveicollis* (Lucas) in cucumber genotypes. *Indian Journal of Entomology*, 75(1), 90–93.

Kumar, A., Simon, S., and Yogi, K. 2013. Studies on the management of imported pests of water melon, *Citrullus* (*vulgaris*) *lanatus* with botanical pesticides. *Pestology*, 37(4), 42–45.

Kumar, B. and Agarwal, M.L. 2005. Efficacy of different attractants and bait combinations against *Batrocera cucurbitae* (Coquillett). *Indian Journal of Plant Protection*, 33, 194–196.

Kumar, S. and Kumar, J. 1998. Laboratory evaluation of certain insecticides against strains of hadda beetle, *Epilachna vigintioctopunctata* Fab. resistant to malathion and endosulfan. *Pest Management and Economic Zoology*, 6, 133–137.

Lakshmi, M.V., Rao, G.R., and Rao, P.A. 2005. Efficacy of different insecticides against red pumpkin beetle, *Raphidopalpa foveicollis* Lucas on pumpkin, *Cucurbita maxima* Duchesne. *Journal of Applied Zoological Researches*, 16(1), 73–74.

Lima, C.H.O., Sarmento, R.A., Rosado, J.F., Silveira, M.C.A.C., Santos, G.R., Pedro Neto, M., Erasmo, E.A.L., Nascimento, I.R., and Picanco, M.C. 2014. Efficiency and economic feasibility of pest control systems in water melon cropping. *Journal of Economic Entomology*, 107(3), 1118–1126.

Luna, R.K., Sharma, A., Sehgal, R.N., Kumar, R., and Gupta, R. 2008. Bioefficacy of drek (*Melia azedarach*) seeds against red pumpkin beetle, *Aulacophora foveicollis* Lucas (Coleoptera: Chrysomelidae). *Indian Journal of Forestry*, 31(3), 357–360.

Mahmood, T., Khokhar, K.M., and Shakeel, M. 2006. Comparative effect of different control measures on red pumpkin beetle, *Aulacophora foveicollis* on cucumber. *Sarhad Journal of Agriculture*, 22, 473–475.

Mahmood, T., Tariq, M.S., Khokhar, K.M., Hidayatullah, and Hussain, S.I. 2010. Comparative effect of different plant extracts and insecticide application as dust to control the attack of red pumpkin beetle on cucumber. *Pakistan Journal of Agricultural Research*, 23(3–4), 196–199.

Matin, T., Kamal, A., Gogo, E., Saidi, M., Deletre, E., Bonafos, R., Simon, S., and Ngouajio, M. 2014. Repellent effect of alphacypermethrin treated netting against *Bemisia tabaci* (Hemiptera: Aleyrodidae). *Journal of Economic Entomology*, 107(2), 684–690.

Mehta, R.K., Chandel, R.S., and Kashap, N.P. 2002. Control of fruit fly, *Batrocera cucurbitae* (Coquillett) on cucumber in Himachal Pradesh. *Pestology*, 26(10), 53–55.

Mohamed, M.A. 2012. Impact of planting dates, spaces and varieties on infestation of cucumber plants with white fly, *Bemisia tabaci* (Genn.). *Journal of Basic and Applied Zoology*, 65(1), 17–20.

Mondal, S. and Chatak, S.S. 2009. Bioefficacy of some indigenous plant extracts against epilachna beetle (*Henosepilachna vigintioctopunctata* Fab.) infesting cucumber. *Journal of Plant Protection Sciences*, 1(1), 71–75.

Mukherjee, A., Sarkar, N., and Barik, A. 2014. Long-chain free fatty acids from *Momordica cochinchinensis* leaves as attractants to its insect pest, *Aulacophora foveicollis* Lucas (Coleoptera: Chrysomeliade). *Journal of Asia-Pacific Entomology*, 17(3), 229–234.

Nath, D. and Ray, D.C. 2012. Traditional management of red pumpkin beetle, *Raphidopalpa foveicollis* Lucas in Cachar district, Assam. *Indian Journal of Traditional Knowledge*, 11(2), 346–350.

NYS. 2014. *Production Guide for Organic Cucumbers and Squash*. IPM Publication No. 135. New York State Department of Agriculture and Markets, New York.

Osman, M.S., Uddin, M.M., and Adnan, S.M. 2013. Assessment of the performance of different botanicals and chemical insecticides in controlling red pumpkin beetle, *Aulacophora foveicollis* (Lucas). *Persian Gulf Crop Protection*, 2(3), 76–84.

Pair, S.D. 1997. Evaluation of systematically treated squash trap plants and attracticidal baits for early-season control of striped and spotted cucumber beetles (Coleoptera: Chrysomelidae) and squash bug (Hemiptera: Coreidae) in cucurbit crops. *Journal of Economic Entomology*, 90(5), 1307–1314.

Patel, L.C. and Mondal, C.K. 2011. Effect of bait and lure based fruit fly management in cucumber. *Pestology*, 36(4), 15–19.

Pedersen, A.B. and Godfrey, L.D. 2011. Evaluation of cucurbitacin-based gustatory stimulant to facilitate cucumber beetle (Coleoptera: Chrysomelidae) management with foliar insecticides in melon. *Journal of Economic Entomology*, 104(3), 1294–1300.

Peng, J.C., Yeh, S.D., Huang, L.H., Li, J.T., Cheng, Y.F., and Chen, T.C. 2011. Emerging threat of thrips-borne melon yellow spot virus on melon and watermelon in Taiwan. *European Journal of Plant Pathology*, 130(2), 205–214.

Pitan, O.O.R. and Esan, E.O. 2014. Intercropping cucumber with amaranthus (*Amaranthus cruentus* L.) to suppress populations of major insect pests of cucumber (*Cucumis sativus* L.). *Archives of Phytopathology & Plant Protection*, 47(9), 1112–1119.

Prokopy, R.J., Miller, N.W., Pinero, J.C., Barry, J.D., Tran, L.C., Oride, L.K., and Vargas, R.I. 2003. Effectiveness of GF-120 fruit fly bait spray applied to border area plants for control of melon flies (Diptera: Tephritidae). *Journal of Economic Entomology*, 96, 1485–1493.

Quang, L.D., Gee, Y.L., and Yong, H.C. 2010. Insecticidal activities of crude extracts and phospholipids from *Chenopodium ficifolium* against melon and cotton aphid, *Aphis gossypii*. *Crop Protection*, 29(10), 1124–1129.

Rahaman, M.A. and Prodhan, M.D.H. 2007. Effect of net barrier and synthetic pesticides on red pumpkin beetle and yield of cucumber. *International Journal of Sustainable Crop Production*, 2(3), 30–34.

Rahaman, M.A., Prodhan, M.D.H., and Maula, A.K.M. 2008. Effect of botanical and synthetic pesticides in controlling epilachna beetle and the yield of bitter gourd. *International Journal of Sustainable Crop Production*, 3(5), 23–26.

Rai, D., Singh, R.K., Ram, H.H., and Singh, D.K. 2004. Evaluation of cucumber (*Cucumis sativus* L.) genotypes for field resistance to red pumpkin beetle (*Aulacophora foveicollis* Lucas). *Indian Journal of Plant Genetic Resources*, 17(1), 8–9.

Rai, N. 2008. *Handbook of Vegetable Science*. NIPA Publications, New Delhi, India, 765pp.

Rajak, D.C. and Singh, H.M. 2002. Comparative efficacy of pesticides against red pumpkin beetle, *Aulacophora foveicollis* on musk melon. *Annals of Plant Protection Sciences*, 10, 147–148.

Rathod, S.T., Borad, P.K., and Bhatt, N.A. 2009. Bioefficacy of neem-based and synthetic insecticides against red pumpkin beetle, *Aulacophora foveicollis* (Lucas) on bottle gourd. *Pest Management in Horticultural Ecosystems*, 15(2), 150–154.

Razmjou, J., Mohammadi, M., and Hassanbour, M. 2011. Effect of vermicompost and cucumber cultivation on population growth attributes of the melon aphid (Hemiptera: Aphididae). *Journal of Economic Entomology*, 104(3), 1379–1383.

Said, M.K. and Muhammad, J. 2000. Efficacy of different concentrations of sevin dust against red pumpkin beetle (*Aulacophora foveicollis* Lucas) causing damage to musk melon (*Cucumis melo* L.). *Pakistan Journal of Biological Sciences*, 3(1), 183–185.

Saljoqi, A.U.R. and Khan, S. 2007. Relative abundance of the red pumpkin beetle, *Aulacophora foveicollis* Lucas on different cucurbitaceous vegetables. *Sarhad Journal of Agriculture*, 23(1), 109–114.

Sankaraiyah, K. 2010. Ovipositional deterrence activity of select plants against *Henosepilachna vigintiocto-punctata* (Fabr.) Coleoptera: Coccinellidae. *Indian Journal of Entomology*, 72(3), 273–275.

Sapkota, R., Dahal, K.C., and Thapa, R.B. 2010. Damage assessment and management of cucurbit fruit flies in spring-summer squash. *Journal of Entomology and Nematology*, 2(1), 7–12.

Satpathy, S. and Rai, S. 2002. Luring ability of indigenous food baits for fruit fly, *Batrocera cucurbitae* (Coq.). *Indian Journal of Entomology*, 26(3), 249–252.

Shivalingaswamy, T.M., Kumar, A., Satpathy, S., Bhardwaj, D.R., and Rai, A.B. 2008. Relative susceptibility of bottle gourd cultivars to red pumpkin beetle, *Aulacophora foveicollis* Lucas. *Journal of Vegetable Science*, 15(1), 97.

Shooker, P., Khayrattee, F., and Permalloo, S. 2006. Use of maize as trap crop for the control of melon fruit fly, *B. cucurbitae* (Diptera: Tephritidae) with GF-120. Biocontrol and other control methods, 354pp. http://www.fela.edu/FlaEnt/fe87.

Sidahmed, O.A.A., Taha, A.K., Hassan, G.A., and Abdalla, I.F. 2014. Evaluation of pheromone dispenser units in methyl eugenol trap against *Batrocera invadens* Drew, Tsuruta & White (Diptera: Tephritidae) in Sudan. *Sky Journal of Agricultural Research*, 3(8), 148–151.

Smyth, R.R. and Hoffman, M.P. 2010. Seasonal incidence of two co-occurring adult parasitoids of *Acalymma vittatum* in New York state: *Centistes (Syrrhizus) diabroticae* and *Celatoria setusa*. *Biocontrol*, 55, 219–228.

Snyder, W.E. and Wise, D.H. 2001. Contrasting trophic cascades generated by a community of generalist predators. *Ecology*, 82, 1571–1583.

Sood, N. and Sharma, D.C. 2004. Bioefficacy and persistent toxicity of different insecticides and neem derivatives against cucurbit fruit fly, *Batrocera cucurbitae* Coq. on summer squash. *Pesticide Research Journal*, 16(2), 22–25.

Stonehouse, J.M., Afzal, M., Zia, Q., Mumford, J.D., Poswal, A., and Mahmood, R. 2002. "Single-killing-point" field assessment of bait and lure control of fruit flies (Diptera: Tephritidae) in Pakistan. *Crop Protection*, 21(8), 651–659.

Stonehouse, J.M., Mumford, J.D., Verghese, A., Shukla, R.P., Satpathy, S., Singh, H.S., Thomas, J. et al. 2007. Village level area-wide fruit fly suppression in India: Bait application and male annihilation at village level and farm level. *Crop Protection*, 26, 788–793.

Tandon, P. and Sirohi, A. 2009. Laboratory assessment of the repellent properties of the ethanolic extracts of four plants against *Raphidopalpa foveicollis* Lucas (Coleoptera: Chrysomelidae). *International Journal of Sustainable Crop Production*, 4(2), 1–5.

University of Delaware, 2012. *IPM—Cucurbit Scouting Guidelines*. College of Agriculture and Natural Resources, Newark, DE.

Vargas, R.I., Burns, R.E., Mau, R.F.L., Stark, J.D., Cook, P., and Pinero, J.C. 2009. Captures in methyl eugenol and cue-lure detection traps with and without insecticides and with a farma tech solid lure and insecticide dispenser. *Journal of Economic Entomology*, 102(2), 552–557.

Vargas, R.I., Leblanc, L., Putoa, R., and Eitam, A. 2007. Impact of introduction of *Batrocera dorsalis* (Diptera: Tephritidae) and classical biological control releases of *Fopius arisanus* (Hymenoptera: Braconidae) on economically important fruit flies in French Polynesia. *Journal of Economic Entomology*, 100(3), 670–679.

Vargas, R.I., Pinero, J.C., Jang, E.B., Mau, R.F.L., Gomez, L., Stoltman, L., and Mafra-Neto, A. 2010. Response of melon fly (Diptera: Tephritidae) to weathered SPLAT-spinosad-Cue-Lure. *Journal of Economic Entomology*, 103(5), 1594–1602.

Vargas, R.I., Souder, S.K., Hoffman, K., Mercoghiano, J., Smith, T.R., Hammond, J., Davis, B.J., Brodie, M., and Dripps, J.E. 2014. Attraction and mortality of *Batrocera dorsalis* (Diptera: Tephritidae) to STATIC spinosadME weathered under operational conditions in California and Florida: A reduced-risk male annihilation treatment. *Journal of Economic Entomology*, 107(4), 1362–1369.

Vargas, R.I., Souder, S.K., Mackey, B., Cook, P., and Stark, J.D. 2012. Field trials of solid triple lure (Trimedlure, methyl eugenol, raspberry ketone and DDVP) dispensers for detection and male annihilation of *Ceratitis capitata*, *Batrocera dorsalis* and *Batrocera cucurbitae* (Diptera: Tephritidae) in Hawaii. *Journal of Economic Entomology*, 105(5), 1557–1565.

Vargas, R.I., Stark, I.D., Hertlein, M., Mafra-Neto, A.C.R., and Pinero, J.C. 2008. Evaluation of SPLAT with spinosad used methyl eugenol or Cue-lure for "attract and kill" of oriental and melon fruit flies (Diptera: Tephritidae) in Hawaii. *Journal of Economic Entomology*, 101(3), 759–768.

Verma, H., Singh, S., and Ahuja, D.B. 2013. Fruit characters of round gourd in relation to fruit fly, *Batrocera cucurbitae* Coquillett. *Indian Journal of Entomology*, 72, 169–171.

Verma, H., Singh, S., Jat, B.L., and Ahuja, D.B. 2010. Evaluation of ecofriendly IPM modules against fruit fly, *Batrocera cucurbitae* (Coquillett) on round gourd. *Indian Journal of Entomology*, 72(2), 185–187.

Verma, S.C. 2012. Management of red pumpkin beetle, *Raphidopalpa foveicollis* Lucas and serpentine leaf miner, *Liriomyza trifolii* Burgess on cucumber under mid-hill conditions of Himachal Pradesh. *Journal of Insect Science*, 25(3), 311–313.

Vishwakarma, R., Chand, P., and Ghatak, S.S. 2011b. Potential plant extracts and entomopathogenic fungi against red pumpkin beetle, *Raphidopalpa foveicollis* (Lucas). *Annals of Plant Protection Science*, 19(1), 84–87.

Vishwakarma, R., Prasad, P.H., Ghatak, S.S., and Mondal, S. 2011a. Bioefficacy of plant extracts and entomopathogenic fungi against epilachna beetle, *Henosepilachna vigintioctopunctata* (Fabricius) infesting bottle gourd. *Journal of Insect Science*, 24, 65–70.

Webb, S.E. 2013. *Insect Management for Cucurbits*. Florida Cooperative Extension Service, University of Florida, Gainesville, FL, Doc. No. ENY-460.

Williamson, J. 2014. *Cucumber, Squash, Melon & Other Cucurbit Insect Pests.* Clemson University Cooperative Extension Service, Clemson, SC.

Xue, H.W. and Wu, W.J. 2013. Preference of *Batrocera cucurbitae* (Diptera: Tephritidae) to different colors: A quantitative investigation using virtual wavelength. *Acta Entomologica Sinica*, 56(2), 161–166.

Yardim, E.N.N., Arancon, N.Q., Edwards, C.A., Oliver, T.J., and Byrne, R.J. 2006. Suppression of tomato hornworm (*Manduca quinquemaculata*) and cucumber beetles (*Acalymma vittatum* and *Diabrotica undecimpunctata*) populations and damage by vermicomposts. *Pedobiologia*, 50, 23–29.

Zhang, R., He, S., and Chen, J. 2014. Monitoring of *Batrocera dorsalis* (Diptera: Tephritidae) resistance to cyantraniliprole in the south of China. *Journal of Economic Entomology*, 107(3), 1233–1236.

21 Cucurbit Insect and Related Pests

Paul J. McLeod and Tahir Rashid

CONTENTS

21.1   INTRODUCTION

The array of cucurbit insects is as diverse as the types of cucurbits described elsewhere in this book. Insects affect all cucurbits and range from the smallest whiteflies, aphids, and thrips to lepidopterous caterpillars that can reach 40 mm at maturity. Insects feed directly on cucurbit foliage, stems, roots, and fruits. Plant injury can vary from insignificant to loss of fruit or plant death. Not only is direct feeding injurious to the fruit, making it unmarketable, but feeding on foliage may also weaken plants to the point of reducing or preventing harvest. Among the most significant injury, transmission of plant pathogens is often considered the greatest impact that insects have on cucurbit production. Among the most damaging insects is the cucumber beetle complex. Although these beetles can damage and occasionally kill seedling cucurbits emerging from the soil, their greatest impact is the transmission of bacterial wilt that can cause a sudden wilt and plant decline. Aphids generally have little direct impact on cucurbits but can transmit potyviruses. These plant diseases have devastated squash production in many areas of the United States and throughout the world in recent years. Despite many efforts at breeding cucurbits for disease resistance and intensive insect management, these two diseases continue to plague both the home gardener and the commercial producer of cucurbits. Attempting to reduce cucurbit diseases by managing cucumber beetles and aphids is not likely to prove successful by the home gardener. The third insect often causing significant problems is the squash bug. Many additional insects attack cucurbits but are not common and their impact is usually limited (Capinera 2001 and Mekinlay 1992).

21.2   INSECTS AND RELATED PESTS OF CUCURBITS

21.2.1   Aphids, including the Melon Aphid *Aphis gossypii* Glover (Hemiptera: Aphididae)

21.2.1.1   Identification and Biology

Several aphids can be detected in cucurbit production (Photo 21.1). Perhaps the most common is the melon aphid, also known as the cotton aphid. Melon aphids occur worldwide and can be detected on many host plants, both domestic and wild. Blackman and Eastop (1984) stated that the melon aphid occurred on over 700 plants. Adults are small (about 2 mm in length), soft-bodied insects that feed on plants by inserting their stylet into the plant and removing large amounts of plant sap. In warm climates, the most common forms are wingless (apterous) females that reproduce parthenogenetically.

PHOTO 21.1   Adult and immature aphids.

In cooler environments, the melon aphid may have wings (alate) and reproduce sexually and lay eggs. Color varies from yellow to dark green. Feeding commonly occurs on the underside of foliage, causing a downward cupping, and along new growth terminals. Not only do aphids adversely affect plants by feeding, they produce honeydew upon which black sooty mold grows. Perhaps the greatest impact of aphids on cucurbits is the transmission of viral diseases that can result in total loss of harvest. Much of the U.S. cucurbit production area was devastated by potyvirus diseases during the 1980s. These diseases are still prevalent. Aphid reproduction is rapid, and newly born aphids can reach maturity and reproduce in as little as 1 week. Thus, numerous generations occur each year. As a result, population increases can be dramatic. But just as dramatic, populations often crash due to the action of natural enemies, including fungal pathogens and parasitic and predatory insects. Insecticide application often decreases the effects of beneficial organisms and large increases in aphid populations may result.

21.2.1.2 Management

In many areas, aphids on cucurbits may be managed by natural enemies. These include predatory insects such as lady beetles, syrphid fly larvae, and ant lion larvae; parasitic wasps; and fungi pathogens. Avoidance or delayed use of insecticides may promote the impact of natural enemies on cucurbit aphids. Additional tactics include planting after temperatures are warm and appropriate use of fertilization and irrigation. This will promote vigorous plant growth that may reduce aphid impact. In subtropical areas of commercial cucurbit production such as the southeastern United States, the use of foliar insecticide sprays may be required. Spray coverage is critical as aphids prefer the underside of foliage. Reflective mulches and application of oil have been reported to slow the spread of viral diseases.

21.2.2 Cucumber Beetles, including the Spotted Cucumber Beetle *Diabrotica undecimpunctata* Mannerheim, the Striped Cucumber Beetle *Acalymma vittatum* (Fabricius), the Western Striped Cucumber Beetle *Acalymma trivittatum* (Mannerheim), and the Banded Cucumber Beetle *Diabrotica balteata* LeConte (Coleoptera: Chrysomelidae)

21.2.2.1 Identification and Biology

Soon after cucurbit seedlings emerge, cucumber beetles attack the new plants. Adult beetles pass the winter and early spring months in weeds and brush near production fields (McLeod 2006). Although beetles can be found feeding on several noncucurbit hosts, they prefer cucurbits and are capable of inflicting severe damage to emerging seedlings. When numbers are high, as often occurs in the southern United States, their feeding may result in death of cucurbit seedlings. Adult beetles are about 5 mm long and emerge from pupa within the soil, mate and deposit up to a few hundred eggs near plants but in the soil. Larvae develop within the soil by feeding on plant roots but damage from feeding by the larvae is usually minimal. The spotted cucumber beetle adult (Photo 21.2) occurs throughout the United States and has six black spots on each outer wing, while the striped cucumber beetle (Photo 21.3) occurs east of the Rocky Mountains and is slightly smaller and has three black lines on the upper abdomen. Adult spotted cucumber beetles feed on numerous host plants including cucurbits, beans, peas, corn, and many others. Cucurbits, however, are their preferred host plant along with the striped cucumber beetle. The western striped cucumber beetle is similar to the striped cucumber beetle but occurs west of the Rocky Mountains (Davidson and Lyon 1979). The banded cucumber beetle (Photo 21.4) occurs in warmer regions of Central America and only in the southern United States. This species has three transverse bands or spots across the elytra. When adult populations are high, damage to cucurbit seedlings may be excessive and plants may be killed (Photos 21.5 and 21.6). This often requires management of the adult beetles. Once plants develop a few true leaves, plants can generally tolerate additional damage and management is

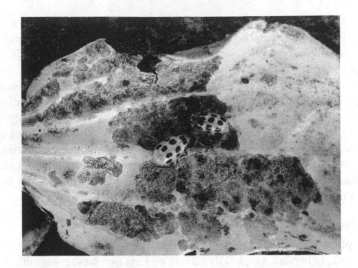

PHOTO 21.2   (See color insert.) Spotted cucumber beetle.

PHOTO 21.3   Striped cucumber beetle.

PHOTO 21.4   (See color insert.) Banded cucumber beetle.

PHOTO 21.5   Foliar damage by cucumber beetles.

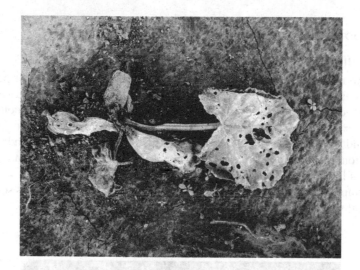

PHOTO 21.6   Seedling damage by cucumber beetles.

usually not required. Cucumber beetles are also responsible for the transmission of bacterial wilt, a disease that results in a very rapid wilt of cucurbit plants. Adult beetles may also feed on the outer skin of fruit, but this is generally minimal (Photo 21.7). Multiple generations occur each year.

21.2.2.2 Management

The most susceptible stage of cucurbits to cucumber beetle attack is just after plants emerge from the soil, if directly seeded, and immediately after transplants are set into the field. Planting when temperatures are warm and the use of irrigation and fertilization will promote rapid plant growth and the resulting damage from adult beetles will be minimized. If beetle numbers increase beyond acceptable levels, the use of insecticides may be warranted. In areas with a history of high cucumber beetle populations, treating the soil or seed with neonicotinoid insecticides can protect emerging seedlings. Once plants have a few true leaves, low numbers of adult beetles may be tolerated with a minimal effect on yield. The use of insecticides for reduction of bacterial wilt is questionable.

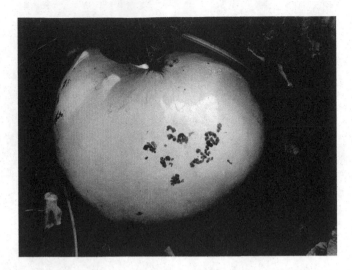

PHOTO 21.7   Cucumber beetle feeding on fruit.

21.2.3   Leaffooted Bug *Leptoglossus* spp. (Hemiptera: Coreidae)

21.2.3.1   Identification and Biology

In much of the world, especially Africa and Asia, the leaffooted bug is a major pest of cucurbits. In the United States, it is sporadic in abundance. Adults are up to 25 mm in length and dark grey in color. A white band may occur across the elytra (Photo 21.8). The name "leaffooted" comes from the flat leaf-like expansion on the hind tibia. Adults along with nymphs (Photo 21.9) feed on the foliage, stems, and fruits by inserting their stylet into the plant tissue. Sap is withdrawn resulting in wilt, leaf distortion, and fruit deformation. Overwintering is on plant debris, and in spring, adults mate and females lay dark eggs in a row on plant material. Nymphs emerge in about 1 week and complete five nymphal instars in approximately 30 days. The host range includes several vegetables along with several weeds including thistles.

PHOTO 21.8   Adult leaffooted bug.

PHOTO 21.9 Immature leaffooted bug.

21.2.3.2 Management

Little is known about the impact of beneficial organisms on leaffooted bugs. In small gardens, adults and nymphs can be picked and destroyed. In commercial cucurbit production, low numbers may be tolerated with little effect on yield. In fields with greater numbers, insecticide use may be warranted.

21.2.4 PICKLEWORM *DIAPHANIA NITIDALIS* (STOLL) AND MELONWORM *DIAPHANIA HYALINATA* L. (LEPIDOPTERA: PYRALIDAE)

21.2.4.1 Identification and Biology

Both pickleworm and melonworm occur throughout much of the world including the United States, Asia, and on the Caribbean Islands. Adult pickleworm and melonworm moths lay eggs individually or in small groups on the leaf surface of cucurbits (Photo 21.10). Eggs generally hatch in 3–4 days, and emerging larvae initially feed on cucurbit foliage. While melonworm larvae

PHOTO 21.10 Adult melonworm.

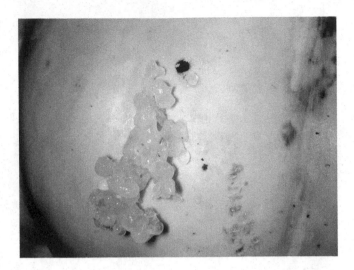

PHOTO 21.11   Fruit damage by pickleworm.

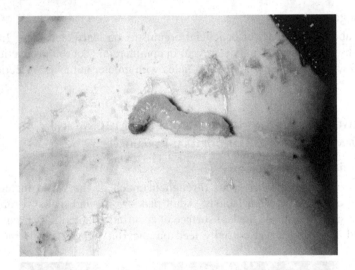

PHOTO 21.12   Pickleworm larvae.

continue to develop on foliage, pickleworm larvae move to the fruit after about 1 week. Here, the 1.3 cm long larvae chew an entrance hole and enter the fruit (Photo 21.11). Not only does the tunneling cause fruit injury, but fungi are introduced that further destroy the fruit. Infested or damaged fruit is unmarketable. In areas with high pickleworm populations, entire fields can be lost. Melonworm larvae generally feed only on foliage where development is completed in about 2 weeks. Upon reaching maturity, larvae approach 30 mm in length. Mature pickleworm larvae are light green and without stripes (Photo 21.12). Mature melonworm larvae are dark green and possess two lateral white stripes (Photo 21.13). Both species usually pupate on the foliage within a chamber composed of rolled leaves.

21.2.4.2 Management

Diaphania populations rarely reach high levels in home gardens and attempts at management by home gardeners are not advisable. If management cannot be avoided, the use of "hard insecticides" should be delayed as long as possible. *Bacillus thuringiensis* (BT) likely offers the best alternative

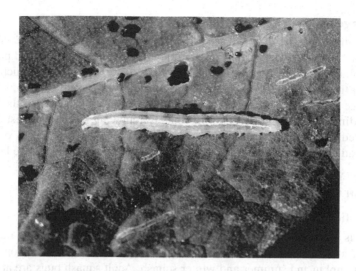

PHOTO 21.13   (See color insert.) Melonworm larvae.

to the "hard insecticide." In small fields, searching for damaged fruit and removal may reduce the insect population to an acceptable level. Sanitation, including destruction of crops immediately following the last harvest, will likely reduce damage in subsequent plantings.

21.2.5   Spider Mite *Tetranychus urticae* Koch (Acari: Tetranychidae)

21.2.5.1   Biology

Although not true insects, mites impact cucurbit production in a similar manner as insects and are thus included in this book. Mites are minute (<0.5 mm in length), have a cosmopolitan distribution, and feed by extracting sap from leaves. The result of this feeding is a whitish or bronze appearance on the leaves (Foster and Flood 1995). When large numbers of mites are present, silk webbing can be seen on leaves and stems (Photo 21.14). Adult females deposit up to about 100 eggs on the foliage. Larvae emerge in as few as 3 days and begin to feed. Within about 1 week, adults may appear, and the cycle is repeated. Multiple generations occur each year. Hot, dry weather promotes mite increase.

PHOTO 21.14   (See color insert.) Spider mites with silk webbing.

21.2.5.2　Management

Low numbers of mites may be tolerated with a minimal effect on yield. If mite numbers increase beyond acceptable levels, the use of an acaricide may be warranted. Mites feed predominately on the bottoms of foliage. Thus, acaricides must be applied in such a manner in which the mite comes into contact with the insecticide, that is, apply the acaricide to the leaf bottom. The use of a surfactant and a large volume of water should assist the spread of the material to the bottom surface. Sanitation, including destruction of crops immediately following the last harvest, may have some effectiveness in reducing damage in subsequent plantings. In areas with a history of high mite numbers, removal of adjacent weeds several weeks prior to transplanting may reduce later infestations.

21.2.6　Squash Bug *Anasa tristis* (De Geer) (Hemiptera: Coreidae)

21.2.6.1　Identification and Biology

The squash bug is often a major pest of cucurbits in the United States. It also occurs throughout Central America and in southern Canada. Each year, reports of significant populations surface, particularly on pumpkin and summer and winter squash. Adult squash bugs are about 18 mm long and dark gray (Photo 21.15). Like other true bugs, adults have outer wings with the distal half membranous and the proximal portion hardened. Adults overwinter in old cucurbit fields or in nearby borders. With the arrival of spring, squash bugs migrate into recently planted cucurbit fields. They feed by inserting their piercing mouthparts into the plant, injecting a digestive toxin, and extracting plant fluids. Shortly after mating, an egg mass is laid, typically in a pattern, on the leaf surface (Photo 21.16). Eggs are a shiny bronze and hatch in 6–14 days. Nymphs are wingless and feed gregariously on all parts of the plant (Photo 21.17). Numbers of nymphs and adults may reach several hundred on a single plant. Heavily infested plants are severely weakened and fruit production and quality are greatly reduced. Egg laying by a single adult may continue for several weeks. Thus, all stages of nymphs may be present along with adults. One generation occurs each year.

PHOTO 21.15　Adult squash bug.

PHOTO 21.16   Squash bug eggs.

PHOTO 21.17   Immature squash bugs.

21.2.6.2   Management

Hand picking of adults, nymphs, and egg masses may provide the home gardener with a sufficiently effective form of management. Also, and of utmost importance, is field sanitation. Squash bugs feed only on cucurbits, and reducing the length of time host plants are available greatly impacts population. Thus, destruction of crops immediately following the last harvest will likely reduce damage in subsequent plantings and will reduce overwintering habitat. Also, elimination of cucurbit weeds in adjacent areas will aid in population reduction. As plants develop in the spring and early summer, they should be searched for adult squash bugs that are often detected on the plant base or on soil near the plant. Low numbers of adults may be tolerated, particularly during periods of rapid plant development. Plants may simply outgrow the damage from low squash bug numbers. Proper fertilization and irrigation are important. Impact of beneficial insects on squash bug is minimal. In commercial plantings, the use of insecticides may be required. Bifenthrin currently offers the best level of control (Thompson 2001).

21.2.7 SQUASH VINE BORER *MELITTIA CUCURBITAE* (HARRIS) (LEPIDOPTERA: SESIIDAE)

21.2.7.1 Identification and Biology

The squash vine borer is a noteworthy pest in areas where cucurbits, especially pumpkins, are grown annually. It occurs in east of the Rocky Mountains in the United States and portions of Central America and southern Canada. Adult moths emerge from overwintering larvae or pupae from the soil in early spring, and mate and lay eggs singly on cucurbit stems (Cranshaw 2004). Caterpillars emerge in 4–10 days and enter the stem to feed (Photo 21.18). Initially, feeding has little effect on the plant but as larvae mature within 3–4 weeks, the infested plant or a portion of a plant may suddenly wilt. Close examination of the stem close to the soil will reveal a mass of caterpillar excrement expelled from the stem (Photo 21.19). By splitting the stem with a knife, the larvae can be located. Mature larvae are white with a brown head and reach about 28 mm in length (Photo 21.20). Prior to pupation, larvae exit the plant and burrow into the soil. Little is known of the squash vine borer biology in the United States, but two generations likely occur annually.

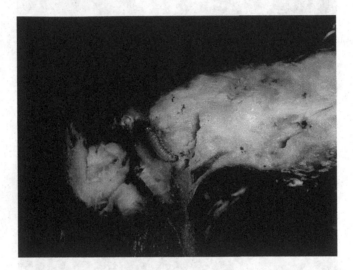

PHOTO 21.18   Young squash vine borer larvae.

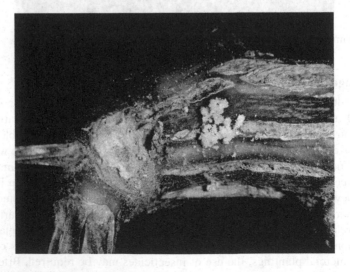

PHOTO 21.19   Squash vine borer excrement on stem.

PHOTO 21.20   Mature squash vine borer larvae.

21.2.7.2   Management

Of utmost importance is field sanitation. Squash vine borers feed only on cucurbits and reducing the length of time host plants are available greatly impacts population. Thus, destruction of crops immediately following the last harvest will likely reduce damage in subsequent plantings. Disking cucurbit fields after harvest and during the fall and winter will expose borers to the adverse effects of winter and kill overwintering stages. Low numbers of borers may be tolerated, particularly during periods of rapid plant development. Plants may simply outgrow the damage from low squash vine borer numbers. Proper fertilization and irrigation are important. In small gardens, damaged stems can be split with a knife and exposed larvae destroyed. By killing the larvae and covering the damaged stalk with soil, the plant may recover sufficiently to produce fruit. The use of synthetic insecticides may not prove beneficial.

21.2.8   Thrips, including *Thrips palmi* (Thysanoptera: Thripidae)

21.2.8.1   Biology

Thrips are minute insects found on the blooms and leaf surfaces of cucurbits. Distribution is worldwide. Adults are dark brown to black and about 1 mm long (Photo 21.21). Two pairs of wings are

PHOTO 21.21   Adult thrips.

present and long hairs occur on the edges of the wings (Photo 21.22). Immature thrips are smaller and lighter in color. Adults and immature stages "rasp" the leaf surface with their mouthparts (Photo 21.23) and feed on the exuding plant sap. Not only can this feeding cause plant injury, but thrips can also transmit plant diseases. Large populations of thrips can cause yellowing of foliage and stunting. After mating females insert eggs into the plant tissue. Larvae emerge and feed by scraping the plant surface and ingesting plant sap. Maturity is reached in less than 1 week and pupation occurs in the soil. Adults emerge in about 4 days, and the cycle is repeated. Many generations occur annually.

PHOTO 21.22   Thrips

PHOTO 21.23   Thrips mouth parts.

21.2.8.2 Management

Low numbers of thrips may be tolerated with a minimal effect on yield. If insecticides are used, systemic insecticides may provide more effective control. These include the neonicotinoids. Sanitation, including destruction of crops immediately following the last harvest, may have some effectiveness in reducing damage in subsequent plantings.

21.2.9 WHITEFLY *BEMISIA* SPP. (HEMIPTERA: ALEYRODIDAE)

21.2.9.1 Identification and Biology

Whiteflies occur worldwide but are most injurious in warm climates. In small, home gardens, whiteflies rarely cause significant problems, probably due to the diversity of plantings within a small area. In larger gardens with large numbers of cucurbits, whiteflies have become increasingly important. In years with hot dry summers, numerous reports of whitefly problems arise from cucurbit producers, especially those who produce pumpkins in the fall (Photo 21.24). Adult whiteflies are minute (about 1 mm in length) and white in color (Photo 21.25). Eggs are inserted into leaf tissue

PHOTO 21.24 **(See color insert.)** Pumpkins damaged by whitefly.

PHOTO 21.25 Adult whitefly.

and nymphs, also known as "crawlers," emerge and feed by extracting sap from plants. Not only does this feeding damage the plant, but whiteflies are also responsible for virus disease transmission. Shortly after hatching, nymphs attach themselves to the underside of the leaf surface where they feed for 1–4 weeks. Developmental time is greatly dependent on temperature. Whiteflies can be found on most vegetable crops and also occur on numerous alternative hosts including weeds (Sorensen and Baker 1994).

21.2.9.2 Management

Most summers, home gardeners should not have to manage their cucurbits for whiteflies and toleration is the best practice. Several additional practices may improve whitefly management. First, location of cucurbit fields in areas of low whitefly populations should be attempted. In arid areas or during periods of drought, whitefly management is difficult and may not be successful. Many weeds, especially velvetleaf, harbor large whitefly populations. Removal of these alternate host plants several weeks prior to the cucurbit production season will reduce the likelihood of high whitefly populations at planting. Sanitation, including destruction of crops immediately following the last harvest, will likely reduce damage in subsequent plantings. In commercial plantings, the use of insecticides may be required. Scouting for whiteflies is important because high populations may be very difficult to manage. When low numbers of whiteflies are detected and plants are young, the use of neonicotinoid insecticides has proven successful.

REFERENCES

Blackman, R.L. and V.F. Eastop. 1984. *Aphids on the World's Crops: An Identification and Information Guide*. Chichester, U.K.: John Wiley & Sons.

Capinera, J.L. 2001. *Handbook of Vegetable Pests*. San Diego, CA: Academic Press.

Cranshaw, W. 2004. *Garden Insects of North America*. Princeton, NJ: Princeton University Press.

Davidson, R.H. and W.F. Lyon. 1979. *Insect Pests of Farm, Garden, and Orchard*. New York: John Wiley & Sons.

Foster, R. and B. Flood. 1995. *Vegetable Insect Management with Emphasis on the Midwest*. Willoughby, OH: Meister Publishing Co.

McKinlay, R.G. 1992. *Vegetable Crop Pests*. Boca Raton, FL: CRC Press, Inc.

McLeod, P. 2006. *Identification, Biology and Management of Insects Attacking Vegetables in Arkansas*. Santa Cruz, Bolivia: Sirena Press.

Sorensen, K.A. and J.R. Baker. 1994. *Insect and Related Pests of Vegetables*. Raleigh, NC: North Carolina State University Press.

Thomson, W.T. 2001. *Agricultural Chemicals: Book I—Insecticides*. Fresno, CA: Thomson Publications.

Section IX

Therapeutic and Medicinal
Values of Cucurbits

22 Cucurbits
Assortment and Therapeutic Values

Krishan Pal Singh, Beena Singh, Prem Chand, and Reena Nair

CONTENTS

22.1 INTRODUCTION

Vegetable crops include a large number of species primarily used as an essential component in our daily diet. These provide vitamins, minerals, fiber, amino acids, and active metabolites, thereby offering health benefits when used in diets. In general, plant genetic resources include plant species that not only provide food, feed, fodder, fiber, and medicine but also are sources of shelter, energy, and other uses that support livelihood.

A plant group with the predominant species used as human food belongs to the family Cucurbitaceae. In spite of the existing marginalization of some of these species, from very remote times, all have contributed as indispensable food products to the diet of rural as well as urban communities of the Asian continent and many other parts of the world. Cucurbits form one of the excellent fruits in nature having a composition of all the essential constituents required for good health of humans. But still the medicinal property of this family is not well thought out. These are used for daily consumption as vegetables and salads because of their availability at low cost.

Cucurbitaceous vegetables form a distinct group among cross-pollinated vegetable crops with respect to the method of improvement. Besides having wide-ranging sex forms and sex expressions that favor outbreeding, they are not completely cross-pollinated as natural self-pollination takes place to some extent. An added distinctive feature of this group is that they do not suffer from inbreeding depression, or do so only negligibly, in contrast to cross-pollinated crops. The phenomenon allows breeding of cultivars through pure line or single-plant selection.

Genetic diversity in a crop signifies adaptation to different environments and growing conditions. Certain genotypes inherently have the ability to withstand drought and poor soil, to resist insects and pests, or to give higher protein yields. Genetic diversity is important in horticulture for breeding plants with characteristics to the ecological conditions, nutritional needs, and other uses by farmers and for conferring at least partial resistance to insects and diseases. There subsists a symbiotic association in the ecological niche in which the crop grows. Diversity plays an important role in nutrient cycling, controlling insect population, and preventing plant disease.

Conservation of plant genetic resources is of crucial importance for the continuous availability and improvement of crops. Wide variation in cucurbit germplasms is prominent with regard to the color of seeds, growth in dryland and/or uplands, climatic effects, and exacting characters in fruits or plants regionwise. Hundreds of such genotypes subsist but are endangered by extinction. These diverse types were developed over the years to suit different ecological conditions.

Cucurbits belong to the family Cucurbitaceae that comprises of 118 genera and 825 species in the world; out of these, nearly 36 genera and 100 species are found in India, although most have other world origins (Whitaker and Davis, 1962). The genetic diversity in cucurbits extends to both vegetative and reproductive characteristics. Cucurbits are mostly annual, monoecious, herbaceous, tendril-bearing vines propagated through seeds unlike the dioecious ones that have perennial habit and are propagated through stem cuttings. This group includes gourds, melons, pumpkins, and squashes that make it the largest group of summer vegetables like ash gourds, bitter gourds, bottle gourds, ridge gourds, round gourds, snake gourds, sponge gourds, little gourds, pointed gourds, pumpkins, muskmelons, watermelons, summer squashes, winter squashes, kakrols, kartolis, and other large number of mostly trailing and climbing vine crops. These are characterized by their fleshy fruits. In this family approximately 38 species are economically important, and its cultivation is done throughout the world from tropical to temperate zones. Out of these species, *Citrullus*, *Cucumis*, *Cucurbita*, and *Lagenaria* are of great economic importance and are cultivated commercially. Some are adapted to humid conditions and others are found in arid areas. Most are susceptible to frost and so they are grown with protection in temperate areas or to coincide with the warm portion of the annual cycle.

The family Cucurbitaceae is well defined and is taxonomically isolated from other plant families. Two subfamilies, namely, Zanonioideae and Cucurbitoideae, are well characterized, the former by possessing small, striate pollen grains and the latter by having the styles united into a single column. Mainly, the food plants fall within the subfamily Cucurbitoideae. Further definition finds cucumber (*Cucumis sativus* L.) and melon (*Cucumis melo* L.) to be within the subtribe Cucumerinae, tribe Melothrieae, and watermelon (*Citrullus lanatus* [Thunb.] Matsum and Nakai) assigned to the tribe Benincaseae, subtribe Benincasinae. Many cultivated and wild species of Cucurbitaceae date back to prehistoric times (Table 22.1).

22.2 SPECIFIC CHARACTERS OF CUCURBITS

There is a substantial range in the monoploid (x) chromosome number (Jeffrey, 1990) in cucurbits, including 7 (*C. sativus*), 11 (*Citrullus* sp., *Momordica* sp., *Lagenaria* sp., *Sechium* sp., and *Trichosanthes* sp.), 12 (*Benincasa hispida*, *Coccinia cordifolia*, *Cucumis* sp., except *C. sativus* and *Praecitrullus fistulosus*), 13 (*Luffa* sp.), and 20 (*Cucurbita* sp.) (Table 22.2).

22.3 GENETIC RESOURCES OF CUCURBITACEOUS VEGETABLES

The plant group with the most number of species used as human food is of the Cucurbitaceae family. Within this family, *Cucurbita* stands out as one of the most important genera.

Cucurbita species: The genus *Cucurbita* comprises of about 27 species (both wild and cultivated) mostly concentrated in the tropical and subtropical regions of the Central and South America. In this genus, few species are of commercial importance, namely, field pumpkin or kashiphal (*Cucurbita moschata* Duch. Ex. Poir.), winter squash or pumpkins (*Cucurbita maxima* Duch. Ex. Lam.), summer squash or squash or chappan kaddu of north India or vegetable marrow (*Cucurbita pepo* L.), and winter squash (*Cucurbita mixta* Pang. syn. *C. angyrosperma*). *C. pepo* is introduced as a crop that withstands cooler climate. *Cucurbita texa* Gray grows wild in Texas (Chadha and Lal, 1993).

TABLE 22.1
Specific Characteristics of the Family Cucurbitaceae

Features	Habit/Characteristics
Plant type (herbaceous)	Plants are mostly climbing/trailing plants; usually with tendrils present at nodes; simple or branched
Leaves	Alternate, palmate, and sometimes pinnately veined; often lobed; sometimes palmate compound and extra floral nectarines often present; stipules absent
Flowers	Mostly unisexual (monoecious or dioecious); usually actinomorphic; mostly in axillary inflorescences or solitary in axils
Calyx	Mostly five-lobed
Corolla	Five-lobed or has five free petals; the corolla of male flowers sometimes unlike that of female flowers
Stamens	Basically five in number; alternate with corolla lobes but often reduced to three, with two having tetrasporangiate dithecal anthers and one having a bisporangiate monothecal anther; stamens free or variously united; staminodes may be present in female flowers
Gynoecium	Two, three, or five carpels unite to form an inferior or half-inferior compound ovary; ovules one to many, anatropous; style commonly solitary, stigmas one to five, leaves entire or lobed
Fruit	A pepo, berry, gourd, or capsule; indehiscent or dehiscing via operculum valves or by splitting irregularly, occasionally explosively dehiscent
Seeds	One to many; large and commonly compressed, sometimes winged, surfaces smooth or variously ornamented; embryo large; endosperm absent or nearly so

Source: More, T.A. and Bhardwaj, D.R., Cucurbit biodiversity, *National Symposium on Vegetable Biodiversity*, Indian Society of Vegetable Science, Indian Institute of Vegetable Research and Jawaharlal Nehru Krishi Vishwa Vidyalaya, Jabalpur, India, April 4–5, 2011, pp. 20–31.

TABLE 22.2
Cucurbits Prevailing in the World and in India

Cucurbits	World	India
Benincasa sp.	1	1
Citrullus sp.	4	2
Lagenaria sp.	6	1
Sechium sp.	8	1
Luffa sp.	9	6
Cucurbita sp.	27	5
Coccinia sp.	30	1
Trichosanthes sp.	44	22
Cucumis sp.	50	5
Momordica sp.	60	7

Source: More, T.A. and Bhardwaj, D.R., Cucurbit biodiversity, *National Symposium on Vegetable Biodiversity*, Indian Society of Vegetable Science, Indian Institute of Vegetable Research and Jawaharlal Nehru Krishi Vishwa Vidyalaya, Jabalpur, India, April 4–5, 2011, pp. 20–31.

Cucumis species: The genus *Cucumis* comprises of about 26 species. The major crops of economic importance are cucumber (*C. sativus*), muskmelon (*C. melo*), snap melon (*C. melo* var. *momordica*), and long melon (*C. melo* var. *utilissimus*). The Indian subcontinent is said to be the center of origin for *C. sativus* and center of diversity for *C. melo* (Zeven and de Wet, 1982). The wild species *Cucumis hardwickii* is found growing in the natural habitats in the foothills of the Himalayas. The free hybridization with cultivated *C. sativus* with no reduction of fertility in F$_2$ generation suggested that *C. hardwickii* is likely the progenitor of cultivated cucumber. Cucumber (*C. sativus* var. *sativus*) has three to five lobed leaves, with an ovary of usually three placentas, with fruits of oblong shape, and with obscurely known species *trigonus* or *cylindrica*. However, *C. sativus* var. *sikkimensis* has seven to nine lobed leaves, ovary of often five placentas, and fruits of ovoid oblong shape, which are adapted in temperate and humid climate.

Muskmelon (*C. melo*) based on the distribution of diversity can be grouped into seven subsets (Munger and Robinson, 1991). These are as follows:

1. *C. melo* var. *agrestis* (*Kachri*): It is a wild type with slender vines and small inedible fruits, probably a synonym of *C. melo* var. *callosus* (*Cucumis callosus*) and *C. melo* var. *trigonus* (*Cucumis trigonus*).
2. *C. melo* var. *cantalupensis* Naud. (*muskmelon or cantaloupe*): The fruit is of medium size with netted, warty, or scaly surface, has usually orange but sometimes green flesh, and has aromatic or musky flavor. It is usually andromonoecious in nature and dehiscent at maturity.
3. *C. melo* var. *flexuosus* Naud. (*C. melo* var. *utilissimus*): It is also known as snake melon, snake cucumber, and Tar-kakari. This fruit is long and slender and consumed at the imma-ture stage; it is monoecious.
4. *C. melo* var. *momordica* (*C. melo* var. *momordica*): Snap melon or phut is a monoecious crop grown in India and other Asian countries. It has white to pale orange, less sweet pulp. The smooth surface of the fruit starts cracking at the time of maturity.
5. *C. melo* var. *conomon* (*sweet or pickling melon*): It is generally andromonoecious in nature and bears small fruits with skin and white flesh.
6. *C. melo* var. *inodorous* Naud. (*winter melon*): The fruit has smooth or wrinkled surface with white or green flesh and no musky odor. It is also andromonoecious in nature and usually requires more time for maturity.
7. *C. melo* var. *dudaim* Naud. (*mango melon, pomegranate melon*): This fruit is small in size, globular in shape, smooth and motted but not netted in surface texture, and acidic in flavor.

Out of the seven melons, *C. melo* var. *agrestis*, *C. melo* var. *cantalupensis* Naud. (Muskmelon), *C. melo* var. *utilissimus*, *C. melo* var. *momordica*, and *C. melo* var. *flexuosus* Naud are available in the Indian subcontinent.

Luffa species: The Indian gene center has rich diversity in genetic resources of *Luffa* spe-cies. The genus comprises of nine species in the world, and out of these, seven species, that is, *Luffa acutangula*, *L. cylindrica*, *L. echinata*, *L. graveolens*, *L. hermaphrodita*, *L. tuberosa*, and *L. umbellata*, are native to India. There is ambiguity with regard to *L. tuberosa* and *L. umbellata* because they are considered synonyms to those species of *Momordica* and *Cucurbita* genera, respectively (Chadha and Lal, 1993). Sponge gourd (*Luffa cylindrica*) and ridge gourd (*L. acutangula*) have a rich diversity throughout India. Rich diversity in vine and fruit morphological characteristics occurs in the north-eastern region of India, which includes Sikkim, West Bengal, and western, central, and southern India. *L. hermaphrodita*, which is considered to have originated from *L. graveolens*, is another potential species distributed in parts of north-central India. *L. acutangula* var. *amara*, grown in peninsular India, is a wild relative of sponge gourd (*L. cylindrica*) and *Luffa echinata* in natural habitats in the western Himalayas, central India, and Gangetic plains.

Lagenaria species: Under the genus *Lagenaria*, six species are reported. Out of these, *L. abyssinica*, *L. siceraria*, and *L. leucantha* are common. Among them, *L. siceraria* is generally cultivated in all tropical parts of the world, especially in India and few African countries. The remaining species are wild, perennial, and dioecious in nature. Two wild species, that is, *L. abyssinica* and *L. breviflora*, are perennial in nature. There are suggestions that *Lagenaria* occurs in wild form in South America and in India.

Citrullus species: The genus *Citrullus* has two species of economic importance, namely, watermelon (*C. lanatus* [Thunb.]) and round gourd (*Citrullus vulgaris* Schrad var. *fistulosus*). Cultivation of large watermelon, *C. lanatus* var. *citroides*, is comparatively recent, and Soviet varieties grown today still have the shape of their African ancestors (*C. lanatus* var. *caffer*). Shimotsuma (1963) reported that *C. vulgaris*, *Citrullus colocynthis*, *Citrullus ecirrhosus*, and *Citrullus naudinianus* are related and cross compatible with each other. Whitaker and Davis (1962) considered *C. colocynthis* as probably an ancestor of watermelon.

Trichosanthes species: *Trichosanthus* is the largest genus of the family cucurbitaceae with about 100 species all over the world (Schaefer et al, 2008) out of which 22 species is reported to be of Indian origin/parts of Tropical Asia and Indo-malayan region. Among them, *Trichosanthes anguina* L. (snake gourd) and *Trichosanthes dioica* Roxb. (pointed gourd) are cultivated throughout the India. The major zone for distribution of diversity for *T. dioica* is north-central and northeastern India including West Bengal, whereas *T. anguina* is distributed throughout the country. However, the rich diversity in *T. anguina* has been observed in northeastern states, West Bengal, Malabar Coast, and Eastern Ghats in low and midhills.

Momordica species: The genus *Momordica* has 7 species reported in India and 60 species in other parts of the world. The cultivated species *Momordica charantia* (bitter gourd) is grown all over the country in tropical and subtropical climate. *Momordica dioica* (kartoli) grows all over West Bengal, Assam, parts of Bihar, and adjoining areas; *Momordica cochinchinensis* (kakrol or sweet gourd or golkakora) is most popular particularly in Tripura, Assam, and West Bengal.

Benincasa species: *B. hispida* (syn. *Benincasa cerifera*), known as ash gourd, white gourd, white pumpkin, hairy melon, Chinese preserving melon, winter melon, or wax gourd, is reported to be native of Java and Japan. It was domesticated in India during prehistoric time. It is widely grown all over the country in tropical and subtropical climates and possesses variability in fruit morphological characteristics and quality.

Coccinia species: *Coccinia* has about 35 species distributed in tropical Africa and Asia. Out of these, only one species, that is, *Coccinia grandis*, is under cultivation. The related species, *Coccinia histella and Coccinia sessilifolia,* are widely cultivated in several countries like Africa, Central America, China, Malaysia, Australia, and other tropical Asian countries. The diversity of cultivar in China suggested that this crop may be indigenous to southern China (Yang and Walters, 1992).

Sechium species: The genus *Sechium* is said to have originated in the mountainous region of America and Mexico. It includes *Microsechium compositum*, *Microsechium gintonii*, *Sechium compositum*, *Sechium edule*, *Sechium hintonii*, and *Sechium jamaicense*. Principally, it was confined in tropical and subtropical climate and particularly in midhill conditions. Maximum variability occurs in Sikkim. In Meghalaya and Mizoram, it is grown on a commercial scale and is popular in Darjeeling Hills as well.

22.4 SPECIFIC ADAPTABILITY OF CUCURBITS IN INDIA

There are several cucurbitaceous vegetables that have specific adaptability in a particular area where they subsist. For example, chow-chow (*S. edule*) has specific adoption in Mizoram and *M. cochinchinensis* in Tripura, Assam, and West Bengal and *T. dioica* in eastern U.P.,

TABLE 22.3
Specific Adaptability of Some Cucurbitaceous Vegetables in India

Crop	Area
Bitter gourd	Tamil Nadu, Kerala, Uttar Pradesh, Bihar, West Bengal, Maharashtra
Chow-chow	Mizoram, Karnataka, Maharashtra
Kakrol	Mizoram, Tripura, West Bengal, Bihar, Vindhya Hills of Uttar Pradesh
Muskmelon	Rajasthan, Eastern Uttar Pradesh, Punjab
Watermelon	Rajasthan, Punjab, Haryana, Western Uttar Pradesh, Karnataka, Madhya Pradesh

Bihar, and West Bengal. A detailed list of some vegetables that have adoption in specific environment is given in Table 22.3.

22.5 REASONS FOR LOSS IN CUCURBIT GENETIC DIVERSITY

Overexploitation and displacement by exotic and improved varieties appear to have taken a heavy toll on cucurbitaceous genetic resources. The main causes for losses of genetic diversity are genetic erosion, genetic vulnerability, and genetic wipeout. These are not mutually exclusive but are, in fact, interlocked by the demand of increasing population and rising expectations (Table 22.4).

22.5.1 ALLOCATION OF GENETIC DIVERSITY

India is situated in the northern hemisphere between 8°4′ to 37°6′N latitude and 68°7′ to 97°25′E longitude, stretched about 3214 km from north to south and about 2933 km from east to west (Research, Reference and Training Division, 2007) covering an area of 32,87,263 km². There are three geographical regions: the Himalayas and eastern hills, the Indo-Gangetic plains (alluvial tract), and the peninsular shield (metamorphosed rocks). Geographically, six regions can be outlined: the Great Plains, the mountain zone, the northwestern Gangetic, the northeastern region

TABLE 22.4
Crop Variability and Gene Erosion in Indian Region

Crop	CS	DS	GVS	GES	GCP
Ash gourd	C	W	H	M	M
Bitter gourd	C	W	H	M	H
Bottle gourd	C	W	H	M	N
Cucumber	C	W	H	M	M
Luffa	C	W	H	H	H
Pointed gourd	C	L	H	H	M
Pumpkin	C	W	H	M	H
Snake gourd	C	W	H	M	H
Tinda	C	L	M	M	M
Watermelon	C	R	H	M	M

Notes: CS, crop status (C, cultivated; W, wild); DS, distribution status (W, widespread distribution; R, regional distribution; L, localized distribution); GVS, germplasm variability status (H, high; M, medium; L, low); GES, genetic erosion status (H, high; M, medium; L, low); GCP, general crop priorities (H, high; M, medium).

TABLE 22.5

Center of Diversity of Major Cucurbitaceous Crops

Gene Center	Primary Center of Diversity	Secondary Center of Diversity
African	Watermelon, melon, bottle gourd	—
Central America and Mexican region	Pumpkin, squash	—
Central Asia	—	Watermelon and muskmelon
Chinese-Japanese	Wax gourd	Watermelon
Hindustan center	Wax gourd, cucumber, ridge gourd, bitter gourd, sponge gourd	Watermelon, bottle gourd
Indo-Chinese	Wax gourd, sponge gourd, ridge gourd, bitter gourd, chayote, cucumber, bottle gourd	Bottle gourd, cucumber
Mediterranean	Watermelon	—
North American region	—	Muskmelon, watermelon, squash, pumpkin
South American region	Pumpkin, chayote	—

Source: More, T.A. and Bhardwaj, D.R., Cucurbit biodiversity, *National Symposium on Vegetable Biodiversity*, Indian Society of Vegetable Science, Indian Institute of Vegetable Research and Jawaharlal Nehru Krishi Vishwa Vidyalaya, Jabalpur, India, April 4–5, 2011, pp. 20–31.

comprising of Brahmaputra and Burma valleys, the desert region, the central and southern plateau region, and the western and eastern peninsular region (Table 22.5).

22.6 DOMESTICATION OF CUCURBITS

In India, diversity in cucurbits is observed in several wild relatives/taxa, namely, *Luffa*, *Momordica*, *Citrullus*, *Cucumis*, *Coccinia*, and *Trichosanthes*. Among the cucurbitaceous types, in *Luffa*, most of the species occur in disturbed sites, forest openings, etc. *L. acutangula* var. *amara* occurs in peninsular India and is the wild relative of the cultivated smooth gourd. *L. echinata* occurs in the western Himalayas, central India, and upper Gangetic plains, while *L. graveolens* (considered a wild progenitor of *L. hermaphrodita*) occurs in Bihar, Sikkim, and Tamil Nadu. *L. umbellata* is confined to the eastern coast. In *Momordica, M. balsamina* occurs in the semidry northwestern plains and only sporadically elsewhere in the upper Gangetic region and in the northern parts of Western and Eastern Ghats. *M. dioica* and *M. cochinchinensis* occur wild/semiwild in the Gangetic plains extending eastward. *Momordica cymbalaria* is restricted to the Western Ghats and Maharashtra, with only sporadic occurrence in the eastern peninsular region. *Momordica subangulata* and *M. macrophylla* occur largely in the northeastern region; *M. subangulata* also exhibits sporadic distribution in the Deccan plateau, extending to the Eastern Ghats. *Momordica denudata* is largely confined to the eastern peninsular tract. *Trichosanthes* has 21 species in India. A widely distributed species is *Trichosanthes bracteata* occurring in eastern India, extending to the south, and sporadically in the Himalayas (1500 m). *Trichosanthes cordata* (related to *T. anguina*) occurs in the peninsular region extending to the northeastern plains and hills (Table 22.6).

22.7 PRIORITY CROPS FOR COLLECTION IN SPECIFIC AREAS

Important vegetable crops for which native diversity still needs more emphasis for collection include *Cucumis* species, *Trichosanthes cucumerina, T. dioica, T. bracteata, C. melo, C. melo* var. *utilissimus, C. hardwickii, C. hystrix, C. setosus, C. cordifolia, Luffa* species, *L. siceraria, Citrullus*

TABLE 22.6
Domestication of Cucurbitaceous Vegetable Crop

Species	Origin	Diversity/Domestication in India
Benincasa hispida	Southeast Asia, Indo-China	Japan to India, tropical/subtropical regions
Citrullus lanatus	West Africa	Uttar Pradesh, Maharashtra
Cucumis melo	South Africa	Domestication in India, northwestern and Indo-Gangetic plains
Cucumis sativus	India (the Himalayas)	Throughout India
Cucurbita maxima	North and Central America	Secondary center of diversity, concentrated in North Eastern Himalayan (NEH) region
Cucurbita moschata	Mexico, Central America	Tropical and subtropical region
Cucurbita pepo	North America, Mexico	Secondary center of diversity
Lagenaria siceraria	Latin America, Africa	Throughout India
Luffa acutangula	India	Western Ghats and south India
Luffa cylindrica	Japan, Brazil, India	Eastern peninsular India
Praecitrullus fistulosus	India	Northwest, Indo-Gangetic plains
Trichosanthes anguina	South Asia, India	Malabar coast, Eastern Ghats, NEH region
Trichosanthes dioica	India	U.P., Bihar, West Bengal, Assam

Source: More, T.A. and Bhardwaj, D.R., Cucurbit biodiversity, *National Symposium on Vegetable Biodiversity*, Indian Society of Vegetable Science, Indian Institute of Vegetable Research and Jawaharlal Nehru Krishi Vishwa Vidyalaya, Jabalpur, India, April 4–5, 2011, pp. 20–31.

colocynths, *M. dioica, M. cochinchinensis, M. charantia, Praecitrullus fistulosus*, and *B. hispida* on priority basis. A large number of collections have already been made through:

1. *Introduction from abroad*: A number of accessions of cucurbitaceous vegetable crops and their wild relatives were introduced from different countries. This includes *C. lanatus* (8 from the United States), *C. melo* (15 from France and Japan), *C. sativus* (15 from Japan and the United States), *Cucumis heptadactylus* (1 from the United States), *Cucumis metuliferus* (16 from the United States), *Cucumis anguria* (94 from the United States), *C. pepo* (2 from the United States), *Cucurbita* species (4 from Algeria and Japan), *L. cylindrica* (1 from Japan), and *L. siceraria* (1 from Japan).

2. *Collection through exploration*: In a number of specific crops or with other crops, several explorations were undertaken in all agroecological zones to collect different cultivated/ landrace/wild/weedy cucurbitaceous vegetables. In the year 2001–2002 under NATP on a plant biodiversity program, several lines in different crops, namely, ash gourd (44), bitter gourd (146), bottle gourd (147), *Citrullus* (39), cucumber (98), *C. moschata* (4), other *Cucumis* species (487), ivy gourd (9), Kachari (6), long melon (5), muskmelon (48), pointed gourd (20), wild pumpkin (1), ridge gourd (121), satputia (8), snap melon (41), snake gourd (51), spine gourd (1), sponge gourd (133), summer squash (5), tinda (14), and watermelon (16), were collected by NBPGR, IIVR, SAUs, and NGOs in collaboration. The area surveyed included parts of Arunachal Pradesh, Meghalaya, Assam, Uttaranchal, Orissa, Uttar Pradesh, Andhra Pradesh, Kerala, Rajasthan, Punjab, Himachal Pradesh, Jharkhand, Jammu and Kashmir, Haryana, Bihar, Madhya Pradesh, Gujarat, Karnataka, and Tamil Nadu (Table 22.7).

22.7.1 EVALUATION AND CHARACTERIZATION

The main objective of vegetable breeding programs is to develop varieties that are superior to the existing ones. Evaluation and characterization activities should go simultaneously. Evaluation of PGR is often carried out under field conditions but a well-organized field evaluation system must

TABLE 22.7

Priority Crops for Collection in Specific Areas

Crop	Region to Be Explored
Benincasa hispida	NEH region
Citrullus colocynthis	Northwestern plains, Rajasthan, Gujarat
Coccinia cordifolia	Eastern Uttar Pradesh, Madhya Pradesh, West Bengal, and NEH region
Cucumis hardwickii	Western Himalayan foothills
Cucumis melo and related wild species	Jammu and Kashmir, H.P. hills, Uttar Pradesh, Rajasthan, Karnataka, Andhra Pradesh
Cucumis prophetarum	Northwestern plains, Sirohi and Abu areas in Rajasthan
Cucumis sativus	Indo-Gangetic plains, sub-Himalayan tract, Western Ghats, eastern peninsular tract
Cucumis setosus	Eastern India and upper Gangetic plains
Cucurbita species	NEH region
Lagenaria	Indo-Gangetic plains, Tarai region, northeastern plains
Luffa species	Indo-Gangetic plains, Tarai region, northeastern plains
Momordica	Central peninsular tract
Momordica cochinchinensis	Peninsular region, West Bengal, NEH region
Sechium edule	NEH region
Trichosanthes bracteata	Himalayan ranges, eastern India, A&N region
Trichosanthes cucumerina	Southern peninsular tract, Kerala
Trichosanthes dioica	Bihar, West Bengal, riverbank of Ganga and plains, Assam Valley

be given priority because many time fields for experimentation suffer various soil stress problems. So, it is essential to develop polyhouse, glasshouse, greenhouse, and net house facilities for evaluation. Such conditions are always more conferral for evaluation of PGR, as under such environments, there are less chances of loss of germplasm lines. Such facilities, along with epiphytotic arrangements, are suitable for evaluation of lines for biotic stress resistance. Screening or evaluation for abiotic stresses like extreme temperature, salinity, sodic soil, and excessive moisture resistance is conducted under controlled laboratory conditions. Utilization of PGR depends only on the quality of evaluated data.

22.7.2 UTILIZATION OF GENETIC RESOURCES

In cucurbitaceous vegetables, several varieties/hybrids have been developed by utilization of genetic resources for commercial purposes.

22.7.2.1 Release of Varieties and Hybrids through Utilization of Genetic Resources

The evaluation of indigenous and exotic germplasm introductions and their hybridization resulted in the selection of several superior varieties/hybrids of different cucurbitaceous vegetables. As a result, multidisciplinary, multilocation testing of new research materials has given rays of hope in identification and recommendation of varieties/hybrids for cultivation in various agroclimatic regions of the country.

22.8 CONSERVATION OF CUCURBIT BIODIVERSITY

The rich heritage and ethnic culture has favored the preservation of the richest diversity including rare landraces/primitive types of useful vegetables like eggplant, cucumber, ridge and sponge gourd, and a number of root and tuber crop species. A number of vegetable crops were brought to India from other regions by travelers, invaders like Persians, Turkish, Mughals, Portuguese, Dutch, French, and British that acclimatized and developed good amount of diversity. However, diversity for valuable genetic resources is threatened in recent times. Therefore, conservation of the vegetable genetic wealth, particularly their wild relative, is thus essentially required for future utilization. Germplasm conservation can be done *in situ*, *ex situ*, and *in vitro*.

1. *In situ germplasm conservation*: Conservation of germplasm in its natural habitat or in the area where it grows naturally is known as *in situ* germplasm conservation. This is achieved by protecting this area from human interference; such an area is often called natural park, biosphere reserve, or gene sanctuary. A gene sanctuary is best located within the center of origin of crop species concerned, preferably covering the microcenter within the center of origin. Mostly, *in situ* conservation is an ideal method of conserving wild plant genetic resources and perennial vegetables that either do not set or set recalcitrant seed and do not produce plants true to type.
2. *Ex situ germplasm conservation*: Conservation of germplasms away from its natural habitat is called *ex situ* conservation. *Ex situ* conservation requires collection and systematic storage of seeds/propagules for short, medium, and long term. NBPGR, New Delhi, the nodal agency for *ex situ* conservation of PGR for food and agriculture, is maintaining active and base collections of various crop species and their wild relatives including vegetables in a network of gene banks in the country.
3. *In vitro conservation*: It includes (a) conservation of cells, tissues, and organs in glass or plastic containers under aseptic conditions through slow growth of cultures and (b) cryopreservation of cultures (tissues, organs, pollens, somatic/zygotic embryos, or embryogenic cell cultures in liquid nitrogen at −150°C to −196°C). It may be called cell and organ bank.

22.9 THERAPEUTIC VALUES IN CUCURBITS

Cucurbits are actually useful for human health and a rich source of water, protein, fats, minerals, fiber, carbohydrates, energy, calcium, magnesium, potassium, phosphorus, iron, sodium, copper, sulfur, chlorine, etc. It has also got vitamin A, thiamine, riboflavin, vitamin C, and nicotinic and oxalic acid (Table 22.8).

White gourd (B. hispida): The fruits are edible. Sweets are prepared from the pulp of fruits. The fruits are used for making jam, jelly, murabba, and cake. Fruit is tonic, nutritive, diuretic, and antiperiodic, which is used specifically for hemorrhages from internal organs. It is also useful in case of insanity, epilepsy, and other nerve diseases. Seeds are vermifuge against tapeworm and diuretic. It is also beneficial in case of constipation, heart disease, tuberculosis, and colic pain and used as an aphrodisiac.

Watermelon (C. lanatus): The ripe fruits are edible and largely used for making confectionary. Its nutritive values are also useful to the human health. The fruit is used in cooling and strengthening and as an aphrodisiac, astringent to the bowels, indigestible, expectorant, diuretic, and stomachic, which purifies the blood, allays thirst, cures biliousness, and treats sore eyes, scabies, and itches. The seeds are tonic to the brain.

Cucumber (C. sativus): It lowers human constipation and is good for digestion. The fruits are much used during summer as a cooling food. They are used as salads and for cooking curries. The tender fruits are preferred for pickling kernels of the seeds and also used in confectionary.

Muskmelon (C. melo): Its ripe fruits are very useful against kidney disease. The fruits are extensively used as dessert fruits and are highly esteemed in the summer months. The seeds are diuretic, cooling, nutritive, and beneficial to the enlargement of prostate gland. The pulp is diuretic and beneficial in chronic or acute eczema.

TABLE 22.8
Nutrient Components of the Fruits of Cultivated Cucurbits (per 100 g of Edible Portion)

Nutrient Components	1	2	3	4	5	6	7	8	9	10	11	12	13	14
Water moisture	92.5	92	96.3	92.7	92.6	94	92	96.1	95.2	93	80	92.4	92	94.6
Protein (g)	0.4	1	0.4	0.6	1.4	1	1.5	0.2	0.5	1.2	2.1	1.6	2	0.5
Fat (g)	0.1	—	0.1	0.2	0.1	0.1	0.1	1	0.1	0.2	—	0.2	0.3	0.3
Minerals (g)	0.3	—	0.3	0.3	0.6	0.6	0.6	0.5	0.3	—	—	0.8	0.5	0.5
Fiber (g)	0.8	—	0.4	0.5	0.7	0.7	0.7	0.6	0.5	2	—	0.8	3	0.8
Carbohydrate (g)	1.9	6.5	2.5	5.9	4.6	4.6	4.8	2.5	3	5	17.4	4.2	4.2	3.3
Energy (cal)	10	35	13	125	13	21	25	12	20	22	80	25	20	18
Calcium (mg)	30	7	10	0.17	10	23	15	120	40	25	36	20	30	50
Magnesium (mg)	—	—	11	—	14	14	15	5	11	—	—	17	9	53
Phosphorus (mg)	20	7	25	—	30	30	32	10	40	—	—	70	40	20
Iron (mg)	0.8	—	1.5	0.04	0.7	0.3	0.8	0.7	1.6	1	—	1.3	1.7	1.1
Sodium (mg)	—	—	10.2	—	5.6	5.6	5.9	1.8	2.9	—	—	17.8	2.6	25.4
Potassium (mg)	—	—	50	—	139	139	150	87	50	—	—	152	83	34
Copper (mg)	—	—	0.1	—	0.2	0.2	0.25	0.3	0.16	—	—	0.18	1.11	0.11
Sulfur (mg)	—	—	17	—	16	16	18	10	13	—	—	15	17	35
Chlorine (mg)	—	—	15	—	4	4	5	—	7	—	—	8	4	21
Vitamin (I.U.)	20	599	40	190	1840	2000	1700	60	56	84	125	210	255	160
Thiamine (mg)	0.06	0.05	0.03	0.06	0.06	0.04	0.07	0.03	0.07	0.03	0.08	0.07	0.05	0.04
Riboflavin (mg)	0.1	0.05	0.01	—	0.04	0.02	0.03	0.01	0.01	0.03	0.06	0.09	0.06	0.06
Nicotinic acid (mg)	0.4	—	0.2	0.4	0.5	0.5	0.6	0.2	0.2	0.3	—	0.5	0.5	0.3
Vitamin C (mg)	1	6	7	35	2	15	20	6	5	7	—	88	29	5
Oxalic acid (mg)	—	—	15	—	—	—	—	—	27	—	—	—	7	34

Source: Rahman, A.H.M.M. et al., *J. Appl. Sci. Res.*, 4(5), 555, 2008, http://www.aensiweb.com/old/jasr/jasr/2008/555–558.pdf

Notes: 1, *Benincasa hispida*; 2, *Citrullus lanatus*; 3, *Cucumis sativus*; 4, *Cucumis melo*; 5, *Cucurbita maxima*; 6, *Cucurbita moschata*; 7, *Cucurbita pepo*; 8, *Lagenaria siceraria*; 9, *Luffa acutangula*; 10, *Luffa cylindrica*; 11, *Momordica cochinchinensis*; 12, *Momordica charantia*; 13, *Trichosanthes dioica*; 14, *Trichosanthes anguina*.

Pumpkin (*C. maxima*): The fruits are useful in human blindness. Matured fruits of pumpkin are used as a table vegetable for baking pies and for making jam; they are also used as a livestock feed. The young fruit resembles the vegetable marrow in flavor, but the full grown fruit is much liked. The seeds are anthelmintic and used as diuretic and tonic. The fruit pulp is often used against inflammations and boils.

Squash gourd (*C. moschata*): Matured fruits of squash gourd are used as a table vegetable for baking pies and for making jam; they are also used for livestock feed. The flesh is usually fine grained and mild flavored and is thus suitable for baking.

Squash (*C. pepo*): The leaves are digestible, hematinic, and analgesic and help in removing biliousness. They are also used as an external applicant for burns. The seeds are diuretic, tonic, and fattening; cure sore chests, bronchitis, and fever; allay thirst, and are good for the kidney and the brain.

Bottle gourd (*L. siceraria*): White pulp of fruit is cool, emetic, purgative, diuretic, and antibilious. Oil from the seeds is used for cooling and relieving headache. Seeds are nutritive and diuretic. Decoction of leaves mixed with sugar is given in jaundice. The warmth of tender stem relieves earache. The fruit is used against cholera.

Ridge gourd (*L. acutangula*): The plant fruit is demulcent, diuretic, and nutritive. The seed possesses purgative and emetic properties. The pounded leaves are applied locally against splenitis, hemorrhoids, and leprosy. The juice of fresh leaves is dropped into the eyes of children during granular conjunctivitis, also to prevent the eyelids adhering at night from excessive meibomian secretion.

Sponge gourd (*L. cylindrica*): The dried fruits yield a spongy substance that is used as a bath sponge. The seeds are emetic and cathartic. Young fruits are cool, demulcent, and produce loss of appetite.

Teasel gourd (*M. cochinchinensis*): Fruits and leaves are used in external application for lumbago, ulceration, and fracture of bones. The seeds are used as aperients and in the treatment of ulcers, sores, and obstructions of liver and spleen.

Bitter gourd (*M. charantia*): The fruits are considered tonic, stomachic, carminative, and cool. The fruits are very much useful against human diabetes. The fruits are also used as febrifuge and in rheumatism, gout, and disease of liver and spleen. The fruits and leaves are anthelmintic and useful in piles, leprosy, and jaundice and as a vermifuge.

Pointed gourd (*T. dioica*): Fresh juice of unripe fruit is used as a cooling agent and laxative. The fruit is also used in spermatorrhoea. The leaf is an aperient and also used as tonic and febrifuge. This is also used as part of the diet of patients with subacute cases of enlarged liver and spleen. The fruit is febrifuge, laxative, and antibilious.

Snake gourd (*T. anguina*): The fruits are used as tonic and laxative. The seeds are anthelmintic and have antidiarrheal properties; these are used against biliousness and in syphilis. The seeds are also used as a coolant.

REFERENCES

Chadda, M.L. and Lal, T. 1993. Improvement of cucurbits. In: K.L. Chadda and G. Kalloo (eds.), *Advances in Horticulture*, Vol. V—*Vegetable Crops: Part 1*. pp. 137–179. Malhotra Publishing House, New Delhi, India.

Jeffrey, C. 1990. Systematics of the Cucurbitaceae: An overview. In: D.M. Bates, R.W. Robinson, and C. Jeffrey (eds.), *Biology and Utilization of the Cucurbitaceae*, pp. 3–28. Cornell University Press, Ithaca, NY.

More, T.A. and Bhardwaj, D.R. 2011. Cucurbit biodiversity. In *National Symposium on Vegetable Biodiversity*. Indian Society of Vegetable Science, Indian Institute of Vegetable Research and Jawaharlal Nehru Krishi Vishwa Vidyalaya, Jabalpur, India, April 4–5, 2011, pp. 20–31.

Munger, H.M. and Robinson, R.W. 1991. Nomenclature of *Cucumis melo* L. *Cucurbit Genet. Coop. Rep.* 14: 43–44.

Rahman, A.H.M.M., Anisuzzaman, M., Ahmed, F., Rafiul Islam, A.K.M., and Naderuzzaman, A.T.M. 2008. Study of nutritive value and medicinal uses of cultivated cucurbits. *J. Appl. Sci. Res.* 4(5): 555–558. http://www.aensiweb.com/old/jasr/jasr/2008/555–558.pdf.

Research, Reference and Training Division. 2007. *India Yearbook*, p. 1. Publications Division, Ministry of Information & Broadcasting, Government of India, Delhi, India.

Shimotsuma, M. 1963. Cytogenetical studies in the genus *Citrullus*. VII. Inheritance of several characters in watermelons. *Jpn. J. Breed.* 13: 235–240.

Whitaker, T.W. and Davis, G.N. 1962. *Cucurbits—Botany, Cultivation and Utilization*. Leonard Hill Ltd., London, U.K.

Yang, S.L. and Walters, T.W. 1992. Ethnobotany and the economic role of the Cucurbitaceae of China. *Econ. Bot.* 46: 349–367.

Zeven, A.C. and de Wet, J.M.J. 1982. *Dictionary for Cultivated Plants and Their Regions of Diversity: Excluding Most Ornamentals Forest Trees and Lower Plants*, 227pp. Center for Agricultural Publishing and Documentary, Wageningen, the Netherlands.

Section X

Growth Responses of Cucurbits
under Stressful Conditions
(Abiotic and Biotic Stresses)

23 Soil Salinity
Causes, Effects, and Management in Cucurbits

Akhilesh Sharma, Chanchal Rana, Saurabh Singh, and Viveka Katoch

CONTENTS

23.1 INTRODUCTION

23.1.1 What Is Soil Salinity?

Soil salinity is a major limiting factor that endangers the capacity of agricultural crops to sustain the growing human population. It is characterized by a high concentration of soluble salts that significantly reduces the yield of most crops. Soils with an electrical conductivity (EC) of the

419

saturation soil extract of more than 4 dS m^{-1} at 25°C are called saline soils, which are equivalent to approximately 40 mM NaCl and generate an osmotic pressure of approximately 0.2 MPa. Salts generally found in saline soils include chloride and sulfates of Na, Ca, Mg, and K. Calcium and magnesium salts are at a high enough concentration to offset the negative soil effects of sodium salts. The pH of saline soils is generally below 8.5. The normal desired range is 6.0–7.0.

23.1.2 CHARACTERISTICS OF SALINE SOILS

1. The soluble salt concentration in the soil solution is very high, which also results in high osmotic pressure of the soil solution. Osmotic pressure is closely related to the rate of water uptake and growth of plants. This causes wilting of plants and nutrient deficiency. A salt content of more than 0.1% is injurious for plant growth (Table 23.1).
2. EC of the soil saturation extract is important as a measure for the assessment of saline soil for the plant growth and is expressed as dS m^{-1} (earlier mmhos cm^{-1}). Salinity effects are negligible below 2 dS m^{-1}. However, yields of very sensitive crops may be restricted between 2 and 4 dS m^{-1}, while yields of many crops may be restricted between 4 and 8 dS m^{-1}. On the other hand, only tolerant crops yield satisfactorily between 8 and 16 dS m^{-1}, whereas above 16 dS m^{-1}, only high-tolerant crops grow (Table 23.2).
3. Determination of water-soluble boron concentration is also an important parameter for characterization of saline soils. Boron concentration above 1.5 ppm is unsafe for plant growth.
4. Soil texture is also an important criterion to characterize saline soils. Sandy soils with 0.1% salt concentration cause injury to the growth of common crops, while the crops grow normally in clayey soils with the same salt content. Saturation percentage is considered as a characteristic property of every soil. For salinity appraisal, soil texture and EC of saturation extract are considered simultaneously.

TABLE 23.1
Characteristics of Salt-Affected Soils

Characteristics	Saline Soils	Sodic Soils
Content in soil	Excess of neutral salts	Excess of sodium salts
pH	<8.5	>8.5
EC (dS m^{-1})	>4	<4
Exchangeable sodium percentage (%)	<15	>15
Physical condition of soil	Flocculated	Deflocculated
Color	White	Black
Organic matter	Slightly less than normal soils	Low
SAR	<13	>13
Total soluble salt contents (%)	>0.1	<0.1

TABLE 23.2
Classification of Saline Soils

Salt Concentration of the Soil Water (Saturation Extract)

(g/L)	(dS m^{-1})	Salinity
0–3	0–4.5	Nonsaline
3–6	4.5–9	Slightly saline
6–12	9–18	Moderately saline
>12	>18	Highly saline

23.1.3 Problems of Salt-Affected Soils

The various problems associated with saline soils that interfere with plant growth are as follows:

1. Soils are generally barren but potentially productive.
2. Saline soils have a high wilting point and low amount of available moisture.
3. Excessive salts in the soil solution increase the osmotic pressure of soil solution compared to cell sap, which makes it difficult for plant roots to extract moisture due to increased potential force that holds water. If salt concentration in the soil is greater than that of the plant, water moves from the plant into the soil, that is, plasmolysis, which leads to wilting/death of the plant.
4. High concentration of soluble salts cause toxicity to the plant, for example, root injury and inhibition of seed germination.

23.1.4 Present Status and Causes of Salinity

Salinization is a process that results in an increased concentration of salts in soil and water. Of these salts, sodium chloride is the most common. With an increase in concentration of soluble salts, it becomes more difficult for plants to extract water from the soil. Higher salt concentrations can be created by poor soil drainage, improper irrigation, irrigation water with high levels of salts, and excessive use of manure or compost as fertilizer. Salinization affects many irrigated areas mainly due to the use of brackish water. Salt-affected soils cover about 800 million ha of land, which accounts for more than 6% of the total land area in the world. There are two kinds of soil salinity, namely primary (natural) and secondary (due to human activity, i.e., dry land and irrigated land salinity). A majority of saline soils have emerged due to natural causes such as accumulation of salts over long periods of time in arid and semiarid zones (Munns and Tester 2008). This is because of the fact that the parent rock from which it formed contains salts, mainly chlorides of sodium, calcium, and magnesium, and to some extent, also contains sulfates and carbonates. Sea water is another source of salts in low-lying areas along the coast. Besides natural salinity, a significant proportion of cultivated land has become saline due to land clearing or irrigation.

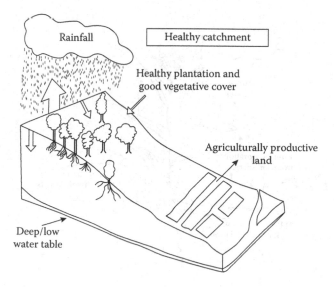

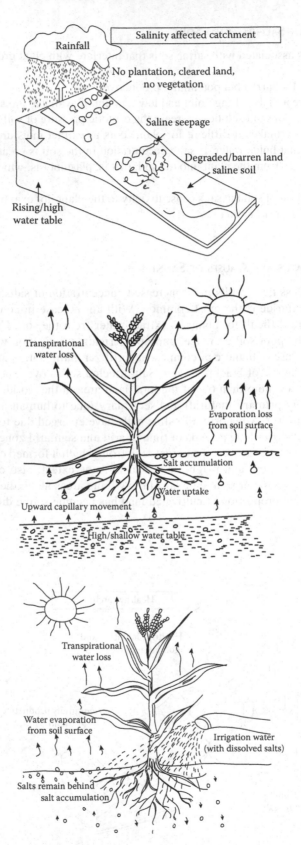

These factors raise the water table and cause the accumulation of salts in the root zone. Presently, out of 230 million ha of irrigated land, around 45 million ha are salt-affected worldwide, which accounts for 20% of the irrigated area. About 1.5 million ha of land is taken out of production every year as a result of high salinity levels in the soil. The irrigation water contains calcium (Ca^{2+}), magnesium (Mg^{2+}), and sodium (Na^+). After irrigation, the water applied to the soil is used by the crop or evaporates directly from the moist soil. Ca^{2+} and Mg^{2+} often precipitate into carbonates, leaving Na^+ dominant in the soil (Serrano et al. 1999). The salt, however, is left behind in the soil. As a result, Na^+ concentrations often exceed those of most macronutrients by one or two orders of magnitude, and by even more in the case of micronutrients. High concentrations of Na^+ in the soil solution may depress nutrient-ion activities and produce extreme ratios of Na^+/Ca^{2+} or Na^+/K^+. The increase in cations and their salts, particularly NaCl, in the soil generates external osmotic potential, which can prevent or reduce the influx of water into the root. The resulting water deficit is similar to drought conditions and additionally compounded by the presence of Na^+ ions (Bohnert 2007). Highly saline soils are sometimes recognizable by a white layer of dry salt on the soil surface. Irrigated land though covers only 15% of the total cultivated land but has high productivity; as a result, they produce one-third of the world's total food.

23.1.5 SALINITY STRESS AND PLANT GROWTH

Plants are stressed in two ways in a high-salt environment. In addition to the water stress imposed by the increase in osmotic potential of the rooting medium as a result of high-solute content, there is the toxic effect of high concentration of ions. Few plant species have adapted to saline stress, but the majority of crop plants are susceptible (they may not survive or survive but with low yield). Soil salinity leads to reduction in biomass production by affecting important physiological and biochemical processes of the plant (Ahmad and John 2005; Ahmad 2010; Ahmad and Sharma 2010). At low salt concentrations, yields are either mildly affected or not affected at all (Maggio et al. 2001). With the increase in salt concentration, the yield reduction is drastic as most crop plants are not able to grow at high concentrations of salt. On the contrary, halophytes can survive salinity and have the capability to grow on saline soils of coastal and arid regions due to specific mechanisms of salt tolerance developed during their phylogenetic adaptation. High salinity affects plants in several ways like water stress, ion toxicity, nutritional disorders, oxidative stress, alteration of metabolic processes, membrane disorganization, reduction of cell division and expansion, and genotoxicity (Munns 2002b; Zhu 2007). These factors together hamper growth and development of the plant that may affect plant survival. During the onset and development of salt stress within a plant, all the major processes such as photosynthesis, protein synthesis, enzyme activity and energy, and lipid metabolism are affected (Parida and Das 2005). Therefore, as a result, premature senescence of older leaves and toxicity symptoms (chlorosis, necrosis) on mature leaves may occur (Hasegawa et al. 2000). In the initial stages, plants experience water stress that causes reduction of leaf expansion. The osmotic effects of salinity stress can be observed immediately after salt application and continue for the duration of salt exposure, which results in the inhibition of cell expansion and cell division along with stomatal closure (Flowers 2004). During long-term exposure to salinity, plants experience ionic stress, which can lead to premature senescence of adult leaves and thus a reduction in the photosynthetic area available to support further growth (Cramer and Nowak 1992). High salinity affects rhizosphere, which is bioenergetically taxing as microorganisms need to maintain an osmotic balance between their cytoplasm and the surrounding medium while excluding sodium ions from the cell interior, and as a result, sufficient energy is required for osmoadaptation (Oren 2002; Jiang et al. 2007).

Munns (2002a) described characteristic changes over different time scales in the plant's development, that is, from the imposition of salinity stress till maturity. Moments after salinization, cells dehydrate and shrink but regain their original volume hours later. Despite this recovery, cell elongation and, to a lesser extent, cell division are reduced, leading to lower rates of leaf and root growth.

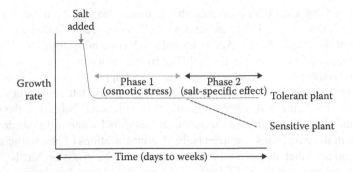

FIGURE 23.1 Two-phase growth response to salinity for genotypes differing in the rate of salt toxicity in leaves. (From Munns, R., *New Phytol.*, 167, 645, 2005.)

Over the next days, reduction in cell division and elongation translates into slower leaf appearance and size. Plants that are severely salt-stressed often develop visual injury due to excessive salt uptake. After a few weeks, lateral shoot development is affected, and after some months, clear differences in overall growth and injury are observed between salt-stressed plants and their nonstressed controls. Based on these sequential differences in response to salinity, a two-phase model describing the osmotic and ionic effects of salt stress (Figure 23.1) was proposed by Munns (2002a, 2005). Identification of plant genotypes capable of increased tolerance to salt and incorporation of these desirable traits into economically useful crop plants may reduce the effect of salinity on productivity. Plants sensitive or tolerant to salinity differ in the rate at which the salt reaches toxic levels in leaves. Timescale is days or weeks or months, depending on the species and the salinity level. During phase 1, growth of both types of plants is reduced because of the osmotic effect of the saline solution outside the roots. During phase 2, old leaves in the sensitive plant die and reduce the photosynthetic capacity of the plant. This exerts an additional effect on growth. However, the physiological, biochemical, and molecular mechanisms of salt tolerance in plants are not yet sufficiently understood, and hence progress in developing salt-tolerant crops has been slow (Lauchli and Grattan 2007).

23.2 STATUS OF CUCURBITS IN RELATION TO SALINITY

The Cucurbitaceae family ranks among the highest of plant families for number and percentage of species used as human food. It consists of 98 proposed genera with 975 species mainly in tropical and subtropical regions, the most important of which are *Cucumis* (cucumber, musk melon), *Cucurbita* (squash, pumpkin, zucchini, some gourds), *Lagenaria* (bottle gourd), *Citrullus* (watermelon), and many others. Cucurbits are grown around the tropics and in temperate areas and are sensitive to frost. Cucurbit crops are, in general, similar in their appearance and requirements for growth. A majority of them are annual bearing vines with trailing growth habit, grown during summer season. They require congenial environmental conditions for better growth to harness maximum productivity per unit area.

Cucurbits differ considerably in their ability to tolerate salinity stress. A majority of the cucurbits are moderately sensitive to salt stress. Cucumber (*Cucumis sativus* L.) and musk melon (*Cucumis melo* L.) are moderately sensitive to salinity. Salinity improves the musk melon fruit quality by increasing the dry matter, total sugars, total soluble solids, and pulp firmness. Squash (*Cucurbita pepo*) is moderately salt tolerant. Villora et al. (1999) reported that salinity improves zucchini fruit quality through enhancement of physical (fruit firmness) and chemical properties (total soluble solids). Similarly, bitter gourd or balsam pear (*Momordica charantia* L.), widely recognized for its hypoglycemic properties, is one of the extensively grown cucurbitaceous vegetables reported to have salinity tolerance. Trajkova et al. (2006) reported that cucumber is more susceptible to NaCl than CaCl$_2$, which points to Na-specific salinity effects. This may be attributed to inefficient compartmentation of Na within the cell, which forces the plant to exclude Na from the leaf. To exclude Na, the plant must expend energy for osmotic adjustment. When Na exclusion breaks down, the plant suffers directly from Na toxicity at a biochemical level.

However, research on crop improvement under a stress situation on cucurbits is very meager. Watermelon (*Citrullus lunatus* L.) is moderately sensitive to salinity, and the reduction in yield due to salinity ranged from 0% at 2.5 mmhos cm^{-1} to 10% at 3.3 mmhos cm^{-1}, 25% at 4.4 mmhos cm^{-1}, 50% at 6.3 mmhos cm^{-1}, and 100% at 10 mmhos cm^{-1} (Yetisir and Uygur 2010). Dhillon et al. (2012) reported that Indian germplasm of cucurbits exhibit few salt-tolerant lines and also, it has been mentioned that bottle gourd withstands salt stress better than watermelon and winter squash.

Salinity Reaction		
Low Tolerant/Sensitive	**Medium-Tolerant/Sensitive**	**Highly Tolerant/Sensitive**
Nil	Cucumber (*C. sativus* L.), musk melon (*C. melo* L.), squash (*C. pepo* L.), pumpkin (*Cucurbita maxima* Duch.), bottle gourd (*Lagenaria siceraria* Standl.), watermelon (*C. lunatus* Thunb.), winter squash (*Cucurbita moschata* Duch.)	Ash gourd (*Benincasa hispida* Thunb.), bitter gourd (*M. charantia* L.)

23.3 STRATEGIES TO COMBAT SOIL SALINITY IN DIFFERENT CUCURBITS

Plant growth under salt stress conditions is a complex mechanism, and the way it is affected by the stress is not fully understood because the response of plants to excessive salinity is multifaceted and involves changes in plant morphology, physiology, and metabolism (Ali et al. 2012). Identification of plant genotypes capable of increased tolerance to salt and incorporation of these desirable traits into economically useful crop plants may reduce the effect of salinity on productivity. The improvement of salinity tolerance in the crops through conventional breeding has very limited success due to genetic and physiological complexity of this trait (Flowers 2004). In addition, tolerance to saline condition is a developmentally regulated and stage-specific phenomenon, that is, tolerance at one stage of plant development does not always correlate with tolerance at other stages (Foolad 2004). Breeding for salt tolerance requires effective screening methods, existence of genetic variability, and ability to transfer the genes to the species of interest. In addition to tolerant cultivars, several cultural practices needs to be applied with an aim that each contribute to a certain extent to allow plants to better withstand the deleterious effects of salt. Some of the proposed practices, like the application of chemical fertilizers at levels somewhat above the optimum in freshwater irrigation and the application of chemical amendments or leaching salts to deeper soil layers, are hardly compatible with the urgent need to preserve the environment (Cuartero et al. 2006).

23.3.1 SOURCES OF TOLERANCE AND DEVELOPMENT OF GENOTYPES TOLERANT TO SALINITY

23.3.1.1 Conventional Breeding Approaches

To ensure sustainable productivity of agricultural crops in the future, there is a need to select and characterize salt-tolerant plants. In order to improve salt tolerance through breeding, genetic variability for the trait is required. The genetic pool of cucurbits possess only partial degree of tolerance to salinity along with their nonstable nature toward salt tolerance (Hasegawa et al. 1980), which makes it even more difficult to cope with. The results of most studies have shown that the resistance to salt stress is usually correlated with a more efficient antioxidant system. The quantitative nature of salt tolerance has its roots in the physiological processes that involve multiple genes, each with a small and unknown effect (Quesada et al. 2000). Major efforts are being directed toward the genetic transformation of plants in order to raise their tolerance (Borsani et al. 2003), and in spite of the complexity of the trait, the transfer of a single gene or a few genes has led to improvement in salt tolerance. Evaluation of salt-tolerant lines through screening and then hybridizing them with high-yielding lines to incorporate salt-tolerant genes by backcrossing has been suggested by many researchers (Munns et al. 2006). Salt-tolerant lines are not as such available in abundance, and their identification is very cumbersome. It is complicated to improve salinity tolerance of cucurbits through simple selection procedures or pedigree breeding due to the presence of dominance gene action (Kere et al. 2013). Yeo et al. (1988) and Cuartero and Fernandez-Munoz (1999) suggested pyramiding of desired genes in a single genotype.

Cucumber is one of the most important cucurbits grown throughout the world and is sensitive to salinity (Dorota 1997), though certain reports categorized it as moderately sensitive to salt stress (Maas 1993), indicating genotypic variation for salt tolerance. Salt stress in cucumber involves both osmotic stress, by limiting absorption of water from soil, and ionic stress, resulting from high concentrations of potentially toxic salt ions within plant cells (Savvas et al. 2005). Therefore, cucumber production in saline soils requires salt-tolerant varieties. Tiwari et al. (2011) were of the opinion that the most feasible alternative to grow crops under salinity prone environments is through genetic improvement. For breeding salt-tolerant cucumber, an understanding of the mechanism of inheritance pattern involved in salinity tolerance is required. Being native to India and despite its wide genetic variability, no information is available for salinity tolerance in cucumber (Malik et al. 2010). Cucumber lines 'CRC 8', 'CHC 2', 'G 338', 'CH 20', and '11411Sare' are known to be tolerant to soil salinity (Kere et al. 2013). The musk melon line Calif-525 is salt tolerant (Shannon et al. 1984; Whitaker 1979).

Of late, Munns et al. (2012) developed tolerant wheat through conventional breeding. This provides hope for genetic improvement of salt-tolerant cucumber if accurate selection and screening methods are identified. However, the genetics of salt tolerance in cucumber is poorly understood (Tiwari et al. 2011) due to the complexity of salt tolerance (Munns and Tester 2008). Therefore, it is necessary to investigate viable selection traits that can predict salinity tolerance of cucumber at the seedling stage. The potential for genetic improvement of salt tolerance in cucumber is feasible if the gene action of superior parents is fully understood and a suitable breeding program is employed (Dashti et al. 2012). Munns et al. (2006) suggested that exploitation of naturally occurring inter and intra-specific genetic variability by hybridization of selected salt-tolerant genotypes with high-yielding genotypes adapted to a specific environment is a descent approach to develop salt-tolerant varieties.

23.3.1.2 Screening Techniques Against Salt Stress

To identify salt-tolerant plants, they have to be screened in saline medium/conditions. Plants do not develop salt tolerance unless they are grown in saline conditions, which means that they must be hardened to salt stress (Levitt 1980). Various methods are used for screening segregating material for salt resistance. The commonly used methods are lysimeter microplots, sand culture, and solution culture tanks. Replicated experiments are conducted over seasons to get more reliable results. Genotypes that survive under salinity conditions are considered as tolerant and screened further. In cucumber, cultivars 'Keyan' and 'Danito' were identified as salinity-tolerant cultivars (Baghbani et al. 2013). Na+ exclusion is widely accepted as an efficient salt selection criterion for cereals (Munns and Tester 2008; Munns et al. 2012).

Malik et al. (2010) have partly explained the mechanism by identifying the parameters as an index for *in vitro* screening of salt tolerance in cucumber genotypes and found that the salt-tolerant genotype (Hazerd) successfully tolerated highest salinity level (120 mM) by accumulating significantly higher levels of free proline and exhibited higher antioxidant enzyme (superoxide dismutase [SOD] and peroxidase [POD]) activities besides showing low lipid peroxidation and electrolyte leakage with slight reduction in photosynthetic pigment. Furthermore, higher salinity tolerance was also correlated to limited translocation of Na+ ions to leaves, resulting in the maintenance of high K+/Na+ ratio.

23.3.1.3 Nonconventional Breeding Methods for Salinity Stress Resistance

In watermelon, transgenic plants have been produced expressing the *HAL1* gene under the control of 35S promoter with a double enhancer sequence from the cauliflower mosaic virus and RNA4 leader sequence of alfalfa mosaic virus (Ellul et al. 2003). The constitutive expression of HAL1 gene showed a beneficial effect on rooting of plants grown under *in vitro* saline conditions.

Since salt concentration in soil is highly variable, it is necessary to frequently test genotypes/plants tolerance to salinity in several salt concentrations applied to root system. Genotype × salt treatment interaction has been found in several occasions and species (Lee et al. 2004). When appropriate segregant populations (RIL or DH) are grown in at least two salinity conditions (control and saline), more quantitative trait loci (QTL) have been identified in saline than in control conditions, and significant QTL × E interaction has been found in all the experiments designed to detect the interaction. It is necessary to assess that QTL detected under saline conditions are expressed in different salt concentrations, otherwise QTL should be found for each specific salt concentration on which tolerant genotypes are to be grown.

Furthermore, in salt-tolerant lines, identifying molecular markers tightly linked to the gene of interest can give us a lead to reduce the otherwise significant influence of environmental factors as suggested by Ashraf et al. (2008). The halo-tolerance due to HAL1 gene of *Saccharomyces cerevisiae* has been identified and confirmed as a molecular tool for genetic engineering for salt-stress protection in watermelon and other crop species (Ellul et al. 2003). In another study by Kere et al. (2013), a strong, positive correlation was opined between the RLN14 (relative leaf number) and TOL (tolerance), whereas VL (vine length) and TOL indicated a weak negative correlation but notably RLN14 and VL showed a strong negative correlation.

23.3.2 GRAFTING AS A TOOL TO MANAGE SALINITY STRESS

Salt tolerance is a complex characteristic both genetically and physiologically (Flowers 2004), which ultimately provides limited success through conventional breeding methods. Therefore, grafting can represent an interesting alternative to avoid or reduce yield losses caused by salinity stress in high-yielding genotypes belonging to Cucurbitaceae family. Grafting as a tool for enhancing the plant characteristics is well known, and investigations have indicated that grafting may limit nutrient and heavy metal toxicity (Edelstein et al. 2005; Arao et al. 2008; Rouphael et al. 2008a; Savvas et al. 2009). Grafting is an integrative reciprocal process and, therefore, both scion and rootstock can influence salt tolerance of the grafted plants. Biochemical mechanisms of uptake in the roots are governed by the demand in the sink, that is, shoot (Marschner 1995). However, the uptake efficiency depends upon the rootstock. So, grafting can serve as an important tool to prevent salt stress by inhibiting Na and Cl uptake. Grafted plants grown under saline conditions often exhibited better growth and yield, higher photosynthesis and leaf water content, greater root-to-shoot ratio, higher accumulation of compatible osmolytes, abscisic acid and polyamines in leaves, greater antioxidant capacity in leaves, and lower accumulation of Na^+ and/or Cl^- in shoots than ungrafted or self-grafted plants.

Grafting of bottle gourd rootstock affects nitrogen metabolism in NaCl-*stressed* watermelon leaves and enhances short-term *salt* tolerance. The plant growth, nitrogen absorption, and assimilation in watermelon were investigated in self-grafted and grafted seedlings using the salt-tolerant bottle gourd rootstock 'Chaofeng Kangshengwang' exposed to 100 mM NaCl for 3 days. Biomass and NO^{3-} uptake rate were significantly increased by rootstock, while these values were remarkably decreased by salt stress. However, compared with self-grafted plants, rootstock-grafted plants showed higher salt tolerance with higher biomass and NO^{3-} uptake rate under salt stress. These results indicated that the salt tolerance of rootstock-grafted seedlings might be enhanced owing to the higher nitrogen absorption and the higher activities of enzymes for nitrogen assimilation induced by the rootstock (Yang et al. 2013). As reported by Goreta et al. (2008), when watermelon

('Fantasy') was grafted onto 'Strongtosa' rootstock (*C. maxima* Duch. × *C. moschata* Duch.), the reductions in shoot weight and leaf area due to exposure to salinity were lower in ungrafted plants. Similarly, Yetisir and Uygur (2010) demonstrated that grafted 'Crimson Tide' watermelon onto rootstocks of *C. maxima* and *L. siceraria* resulted in higher growth than ungrafted plants under saline conditions (8.0 dS m^{-1}).

Romero et al. (1997) compared the effect of salinity (4.6 dS m^{-1}) on two varieties of melon (*C. melo* L.) grafted onto three hybrids of squash (*C. maxima* Duch. × *C. moschata* Duch.) with its effects on ungrafted melons and found that grafted melons were more tolerant to salinity and gave higher yields than ungrafted ones. Huang et al. (2009b) determined the fruit yield response of cucumber variety, 'Jinchun No. 2' either self-grafted or grafted onto the commercial salt-tolerant rootstock 'Black Seeded' fig leaf gourd (*Cucurbita ficifolia* Bouche) and Chaofeng Kangshengwang (*L. siceraria* Standl.) and studied the response under different saline conditions (0, 30, or 60 mM NaCl). Plants grafted onto 'Figleaf Gourd' and 'Chaofeng Kangshengwang' had higher fruit number and marketable fruit yield compared to self-grafted plants at all salt levels. The total fruit yield of plants grafted onto 'Figleaf Gourd' increased by 15%, 28%, and 73% under 0, 30, and 60 mM NaCl stress, respectively, whereas the respective values were 14%, 33%, and 83% in the plants grafted onto 'Chaofeng Kangshengwang' over self-grafted plants. They concluded that grafting cucumber onto 'Black Seeded' figleaf gourd increased plant tolerance to salinity induced by major nutrients.

Rouphael et al. (2008b) attributed the improved crop performance of grafted cucumber plants on pumpkin rootstock to the ability of the pumpkin rootstock to check the accumulation of Cu. Pumpkin plants have a deep and sturdy root system, which gives them an advantage to adapt to saline conditions. Melon plants grafted onto the commercial rootstock 'TZ-148' (*C. maxima* Duchesne × *C. moschata* Duchesne) prevented boron toxicity (Edelstein et al. 2005, 2007). Rouphael et al. (2012) suggested that the use of salt-tolerant *Cucurbita* hybrid rootstocks (*C. maxima* Duch. × *C. moschata* Duch.) 'P360' and 'PS1313', respectively, can improve melon and cucumber photosynthetic capacity under salt stress and consequently crop performance.

El-Shraiy et al. (2011) highlighted the significance of grafting cucumber on salt tolerance rootstock (Shintosa Supreme pumpkin), which increased fruit yield, fruit number, fruit weight, fresh and dry weight, plant height, leaf area, and leaf and relative water content (LRWC) compared to ungrafted plants under saline conditions. These positive effects of grafting cucumber significantly increased chlorophyll, carotenoid, proline, and total soluble protein concentrations although there was a reduction in titratable acidity, total soluble solids, and EC in fruit juice compared to ungrafted plants.

The use of rootstocks to combat salinity has also resulted in improved plant vigor through efficient use of nutrients and water, disease tolerance, cold tolerance, heat tolerance, and tolerance to wet soil conditions (Lee et al. 2010). Colla et al. (2005) ascribed that different rootstocks show variable results. The rootstocks of *Cucurbita* spp. and *Lagenaria* spp. serve as the best as they have the potential of improving the salt tolerance of scion by reducing the Na uptake better than *Citrullus* spp. and *Cucurbita* spp. In contrast, Taffouo et al. (2008) concluded that *Lagenaria* spp. is best suited to tolerate saline conditions. However, in an argument report stating similar sensitivity for grafted and nongrafted watermelon plants, the increased yield was due to grafting per se (Colla et al. 2006). Many reports suggest grafting to be an efficient way out for saline soils (Lauchli and Epstein 1970; Xiang et al. 2009; Lee et al. 2010). Different research reports revealed that the losses due to salinity in cucurbitaceae family can be avoided or negated by grafting cucurbits onto rootstocks capable of ameliorating salt-induced damage to the shoot (Estan et al. 2005; Wei et al. 2007; Zhu et al. 2008a,b; He et al. 2009; Huang et al. 2010; Yetisir and Uygur 2010; Zhen et al. 2010). Thus, the identification of compatible rootstocks with tolerance to other types of salinity is a basic requirement for the continued success of grafting (Colla et al. 2010) (Table 23.3).

TABLE 23.3

Exclusion and/or Inclusion of Na+ and Cl− in Grafted Vegetables under Saline Conditions

Scion Species	Rootstock Species	Ion Exclusion and/or Inclusion in the Scion	Ion Exclusion and/or Inclusion in the Rootstock	References
Cucumis sativus L.	Cucurbita moschata	Na+ exclusion		Chen and Wang (2008)
Cucumis sativus L.	Cucurbita ficifolia	Na+ exclusion		Chen and Wang (2008)
Cucumis sativus L.	Cucurbita moschata	Na+ exclusion and Cl− inclusion	Na+ and Cl− inclusion	Zhu et al. (2008a)
Cucumis sativus L.	Lagenaria siceraria	Na+ and Cl− exclusion	Similar Na+ and Cl−	Huang et al. (2009a,c)
Citrullus lanatus	Lagenaria siceraria	Na+ exclusion	Na+ inclusion	Zhu and Guo (2009)
Citrullus lanatus	Cucurbita maxima × C. moschata	Na+ exclusion and Cl− inclusion		Colla et al. (2006)
Cucumis melo L.	Cucurbita maxima × C. moschata	Na+ and Cl− exclusion		Romero et al. (1997)

Source: Colla, G. et al., Sci. Hortic., 127, 147, 2010.

23.3.3 Chemical Amendments to Ameliorate Soil Salinity Stress

Soil amendments are the materials applied to the soil with an objective to make the soil suitable for plant growth and development. Salinity is considered as an undesirable chemical property of the soil, which causes poor and scattered seed germination besides difficulty in seedling establishment and burning of plant tissues. This reduces crop yield due to poor plant growth and in extreme cases, may lead to death of plants. The most effective way to handle this problem is to change the old soil by replacing it with soil containing desirable characteristics. This is practically not feasible. However, the chemical properties of saline soils can be improved with desirable characteristics to some extent by the addition of organic matter and different chemical amendments. In addition, leaching or removal of soluble salts can also reduce salinity in soils. Some of the proposed practices, viz., application of chemical amendments or leaching salts to deeper soil layers, seed priming, etc., have been advocated. All chemical amendments are not suitable for all soil conditions, for example, gypsum is suitable in saline soil having a pH range up to 9, while limestone is suitable in saline soil with a pH less than 8.

Saline soils can be ameliorated by replacing excess sodium (Na^+) from the cation exchange sites, by providing calcium (Ca^{2+}) as a source. However, worldwide, the cost of chemical amendments has, in general, increased because of the reduction in subsidies for their purchase. The reclamation of saline soils can be done by using different methods such as physical amelioration (deep ploughing, subsoiling, sanding, profile inversion), chemical amelioration (use of gypsum, calcium chloride, limestone, sulfur, and iron sulfate), and electroreclamation through treatment with electric current (Mahdy 2011). The most effective methods are based on the removal of soluble sodium and changing the ionic composition of soils through applied chemicals and simultaneous leaching of sodium salts from the soil profile (Chhabra 1994). The use of organic matter improves the soil structure and permeability. This enhances leaching of salt, reduces surface evaporation, and inhibits salt accumulation in the surface layers. In addition, it also increases water infiltration, water-holding capacity, and aggregate stability and reduces EC (Qadir et al. 2001). The organic matter of high cation exchange capacity (CEC) can adsorb some soluble salts, decrease pH, and promote aggregation. The decrease in pH below 8 can lead to charging of clay minerals and electrostatic adsorption of the organic compounds.

23.3.3.1 Calcium

Application of Ca^{2+} has been shown to ameliorate the adverse effects of salinity in a variety of plant species (Caines and Shannon 1999; Shabala et al. 2003; Arshi et al. 2005; Renault 2005). Calcium (Ca) plays an essential role in processes that preserve the structural and functional integrity of plant membranes, stabilize cell wall structure, regulate ion transport and selectivity, and control ion-exchange behavior as well as enzyme activities (Esringu et al. 2011). It has been hypothesized that a high concentration of Ca can protect the cell membrane from the adverse effects of salinity (Kaya et al. 2002). Maintaining an adequate supply of calcium in saline soil solutions is an important factor in controlling the severity of specific ion toxicities, particularly in crops that are susceptible to sodium and chloride injury (Grattan and Grieve 1999).

Cerda and Martinez (1988) reported calcium deficiency in addition to sodium toxicity under saline conditions in cucumber. Application of Ca ameliorates the adverse effect of salinity in plants by facilitating greater potassium (K) in Na selectivity (Hasegawa et al. 2000). Dabuxilatu and Ikeda (2005) reported that an increase in Ca concentration had an ameliorative effect on cucumber plants where the growth was inhibited due to salinity. The cation imbalance due to increased Na^+ concentration and decreased K^+ and Mg^{2+} concentrations might be responsible for growth inhibition in cucumber plants. They were of the view that the beneficial effect of a high Ca concentration in a saline environment would be due to the maintenance of K/Na selectivity and adequate Ca status in roots.

Lei et al. (2014) found that supplementary $Ca(NO_3)_2$ at 10 mM ameliorated the negative effects of NaCl on plant dry mass, relative growth rate, as well as Ca^{2+}, K^+, and Na^+ content, especially for pumpkin rootstock-grafted cucumber plants. The addition of Ca^{2+} in combination with pumpkin rootstock grafting is a powerful way to increase cucumber salt tolerance as supplementary $Ca(NO_3)_2$ distinctly stimulated the plasma membrane H^+-ATPase gene (PMA) expression as well as higher plasma membrane Na^+/H^+ antiporter encoding gene (SOS_1) expressions than the self-grafted plants under NaCl + Ca treatment. Ca and N supplemented into the soil in the form of $Ca(NO_3)_2$ can significantly improve plant growth, fruit yield, and membrane permeability affected by high salinity along with correcting both Ca and N deficiencies. Calcium nitrate applied at 1 g/kg to the soil offers an economical and simple solution to crop production problems caused by high salinity in the soils of arid and semiarid regions of the world (Kaya and Higgs 2002).

23.3.3.2 Silica

Silica (Si) is beneficial for the growth of many plants under various abiotic (e.g., salt, drought, and so on.) and biotic (diseases and insect-pests) stresses (Ali et al. 2012). There are different mechanisms by which Si mediates salinity tolerance in plants (Liang et al. 2005) including increased plant water status (Chinnusamy et al. 2005), salt stress due to ion toxicity (Romero-Aranda et al. 2006), enhanced photosynthetic activity and maintenance of ultra structure of leaf organelles (Shu and Liu 2001), stimulation of scavenging system of reactive oxygen species (Zhu et al. 2004), immobilization of toxic Na ion (Liang et al. 2003), reduced Na uptake in plants and enhanced K uptake (Liang et al. 2005), and higher K:Na selectivity (Hasegawa et al. 2000). Its application helps to improve the defensive system of the plants by producing antioxidants, which in turn detoxify reactive oxygen species. Morphological and physiological improvement in plants was observed due to Si deposition within the plant body under salt stress conditions. Silicon improves growth and dry matter production under salt stress conditions. Silicate has increased resistance against salinity in cucumber (Amirossadat et al. 2012). Si may be involved in the metabolic or physiological changes in plants, and its addition may protect the plant tissues from membrane oxidative damage under salt stress, thus mitigating salt toxicity and improving the growth of cucumber plants (Zhu et al. 2004). It is, therefore, suggested that supplemental application of Si must be included in salt stress alleviation management techniques.

23.3.3.3 Other Nutrients

Salt stress also has significant effects on nitrogen (N) nutrition in plants. Salinity reduces the uptake of NO_3 in many plant species, mostly due to the high Cl content of saline soil (Esringu et al. 2011). The effectiveness of N application under salinity stress conditions has been observed in cucumber (Cerda and Martinez 1988) and melon (Feigin et al. 1987). The form in which N is supplied to salt-stressed plants can influence salinity-N relations as well as affect the relation of salinity with other nutrients (Martinez and Cerda 1989). Melon plants supplemented with ammonia were more sensitive to salinity than nitrate-fed plants when grown in solution cultures (Feigin 1990). In addition, Martinez and Cerda (1989) found that Cl^- uptake was reduced in cucumber when only nitrate was added to the solution but when half the nitrate in the solution was replaced by ammonia, Cl accumulation was enhanced. They further observed that accumulation of K in the plant increased with nitrate as the only N-source, whereas there was reduction in K when both nitrate and ammonia were used. Similar effects were found in salt-stressed melon (Feigin 1990; Adler and Wilcox 1995). The relationship between salinity and N is complex. The majority of studies indicate reduction of N accumulation in the shoot. In contrast, increased accumulation or no effect of N has also been reported. There is no supporting evidence in favor of reports stating growth limiting effects of reduced N accumulation under saline conditions (Grattan and Grieve 1999).

The effects of phosphorus (P) have been found to be very complex and are known to vary with growing conditions of the plant, plant type, and even cultivar (Grattan and Grieve 1994). Salinity decreases the concentration of P in the plant tissue (Sharpley et al. 1992), but differences among studies may be ascribed to variation in P concentration in different experiments and also due to the occurrence of interactions among other nutrients simultaneously. The reduction in availability of phosphate in saline soils may be attributed to ionic strength effects decreasing phosphate activity and sorption processes controlling phosphate concentrations in soil solution and due to low-solubility of Ca–P minerals. Therefore, phosphate concentrations decreased with increase in salinity ($NaCl + CaCl_2$).

Potassium (K) is also a major plant macronutrient that plays an important role in stomatal behavior, osmoregulation, enzyme activity, cell expansion, neutralization of nondiffusible negatively charged ions, and membrane polarization (Elamalai et al. 2002). Metabolic toxicity of Na is largely due to its ability to compete with K for binding sites essential for cellular function (Bhandal and Malik 1988). High NaCl has shown to induce K deficiency in melon. The supplementation with KNO_3 and proline significantly ameliorated the adverse effects of salinity on plant growth, fruit yield, and other physiological parameters by maintaining membrane permeability and increasing concentrations of Ca^{2+}, N, and K^+ in the leaves of plants subjected to salt stress (Kaya et al. 2007). Foliar application of KNO_3 at 250 ppm induced increased fruit formation and fruit weight in bottle gourd grown under saline conditions (Ahmad and Jabeen 2005). Similarly, Al-Hamzawi (2010) observed the efficacy of foliar application of KNO_3 at 15 mM in enhancing total yield along with different growth and yield parameters and simultaneous increase in storage life.

Similarly, magnesium (Mg) is an important component of chlorophyll and plays a vital role in photosynthesis; in addition, it assists in phosphate metabolism, plant respiration, protein synthesis, and activation of several enzyme systems in the plant (Marschner 1995). Ibekwe et al. (2010) reported that the effects of salinity, boron, and pH were more severe on the rhizosphere bacterial population during the first week of growing cucumber and further salinity impact decreases with plant growth. Thus, early detection of stress may provide some remedial action to improve soil quality and crop performance.

23.3.3.4 Proline in Alleviating Salt Stress

Salinity stress decreases amino acids in plants, viz., cysteine, arginine, and methionine, whereas proline concentration rises in response to salinity stress. Proline accumulation is a well-known measure adopted for the alleviation of salinity stress. Intracellular proline, which is accumulated during salinity stress, not only provides tolerance toward stress but also serves as an organic nitrogen reserve during stress recovery. Huang et al. (2009a) reported that foliar application of proline on the salt-sensitive cucumber cultivar 'Jinchun No. 2' significantly alleviated the growth inhibition of plants induced by NaCl, which could be partially attributed to higher leaf relative water content and POD activity, higher proline and Cl⁻ contents, and lower malondialdehyde content. Similar effects of proline on alleviation of salinity-induced damage were reported by Yan et al. (2012) in melon cultivars, namely 'Yuhuang' and 'Xuemei'.

23.3.4 SEED PRIMING

Priming is one of the physiological methods that improves seed performance and provides faster and synchronized germination (Sivritepe et al. 2003). Salinity has an adverse effect on seed germination of several vegetable crops by creating an osmotic potential outside the seed inhibiting the absorption of water or by the toxic effect of Na^+ and Cl^- (Khajeh-Hosseini et al. 2003). Osmotic and saline stresses are responsible for the inhibition and delay of germination and plant growth (Almansouri et al. 2001). Water uptake during the imbibition phase decreases and salinity induces an excessive absorption of toxic ions in the seed (Murillo-Amador et al. 2002). Priming of seeds with water (Casenave and Toseli 2007), inorganic salts (Patade et al. 2009), osmolytes, and hormones (Iqbal and Asraf 2007) has been demonstrated as a successful cost-effective strategy for improving seed vigor and seedling growth under saline conditions (Foti et al. 2008). Priming improves germination, emergence, and establishment of several seed species (Singh 1995; Basra et al. 2005). NaCl priming could be used as an adaptation method to improve salt tolerance of seeds. Higher salt tolerance of plants from primed seeds seems to be the result of a higher capacity for osmotic adjustments since plants from primed seeds have more Na and Cl in roots and more sugar and organic acids in leaves in comparison to nonprimed seeds (Cayuela et al. 1996). Passam and Kakourioitis (1994) found that seed priming enhances germination, emergence, and growth under saline conditions in the cucumber, but benefits of NaCl priming did not persist beyond the seedling stage. On the other hand, Franco et al. (1993) observed that melon is salt tolerant between the fruit development and harvest stage, but it is sensitive during germination and seedling growth stage. Priming of melon seeds with NaCl resulted in increased salt tolerance in seedlings of melon cultivars 'Kirkagac' and 'Hasanbey' (Sivritepe et al. 1999). In other studies, priming with NaCl increased salt tolerance of seedlings by promoting K and Ca accumulation besides inducing osmoregulation by the accumulation of organic solutes, enhancing yields and quality of melon seeds (Sivritepe et al. 2003) and cucumber (Esmaielpour et al. 2006). Joshi et al. (2013) reported the beneficial effects of presoaking (priming) treatment with 2 mM $CaCl_2$ solution for 24 h under saline conditions on germination and seedling growth of cucumber, which could be ascribed to its role in activation of antioxidant system and accumulation of proline. Balouchi (2014) observed that osmopriming by PEG (−5 bar) and hydropriming on *C. pepo* was good for germination and seedling growth under saline conditions. Primed seeds should be sown immediately after priming since they lose their storage life instantly (Basra et al. 2003).

23.3.5 ADDITIONAL PRACTICES FOR AMELIORATION OF SALINE SOILS

It is essential to consider the following points before reclamation of saline soils:

1. Good quality irrigation water with low salt content. It is important to determine total soluble salt, sodium adsorption ratio (SAR), and sometimes boron also.
2. Degree of salinity.
3. Nature and distribution of salts in the root zone.
4. Level of subsoil water (water table).

As mentioned earlier, a majority of crop plants are susceptible to salt injury during germination or in the early seedling stages. Any management practice that helps in reducing salt concentration during these stages would benefit the crop by promoting plant growth and development. Salts in the root zone are dynamic and often vary with climate. In case of well-drained soils, rains tend to push salts below the root zone, while dry periods bring salts near the soil surface. In contrast, the water table rises close to the surface during the rainy period in poorly drained soils and as a result salts move upward with the water, causing high salinity than during the dry period. Hence, the selection of crops for planting according to their salt tolerance can be done according to climate cycles and soil drainage classification. This indicates that planning for cultivation of annual crops like a majority of cucurbits is easier than for perennials. The following technical requirements are necessary for the reclamation of saline soils.

23.3.5.1 Adequate Drainage

Salt problems often occur in soils with poor internal drainage. Low-permeability soil layers may restrict the flow of water from deeper layers much slower than evapotranspiration (ET) from the soil surface. In such situations, select those crops like bitter gourd and ash gourd that can tolerate salinity without much reduction in yield. Artificial drains can also be provided to allow the removal of leaching water and salts from soils.

23.3.5.2 Quality of Water and Irrigation Frequency

Irrigation management can be used to decrease the level of salts in the root zone of the crop. The management of saline soil becomes difficult if the irrigation water contains high salt concentration and the situation becomes worse if the seasonal rainfall is less. Permeability of the subsurface soils is important for salt management. Leaching of sufficient salts down (beyond the root zone) has an indirect correlation with evapotranspiration (bringing water and salts back toward the surface). So, it is essential to apply quality irrigation water; free/low in salt contents (1500–2000 ppm total salts), particularly sodium, by leaching of salts below the root zone to ensure optimal soil conditions for better plant growth. Leaching works well on saline soils that have good structure and internal drainage. Frequent irrigations leach salts more efficiently from the soil profile by diluting salt accumulation in the root zone and thereby reducing salinity to a certain extent. The salts cannot be dissolved and leached out of the soil if water infiltration does not take place in the soil.

Preplant irrigation helps in flushing out/leaching the salts concentrated on the surface of saline soils. Cucurbits are more susceptible to salt injury during germination or in the early seedling stages. An early-season application of good quality water at frequent intervals may provide good conditions for the crop to grow through its most injury-prone stages.

23.3.5.3 Appropriate Planting Method

The salinity impact may be reduced if the planting of seeds/seedlings is done at appropriate positions on the ridges. Depending upon the irrigation system, the furrows, ridges, and planting of seed/seedlings can be planned. In general, planting should be done on the shoulder of the ridge rather than on the top or center, while irrigation is applied through furrows on both sides of the ridge because evaporation will cause accumulation of more salts at the ridge top or center. On the other hand, if irrigation is applied in alternate furrows, then plant only on one shoulder of the ridge, closer to the irrigated furrow.

23.3.5.4 Mulching

Mulching with crop residues helps in reducing evaporation from the soil surface. This, in turn, reduces the upward movement of salts and lessens the accumulation of salts. Inorganic mulches integrated with the drip irrigation system effectively reduce salt concentration. Subsurface drip irrigation pushes salts to the edge of the soil wetting front, reducing harmful effects on seedlings and plant roots.

23.3.5.5   Deep Tillage

In saline soils, accumulation of salts is close to the surface. Deep tillage would mix the salts present in the surface zone and would reduce the concentration of the salts near the surface. In the impervious hard pan soils, salt leaching process does not occur. Chiseling would be effective to improve water infiltration and downward movement of salts.

23.3.5.6   Bioamelioration of Saline Soils

The beneficial effects of arbuscular mycorrhizal fungi (AMF) to promote plant growth and salinity tolerance have been reported by many research workers. They promote salinity tolerance by employing various mechanisms by enhancing and improving nutrient acquisition (Al-Karaki and Al-Raddad 1997), plant growth hormones, rhizospheric and soil conditions (Lindermann 1994), physiological and biochemical properties of the host (Smith and Read 1997), and defending roots against soil-borne pathogens (Dehne 1982). In addition, AMF can improve water absorption capacity of plants by increasing root hydraulic conductivity and favorably adjusting the osmotic balance and composition of carbohydrates, which may lead to an increase in plant growth and subsequent dilution of toxic ion effect (Evelin et al. 2009). These benefits of AMF may provide an opportunity for bioamelioration of saline soils. In addition, potential salt-tolerant bacteria isolated from the soil or plant tissues help to alleviate salt stress by promoting seedling growth and increase biomass of crop plants grown under salinity stress (Chakraborty et al. 2011).

23.4   CONCLUSIONS

Salinity would be a major threat to the agriculture sector in the near future. A lot of research work has been done on cucurbits with respect to salinity management. There is a need to concentrate more on the development of salinity-resistant/salinity-tolerant cultivars for sustainable production. In addition to tolerant cultivars, several economic, cultural practices need to be standardized with the aim that each contributes to a certain extent to allow plants to better withstand the deleterious effects of salt. The most practical approach would be to integrate traditional breeding approaches with genetic manipulation and soil amelioration practices to cope with the increasing soil salinity constraints.

REFERENCES

Adler, P.R. and G.E. Wilcox. 1995. Ammonium increases the net rate of sodium influx and partitioning to the leaf of muskmelon. *J. Plant Nutr.* 18: 1951–1962.

Ahmad, P. 2010. Growth and antioxidant responses in mustard (*Brassica juncea* L.) plants subjected to combined effect of gibberellic acid and salinity. *Arch. Agro. Soil Sci.* 56(5): 575–588.

Ahmad, P. and R. John. 2005. Effect of salt stress on growth and biochemical parameters of *Pisum sativum* L. *Arch. Agron. Soil Sci.* 51: 665–672.

Ahmad, P. and S. Sharma. 2010. Physico-biochemical attributes in two cultivars of mulberry (*M. alba*) under NaHCO3 stress. *Int. J. Plant Prod.* 4(2): 79–86.

Ahmad, R. and R. Jabeen. 2005. Foliar spray of mineral elements antagonistic to sodium—A technique to induce salt tolerance in plants growing under saline conditions. *Pak. J. Bot.* 37(4): 913–920.

Al-Hamzawi, M.K.A. 2010. Effect of calcium nitrate, potassium nitrate and Anfaton on growth and storability of plastic houses cucumber (*Cucumis sativus* L. cv. Al-Hytham). *Am. J. Plant Physiol.* 5(5): 278–290.

Ali, A., S.M.A. Basra, S. Hussain, J. Iqbal, M. Ahmad, and M. Sarwar. 2012. Salt stress alleviation in field crops through nutritional supplementation of silicon. *Pak. J. Nutr.* 11(8): 637–655.

Al-Karaki, G.N. and A. Al-Raddad. 1997. Effect of arbuscularmycorrhizal fungi and drought stress on growth and nutrient uptake of two wheat genotypes differing in drought resistance. *Mycorrhiza.* 7: 83–88.

Almansouri, M., J.M. Kinet, and S. Lutts. 2001. Effect of salt and osmotic stresses on germination in durum wheat (*Triticum durum*). *Plant Soil.* 231: 243–254.

Amirossadat, Z., A.M. Ghehsareh, and A. Mojiri. 2012. Impact of silicon on decreasing of salinity stress in greenhouse cucumber (*Cucumis sativus* L.) in soilless culture. *J. Biol. Environ. Sci.* 6(17): 171–174.

Arao, T., H. Takeda, and E. Nishihara. 2008. Reduction of cadmium translocation from roots to shoots in eggplant (*Solanum melongena*) by grafting onto *Solanum torvum* rootstock. *Soil Sci. Plant Nutr.* 54: 555–559.

Arshi, A., M.Z. Abdin, and M. Iqbal. 2005. Ameliorative effect of CaCl$_2$ on growth, ionic relations and proline content of senna under salinity stress. *J. Plant Nutr.* 28: 101–125.

Ashraf, H., H.R. Athar, P.J.C. Harris, and T.R. Kwon. 2008. Some prospective strategies for improving crap salt tolerance. *Adv. Agron.* 97: 45–110.

Baghbani, A., A.H. Forghani, and A. Kadkhodaie. 2013. Study of salinity stress on germination and seedling growth in greenhouse cucumber cultivars. *J. Basic Appl. Sci. Res.* 3(3): 1137–1140.

Balouchi, H. 2014. Effects of seed priming on germination and seedling growth of cucurbit (*Cucurbita pepo*) medical plants under salinity stress. *J. Crop Prod. Process.* 3(10): 165–180.

Basra, S.M.A., I. Afzal, R.A. Rashid, and A. Hameed. 2005. Inducing salt tolerance in wheat by seed vigor enhancement techniques. *Int. J. Biotechnol. Biol.* 1: 173–179.

Basra, S.M.A., E. Ulah, E.A. Waraich, M.A. Chema, and I. Afzal. 2003. Effect of storage on growth and yield of primed canola (*Brasica napus*) seeds. *Int. J. Agric. Biol.* 2: 17–120.

Bhandal, I.S. and C.P. Malik. 1988. Potassium estimation, uptake, and its role in the physiology and metabolism of flowering plants. *Int. Rev. Cytol.* 110: 205–254.

Bohnert, H.J. 2007. *Abiotic Stress.* John Wiley & Sons, Ltd., Hoboken, NJ.

Borsani, O., V. Valpuesta, and M.A. Botella. 2003. Developing salt-tolerant plants in a new century: A molecular biology approach. *Plant Cell Tiss. Org. Cult.* 73: 101–115.

Caines, A.M. and C. Shannon. 1999. Interactive effect of Ca and NaCl salinity on the growth of two tomato genotypes differing in Ca use efficiency. *Plant Physiol. Biochem.* 37(7/8): 569–576.

Casenave, E.C. and M.E. Toselli. 2007. Hydropriming as a pre-treatment for cotton germination under thermal and water stress conditions. *Seed Sci. Technol.* 35: 88–98.

Cayuela, E., F. Prez-Alfocea, M. Caro, and M.C. Bolarin. 1996. Priming of seeds with NaCl induces physiological changes in tomato plants grown under salt stress. *Physiol. Plant.* 96: 231–236.

Cerda, A. and V. Martinez. 1988. Nitrogen fertilization under saline conditions in tomato and cucumber plants. *J. Hortic. Sci.* 63(3): 451–458.

Chakraborty, A.P., P. Dey, B. Chakraborty, U. Chakraborty, and S. Roy. 2011. Plant growth promotion and amelioration of salinity stress in crop plants by a salt-tolerant bacterium. *Recent Res. Sci. Technol.* 3: 61–70.

Chen, G. and R. Wang. 2008. Effects of salinity on growth and concentrations of sodium, potassium, and calcium in grafted cucumber seedlings. *Acta Horticulture.* 771: 217–224.

Chhabra, R. 1994. *Soil Salinity and Water Quality.* Oxford & IBH Publishing Co., New Delhi, India.

Chinnusamy, V., A. Jagendorf, and J.K. Zhu. 2005. Understanding and improving salt tolerance in plants. *Crop Sci.* 45: 437–448.

Colla, G., S. Fanasca, M. Cardarelli, Y. Rouphael, F. Saccardo, A. Graifenberg, and M. Curadi. 2005. Evaluation of salt tolerance in rootstocks of Cucurbitaceae. *Proc. Int. Symp. Soilless Cult. Hydroponics, Acta Hortic.* 697: 469–474.

Colla, G., Y. Roupaphel, M. Cardarelli, and E. Rea. 2006. Effect of salinity on yield, fruit quality, leaf gas exchange and mineral composition of grafted watermelon plants. *Hortic. Sci.* 3: 622–627.

Colla, G., Y. Roupaphel, C. Leonardic, and Z. Bied. 2010. Role of grafting in vegetable crops grown under saline conditions. *Sci. Hortic.* 127: 147–155.

Cramer, G.R. and R.S. Nowak. 1992. Supplemental manganese improves the relative growth, net assimilation and photosynthetic rates of salt-stressed barley. *Physiol. Plant.* 84: 600–605.

Cuartero, J., M.C. Bolarin, M.J. Asins, and V. Moreno. 2006. Increasing salt tolerance in tomato. *J. Exp. Bot.* 57: 1045–1058.

Cuartero, J. and R. Fernandez-Munoz. 1999. Tomato and salinity. *Sci. Hortic.* 78: 83–125.

Dabuxilatu, G. and M. Ikeda. 2005. Interactive effect of salinity and supplemental calcium application on growth and ionic concentration of soybean and cucumber plants. *Soil Sci. Plant Nutr.* 61(4): 549–555.

Dashti, H., M.R. Bihamta, H. Shirani, and M.M. Majidi. 2012. Genetic analysis of salttolerance in vegetative stage in wheat (*Triticum aestivum*). *Plant Omics.* 5: 19–23.

Dehne, H.W. 1982. Interaction between vesicular-arbuscularmycorrhizal fungi and plant pathogens. *Phytopathology.* 72: 1115–1119.

Dhillon, N.P.S., A.J. Monforte, M. Pitrat, S. Pandey, P.K. Singh, K.R. Reitsma, J. Garcia-Mas, A. Sharma, and J.D. McCreight. 2012. Melon landraces of India: Contributions and importance. *Plant Breed. Rev.* 35: 85–150.

Dorota, Z. 1997. *Irrigating with High Salinity Water.* Florida Cooperative Extension Service, Institute of Food and Agriculture Sciences, University of Florida, Gainesville, FL; *Bulletin* 322: 1–5.

Edelstein, M., M. Ben-Hur, R. Cohen, Y. Burger, and I. Ravina. 2005. Boron and salinity effects on grafted and non-grafted melon plants. *Plant Soil.* 269: 273–284.

Edelstein, M., M. Ben-Hur, and Z. Plaut. 2007. Grafted melons irrigated with fresh or effluent water tolerate excess boron. *J. Am. Soc. Hortic. Sci.* 132: 484–491.

Elamalai, R.P., P. Nagpal, and J.W. Reed. 2002. A mutation in the *Arabidopsis* KT2/KUP2 potassium transporter gene affects shoot cell expansion. *Plant Cell.* 14: 119–131.

Ellul, P., G. Ros, A.S. Atar, L.A. Roig, R. Serrano, and V. Moreno. 2003. The expression of the *Saccharomyces cerevisiae* HAL1 gene increases salt tolerance in transgenic watermelon [*Citrullus lanatus* (Thunb.) Matsun. & Nakai.]. *Theor. Appl. Genet.* 107: 462–469.

El-Shraiy, A.M., M.A. Mostafa, S.A. Zaghlool, and S.A.M. Shehata. 2011. Alleviation of salt injury of cucumber plant by grafting onto salt tolerance rootstock. *Aust. J. Basic Appl. Sci.* 5(10): 1414–1423.

Esmaielpour, B., K. Ghassemi-Golezani, F.R. Khoei, V. Gregoorian, and M. Toorchi. 2006. The effect of NaCl priming on cucumber seedling growth under salinity stress. *J. Food Agric. Environ.* 4(2): 347–349.

Esringu, A., C. Kant, E. Yildirim, H. Karlidag, and M. Turan. 2011. Ameliorative effect of foliar nutrient supply on growth, inorganic ions, membrane permeability, and leaf relative water content of physalis plants under salinity stress. *Commun. Soil Sci. Plant Anal.* 42: 408–423.

Estan, M.T., M.M. Martinez-Rodriguez, F. Perez-Alfocea, T.J. Flowers, and M.C. Bolarin. 2005. Grafting raises the salt tolerance of tomato through limiting the transport of sodium and chloride to the shoot. *J. Exp. Bot.* 56: 703–712.

Evelin, H., R. Kapoor, and B. Giri. 2009. Arbuscular mycorrhizal fungi in alleviation of salt stress: A review. *Ann. Bot.* 104(7): 1263–1280.

Feigin, A. 1990. Interactive effects of salinity and ammonium/nitrate ratio on growth and chemical composition of melon plants. *J. Plant Nutr.* 13: 1257–1269.

Feigin, A., I. Rylski, A. Meiri, and J. Shalhevet. 1987. Response of melon and tomato plants to chloride-nitrate ratios in saline nutrient solutions. *J. Plant Nutr.* 10: 1787–1794.

Flowers, T.J. 2004. Improving crop salt tolerance. *J. Exp. Bot.* 55: 307–319.

Foolad, M.R. 2004. Recent advances in genetics of salt tolerance in tomato. *Plant Cell Tiss. Org.* 76: 101–119.

Foti, R., K. Abureni, A. Tigere, J. Gotos, and J. Gere. 2008. The efficacy of different seed priming osmotica on the establishment of maize (*Zea mays* L.) caryopses. *J. Arid Environ.* 72: 1127–1130.

Franco, J.A., C. Esteban, and C. Rodriguez. 1993. Effects of salinity on various growth stages of muskmelon cv. Revigal. *J. Hortic. Sci.* 68: 899–904.

Goreta, S., V. Bucevic-Popovic, G.V. Selak, M. Pavela-Vrancic, and S. Perica. 2008. Vegetative growth, superoxide dismutase activity and ion concentration of salt stressed watermelon as influenced by rootstock. *J. Agric. Sci.* 146: 695–704.

Grattan, S.R. and C.M. Grieve. 1994. Mineral nutrient acquisition and response by plants grown in saline environments. In: *Handbook of Plant and Crop Stress*, Ed. Pessarakli, M. Marcel Dekker, New York, pp. 203–226.

Grattan, S.R. and C.M. Grieve. 1999. Salinity-mineral nutrient relations in horticultural crops. *Sci. Hortic.* 78: 127–157.

Hasegawa, P., R.A. Bressan, J.K. Zhu, and H.J. Bohnert. 2000. Plant cellular and molecular responses to high salinity. *Annu. Rev. Plant Mol. Biol.* 51: 463–499.

Hasegawa, P.M., R.A. Bressan, and A.K. Handa. 1980. Growth characteristics of NaCl selected and non-selected cell lines of *Nicotiana tabacum* L. *Plant Cell Physiol.* 21: 1347.

He, Y., Z.J. Zhu, J. Yang, X.L. Ni, and B. Zhu. 2009. Grafting increases the salt tolerance of tomato by improvement of photosynthesis and enhancement of antioxidant enzymes activity. *Environ. Exp. Bot.* 66: 270–278.

Huang, Y., Z.L. Bie, S. He, B. Hua, A. Zhen, and Z. Liu. 2010. Improving cucumber tolerance to major nutrients induced salinity by grafting onto *Cucurbita ficifolia*. *Environ. Exp. Bot.* 69: 32–38.

Huang, Y., Z.L. Bie, Z.X. Liu, A. Zhen, and W.J. Wang. 2009a. Exogenous proline increases the salt tolerance of cucumber by enhancing water status and peroxidase enzyme activity. *Soil Sci. Plant Nutr.* 55: 698–704.

Huang, Y., R. Tang, Q.L. Cao, and Z.L. Bie. 2009b. Improving the fruit yield and quality of cucumber by grafting onto the salt tolerant rootstock under NaCl stress. *Sci. Hortic.* 122: 26–31.

Huang, Y., J. Zhu, A. Zhen, L. Chen, and Z.L. Bie. 2009c. Organic and inorganic solutes accumulation in the leaves and roots of grafted and ungrafted cucumber plants in response to NaCl stress. *J. Food Agric. Environ.* 7: 703–708.

Ibekwe, A.M., J.A. Poss, S.R. Grattan, C.M. Grieve, and D. Suarez. 2010. Bacterial diversity in cucumber (*Cucumis sativus*) rhizosphere in response to salinity, soil pH, and boron. *J. Soil Biol. Biochem.* 42: 567–575.

Iqbal, M. and M. Ashraf. 2007. Seed treatment with auxins modulates growth and ion partitioning in salt-stressed wheat plants. *J. Int. Plant Biol.* 49: 1003–1015.

Jiang, Y., B. Yang, N.S. Harris, and M.K. Deyholos. 2007. Comparative proteomic analysis of NaCl stress-responsive proteins in *Arabidopsis* roots. *J. Exp. Bot.* 58: 3591–3607.

Joshi, N., A. Jain, and K. Arya. 2013. Alleviation of salt stress in *Cucumis sativus* L. through seed priming with calcium chloride. *Indian J. Appl. Res.* 3: 22–25.

Kaya C., B.E. Ak, D. Higgs, and B. Murillo-Amador. 2002. Influence of foliar-applied calcium nitrate on strawberry plants grown under salt-stressed conditions. *Aust. J. Exp. Agric.* 42: 631–636.

Kaya, C. and D. Higgs. 2002. Calcium nitrate as a remedy for salt-stressed cucumber plants. *J. Plant Nutr.* 25(4): 861–871. http://dx.doi.org/10.1081/PLN-120002965.

Kaya, C., A.L. Tuna, M. Ashraf, and H. Altunlu. 2007. Improved salt tolerance of melon (*Cucumis melo* L.) by the addition of proline and potassium nitrate. *Environ. Exp. Bot.* 60: 397–403.

Kere, G.M., Q. Guo, J. Shen, J. Xu, and J. Chen. 2013. Heritability and gene effects for salinity tolerance in cucumber (*Cucumis sativus* L.) estimated by generation mean analysis. *Sci. Hortic.* 159: 122–127.

Khajeh-Hosseini, M., A.A. Powell, and I.J. Bimgham. 2003. The interaction between salinity stress and seed vigor during germination of soybean seeds. *Seed Sci. Technol.* 31: 715–725.

Lauchli, A. and E. Epstein. 1970. Transport of potassium and rubidium in plant roots. The significance of calcium. *Plant Physiol.* 45: 639–641.

Lauchli, A. and S.R. Grattan. 2007. Plant growth and development under salinity stress. In: *Advances in Molecular Breeding toward Drought and Salt Tolerant Cropseds*, Eds. Jenks, M.A. et al. Springer, Heidelberg, Germany, pp. 1–32.

Lee, G.J., H.R. Boerma, M.R. Villaprcia, X. Zhou, T.E. Carter Jr., and Z. Ll. 2004. A major QTL condi-tioning salt tolerance in S-100 soybean and descendent cultivars. *Theor. Appl. Genet.* 109: 1610–1619.

Lee, J.M., C. Kubota, S.J. Tsao, Z. Bie, P.H. Echevarria, L. Morra, and M. Oda. 2010. Current status of vegetable grafting: Diffusion, grafting techniques, automation. *Sci. Hortic.* 127: 93–105.

Lei, B., Y. Huang, J.J. Xie, Z.X. Liu, A. Zhen, M.L. Fan, and Z.L. Bie. 2014. Increased cucumber salt tolerance by grafting on pumpkin rootstock and after application of calcium. *Biol. Plant.* 58(1): 179–184.

Levitt, J. 1980. *Responses of Plants to Environmental Stresses*, Vol. II, 2nd edn. Academic Press, New York, p. 607.

Liang, Y.C., Q. Chen, Q. Liu, W. Zhang, and R. Ding. 2003. Effects of silicon on salinity tolerance of two barley cultivars. *J. Plant Physiol.* 160: 1157–1164.

Liang, Y.C., W.Q. Zhang, J. Chen, and R. Ding. 2005. Effect of silicon on H^+-ATPase and H^+-PPase activity, fatty acid composition and fluidity of tonoplast vesicles from roots of salt-stressed barley (*Hordeum vulgare* L.). *J. Environ. Exp. Bot.* 53: 29–37.

Lindermann, R.G. 1994. Role of VAM in biocontrol. In: *Mycorrhizae and Plant Health*, Eds. Pfleger, F.L. and R.G. Linderman. American Phytopathological Society, St. Paul, MN, pp. 1–26.

Maas, E.V. 1993. Testing crops for salinity tolerance. In: *Proceedings of the Workshop on Adaptation of Plants to Soil Stresses*, Eds. Maranville, J.W., B.V. Baligar, R.R. Duncan, and J.M. Yohe, University of Nebraska, Lincoln, NE, August 1–4, 1993. INTSORMIL. Publication No. 94-2, pp. 234–247.

Maggio, A., P.M. Hasegawa, R.A. Bressan, M.F. Consiglio, and R.J. Joly. 2001. Unraveling the functional relationship between root anatomy and stress tolerance. *Aust. J. Plant Physiol.* 28: 999–1004.

Mahdy, A.M. 2011. Comparative effects of different soil amendments on amelioration of saline-sodic soils. *Soil Water Res.* 6(4): 205–216.

Malik, A.A., W.G. Li, L.N. Lou, J.H. Weng, and J.F. Chen. 2010. Biochemical/physiological characterization and evaluation of in vitro salt tolerance in cucumber. *Afr. J. Biotechnol.* 9: 3284–3292.

Marschner, H. 1995. *Mineral Nutrition of Higher Plants*. Academic Press Inc., London, U.K., 889pp.

Martinez, V. and A. Cerda. 1989. Influence of N source on rate of Cl, N, Na, and K uptake by cucumber seedlings grown in saline conditions. *J. Plant Nutr.* 12: 971–983.

Munns, R. 2002a. Comparative physiology of salt and water stress. *Plant Cell Environ.* 25: 239–250.

Munns, R. 2002b. Salinity, growth and phytohormones. In: *Salinity: Environment—Plants—Molecules*. Eds. Lauchli, A. and Luttge, U. Kluwer Academic Publishers, Dordrecht, the Netherlands, pp. 271–290.

Munns, R. 2005. Genes and salt tolerance: Bringing them together. *New Phytol.* 167: 645–663.

Munns R., R.A. James, and A. Lauchli 2006. Approaches to increasing the salt tolerance of wheat and other cereals. *J. Exp. Bot.* 57: 1025–1043.

Munns, R., R.A. James, B. Xu, A. Athman, S.J. Conn, C. Jordans, C.S. Byrt et al. 2012. Wheat grain yield on saline soils is improved by an ancestral Na+ transporter gene. *Nat Biotechnol.* 30(4): 360–364.

Munns, R. and M. Tester. 2008. Mechanisms of salinity tolerance. *Annu. Rev. Plant Biol.* 59: 651–681.

Murillo-Amador, B., R. Lopez-Aguilar, C. Kaya, J. Larrinaga-Mayoral, and A. Flores-Hernandez. 2002. Comparative effects of NaCl and polyethylene glycol on germination, emergence and seedling growth of cowpea. *J. Agron. Crop Sci.* 188: 235–247.

Oren, A. 2002. Diversity of halophilic microorganisms: Environments, phylogeny, physiology, and applications. *J. Ind. Microbiol. Biotechnol.* 28: 56–63.

Parida, A.K. and A.B. Das. 2005. Salt tolerance and salinity effects on plants: A review. *Ecotoxicol. Environ. Saf.* 60(3): 324–349.

Passam, H.C. and D. Kakouriotis. 1994. The effects of osmo-conditioning on the germination, emergence and early plant growth of cucumber under saline conditions. *Sci. Hortic.* 57(3): 233–240.

Patade, V.Y., S. Bhargava, and P. Suprasanna. 2009. Halopriming imparts tolerance to salt and PEG induced drought stress in sugarcane. *Agric. Ecosyst. Environ.* 134: 24–28.

Qadir, M., S. Schubert, A. Ghafoor, and G. Murtaza. 2001. Amelioration strategies for sodic soils: A review. *Land Degrad. Dev.* 12: 357–386.

Quesada, V., M.R. Ponce, and J.L. Micol. 2000. Genetic analysis of salt-tolerant mutants in *Arabidobsis thaliana. Genetics.* 154: 421–436.

Renault, S. 2005. Response of red-oiser dogwood (*Cornus stolonifera*) seedlings to sodium sulphate salinity: Effects of supplemental calcium. *Physiol. Plant.* 123: 75–81.

Romero, L., A. Belakbir, L. Ragala, and J.M. Ruiz. 1997. Response of plant yield and leaf pigments to saline conditions: Effectiveness of different rootstocks in melon plants (*Cucumis melo* L.). *Soil Sci. Plant Nutr.* 43: 855–862.

Romero-Aranda, M.R., O. Jurado, and J. Cuartero. 2006. Silicon alleviates the deleterious salt effect on tomato plant growth by improving plant water status. *J. Plant Physiol.* 163: 847–855.

Rouphael, Y., M. Cardarelli, G. Colla, and E. Rea. 2008a. Yield, mineral composition, water relations, and water use efficiency of grafted mini-watermelon plants under deficit irrigation. *HortScience.* 43: 730–736.

Rouphael, Y., M. Cardarelli, E. Rea, and G. Colla. 2008b. Grafting of cucumber as a means to minimize copper toxicity. *Environ. Exp. Bot.* 63: 49–58.

Rouphael, Y., M. Cardarelli, E. Rea, and G. Colla. 2012. Improving melon and cucumber photosynthetic activity, mineral composition, and growth performance under salinity stress by grafting onto Cucurbita hybrid rootstocks. *Photosynthetica.* 50: 180–188.

Savvas, D., D. Papastavrou, G. Ntatsi, A. Ropokis, C. Olympios, H. Hartmann, and D. Schwarz. 2009. Interactive effects of grafting and Mn-supply level on growth, yield and nutrient uptake by tomato. *HortScience.* 44: 1978–1982.

Savvas, D., V.A. Pappa, G. Gizas, and A. Kotsiras. 2005. NaCl accumulation in a cucumber crop grown in a completely closed hydroponic system as influenced by NaCl concentration in irrigation water. *Euro. J. Hortic. Sci.* 70: 217–223.

Serrano, R., A. Culianz-Macia, and V. Moreno. 1999. Genetic engineering of salt and drought tolerance with yeast regulatory genes. *Sci. Hortic.* 78: 261–269.

Shabala, S.N., L. Shabala, and E. Volkenburgh van. 2003. Effect of calcium on root development and root ion fluxes in stalinized barley seedlings. *Funct. Plant Biol.* 30: 507–514. doi:101071/FP03016.

Shannon, M.C., G.W. Bohn, and J.D. McCreight. 1984. Salt tolerance among muskmelon genotypes during seed emergence and seedling growth. *HortScience.* 19: 828–830.

Sharpley, A.N., J.J. Meisinger, J.F. Power, and D.L. Suarez. 1992. Root extraction of nutrients associated with long-term soil management. In: *Advances in Soil Science*, Ed. Halfield, J.L. and B.A. Stewart. Springer-Verlag, New York. pp. 151–217, Vol. 19.

Shu, L.Z. and Y.H. Liu, 2001. Effects of silicon on growth of maize seedlings under salt stress. *Agro-Environ. Prot.* 20: 38–40.

Singh, B.G. 1995. Effect of hydration-dehydration seed treatments on vigor and yield of sunflower. *Indian J. Plant Physiol.* 38: 66–68.

Sivritepe, H.O., A. Eris, and N. Sivritepe. 1999. The effect of NaCl priming on salt tolerance in melon seedlings. *Acta Hortic.* 492: 77–84.

Sivritepe, N., H.O. Sivritepe, and A. Eris, 2003. The effects of NaCl priming on salt tolerance in melon seedlings grown under saline conditions. *Sci. Hortic.* 97: 229–237.

Smith, S.E. and D.W. Read. 1997. *Mycorrhizal Symbiosis*, 2nd edn. Academic, London, U.K.

Taffouo, V.D., N.L. Djiotie, M. Kenne, N. Din, J.R. Priso, S. Dibong, and A. Akoa. 2008. Effects of salt stress on physiological and agronomic characteristics of three tropical cucurbit species. *J. Appl. Biosci.* 10: 434–441.

Tiwari, J.K., A.D. Munshi, R. Kumar, R.K. Sharma, and A.K. Sureja. 2011. Inheritance of salt tolerance in cucumber (*Cucumis sativus* L.). *Indian J. Agric. Sci.* 81(5): 398–401.

Trajkova, F., N. Papadantonakis, and D. Sanas. 2006. Comparative effects of NaCl and CaCl$_2$ salinity on cucumber grown in a closed hydroponic system. *HortScience.* 41(2): 437–441.

Villora, G., D.A. Moreno, G. Pulgar, and L. Romero. 1999. Zucchini growth, yield, and fruit quality in response to sodium chloride stress. *Plant Nutr.* 22: 855–861.

Wei, G.P., Y.L. Zhu, Z.L. Liu, L.F. Yang, and G.W. Zhang. 2007. Growth and ionic distribution of grafted eggplant seedlings with NaCl stress. *Acta Bot. Boreal-Occid. Sin.* 27: 172–178.

Whitaker, T.W. 1979. The breeding of vegetable crops: Highlights of the past seventy-five years. *HortScience.* 14: 359–363.

Xiang, L., G. Shi-rong, T. Jing, and D. Jiu-ju, 2009. Effects of grafting on the growth and the salt-tolerance of the watermelon under NaCl stress. *Jiangsu J. Agric. Sci.* 25(03).

Yan, K., P. Chen, H. Shao, S. Zhao, L. Zhang, L. Zhang, G. Xu, and J. Sun. 2012. Responses of photosynthesis and photosystem II to higher temperature and salt stress in sorghum. *J. Agron. Crop Sci.* 198(3): 218–225.

Yang, Y., X. Lu, B. Yan, B. Li, J. Sun, S. Guo, and T. Tezuka. 2013. Bottle gourd rootstock-grafting affects nitrogen metabolism in NaCl-stressed watermelon leaves and enhances short-term salt tolerance. *J. Plant Physiol.* 170(7): 653–661.

Yeo, A.R., M.E. Yeo, and T.J. Flowers. 1988. Selection of lines with high and low sodium transport from within varieties of an inbreeding species: Rice (*Oryza sativa*). *New Phytol.* 110: 13–19.

Yetisir, H. and V. Uygur. 2010. Responses of grafted watermelon onto different gourd species to salinity stress. *J. Plant Nutr.* 33: 315–327.

Zhen, A., Z.L. Bie, Y. Huang, Z.X. Liu, and Q. Li. 2010. Effects of scion and rootstock genotypes on the antioxidant defense systems of grafted cucumber seedlings under NaCl stress. *Soil Sci. Plant Nutr.* 56: 263–271.

Zhu, J., Z.L. Bie, Y. Huang, and X.Y. Han. 2008a. Effect of grafting on the growth and ion contents of cucumber seedlings under NaCl stress. *Soil Sci. Plant Nutr.* 54: 895–902.

Zhu, J.K. 2007. *Plant Salt Stress.* John Wiley & Sons, Ltd., Hoboken, NJ.

Zhu, S.N. and S.R. Guo. 2009. Effects of grafting on K$^+$, Na$^+$ contents and distribution of watermelon (*Citrullus vulgaris* Schrad.) seedlings under NaCl stress. *Acta Horticulture.* 36: 814–820.

Zhu, S.N., S.R. Guo, G.H. Zhang, and J. Li. 2008b. Activities of antioxidant enzymes and photosynthetic characteristics in grafted watermelon seedlings under NaCl stress. *Acta Bot. Boreal-Occident. Sin.* 28: 2285–2291.

Zhu, Z., G. Wei, J. Li, Q. Qian, and J. Yu. 2004. Silicon alleviates salt stress and increases antioxidant enzymes activity in leaves of salt-stressed cucumber (*Cucumis sativus* L.). *Plant Sci.* 167: 527–533.

24 Physiological and Biochemical Responses of Cucurbits to Drought Stress

Amir Hossein Saeidnejad

CONTENTS

24.1 INTRODUCTION

Plant growth and productivity is adversely affected by nature's wrath in the form of various abiotic and biotic stress factors. Plants are frequently exposed to a plethora of stress conditions and various anthropogenic activities have accentuated the existing stress factors. All these stress factors are a menace for plants and prevent them from reaching their full genetic potential and limit crop productivity worldwide (Bray et al., 2000). Hence, these stresses threaten the sustainability of the agricultural industry.

Due to the growth of population and expansion of agricultural, energy, and industrial sectors, the demand for water has extensively increased and even water scarcity has been occurring almost every year in many parts of the world (Saeidnejad et al., 2013). Drought stress has been known to induce a sequence of morphological, biochemical, and molecular alterations that negatively affect plant growth and productivity. The occurrence of drought is widespread across many parts of the world. It seems worldwide losses in crop yields from water deficit probably exceed the cumulative loss of all other stresses (Thapa et al., 2011). Considering these facts, understanding the genetic and molecular mechanisms controlling drought response has become vital in evolving strategies to enhance drought tolerance in crop plants. Although the general effects of drought on plant growth are fairly well known, the primary effects of water deficit at the biochemical and the molecular levels are not well understood (Bhatnagar-Mathur et al., 2009).

Cucurbits are among the economically most important vegetable crops worldwide and are grown in both temperate and tropical regions (Sanjur et al., 2002). Cucurbits cultivated for seed consumption are reported to be rich in nutrients (de Mello et al., 2000, 2001). The fruits of cucurbits are very useful for human health, that is, blood purification, digestion, and constipation treatment (Rahman et al., 2008).

Despite their agronomic, cultural, and culinary importance, these plants lack attention from research and development so that they are categorized as orphan crops (IPGRI, 2002). The limit of proper knowledge, such as methods of production, preservation, and mechanisms of tolerance to abiotic stresses, especially drought, is an important deterrent to their wider production, which should result in food security and increased incomes for peasants.

24.2 CUCUMBER

As regards the importance of cucumber in agriculture, it is remarkable to know that cucumber is the fourth most important vegetable crop worldwide. As the first in cucurbits, the cucumber genome sequence explicated chromosomal evolution in the genus *Cucumis* and afforded novel insights into several important biological processes such as biosynthesis of cucurbitacin and "fresh green" odor (Lv et al., 2012). Cucumber is being developed as a new model species in plant biology due to its small number of genes, rich diversity of sex expression, suitability for vascular biology studies, short life cycle (3 months from seed to seed), and accumulating resources in genetics and genomics (Guo et al., 2010). Cucumber is mostly consumed as a raw fruit, and it also has a variety of medicinal characteristics (Shah et al., 2013). Considering the structural properties of the plant with large leaves and shallow root system, it is susceptible to drought conditions.

Lack of precipitation and limited water supply during crop growing season are important considerations that need careful attention for crop production. Therefore, the relationships between crop water consumption, crop water stress, and yield have to be determined for better water conservation.

24.2.1 DROUGHT STRESS AND YIELD RELATION

Yield is known as one of the most important factors that is affected by drought conditions. In a study carried out under field conditions, different drip irrigation regimes were evaluated to determine a threshold value for crop water stress index (CWSI) based on irrigation programming. Four different irrigation treatments such as 50% (T-50), 75% (T-75), 100% (T-100), and 125% (T-125) of irrigation water applied/cumulative pan evaporation (IW/CPE) ratio with a 3-day period were studied. Maximum marketable fruit yield was from T-100 treatment with 76.65 t ha^{-1} in 2002 and 68.13 t ha^{-1} in 2003. Fruit yield was reduced significantly as irrigation rate was decreased (Simsek et al., 2005).

24.2.2 BIOCHEMICAL RESPONSES OF CUCUMBER TO DROUGHT STRESS

24.2.2.1 Photosynthesis Properties

Photosynthesis is one of the key processes to be affected by water deficits via decreased CO_2 diffusion to the chloroplast and metabolic constraints. The relative impact of those limitations varies with the intensity of the stress, the occurrence (or not) of superimposed stresses, and the species we are dealing with (Pinheiro and Chaves, 2011). Studies have shown that stomatal closure is one of the first line of responses to dehydration condition and an efficient method to reduce water loss through the decrement of leaf transpiration; however, this process limits CO_2 diffusion into the mesophyll for photosynthesis and leads to low mesophyll conductance during mild water deficit (Cornic, 2000; Grassi and Magnani, 2005; Flexas et al., 2006). In mild water deficit, stomatal limitation is the major reason for causing reduced photosynthesis. Photosynthesis significantly declined in some plants during severe water deficit due to the inhibition of biochemical or metabolic (nonstomatal) limitations. Some studies have shown that water deficit induces metabolic impairment, particularly by reductions in Rubisco activity and protein content (Lawlor and Cornic, 2002; Bota et al., 2004; Jorge et al., 2006; Dias and Brüggemann, 2007; Feller et al., 2008).

In a very recent study by Zhang et al. (2013), cucumber seedlings with six to seven leaves were deprived of water. Results indicated that a decrease in stomatal conductance occurred early in response to water deficit and may prevent excessive dehydration and improve water use efficiency in cucumber plants. The CO_2 assimilation rate declined gradually during dehydration and was significantly affected after 7–8 days of water deficit in relation to well-watered plants. Meanwhile, relative expressions of rbcL and rbcS were increased at first and then declined rapidly to reach the minimum at the eighth day of dehydration during water deficit. After 2 days of rewatering, the mentioned parameters increased rapidly and nearly reached the level of the control. These results indicated that in addition to the lower CO_2 available for carboxylation (in consequence of the decreased stomatal conductance), reduced amount and activity of Rubisco limited photosynthesis in dehydrated cucumber plants. The observed down-regulation of stomatal conductance (g_s) after 1 day of rewatering imposed a substantial limitation on photosynthesis, and it is consistent with the results of other studies (Gallé et al., 2007; Galmés et al., 2007).

24.2.2.2 Antioxidant Activity

When the effect of short-term polyethylene glycol (PEG)-induced water stress on antioxidant activity of cucumber seedlings was evaluated, results showed that it caused excessive generation of reactive oxygen species (ROS) including superoxide and hydrogen peroxide. Meanwhile, malondialdehyde (MDA) content increased. Antioxidant enzymes including superoxide dismutases (SOD), peroxidases (POD), catalase (CAT), and ascorbate peroxidase (APX) activities increased at different times and to different extents under water stress, while ascorbate (AsA) and glutathione (GSH) content, glutathione reductase (GR), dehydroascorbate reductase (DHAR), and monodehydroascorbate reductase (MDHAR) activities all decreased when compared to the control. According to this, it can be concluded that water stress strongly disrupted the normal metabolism of roots and restrained water absorption, and seemingly the enzymatic system played a more important role in protecting cucumber seedling roots against oxidative damage than the nonenzymatic system in short-term water deficit stress (Fan et al., 2014).

The combination of plant growth-promoting rhizobacterium strains and drought stress induction significantly enhanced SOD activity and mitigated the drought-triggered down-regulation of the expression of the genes cAPX, rbcL, and rbcS encoding cytosolic ascorbate peroxidase, and ribulose-1,5-bisphosphate carboxy/oxygenase (Rubisco) large and small subunits, respectively (Wang et al., 2012).

24.2.3 Role of Plant Growth Regulators in Drought Stress Regulation

24.2.3.1 Abscisic Acid

It has been well documented that abscisic acid (ABA) is involved in the drought response of many species in higher plants (Reddy et al., 2004; Shinozaki and Yamaguchi-Shinozaki, 2007). ABA is suggested to participate in the transduction of a stress-induced signal and the expression of drought-resistance genes. The onset of this phytohormone accumulation coincides with the loss in plant turgor (Shinozaki and Yamaguchi-Shinozaki, 1997). However, the interrelation between the changes in the ABA content and the development of adaptation to water deficit is still insufficiently followed. In the study of changes in the ABA content in cucumber (*Cucumis sativus* L.) leaves during adaptation to drought, it was observed that there were increases of the ABA content in cucumber leaves twice during adaption; one appeared at the initial stage of adaption and another was observed before the onset of leaf permanent wilting, and they proved that ABA was involved in the development of cucumber defense responses during the adaption to drought (Pustovoitova et al., 2004).

24.2.3.2 Indoleacetic Acid

The evidence on the role of indoleacetic acid (IAA) in drought tolerance induction seems to be contradictory. It was assumed for a long time that drought leads to IAA content reduction. However, it became more clear that the adaptation to drought is accompanied by an increase in IAA content (Zholkevich and Pustovoitova, 1993). IAA content increased in various plant organs subjected to water deficit. But still a complete pattern of changes in the content of IAA during drought adaptation is not available. Pustovoitova et al. (2004) reported the changes in IAA content in cucumber leaves exposed to drought stress and according to their results, an increased IAA content was maintained for a longer period, throughout about two-thirds of the adaptation period. A second increase in ABA content was observed before the onset of leaf permanent wilting, when IAA content already decreased. These data suggest that not only ABA but also IAA are involved in the development of defense responses during the adaptation to drought.

24.2.4 ALLEVIATION OF DROUGHT STRESS BY EXOGENOUS APPLIED TREATMENTS

Exogenous caffeic acid (CA, 3,4-dihydroxycinnamic acid) disrupts plant–water relations, but at certain concentrations and exposure durations, CA is an effective antioxidant (Gulcin, 2006), and it increases oxidation resistance in many different systems (Fukumoto and Mazza, 2000). In a study, CA pretreatment alleviates growth inhibition, increases RWC in leaves, and decreases the osmotic potential of cucumber under PEG-induced dehydration stress. This pretreatment also reduces the levels of O_2^-, H_2O_2, and MDA, increases the endogenous CA content, and increases the activities of SOD, CAT, GPX, GSH-Px, APX, DHAR, MDHAR, and GR in dehydration-stressed leaves (Wan et al., 2014).

5-Aminolevulinic acid (ALA) at low concentrations enhances chlorophyll biosynthesis and photosynthesis of crops, thus regulating the growth and development of plants and increasing the yield of crops. Meanwhile, exogenous application of ALA regulates antioxidant enzyme activities (Memon et al., 2009) and thereby increases the resistance of plants to a wide range of stresses (Liu et al., 2011; Naeem et al., 2011). When drought stress and ALA pretreatment effects on cucumber was investigated, results implied that drought inhibited plant growth, elevated the percentage of leaf withering, and increased the levels of ROS, while the combination of ALA pretreatment and PEG improved growth inhibition and reduced the leaf withering percentage and ROS levels. It also increased the activities of SOD, CAT, APX, GSH-Px, GPX, DHAR, MDHAR, and GR and elevated the contents of AsA and GSH, but the combined effects of ALA pretreatment and PEG increased the antioxidant activities more than PEG treatment alone. Therefore, ALA pretreatment can induce tolerance to drought stress in cucumbers by increasing the activities of antioxidants (Li et al., 2011).

It is well known that sugars trigger many stress-responsive genes in *Arabidopsis* and play a role in environmental responses (Price et al., 2004). Based on a hypothesis, alleviative effects of exogenous glucose on dehydration stress of cucumber seedlings was examined, and it was observed that it changes levels of antioxidants such as SOD, CAT GSH-Px, and GR, carbohydrate metabolism–related enzymes, and soluble sugars, while reducing dehydration stress of cucumber. These results indicate that exogenous glucose may have application possibility for a future practical trial of stress reduction (Huang et al., 2013).

There is a growing interest in the possible involvement of polyamines (PAs) in the defense reaction of plants to various environmental stresses (Bouchereau et al., 1999). When cucumber seedlings were treated with spermidine (Spd) prior to dehydration, it influenced enzymes of the antioxidative system under stress conditions; and an increase of guaiacol peroxidase activity, and, to a lesser degree, a reduction of SOD and catalase activities in Spd-treated plants in comparison to untreated stressed plants was recorded. Hydrogen peroxide and superoxide radical contents were also reduced in stressed plants after Spd pretreatment (Kubis, 2008).

Plant growth–promoting rhizobacterium (PGPR) induces physical and chemical changes in plants, resulting in enhanced tolerance to abiotic stresses termed as induced systemic tolerance. In an experiment, drought-stressed cucumber plants treated with three PGPR strains (*Bacillus cereus* AR156, *Bacillus subtilis* SM21, and *Serratia* sp. XY21) showed enhanced leaf monodehydroascorbate (MDA) content and relative electrical conductivity by 40% and 15%, respectively; increased leaf proline content and root recovery intension by 3.45-fold and 50%, respectively; and also maintained leaf chlorophyll content in cucumber plants under drought stress (Wang et al., 2012). It seems that it could be an efficient approach for drought stress tolerance induction.

24.3 MELONS

Melons (*Cucumis melo* L.) are important horticultural crops with a worldwide production of 27.3 million MT, with China, Iran, Turkey, Egypt, and the United States accounting for 68% of the world production (FAO, 2013). Usually all melon types are cultivated with similar cultural practices, particularly irrigation, but the acclimatization response to deficit irrigation varies among the cultivars or genetic makeup (Leskovar et al., 2004).

24.3.1 GROWTH AND YIELD UNDER DROUGHT STRESS

Melon plants are highly productive under adequate water conditions, but water scarcity is a major constraint to horticultural production in arid and semiarid regions around the world. The relatively shallow depth of melon roots requires soil water to be maintained at a minimum of 65% of capacity in order to avoid water stress (Anonymous, 2005). Roots are mainly located within the top 40–50 cm of soil and develop rapidly. As the plant grows, evapotranspiration increases (Allen et al., 1998). Therefore, irrigation must be scheduled to avoid excessive moisture or water stress that can lead to reduced yield, lower quality, and plant disease.

Available information about the role of water scarcity in melon growth and productivity is often conflicting in the literature. But it could generally mention that melon has critical periods of growth when water is a necessity for optimal yield and quality. Considering yield responses, the crop showed a negative response to irrigation deficit (40% ETc), which was observed clearly in the marketable yield (Cabello et al., 2009). These results suggest that the crop is not very sensitive to moderate water deficits up to 25% of the ETc, but a water deficit of 40% of the ETc can reduce melon yield by 22%. It was also shown that moderate water stress reduced marketable fruit yield by about 14%–17%. Severe water stress had a much more marked effect, reducing marketable fruit yield by 55%–59% (Kirnak et al., 2005).

Based on the evaluation of various researches, it is obvious that fruit weight in melon is more sensitive to water stress than fruit number. In fact, reduced melon yield, which is a common consequence of water stress condition, is largely attributed to decreased fruit weight, than to fruit number (Long et al., 2006; Dogan et al., 2008).

24.3.2 BIOCHEMICAL RESPONSE OF MELONS TO DROUGHT STRESS

Similar to most other members of the Cucurbitaceae family, the biochemical response of melon to drought stress was not comprehensively discussed. As regards the antioxidative properties, by exposing melon seedlings to PEG-induced osmotic stress, not only were the fresh and dry weights of shoot and root tissues reduced, but the antioxidative system positively responded to the stress condition. These results suggest that drought tolerance in melon cultivars might be closely related to an increase in capacity for antioxidant enzyme activity and the osmoprotective function of proline with a considerable accumulation (Kavas et al., 2013).

Chlorophyll fluorescence parameters provide a quick and accurate technique for quantifying the ability of individual species to tolerate water stress. The ability of a plant to tolerate environmental stresses and the extent to which those stresses damage the photosynthetic apparatus can be examined by measuring chlorophyll fluorescence (Maxwell and Johnson, 2000). According to previous studies, there is a strong correlation between chlorophyll fluorescence and environmental stress tolerance, and these can serve as reliable indicators of how well a plant is coping with stress (Mishra et al., 2012). However, remarkable resistance of the photosynthetic apparatus to water shortages has been reported; a 30% leaf water deficit has been estimated as the limit above which photosynthetic biochemistry is significantly affected (Cornic and Fresneau, 2002). In melon, it was reported that no significant variation in the Fv/Fm ratio was observed in either cultivar at both PEG concentrations (Kavas et al., 2013). Stable Fv/Fm ratios also confirm previous observations that the photosynthetic machinery is resistant to a certain level of water deficit (Kocheva et al., 2005).

24.3.3 DROUGHT STRESS AND QUALITATIVE PROPERTIES OF MELON

Considering total soluble solids (TSS) as an important qualitative parameter, different results were observed by researchers. For instance, Long et al. (2006) obtained a reduction in TSS when water deficit was applied during the harvest period, before harvest, or during both periods. They reported that water stress during the critical period of sugar accumulation in the fruit was likely to have reduced assimilate supply to the fruit by slowing the rate of photosynthesis in the source leaves, reducing sugar accumulation. The results were also observed by some other researchers (Lester et al., 1994; Fabeiro et al., 2002) showing that fruit sugar content is affected positively by water stress. It is also necessary to note that there are some reports that mentioned that there was no effect of water scarcity on TSS (Hartz, 1997).

24.4 WATERMELON

Watermelon (*Citrullus lanatus* L.) is an important crop accounting for approximately 7% of the world agricultural area devoted to vegetable crops. China is the largest producer and consumer with an annual production of about 68 million tons (http://faostat.fao.org).

Most crops, including watermelon, respond positively to irrigation with respect to growth and yield. For watermelon, the amount and timing of irrigation are important for efficient use of applied water and for maximizing crop yields. Relatively little data of this nature are available for watermelon, particularly for recently developed hybrids. It was shown that watermelon is more sensitive to water deficit during flowering, fruit growth, and late vegetative periods than during early and total vegetative periods.

24.4.1 YIELD AND GROWTH OF WATERMELON AND DROUGHT STRESS RELATION

When the yield response of watermelon to irrigation shortage was investigated by Erdem and Yuksel (2003), they found that if 50% water deficit occur during the total growing period, relative yield reduction would be 64% and if the same amount of water deficit occurs separately for early vegetative, late vegetative, total vegetative, flowering, and fruit growth periods, relative yield reductions would be 15%, 30%, 24%, 34%, 32%, respectively. In fact, the results of the researches in this field have a similar output that when watermelon is grown under limited water supply, supplemental irrigation must be programmed to ensure sufficient water is available in the soil during the most sensitive periods, flowering and fruit growth. For instance, it was shown that supplemental irrigation significantly increased watermelon yields and water use efficiencies of *C. lanatus* and this effect was more bold where a 100 mm deep surface mulch of a mixture of gravel and sand was applied (Wang et al., 2004).

24.4.2 Role of Plant Growth Regulators in Drought Stress Regulation

24.4.2.1 Abscisic Acid

The role of ABA in watermelon response to drought conditions has also been investigated. When watermelon seedlings were subjected to drought stress, expressions of ClBG1 and ClCYP707A1 were significantly down-regulated, while expressions of ClBG2 and ClNCED4 were up-regulated in the roots, stems, and leaves. The expression of ClBG3 was down-regulated in the root tissue but was up-regulated in stems and leaves. In conclusion, endogenous ABA content was modulated by a dynamic balance between biosynthesis and catabolism (Li et al., 2012).

24.4.3 Alleviative Mechanisms for Drought Stress Conditions

When water supply is limited, crop management practices that enhance drought resistance, plant water-use efficiency, and plant growth are particularly beneficial (Kirnak et al., 2001). Among the strategies that are needed to overcome this problem, inoculation with arbuscular mycorrhizal (AM) fungi is considered as a sustainable mitigation practice.

There are several studies showing that AM fungi have a positive effect on plant growth and productivity (Ruiz-Lozano et al., 1995; Subramanian and Charest, 1999; Kaya et al., 2003; Ruiz-Lozano, 2003; Al-Karaki et al., 2004). Improved adaptation of plants inoculated with AM fungi to water stress conditions has been linked to several interrelated observations and mechanisms including enhanced nutrient uptake, transpiration, increased root length, development of external hyphae, improved leaf water potential, and changes in leaf elasticity (Ellis et al., 1985; Auge et al., 1987; Davies et al., 1992). Mycorrhizal inoculation also improves root P uptake, particularly under dry soil conditions. Mycorrhizae, therefore, are likely to be important for increasing P acquisition during drought periods (Bryla and Duniway, 1997). As regards the watermelon, it was observed that when plants were exposed to limited water conditions and also inoculated with AM, mycorrhizal plants had significantly higher biomass and fruit yield compared to nonmycorrhizal plants, whether plants were water stressed or not. AM colonization increased water use efficiency (WUE) in both well-watered and water-stressed plants. Macro- (N, P, K, Ca, and Mg) and micro- (Zn, Fe, and Mn) nutrient concentrations in the leaves were significantly reduced by water stress. The necessary point to mention in this report was that mycorrhizal colonization of stressed plants restored leaf nutrient concentrations to levels in nonstressed plants in most cases (Kaya et al., 2003).

The extent and mode of plant response to drought depends on the AM fungal species involved, the interaction between the plant and the introduced fungi, and the degree of water stress (Ruiz-Lozano et al., 1995). The majority of the available data regarding the response of mycorrhizal plants to water stress are derived from experiments conducted under controlled conditions (sterilized soils) without taking into account native populations of AM fungi (Davies et al., 1992; Ruiz-Lozano et al., 1995; Kaya et al., 2003; Subramanian et al., 2006). Little is known about the participation of mycorrhizal inoculation on crop productivity under field conditions, where introduced AM fungi must coexist and compete with the AM population of native fungi.

When in a study the focus was on the impact of watering level and inoculation with allochthonous AM fungi on the diversity and presence of AM fungi in watermelon roots by using molecular techniques, it was shown that the diversity of AM fungal colonizers was strongly affected by inoculation and also by water stress in the inoculated plants. Inoculation affected fungal presence under water limitation conditions only. The latter was in line with the significant beneficial effect of inoculation on both WUE and yield only under water limitation (Omirou et al., 2013).

24.4.4 QUALITATIVE PROPERTIES OF WATERMELON AND DROUGHT STRESS

One of the reasons that make watermelon a healthy fruit is its lycopene content. Lycopene is a type of carotenoid that is also found in ripe tomato and pink grape fruit and gives them a characteristic red color and makes fundamental contributions to human health. So, it is important to know the trend of changes in lycopene content under drought conditions, although the available data in this field are limited. When some diploid and triploid cultivars were exposed to drought conditions that were applied by evapotranspiration rates, lycopene content was significantly higher at 0.75 and 1.0 ET. Furthermore, lycopene and vitamin C content did not decrease with deficit irrigation at 0.75 ET. With respect to the fact that this is the result of a single experiment, the judgment of how lycopene responds to drought conditions in watermelon needs more investigations to be done (Leskovar et al., 2004).

24.5 SQUASH

Squash (Cucurbita) is an important commercial crop that has gained popularity for both open-field and greenhouse in the Mediterranean region (Rouphael and Colla, 2005). These are grown 1.5 million ha annually and in 2007, the yield exceeded 20 million MT worldwide. China accounts for about 22% of the production area and 31% of worldwide production (Liu et al., 2008).

Squash is known as a sensitive plant to water scarcity and excessiveness and may be damaged. Since squash rooting depth is relatively shallow, soil water has to be maintained above 50% of the available soil water in order to avoid detrimental water deficit (Hess et al., 1997). Squash roots, most of which are in the top 40–50 cm of soil, develop rapidly. Irrigation should be scheduled to avoid excessive moisture or water stress. Lack of adequate soil water at harvest can result in misshapen fruits, but too much soil water can aggravate root and stem rot diseases (Richard et al., 2002).

The available data with the focus on the effects of water deficit and drought conditions on squash are limited. For instance, Richard et al. (2002) reported that the yield of squash plants reduced significantly when plants were exposed to drought condition by deficit irrigation. Another research by Ahmet et al. (2004) showed that reducing applied water by deficit irrigation in squash led to decreased yield, significantly. Amer (2011) indicated that when drought conditions was applied by irrigating as a fraction of crop evapotranspiration, it was observed that squash fruit yield and quality were significantly affected. In fact, both fruit and seed yields were significantly affected in a linear relationship ($r_2 \geq 0.91$) by drought stress condition. It is necessary to point out that not only the yield but also the seed quality was reduced by applied drought stress in treated plants.

Besides quantitative properties of squash that are affected by drought conditions, qualitative properties are also proved to be influenced by drought stress. It was shown that the mineral content of fruits and antioxidant activity are affected by exposing squash plants to drought stress conditions. In fact, when different irrigation quantities were applied, it was indicated that total phenolic content and antioxidant activity were affected along with reducing applied water level, although the difference between the treatments was not significant (Kuslu et al., 2011).

24.6 *LAGENARIA SICERARIA*

The plant *Lagenaria siceraria*, known as bottle gourd, is an annual, herbaceous, climbing plant with a long history of traditional medicinal uses in many countries, especially in tropical and subtropical regions. Since ancient times, the climber has been known for its curative properties and has been utilized for treatment of various ailments, including jaundice, diabetes, ulcer, and piles (Prajapati et al., 2010). Although the plant is identified as an excellent model crop for improving food security and helping economic prosperity of rural communities, in-depth investigations on the crop are scant and it is difficult to use reports about other cucurbit species due to variations in morphology and phenology, as well as to describe its tolerance to water stress (Chimonyo and Modi, 2013).

A very interesting research by Sithole and Modi (2015) was found to investigate the drought tolerance of *L. siceraria* compared with pumpkin (*Cucurbita maxima*) and cucumber (*Cucurbita pepo*), the closest relatives. As a general conclusion, all of the landraces and hybrids showed lower chlorophyll content index, stomatal conductance, and plant growth in response to decreasing water availability. However, the landraces showed better stomatal regulation indicating better acclimation to water stress. The comparable performance of landraces to the exotic pumpkin and cucumber varieties and their water stress tolerance suggests that bottle gourd could be promoted as an alternative food security crop in rural communities.

An unpublished research also focused on the water stress effects on physiological performance of *L. siceraria* and significant negative effects on vine length, branch number, stomatal conductance, and stem and total fresh mass. Water stress also caused a significant reduction in plant vine length of 8.2% and lateral branch number by 25%. This is because water stress has a greater effect on lateral branch production than vine length (Chimonyo and Modi, 2013).

REFERENCES

Ahmet, E., S. Sensoy, C. Kucukyumuk, and I. Gedik. 2004. Irrigation frequency and quantity affect yield components of spring squash (*Cucurbita pepo* L.). *Agric Water Manage* 67, 63–76.

Al-Karaki, G., B. McMichael, and J. Zak. 2004. Field response of wheat to arbuscular mycorrhizal fungi and drought stress. *Mycorrhiza* 14, 263–269.

Allen, R.G., L.S. Pereira, D. Raes, and M. Smith. 1998. *Crop Evapotranspiration: Guidelines for Computing Crop Water Requirements.* Irrigation and Drainage Paper No. 56. Food and Agriculture Organization, Rome, Italy.

Amer, K.H. 2011. Effect of irrigation method and quantity on squash yield and quality. *Agric Water Manage* 98, 1197–1206.

Augé, R.M., K.A. Schekel, R.L. Wample. 1987. Leaf water and carbohydrate status of VA mycorrhizal rose exposed to water deficit stress. *Plant and Soil* 99: 291–302.

Bhatnagar-Mathur, P., M.J. Devi, V. Vadez, and H.K. Sharma. 2009. Differential antioxidative responses in transgenic peanut bearno relationship to their superior transpiration efficiency underdrought stress. *J Plant Physiol* 166, 1207–1217.

Bota, J., H. Medrano, and J. Flexas. 2004. Is photosynthesis limited by decreased Rubisco activity and RuBP content under progressive water stress? *New Phytol* 162, 671–681.

Bouchereau, A., A. Aziz, F. Larher, and J. Martin-Tanguy. 1999. Polyamines and environmental challenges: Recent development. *Plant Sci* 140, 103–125.

Bray, E.A., J. Bailey-Serres, and E. Weretilnyk. 2000. Responses to abiotic stresses. In: Gruissem, W. (ed.). *Biochemistry and Molecular Biology of Plants*, pp. 1158–1249. American Society of Plant Physiologists, Rockville, MD.

Bryla, D.R. and J.M. Duniway. 1997. Growth, phosphorus uptake, and water relations of safflower and wheat infected with anarbuscular mycorrhizal fungus. *New Phytol* 136, 581–590.

Cabello, M.J., M.T. Castellanos, F. Romojaro, C. Martínez-Madrid, and F. Ribas. 2009. Yield and quality of melon grown under different irrigation and nitrogen rates. *Agric Water Manage* 96, 866–874.

Chimonyo, V.G.P. and A.T. Modi. 2013. Seed performance of selected bottle gourd (*Lagenaria siceraria* (Molina) Standl.). *Am J Exp Agric* 3, 740–766.

Chimonyo, V.G.P. and A.T. Modi. 2013. Growth responses of selected bottle gourd landraces to water stress under controlled environment conditions. The 3rd International Conference on Neglected and Underutilized Species. Accra, Ghana.

Cornic, G. 2000. Drought stress inhibits photosynthesis by decreasing stomatalaperture—Not by affecting ATP synthesis. *Trends Plant Sci* 5, 187–188.

Cornic, G. and C. Fresneau. 2002. Photosynthetic carbon reduction andoxidation cycles are the main electron sinks for photosystem II activity during a mild drought. *Ann Bot* 89, 887–894.

Davies, F.T., J.R. Potter, and R.G. Linderman. 1992. Mycorrhiza and repeated drought exposure affect drought resistance and extraradical hyphae development of pepper plants independent of plant size and nutrient content. *J Plant Physiol* 139, 289–294.

de Mello, M.L.S., P.S. Bora, and N. Narain. 2001. Fatty and amino acids composition of melon (*Cucumis melo* var. saccharinus) seeds. *J Food Comp Anal* 14, 69–74.

de Mello, M.L.S., N. Narain, and P.S. Bora. 2000. Characterisation of some nutritional constituents of melon (*Cucumis melo* hybrid AF-522) seeds. *Food Chem* 68, 411–414.

Dias, M.C. and W. Brüggeman. 2007. Differential inhibition of photosynthesis underdrought stress in Flaveria species with different degrees of development of the C_4 syndrome. *Photosynthetica* 45, 75–84.

Dogan, E., H. Kirnak, K. Berekatoglu, L. Bilgel, and A. Surucu. 2008. Water stress imposed on muskmelon (*Cucumis melo* L.) with subsurface and surface drip irrigation systems under semiarid climatic conditions. *Irrig Sci* 26(2), 131–138.

Ellis, J.R., H.J. Larsen, and M.G. Boosalis. 1985. Drought resistance of wheat plants inoculated with vesicular–arbuscular mycorrhizae. *Plant Soil* 86, 369–378.

Erdem, Y. and A.N. Yuksel. 2003. Yield response of watermelon to irrigation shortage. *Sci Hortic* 98, 365–383.

Fabeiro, C., F. Martín, and J.A. de Juan. 2002. Production of muskmelon (*Cucumis melo* L.) under controlled deficit irrigation in a semi-arid climate. *Agric Water Manage* 54, 93–105.

Fan, H., L. Ding, C. Du, and X. Wu. 2014. Effect of short-term water deficit stress on antioxidative systems in cucumber seedling roots. *Bot Stud* 55, 46.

FAO, 2013. FAOSTAT. Available at http://faostat3.fao.org/home/index.html. accessed on 12/8/2014.

Feller, U., I. Anders, and K. Demirevsk. 2008. Degradation of Rubisco and other chloroplast proteins under abiotic stress. *Gen Appl Plant Physiol* 34: 5–18.

Flexas, J., M. Ribas-Carbó, J. Bota, J. Galmés, M. Henkle, S. Martínez-Cãnellas, and H. Medrano. 2006. Decreased Rubisco activity during water stress is not induced by decreased relative water content but related to conditions of low stomatal conductance and chloroplast CO_2 concentration. *New Phytol* 172, 73–82.

Fukumoto, L.R. and G. Mazza. 2000. Assessing antioxidant and prooxidant activities of phenolic compounds. *J Agric Food Chem* 48, 3597–3604.

Gallé, A., P. Haldimann, and U. Feller. 2007. Photosynthetic performance and water relations in young pubescent oak (*Quercus pubescens*) trees during drought stress and recovery. *New Phytol* 174, 799–810.

Galmés, J., H. Medrano, and J. Flexas. 2007. Photosynthetic limitations in response to water stress and recovery in mediterranean plants with different growth forms. *New Phytol* 175, 81–93.

Grassi, G. and F. Magnani. 2005. Stomatal mesophyll conductance and biochemical limitations to photosynthesis as affected by drought and leaf ontogeny in ash and oak trees. *Plant Cell Environ* 28, 834–849.

Gulcin, I. 2006. Antioxidant activity of caffeic acid (3,4-dihydroxycinnamic acid). *Toxicology* 217, 213–220.

Guo, S., Y. Zheng, J. Joung, S. Liu, Z. Zhang, O. Crasta, B.W. Sobral, W. Xu, S. Huang, and Z. Fei. 2010. Transcriptome sequencing and comparative analysis of cucumber flowers with different sex types. *BMC Genomics* 11: 384.

Hartz, T.K. 1997. Effects of drip irrigation scheduling on muskmelon yield and quality. *Sci Hortic* 69(1), 117–122.

Hess, M., M. Bill, S. Jason, and S. John. 1997. *Oregon State University Western Oregon Squash Irrigation Guide*, vol. 541. Department of Bioresource Engineering, Corvallis, OR, pp. 737–6304.

Huang, Y., Y. Nie, Y. Wan, S.H. Chen, Y. Sun, X. Wang, and J. Bai. 2013. Exogenous glucose regulates activities of antioxidant enzyme, soluble acid invertase and neutral invertase and alleviates dehydration stress of cucumber seedlings. *Sci Hortic* 162, 20–30.

IPGRI. 2002. Neglected and underutilized plant species: Strategic action plan of the International Plant Genetic Resources Institute (IPGRI). IPGRI, Rome, Italy.

Jorge, I., R.M. Navarro, C. Lenz, D. Ariza, and J. Jorrín. 2006. Variation in the holm oak leaf proteome at different plant developmental stages, between provenances and in response to drought stress. *Proteomics* 6, 207–214.

Kavas, M., M.C. Baloğlu, O. Akça, F.S. Köse, and D. Gökçay. 2013. Effect of drought stress on oxidative damage and antioxidant enzyme activity in melon seedlings. *Turk J Biol* 37, 491–498.

Kaya, C., D. Higgs, H. Kirnak, and I. Tas. 2003. Mycorrhizal colonization improves fruit yield and water use efficiency in watermelon (*Citrullus lanatus* Thumb) grown under well-watered and water-stressed conditions. *Plant Soil* 253, 287–292.

Kirnak, H., D. Higgs, C. Kaya, and I. Tas. 2005. Effects of irrigation and nitrogen rates on growth, yield, and quality of muskmelon in semiarid regions. *J Plant Nutr* 28, 621–638.

Kirnak, H., C. Kaya, D. Higgs, and S. Gercek. 2001. A long-term experiment to study the role of mulches in physiology and macro-nutrition of strawberry grown under water stress. *Aust J Agric Res* 52, 937–943.

Kocheva, K.V., M.C. Busheva, G.I. Georgiev, P.H. Lambrev, and V.N. Goltsev. 2005. Influence of short-term osmotic stress on the photosynthetic activity of barley seedlings. *Biol Plant* 49, 145–148.

Kubis, J. 2008. Exogenous spermidine differentially alters activities of some scavenging system enzymes, H_2O_2 and superoxide radical levels in water-stressed cucumber leaves. *J Plant Physiol* 165, 397–406.

Kuslu, Y., U. Sahin, F.M. Kiziloglu, and S. Memis. 2011. Fruit yield and quality, and irrigation water use efficiency of summer squash drip-irrigated with different irrigation quantities in a semi-arid agricultural area. *J Integr Agric* 13(11), 2518–2526.

Lawlor, D.W. and G. Cornic. 2002. Photosynthetic carbon assimilation and associated metabolism in relation to water deficits in higher plants. *Plant Cell Environ* 25, 275–294.

Leskovar, D.I., H. Bang, K.M. Crosby, N. Maness, J.A. Franco, and P. Perkins-Veazie. 2004. Lycopene, carbohydrates, ascorbic acid and yield components of diploid and triploid watermelon cultivars are affected by deficit irrigation. *J Hortic Sci Biotechnol* 79, 75–81.

Lester, G.E., N.F. Oebker, and J. Coons. 1994. Preharvest furrow and drip irrigation schedule effects on postharvest muskmelon quality. *Postharvest Biol Technol* 4, 57–63.

Li, D., J. Zhang, W. Sun, Q. Li, A. Dai, and J. Bai. 2011. 5-Aminolevulinic acid pretreatment mitigates drought stress of cucumber leaves through altering antioxidant enzyme activity. *Sci Hortic* 130, 820–828.

Li, Q., P. Li, L. Sun, Y. Wang, K. Ji, Y. Sun, S.H. Dai, P. Chen, C.H. Duan, and P. Leng. 2012. Expression analysis of β-glucosidase genes that regulate abscisic acid homeostasis during watermelon (*Citrullus lanatus*) development and under stress conditions. *J Plant Physiol* 169, 78–85.

Liu, D., Z.F. Pei, M.S. Naeem, D.F. Ming, H.B. Liu, F. Khan, and W.J. Zhou. 2011. 5 Aminolevulinic acid activates antioxidative defense system and seedling growth in *Brassica napus* L. under water-deficit stress. *J Agron Crop Sci* 197, 284–295.

Liu, Y.S., D.P. Lin, X.W. Sun, and C.L. Wang. 2008. Advances of the cucurbit industry and cucurbit science and technology in China. *China Cucurbits Veg* 6, 4–9.

Long, R.L., K.B. Walsh, and D.J. Midmore. 2006. Irrigation scheduling to increase muskmelon fruit biomass and soluble solids concentration. *HortScience* 41(2), 367–369.

Lv, J., J. Qi, Q. Shi, D. Shen, S. Zhang, G. Shao, H. Li et al. 2012. Genetic diversity and population structure of cucumber (*Cucumis sativus* L.). *PLoS One* 7, e46919.

Maxwell, K. and G.N. Johnson. 2000. Chlorophyll fluorescence, apractical guide. *J Exp Biol* 51, 659–668.

Mcmon, S.A., X. Hou, L. Wang, and Y. Li. 2009. Promotive effect of 5-aminolevulinic acidon chlorophyll, antioxidative enzymes and photosynthesis of pakchoi (*Brassica campestris* ssp. *chinensis* var. *communis* Tsenet Lee). *Acta Physiol Plant* 31, 51–57.

Mishra, K.B., R. Iannacone, A. Petrozza, A. Mishra, N. Armentano, G. LaVecchia, M. Trtilek, F. Cellini, and L. Nedbal. 2012. Engineered drought tolerance in tomato plants is reflected in chlorophyll fluorescence emission. *Plant Sci* 182, 79–86.

Naeem, M.S., M. Rasheed, D. Liu, Z.L. Jin, D.F. Ming, K. Yoneyama, Y. Takeuchi, and W.J. Zhou. 2011. 5-Aminolevulinic acid ameliorates salinity-induced metabolic, water-related and biochemical changes in *Brassica napus* L. *Acta Physiol Plant* 33, 517–528.

Omirou, M., I.M. Ioannides, and C. Ehaliotis. 2013. Mycorrhizal inoculation affects arbuscular mycorrhizal diversity in watermelon roots, but leads to improved colonization and plant response under water stress only. *Appl Soil Ecol* 63, 112–119.

Pinheiro, C. and M.M. Chaves. 2011. Photosynthesis and drought: Can we make metabolic connections from available data? *J Exp Bot* 62, 869–882.

Prajapati, R.P., M. Kalariya, S.K. Parmar, and N.R. Sheth. 2010. Phytochemical and pharmacological review of *Lagenaria sicereria*. *J Ayurveda Integr Med* 1(4), 266–272.

Price, J., A. Laxmi, S.K. St Martin, and J.C. Jang. 2004. Global transcription profiling reveals multiple sugar signal transduction mechanisms in *Arabidopsis*. *Plant Cell* 16, 2128–2150.

Pustovoitova, T.N., N.E. Zhdanova, and V.N. Zholkevich. 2004. Changes in the levels of IAA and ABA in cucumber leaves under progressive soil drought. *Russ J Plant Physiol* 51: 513–517.

Rahman, A.H.M.M., M. Anisuzzaman, F. Ahmed, A.K.M. Rafiul Islam, and A.T.M. Naderuzzaman. 2008. Study of nutritive value and medicinal uses of cultivated cucurbits. *J Appl Sci Res* 4(5), 555–558.

Reddy, A.R., K.V. Chaitanya, and M. Vivekanandan. 2004. Drought-induced responses of photosynthesis and antioxidant metabolism in higher plants. *J Plant Physiol* 161, 1189–1202.

Richard, M., A. Jose, G. Mark, and M. Keith. 2002. Spring squash production in California. Vegetable Reproduction Series Publication 7245, Vegetable Research and Information Center, Davis, CA.

Rouphael, Y. and G. Colla. 2005. Growth, yield, fruit quality and nutrient uptake of hydroponically cultivated zucchini squash as affected by irrigation systems and growing seasons. *Sci Hortic* 105, 177–195.

Ruiz-Lozano, J.M. 2003. Arbuscular mycorrhizal symbiosis and alleviation of osmotic stress new perspectives for molecular studies. *Mycorrhiza* 13, 309–317.

Ruiz-Lozano, J.M., M. Gomez, and R. Azcon. 1995. Influence of different *Glomus* species on the time-course of physiological responses of lettuce to progressive drought stress periods. *Plant Sci* 110, 37–44.

Saeidnejad, A.H., M. Kafi, H.R. Khazaei, and M. Pessarakli. 2013. Effects of drought stress on quantitative and qualitative yield and antioxidative activity of *Bunium persicum. Turk J Bot* 37, 930–939.

Sanjur, O.I., D.R. Piperno, T.C. Andres, and L. Wessel-Beaver. 2002. Phylogenetic relationships among domesticated and wild species o *Cucurbita* (Cucurbitaceae) inferred from a mitochondrial gene: Implications for crop plant evolution and areas of origin. *Proc Natl Acad Sci USA* 99, 535–540.

Sensoy, S., A. Ertek, I. Gedik, and C. Kucukyumuk. 2007. Irrigation frequency and amount affect yield and quality of field-grown melon (*Cucumis melo* L.). *Agric Water Manage*, 88: 269–274.

Shah, P., S. Dhande, Y. Joshi, and V. Kadam. 2013. A review on *Cucumis sativus* (Cucum-ber). *Res J Pharm Phytochem* 5, 49–53.

Shinozaki, K. and K. Yamaguchi-Shinozaki. 1997. Gene expression and signal transduction in water-stress response. *Plant Physiol* 115(2), 327–334.

Shinozaki, K. and K. Yamaguchi-Shinozaki. 2007. Gene networks involved in drought stress response and tolerance. *J Exp Bot* 58, 221–227.

Simsek, M., T. Tonkaza, M. Kacırab, N. Comlekcioglu, and Z. Dogan. 2005. The effects of different irrigation regimes on cucumber (*Cucumbis sativus* L.) yield and yield characteristics under open field conditions. *Agric Water Manage* 73: 173–191.

Sithole, N., and A. Modi. 2015. Responses of selected bottle gourd [*Lagenaria siceraria* (Molina Standly)] landraces to water stress. *Acta Agr Scand B-S P.* 65(4): 350–356.

Subramanian, K., P. Santhanakrishnan, and P. Balasubramanian. 2006. Responses of field grown tomato plants to arbuscular mycorrhizal fungal colonization under varying intensities of drought stress. *Sci Hortic* 107, 245–253.

Subramanian, K.S. and C. Charest. 1999. Acquisition of N by external hyphae of an arbuscular mycorrhizal fungus and its impact on physiological responses inmaize under drought-stressed and well-watered conditions. *Mycorrhiza* 9: 69–75.

Thapa, G., M. Dey, L. Sahoo, and S.K. Panda. 2011. An insight into the drought stress induced alterations in plants. *Biol Plant* 55, 603–613.

Wan, Y., S.H. Chen, Y. Huang, X. Li, Y. Zhang, X. Wang, and J. Bai. 2014. Caffeic acid pretreatment enhances dehydration tolerance in cucumber seedlings by increasing antioxidant enzyme activity and proline and soluble sugar contents. *Sci Hortic* 173, 54–64.

Wang, C.J., W. Yang, C. Wang, C. Gu, and D.D. Niu. 2012. Induction of drought tolerance in cucumber plants by a consortium of three plant growth promoting rhizobacterium strains. *PLoS One* 7(12), e52565.

Wang, Y., Z.K. Xie, F. Li, and Z.H. Zhang. 2004. The effect of supplemental irrigation on watermelon (*Citrullus lanatus*) production in gravel and sand mulched fields in the Loess Plateau of northwest China. *Agric Water Manage* 69, 29–41.

Zhang, L., L. Zhang, J. Sun, Z. Zhang, H. Rena, and X. Sui. 2013. Rubisco gene expression and photosynthetic characteristics of cucumber seedlings in response to water deficit. *Sci Hortic* 161, 81–87.

Zholkevich, V.N. and T.N. Pustovoitova. 1993. Growth and phytohormone content in *Cucumis sativus* L. leaves under water deficiency. *Russ J Plant Physiol* 40, 676–680.

25 Growth Responses of Watermelon to Biotic and Abiotic Stresses

Satya S.S. Narina

CONTENTS

25.1 INTRODUCTION

Watermelon (*Citrullus lanatus* var. *lanatus*), a C_3 xerophyte, belongs to the family Cucurbitaceae, order Cucurbitales, and tribe Benincasinae. It originated from South Africa. The cultivation of watermelon slowly extended from tropical Africa to India and China by the tenth century to European countries by the seventeenth century. The European colonists introduced watermelon in the United States where Spanish settlers started growing it initially in Florida during the sixteenth century. Its cultivation was quickly accepted by other states and was spread to Hawaii and pacific islands (David and Donald, 2012). China is the lead producer (70,000,000 t) of watermelons globally (2012 census, UN_FAOSTAT).

The largest watermelon-producing states among the 44 watermelon-growing states in the United States (total area 133,700 acres, total production 1,771,720 Mt) are Georgia, Florida, Texas, California, and Arizona (USDA, 2014). Per capita consumption of watermelons in the United States is 7.03 kg (15.5lb) in 2010. Watermelons accounted for more than half of U.S. melon export (mainly to Canada) averaging over 300 million lb, nearly 1/10th of the world's total (ERS.USDA.Gov).

Watermelon is grown throughout the world and is the most commonly consumed vegetable as a fresh fruit. A slice of 280 g provides 80 cal of energy and 0 cal from fat, with various potential nutritional values (Table 25.1). It is used to make juice, desserts, pickles, candies, and baskets in various parts of the world. The fruit is a good resource for carbohydrates (6.4 g/100 g), vitamins A (590 IU) and C, lycopene (4100 µg/100 g, range 2300–7200), potassium, amino acid, and citrulline. It is used to build the immune system and treat cardiovascular problems in humans and seed oil has anthelmintic properties.

Watermelon (*C. lanatus* [Thumb.] Matsum and Nakai) was formerly *Citrullus vulgaris* (2n = 22, x = 11). Bailey (1930) classified commercial cultivars as *C. lanatus* var. *lanatus*, and wild accessions as *C. lanatus* var. *citroides* (preserving melon). The species were also classified based on cucurbitacin content into two groups: (1) species with cucurbitacin E, which include *C. lanatus*, *C. colocynthis*, and *C. ecirrhosus*, and (2) the species with cucurbitacins E and B, which includes *C. naudinianus*. These four species were cross-compatible, with distinct differences in flowering habit, genetic and structural changes in chromosomes, and were geographically isolated (Wehner, 2003).

Citrullus colocynthis is the wild ancestor for the cultivated *C. lanatus*. A wild watermelon species, *Citrullus L. lanatus* var. *caffer*, was observed in the Kalahari Desert in Africa. There were two other closely related species: *Praecitrullus fistulosus* (India and Pakistan) and *Acanthosicyos naudinianus* (southern Africa). The genus *Citrullus* now include *C. lanatus* (*var lanatus, va. caffer, var. citroides*), *C. ecirrhosus*, *C. colocynthis*, *C. naudinianus*, and *C. rehmii*. The cultivated and wild *Citrullus* species originated from *C. ecirrhosus* from Namibia (Dane and Liu, 2006). *C. ecirrhosus* is more closely related to *C. lanatus* than either is to *C. colocynthis*.

TABLE 25.1

Nutritional Value of a Slice (280 g) of Watermelon Fruit

Nutrient	Quantity (g)	Daily Value (%)
Total fat	0	0
Saturated fat	0	0
Trans fat	0	0
Cholesterol	0	0
Sodium	0	1
Potassium		
Total carbohydrates	21	7
Dietary fiber	1	
Sugars	20	
Protein	1	
Vitamin A		30
Calcium		2
Vitamin C		25
Iron		4

25.2 PHYSIOLOGICAL STAGES OF GROWTH AND DEVELOPMENT I

25.2.1 GERMINATION, EMERGENCE, AND SEEDLING GROWTH

Seed sowing is usually done at 6–8 ft × 3–4 ft in March–April. The seeds of cucurbitaceous crops are nonendospermic and germination is epigeal. If seeds are left too long in the fruit, they will germinate in situ. The harvested seeds can be planted the next day as there is no dormancy in watermelon seeds. The germination of seeds takes 2 days to 2 weeks and is dependent on the temperature, water activity of the medium, and age of the seed. The optimum germination temperature of triploid watermelon seed is 29°C–32°C. Maximum seed germination in watermelon was reported between 18°C and 22°C (Milotay et al., 1991; Singh et al., 2001) and at an optimum temperature of 25°C (Jennings and Saltveit, 1994).

Seeds will not germinate below 15°C and will show reduced emergence under prolonged low-temperature conditions though there is available soil moisture due to the effect of pathogenic microorganisms on germinating seeds (Singh et al., 2001). Germination of watermelon seeds improved at 15°C after priming with inorganic osmotica (Sachs, 1977).

Germination was positively correlated with water activity at a specific temperature. Seed treatment with fungicides and bactericides is beneficial and improves seed germination and emergence. The emergence of seedlings was high in seeds grown in sterilized soils.

Tetraploid watermelon has poor germination and low seedling vigor mainly due to thick seed coat and seed coat adherence to emerged cotyledons. Seed treatments such as nicking and presoaking in distilled water or H_2O_2 enhanced the germination of tetraploid seeds but showed genotypic variation (Muhammad et al., 2006). While planting, seeds positioned with the radicle up reduced seed coat adherence to cotyledons. Sowing of water-imbibed watermelon seeds improved seed germination percentage (Hall et al., 1989).

25.2.2 PHOTOSYNTHESIS

The 5-aminolevulinic acid (ALA) treatment (50–200 mg/L) enhanced the net photosynthetic rate, stomatal conductance and transpiration rate of the leaves of watermelon seedlings (Kang et al., 2006) and significantly increased the maximum fluorescence (Fm), the variable fluorescence (Fv), PS II maximal photochemical efficiency (Fv/Fm), potential PS II photochemical efficiency (Fv/F_o), and the ability of PS II reaction center to trap energy from antenna pigment (1/F_o–1/Fm) in the dark-adapted leaves of watermelon seedlings. The measurements of light response curves showed that PS II photochemical efficiency (Fv'/Fm'), PS II actual photochemical efficiency (ΦPS II), photochemical quench (qP), and photochemistry (P) of the light-adapted leaves decreased as actinic light intensity increased. The nonphotochemical quench (NPQ), electronic transfer rate (ETR), photochemistry rate (PCR), antenna heat dissipation (D), and nonphotochemistry in PS II reaction center (E) increased as the actinic light intensity increased. At a actinic light intensity of about 40 µmol/m²/s, an inflection was found in the light response curves of chlorophyll fluorescence parameters including Fv'/Fm', ΦPS II, qP, NPQ, P, D, and E, which was possibly related to the light compensation point of photosynthesis. Additionally, exogenous ALA treatments significantly promoted Fv'/Fm', ΦPS II, P, PCR, and ETR. When the actinic light intensity was higher than about 1500 µmol/m²/s, NPQ was higher in ALA-treated leaves than that of the control, suggesting that photoinhibition induced by ALA treatment was beneficial for energy dissipation to protect leaf photosynthesis. An analysis of the energy distribution of PS II reaction center showed that ALA treatment led to lower levels of D and higher levels of E and P, suggesting that ALA enhanced energy harvested by antenna pigments into PS II reaction center, where part of the energy dissipated nonphotochemically, which was important for maintaining higher level of photochemistry of PS II (Kang and Wang, 2008).

A protective mechanism against photoinjury during photosynthesis in watermelon was induced by ALA treatments by decreasing the minimum fluorescence (F_o) and increasing the Fm, resulting in higher levels of the Fv/Fm of the dark-adapted leaves, and increasing the actual photochemical efficiency of the photosystem (phiPS II), ETR, qP, and NPQ in light-adapted leaves. Thus, the photon energy irradiated on watermelon leaves was more easily harvested by PS II and was transformed into bioelectric energy in the reaction center and transferred by the photosynthetic electronic chains in the ALA-treated plants (Wang et al., 2010).

ALA treatments raised the activities of anti-oxidative enzymes as SOD and POD, but exerted little influence on the CAT and APX activities in the leaves of watermelon seedlings, which implied that the ALA treatments enhanced the metabolism of active oxygen species by raising the SOD and POD activities in watermelon plants, thus maintaining the cellular photosynthetic capacity. Thirty days after the ALA treatments, the dry weights, starch and solvable sugar contents of treated watermelon plants increased separately by 42%–54%, 62.2%–207.0% and 32.0%–87.1% respectively. Thus, ALA treatments could improve the activities of anti-oxidative enzymes in watermelon seedling at low temperature under weak light, and promote the photosynthesis and photosynthate accumulations as well as the plant growth (Kang et. al., 2006).

25.2.3 VEGETATIVE GROWTH

Watermelon is a trailing vine with coarse, hairy, and pinnately lobed leaves and white to yellow flowers. The watermelon is grown for its fruit, a kind of berry known as PEPO botanically. The fruit has a smooth, hard rind, usually green with dark-green stripes or yellow spots, and a juicy, sweet interior flesh, usually deep red to pink, but sometimes orange, yellow, or white, with many seeds. The stems are thin, hairy, angular, and grooved and have branched tendrils at each node. The stems are highly branched and up to 30 ft long in regular cultivars and are shorter and less branched in dwarf types. Roots are extensive but shallow, with a taproot and many lateral roots.

Inoculating the arbuscular mycorrhizal fungus *Glomus versiforme* could activate the defensive enzyme (root phenylalanine ammonialyase [PAL], catalase [CAT], peroxidase [POD], β-1,3-glucanase, and chitinase) activities of nongrafted and grafted watermelon seedlings, enable the seedling roots to produce rapid response to adversity, and improve the capability of watermelon seedling against continuous cropping obstacle (Chen et al., 2013).

25.3 PHYSIOLOGICAL STAGES OF GROWTH II

Watermelon is a warm-season crop. It requires a long growing season. Flowering and fruit development are promoted by high light intensity and high temperature.

Watermelon is monoecious. Flowering begins about 8 weeks after seeding. Flowers are born in the order of staminate (male)>, perfect (hermaphroditic)>, or pistillate (female). There are andromonoecious (staminate and perfect) types observed mainly in the older varieties or accessions collected from the wild. The pistillate flowers have an inferior ovary, and the size and shape of the ovary is correlated with final fruit size and shape. In many varieties, the pistillate or perfect flowers are borne at every seventh node, with staminate flowers at the intervening nodes. The flower ratio of typical watermelon varieties is 7:1 staminate/pistillate, but the ratio ranges from 4:1 to 15:1.

The fruit of watermelon has a round to cylindrical shape, is up to 24 in. long, and has a rind 0.4–1.5 in. thick. The edible part of the fruit is the endocarp (placenta). Fruits usually weigh 8–35 lb with exceptions as large as 262 lb and smaller, in the range of 2–8 lb. Fruit rind varies from thin to thick and brittle to tough. Seeds continue to mature as the fruit ripens and the rind lightens in color.

Harvesting starts from July through August. The harvesting indices are a combination of the following that may vary based on location, variety, and plant growth (Crop profile, VA2003):

1. Tendrils or pigtails on vines nearest the fruit wilt and change color from green to brown.
2. The ground spot on the belly of the melon changes from white to light yellow.

3. The thumping sound changes from a metallic ringing when immature to a soft hollow sound when mature.
4. The green bands (striped varieties) gradually break up as they intersect at the blossom end of the melon.

Seeds will be easier to extract from the fruit if the fruit is held in storage (in the shade or in the seed processing room) for a few days after harvest. The optimum temperature for storage is 60°F.

25.4 GROWTH RESPONSES OF WATERMELON UNDER VARIOUS STRESSFUL CONDITIONS

The current section of the chapter attempts to overview recent advances in research to find genetic, physiological, and biochemical mechanisms that were responsive to various stresses and were manipulated to improve watermelon ability to withstand longer periods of stress during critical stages of growth that determines watermelon yield. Scientific breeding of crop plants for adaptation to biotic and abiotic stresses is necessary to improve their productivity and superior quality for food, feed, fiber, and other industrial uses.

25.4.1 BIOTIC STRESS

Watermelons are susceptible to several diseases that attack the roots, foliage, and fruit. The most common are *fusarium* wilt, anthracnose, downy mildew, and virus diseases followed by *Cercospora* leaf spot, gummy stem blight, powdery mildew, bacterial fruit blotch, damping-off, root rot/vine decline, and root knot nematode (Parris, 1949).

A preventive method of using cultural practices and resistant varieties will be helpful. None of the varieties were resistant to nematodes or insects, but were showing variable resistance to major pathogens *fusarium* and anthracnose. Resistance to either anthracnose, downy mildew, powdery mildew, or watermelon mosaic virus (WMV) was found among all major groups of *Citrullus* plant introductions (PIs), and resistance to gummy stem blight or *fusarium* wilt may exist among *C. lanatus* var. *citroides* PIs (Levi et al., 2001).

Plant defense mechanisms include (1) structural characteristics like thick, waxy coating and hairs, thick cuticle and cell wall, and stomatal structure and behavior; (2) chemical defenses like inhibitors present in plant cell before infection and released in response to its environment, lack of recognition between host and pathogen, and lack of host receptors and sensitive sites for toxins; and (3) induced defenses like hypersensitive responses (HRs) and systemic acquired resistance (SAR).

HR is the most common resistance response to viruses, bacteria, fungi, nematodes, and insects. SAR is the final line of defense initiated by signal molecules (salicylic acid, SA; jasmonic acid-methyl jasmonate, JA; ethylene, and saponins) that involve defense-related proteins, R-genes (resistance), and phytochemicals. It is important to understand the defense-related biochemical responses, mainly complex metabolic changes that occur due to pathogen and pest attack in watermelon.

Major pathogens and insect pests that cause economic yield loss were discussed below along with the structural and biochemical changes during biotic stresses, including cultivar improvements in the watermelon crop.

25.4.1.1 Fusarium Wilt

A soilborne disease caused by *Fusarium oxysporum* f. sp. niveum (FON) is the major production problem worldwide and in the United States. There were three races (0, 1, and 2); race 1 is the most commonly occurring race throughout the United States and worldwide, and race 2 is aggressive and was present in Delaware, Florida, Indiana, Oklahoma, Maryland, and Texas in the United States. The disease is severe at temperatures between 77°F and 81°F (25°C–27°C). Damping-off

FIGURE 25.1 Fusarium wilt of watermelon. (Courtesy of D. S. Egel and R. D. Martyn. Fusarium wilt of watermelon and other cucurbits. The Plant Health Instructor. 2007.)

may occur in young seedlings and rot in the soil, where the hypocotyls are surrounded by a watery and soft rot causing the plants to become stunted. Later, wilting occurs in more mature plants, causing the plant to die. The infected plants were noticeably wilted on one side with other shoots being healthy, observed with flaccid, withered, and had brown leaves with discolored brown vascular tissues (Figure 25.1).

The watermelon cultivars, Afternoon Delight, Crimson Sweet, Indiana, Calhoun Gray, Smokylee, and Summit, were highly resistant and Sweet Princess, Jubilee, Charleston 76, Klondike R7, and Summerfield are slightly resistant to FON (Elmstrom and Hopkins, 1981). Long crop rotations with grass crops, delaying final thinning and resistant cultivars, are the best for controlling FON. The combined use of green manure, hairy vetch (*Vicia villosa* Roth) organic mulch, and partial cultivar resistance can provide adequate suppression of FON (Zhou and Everts, 2006). This combination also reduced the severity of powdery mildew (caused by *Podosphaera xanthii*) and Plectosporium blight (caused by *Plectosporium tabacinum*) on pumpkin, a vegetable from cucrbitaceous family (Everts, 2002).

The capacities of self-regulating and returning to normal status of the resistant watermelon cultivar 'Kelunsheng' were stronger than the susceptible 'Zaohua', in the content of malondialdehyde (MDA), relative conductivity, and activities of superoxide dismutase (SOD) and CAT, when infected by FON at seedling stage. The capacities of resistant cultivar in inhibiting chlorophyll deterioration and maintaining higher carotenoid content were significantly stronger than those of susceptible cultivar. The resistant cultivar maintained higher activity of dehydrogenase, relatively higher content of vitamin C, and relatively lower content of soluble sugar with high content of soluble protein than the susceptible cultivar (Wang et al., 2002).

The procedure that combines root dipping inoculation with spore-culturing method at the seedling stage and further tests for resistant materials in open fields was recommended for accurate and consistent results in comparatively shorter period and accelerated the process of watermelon breeding resistant to FON (Zhang et al., 1991). The FON incidence rates of highly resistant varieties Tianniu and Zaochunhongyu were 19.67% and 18.45%, respectively, while the incidence rates of highly susceptible varieties Tianshi and Heimeiren were 88.22% and 93.33%, respectively, and showed clearly the physiological race that infected watermelon, provided the basis for watermelon-resistant variety breeding and selection of production cultivar in Heilongjiang (An et al., 2009).

Development of genetic populations segregating for FON race 2 is in progress. Three single nucleotide polymorphisms (SNPs) and a major quantitative trait locus (QTL) were identified as associated with race 1 resistance (Amnon, 2014) and quantitative trait loci (QTL) markers for fruit, seed, and flower traits as well as rind thickness (Cecilia, 2012).

FIGURE 25.2 **(See color insert.)** Anthracnose of watermelon. (Credit to Dr. D.M. Ferrin, LSU AgCenter, Baton Rouge, LA.)

25.4.1.2 Anthracnose

Two races of the anthracnose fungus (*Colletotrichum orbiculare*) are common on cucurbit crops. Warm, wet weather favors the infection and spread of the disease. Most cultivars of watermelon are resistant to race 1 and none have race 2 resistance. Race 2 fungus occurs on aboveground parts of the watermelon. The severity of the infection leads to numerous lesions on the leaves (Figure 25.2), and vine defoliation leads to yield loss and lower fruit quality.

The best time to investigate anthracnose resistance was 7–10 days after inoculation with an inoculation of 1×10^6 pfu/mL at 2-true-leaf stage (Zhao-Ming et al., 2009). The antifungal protein isolated from a bacterial strain, MET0908, acted on the cell wall of *Colletotrichum lagenarium* isolated from anthracnose infected fruits, leaves, and stems of watermelon and exhibited β-1, 3-glucanase activity (Kim and Chung, 2004).

The activity of POD and polyphenol oxidase (PPO) and the contents of chlorophyll were higher in anthracnose resistant cultivars (cv. Xinongbahao and Hongxiaoyu) compared to susceptible cultivars (Linglongwang), and the peak value of soluble protein content was observed earlier in cultivars with resistance compared to susceptible ones when seedlings were inoculated with 1×10^6 pfu/mL spore solution of *C. orbiculare* (Ting et al., 2009).

25.4.1.3 Downy Mildew

Pseudoperonospora cubensis a fungus that causes small, circular, irregular, chlorotic spots on the upper leaf surface that later turn brown, necrotic, and blighted leaf (Figure 25.3) with purple to gray downy growth on underside of lesions. The lesions are not restricted by the veins (Lebeda and Cohen, 2011). Infection is severe during winter and infected leaves tend to curl upward from the margins. Higher temperatures and relative humidity prevailing under closed covers in winter normally favor the development of downy mildew in watermelon. The disease is prevalent in warm humid climate of temperate areas such as Americas, Europe, Japan, Australia and South Africa, tropical regions and semi-arid regions, such as the Middle east.

Reduction in photosynthetic activity early in plant development, stunted plants, sunscald, reduced fruit quality and yield due to premature defoliation are some of the symptoms associated with growth and development due to the disease (Colucci and Holmes, 2010; Savory et al., 2011).

25.4.1.4 Gummy Stem Blight (Black Rot)

Didymella bryoniae (*Phoma cucurbitacearum*) causes crown blight, extensive defoliation, (Figure 25.4), and fruit rot of watermelon in southeastern United States. The fungus is airborne,

FIGURE 25.3 **(See color insert.)** Downy mildew on watermelon. (Credit to Dr. D.M. Ferrin, LSU AgCenter, Baton Rouge, LA.)

FIGURE 25.4 **(See color insert.)** Gummy stem blight on watermelon. (Credit to Dr. D.M. Ferrin, LSU AgCenter, Baton Rouge, LA.)

seedborne, and soilborne. The infection increases due to wet plant surface and high relative humidity. Crop rotation, various irrigation methods, and solarization are found effective in controlling disease spread. Plant stand, vine length, and fruit set were increased in cabbage-amended, solarized plots with reduced area under disease progress compared to annual cropping of watermelon alone (Keinath, 1996).

Among the USDA collections, most resistant PIs identified are 279461, 254744, 482379, 244019, 526233, 482276, 164248, 482284, 296332, 490383, 271771, and 379243 and most susceptible are 226445, 534597, 525084, 223764, 169286, and 183398 (Song et al., 2004).

25.4.1.5 Powdery Mildew

The symptoms *Podosphaera xanthii* infection on watermelon include circular, tan to dark brown spots that begin at the leaf margins to the entire leaf (Figure 25.5), the formation of blights, stem cankers, extensive defoliation, and affect the cotyledons, hypocotyls, and fruits (Antonia, 2008; Maynard and Hopkins, 1999). Exposed vines and fruits may sunscald and shrivel. The most resistant accessions identified were PI 632755, PI 386015, PI 346082, PI 525082, PI 432337, PI 386024, PI 269365, and PI 189225 (Antonia, 2008).

FIGURE 25.5 Powdery mildew on watermelon. (Copyright to Tom Isakiet, Taken from Aggie-horticulture. tamu.edu, College Station, TX.)

25.4.1.6 Phytophthora Blight

The phytophthora blight and rot were caused by *Phytophthora capsici* Leon and *P. drechsleri* Tuck and were first reported in Colorado and California followed by the north and southeastern states of the United States and China. *P. capsici* causes seedling damping-off, leaf spot, foliar blight, root and crown rot, stem lesion, and fruit rot (Figure 25.6). Crown rot, an initial symptom on the root in the field, causes the entire plant to completely collapse and die in a short period of time. The affected vine tissue turns brown in color, appears water soaked, and often collapses. The infected fruit develops dark, water-soaked lesions and is commonly covered with white mold. Two resistant accessions of *P. capsici* (IT185446 and IT187904) identified through germplasm characterization could be used as a rootstock and as a source of resistance in breeding resistant watermelon varieties against *Phytophthora* (Kim et al., 2013).

25.4.1.7 Bacterial fruit blotch

The disease is caused by *Acidovorax avenae subsp. citrulli* and can destroy 100 % of marketable watermelon if the infection develops early (Lewis et al., 2015). Symptoms include dark angular patches on seedling transplants, mature leaves and fruits. Severe infestation results in small, dark green stain on the fruits upper surface causing rupture of the fruit and fruit rot (Figure 25.7); though the disease is confined to the rind, causes severe economic losses. No resistance identified yet.

FIGURE 25.6 Blight and rot on watermelon. (Credit to K. Everts, from the University of Maryland, College Park, MD.)

FIGURE 25.7 Bacterial fruit blotch. (Courtesy of Lewis et al., 2015, from MU, MI.)

25.4.1.8 Aphids, Thrips and Mites

Clear polyethylene (PE) films were observed beneficial in reducing winged aphid populations and are recommended mulches for early spring planting of watermelon under Mediterranean climate. Watermelon vegetative growth (main vine leaf number, vine length, vine diameter, and branch number) and early yield were high and positively correlated with clear PE mulch (Ban et al., 2009). No melon aphid or aphid borne virus resistance was identified in any commercially available varieties. Broad mites (*Polyphagotarsonemus latus*) were severe pests on watermelon (Zang et al., 2003) in western Indiana under drought conditions, wherein mites sucked the sap from the underside of the leaf, resulting in reduced photosynthetic area (Figure 25.8). Photosynthetic assimilates for plant growth complete defoliation of vine and unmarketable fruits.

Mite injury occurred mainly on the growing terminals (Figure 25.9) and tender apical leaves which turned bronze color, distorted and curled upwards. PIs in accessions belonging to *Citrullus lanatus var. lanatus* (PI 357708), *Citrullus lanatus var. citroides* (PI 500354), *Citrullus colocynthis* (PI 386015, PI 386016, PI 525082), and *Parecitrullus fistulosus* (PI 449332) had significantly lower broad mite injury ratings and counts compared to commercial cultivar, 'Mickey Lee' (Chandrashekar et al., 2007). Melon thrips (*Thrips palmi*) is a problem where the melon crop is planted near to the other susceptible crops like tobacco and can be controlled by shaking the vine and avoiding planting near the infested crops (Webb et.al., 2001).

25.4.1.9 Viral Diseases

Among various cucurbit viruses, WMV-1 and WMV-2, watermelon vine decline (WVD), potato ringspot virus (PRSV-W), zucchini yellow mosaic virus (ZYMV), tobacco ring spot virus (TRSV),

FIGURE 25.8 Spider mite damage on watermelon leaves. (Copyright to Jerry Brust, taken from the University of Maryland, College Park, MD.)

FIGURE 25.9 Stunted tip by broad mite infestation. (Copyright to Chandrasekhar et al., 2007.)

squash leaf curl virus (SLCV), and squash mosaic virus (SqMV) are most common in southern United States (Ali, 2011).

PRSV-W causes mosaic, blistering, puckering, and chlorosis of infected watermelon leaves. PRSV-W and WMV cause mottling, leaf deformation, and shoe strings in watermelon. Yellowing and mosaic are caused by ZYMV. WMV-2 infects various cucurbit hosts (muskmelon, bottle gourd), leading to physiological and metabolic changes in chlorophyll and carotenoid synthesis, cellular nutrient (phosphorous, nitrogen) availability in the leaf, and reduced photosynthetic activity (Muqit et al., 2007).

WMV-1 induced changes in leaf chlorophyll (14.37%–54.93%) and carotenoids (30.66%–70.41%) with fewer yields, and poor-quality fruits (Sandhu et al., 2007). The virus also caused disintegration of infected leaf tissue, hypertrophy, compactness of tissue, and decrease in the number of chloroplasts in the tissue depending on the virulence. The virus also induced reduced vigor of vine and its length and reduced the number of branches per vine. WVD was transmitted by whiteflies, resulting in severe economic losses in Florida due to yellowing and browning and collapse of the entire vine (Baker et al., 2008).

Viruses are transmitted by aphids and whiteflies. Control measures of insect vector, use of silver plastic mulch and insect-/virus-resistant varieties would be beneficial. High resistance to mosaic virus was observed in watermelon collections PI 595203 (ZYMV-CH and WMV) and PI 482261 (ZYMV-FL) and was controlled by a single recessive gene (Xu et al., 2004).

Resistance to PRSV-W was reported in accessions PI 244017, PI 244019, PI 485583 (Guner et. al., 2008) and PI 595203 (Strange et al., 2002). It was controlled by a single recessive gene, prv (Guner et al., 2008). Resistance to ZYMV was controlled by a single recessive gene zym-FL- in Citrullus lanatus and zym-FL-2 (Wehner, 2012) and zym-CH (Xu et al. 2004).

25.4.2 Abiotic Stress

Abiotic stresses include drought, flooding, salinity, heat, chilling, freezing (high or low temperatures), radiation, nutrient deficiency, and heavy metal stress. The changes in the duration of predominant rainy season due to climate change and land degradation due to heavy use of fertilizers to boost the productivity per unit area were the major causes for these stresses. Severities of these stresses induce severe cellular damage, leading to impaired growth, development, reproduction, and fruit set with significant yield reductions in watermelon.

Reactive oxygen species (ROS) production will be enhanced during abiotic stresses and include free radicals such as superoxide anion ($O_2^{\cdot-}$), hydroxyl radical ($\cdot OH$), as well as nonradical molecules like hydrogen peroxide (H_2O_2) and singlet oxygen ($^1O_2$). ROS are formed by the inevitable leakage of electrons onto O_2 from the electron transport activities of chloroplasts, mitochondria, cell wall, and plasma membranes or as a byproduct of various metabolic pathways localized in different cellular compartments (Sharma et al., 2012).

Excessive ROS result in peroxidation of lipids, irreversible changes in amino acid sequence of proteins, mutations in nucleic acids, enzyme inhibition, and activation of programmed cell death (PCD) pathway, ultimately leading to death of the cells. Plant defensive mechanisms against these environmental stresses include SOD that catalyze the dismutation of superoxide to hydrogen peroxide and oxygen; CAT, a heme-containing enzyme that catalyzes the dismutation of hydrogen peroxide into water and oxygen; and ascorbate, tocopherol, glutathione (GSH), and carotenoids that act as antioxidants, scavenging oxygen free radicals (McKersie, 1996). The predominant abiotic stress response of watermelon were incorporated below.

25.4.2.1 Nutrient Deficiency Stress

Productivity is optimum on well-drained, irrigated soils with balanced nutrition (lb/acre of 120–180 N, 150 P_2O_5, 120–150 K_2O). Nitrogen deficiency in watermelon is usually observed in Arizona soils. If N deficiency is revealed after fruit set, it cannot be corrected. Preplant soil analysis is more valuable as nitrogen deficiency affects the growth, vigor, and yield of watermelon. Petiole nitrate levels are normally high during early growth stage and gradually reduce toward fruit set. The nitrate concentration of petiole should be maintained above 4000 ppm NO_3–N throughout the season. If there is a decline in the level to 2000 ppm NO_3–N, visual symptoms such as pale green foliage or reduced vine growth occur (Anonymous, 2015).

Early yields were positively correlated with phosphorous (P) fertilization and maximum yield was realized with little P fertilizers (Hoschmuth et al., 1993). K^+ is required for flowering, fruit development, and better fruit quality. K deficiency and associated blossom end rot due to calcium deficiency were often observed in watermelon grown on soils where previously long-fruited cultivars were grown and/or there is less irrigation such as those in Florida. Higher levels of N and K enhanced the demand for absorption of banded P, and the yield was not affected by P absorption (Hochmuth and Hanlon, 2015).

The watermelon genotype 'YS' was more tolerant to K^+ deficiency and displayed less inhibited root growth than the genotype '8424', and the results from transcriptome analysis revealed that repressed defense and stress response can save energy for better root growth in YS, which can facilitate K^+ uptake and increase K efficiency and tolerance to K^+ deficiency (Fan et al., 2013).

Copper is vital for the metabolism of higher plants and physiological and biochemical processes, especially electron transport during photosynthesis (component of plastocyanin) and carbohydrate and protein accumulation. On a very acidic, Leon fine sandy soil, at Gainesville, Florida, copper (Cu) deficiency resulted in the reduced and stunned growth, with regular shaped leaves cupped upward and marginal burning of young, expanding leaves and death occurring at the growing points. Fruit set was absent on plants with severe Cu deficiency. Increased yield due to Cu fertilization increased the Cu levels in the leaf tissues (Locascio et al., 1964). Organic nitrogen fertilization made Cu and Zinc (Zn) available to the growing watermelon (Fiskell et al., 1964).

Manganese (Mn) deficiency starts before the watermelon begins to flower, characterized by yellowing of leaves with interveinal chlorosis with white spots after fruit set, and spreads from lower leaves to upper leaves over time. Soils with a high pH and a level of easily reductive manganese of less than 100 mg Mn/kg (extracted by 2% hydroquinone + 1 M ammonium acetate) are usually deficient in Mn. Low-molecular-weight organic compounds, green manure, and biosolids added to a soil with a high content of Mn increased Mn phytotoxicity.

Watermelon is sensitive to molybdenum (Mo) deficiency. Mo deficiency is characterized by stunted plants with whitish-tan interveinal chlorosis and marginal leaf burn in severe cases. Older leaves are affected first.

25.4.2.2 Salinity Stress

Watermelons grow best on sandy loam soils, tolerant to slight acidity (pH 6–7). Inhibition of shoot and root growth development is the primary response to salinity stress. Root extension growth is severely inhibited by high concentrations of NaCl in the growth medium compared to lateral root formation (Bernstein and Kafkafi, 2002).

The stand establishment rate of transplanted seedlings was significantly reduced due to the increase of nutrient solution salinity, and significantly higher values of relative growth rate (RGR), net assimilation rate (NAR), and leaf area ratio (LAR) were recorded in case of root-pruned, splice-grafted seedlings (Balliu et al., 2014).

25.4.2.3 Drought Stress

Agriculture is mostly dependent on irrigation. Drought is the most devastating abiotic stress, leading to economic yield loss. During normal growth conditions, roots absorb water and nutrients from the soil to support plant function well with balanced system. Under water stress, the entire plant system will be altered, resulting in various structural, molecular, cellular, metabolic, functional, and phenotypic changes like hardening of cell wall and reduced or elongated root to adapt to changing surrounding environmental conditions (Atkinson and Urwin, 2012; Ghosh and Xu, 2014).

At the early stage of drought stress, root development of wild watermelon (*C. lanatus* sp.) was enhanced, indicating the activation of a drought avoidance mechanism for absorbing water from deep soil layers, while the enhancement of physical desiccation tolerance at later stages of drought stress, by regulating its root proteome (Yoshimura et al., 2008). Wild watermelon showed vigorous root growth and a domesticated cultivar showed retarded root growth under drought stress (Kajikawa et al., 2010). Leaves with decreased transpiration and a concomitant increase in leaf temperature were reported in wild watermelon under water deficit conditions (Akashi et al., 2011).

Under drought conditions, in the presence of strong light, wild watermelon accumulates high concentrations of citrulline (Kawasaki et al., 2000), glutamate, arginine, and DRIP-1, a drought-induced peptide (DRIP) in its leaves (Yokota et al., 2002). A comparison of the proteome of stressed plants with that of unstressed plants revealed 23 stress-induced and 6 stress-repressed proteins and confirmed that 15 out of 23 upregulated proteins were heat shock proteins (HSPs) and 10 out of 15 HSPs were small HSPs (sHSP). The defense response of wild watermelon involves several functional proteins related to cellular defense and metabolism to provide protection against water deficit stress (Akahsi et al., 2011), and these traits can be exploited to develop new cultivars of cultivated watermelon with drought tolerance.

25.4.2.4 Temperature Stress

Optimum temperature for watermelon growth ranges from 20°C to 32°C (Bates and Robinson, 1995). Low-temperature stresses (<5°C) are most common in temperate regions like North America and Europe, and high-temperature stresses (>45°C) are extreme in tropical and subtropical regions of the world like China, India, Africa, and the southern and central regions of temperate countries like Europe and the Americas, resulting in famine due to complete crop failure.

Plant height, relative chlorophyll content, and leaf and stem dry weight of watermelon decreased linearly with the decrease in day temperature from 25°C to 15°C (Inthichack et al., 2014). High temperatures usually impact the reproductive growth of plants rather than the vegetative growth due to pollen infertility, which eventually reduces the yield (Bita and Gerats, 2013).

25.4.2.4.1 High-Temperature Stress

High-temperature stresses usually slow down or totally inhibit seed germination, depending on the cultivar and the intensity of the stress (Singh et al., 2001). During the later stages of growth, photosynthesis, respiration, water relations and membrane stability, levels of hormones primary and secondary metabolites were affected (Bita and Gerats, 2013).

Various physiological injuries include scorching of leaves and stems, leaf abscission and senescence, shoot and root growth inhibition, or fruit damage (Vollenweider and Günthardt-Goerg, 2005).

Thermal stress induced the accumulation of phenolic compounds by activating their biosynthesis and inhibiting their oxidation (Rivero et al., 2001).

Heat stress resulted in the enhanced expression of a variety of HSPs, protein kinases, and other stress-related proteins and production of antioxidants and ROS that constitute major plant responses to heat stress (Wahid et al., 2007). The peroxisomal SOD from watermelon cotyledons has similarities with chloroplast and cytosolic SODs and has high thermal stability and resistance to inactivation by hydrogen peroxide that protects toxic effects of superoxide radicals produced within the cell during stress (Bueno et al., 1995). Light-green and gray-green watermelons are less subject to sunburn injury than dark-green and striped varieties (Roberts et al., 1914).

Information on important traits for high temperature tolerance, like membrane lipid saturation and accumulation of osmoprotectants (proline, glycine betaine, soluble sugars), was lacking for watermelon crop.

25.4.2.4.2   Low-Temperature/Chilling Stress

Crop failures in tropical regions with extreme temperature stress, made possible expansion of cultivation of new crops in nontraditional areas with low temperatures like northern America and Canada. Though the degree of stress resistance varied from plant to plant of each cultivar and or genotype, scientists made extensive progress in identifying low-temperature stress responses to develop watermelon cultivars with chilling tolerance.

Due to fluctuations in spring temperatures, early planted watermelons were suffering from germination and emergence (Singh et al., 2001). Cultivar variability for germination and cold tolerance traits in 'Blackstone' and 'Starbrite' cultivars due to their highest germination percentage (90%–95%) at 14°C can be used to breeding cultivars with better germinability and earliness (Singh et al., 2001).

Carnival variety can be tolerant to cold stress (2°C) during transplanting and yield without any long-term detrimental effects in South Carolina, United States (Korkmaz and Dufault, 2002). The most resistant PI 244018 and the most susceptible NH Midget and Golden were identified from studies conducted with controlled light intensity of 500 mmol/m²/s and photosynthetic photon flux density (PPFD) and optimal conditions for chilling treatment of 36 h at 4°C or 24 h at 2°C (Kozik and Wehner, 2014).

Chilling stress induced the accumulation of soluble phenolics and the highest phenylalanine ammonialyase activity and the lowest POD and PPO activities and reduced shoot weight (Rivero et al., 2001).

Seed soaking with 0.5 mM salicylic acid (SA) was effective in protecting watermelon seedlings from the damaging effects of chilling stress at the early stages of growth. Due to inhibition of proline accumulation and leaf electrolyte leakage (Mohammad et al., 2013), and the concentration of SA at 1.0 mmol/L showed the most effective treatment for chilling tolerance (up to 5°) at 3-leaf stage (Lu and Yu, 2004).

The activities of three protective enzymes, such as POD, CAT, and adenosine triphosphatase, increased, while the MDA content decreased in the leaves of SA-treated plants. The decline of chlorophyll content of the leaves pretreated by SA under chilling temperature was reduced (Lu and Yu, 2004).

When seedlings are at four leaf stage, the activities of POD and APX (ascorbate peroxidase) were increasing at 15°C stress and declining at 10°C stress while APX changed less than POD (Guang et al., 2006). Watermelon varieties Zao-Jia 8424, variety No. 4 and No. 6 are cold-sensitive type, Wan-Fu-Lai was moderately cold tolerant type, and wild watermelon, variety No. 2 and No. 5 are cold tolerant type. The leakage of electrolytes, the content of Proline and the activities of POD were increased with the extension of low temperature treatment. The activities of SOD of Wan-Fu-Lai and Wild watermelon increased, and the soluble sugar of wild watermelon displayed the trend to declining-rising. The Zao-Jia 8424 and WanFu-Lai cultivars of watermelon showed the trend to rising with the extension of low temperature treatment (Peng et al., 2006).

The auto-tetraploid and triploid had lower cold tolerance than that of diploid cultivars under the condition of $(12 \pm 1)°C$ temperature and 1000lx. With the increase of stress time, the SOD and POD activity of triploid cultivars were gradually increased. The SOD and POD activities of diploid were increased initially and declined gradually while the MDA contents were decreased initially and increased thereafter with increase of stress. The initial increase in activities of SOD and POD were greater in diploids than in triploid and tetraploid cultivars. The soluble protein (PRO) content of diploid and triploid was increased at first, and then decreased (Liu et al, 2003).

The ALA treatment improved the chilling tolerance in watermelon by regulating assimilatory accumulation, resulting in higher net photosynthetic rate (34%–44%) and soluble sugar (32%–87%) and starch (70%–227%) content of leaves under low light and chilling stresses (Kang et al., 2006; Wang et al., 2010).

25.4.2.5 Radiation of Ultraviolet B Stress

The flux of photochemically active ultraviolet B (UV-B) photons (wavelength, λ <315 nm) into the troposphere is limited by the amount of stratospheric O_3 (Cicerone, 1987). The decrease in stratospheric ozone (O_3) and the consequent possible increase in UV-B is a critical issue. The effects of UV-B, CO_2, and O_3 on plants have been studied under growth chamber, greenhouse, and field conditions. Biomass responses of plants to enhanced UV-B can be negative (adverse effect), positive (stimulator effect), or neutral (tolerant). There are inconsistencies between the results obtained under controlled conditions and those obtained from field observations. Based on measurements of biomass accumulation, watermelon was highly sensitive to increased UV-B radiation (Krupa and Kickert, 1989)

25.4.2.6 Heavy Metal Stress

Heavy metal (HM) toxicity is one of the major abiotic stresses, leading to hazardous effects in plants. A common consequence of HM toxicity is the excessive accumulation of ROS and methylglyoxal (MG), both of which can cause peroxidation of lipids, oxidation of protein, inactivation of enzymes, DNA damage, and/or interaction with other vital constituents of plant cells. Higher plants have evolved a sophisticated antioxidant defense system and a glyoxalase system to scavenge ROS and MG. In addition, HMs that enter the cell may be sequestered by amino acids, organic acids, and GSH or by specific metal-binding ligands.

Being a central molecule of both the antioxidant defense system and the glyoxalase system, GSH is involved in both direct and indirect controls of ROS and MG and their reaction products in plant cells, thus protecting the plant from HM-induced oxidative damage. Recent plant molecular studies have shown that GSH by itself and its metabolizing enzymes—notably glutathione S-transferase, glutathione peroxidase, dehydroascorbate reductase, glutathione reductase, and glyoxalases I and II—act additively and coordinately for efficient protection against ROS- and MG-induced damage, in addition to detoxification, complexation, chelation, and compartmentation of HMs.

A copper level of 50 mg/kg in the soil was beneficial for the growth of watermelon plants, while that above 100 mg/kg proved toxic. The phytoremediating role of watermelon on copper-polluted soil is low (Vijayarengan and Jose, 2014). The Mt2 and PCS genes were shown to act as potent scavengers of hydroxyl radicals in watermelon (*C. lanatus*) when exposed to HMs (Akashi et al., 2004). Several studies have indicated that some rootstocks are capable of restricting the uptake and/or the transport of heavy metals (e.g., Cd, Ni, Cr) and micronutrients (e.g., Cu, B, and Mn) to the shoot, thereby mitigating the stress caused by their excessive external concentrations (Savvas et al., 2010).

25.4.2.7 Pollution Stress

Ozone is produced in air during the daylight hours from chemicals released from automobile exhausts and the combustion of fossil fuels (oil and coal). Ozone or the ozone layer in the stratosphere (mostly 90% O_3) serves as a shield against biologically harmful solar UV radiation, initiates key stratospheric chemical reactions, and transforms solar radiation into heat and the mechanical

FIGURE 25.10   Advanced stage of ozone injury in watermelon. (Copyright to Simon et al., 1988 from PU, IN.)

energy of atmospheric winds. The remaining 10% O_3 that is in the troposphere is harmful to humans, animals, and vegetation (Krupa and Kickert, 1989; Cicerone, 1987).

Pollution injury was observed first on the pollinizer plants. Ozone is the most phytotoxic air pollutant that damages the leaves of field-grown melons and watermelons in several midwestern and eastern states of the United States and Europe. Injury on watermelon leaves consists of premature chlorosis (yellowing) on older leaves. Leaves subsequently develop brown or black spots with white patches. Watermelons are generally more susceptible than other cucurbits to ozone damage (Figure 25.10). Damage is more prevalent when fruits are maturing or when plants are under stress. Injury is seen on crown leaves first and then progresses outward.

An autumn crop of watermelon (cv. Sugar Baby) in open-top chambers in Indiana treated with either charcoal-filtered or nonfiltered air showed a significant decrease in marketable yield by weight and number (21%) compared with those grown in clean air (Synder et al., 1991). In two studies using open-top chambers in commercial fields in Spain, the soluble solid content of watermelon was decreased by 4%–8% due to seasonal ambient ozone levels (Gimeno et al., 1999).

Symptoms due to injury by air pollution are highly similar to those due to biotic or nutrient stresses. Therefore, leaf tissues should be checked for Mn toxicity; N, P, Mg, B, and Fe deficiency; moisture stress; insect damage by mites and thrips; pesticide toxicity; and virus infection if checking for air pollution damage caused by ozone. If the pollution injury is caused by sulfur dioxide, leaf tissues should be checked for N, K, Mn, Mg, and Ca deficiency, moisture stress, high temperature, pesticide toxicity, and other leaf disease-causing organisms (Lacasse and Treshow, 1976; Simon et al., 1988).

The key to avoiding air pollution injury is to plant tolerant varieties to limit plant stresses. Seedless watermelon varieties tend to be more resistant to air pollution injury than seeded varieties and "ice box" types are the most susceptible. Fungicides, thiophanate methyl (Topsin), offer some protection against ozone damage. Watermelon cultivars are classified as tolerant (Charleston Gray, Jubilee, picnic and super sweet), intermediate (Chilean Black, Mirage, Tender Sweet, and Orange Flesh), and susceptible (Crimson Sweet, Madera, Sugar Baby, Moran, Petite Sweet, and Blue Belle) to pollution injury based on evaluations at southwest Purdue Agricultural Research Center, Vincennes, IN, in 1984–1985.

Conclusively, watermelon is economically viable and nutritionally qualitative consumer preferred fruit commodity in the world. The water withdrawal requirements for agriculture were expected to be increased by 2030 and so as growing challenges for watermelon under abiotic stress. Stress tolerance research to transform melons and create variability and approaches that would improve the crops ability to stand biotic and abiotic stresses is in progress (Table 25.2). Lot of work done

TABLE 25.2
Summary on Biotic and Abiotic stress, Resistance, and Responses in Watermelon

Common Name of the Stress	Causal Agent (Scientific Name)	Symptoms	Sources of Resistance (Cv./PIs/ Genes)	Response to Stress
Fusarium wilt	*Fusarium oxysporium* f. sp. *Niveum* (FON)	Damping off and rot Fusarium wilt of watermelon (Courtesy D.S. Egel)	Race 1 of FON is controlled by a single dominant gene *Fo-1* (Henderson et al., 1970; Netzer and Weintall, 1980) Cv. Tianniu and Zaochunhongyu (An et al., 2009) Cv. Kelunsheng (Wang et al., 2002)	Self-regulation, less incidence of the disease in highly resistant cultivars with high contents and activities of SOD, CAT, and MDA
Anthracnose	*Colletotrichum orbiculare*	Numerous lesions on leaves and defoliation Anthracnose of watermelon (Credit to Dr. DM Ferrin, LSU AgCenter, Baton Rouge, LA)	Races 1 and 3 are controlled by a single dominant gene *Ar-1* (Layton, 1937) Race 2 is controlled by a single dominant gene *Ar-2-1* (Winstead et al., 1959) Sources for *Ar-2-1* are W695 citron and PIs 189225, 271775, 271779, and 299379 Cv. Xinongbabahao and Hongxiaoyu (Ting et al., 2009) *C. lagenarium*—antifungal protein "MET0908"	Increased activities of POD and PPO in resistant cultivars
Gummy stem blight	*Phoma cucurbitacearum*	Crown blight, defoliation and fruit rot Gummy stem blight on watermelon (Credit to Dr. DM Ferrin, LSU AgCenter, Baton Rouge, LA)	PIs from USDA showed resistance (Song et al., 2004) Inherited by a recessive gene *db* (Norton, 1979)	Reduced incidence of blight, rot, and defoliation in resistant cultivars
Downy mildew	*Pseudoperonospora cubensis*	Irregular chlorotic spots, brown, blighted, curled, necrotic leaf and stunted plants Downy mildew on watermelon (Credit to Dr. DM Ferrin, LSU AgCenter, Baton Rouge, LA)	—	—

(Continued)

TABLE 25.2 (Continued)
Summary on Biotic and Abiotic stress, Resistance, and Responses in Watermelon

Common Name of the Stress	Causal Agent (Scientific Name)	Symptoms	Sources of Resistance (Cv./PIs/Genes)	Response to Stress
Powdery mildew	*Podosphaera xanthii*	Leaf blight, stem cankers; Powdery mildew on watermelon (Copy right to Tom Isakiet, Taken from Aggie-horticulture.tamu.edu, College Station, TX)	PIs from USDA (Antonia, 2008)	—
Phytophthora blight	*Phytophthora capsici* and *P. drechsleri*	Damping off, leaf spot, root and crown rot, fruit rot and stem lesions. Blight and rot on watermelon (Credit to K. Everts, from the University of Maryland, College Park, MD)	Root stocks of IT 185446 and IT 187904 (Kim et al., 2013)	—
Bacterial fruit blotch	*Acidovorax avenae* subsp. *citrulli*	Leaf and fruit spots, fruit rupture; Bacterial fruit blotch (Courtesy of Lewis et al., 2015, from MU, MI)	—	—
Aphids	*Aphis gossypii*	Distorted curled leaves, fruits coated with sticky secretion and sooty mold, loss of vigor, and stunted, dead plants	For pathogen WMV1; Host resistance—PI180283—single dominant gene; Host—*Cucuumis metuliferus*—*wmv* gene; Host—PI 180280—*wmv-1* gene	Vector for WMV
Mites	*Polyphagotarsonemus latus*—broad mite; *Acari family*—spider mites	Stunted tips, bronze, discolored curled leaves, defoliation and unmarketable fruits; Stunted tip by broad mite infestation (Copy right to Chandrasekhar et al., 2007); Spider mite damage on watermelon leaves (Copy right to Jerry Brust, taken from the University of Maryland, College Park, MD)	PIs from USDA selected for broad mite resistance (Chandrasekhar et al., 2007)	—

(Continued)

TABLE 25.2 (Continued)

Summary on Biotic and Abiotic stress, Resistance, and Responses in Watermelon

Common Name of the Stress	Causal Agent (Scientific Name)	Symptoms	Sources of Resistance (Cv./PIs/ Genes)	Response to Stress
Viral diseases	WMV 1 and 2, WVD, PRSV-W, ZYMV, TRSV, SLCV, SqMV	Mosaic, leaf curl, and ringspot virus, causing leaf chlorosis and poor fruit quality	Single recessive gene—PI 595203 and PI 482261 (Xu et al., 2004)	Insect vector population increases and reduced photosynthates
Nutrient deficiency stress	Major: N, P, and K Miner: Cu, Zn, and Mn, Mo	Varies with nutrient in deficiency—reduced vine, fruit and root growth, pale green leaves with marginal leaf burn, and yellowing	Genotype "YS"—tolerant to K deficiency (Fan et al., 2013)	
Salinity stress	High concentration of NaCl	Root growth inhibition	Root pruned slice grafting (Balliu et al., 2014)	—
Drought stress	Reduced water availability to growing seedling	Retarded root growth, increased transpiration, and heat stress in susceptible varieties	Drought avoidance—wild watermelon	Accumulation of citrulline, glutamate, arginine, DRIP-1, and HSPs
				Elongated roots, reduced transpiration, leaf area, and hardening of cell walls to adapt to drought stress
High temperature stress	At >45°C	Inhibition of seed germination, leaf scorching, abscission, senescence, inhibition shoot and root growth, and fruit damage	Light green and gray green watermelons are less susceptible (Roberts et al., 1914)	Accumulation of phenolic compounds, enhanced expression of HSPs, protein kinases, production of antioxidants and ROS

(Continued)

TABLE 25.2 (Continued)
Summary on Biotic and Abiotic stress, Resistance, and Responses in Watermelon

Common Name of the Stress	Causal Agent (Scientific Name)	Symptoms	Sources of Resistance (Cv./PIs/ Genes)	Response to Stress
Chilling stress	Starts at 15°C and severe at <5°C	Stunted plant growth, wilted leaves with necrotic/chlorotic lesions, and reduced leaf expansion	Blackstone, starbrite (Singh et al., 2001) Carnival, PI 244018 (Kozik and Wehner, 2014) The single dominant gene *Ctr* was provided cool temperature resistance (Provvidenti, 1992, 2003)	Accumulation of soluble phenolics, increased activity of polyphenol oxidase, phenylalanine ammonia lyase, and reduced POD and shoot weight Increased susceptibility to diseases
Heavy metal stress	Excessive accumulation of ROS and MG, causing DNA damage	Death of the plant due to oxidative damage	Mt2 and PCS genes (Akashi et al., 2004)	—
Radiation stress	Photochemically active UV-B photons	Highly sensitive to UV-B	—	—
Pollution stress	Ozone, SO$_2$	Premature chlorosis Advanced stage of ozone injury in watermelon (Taken from PU, IN)	Seedless type	Mineral deficiency or toxicity and leaf diseases

in breeding for fruit quality, morphological trait improvement to date and biotic stress resistance (Wehner, 2012). Intensive crop improvement studies of metabolic, biochemical and physiological traits that are identified for major biotic and abiotic stresses including postharvest storage quality of fruit, and that were currently lacking, would be beneficial.

REFERENCES

Akashi K, Nishimura N, Ishida Y, Yokota A (2004). Potent hydroxyl radical-scavenging activity of drought-induced type-2 metallothionein in wild watermelon. *Biochemical and Biophysical Research Communications* 323(1):72–78.

Akashi K, Yoshida K, Kuwano M, Kajikawa M, Yoshimura K, Hoshiyasu S, Inagaki N, Yokota A (2011). Dynamic changes in the leaf proteome of a C3 xerophyte, *Citrullus lanatus* (wild watermelon), in response to water deficit. *Planta* 233(5):947–960. doi: 10.1007/s00425-010-1341-4.

Ali A (2011). Symptoms produced by various viruses on watermelon. Report with pictures from The University of Tulsa, Tulsa, OK. Accessed on January 27, 2015. http://www.nationalwatermelonassociation.com/pdfs/Symptoms%20produced%20by%20various%20viruses%20on%20watermelon.pdf.

Amnon L (2014). Genetic enhancement of watermelon, broccoli, and leafy brassicas for economically important traits. Vegetable Research USDA_ARS, 2700 Savannah highway, Charleston, SC, Annual Report.

An M-J, Wu F-Z, Liu B (2009). Study on the differentiation of physiological race from *Fusariun oxysporum* f. sp. *niveurn* and the resistance of different watermelon cultivars in Heilongjiang. *Journal of Shanghai Jiaotong University—Agricultural Science* 27(5):494–500.

Anonymous (2015). Watermelon. Accessed on February 13, 2015. http://ucanr.org/sites/nm/files/76631.pdf.

Antonia TY (2008). Breeding for resistance to powdery mildew race 2W in watermelon [*Citrullus lanatus* (Thunb.) Matsum. and Nakai]. A dissertation submitted to the Graduate Faculty of North Carolina State University, Raleigh, NC. In partial fulfillment of the requirements for the degree of Doctor of Philosophy, North Carolina State University, Raleigh, NC.

Atkinson NJ, Urwin PE (2012). The interaction of plant biotic and abiotic stresses: From genes to the field. *Journal of Experimental Botany* 63:3523–3544. doi: 10.1093/jxb/ers100.

Baker C, Webb S, Adkins S (2008). *Squash Vein Yellowing Virus, Causal Agent of Watermelon Vine Decline in Florida*. Department of Agriculture & Consumer Services, Gainesville, FL, pp. 1–4.

Balliu A, Sallaku G, Islami E (2014). Root pruning effects on seedlings' growth and plant stand establishment rate of watermelon grafted seedlings. *Proceedings of the Eurasian Symposium on Vegetable Greens: Acta Horticulturae* 1033:19–24.

Ban D, Žanić K, Dumičić G, Čuljak TG, Ban SG (2009). The type of polyethylene mulch impacts vegetative growth, yield, and aphid populations in watermelon production. *Journal of Food, Agriculture and Environment* 7(3,4):543–550.

Bates MD, Robinson RW (1995). Cucumbers, melons, and watermelons. In: J. Smartt and N.W. Simmonds (eds.), *Evolution of Crop Plants*. Longman Scientific & Technical, Essex, U.K., pp. 89–97.

Bernstein N, Kafkafi U (2002). Root growth under salinity stress. In: Y Waisel, A Eshel, and U Kafkafi (eds.), *Plant Roots: The Hidden Half*. CRC Press, Boca Raton, FL, pp. 787–805.

Bita CE, Gerats T (2013). Plant tolerance to high temperature in a changing environment: Scientific fundamentals and production of heat stress-tolerant crops. *Frontiers in Plant Science* 4:273. doi: 10.3389/fpls.2013.00273.

Bueno P, Varela J, Gimenez-Gallego G, Del Rio LA (1995). Peroxisomal copper, zinc superoxide dismutase. Characterization of the isoenzyme from watermelon cotyledons. *Plant Physiology* 108(3):1151–1160.

Cecilia M (2012). Quantitative trait loci (QTL) mapping of important traits in watermelon. College of Agriculture and Environmental Sciences, University of Georgia, Clark County, GA.

Chandrasekar SK, Shepard BM, Hassell R, levi A, Simmons, AM (2007). Potential sources of resistance to broad mites (*Polyphagotarsonemus latus*) in watermelon. *Germplasm Hort Science* 42(7):1539–1544.

Chen K, Sun JQ, Liu RJ, Li M (2013). Effect of arbuscular mycorrhizal fungus on the seedling growth of grafted watermelon and the defensive enzyme activities in the seedling roots. *Ying Yong Sheng Tai Xue Bao* 24(1):135–141. http://www.ncbi.nlm.nih.gov/pubmed/23718001.

Cicerone RJ (1987). Changes in stratospheric ozone. *Science* 237:35–42.

Colucci SJ, Holmes GJ (2010). Downy mildew of *cucurbits*. *The Plant Health Instructor*. APS, Valent USA corporation, NC, p. 1. doi: 10.1094/PHI-I-2010-0825-01.

Crop Profile, VA 2003 Crop Profile for Watermelon in Virginia. General Production Information. Prepared in April 2003, Accessed on January 29, 2015. http://www.ipmcenters.org/cropprofiles/docs/VAwatermelon.pdf.

Dane F, Liu J (2006). Diversity and origin of cultivated and citron type watermelon (*Citrullus lanatus*). *Genetic Resources and Crop Evolution* 54(6):1255. doi:10.1007/s10722-006-9107-3.

David M, Donald NM (2012). Chapter 6: Cucumbers, melons and watermelons. In: K.F. Kiple and K.C. Ornelas (eds.), *The Cambridge World History of Food, Part 2*. Cambridge University Press, Cambridge, U.K. doi: 10.1017/CHOL9780521402156. ISBN 9780521402156.

Egel DS, Martyn RD (2007). Fusarium wilt of watermelon and other cucurbits. The Plant Health Instructor. Department of Botany and Plant Pathology Purdue University, West Lafayette, IN. DOI: 10.1094/PHI-I-2007-0122-01. Updated 2013. (Accessed on January 30, 2015.)

Elmstrom GW, Hopkins DL (1981). Resistance of watermelon cultivars to *Fusarium* wilt. *Plant Disease* 65:825–827.

Everts KL (2002). Reduced fungicide applications and host resistance for managing three diseases in pumpkin grown on a no-till cover crop. *Plant Disease* 86:1134–1141.

Fan M, Huang Y, Zhong Y, Kong Q, Xie J, Niu M, Xu Y, Bie Z (2013). Comparative transcriptome profiling of potassium starvation responsiveness in two contrasting watermelon genotypes. *Planta* 239(2):397–410. doi: 10.1007/s00425-013-1976-z. Epub November 2, 2013.

Fiskell JG, Everett PH, Locascio SJ (1964). Minor element release from organic N fertilizer materials in laboratory and field studies. *Agriculture and Food Chemistry* 12:363–367.

Ghosh D, Xu J (2014). Abiotic stress responses in plant roots: A proteomic perspective. *Frontiers in Plant Science* 5:6.

Gimeno BS, Bermejo V, Reinert RA, Zheng Y, Barnes JD (1999). Adverse effects of ambient ozone on watermelon yield and physiology at a rural site in eastern Spain. *New Phytologist* 144:245–260.

Guang P, Hong S, Hong S, Sheng XZ (2006). Comparative analyses of some physiological characters of watermelon seedlings under 10°C and 15°C low temperature treatments. *Journal of Wuhan Botany Research* 24(5):441–445.

Guner N, PesicVan-Esbroeck Z, Wehner TC. (2008). Inheritance of resistance to Papaya ringspot virus-watermelon strain in watermelon. *J. Hered.* (in review).

Hall MR, Ghate SR, Phatak SC (1989). Germinated seeds for field-establishment of watermelon. *Horticultural Science* 24:236–238.

Henderson WR, Jenkins SF, and Rawlings JO (1970). The inheritance of Fusarium wilt resistance in watermelon, Citrullus lanatus (Thunb.) *Mansf. J. Amer. Soc. Hort. Sci* 95:276–282.

Hochmuth G, Hanlon Ed A (2015). A summary of N, P and K research with watermelon in Florida. Publication # 325/CV232, pp. 1–20. Accessed on January 30, 2015. http://edis.ifas.ufl.edu.

Hoscmuth GJ, Hanlon Ed A, Cornell J (1993). Watermelon phosphorous requirements in soils with low Mehlich-I-extractable phosphorus. *Horticultural Science* 28(6):630–632.

Inthichack P, Yasuyo Nishimura Y, Fukumoto Y (2014). Effect of diurnal temperature alternations on plant growth and mineral composition in cucumber, melon and watermelon. *Pakistan Journal of Biological Sciences* 17:1030–1036.

Jennings P, Saltveit ME (1994). Temperature effects on imbibition and germination of cucumber (*Cucumis sativus*) seeds. *Journal of the American Society of Horticultural Science* 119:464–467.

Kajikawa M, Morikawa K, Abe Y, Yokota A, Akashi K (2010). Establishment of a transgenic hairy root system in wild and domesticated watermelon (*Citrullus lanatus*) for studying root vigor under drought. *Plant Cell Report* 29(7):771–778. doi: 10.1007/s00299-010-0863-3.

Kang L, Cheng Y, Wang L-J (2006). Effect of 5-aminolevulinic acid (ALA) on the photosynthesis and antioxidative enzymes activities of the leaves of greenhouse watermelon in summer and winter. *Acta Botanica Boreali-Occidentalia Sinica* 26(11):2297–2301.

Kang L, Wang L-J (2008). Effects of ALA treatments on light response curves of chlorophyll fluorescence of watermelon leaves. *Journal of Nanjing Agricultural University* 31:31–36.

Kawasaki S, Miyake C, Kohchi T, Fujii S, Uchida M, Yokota A (2000). Responses of wild watermelon to drought stress: Accumulation of an ArgE homologue and citrulline in leaves during water deficits. *Plant and Cell Physiology* 41(7):864–873.

Keinath AP (1996). Soil amendment with cabbage residue and crop rotation to reduce gummy stem blight and increase growth and yield of watermelon. *Plant Disease* 80:564–570.

Kim M, Shim Ch, Kim YK, Jee HJ, Hong SJ, Park JH, Han EJ (2013). Evaluation of watermelon germplasm for resistance to phytophthora blight caused by *Phytophthora capsici*. *Plant Pathology Journal* 29(1):87–92. doi: 10.5423/PPJ.OA.02.2012.0031.

Kim PI, Chung, K (2004). Production of an antifungal protein for control of *Colletotrichum lagenarium* by *Bacillus amyloliquefaciens*. *FEMS Microbiology Letters* 234(1):177–183.

Korkmaz A, Dufault RJ (2002). Short-term cyclic cold temperature stress on watermelon. *Horticultural Science* 37(3):487–489.

Kozik EU, Wehner TC (2014). Tolerance of watermelon seedlings to low-temperature chilling injury. *Horticultural Science* 49(3):240–243.

Krupa SV, Kickert SN (1989). The greenhouse effect: Impacts of ultraviolet-B (UV-B) radiation, carbon dioxide (CO_2), and ozone (O_3) on vegetation. *Environmental Pollution* 61:263–393.

Lacasse NL, Treshow M (1976). *Diagnosing Vegetation Injury Caused by Air Pollution*. Air Pollution Training Institute Applied Science Association Inc., Research Triangle Park, NC.

Lebeda A, Cohen Y (2011). Cucurbit downy mildew (*Pseudoperonospora cubensis*)—Biology, ecology, epidemiology, host-pathogen interaction and control. *European Journal of Plant Pathology* 129:157–192. doi 10.1007/s10658-010-9658-1.

Levi A, Thomas CE, Keinath AP, Wehner TC (2001). Genetic diversity among watermelon (*Citrullus lanatus* and *Citrullus colocynthis*) accessions. *Genetic Resources and Crop Evolution* 48:559–566.

Liu W, Wang M, Yan Z (2003). Studies on physiological and biochemical characteristics of seedlings of different ploidy watermelons under cold-stress. *Journal of Fruit Science* 20(1):44–48.

Lewis WJ, Timothy PB, Barbara C. (2015). Watermelon Bacterial Fruit Blotch, Adapted from Latin, RX and D Hopkins 1995 Bacterial fruit blotch of watermelon. The hypothetical exam question becomes reality. *Plant Disease* 79(8):761–764.

Locascio SJ, Everett PH, Fiskell JG (1964). Copper as a factor in watermelon fertilization. *Florida State Horticultural Society* 77:190–194.

Lu J-F, Yu J-H (2004). Effects of SA on physiological indexes of chilling-tolerance in watermelon seedlings. *Journal of Gansu Agricultural University.*

Maynard DN, Hopkins DL (1999). Watermelon fruit disorders. *Horticultural Technology* 9(2):155–161.

McKersie BD (1996). Oxidative stress. Department of Crop Science, University of Guelph, Guelph, Ontario, Canada. Accessed on February 10, 2015. http://www.plantstress.com/articles/oxidative%20stress.htm.

Milotay PL, Kovac, Barta A (1991). The effect of suboptimum temperatures on cucumber seed germination and seedling growth. *Zoldsetermesztesi Kuutato Intezer Bullentinje* 24:33–45.

Mohammad S, Fardin G, Sajad F, Fatemeh B (2013). Chilling tolerance improving of watermelon seedling by salicylic acid seed and foliar application. *Notulae Scientia Biologicae* 5(1):67–73.

Muhammad JJ, Sung WK, Kim DH, Haider A (2006). Seed treatment and orientations affects germination and seedling emergence in tetraploid watermelon. *Pakistan Journal of Botany* 38(1):89–98.

Muqit A, Akanda AM, Kader KA (2007). Biochemical alteration of cellular components of ash gourd due to infection of three different viruses. *International Journal of Sustainable Crop Production* 2(5):40–42.

Netzer D, Weintall C. (1980). Inheritance of resistance to race 1 of Fusarium oxysporum f. sp. niveum. *Plant Dis.* 64:863–854.

Norton JD. (1979). Inheritance of resistance to gummy stem blight in watermelon. *HortScience* 14:630–632.

Parris GK (1949). Diseases of watermelon in Florida. Watermelon and Grape Investigations Laboratory, Bulletin 459, Florida Agricultural Experimental Station, Gainesville, FL. http://fshs.org/proceedings-o/1949-vol-62/146–148(PARRIS).pdf.

Peng J, Sun Y, Shi R, Xie G (2006). Effects of 10°C low temperature on physiological index of hardness in watermelon seedlings. *Anhui Agricultural Science Bulletin*, p. 10.

Provvidenti R. (1992). Cold resistance in accessions of watermelon from Zimbabwe. *Cucurbit Genet. Coop. Rpt.* 15:67–69.

Provvidenti R. (2003). Naming the gene conferring resistance to cool temperatures in watermelon. *Cucurbit Genet. Coop. Rpt.* 26 (in press).

Rivero RM, Ruiz JM, García PC, López-Lefebre LR, Sánchez E, Romero L (2001). Resistance to cold and heat stress: Accumulation of phenolic compounds in tomato and watermelon plants. *Plant Science* 160(2):315–321.

Roberts W, Motes J, Edelson J (1914). Watermelon production. Oklahoma Cooperative Extension Service Division of Agricultural Sciences and Natural Resources, Oklahoma State University, USA, HLA 6236, pp. 1–4. http://pods.dasnr.okstate.edu/docushare/dsweb/Get/Document-1110/F-6236web.pdf. (Accessed on January 15, 2015.)

Sachs M (1977). Priming of watermelon seeds for low temperature germination. *Horticultural Science* 102:175–178.

Sandhu PS, Kang SS, Kaul VK (2007). Biochemical and histological deviations induced in muskmelon by mosaic disease and its impact on yield and quality. *Indian Journal of Virology* 18(2):79–82.

Savory EA, Granke LL, Quesada-Ocampo LM, Varbanova M, Hausbeck MK, Day B (2011). The cucurbit downy mildew pathogen *Pseudoperonospora cubensis*. *Molecular Plant Pathology* 12(3):217–226.

Savvas D, Colla G, Rouphael Y, Schwarz D (2010). Review: Amelioration of heavy metal and nutrient stress in fruit vegetables by grafting. *Scientia Horticulturae* 127(2010):156–161.

Sharma P, Jha AB, Dubey RS, Pessarakli M (2012). Reactive oxygen species, oxidative damage, and antioxidative defense mechanism in plants under stressful conditions. *Journal of Botany* 26:Article ID 217037. doi:10.1155/2012/217037.

Simon JE, Decoteau DR, Simini M (1988). Identifying air pollution damage on melons. HO-192, Purdue University, Cooperative Extension Service, West Lafayette, IN. https://www.extension.purdue.edu/extmedia/HO/HO-192.html.

Singh S, Singh P, Sanders DC, Wehner TC (2001). Germination of watermelon seeds at low temperature. *Cucurbit Genetics Cooperative Report* 24:59–64.

Song R, Gusmini G, Wehner TC (2004). Screening the watermelon germplasm collection for resistance to gummy stem blight. *Acta Horticultural* 637:63–68.

Strange EB, Guner N, Pesic-VanEsbroeck Z, Wehner TC. (2002). Screening the watermelon germplasm collection for resistance to Papaya ringspot virus type-W. *Crop Sci*. 42:1324–1330.

Synder RG, Simon JE, Reinert RA, Simini M, Wilcox GE (1991). Effects of air quality on growth, yield and quality of watermelon. *HortScience* 26:1045.

Ting L, Xian Z, Yonxin W (2009). Biochemical and physiological changes of different watermelon cultivars inoculated by *Colletotrichum orbiculare*. Institute of Agricultural Information, Chinese Academy of Agricultural Sciences, AGRIS, Food and Agricultural Organization of United Nations, 37(2). http://agris.fao.org/agris-search/search.do?recordID=CN2010000062. (Accessed on January 18, 2015.)

USDA (2014). Crop Production. Todays reports from National agricultural statistics service. Accessed on October 27, 2014. http://www.nass.usda.gov/Publications/Todays_Reports/reports/vgan0314.pdf.

Vijayarengan P, Jose MD (2014). Changes in growth, pigments and phytoremediating capability of four plant species under copper stress. *International Journal of Environmental Biology* 4(2):119–126.

Vollenweider P, Günthardt-Goerg MS (2005). Diagnosis of abiotic and biotic stress factors using the visible symptoms in foliage. *Environmental Pollution* 137:455–465.

Wahid A, Gelani S, Ashraf M, Foolad MR (2007). Review: Heat tolerance in plants: An overview. *Environmental and Experimental Botany* 61:199–223.

Wang JM, Guo CR, Zhang ZG, He YC, Li WY (2002). Biochemical and physiological changes of different watermelon cultivars Infected by *Fusarium oxysporum*. *Scientia Agricultura Sinica*, p. 11.

Wang LJ, Sun YP, Zhang ZP, Kang L (2010). Effects of 5-Aminolevulinic acid (ALA) on photosynthesis and chlorophyll fluorescence of watermelon seedlings grown under low light and low temperature conditions. *Acta Horticultural (ISHS)* 856:159–166. http://www.actahort.org/books/856/856_21.html.

Wehner TC (2003). Watermelon. *World Book Encyclopedia*, Chicago, IL, Vol. 146, pp. 368–405. http://cuke.hort.ncsu.edu/cucurbit/wehner/articles/book16.pdf. (Accessed on January 28, 2015.)

Wehner TC. (2012). Gene list for watermelon. Cucurbit Genetics Cooperative, Department of Horticulture, North Carolina State University, Raleigh, NC. http://cuke.hort.ncsu.edu/cgc/cgcgenes/wmgenes/gene12wmelon.html (Accessed on October 12, 2015.)

Webb SE, Riley DG, Brust GE. (2001). in *Watermelons*. Characteristics, production, and marketing, Insect and mite pests, ed Maynard D.N. ASHS Press, Alexandria, VA, pp. 131–149.

Xu Y, Kang D, Shi Z, Shen H, Wehner T (2004). Inheritance of resistance to zucchini yellow mosaic virus and watermelon mosaic virus in watermelon. *Journal of Heredity* 95(6):498–502.

Yokota Y, Kawasaki S, Iwano M, Nakamura C, Miyake C, Akashi K (2002). Citrulline and DRIP-1 protein (ArgE homologue) in drought tolerance of wild watermelon. *Annals of Botany* 89(7):825–832. doi: 10.1093/aob/mcf074.

Yoshimura K, Masuda A, Kuwano M, Yokota A, Akashi K (2008). Programmed proteome response for drought avoidance/tolerance in the root of a C3 xerophyte (wild watermelon) under water deficits. *Plant and Cell Physiology* 49(2):226–241. doi: 10.1093/pcp/pcm180.

Zhang X, Qian X, Gu Q (1991). Studies on the methods for identifying watermelon on resistances to Fusarium wilt, Institute of Agricultural Information, Chinese Academy of Agricultural Sciences, 8:4.

Zhang, Z.-Q. (2003) Mites of greenhouses, identification, biology and control. CABI Publ. Internat, Wallingford, UK.

Zhao-Ming C et al. (2009). Study on different identification methods of anthracnose resistance of watermelon at seeding stage. *Anhui Agricultural Science Bulletin* 15. http://en.cnki.com.cn/Article_en/CJFDTotal-AHNB200915073.htm.

Zhou XG, Everts KL (2006). Suppression of Fusarium wilt of watermelon enhanced by hairy vetch green manure and partial cultivar resistance. Online *Plant Health Progress*. doi: 10.1094/PHP-2006-0405-01-RS, The American Phytopathological Society, St. Paul, MN. pp. 99–100.

Section XI

Examples of Cucurbit Crop Plants Growth and Development and Cultural Practices

26 Physiological Stages of Growth and Development in Bitter Melon

Satya S. Narina

CONTENTS

26.1 INTRODUCTION

Bitter melon, *Momordica charantia* (2n = 22), cucurbitaceous vegetable crop. It is also called bitter gourd, balsam pear, fukwa, karela, nigai uri, and ampalaya. It is normally grown in hot, humid areas as an annual crop and as a perennial in mild areas and during frost-free winters. Its cultivation initially started in South and East Asia and later spread to nontraditional climatic zones of other countries. California and Florida are the main producers of bitter melon in the United States, with an average yield of 5–7 t/acre. A considerable volume of bitter melon is also imported from Mexico, Dominican Republic, and Honduras.

The bittermelon fruit has vitamins A, B, and C, iron (Fe), minerals, phosphorous (P), and dietary fiber. Every 100 g edible portion of the plant contains 83–92 g water, 1.5–2 g protein, 0.2–1 g fat, 4–10.5 g carbohydrates, and 0.8–1.7 g fiber. Its energy value is 105–250 kJ/100 g. It is high in minerals such as Calcium (20–23 mg), Fe (1.8–2 mg), P (38–70 mg), and vitamin C (88–96 mg) (Tolentino and Cadiz, 2005).

Different parts of bittermelon extracts are reported to have many pharmacological activities with potential medical components, such as the ribosome inactivating protein (RIP); MAP30 (Momordica anti-HIV protein), which suppresses human immunodeficiency virus (HIV) activity;

M. charantia lectin (MCL); *M. charantia* inhibitor (MCI); and momordicosides A and B, which can inhibit tumor growth. The biologically active proteins of momordin, α- and β-momorcharin, and momordicine have highly effective antidiabetic, antiviral, antibacterial, anti-inflammatory, and anti-rheumatic properties to function as febrifuge medicine for several human diseases. Leaf and fruit extracts from bitter melon were used as attractants for insect pests like *Epilachna dodecastigma* and to induce allelopathic stress in other vegetable crops.

Another cultivated species, *Momordica dioica* (bitter-less bitter gourd, kakrol, teasle gourd, 2n = 28) is a perennial, dioecious climber with thickened roots. Bitter-less bitter gourd produces relatively small and oval shaped edible fruit vegetable rich in Ca, Fe, P and carotenoids. This fruit is in demand for internal as well as external markets due to its medicinal properties (Jeffrey, 1990).

26.1.1 History, Nomenclature, and Taxonomy

The genus *Momordica* belongs to the subtribe Thladianthinae, tribe Joliffieae, subfamily Cucurbitoideae, family Cucurbitaceae. The genus *Momordica* has 45 species domesticated in Asia and Africa (Robinson and Decker-Walters, 1997). The genus *Momordica* was grouped into monoecious (*M. charantia* L. and *M. balsamina* L.) and dioecious (*M. dioica* Roxb., *M. sahyadrica* Joseph and Antony, *M. cochinchinensis* [Lour.] Spreng., and *M. subangulata* Blume ssp. renigem [G. Don] W.J.J. deWilde) groups in India (Behra et al., 2015).

Indian bitter gourd was again classified into two botanical varieties based on fruit size, shape, color, and surface texture: (1) *M. charantia* var. *charantia* has large fusiform fruits, which do not taper at both ends and possess numerous triangular tubercles giving the appearance of a "crocodile's back"; and (2) *M. charantia* var. *muricato* (wild), which develops small and round fruits with tubercles, more or less tapering at each end (Chakravarty, 1990). Both varieties are widely cultivated throughout tropical and subtropical regions of India.

Yang and Walters (1992) classified bitter gourd in China into three horticultural groups or types:

1. *A small-fruited type*: Fruits are 10–20 cm long, 0.1–0.3 kg in weight, usually dark green, and very bitter.
2. *A long-fruited type*: Fruits are 30–60 cm long, 0.2–0.6 kg in weight, light green in color with medium-size protuberances, and only slightly bitter.
3. *A triangular-fruited type*: Cone-shaped fruits are 9–12 cm long, 0.3–0.6 kg in weight, light to dark green with prominent tubercles, and moderately to strongly bitter.

Reyes et al. (1994) reclassified Indian and southeast Asian *M. charantia* as var. *minima* (<5 cm) and maxima (>5 cm) based on fruit diameter. Efforts made in crossing between various species lead to either sexually incompatible interspecific crosses (Singh, 1990) or failed to set fruit (Vahab and Peter, 1993).

26.2 PHYSIOLOGICAL STAGES OF GROWTH AND DEVELOPMENT I

Bittermelon is a rapidly growing herbaceous vine with thin stems and tendrils which needs to be trellised, to provide support for the climbing vine. The trellis should be 6 ft high, 4–6 ft apart. The seed can be planted directly or grown as seedling at 1.5–2 ft spacing and between row spacing of 3–5. The summer season crop is sown from January to June in the plains (Singh et al., 2006).

26.2.1 Germination and Emergence

An optimum temperature of 25°C–28°C is required for the seed germination (Peter et al., 1998). Field emergence is always a problem even with seeds of high germinability due to thick seed coat. To overcome this problem and reduce the seed rate, presowing seed treatments (soaking or priming) are in practice. Pregerminated seeds are proven superior in emergence and establishment. Seed soaking is successful in seedling establishment under suboptimal temperature conditions (Wang et al., 2002).

Seed germination was high (80%–90%) in household vinegar (pH 3.7) and potassium nitrate (KNO_3, 0.2%–0.3%) and was negatively associated with storage time (Fonska and Fonska, 2011). The pregermination was high (100%) when seeds were soaked in panchakavya (@3% for 9 h) for 7 days compared to 2% KNO_3 (Thirusenduraselvi and Jerlin, 2007). The highest germination (85.18%), number of branches/plant (8.64%), and number of fruits/plant (20.70%) were obtained when the bitter gourd seeds were soaked for 12 h. Earlier emergence (6.28%) and earlier flowering (39.40%) were recorded in plants where seeds are presoaked for 16 h. The cultivar 'Palee' significantly enhanced the germination (85.56%), days to flowering (39.55%), number of branches/plant (8.86%), and fruits/plant (21.09%). Seed soaking in water for 12 h has the potential to improve germination and seedling growth of bitter gourd cultivars (Saleem et al., 2014). An efficient system for somatic embryogenesis and organogenesis using leaf callus and petiole in *M. charantia* was reported (Thiruvengadam et al., 2012).

26.2.2 SEEDLING GROWTH AND PHOTOSYNTHESIS

Seed treatment with a carbon-based nanoparticle, fullerol [$C_{60}(OH)_{20}$], resulted in increase in biomass yield (54%) and water content (24%). Increase in fruit length (20%), fruit number (59%), and fruit weight (70%) led to an improvement of up to 128% in fruit yield. Contents of two anticancer phytomedicines, cucurbitacin-B (74%) and lycopene (82%), were observed enhanced the contents of two antidiabetic phytomedicines, charantin and insulin, were augmented up to 20% and 91%, respectively (Kole et al., 2013). Accumulation of root sugars is independent of leaf sugar in flooding-tolerant bitter melon (Cv. New Known You #3) under flooded conditions (Su et al., 1998).

26.2.3 VEGETATIVE GROWTH

Exogenous application of Napthalene acetic acid (NAA) and Gibberellic acid (GA_3) slightly increased the alkaloid content of bittermelon fruits but slightly decreased their foliar alkaloid content. NAA treatment slightly decreased foliar chlorophyll, and GA3 slightly increased foliar chlorophyll (Tolentino and Cadiz, 2005).

Fatty acids was observed in young, mature, and senescent leaves of bitter melon representing total fatty acids of 87.30%, 95.25%, and 83.11% respectively. The proportion of saturated fatty acids was highest in senescent leaves (78.60%), followed by young (69.42%) and mature leaves (48.92%), with the balance accounted for by unsaturated fatty acids. Palmitic acid was the predominant saturated fatty acid in the three types of leaves, whereas alpha-linolenic acid was the predominant unsaturated fatty acid (Sarkar et al., 2013).

Larger varieties (Big Top Medium, Hanuman, Jade and White) were more productive than the small varieties (Indra and Niddhi) in terms of total fruit weight and yield per flower pollinated in bittermelon when grown in greenhouse. The bioactivity (total phenolic and saponin compounds and antioxidant activity) of the two small varieties and Big Top Medium was significantly higher than that of the other three large varieties (Tan et al., 2014).

26.3 PHYSIOLOGICAL STAGES OF GROWTH II

26.3.1 REPRODUCTIVE GROWTH

26.3.1.1 Flowering

Flowering starts from 45–55 days after sowing, and will continue for about 6 months. In the Northern Hemisphere, flowering occurs during June to July. Long days cause male flowers to bloom up to 2 weeks before female flowers, while short days have the opposite effect. Spraying vines with flowering hormones after they have six to eight true leaves will increase the number of female flowers and can double the number of fruits. An application of gibberellic acid at 25–100 ppm increases female flowers by 50% and can work for up to 80 days (Palada and Chang, 2003).

26.3.1.2 Fruit Setting

Flowers are cross pollinated by insects in the Northern Hemisphere, fruiting occurs during August to November. The fruit has a distinct warty exterior and an oblong shape. It is hollow in crosssection, with a relatively thin layer of flesh surrounding a central seed cavity filled with large, flat seeds and pith. The fruit is most often eaten green or as it is beginning to turn yellow and turns orange when fully ripe.

26.3.1.2.1 Biochemical changes during various stages of fruit development

Fruits were studied for physiological and biochemical changes during seven sequential developmental stages, starting from very young stage to postripened stage. A gradual decrease of chlorophyll-a (5.25-fold), chlorophyll-b (13.0-fold), total chlorophyll (8.23-fold), starch (6.5-fold), and free amino acids (14.4-fold), total proteins (67.2%) and RNA (55.1%) from premature stage to postripened stage. However, a gradual increase was observed in carotenoids (2.5-fold) and total sugars (209%), nonreducing sugars (317%), and phenol (2.9-fold), from very young stage to preripened stage.

The quantities of anthocyanins, reducing sugars, and DNA was observed unstable.

26.3.1.2.1.1 Hydrolytic enzymes (amylase, invertase, and peroxidase) The activity of amylase decreased from mature stage to postripened stage (86.73%). The activities of invertase and peroxidase were decreased (94.3%) at very young stage and increased (44.64 %) at mature stage.

Cell wall degrading enzymes (such as cellulase, polygalacturonase [PG], and pectin methyl esterase [PME]): 6.38-fold increased activity of cellulase was observed from very young stage to ripened stage. The PG activity gradually increased 2.37-fold, and an inconsistent activity of PME was noticed during ripening. These biochemical changes also reflect cellular changes such as increasing cell number, enlargement of cell, decreasing cell content, and separation of middle lamella, which were observed during the mature stage to postripened stage (Shah and Rao, 2013).

The influence of ripening stages on the phenolic bioactive substances and the corresponding antioxidant activity of bitter melon (*M. charantia*) showed increased ferric reducing antioxidant power (FRAP), chroma (lightness and yellowish), total phenolic content, and decreased 2,2-diphenyl-1-picrylhydrazyl (DPPH) activity as the ripening advanced (Amina and Anna, 2011). Chlorophyll, protein, and starch content were decreased, and carotenoids, anthocyanins, and sugars increased during ripening. The bittermelon fruit treated with 1-methylcyclopropene (1-MCP) for 12 h and 70 µM calcium pectate has showed delayed ripening (Anbarasan and Tamilmani, 2013a,b).

Exogenous application of NAA and GA$_3$ at the five-leaf stage induced parthenocarpy in bittermelon (Tolentino and Cadiz, 2005). It decreased fruit length and fresh weight of plants when treated with 20 ppm NAA and increased fruit diameter in plants when treated with 100 ppm NAA was observed. GA$_3$ treatment had the decreased fresh weight and diameter in plants treated with 20 ppm GA$_3$ (Tolentino and Cadiz, 2005). Seedless condition was the reason for reduced fresh weight and diameter of the fruit.

26.3.1.3 Harvesting

Fruits can be harvested at any stage of development but are typically harvested full sized but green, about 2 weeks after anthesis. At this stage of maturity, the pulp and seed color is white to creamy-white. The bitter gourd can yield up to 40 t/ha, with an average of 8–10 t/ha. The fruits are normally harvested every 2–3 days manually by cutting the stem vine with a knife or shear. Pulling the fruit vine stem could incur root damage leading to senescence of the leaves, flowers, and developing fruits. The bitter gourd fruit has high moisture content, a large surface:volume ratio, and a relatively thin cuticle, which makes it very susceptible to moisture loss and physical injury.

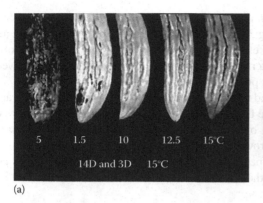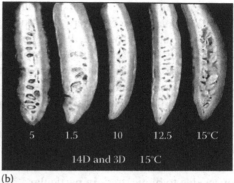

(a)   (b)

FIGURE 26.1   (See color insert.) Visual quality of bitter melon (a, b), after storage at different temperatures for 14 days. Fruits were evaluated 3 days after transfer to 15°C (59°F). (Zong et al., 1993 from UC, CA.)

26.3.1.4   Postharvesting

Bittermelon fruits are climacteric, have a relatively high respiration rate and therefore prompt removal of the field heat via hydrocooling or room cooling is recommended. Respiration rates of fruits at 20°C and 10°C were approximately 40 and 15 L CO_2/g/h, respectively. Ethylene production rates at these temperatures were 0.1–0.3 nL/g/h (Zong et al., 1995). Bittermelon fruits have a marketable shelf life of 7–12 days when stored at 10°C–12.5°C temperature (Zong et al., 1993) and 85%–90% relative humidity (RH).

Fruits stored at 15°C continued to develop, showing undesirable changes including seed development, loss of green color, and fruit splitting. The immature fruit maintained postharvest quality better than fruits harvested at the fully developed green stage. Fruits stored at 5°C–7°C developed visible symptoms of pitting, which eventually coalesced to form larger sunken dark brown pits in the ribbed regions. Other chilling injury symptoms include surface discoloration from green to dark brown, secondary infections, russetting, and internal tissue breakdown (Figure 26.1).

Bitter melons stored at 15°C in controlled atmospheres (21%, 5%, or 2.5% O_2 in combination with 0%, 2.5%, 5%, or 10% CO_2) were not different in quality from air-stored fruits at 2 weeks. Fruits stored 3 weeks in 2.5% or 5% CO_2 in combination with 2.5% O_2 showed greater retention of green color and had less decay and splitting than air-stored fruits (Zong et al., 1995).

26.3.2   Senescence

Female plants senesced earlier than male plants. The chlorophyll levels of defruited and male plants remained higher than those of female and monoecious plants, indicating the senescence pattern as follows: female > monoecious > male > defruited. Maximum deferment of senescence in both defruited and male plants was due to the absence of fruits. Fruiting, may be the initiator of the senescence signal. The senescence signal developed in the fruits migrated downward for the induction of leaf senescence. Protein content in the leaves and dry weight of aerial plant parts remained higher compared to defruited and monoecious plants (Ghosh, 2005).

26.4   GROWTH RESPONSES OF BITTER MELON UNDER PHYSIOLOGICAL STRESSES

26.4.1   Biotic Stress

Biotic stress in crop plants is caused due to infestation with insect pests, fungal, bacterial and viral diseases. Bittermelon is infected by watermelon mosaic potyvirus, downy mildew, *Cercospora* leaf spot, bacterial wilt, *fusarium* wilt, and root knot nematode. The melon fruit fly *Bactrocera cucrbitaceae* (Coquillett) is widely distributed in temperate, tropical, and subtropical regions of the world and is a major pest of bittermelon causing 30 to 100% loss depending on the infestation and environment (Gupta and Verma, 1992; Sapkota et al., 2010). Beetles, thrips, cutworms, bollworm, aphids, and mites are other common pests of bittermelon. Temperature below 32°C and RH between 60% and 70% favor the abundance of the insect population.

Early primary resistance mechanism observed to combat nematode attack in bittermelon is rapid augmentation in the protein concentration in the first week and total sugar level in the fourth week of inoculation in infected roots. Greater influx of sugars was observed in the attacked tissue for providing nutrition to the nematodes for their growth and survival (Surendra and Aditi, 2014).

No published data are available on major pest or disease incidence or resistance developed in response to biotic stress in bitter melon in the United States and information on biochemical responses due to biotic stress in particular.

26.4.2 ABIOTIC STRESS

Abiotic stress is caused due to temperature fluctuations (low, high and freezing), flooding, drought, salinity and pollution. Bittermelon fruits can be classified as moderately sensitive to chilling injury. Bitter melon can be successfully marketed for 1–2 weeks if kept at 10°C–12.5°C (50°F–55°F) during postharvest handling to reduce deterioration and to avoid chilling injury (Zong et al., 1993).

Bittermelon exhibited protection mechanism against oxidative damage by maintaining a highly induced antioxidant system under three stresses, (Sodium Chloride (NaCl), Ultraviolet radiation (UV-B) and water stress) at three different stages (pre-flowering, flowering and post flowering) of plant growth (Agarwal and Shaheen, 2007). Except for peroxidase (POX), all enzyme activities including superoxide dismutase (SOD), catalase (CAT), polyphenol oxidase (PPO), glutathione reductase (GR), and concentrations of ascorbate (ASA), hydrogen peroxide (H_2O_2), and thiobarbituric acid reactive substances (TBARS) were maximum at the flowering stage under all three stresses. All the enzyme activities, SOD, CAT, POX, PPO, GR, and the concentrations of ASA, H_2O_2, and TBARS were elevated under NaCl and UV-B stresses at all growth stages with the exception of H_2O_2 concentration at the postflowering stage under UV-B radiation. Greater quantities of the inorganic ions Na+ and Cl− were accumulated at all growth stages under salt stress.

Drought led to decreases in the concentrations of H_2O_2 and ASA and activities of PPO and GR and elevated concentrations of TBARS and activities of SOD, CAT, and POX at all three stages in comparison with control values (Agarwal and Shaheen, 2007). The photosynthetic pigments decreased at all stages under all stresses. The chlorophyll stability index decreased under NaCl stress, accelerated only at the postflowering stage under UV-B radiation, and significantly increased at pre- and postflowering stages under water stress. Protein concentration was reduced under NaCl stress (except at the pre-flowering stage), and increased under UV-B and water stresses.

Exogenous silicon application may increase germination rate, germination index, and vitality index, and contributes reduced melanaldehyde contents and increased antioxidant enzyme (SOD, POD, and CAT) activities under NaCl stress (Wang et al., 2010).

Application of Zinc Sulphate and Sodium Borate ($ZnSO_4 \cdot 7H_2O$ and $Na_2B_4O_7 \cdot 10H_2O$) on Zinc (Zn) and Boron (B)-deficient soils could increase bittermelon yield and its protein, vitamin, and amino acid contents, and decrease the $NO_3(-)$–N content and delayed senescence. The nutrients Zn and B increased polyamines (PAs), putrescine (Put), spermidine (Spd), spermine (Spm), indole acetic acid (IAA), gibberellic acid (GA3), and ascorbic acid (ASA) contents and SOD, POD, and CAT activities, and reduce malonaldehyde (MDA) and abscisic acid (ABA) contents in leaves, which inhibited membrane lipid peroxidation (Shi and Cheng, 2004).

Flooding treatment reduced leaf photosynthetic rate, stomatal conductance, transpiration, soluble protein, and activity of ribulose-1,5-bisphosphate carboxylase/oxygenase (rubisco) and these flooding effects were milder in grafted than those on ungrafted bitter melon. No significant changes were observed for leaf internal CO_2 concentration, percent activation of rubisco, starch content, and leaf dark respiration rate (Liao, 1996). In Taiwan, the yield of bittermelon is increased by grafting with luffa (*Luffa* spp.) as luffa resists fusarium wilt and is more tolerant to flooding (Liao and Lin, 1996; Lin et al., 1998).

26.5 CONCLUSION

Majority of the research was concentrated around tissue culture for regeneration and for studying biomedical properties of the fruit, being a medicinally important fruit vegetable. Research done in the 1900s was on postharvest storage studies, and this could be due to production difficulties during this period in the United States. The bitter melon crop cultivation in southern United States, California, and Arizona in field climate is producing higher yields with better crop growth. There were very less or no research studies on stress responses that would benefit for crop improvement and their production in resource-free climatic zones for wide adaptability.

Most of the bittermelon production in temperate zones like the United States and Australia were confined to greenhouse and high tunnels, mainly for fruit quality, and this might be due to less preference for it as a fresh vegetable compared to other industrial uses of the fruit. There is no publicly available literature on structural and biochemical changes during growth and development of bitter melon under various stresses in particular and not much genetic improvement done for resistance to insect (fruit fly), nematode and other diseases. Research efforts toward identification of cultivars with various abiotic stresses specifically high/freezing/low temperature/water stress tolerances would be helpful for this economically viable crop as temperature induces several plant metabolic changes due to oxidative stress, which leads to changes in accumulation of nutrients and active ingredients in the fruit.

REFERENCES

Agarwal S, Shaheen R (2007). Stimulation of antioxidant system and lipid peroxidation by abiotic stresses in leaves of *Momordica charantia*. *Brazilian Journal of Plant Physiology* 19(2):149–161. doi.org/10.1590/S1677-04202007000200007.

Aminah A, Anna PK (2011). Influence of ripening stages on physicochemical characteristics and antioxidant properties of bitter gourd (*Momordica charantia*). *International Food Research Journal* 18(3):895–900.

Anbarasan A, Tamilmani C (2013a). Effect of calcium pectate on the biochemical and pigment changes during the ripening of bitter gourd fruit (*Momordica charantia* L. var-Co-1). *Asian Journal of Plant Science and Research* 3(4):51–58.

Anbarasan A, Tamilmani C (2013b). Effect of 1-Methylcyclopropene (1-MCP) on the pigment changes during ripening of bitter gourd fruit (*Momordica charantia* L. var. Co-1). *International Journal of Agricultural and Food Science* 3(2):44–49. http://www.urpjournals.com.

Behra TK, Behra S, Bharathi LK, John KJ, Simon PW, Staub JE (2015). Bitter gourd: Botany, horticulture, breeding. In: J. Janick (ed.), *Horticultural Reviews*, Vol. 37, pp. 101–141. John Wiley & Sons, Inc., Hoboken, NJ. http://naldc.nal.usda.gov/download/42264/PDF. Accessed on January 28, 2015.

Chakravarty HL (1990). Gucurbits of India and their role in the development of vegetable crops. In: D.M. Bates, R.W. Robinson and C. Jeffrey (eds.), *Biology and Utilization of CucvrbiLoceao*, pp. 325–334, Cornell University Press, Ithaca, NY.

Fonseka HH, Fonseka RM (2011). Studies on deterioration and germination of bitter gourd seed (*Momordica charantia* L.) during storage. *Acta Horticulturae* (*ISHS*) 898:31–38. http://www.actahort.org/books/898/898_2.htm.

Ghosh A (2005). Mechanism of monocarpic senescence of *Momordica dioica*: Source-sink regulation by reproductive organs. *Pakistan Journal of Scientific and Industrial Research* 48(1):55–56.

Gupta D and Verma AK (1992). Population fluctuations of the maggots of fruitflies Daucus cucrbitae Coquillett and D. tau (walker) infesting cucrbitaceous crops. *Adv.Pl.Sci.* 5:518–523.

Jeffrey C (1990). An outline classification of the Cucurbitaceae (Appendix). In: D.M. Bates, W.R. Robinson, and C. Jeffrey (eds.), *Biology and Utilization of the Cucurbitaceae*, pp. 449–463, Cornell University Press, Ithaca, NY.

Kole C, Kole P, Randunu KM, Choudhary P, Podila R, Ke PC, Rao AP, Marcus RK (2013). Nanobiotechnology can boost crop production and quality: First evidence from increased plant biomass, fruit yield and phytomedicine content in bitter melon (*Momordica charantia*). *BMC Biotechnology* 13:37. doi: 10.1186/1472-6750-13-37.

Liao ChT (1996). Photosynthetic responses of grafted bitter melon seedlings to flood stress. *Environmental and Experimental Botany* 36(2):167–172. doi:10.1016/0098-8472(96)01009-X.

Liao ChT, Lin CH (1996). Photosynthetic responses of grafted bitter melon seedlings to flood stress. *Environmental and Experimental Botany* 36(2):167–172.

Lin YS, Hwang CH, Soong SC (1998). Resistance of bitter gourd-luffa grafts to *Fusarium oxysporum* f.sp. *momordicae* and their yield. [Chinese] *Chih Wu Pao Hu Hsueh Hui Hui K'An* 40(2):121–132.

Palada MC, Chang LC (2003). Suggested cultural practices for bittergourd. International Cooperators Guide, AVRDC Pub#03-547. http://203.64.245.61/web_crops/cucurbits/bittergourd.pdf. Accessed on January 28, 2015.

Peter KV, Sadhu MK, Raj M, Prasanna KP (1998). Improvement and cultivation of bitter gourd, snake gourd, pointed gourd and ivy gourd. In: NM Nayar and A More (eds.), *Cucurbits*, pp. 187–195, Science Publishers Inc., Enfield, NH.

Reyes MEC, Gildemacher BH, Jansen GJ (1994). *Momordica* L. In: JS Siemonsma and K Piluek (eds.), *Plant Resources of South-East Asia: Vegetables*, pp. 206–210, Pudoc Scientific Publishers, Wageningen, the Netherlands.

Robinson RW, Decker-Walters DS (1997). *Cucurbits*, CAB International, Oxford, U.K.

Saleem MS, Sajid M, Ahmed Z, Ahmed S, Ahmed N, Islam, SMUI (2014) Effect of seed soaking on seed germination and growth of bitter gourd cultivars. *IOSR Journal of Agriculture and Veterinary Science* 6(6):7–11.

Sapkota R, Dahal KC, and Thapa RB (2010). Damage assessment and management of cucurbit fruit flies in spring-summer squash. *Journal of Entomology and Nematology* 2(1):007–012.

Sarkar N, Mukherjee A, Barik A (2013). Long-chain alkanes: Allelochemicals for host location by the insect pest, *Epilachna dodecastigma* (Coleoptera: Coccinellidae). *Applied Entomology and Zoology* 48(2):171–179.

Shah PT, Rao VR (2013). Physiological, biochemical and cellular changes associated with the ripening of bitter less bitter gourd (*Momordica dioica* Roxb. Ex Willd) fruits. *The International Journal of Engineering and Science* 2(7):1–5.

Shi M, Cheng R (2004). Effects of zinc and boron nutrition on balsam pear (*Momordica charantia*) yield and quality, and polyamines, hormone, and senescence of its leaves. *Ying Yong Sheng Tai Xue Bao* 15(1):77–80.

Singh AK (1990). Cytogenetics and evolution in the cucurbitaceae. In: DM Bates, RW Robinson, and C Jeffrey (eds.), *Biology and Utilization of Cucurbitaceae*, pp. 11–28, Cornell University Press, Ithaca, NY.

Singh NP, Singh DK, Singh YK, Kumar V (2006). *Vegetables Seed Production Technology*, 1st edn., pp. 143–145, International Book Distributing Co., Lucknow, India.

Su PH, Wu TH, Lin CH (1998). Root sugar level in luffa and bitter melon is not referential to their flooding tolerance. *Botanical Bulletin of Academia Sinica* 39:175–217.

Surendra KG, Aditi NP (2014). Study on protein and sugar content in *Meloidogyne incognita* infested roots of bitter gourd. *International Journal of Current Microbiology and Applied Sciences* 3(5):470–478.

Tan SP, Parks SE, Stathopoulos CE, Roach PD (2014). Greenhouse-grown bitter melon: Production and quality characteristics. *Journal of the Science of Food and Agriculture* 94(9):1896–903. doi:10.1002/jsfa.6509.

Thirusenduraselvi D, Jerlin R (2007). Effect of pre-germination treatments on the emergence percentage of bitter gourd Cv. CO-1 seeds. *Tropical Agricultural Research and Extension* 10: 88–89. doi:10.4038/tare.v10i0.1877.

Thiruvengadam M, Chung IM, Chun SC (2012). Influence of polyamines on in vitro organogenesis in bitter melon (*Momordica charantia* L.). *Journal of Medicinal Plants Research* 6(19):3579–3585.

Tolentino MF, Cadiz NM (2005). Effects of naphthaleneacetic acid (NAA) and gibberellic acid (GA3) on fruit morphology, parthenocarpy, alkaloid content and chlorophyll content in bittergourd (*Momordica charantia* L. "Makiling"). *The Philippine Agricultural Scientist* 88(1):35–39.

Vahab MA, Peter KV (1993). Crossability between *Mormordica charantia* and *Moniordica dioica*. Cucurbit Genetics Cooperative Report No. 16. http://cuke.hort.ncsu.edu/cgd/cgcls/c8c16_32,html.

Wang HY, Chen CL, Sung JM (2002). Both warm water soaking and solid priming treatments enhance anti-oxidation of bitter gourd seeds germinated at sub-optimal temperature. *Seed Science and Technology* 31:47–56.

Wang X, Ou-yang Ch, Fan Z, Gao S, Chen F, Tang L (2010). Effects of exogenous silicon on seed germination and antioxidant enzyme activities of *Momordica charantia* under salt stress. *Journal of Animal and Plant Sciences* 6(3):700–708. http://www.biosciences.elewa.org/JAPS.

Yang SL, Walters TW (1992). Ethnobotany and the economic role of the cucurbitaceae of China. *Economy Botany* 46:349–367.

Zong RJ, Cantwell M, Morris LL (1993). Angled luffa, bitter melon, fuzzy melon, yard-long bean. Postharvest handling of Asian specialty vegetables under study California. *Agriculture* 47(2):27–29. doi:10.3733/ca.v047n02p27.

Zong RJ, Morris L, Cantwell M (1995). Postharvest physiology and quality of bitter melon (*Momordica charantia* L.). *Postharvest Biology and Technology* 6:65–72.

27 Snapmelon

Rakesh Kr. Dubey, Vikas Singh,
Garima Upadhyay, and Hari Har Ram

CONTENTS

27.1 INTRODUCTION

Taxonomy

Domain:	Eukaryota
Kingdom:	Plantae
Subkingdom:	Viridaeplantae
Phylum:	Tracheophyta

Snapmelon is native to India. It was intensively grown in the nineteenth century in northern India (Duthie, 1905) where it is commonly known as "phut," which means "to split." Immature fruits are cooked or pickled; the low-sugared matured fruits are eaten raw (Pandit et al., 2005). Ripe fruits usually crack. It is grown on a small scale in north India mainly in Rajasthan, Punjab, Haryana, Uttar Pradesh, and Bihar (Dhillon et al., 2009). Sometimes, it is cultivated as an intercrop with sorghum, maize, and cotton. Its fruits are tender at the young stage and are eaten either raw or cooked. They give out a musky flavor and are eaten as a dessert. They are rich in minerals and vitamins and are useful for lowering blood sugar level. Its seeds are rich in oil with a nutty flavor but are very fiddly to use because of its small size and a fibrous coat covering. Local landraces of snapmelon have been reported to be a good source of disease and insect resistance (Dhillon et al., 2007). Although snapmelon originated in India and also has various important uses, it is grown on a very limited area. The main reason for low area production and productivity of snapmelon in comparison to other vegetable crops is the lack of knowledge and use of unidentified local varieties/landraces with

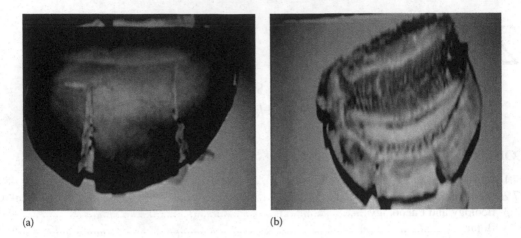

(a)   (b)

FIGURE 27.1 **(See color insert.)** Fruit and seeds of snapmelon. (a) Cracked fruit of snapmelon and (b) snapmelon flesh with seeds.

poor and inconsistent performance. Authors have found a good amount of variation with respect to earliness, number, weight, color, shape, size, flavor, sweetness, and disease resistance in crops grown at their experimental farm, which offers great scope for the improvement of this crop through the application of knowledge of genetics and plant breeding (Figure 27.1).

27.2 MEDICINAL USES

The flowers of snapmelon are expectorant and emetic. The fruit is stomachic. The seed is antitussive, digestive, febrifuge, and vermifuge. The whole seed along with the seed coat is ground into fine flour to make an emulsion with water and is consumed as a vermifuge. It is necessary to take a purge in order to expel the tapeworms or other parasites from the intestine. The root is diuretic and emetic. The fruits can be used as a cooling, light cleanser or moisturizer for the skin. They are also used as a first aid for burns and abrasions.

27.3 ECOLOGY AND PHENOLOGY

Snapmelon requires a rich, well-drained, moisture-retentive soil and grows well in warm, sunny conditions. Being a frost-tender annual plant, the snapmelon is occasionally cultivated especially in warmer climates for its edible fruit. The fruits of snapmelon have a yellow, smooth rind; fluffy pale white or orange flesh colors are mealy and taste sourish with low sugar content. The fruit cracks and disintegrates on full maturity. Half-ripe fruits that turn yellow in color are harvested to be eaten as a salad. Fruits have very poor keeping quality and cannot be stored more than 2–3 days after harvest. The fruits are consumed locally or in nearby markets. Ripe fruits are used for dessert. Raw seeds are rich in oil with a nutty flavor but are very fiddly to use because of its small size and a fibrous coat covering. The seed contains oil between 12.5% and 39.1%.

27.4 BOTANY

Snapmelon plants have shallow, tap root system and a soft, woody stem. Leaves are simple or somewhat palmately lobed, arranged alternately. Flowers are solitary. The staminate and pistillate flowers are produced on different nodes (monoecious). Fruits are true pepo type. *Cucumis melo* var. *momordica* is an annual climber growing to 1.5 m. Staminate flowers are in clusters and the pistillate flowers are usually solitary. The calyx and the corolla of staminate and pistillate flowers are five lobed. It flowers from July to September and the seeds ripen from August to October. The flowers are pollinated by insects and the plant is self-fertile.

27.5 BREEDING METHODS AND IMPROVED CULTIVARS

The breeding methods for improvement of snapmelon are the same as applied in other cucurbits that are cross-pollinated but without inbreeding depression as in self-pollinated crops. The commonly practiced breeding methods in snapmelon include pure line selection, mass selection, recurrent selection, pedigree method, backcross, and heterosis breeding. Pure line or individual plant selection with inbreeding followed in a self-pollinated crop is also applicable in snapmelon, particularly for selection in heterogeneous local cultivars. Improved cultivars are maintained by mass selection. Both mass selection and recurrent selection procedures are effective in population improvement and development of inbred to be utilized in heterosis breeding. Pedigree selection in segregating progenies following hybridization is suitable for obtaining genotypes having combinations of desired attributes from the parents. The backcross method of breeding can be practiced in snapmelon for transferring desired genes, especially for disease resistance, from donor parents. A classical example is the development of the powdery mildew resistant lines PMR-45 in muskmelon and others in the United States, incorporating resistance genes from Indian cultivars. The disease-resistant melon cultivars developed in India have been mentioned under disease resistance. Sweetness in snapmelon is governed by genetic as well as environmental factors. Hence, sweetness in marketed fruits is often unreliable due to natural cross-pollination occurring in the sweet-fruited cultivar grown by the farmer. Random amplified polymorphic DNA (RAPD)-based grouping analysis revealed that Indian snapmelon is rich in genetic variation, and region and subregion approach should be followed across India for acquisition of additional melon landraces. Comparative analysis of the genetic variability among Indian snapmelon and an array of previously characterized reference accessions of melon from Spain, Israel, Korea, Japan, Maldives, Iraq, Pakistan, and India using SSRs showed that Indian snapmelon germplasm contained a high degree of unique genetic variability that is needed to be preserved to broaden the genetic base of melon germplasm available with the scientific community (Staub et al., 2004; Dhillon et al., 2007). Some improved snapmelon varieties were developed in India by selection from local types such as the following:

V. 1. AHS-10: Fruits can be harvested 65–70 days after sowing; fruits are oblong and medium in size (900 g), flesh whitish pink, and sweet in taste, having 4.5%–5.0% TSS. This cultivar bears 4.0–4.5 fruits per vine, giving a yield of 225–230 q/ha under arid conditions.

V. 2. AHS-82: Fruit harvest starts 67–70 days after sowing; each vine bears 4.5–5.0 fruits, giving a yield of 245–250 q/ha. The flesh is light pink and sweet in taste, having 4.3%–4.9% TSS.

27.6 PRODUCTION TECHNOLOGY

The crop is grown on a wide range of soil. Light, sandy loam soil with a good content of organic matter is ideal for growing snapmelon. Generally, the seeds are sown from June to July but in Rajasthan and Bihar states of northern India, seed sowing is done from January to March. Seeds are sown in hills or pits in a row. The distance between rows is 1.5–2.0 m and hills in the rows are 75–100 cm apart. About 2–3 seeds are sown in each hill. About 2–3 kg of seeds is required for 1 ha. The crop grown in the rainy season is not irrigated unless there is a long dry spell. Light hoeing and weeding are done during the early stages of vine growth. The fruits are harvested at the half-ripe stage. At fully ripening, the fruits burst and are not fit for selling in the market. The average yield of snapmelon is about 15–20 tonnes/ha and about 6–8 tonnes/ha when it is grown as an intercrop.

27.7 STRESS TOLERANCE

Snapmelon germplasm has been found to be a very good source of disease and insect resistance. California melon breeders came to India in February 1929, and a powdery mildew (*Podosphaera xanthii* (Castagne) Braun et Shishkoff and *Golovinomyces cichoracearum* (DC) V.P. Heluta) resistant snapmelon collection designated PI 79376, originating from Kathiawar region of Gujarat, India, was presented to them by D.N. Mehta, Second Economic Botanist, Nagpur, Central Provinces, India (Swarup, 2000). The present-day varieties of muskmelon resistant to race 2 of *Podosphaera xanthii* and to *Golovinomyces cichoracearum* owe their origin to this genetic stock. Another snapmelon accession, PI 124111, collected from Kolkata, India, in 1937, is known for its resistance to powdery mildew (Harwood and Markarian, 1968) and downy mildew (*Pseudoperonospora cubensis* [Berk et Curtis] Rostovzev) (Thomas et al., 1988). Subsequently, Indian snapmelon accessions PI 124112, PI 134192, and PI 414723 provided resistance to various diseases like Fusarium wilt (*Fusarium oxysporum*), mildews, Zucchini yellow mosaic virus (ZYMV), papaya ring spot virus (PRSV), Cucurbit aphid-borne yellow mosaic virus (CABYV), and one insect-pest *Aphis gossypii* Glover (Dhiman et al., 1997; Pitrat et al., 2000). This been exploited by muskmelon breeders in developed countries. Recently, PI 124111F (F$_7$ derivatives of snapmelon line PI 124111), which was originally found to be resistant to the five pathotypes of *Pseudoperonospora cubensis*, was reported to be resistant to the newly discovered pathotype 6 in Israel (Cohen et al., 2003). In India, during screenings under natural epiphytotic conditions, More (2002) observed five Indian snapmelon genotypes, viz. 55-1, 55-2, 77, 113, and 114, resistant to downy mildew. Cucumber green mottle mosaic virus (CGMMV) resistance from Indian snapmelon has been incorporated into Indian muskmelon cultivars (Pan and More, 1996). An improved muskmelon cultivar, Punjab Rasila, was developed by crossing snapmelon with a melon variety that is resistant to downy mildew.

27.8 PLANT PROTECTION

27.8.1 Fruit Fly

Fruit fly damage is the major limiting factor in obtaining good quality fruits and high yield. It prefers young, green, and tender fruits for egg laying. The females lay the eggs 2–4 mm deep in the fruit pulp, and the maggots feed inside the developing fruits. At times, the eggs are also laid in the corolla of the flower and the maggots feed on the flowers. A few maggots have also been observed to feed on the stems (Narayanan, 1953). The fruits attacked in early stages fail to develop properly and drop or rot on the plant. Since the maggots damage the fruits internally, it is difficult to control this pest with insecticides. Therefore, there is a need to explore alternative methods of control and develop an integrated pest control strategy for effective management of this pest. The available information on the fruit fly has been reviewed in this chapter to explore the possibilities for successful management of this pest.

27.8.1.1 Distribution of Fruit Fly

The melon fruit fly is distributed all over the world, but India is considered as its native home. It was discovered in Solomon Islands in 1984 and is now widespread in all the provinces, except Makira, Rennell-Bellona, and Temotu (Eta, 1985). In the Commonwealth of the Northern Mariana Islands, it was detected in 1943 and eradicated by sterile-insect release in 1963 (Steiner et al., 1965; Mitchell, 1980) but reestablished from the neighboring Guam in 1981 (Wong et al., 1989). It was detected in Nauru in 1982 and eradicated in 1999 by male annihilation and protein bait spraying but was reintroduced in 2001 (Hollingsworth and Allwood, 2002).

27.8.1.2 Host Range

Fruit fly damages over 81 plant species. Based on the extensive surveys carried out in Asia and Hawaii, plants belonging to the family Cucurbitaceae are preferred most (Dhillon et al., 2005). Doharey (1983) reported that it infests over 70 host plants, among which fruits of bitter gourd (*Momordica charantia*), muskmelon (*Cucumis melo*), snapmelon (*Cucumis melo* var. *momordica*), and snake gourd (*Trichosanthes anguina* and *Trichosanthes cucumeria*) are the most preferred hosts. However, White and Elson-Harris (1994) stated that many of the host records might be based on casual observations of adults resting on plants or caught in traps set in nonhost plant species. In the Hawaiian Islands, melon fruit fly has been observed feeding on the flowers of the sunflower, Chinese bananas, and the juice exuding from sweet corn. Under induced oviposition, broccoli, dry onion (*Allium cepa*), blue field banana (*Musa paradisiaca* sp. *sapientum*), tangerine (*Citrus reticulata*), and longan (*Euphoria longan*) are doubtful hosts of *B. cucurbitae*. The melon fly has a mutually beneficial association with the orchid *Bulbophyllum patens*, which produces zingerone. The males pollinate the flowers and acquire the floral essence and store it in the pheromone glands to attract conspecific females (Hong and Nishida, 2000).

27.8.1.3 Nature and Extent of Damage

Maggots feed inside the fruits but, at times, also feed on flowers and stems. Generally, the females prefer to lay the eggs in soft, tender fruit tissues and have been reported to infest 95% of bitter gourd fruits in Papua New Guinea and 90% snake gourd and 60%–87% pumpkin fruits in Solomon Islands (Hollingsworth et al., 1997). Singh et al. (2000) reported 31.27% damage in bitter gourd and 28.55% in watermelon in India.

27.8.1.4 Life Cycle

The fruit fly remains active throughout the year on one or the other host. During the severe winter months, they hide and huddle together under dried leaves of bushes and trees. During the hot and dry season, the flies take shelter under humid and shady places and feed on honeydew of aphids infesting the fruit trees. This species actively breeds when the temperature falls below 32.2°C and the relative humidity ranges between 60% and 70%. Fukai (1938) reported the survival of adults for a year at room temperature if fed on fruit juices. In general, its life cycle lasts from 21 to 179 days (Narayanan and Batra, 1960). The full-grown larvae come out of the fruit by making one or two exit holes for pupation in the soil. The larvae pupate in the soil at a depth of 0.5–15 cm. The depth up to which the larvae move in the soil for pupation and its survival depends on soil texture and moisture (Pandey and Misra, 1999). The males of *Bactrocera cucurbitae* mate with females for 10 or more hours, and sperm transfer increases with the increase in copulation time. Egg hatchability is not influenced by mating duration (Tsubaki and Sokei, 1988).

27.8.1.5 Strategies for Integrated Management of Fruit Fly

The fruits of cucurbits, of which the melon fly is a serious pest, are picked up at short intervals for marketing and self-consumption. Therefore, it is difficult to rely on insecticides as a means of controlling this pest. In situations where chemical control of melon fruit fly becomes necessary, one has to rely on soft insecticides with low residual toxicity and short waiting periods. Therefore, keeping in view the importance of the pest and crop, melon fruit fly management could be done using local area management or wide area management.

27.8.1.5.1 Local Area Management

Local area management means the minimum scale of pest management over a restricted area such as at the field level/crop level/village level that has no natural protection against reinvasion. The aim of local area management is to suppress the pest rather than eradicate it. Under this management option, a number of methods such as bagging of fruits, field sanitation, protein baits and cue-lure traps, host plant resistance, biological control, and mild insecticides can be employed to keep the pest population below economic threshold level in a particular crop over a period of time to avoid crop losses without health and environmental hazards, which is the immediate concern of the farmers (Pandit et al., 2003).

27.8.1.5.1.1 Bagging of Fruit   Bagging of fruits on the vine (3–4 cm long) with two layers of paper bags at 2–3-day intervals minimizes fruit fly infestation and increases net returns by 40%–58% (Jaiswal et al., 1997).

27.8.1.5.1.2 Field Sanitation   The most effective method in melon fruit fly management is the use of the primary component—field sanitation. To break the reproduction cycle and population increase, growers need to remove all unharvested fruits or vegetables from the field by completely burying them deep into the soil (Pandit et al., 2010). Burying damaged fruits 0.46 m deep into the soil prevents adult fly exclusion and reduces population increase (Klungness et al., 2005).

27.8.1.5.1.3 Monitoring and Control with Parapheromone Lures/Cue-Lure Traps   The principle of this particular technique is the denial of resources needed for laying of eggs by female flies such as protein food (protein bait control) or parapheromone lures that eliminate males. There is a positive correlation between cue-lure trap catches and weather conditions such as minimum temperature, rainfall, and minimum humidity. The sex attractant cue-lure traps are more effective than the food attractant tephritlure traps for monitoring *B. cucurbitae*. Methyl eugenol and cue-lure traps have been reported to attract *B. cucurbitae* males from mid-July to mid-November (Liu and Lin, 1993). A leaf extract of *Ocimum sanctum* that contains eugenol (53.4%), beta-caryophyllene (31.7%), and beta-elemene (6.2%) as the major volatiles attracts flies from a distance of 0.8 km when placed on cotton pads (0.3 mg) (Roomi et al., 1993). Thus, melon fruit fly can also be controlled through the use of *Ocimum sanctum* as the border crop sprayed with a protein bait (protein derived from corn, wheat, or other sources) containing spinosad as a toxicant. A new protein bait, GF-120 Fruit Fly Bait®, containing spinosad as a toxicant, has been found to be effective in the area-wide management of melon fruit fly in Hawaii (Prokopy et al., 2004). The GF-120 Fruit Fly Bait would be highly effective when applied to sorghum plants surrounding cucumbers against protein-hungry melon flies but would be less effective in preventing protein-satiated females from arriving on cucumbers.

27.8.1.5.1.4 Biological Control   Srinivasan (1994) reported *Opius fletcheri* Silv. to be a dominant parasitoid of *B. cucurbitae*. A new parasitoid, *Fopius arisanus*, has also been included in the IPM program of *B. cucurbitae* in Hawaii (Wood, 2001). A Mexican strain of the nematode *Steinernema carpocapsae* Weiser (*Neoaplectana carpocapsae*) has been reported to cause 0%–86% mortality to melon fruit fly after an exposure of 6 days to 5,000–5,000,000 nematodes/cup in the laboratory and an average of 87.1% mortality under field conditions when applied at 500 infective juveniles/cm^2 soil (Lindegren, 1990). Sinha (1997) reported that culture filtrate of the fungus *Rhizoctonia solani* Kuhn is an effective bioagent against *B. cucurbitae* larvae, while the fungus *Gliocladium virens* Origen has been reported to be effective against *B. cucurbitae*. Culture filtrates of the fungi *Rhizoctonia solani*, *Trichoderma viridae* Pers., and *Gliocladium virens* affected the oviposition and the development of *B. cucurbitae* adversely (Sinha and Saxena, 1999).

27.8.1.5.1.5 Chemical Control   Chemical control of the melon fruit fly is relatively ineffective. However, insecticides such as malathion, dichlorvos, phosphamidon, and endosulfan are moderately effective against the melon fly (Agarwal et al., 1987). The application of molasses + malathion (Limithion 50 EC) and water in the ratio of 1:0.1:100 provides good control of melon fly (Akhtaruzzaman et al., 2000). Application of either 0.05% fenthion or 0.1% carbaryl at 50% appearance of male flowers and again 3 days after fertilization is helpful in reducing melon fly damage (Srinivasan, 1991).

27.8.1.5.2 Wide Area Management

Wide area management is not a unitary concept but incorporates a number of related but distinct methods including local area management. The methods used for a wide area management approach include sterile-male insect release, insect transgenesis, and quarantine control techniques in combination with available local area management options. The aim of wide area management is to coordinate and combine different characteristics of an insect eradication program over an entire area within a defensible perimeter. The area must be subsequently protected against reinvasion by quarantine controls, for example, by pest eradication on isolated islands. It has proved to be economically viable, environmentally sensitive, sustainable, and has suppressed fruit flies below economic thresholds with the minimum use of organophosphate and carbamate insecticides. An IPM program that used field sanitation, protein bait applications, male annihilation, and release of sterile flies and parasites reduced fruit fly infestation from 30% to 40% to less than 5% and cut organophosphate pesticide use by 75%–90% (Vargas, 2004). The recent wide area management eradication of *B. cucurbitae* in Seychelles demonstrated a three-tier model including (1) initial population reduction using bait sprays, (2) elimination of reproduction using parapheromone lure blocks to eradicate males and thus prevent oviposition by females, and (3) intensive surveying by traps and fruit inspection until certain that the pest is entirely eradicated (Mumford, 2004). Although the sterile insect technique has been successfully used in area-wide approaches, wide area management needs more sophisticated and powerful technologies in their eradication program, such as insect transgenesis that could be deployed over a wide area and is less susceptible to immigrants. Above all, the use of the geographical information system has been used as a tool to mark site-specific locations of traps, host plants roads, land use areas, and fruit fly populations within a specified operational grid (Mau et al., 2003a).

27.8.1.5.2.1 Sterile-Male Technique

In this technique, sterile males are released in the fields for mating with the wild females. Sterilization is accomplished through irradiation, chemosterilization, or genetic manipulation. In sterile insect programs, the terms "sterility" or "sterile insect" refer to the transmission of dominant lethal mutations that kill the progeny. The females either do not lay eggs or lay sterile eggs. Ultimately, the pest population can be eradicated by maintaining a barrier of sterile flies. A sterile insect program is species specific and is considered an ecologically safe procedure and has been successfully used in area-wide approaches to suppress or eradicate pest insects such as the pink bollworm *Pectinophora gossypiella* in California (Walters et al., 2000), the tsetse fly *Glossina austeni* in Zanzibar (Vreysen, 2001), the New World screwworm *Cochliomyia hominivorax* in North and Central America (Wyss, 2000), and various tephritid fruit fly species in different parts of several continents (Klassen et al., 1994). Chemosterilization (by exposing the flies to 0.5 g tepa in drinking water for 24 h) and gamma irradiation are the only widely tested and accepted sterile-male techniques against melon fly (Odani et al., 1991). Nakamori et al. (1993) found in Okinawa that frequent and intensive release of sterile flies did not increase the ratio of sterile to wild flies in some areas, suggesting that it is important to identify such areas for eradication of this pest. Eradication of this pest has already been achieved through sterile-male release in Kikaijima Islands in 1985, Amami-oshima in 1987, Tokunoshima, and Okienoerabu-jima and Yoronjima Islands in 1989 (Yoshizawa, 1997). In the Mediterranean fruit fly (medfly) *Ceratitis capitata*, release of sterile males increased the effectiveness of the sterile insect program. The use of sterile-male and male annihilation techniques has successfully eradicated the melon fly from Japan for over 24 years (Liu, 1993). However, the suppression of *B. cucurbitae* reproduction through male annihilation with cue-lure may be problematic. Matsui et al. (1990) reported that no wild tephritids were caught with cue-lure traps after intensification of distribution of cue-lure strings, but the mating rates of mature females did not decrease compared to those in control islands. Conventional sterilization based on ionizing radiation causes chromosome fragmentation without centromeres, where the chromosome fragments will not be transmitted correctly to the progeny, and can have adverse effects on viability and sperm quality, resulting in reduced competitiveness of sterilized individuals (Cayol et al., 1999).

27.8.1.5.2.2 Transgene-Based, Embryo-Specific Lethality System　Although the sterile insect technique can be used successfully to suppress economically important pest species, conventional sterilization by ionizing radiation reduces insect fitness, which can result in reduced competition of the sterilized insects (Horn and Wimmer, 2003). A transgene-based, female-specific expression starfruit, *Averrhoa carambola*, of tephritid eggs and larvae (Armstrong et al., 1995) has been demonstrated.

27.8.1.5.2.3 Quarantine　The import and export of infested plant material from one area or country to other noninfested places is the major mode of the spread of insect pests. The spread of the melon fly can be blocked through tight quarantine and treatment of fruits at the import/export ports. Cold treatment at $1.1°C \pm 0.6°C$ for 12 days disinfected the planting materials.

REFERENCES

Agarwal, M.L., Sharma, D.D., and Rahman, O. 1987. Melon fruit fly and its control. *Indian Horticulture* 32: 10–11.

Akhtaruzzaman, M., Alam, M.Z., and Ali-Sardar, M.M. 2000. Efficiency of different bait sprays for suppressing fruit fly on cucumber. *Bulletin of the Institute of Tropical Agriculture, Kyushu University* 23: 15–26.

Armstrong, J.W., Silvam, S.T., and Shishido, V.M. 1995. Quarantine cold treatment for Hawaiian carambola fruit infested with Mediterranean fruit fly, melon fly, or oriental fruit fly (Diptera: Tephritidae) eggs and larvae. *Journal of Economic Entomology* 88: 683–687.

Cayol, J.P., Vilardi, J., Rial, E., and Vera, M.T. 1999. New indices and method to measure the sexual compatibility and mating performance of *Ceratitis capitata* (Diptera: Tephritidae) laboratory-reared strains under field cage conditions. *Journal of Economic Entomology* 92: 140–145.

Cohen, Y., Meron, I., Mor, N., and Zurial, S. 2003. A new pathotype of *Pseudopernospora cubensis* causing downy mildew in cucurbits in Israel. *Phytoparasitica* 31: 452–466.

Dhillon, M.K., Singh, R., Naresh, J.S., and Sharma, H.C. 2005. The melon fruit fly, *Bactrocera cucurbitae*: A review of its biology and management. *Journal of Insect Science* 5: 40. Available online: insectscience.org/5.40.

Dhillon, N.P.S., Ranjana, R., Singh, K., Eduardo, I., Monforte, A.J., Pitrat, M., Dhillon, N.K., and Singh, P.P. 2007. Diversity among landraces of Indian snapmelon (*Cucumis melo* var. *momordica*). *Genetic Resources and Crop Evolution* 54(6): 1267–1283.

Dhillon, N.P.S., Singh, J., Fergany, M., Monforte, A.J., and Sureja, A.K. 2009. Phenotypic and molecular diversity among landraces of snapmelon (*Cucumis melo* var. *momordica*) adapted to the hot and humid tropics of eastern India. *Plant Genetic Resources: Characterization and Utilization* 7(3): 291–300.

Dhiman, J.S., Lal, T., and Dhaliwal, M.S. 1997. Remove from marked records downy-mildew resistance in snapmelon and its exploitation for muskmelon improvement. *Plant Disease Research* 12(1): 88–90.

Doharey, K.L. 1983. Bionomics of fruit flies (*Dacus* spp.) on some fruits. *Indian Journal of Entomology* 45: 406–413.

Duthie, J.F. 1905. *Flora of Upper Gangetic Plain and the Adjacent Sivalik and Sub-Himalyan Tracts*, Vol. 1: *Ranunculaceae to Companulaceae*. Office of the Superintendent of Government Printing, Kolkata, India.

Eta, C.R. 1985. Eradication of the melon fly from Shortland Islands (special report). Solomon Islands Agricultural Quarantine Service, Annual Report. Ministry of Agriculture and Lands, Honiara, Solomon Islands.

Fukai, K. 1938. Studies on the possibility of life of the Formosa melon fly in Japan. *Nojikairyo-Shiryo* 134: 147–213.

Harwood, R.R. and Markarian, D. 1968. A genetic survey of resistance to powdery mildew in muskmelon. *Heredity* 59: 213–217.

Hollingsworth, R.G. and Allwood, A.J. 2002. Melon fly. In: *Pest Advisory Leaflet No. 31*, Secretariat of the Pacific Community, Plant Protection Service, Nabua (Technical Bulletin).

Hollingsworth, R., Vagalo, M., and Tsatsia, F. 1997. Biology of melon fly with special reference to the Solomon Islands. In: A.J. Allwood and D. Rai (eds.), *Management of Fruit Flies in the Pacific*. A Regional Symposium, Nadi, Fiji, Vol. 76, pp. 140–144.

Hong, K.T. and Nishida, R. 2000. Mutual reproductive benefits between a wild orchid, *Bulbophyllum patens* and *Bactrocera* fruit flies via a floral synomone. *Journal of Chemical Ecology* 26: 533–546.

Horn, C. and Wimmer, E.A. 2003. A transgene-based, embryo-specific lethality system for insect pest management. *Nature Biotechnology* 21: 64–70.

Jaiswal, J.P., Gurung, T.B., and Pandey, R.R. 1997. Findings of melon fruit fly control survey and its integrated management 1996/97. Working Paper 97/53, Lumle Agriculture Research Centre, Kashi, Nepal, pp. 1–12.

Klassen, W., Lindquist, D.A., and Buyckx, E.J. 1994. Overview of the joint FAO/IAEA Division's involvement in fruit fly sterile insect technique programs. In: C.O. Calkins, W. Klassen, and P. Liedo (eds.), *Fruit Flies and the Sterile Insect Technique*, CRC Press, Boca Raton, FL. pp. 3–26.

Klungness, L.M., Jang, E.B., Mau, R.F.L., Vargas, R.I., Sugano, J.S., and Fujitani, E. 2005. New approaches to sanitation in a cropping system susceptible to tephritid fruit flies (Diptera: Tephritidae) in Hawaii. *Journal of Applied Science and Environmental Management* 9: 5–15.

Lindegren, J.E. 1990. Field suppression of three fruit fly species (Diptera: Tephritidae) with Steinernema carpocapsae. In: *Proceedings Fifth International Colloquium on Invertebrate Pathology and Microbial Control*, Adelaide, South Australia, Australia, August 20–24, 1990, pp. 1–223.

Liu, Y.C. 1993. Pre-harvest control of Oriental fruit fly and melon fly. In: *Plant Quarantine in Asia and the Pacific, Report of APO Study Meeting*, Taipei, Taiwan, March 17–26, 1992, pp. 73–76.

Liu, Y.C. and Lin, J.S. 1993. The response of melon fly, *Dacus cucurbitae* Coquillett to the attraction of 10% MC. *Plant Protection Bulletin Taipei* 35: 79–88.

Matsui, M., Nakamori, H., Kohama, T., and Nagamine, Y. 1990. The effect of male annihilation on a population of wild melon flies, *Dacus cucurbitae* Coquillett (Diptera: Tephritidae) in Northern Okinawa. *Japanese Journal of Applied Entomology and Zoology* 34: 315–317.

Mau, R.F.L., Jang, E.B., Vargas, R.I., Chan, C.M., Chou, M., and Sugano, J.S. 2003a. Implementation of a geographical information system with integrated control tactics for area wide fruit fly pest management. *Proceedings of Meeting on Area Wide Control for Fruit Flies* 5: 23–33.

Mitchell, W.C. 1980. Verification of the absence of Oriental fruit and melon fruit fly following an eradication program in the Mariana Islands. *Proceedings of the Hawaiian Entomological Society* 23: 239–243.

More, T.A. 2002. Enhancement of muskmelon resistance to disease via breeding and transformation. *Acta Horticulturae* 588: 205–211.

Mumford, J.D. 2004. Economic analysis of area-wide fruit fly management. In: B. Barnes and M. Addison (eds.), *Proceedings of the Sixth International Symposium on Fruit Flies of Economic Importance, Stellenbosch, South Africa*, FruitecPress, Stellenbosch, South Africa, May 6–10, 2002.

Nakamori, H., Shiga, M., and Kinjo, K. 1993. Characteristics of hot spots of melon fly, *Bactrocera (Dacus) cucurbitae* Coquillett (Diptera: Tephritidae) in sterile fly release areas in Okinawa Island. *Japanese Journal of Applied Entomology and Zoology* 37: 123–128.

Narayanan, E.S. 1953. Seasonal pests of crops. *Indian Farming* 3(4): 8–11.

Narayanan, E.S. and Batra, H.N. 1960. *Fruit Flies and Their Control*. Indian Council of Agricultural Research, New Delhi, India, pp. 1–68.

Odani, Y., Sakurai, H., Teruya, T., Ito, Y., and Takeda, S. 1991. Sterilizing mechanism of gamma-radiation in the melon fly, *Dacus cucurbitae*. *Research Bulletin of Faculty of Agriculture Gifu University* 56: 51–57.

Pan, R.S. and More, T.A. 1996. Screening of melon (*Cucumis melo* L.) germplasm for multiple disease resistance. *Euphytica* 88: 125–128.

Pandey, M.B. and Misra, D.S. 1999. Studies on movement of *Dacus cucurbitae* maggot for pupation. *Shashpa* 6: 137–144.

Pandit, M.K., Pal, P.K., and Das, B.K. 2010. Effect of date of sowing on flowering and incidence and damage of melon fruit fly in snap melon *Cucumis melo* var. *momordica*. genotypes. *Journal of Plant Protection Sciences* 2(1): 86–91.

Pandit, M.K., Saha, A., and Mahato, B. 2005. Evaluation of growth and yield potential of some local snapmelon genotypes in the Gangetic alluvial zone of West Bengal. *Crop Research Hisar* 30(2): 192–195.

Pandit, M.K., Saha, A., and Saha, A. 2003. Influence of thermal environment on phenological development and yield of snap melon (*Cucumis melo* var. *momordica*) genotypes in southern West Bengal. *Vegetable Science* 30(21): 150–154.

Pitrat, M., Hanelt, P., and Hammer, K. 2000. Some comments on interaspecific classification of cultivars of melon. *Acta Horticulturae* 510: 29–36.

Prokopy, R.J., Miller, N.W., Pinero, J.C., Oride, L., Perez, N., Revis, H.C., and Vargas, R.I. 2004. Hoe effective is GF-120 fruit fly bait spray applied to border area sorghum plants for control of melon flies (Diptera: Tephritidae). *Florida Entomologist* 87: 354–360.

Roomi, M.W., Abbas, T., Shah, A.H., Robina, S., Qureshi, A.A., Hussain, S.S., and Nasir, K.A. 1993. Control of fruit flies (*Dacus* spp.) by attractants of plant origin. *Anzeiger fur Schadlingskunde, Aflanzenschutz, Umwdtschutz* 66: 155–157.

Singh, S.V., Mishra, A., Bisan, R.S., Malik, Y.P., and Mishra, A. 2000. Host preference of red pumpkin beetle, *Aulacophora foveicollis* and melon fruit fly, *Dacus cucurbitae*. *Indian Journal of Entomology* 62: 242–246.

Sinha, P. 1997. Effects of culture filtrates of fungi on mortality of larvae of *Dacus cucurbitae*. *Journal of Environmental Biology* 18: 245–248.

Sinha, P. and Saxena, S.K. 1999. Effect of culture filtrates of three fungi in different combinations on the development of the fruit fly, *Dacus cucurbitae* Coq. *Annals of Plant Protection Service* 7: 96–99.

Srinivasan, K. 1991. Pest management in cucurbits—An overview of work done under AICVIP. In: *Group Discussion of Entomologists Working in the Coordinated Projects of Horticultural Crops*, Central Institute of Horticulture for Northern Plains, Lucknow, India, January 28–29, 1991, pp. 44–52.

Srinivasan, K. 1994. Recent trends in insect pest management in vegetable crops. In: G.S. Dhaliwal and R. Arora (eds.), *Trends in Agricultural Insect Pest Management*. Commonwealth Publishers, New Delhi, India, pp. 345–372.

Staub, J.E., L'opez-Ses'el Ana, I., and Fanourakis, N. 2004. Diversity among melon landraces (*Cucumis melo* L.) from Greece and their genetic relationships with other melon germplasm of diverse origins. *Euphytica* 136: 151–166.

Steiner, L.F., Harris, E.J., Mitchell, W.C., Fujimoto, M.S., and Christenson, L.D. 1965. Melon fly eradication by over flooding with sterile flies. *Journal of Economic Entomology* 58: 519–522.

Swarup, V. 2000. Genetic resources in vegetable crops in India. In: K. Gautam and K. Singh (eds.), *Emerging Scenario in Vegetable Research and Development*. Research Periodicals and Book Publishing Home, Houston, TX, pp. 346–355.

Thomas, C.F., Cohen, Y., Mc Creight, J.D., Jourdain, E.L., and Cohen, S. 1988. Inheritance of resistance to downy mildew in *Cucumis melo*. *Plant Disease* 72: 33–35.

Tsubaki, Y. and Sokei, Y. 1988. Prolonged mating in the melon fly, *Dacus cucurbitae* (Diptera: Tephritidae): Competition for fertilization by sperm loading. *Research on Population Ecology* 30: 343–352.

Vargas, R.I. 2004. Area-wide Integrated Pest Management for exotic fruit flies in Hawaii. In: *FLC Awards Program. The FLC-TPWG National Meeting*, San Diego, CA, May 3–6, 2004, p. 12.

Vreysen, M.J. 2001. Principles of area-wide integrated tsetse fly control using the sterile insect technique. *Medicine for Tropics* 61: 397–311.

Walters, M.L., Staten, R.T., and Roberson, R.C. 2000. Pink bollworm integrated management using sterile insects under field trial conditions, Imperial Valley, California. In: K.H. Tan (ed.), *Area-Wide Control of Fruit Flies and Other Insect Pests*. Penerbit University Sains Malaysia, Penang, Malaysia, pp. 201–206.

White, I.M. and Elson-Harris, M.M. 1994. *Fruit Flies of Economic Significance: Their Identification and Bionomics*. Commonwealth Agriculture Bureau International, Oxon, U.K., pp. 1–60.

Wong, T.T.Y., Cunningham, R.T., Mcinnis, D.O., and Gilmore, J.E. 1989. Seasonal distribution and abundance of *Dacus cucurbitae* (Diptera: Tephritidae) in Rota, Commonwealth of the Mariana Islands. *Environmental Entomology* 18: 1079–1082.

Wood, M. 2001. Forcing exotic, invasive insects into retreat: New IPM program targets Hawaii's fruit flies. *Agricultural Research Washington* 49: 11–13.

Wyss, J.H. 2000. Screwworm eradication in the Americas-overview. In: K.H. Tan (ed.), *Area-Wide Control of Fruit Flies and Other Insect Pests*. Penerbit University Sains Malaysia, Penang, Malaysia, pp. 79–86.

Yoshizawa, O.1997. Successful eradication programs on fruit flies in Japan. *Research Bulletin of Plant Protection Service Japan* 33: 10.

28 Kachri (*Cucumis callosus* [Rottler] Cogn.)

Rakesh Kr. Dubey, Vikas Singh,
Garima Upadhyay, and Hari Har Ram

CONTENTS

28.1   INTRODUCTION

Systematic Classification of Kachri

Domain	Eukaryota
Kingdom	Plantae
Subkingdom	Viridaeplantae
Phylum	Tracheophyta
Subphylum	Euphyllophytina
Infraphylum	Radiatopses
Class	Spermatopsida
Subclass	Rosidae
Superorder	Violanae
Order	Cucurbitales
Family	Cucurbitaceae
Subfamily	Cucurbitoideae
Tribe	Melothrieae
Genus	*Cucumis*
Specific epithet	*callosus* Cogn.
Botanical name	*Cucumis callosus* Cogn.

Kachri [(*Cucumis callosus* (Rottler) Cogn.)] is commonly known as bitter cucumber in English, kachri in Hindi, and karkati in Sanskrit and belongs to the family Cucurbitaceae. Kachri, a feral species of India, has attracted the attention of muskmelon breeders as this species is reported to possess genes for resistance to fruit fly and leaf-eating caterpillars, as well as genes for drought and field resistance to a host of pests and diseases. The natural distribution pattern of kachri in India comprising of the Vindhya hills and Aravalli mountain ranges extends northward to the Indo-Gangetic plains and southward to the Deccan plateau, touching rain shadow areas of the Western Ghats and also found in the Eastern Himalayan region of India. Kachri is characterized by its morphological distinction and F_1 and BC_1 fertility with *Cucumis melo*, a subspecific rank within *C. melo*. In the most well-quoted revision of *Cucumis* (Kirkbride 1993), *Cucumis callosus* is treated as a synonym of *Cucumis melo*. He followed Jeffrey (1980) in arriving at such a conclusion. The veteran botanist Clarke (1879) observed that Naudin stressed the perennial nature of the root as a distinguishing character from *C. melo*. However, he merged the entity *Cucumis pubescens* Willd (presently equated with *C. melo* subsp. *agrestis*) with *Cucumis trigonus* Roxb. Kurz (1877) who studied Bengal (India) plants separated *C. trigonus*, characterized by solitary peduncles, from *C. pubescens*, with clustered peduncles, and made the latter a variety of *C. melo*. This concept of Kurz (1877) prevails even to this date in the Indian botanical circles. Verma and Pant (1985) treated *C. trigonus* as a synonym of *C. callosus*. Chakravarthy (1959, 1982) and Matthew (1983) retained its separate species status from that of *C. melo* and *C. melo* subsp. *agrestis*. Dubey and Ram (2006) also illustrated the detailed taxonomy, botany, and distribution of kachri. Narrating the history of taxonomic treatments of *C. callosus*, Nesom (2011) aptly stated that "diversity and ambiguity of interpretation are widespread." Hence, its treatment is synonymous with that of *C. melo* (Kirkbride 1993), without even assigning a sub of its specific status. Sound knowledge of the morphological and biological features and correct taxonomic identities is a prerequisite for successful use of germplasm in cucumber and melon breeding (Kristkova et al. 2003; Renner and Schaefer 2008). Wild species are rich reservoirs of useful genes that are not present in cultivated gene pool (Tanksley and McCouch 1997). Underutilized cucurbits, which form a large and diverse commodity group, assume an important role being globally well dispersed and distributed. This chapter presents an overview of kachri (*C. callosus* (Rottler) Cogn.) with respect to its potential uses, botany, taxonomy, distribution, and production technology for its successful cultivation.

28.2   USES

Immaturely harvested fruits are utilized during vegetable cooking and for *Chutney* preparations. The recipe for kachri *Chutney* uses the following ingredients: 50 g kachri, 100 g garlic paste, 10 g coriander powder, 10 g red chilli powder, a pinch of turmeric powder, 2 g cumin seeds, a pinch of asafoetida, salt to taste, and oil. For preparation of *Chutney*, the method would be as follows:

1. Soak kachri for 5–6 h and grind it to a fine paste.
2. Mix well the kachri paste, garlic, salt, red chilli powder, turmeric, and coriander powder.
3. Take oil in a pan. Add the cumin seeds and let them crackle. Add asafoetida as well as the kachri mixture to it. Add around 60 mL of lukewarm water.
4. Let it simmer for 4–5 min and remove from fire.

Besides *Chutney*, half cut green fruit is dried in the sun for future use with *mustard green* during the winter season. It is also used in place of tomato in vegetable dishes for adding sour taste. In India (western Rajasthan), ripe fruit is eaten raw and used in curries. Green fruit is used as a vegetable. Dried fruit rind and seeds are used in curries (Figure 28.1).

(a)   (b)

FIGURE 28.1 (See color insert.) Fruit and seeds of *Cucumis callosus*. (a) Cut fruits with seeds and (b) ripened fruits of *C. callosus*.

28.3 ORIGIN AND DISTRIBUTION

Chakravarthy (1959) reported its historical distribution between 40°N and 40°S of the equator comprising India, Pakistan, Iran, Afghanistan, and parts of North Africa, running eastward to Malaysia, China, and Australia. In India, it has been found to occur naturally in the wild state in the Deccan plateau and Indo-Gangetic plains comprising the states of Tamil Nadu, Karnataka, Maharastra, Madhya Pradesh, Uttar Pradesh, Jharkhand, Chhattisgarh, Odisha, West Bengal, Punjab, Haryana, Himachal Pradesh, and Assam. According to John et al. (2013), a perusal of the herbarium and passport data indicate that germplasm collection trips for this taxa should be centered around the open sunny localities in the districts of Coimbatore, Salem, Tirunelveli, Ramnad, Chengalpet, Sivagangai, and Rajapalayam in Tamil Nadu; Prakasam, Krishna, Chittoor, Nellore, and Hyderabad in Andhra Pradesh; Rangareddy and Medak in Telangana; Bellary in Karnataka; Dhule, Thana, and Pune in Maharashtra; Ganjam in Odisha; Ranigunj, Jalsuka, and Sunderbans in West Bengal; Indore and Jabalpur in Madhya Pradesh; Raipur in Chhattisgarh; Dehradun in Uttarakhand; Banda, Agra, Jaunpur, Sultanpur, Varanasi, Faizabad, Baharaich, and Saharanpur in Uttar Pradesh; Jodhpur in Rajasthan; Sahibganj in Bihar; Sutlej river bank in Punjab; and Palamau and Ranchi in Jharkhand. The authors have also observed the presence of kachri with different names in northeastern India, such as in East Siang, Upper Siang, and Lohit in Arunachal Pradesh; Shillong in Meghalaya; Imphal in Manipur; Dimapur in Nagaland; Guwahati, Kamrup, and Dhemaji in Assam; Agartala in Tripura; and Gangtok in Sikkim. It is known by several vernacular names such as kachri in Rajasthan, Haryana; chipper in Punjab; and pehtool in eastern Uttar Pradesh, and several other names in other parts of the country. Some of the basic questions to be answered regarding the evolution of cultivated plants are their geographic origin, progenitor taxa, and region of domestication (Vavilov 1935). Cucumber (*Cucumis sativus*) is believed to have originated in India, and there is a unanimous agreement on that count due to the prevalence of its wild progenitor *C. sativus* f. *hardwikii* (Royle) W. J. de Wilde et deifies across the country and also due to linguistic evidences (Fuller 2006). However, when it comes to the origin and domestication of melon, majority opinion favors a sub-Saharan African origin with India as a secondary center (Kerje and Grum 2000; Luan et al. 2008; Perin et al. 2002; Whitaker and Bemis 1976; Whitaker and Davis 1962). As pointed out by Sebastian et al. (2010), the impressive species richness in Africa sharing the 2n = 24 chromosome number, which is the same as melon, prompted many to support this view. However, the greatest diversity in cultivated landraces is to be found in Asia (Akashi et al. 2002; Dwivedi et al. 2010; Tanaka et al. 2007) and Australia. Furthermore, successful recovery of F_1 offsprings could not be achieved from the crosses of *C. melo* with various African species of *Cucumis* (Sebastian et al. 2010). These considerations prompted recent workers to think in favor of an Asian origin of melon, and Sebastian et al. (2010) postulated an Indian origin for *C. melo* from *C. callosus*. In fact, Parthasarathy and Sambandam (1980), based on their observations of free compatibility with *C. melo* and normal behavior of chromosomes during diakinesis, had proposed *C. callosus* of India as the progenitor of *C. melo*. Over the past decade, based on the synthesis of a large quantum of new archaeological and genetic evidences, archaeobotanists have postulated the process of domestication as a protracted and diffuse process, progressing parallel in different locations around the Near East (Allaby et al. 2010; Fuller et al. 2011), as against the earlier notion of a focused, single process in the Near East, where the whole package of "founder crops" emanated from a core area at essentially the same time (Abbo et al. 2010; Lev-Yadun et al. 2000). Archaeological evidence suggests that melons were widespread in Egypt, Arabia, India, and China by 2000 BC, and in all likelihood, its cultivation started in more than one region with at least two domestication events, one in India and the other in the Near East (most likely Egypt) as reported by Zohary and Hopf (2000) and probably three events (Egypt, India/Pakistan, Lower Yangtze) as narrated by Fuller (2012). Of particular interest is the recovery of seed remains (by flotation) of *Cucumis* sp. (comparable to *Cucumis prophetarum*, or perhaps *C. trigonus* = *C. callosus*) from two southern Neolithic sites in the Deccan plateau of South India (1800–1200 BC) (Fuller et al. 2001), though not confirmed by direct AMS dating. However, archaeological reports may be of limited value for pinpointing areas and the probable routes of melon or cucumber domestication, as precise identification of fossil seeds of cucumber and melon is extremely difficult (Sebastian et al. 2010). The cross between *C. callosus* and other taxa of Indian occurring melons is presented in Table 28.1. A perusal of Table 28.1 and Figure 28.2 clearly shows that *C. callosus* falls in the primary gene pool of *C. melo*. However, even with hand pollination at optimum stigmatic receptivity, direct crosses yielded only 15% fruit set and reciprocal only 6% compared to 65% in the case of selfing (assisted pollination). Reciprocal crosses with *C. melo* var. common, var. *momordica*, and var. *cantalupensis* failed to set fruits, but var. *maltensis* produced fully developed mature ripe fruit with 232 healthy seeds. There is no natural hybrid of *C. melo* and *C. callosus* and vice versa when both were grown side by side. Similarly, direct (27 flowers) and reciprocal (97 flowers) crosses with the *C. melo* subsp. *agrestis* failed to set fruits. Pollinator specificity and other barrier mechanisms, if any, need to be further investigated. Contrary to this, the entities *melo* and *agrestis* show more intimate gene transfer in nature. Natural hybrids, intermediate between cultivated and wild/feral, are seen occasionally in farmers' field, and often they are bitter, contaminating the main produce. This again strengthens our stand that *C. callosus*, even while falling within the primary gene pool of *C. melo*, is distinct from both cultivated melon and feral *agrestis*, whereas *melo* and *agrestis* forms are much closer. Crossability studies indicate its placement in the primary gene pool of *C. melo* under the broader biological species concept of *C. melo*. F_1 and BC_1 of *C. melo* var. *conomon* and *C. callosus* were found to be fully fertile, the F_1 being intermediate between parents for quantitative traits (John et al. 2013). Kirkbride (1993) referred to Australian forms of *C. melo* with highly dissected leaves and unusual pubescence on the female flower hypanthium and suggested a need for further biosystematic studies to understand and accommodate the variation. The material mentioned may be *C. callosus* in all probability. Kirkbride (1993) recognized *agrestis* and *melo* as two subspecies of *C. melo*. This similarity indicates the divergence of cultivated *melo* from *C. callosus* with this latter entity as a common ancestor for both *melo* and *agrestis*. Morphological variation parallel to that of cultivated *melo* is observed in wild and weedy *agrestis*. During the course of domestication, wild traits like bitterness, small fruit size, long maturity periods, hard flesh, and resistance to biotic and abiotic stress could have been lost. The low genetic variability in cultivated melon was emphasized by many authors (Neuhausen 1992; Shattuck-Eidens et al. 1990).

TABLE 28.1
Details of Successful Crosses of *Cucumis callosus*

Parents	Flowers Pollinated (No.)	Fruit Set (No.)	Fruit Set (%)	Seeds
C. callosus × *C. callosus*	23	15	65.23	Filled, viable
C. melo × *C. callosus*	95	14	14.73	Filled, viable
C. callosus × *C. melo*[a]	18	1	6	Filled, viable
C. melo × *C. melo*[a]	37	20	54	Filled, viable
C. melo[a] × *C. melo* var. *agrestis*	39	11	28.21	Filled, viable
C. melo var. *agrestis* × *C. melo*[a]	158	59	37.34	Filled, viable
C. melo var. *agrestis* × *C. melo* var. *agrestis*	30	8	27	Filled, viable
C. sativus[a] × *C. callosus*	28	8	28.57	Unfilled, not viable
C. callosus × *C. sativus*[b]	54	0	0	No fruit set

Source: Adapted from John, K.J. et al., *Genet. Resour. Crop Evol.*, 60, 1037, 2013. With permission.
[a] Crosses include subcategories of the species.
[b] Successful fruit set only with var. *maltensis*.

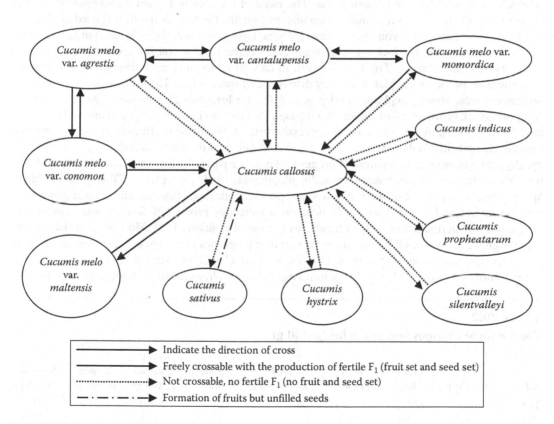

FIGURE 28.2 Crossability polygon of *Cucumis callosus* with other taxa. (Adapted from John, K.J. et al., *Genet. Resour. Crop Evol.*, 60, 1037, 2013. With permission.)

28.4 NUTRITIONAL IMPORTANCE

The biochemical composition of kachri (Table 28.2, Mann et al. 2013) indicates a balanced proximate composition containing a good amount of protein. It can be used as a spasmolytic and anesthetic agent due to the presence of alkaloid content, whereas appreciable amounts of saponins are helpful to boost immune system and lower the risks of various degenerative diseases. The total phenolic content has been reported between 56 and 72 µg GAE/mg extract in kachri. Phenolic compounds contribute to quality and nutritional value in terms of modifying color, taste, aroma, flavor, and also in providing health benefits. In addition, phenolics provide plant defense mechanisms to neutralize reactive oxygen species (ROS) in order to survive and prevent molecular damage and damage by microorganisms, insects, and herbivores. The aqueous seed extract of *C. callosus* (Rottl.) Cogn. possesses significant antioxidant activity. However, the components responsible for antioxidative activity are currently unclear. The study suggested that this plant is a potential source of natural antioxidant that could have great importance as a therapeutic agent in preventing or slowing the progress of aging and age-associated oxidative stresses. Further studies are warranted for the isolation and characterization of antioxidant components and also in vivo studies are needed for understanding its mechanism of action as an antioxidant (Chand et al. 2012).

28.5 MEDICINAL IMPORTANCE

The fruit is traditionally used to prevent insanity, to strengthen memory, and to remove vertigo. The seeds are cooling and astringent and useful in bilious disorder, diabetics, easy bowel syndrome, stomach pain, vomiting, and constipation. The paste of the root is applied on scorpion sting; the decoction of the root is given in indigestion and dropsy; and the pulp of the fruit is used in abortion and to increase menses for women. It is reported to be light, bitter, hot, abortifacient, purgative, blood purifier, and cathartic. The fruit is useful in jaundice, cerebral congestion, colic, constipation, dropsy, fever, worms, and sciatica. The root is given in case of abdominal enlargement, cough, asthma, inflammation of the breast, ulcers, urinary diseases, and rheumatism. The oil from seeds is used for poisonous bites, bowel complaints, epilepsy, and also for blackening of the hair. The details of the compounds are summarized in Tables 28.3 through 28.5 (Soam et al. 2013). Wilfortrine and wilforine have been reported to possess immunosuppressive effects. Wilforine is effective in the treatment of rheumatoid arthritis. Wilfortrine can inhibit leukemia cell growth in mice and shows anti-HIV activity. Pregnane is, indirectly, a parent of progesterone. It is a parent hydrocarbon for two series of steroids stemming from 5α-pregnane (originally allopregnane) and 5β-pregnane (17β-ethyletiocholane). 5β-pregnane is the parent of the progesterones, pregnane alcohols, ketones, and several adrenocortical hormones and is found largely in urine as a metabolic product of 5β-pregnane compounds. Pregnanes are steroid derivatives with carbons present at positions 1–21. Most biologically significant pregnane derivatives fall into one of two groups: pregnenes and pregnadienes. Another class is pregnatrienes. Many compounds related to this, found in *C. callosus,* require further investigations. Astaxanthin is a carotenoid. It belongs to a larger class of phytochemicals known as terpenes and

TABLE 28.2
Biochemical Composition of Kachri (g/100 g)

Ash	Moisture	Crude Fat	Total Protein	Total Carbohydrate	Crude Fiber	Crude Alkaloids (g/100 g)	Tannins (g/100 g)	Saponins (g/100 g)
11.8 ± 0.160	2.5 ± 0.061	6.828 ± 0.165	23.846 ± 0.991	61.214 ± 0.615	0.23 ± 0.165	12.96 ± 0.088	0.177 ± 0.007	7.28 ± 0.077

Source: Adapted from Mann, S. et al., *Int. J. Pharm. Life Sci.*, 4(1), 2335, 2013. With permission.

TABLE 28.3

Alkaloidal Compounds Having Pharmacological Significance in Kachri

SN	Alkaloids	Formula	Molecular Weight	Retention Time (min)	Strength Index	Relative Strength Index
1.	Imidazo[4,5-*E*]1,2,4-triazin-6-one, perhydro-4,5,7-trimethyl-3-thioxo or 4,5,7-trimethyl-3-thioxooctahydro-6*H*-imidazo[4,5-*E*][1,2,4]triazin-6-one	$C_7H_{13}N_5OS$	215	5.46–6.23	250	486
2.	1-Diphenylsilyloxybutane or butoxy(diphenyl)silane	$C_{16}H_{20}OSi$	256	5.46–6.23	250	376
3.	2-α′-Naphthylimino-2,3-dihydro-4*H*-1,3-benzothiazin-4-one or 2-β-naphthylimino-2,3-dihydro-4*H*-1,3-benzothian-4-one	$C_{18}H_{12}N_2OS$	304	13.32–17.94	321	581
4.	3-Methyl-1-phenethyl-4-phenyl(propionyloxy)piperdine	$C_{23}H_{29}NO_2$	351	5.46–6.23	244	314
5.	Palladium, bis(2-butyl-8-quinolinolato)	$C_{26}H_{28}N_2O_2Pd$	506	18.23–19.30	300	600
6.	2,5-Dichloro-4,6-bis-(1-methyl-1*H*-benzoimidazol-2yl)-isophthalic acid, dimethylester	$C_{26}H_{20}Cl_2N_4O_4$	522	15.71–16.23	316	552
7.	5*H*-Indeno(1,2-b)pyrazin-5-one, 6,7,8-tribromo-9-(ethoxycarbonyl)-*N*,*N*′-diethyl-1,2,3,4-tetrahydro-	$C_{18}H_{19}Br_3N_2O_3$	548	11.63–12.27	454	611
8.	Wilforine	$C_{43}H_{49}NO_{18}$	867	18.23–19.30	359	470
9.	Wilfortrine	$C_{41}H_{47}NO_{20}$	873	13.32–17.94	293	503

Source: Adapted from Soam, P.S. et al., *Eur. J. Exp. Biol.*, 3(1), 316, 2013. With permission.

is classified as a xanthophyll. Its primary use for humans is as a food supplement. Research shows that due to astaxanthin's potent antioxidant activity, it may be beneficial in cardiovascular, immune, inflammatory, and neurodegenerative diseases. Tripalmitin is a triglyceride derived from palmitic acid that is used to manufacture soap. Triglycerides have the mechanism for storing unused calories.

28.6 BOTANY

This species of the genus *Cucumis* is distinguishable on the basis of the following key as reviewed by John et al. (2013).

Fruits: Oval in shape; fruit skin covered by light, irregular, greenish-white stripes; fruits remain light green in color after ripening of fruits.

Leaves: Leaves shallowly three-lobed and rough.

Stem: Stem is soft—hairy to glabrous, angled; leaves ovate to reniform. Vine length remains up to 2.0–3.0 m, with many primary branches.

Flowers: Flowers borne solitary on stem, monoecious in nature. Staminate and pistillate flowers are solitary on short, stout pedicels. Under open natural system, it is partly cross and partly self-pollinated by bees. Flowers are in a 7:20 (female-to-male) ratio. Calyx and corolla are five in number and yellow in color. Anther bears abundant yellow color pollen; however, stigma is divided into three lobes. Anthesis starts from 6:00 a.m. and continues up to 10:00 a.m. (Table 28.6).

TABLE 28.4

Steroidal Compounds Having Pharmacological Significance Found in Kachri

SN	Steroids	Formula	Molecular Weight	Retention Time (min)	Strength Index	Relative Strength Index
1.	2-α′-Naphthylimino-2,3-dihydro-4H-1,3-benzothion-4-one	$C_{18}H_{12}N_2OS$	304	13.32–17.94	314	588
2.	Pregan-20-one,3,17,21-tris[(trimethylsilyl)oxy]-O-(phenylmethyl)oxime,(3α′,5α′)	$C_{37}H_{65}NO_4Si_3$	671	11.63–12.27	480	601
3.	5β-pregan-20-one,3α′,11β,17,21-tetrakis(trimethylsiloxy)-, O-methyloxime	$C_{34}H_{69}NO_5Si_4$	683	11.63–12.27	433	578
4.	Silane, [[(3β,5α′,11β,20S)-pregane-3,11,17,20,21-pentayl]pentakis(oxy)]pentakis trimethyl	$C_{36}H_{76}O_5Si_5$	728	15.71–16.23	399	532
5.	5β-Cholan-24-oic acid,4-(23-carboxy-7,12-dioxo-24-nor-5β5-chol-3-en-3-yl)-3,7,12-trioxo-, dimethylester	$C_{50}H_{70}O_9$	814	18.23–19.30	370	488

Source: Adapted from Soam, P.S. et al., *Eur. J. Exp. Biol.*, 3(1), 316, 2013. With permission.

TABLE 28.5

Terpenes, Triacylglycerols, and Similar Compounds Found in Kachri

SN	Terpenes	Formula	Molecular Weight	Retention Time (min)	Strength Index	Relative Strength Index
1.	Astaxanthin	$C_{40}H_{52}O_4$	596	11.63–12.27	459	529
	Triacylglycerols					
2.	Tripalmitin	$C_{51}H_{98}O_6$	806	9.70–11.37	393	493
	Similar compounds					
3.	2-Nonaprenyl-6-methoxyphenol	$C_{52}H_{80}O_2$	736	13.22–18.02	330	410

Source: Adapted from Soam, P.S. et al., *Eur. J. Exp. Biol.*, 3(1), 316, 2013.

28.7 TAXONOMY

C. callosus is a feral species of India. The synonym given for this species is *C. trigonus*. Earlier cytologists have shown that *C. callosus* has 14 somatic chromosomes (Darlington and Wylie 1955; Singh and Ray 1973). The breeding experiments of Naudin (1859) and Sambandam and Chelliah (1972) have given evidence of free compatibility between *C. callosus* and *C. melo*. Hence, to obviate the confusion regarding the chromosome number as well as the relationship with other *Cucumis* species, Parthasarthy and Sambandam (2005) conducted an experiment to ascertain compatibility with other *Cucumis* species. Crosses were made (including reciprocals) with seven *Cucumis* species, namely, *Cucumis metuliferus*, *Cucumis anguria*, *Cucumis longipes* (=*Cucumis anguria* var. *longipes*), *Cucumis zeyheri*, *Cucumis myriocarpus*, *Cucumis dipsaceus*, and *C. melo*. The success of each cross was determined by the percent fruit set and the number of seeds. F_1 fertility was

TABLE 28.6
Comparative Morphology of *Cucumis callosus*, *Cucumis melo*, and *Cucumis melo* ssp. *agrestis*

Character	*Cucumis callosus*	*Cucumis melo*	*Cucumis melo* ssp. *agrestis*
Life span	Perennial	Annual	Annual
Tap root	Tuberous	Nontuberous	Nontuberous
Stem type	Slender	Robust	Moderately slender
Branching	Less, drooping	Branching, procumbent	Highly branching, procumbent
Petiole	Slender	Robust	Moderately robust
Leaf lobing	Deeply lobed	Unlobed	Unlobed
Leaf lamina shape	Suborbicular	More or less reniform	Suborbicular
Leaf pubescence	Hirsute	Villose	Subhirsute
Staminate flowers	Solitary	Fascicles of 4–9	Fascicles of 3–5
Petal color	Light greenish yellow	Bright yellow	Bright yellow
Pistillate flower pedicel	Looped, U-shaped	Slightly curved	Slightly curved
Fruit shape	Round/obovoid	Polymorphous	Oblong or turbinate
Ovary pubescence	Tomentose	Pubescent/puberulent	Pubescent
Seed funicle	Mucronate	Acute	Acute
Seed size	Medium, bulged	Big	Small
Shelf life of fruits	Longer duration (12 months or more)	Short duration (few days)	Short to medium
Fruit pulp taste	Bitter/sour	Sour	Bitter/sour
Dispersal	Difficult as fruit wall is intact	Easy as fruit wall crack or disintegrate	Moderately easy
Leaf length (cm)	6–7.5	6.5–9.5	6.0–7.5
Flower diameter (cm)	2.0	4–6.0	2–4.0
Pedicel length (cm)	0.5–1.0	1–2.5	Up to 1.0
Calyx tube (mm)	2–4.0	6–8.0	3–5.0
Anther length (mm)	2.0	3–4.0	1.5
Fruit length (cm)	3.5–5.0	7–20.0	2.5–6.0
Fruit diameter (cm)	3–5.2	8–15.5	2.8–4.5
Fruit circumstance (cm)	12.4–13.2	22.3–43.0	8–11.3
Single fruit weight (g)	13–21.0	150–1250.0	10–30.0
Flesh thickness (cm)	0.4–0.5	1.4–3.5	0.4–0.6
Seed length (mm)	6.61–7.22	7.8–9.37	4.18–4.94
100 seed weight (g)	1.6–1.7	1.7–1.9	1.3–1.6

Source: Adapted from John, K.J. et al., *Genet. Resour. Crop Evol.*, 60, 1037, 2013. With permission.

determined by pollen fertility and viability (germination). Meiotic studies were carried out in the PMCs of F_1 plants to study the behavior of chromosomes during diakinesis. The metaphase plates revealed a somatic chromosome number of 24 for *C. callosus*. The crosses with seven *Cucumis* species indicated that *C. callosus* was crossable with only *C. melo*. The number of well-developed seeds per fruit in the cross *C. callosus* × *C. melo* was, on an average, 266 seeds, and the seeds showed about 93.0% germination. The pollen mother cells showed normal bivalent formation, indicating cross compatibility between *C. melo* and *C. callosus*. The somatic chromosome number observed by Parthasarathy and Sambandarn (2005) for *C. callosus*, that is, 2n = 24, is in disagreement with that of Singh and Ray (1973), who reported the 2n number to be 14. It is interesting to note the report of Brown et al. (1969) who stated that *Cucumis triganus* from India was mislabeled and in reality it was *Cucumis hardwickii*. *C. triganus* is the synonym for *C. callosus*. But the confusion in the taxonomy of *Cucumis*, especially those found in India, is due to the fact that names have been given based on the morphological differences. Chakravarthy (1959) placed *C. triganus* under *C. callosus* and *C. pubescens* under *C. melo*. Watt (1898) stated that there is much confusion regarding the Indian so-called wild and cultivated species and varieties of *Cucumis*. On the basis of chromosome number and its free compatibility with *C. melo* and the normal behavior of chromosomes during diakinesis, Parthasarthy and Sambandam (2005) concluded that *C. callosus* does not warrant a separate species status and is nothing but a progenitor of *C. melo*. The herb is much branched, commonly prostrate, and perennial. Leaves are cordate, suborbicular, deeply palmate, and five to seven-lobed. Flowers are yellow. Fruits are smooth, obovoid, ellipsoid, with green variegated stripes. Fruiting ranges between August and November.

28.8   BIOSYSTEMATICS

It grows on various types of soils: black cotton soil, clayey loam, gravely soil, and sandy loam with slightly alkaline pH. The crop grows well in an arid environment with low relative humidity (John et al. 2013). It is also grown on sandy soils, prostrate, or climbing on field hedges (Bhandari 1978). Subterranean sprouts emerge from rootstocks with pre-monsoon showers and come to flowering and fruiting from July to December in the Deccan plateau, East and NE India, and Indo-Gangetic plains. Seeds germinate with pre-monsoon showers and take around 60–70 days for flower initiation. Anthesis takes place between 5:00 and 5:30 a.m. The ideal time for pollination for maximum fruit set is between 7:00 and 9:00 a.m. Unlike other *Cucumis* species, the fruits continue to remain on vines for months together without any abscission. However, seeds extracted from 90-day-old fruits were found to be fully viable, indicating a requirement of less than 90 days for physiological maturity. Most of the problems encountered in wide hybridization like nonadaptability of wild taxa, nonsynchronous flowering, and pre- and postfertilization barriers are not observed in *C. callosus* and *C. melo* hybridization programs. However, correct taxonomic delineation and identification is of importance to facilitate the use of these wild species in crop improvement programs. Assigning a distinct taxonomic status is also important from a conservation point of view, lest the entity may not get adequate representation in ex situ gene banks (John et al. 2013).

28.9   PRODUCTION TECHNOLOGY

Kachri is essentially a warm season crop. Generally, a long period of warm and humid climate is required. It can be grown in mild climate but is sensitive to frost. The optimum temperature requirement for successful crop growth is 24°C–30°C. Seed sowing time is from June to July and fruit harvesting remains up to mid November to December. Crop produce (fruit) sold in the market is as high as Rs. 20.0 kg^{-1}. The seed rate for kachri is about 2.5 kg/ha. The seeds are sown on raised beds or in furrows or in pits when it is grown as a sole crop. It is mostly grown as a mixed crop in the *Kharif* season. The spacing followed is 0.5 m × 0.2–0.3 m. Generally, no irrigation is done by farmers when grown as a mixed crop with *Kharif* crops, but it needs at least one or two irrigations for better yield. The fruit takes 8–10 days from anthesis to harvesting stage. Fruits are harvested in immature stage for domestic uses. With proper care, it produces 50–60 q/ha immature fruits.

28.10   TOLERANCE TO BIOTIC AND ABIOTIC STRESSES

The authors have personally observed its potential for tolerance to extreme drought, growing and reproducing for many months, as well as its ability to survive under a severe epidemic of spider mite and its resistance to fruit fly and *Fusarium* wilt. Incorporation of genes from wild species has been advocated for broadening the genetic base of melon/cultivated crops and transfering several useful traits related to abiotic and biotic stress tolerances. A wider distribution across 75,000 km^2 stretch, encompassing diverse agroecological zones touching the Himalayan foothills to Eastern Ghats, Aravalli and Vindhya mountains, and Deccan plateau, indicates the possibility for collection of variability in this wild taxa. Utilization of this taxa in melon improvement will lead to the development of varieties with extreme drought tolerance; resistance to fruit fly, powdery mildew, *Fusarium* wilt, and spider mite; long shelf life; extended harvest time; and noncracking of skin and mesocarp (John et al. 2013).

28.11 PEST MANAGEMENT

28.11.1 FRUIT ROT (*FUSARIUM PALLIDOROSEUM*)

Kachri is a warm season crop that is grown as a wild or cultivated crop on a large scale with pearl-millet, moth-bean, green gram, and cluster bean crops in arid and semi-arid regions. Kachri fruits were found infected with fruit rot damage both quantitatively and qualitatively (Kumawat et al. 2013)

28.11.1.1 Control

The combination of carbendazim 12% + mancozeb 63% at 100, 300, and 500 ppm was found to be the most effective in the inhibition (100%) of mycelial growth of *F. pallidoroseum* followed by the individual application of the two drugs.

28.12 FUTURE THRUST

1. Greater focus on collection, conservation, evaluation, and utilization of landraces
2. Documentation of indigenous technical knowledge
3. Concerted research on breeding and production technology
4. Use as donor for disease resistance breeding in relevant *Cucumis* sp.

REFERENCES

Abbo, S., Lev-Yadun, S., and Gopher, A. 2010. Agricultural origins: Centres and non-centres: A near eastern reappraisal. *Critical Review of Plant Science* 29: 317–328.

Akashi, Y., Fukuda, N., Wako, T., Masuda, M., and Kato, K. 2002. Genetic variation and phylogenetic relationships in East and South Asian melons *Cucumis melo* L. based on the analysis of five isozymes. *Euphytica* 125: 385–396.

Allaby, R.G., Brown, T., and Fuller, D.Q. 2010. A simulation of the effect of inbreeding on crop domestication genetics with comments on the integration of archaeobotany and genetics: A reply to Honne and Heun. *Vegetation History and Archaeobotany* 19: 151–158.

Bhandari, M.M. 1978. *Flora of the Indian Desert*. Scientific Publishers, Jodhpur, India.

Brown, G.B., Deakin, J.R., and Wood, M.B. 1969. Identification of *Cucumis* species by paper chromatography of flavonoids. *Journal of American Society of Horticultural Sciences* 94: 231–234.

Chakravarthy, H.L. 1959. *Monograph of Indian Cucurbitaceae—Taxonomy and Distribution*. Records of the Botanical Survey of India, Delhi, India, Vol. 17, pp. 98–112.

Chakravarthy, H.L. 1982. *Fascicles of the Flora of India: Cucurbitaceae*. Botanical Survey of India, Howrah, India, p. 136.

Chand, T., Bhandari, A., Kumawat, B.K., Sharma, A., Pareek, A., and Bansal, V.K. 2012. *In vitro* antioxidant activity of aqueous extract of seeds of *Cucumis callosus* (Rottl.) Cogn. *Der Pharmacia Letter* 4(3): 840–844.

Clarke, C.B. 1879. Cucurbitaceae. In: *The Flora of British India*, Vol. II: *Sabiaceae to Cornaceae*, ed. J.D. Hooker. L. Reeve & Co., London, U.K., pp. 619–620.

Darlington, C.D. and Wylie, A.P. 1955. *Chromosome Atlas of Flowering Plants*. Allewn & Unwin Ltd., London, U.K.

Dubey, R.K. and Ram, H.H. 2006. Kachri (*Cucumis callosus*) Rottl.) Cogn., taxonomy, distribution and future prospects. In: *Cucurbits Breeding and Production Technology*, eds. P.L. Gautam, H.H. Ram, and H.P. Singh. GBPUAT, Pantnagar, India, pp. 199–203.

Dwivedi, N.K., Dhariwal, O.P., Krishnan, S.G., and Bhandari, D.C. 2010. Distribution and extent of diversity in *Cucumis* species in the Aravalli ranges of India. *Genetic Resources and Crop Evolution* 57: 443–452.

Fuller, D.Q. 2006. Agricultural origins and frontiers in South Asia: A working synthesis. *Journal of World Prehistory* 20: 1–86.

Fuller, D.Q. 2012. New archaeobotanical information on plant domestication from macro-remains: Tracking the evolution of domestication syndrome traits. In: *Biodiversity in Agriculture: Domestication, Evolution, and Sustainability*, eds. P. Gepts, T.R. Famula, R.L. Bettinger, S.B. Brush, A.B. Damania, P.E. McGuire, and C.O. Qualset. Cambridge University Press, Cambridge, U.K., pp. 110–135.

Fuller, D.Q., Corisettar, R., and Venkatasubbiah, P.C. 2001. Southern neolithic cultivation systems: A reconstruction based on archeobotanical evidence. *Journal of Social South Asian Study* 17: 171–187.

Fuller, D.Q., Willcox, G., and Allaby, R.G. 2011. Early agricultural pathways: Moving outside the "core area" hypothesis in Southwest Asia. *Journal of Experimental Botany* 63: 617–633. doi:10.1093/jxb/err307.

Jeffrey, C. 1980. Further notes on Cucurbitaceae. V. The Cucurbitaceae of the Indian subcontinent. *Kew Bulletins* 34: 789–809.

John, K.J., Scariah, S., Nisar, V.A.M., Latha, M., Gopalkrishanan, S., Yadav, S.R., and Bhatt, K.V. 2013. On the occurrence, distribution, taxonomy and genepool relationship of *Cucumis callosus* (Rottler) Cogn., the wild progenitor of *Cucumis melo* L. from India. *Genetic Resources and Crop Evolution* 60: 1037–1046.

Kerje, T. and Grum, M. 2000. The origin of melon *Cucumis melo*: A review of the literature. *Acta Horticulturae* 510: 34–37.

Kirkbride, J.H. 1993. *Biosystematic Monograph of the Genus Cucumis (Cucurbitaceae)*. Parkway Publishers, Boone, NC, p. 159.

Kristkova, E., Lebeda, A., Vinter, V., and Blahousek, O. 2003. Genetic resources of the genus *Cucumis* and their morphological description. *Horticultural Science* 30: 14–42.

Kumawat, R., Jat, R.G., and Kumawat, K. 2013. Efficacy of fungicides against *Fusarium pallidoroseum* causing fruit rot of Kachri (*Cucumis callosus*). *International Journal of Plant Protection* 6: 213–214.

Kurz, S. 1877. Contributions towards knowledge of the Burmese flora. *Journal of Asiatic Society of Bengal* 46(2): 95–106.

Lev-Yadun, S., Gopher, A., and Abbo, S. 2000. The cradle of agriculture. *Science* 288: 1602–1603.

Luan, F., Delannay, I., and Staub, J.E. 2008. Chinese melon (*Cucumis melo* L.) diversity analyses provide strategies for germplasm curation, genetic improvement, and evidentiary support of domestication patterns. *Euphytica* 164: 445–461.

Mann, S., Chaudhary, I., and Gupta, R.K. 2013. Value addition scenario of arid foods of desert area and evaluation of their nutritional and phytochemical potential. *International Journal of Pharmacy and Life Sciences* 4(1): 2335–2339.

Matthew, K.M. 1983. *The Flora of the Tamilnadu Carnatic*, Vol. 3. Diocesan Press, Madras, India.

Naudin, C. 1859. Revue des cucurbitaceaes *cultivees* au Museum en. *Annales des Sciences Naturelles; Botanique, sér* 4(12): 79–164.

Nesom, G.L. 2011. Towards consistency of taxonomic rank in wild/domesticated Cucurbitaceae. *Phytoneuron* 13: 1–33.

Neuhausen, S.L. 1992. Evaluation of restriction length polymorphism in *Cucumis melo*. *Theoretical and Applied Genetics* 83: 379–384.

Parthasarathy, V.A. and Sambandam, C.N. 2005. Taxonomy of *Cucumis callosus* (Rottl.) Cogn., the wild melon of India. *Cucurbits Genetic Cooperative Report* 3: 66–67.

Perin, C., Hagen, L.S., De Conto, V., Katzir, N., Danin-Poleg, Y., Portnoy, V., Baudracco-Arnas, S., Chadoeuf, J., Dogimont, C., and Pitrat, M. 2002. A reference map of *Cucumis melo* based on two recombinant inbred line populations. *Theoretical and Applied Genetics* 104: 1017–1034.

Renner, S.S. and Schaefer, H. 2008. Phylogenetics of *Cucumis* (Cucurbitaceae) as understood. In: M. Pitrat (ed.), *Proceedings of the IXth EUCARPIA Meeting on Genetics and Breeding of Cucurbitaceae*, INRA, Avignon, France.

Sambandam, C.M. and Chelliah, S. 1972. Scheme for the evaluation of cantaloupe and muskmelon varieties for the resistance to the fruitfly, USDA PL-480 Research Project. Annamalai University, Chidambaram, India, 67pp.

Sebastian, P., Schaefer, H., Telford, I.R.H., and Renner, S.S. 2010. Cucumber (*Cucumis sativus*) and melon (*Cucumis melo*) have numerous wild relatives in Asia and Australia, and the sister species of melon is from Australia. *Proceedings of the National Academy of Sciences of the United States of America* 107: 14269–14273.

Shattuck-Eidens, D.M., Bell, R.N., Neuhausen, S.L., and Helentjaris, T. 1990. DNA sequence variation within maize and melon: Observation from polymerase chain reaction, amplification and direct sequencing. *Genetics* 126(1): 207–217.

Singh, A.K. and Ray, R.P. 1973. *Karyological* studies in *Cucumis*. *Caryologia* 27: 153–160.

Soam, P.S., Singh, T., Vijayvergia, R., and Jayabaskaran, C. 2013. Liquid chromatography–mass spectrometry based profile of bioactive compounds of *Cucumis callosus*. *European Journal of Experimental Biology* 3(1): 316–326.

Tanaka, K., Nishitani, A., Akashi, Y., Sakata, Y., Nishida, H., Yoshino, H. and Kato, K. 2007. Molecular characterization of South and East Asian melon, *Cucumis melo* L., and the origin of Group Conomon var. makuwa and var. conomon revealed by RAPD analysis. *Euphytica*, 153: 233–247.

Tanksley, S.D. and McCouch, S.R. 1997. Seed banks and molecular maps: Unlocking genetic potential from the wild. *Science* 277: 1063–1066.

Vavilov, N.I. 1935. Theoretical basis for plant breeding. Origin and geography of cultivated plants. In: *Phytogeographical Basis for Plant Breeding* (D. Love, transl.). Cambridge University Press, Cambridge, U.K., pp. 316–366.

Verma, D.M. and Pant, P.C. 1985. Cucurbitaceae. In: *Flora of India*. Series 3: Flora of Raipur, Durgand and Rajnandgaon. Botanical Survey of India, Kolkata, India, pp. 157–160.

Watt, G. 1898. *Dictionary of Economic Products of India*. Periodical Experts, New Delhi, India, Vol. 2, 626pp.

Whitaker, T.W. and Bemis, W.P. 1976. *Cucurbits, Cucumis, Citrullus, Cucurbita, Lagenaria* (Cucurbitaceae). In: *Evolution of Crop Plants*, ed. N.W. Simmonds. Longman, London, U.K., pp. 64–69.

Whitaker, T.W. and Davis, G.N. 1962. *Cucurbits*. Inter Science Publishers Inc., New York.

Zohary, D. and Hopf, M. 2000. Domestication of plants in the old world. *Genetic Resources and Crop Evolution* 60: 1037–1046.

29 Squashes and Gourds

Sanjeev Kumar, Puja Rattan, and R.K. Samnotra

CONTENTS

29.1   INTRODUCTION

Cucurbits belong to the family Cucurbitaceae. Jeffrey (1990) suggested that it consists of about 118 genera and 825 species. Cucurbits are cultivated throughout the world and are among the most important plant families that supply human with edible products and useful fibers. Cucurbits, which include cucumbers, muskmelons, watermelons, pumpkins, summer squash, winter squash, and gourds, are some of the most popular garden vegetables planted today. Cucurbit crops are similar in their appearance and requirements for growth. They are prostrate, sprawling vines, usually with tendrils. Each vine bears many large, lobed leaves. For all cucurbits, except the bottle gourd, the flowers are bright yellow. Each vine bears two kinds of flowers: pistillate (female) and staminate (male). Although cultivated cucurbits are very similar in aboveground development and root habit, they are extremely diverse in fruit characteristics. Fruits of cucurbits can be eaten raw when immature as in summer squash or mature as in the case of watermelon. The fruits of some of the cucurbits are consumed after being baked (squash), pickled (cucumber), and candied (watermelon). Some of the cucurbits are consumed fresh in salads (cucumber) or as dessert (melon). Cucurbits are also produced for other uses than food. In certain cucurbits like bottle gourd, fruits are used for storage, drinking containers, bottles, utensils, smoking pipes, musical instruments, gourd craft decoration, masks, floats for fish net, and other items. The fiber of a mature loofah fruit can be used as a sponge for personal hygiene, household cleaning, and various other purposes, including filtration. Seeds or fruit parts of some cucurbits are reported to possess purgative, emetic, and antihelminthic properties due to the secondary metabolite cucurbitacin content (Robinson and Decker-Walters, 1997). Chambliss and Jones (1966a) suggested that fruits and roots with high cucurbitacin content can be used as an insect attractant (e.g., cucumber beetle, *Diabrotica* ssp.) or as an insect repellent (e.g., honeybee, *Apis mellifera* L., and yellow jacket wasp, *Vespula* sp.), while Bar-Nun and Mayer (1990) revealed that ectopic application of cucurbitacin can function as a protectant against infection by *Botrytis cinerea*. Therefore, cucurbits are among the largest and the most diverse plant families, have a large range of fruit characteristics, and are cultivated worldwide in a variety of environmental conditions. Cucurbits are associated with the origin of agriculture and human civilizations and are also among the first plant species to be domesticated in both the Old and the New World. Ralf Norrman and Jon Haarberg (1980) examined the symbolic place of cucurbits in Western literature and culture and further extended their analysis to selected non-Western cultural settings. They noted that cucurbits have complex semiotic associations with sex and sexuality, fertility, vitality, moisture, creative power, rapid growth, and sudden death. Cucurbits also figure prominently in the symbolism and cosmologies of many non-Western societies (Table 29.1).

29.2   SUMMER SQUASH

The summer squash, *Cucurbita pepo*, has its origin in areas of Mexico and Central America. Before Columbus arrival in the New World, summer squashes were cultivated in much of what is now the southwestern, midwestern, and eastern United States by Native Americans. Archaeological records of the New World suggest that *Cucurbita* was one of the first plant to be domesticated (Nee, 1990), whereas the first species to be domesticated in the New World was *C. pepo*. It is the most diverse species consisting of wild forms in the United States and Mexico, previously separately classified as *C. texana* Grey and *C. fraterna* Bailey, respectively. It has many botanical synonyms,

TABLE 29.1

Some Defining Characteristics of the Cultivated *Cucurbita* Species

Species	Seed	Leaf	Stem	Peduncle	Fruit Flesh
C. argyrosperma	Large, white, prominent margin that may be scalloped	Moderately lobed, short, soft pubescence	Hard, angular	Hard, corky, sometimes swollen	Very coarse, pale yellow
C. ficifolia	Black or tan, smooth margin	Deeply lobed, smooth margin, round, prickly	Hard, grooved	Hard, angled, slightly expanded at fruit attachment	Coarse, stringy, white
C. maxima	White to brown, oblique seed scar	Almost round, unlobed, prickly	Soft, round	Round, corky, not flared at fruit attachment	Fine, not fibrous, deep orange
C. moschata	White to brown, rough margin, oblique seed scar	Shallow lobes, almost round, soft pubescence	Hard, ridged	Hard, angled, flared at fruit attachment	Fine, not fibrous, deep orange
C. pepo	Light tan, prominent smooth margin, rounded seed scar	Deeply lobed, very prickly	Hard, ridged	Hard, angled, ridged	Coarse, orange

namely, *Cucurbita aurantia* Willd., *Cucurbita courgero* Ser., *Cucurbita elongata* Bean ex Schrad., *Cucurbita esculenta* Gray, *Cucurbita melopepo* L., *Cucurbita ovifera* L., *Cucurbita subverrucosa* Willd., *Cucurbita verrucosa* L., *Cucurbita fastuosa* Salisb., *Cucurbita mammeata* Molina, *Cucurbita maxima* var. *courgero* Ser., *Cucurbita oblonga* (Duch. ex Lam) Link, *Cucurbita venosa* Descourt., *C. pepo* (L.) Dumort, *Pepo citrullus* Sag., *Pepo melopepo* (L.) Moench., *Pepo vulgaris* Moench., and *Pepo verrucosus* (L.) Moench.

There are many forms of *C. pepo* for which different names are used. The term pumpkin is rooted in a Greek and Latin word for a large round of fruits, whereas squash comes from the plural form of a native North American word for something immature and incomplete. Therefore, the term pumpkin is used for the edible *Cucurbita* fruits that are round or nearby round, whereas the term squashes are applied for the edible fruits that deviate in shape from roundness. Another big line of difference is that pumpkins are consumed when they are mature, while squashes are consumed when they are immature. The term "zucchini" is the short plural form of Italian zucca for pumpkin squash or gourd. Zucchini applies to cylindrical shaped fruits, similar to those of the original cultivars bearing the name zucchini (Tapley et al., 1937).

On the basis of genetic relationship, *C. pepo* has been subdivided into two subspecies, *pepo* and *ovifera*, which differ in their origin, the former being Mexican and the latter being from eastern United States (Decker, 1988). Ferriol et al. (2013) also studied polymorphism in *C. pepo*. They grouped the cultivars into eight morphotypes in two subspecies, ssp. *pepo* and ssp. *ovifera*. They reported that in ssp. *ovifera*, the accessions of the different morphotypes were basically grouped according to the fruit color. This may indicate different times of development and also the extent of breeding in the accessions used. The landraces of the spp. *ovifera*, used as ornamental plants in Europe, can be of great interest for preserving the diversity of *C. pepo* (Table 29.2).

29.2.1 Medicinal Value

Summer squash is an economically important plant and is cultivated throughout the world for oil and medical purposes (Fu et al., 2006). Pharmacological effects, comprising antidiabetic, antihypertensive, antitumor, antimutagenic, immunomodulating, antibacterial, antihypercholesterolemic, intestinal antiparasitic, antalgic, and anti-inflammation effects, and utilization possibilities

TABLE 29.2

Summer Squash Grouped into Six Different Extant Groups on the Basis of the Fruit Shape

Group	Fruit Shape
Scallop	Flattened with scalloped margins
Crookneck	Elongated with narrow, long, slightly to very curved neck; broad distal half, convex or pointed distal end
Straightneck	Cylindrical with very short neck or constriction near the stem end and a broad distal half, convex or pointed distal end
Vegetable	Short, tapered cylindrical, narrow at the stem end, broad at the distal end, length to broadest width ratio ranging from 1.5 to 3.0
Cocozelle	Long to extremely long slightly tapered cylindrical, bulbous near blossom end, length to broadest width ration ranging from 3.5 to 8 or even higher
Zucchini	Uniform cylindrical, little or no taper, length to broadest width ration ranging from 3.5 to 4.5

Source: Paris, H.S., *Econ. Bot.*, 43(4), 423, 1986.

of various summer squash species have been reported (Kostalova et al., 2009). As an excellent source of manganese and a very good source of vitamin C, summer squash provides us with a great combination of conventional antioxidant nutrients. But it also contains an unusual amount of other antioxidant nutrients, including carotenoids lutein and zeaxanthin. These antioxidants are especially helpful in antioxidant protection of the eye, including protection against age-related macular degeneration and cataracts. This vegetable also provides us with antioxidant advantages in terms of its antioxidant stability. Recent research has confirmed strong retention of antioxidant activity in summer squash after steaming. Research has also confirmed excellent retention of antioxidant activity in summer squash after freezing. (Anonymous, 2015)

Seeds of summer squash (*C. pepo* L.) have long been used as a medicine for various ailments, particularly as a treatment against worms (Lewis et al., 1997). In Eritrea, Sudan, and Ethiopia, summer squash seeds are used to treat tapeworm, when the dried seeds are eaten on an empty stomach. For many years, particularly in Europe, extracts from summer squash seeds, *C. pepo*, have been used in folk medicine as a remedy for micturition caused by benign prostatic hyperplasia (Younis et al., 2000). Summer squash is used for treating helminth and as a medicine to reduce bad cholesterol.

29.2.2 BOTANY

C. pepo is monoecious (unisexual flowers, with male and female on the same plant) and bears solitary actinomorphic flowers (~10 cm across) that produce nectar. The calyx is campanulate with five free sepals; each sepal is linear and 0.9–3 cm long, smaller on pistillate flowers. The yellow corolla is campanulate and five-parted (occasionally six-parted) with erect to spreading petals that are apically acute, approximately 5–10 cm long, ca. 3 cm broad on smaller flowers. Staminate flowers are borne on a 3–20 cm long peduncle and have three stamens with free filaments (~1.5 cm long) and connivent, twisted anthers (~1 cm long). Pistillate flowers are borne on shorter peduncles, only 2–5 cm long, and have an inferior one (ovary). They have a locular, globose to cylindrical ovary and are of thick style with three bilobed stigmas. The peduncle of the developing fruit is hard and slightly bent, expands at the junction of the fruit, and ranges from 5.6 to 15.1 mm in diameter. Lu and Jeffrey (2011) suggested that *C. pepo* plants are climbing, having stem setose and scabrous-hairy, leaf petiole setose, 6–9 cm; leaf blade triangular or ovate-triangular, both surfaces scabrous-hairy, base cordate, margin irregularly dentate, apex acute. Fruiting pedicel robust, conspicuously angular-sulcate, apex slightly thickened; fruit variable in shape and size with numerous, white, ovate, ca. 20 mm, marginate and obtuse seeds.

29.3 WINTER SQUASH

Cucurbita moschata, which encompasses various cultivars of pumpkin and winter squash, is a plant species in the gourd family (Cucurbitaceae) cultivated in warm areas around the world as food and animal fodder. The names "winter squash" and "pumpkin" are also applied to the cultivars of *C. maxima*, *C. mixta*, and *C. pepo*. Because the common names are used to refer to several different species and those species may also have other common names, it can be difficult to ascertain the species from which each variety is derived. *Cucurbita moschata* is native to the lowlands of tropical and subtropical America (Mexico and South America); it is unique in being spread in two distinct native areas, a major one in Mexico and a minor one in the northern South America (Whitaker and Bemis, 1975). It likely originated in Mexico and Central America and was already widely cultivated in North and South America before the arrival of the Europeans. Archaeologists have found evidence of *C. moschata* in Peruvian sites dated from 4000 to 3000 BC and in Mexican sites from 1440 to 400 BC, suggesting a long history of domestication and cultivation. *C. moschata* is better adapted to hot, humid climates than *C. pepo* and *C. maxima*. *C. moschata* was the most variable and closely related species and nearest to the common ancestor of the genus because of the high interspecific compatibility (Whitaker and Bemis, 1975). An isozyme study showed high allelic diversity in *C. pepo* and *C. moschata*. *C. pepo* shares a common ancestor with *C. moschata* and *C. argyrosperma* but not with *C. maxima* (Decker-Walters et al., 1990). Esteras et al. (2008) revealed in their study that the Spanish accessions of *C. moschata* were highly variable, with the highest variability found in the Canary Islands and in the Mediterranean region within the Iberian Peninsula. Up to nine morphotypes were identified, with variability for fruit size, rind, and flesh color. Junxin Wu et al. (2011) revealed that a total of 1032 landraces of *C. moschata* are maintained in China; little is known about their genetic diversity. The accessions from China were classified into two clusters, which were clearly differentiated from the accessions originating from Mexico, Guatemala, Honduras, and Ecuador. The Chinese group is genetically more closely related to other Asian countries' group (India and Japan). In general, the accessions from the Americas had a greater number of unique loci than those from China. The differences are probably due to a limited number of introductions and genetic drift. A few studies concerning the morphological diversity among the landraces of the squash from different centers of diversity such as Cuba, Korea, and Puerto Rico have revealed a great variability of the squash types (Rios et al., 1997; Chung et al., 1998; Wessel-Beaver, 1998). Recently, Ferriol et al. (2004a,b) also found considerable landraces of *C. moschata* and *C. maxima*, especially in the fruit shape, ribbing, and size. There are also quite a few studies concerning molecular analysis in *Cucurbita* species. Random amplified (RAPD) markers were used to analyze the genetic diversity among *C. moschata* landraces from Korea, southern Africa, and other geographical origins (Youn and Chung, 1998; Baranek et al., 2000; Gwanama et al., 2000). In all the cases, the accessions were grouped according to the agroclimatic regions of origin and not according to the morphological traits. Ferriol et al. (2003) studied the genetic diversity among 19 Spanish accessions of *C. maxima* using two different molecular marker types: sequence-related amplified polymorphism (SRAP) and RAPD. More recently, Ferriol et al. (2004a,b), in addition to morphological analysis, employed SRAP and amplified fragment length polymorphism molecular markers for analyzing the diversity among a large number of *C. moschata* and *C. maxima* landraces. In all the cases, the results from the molecular analysis were correlated with the morphological characteristics. However, very few studies have attempted to compare results derived from the individual versus combined data sets not only for *Cucurbita* species but also for other crop species (Ajmone-Marsan et al., 1992; Franco et al., 1997; Russell et al., 1997). In addition, there are almost very few analyses combining data sets of different nature, that is, qualitative and quantitative morphological data or morphological and molecular marker data (Franco et al., 1997, 1998, 2001). Such kind of analyses could provide evidence whether a better estimate of the genetic diversity is obtainable and also whether the total evidence is within the confidence limits of evidence from the individual data sets. In botanical literature, *C. moschata* is reported as being grown mainly in areas of low altitude with a hot climate and high humidity (Esquinas-Alcazar and Gulick, 1983; Whitaker, 1990). However, while it is true that this species is preferentially grown within these limits, they do not appear to be strictly adhered to, as variants have been found above 2200 m. In general, *C. moschata* is the cultivated *Cucurbita* least tolerant of low temperatures but is relatively drought tolerant. *C. moschata* is cultivated in subtropical and tropical regions worldwide but can also be cultivated sporadically elsewhere.

29.3.1 BOTANY

C. moschata plants are frost-intolerant, monoecious annuals. Stems are hairless or soft hairy and trailing or climbing vines grow to 3 m. Young leaves, flowers, shoot tips, fruits, and seeds are edible. Leaves are simple, alternate, and shallowly lobed, often with white spots along the veins. The peduncle (stem that holds the fruit) is five-angled and flares outward where it is attached to the fruit. *C. moschata* has pentamerous, solitary, axillary flowers. The male flowers have 16–18 cm pedicels and a very short calyx is broadly campanulate to pateriform, expanded, or foliaceous toward the apex, 5–13.5 cm long, with five divisions for up to one-third of their length. The female flowers have thick pedicels of 3–8 cm in length and a globose, ovoid, oblate, cylindrical, piriform, conical, and turbinate ovary. They have a very small calyx and sepals that are more often foliaceous than in the males are measured up to 7.5 cm in length, and are of the thickened style. They have three lobate stigmas (Lira Saade et al., 1995). The fruits (technically referred to as pepos) are relatively large, with shapes ranging from globose to oblong to flat. Seeds are 16–20 mm long. *C. moschata* is self-compatible, but most of its varieties are not self-pollinated due to the spatial differences in the location of staminate and pistillate flowers; thus, they require pollen vectors to effect pollination and thereby facilitate fruit setting (Roubik, 1995).

It produces a variety of fruits that vary considerably in size and shape due to large genetic variations within this species. The fruit is typically orange or yellow and has many creases running from the stem to the bottom. The fruits are usually not harvested when young but are left on the plant to mature for eventual fall harvest as winter squashes. They have a thick shell on the outside, with seeds and pulp on the inside. Pumpkins are monoecious; its female flower is distinguished by the small ovary at the base of the petals. These bright and colorful flowers have an extremely short life and may only open for as short a time as 1 day. The color of pumpkins is derived from the orange pigments abundant in them. The main nutrients are lutein and both alpha and beta carotene; the latter generates vitamin A in the body (Ahamed et al., 2011).

29.3.2 NUTRITIONAL VALUE

Pumpkin has received considerable attention in recent years because of the nutritional and health benefits of the bioactive compounds obtainable from its seeds and fruits. The chemical and pharmacological properties of *C. moschata* extracts from its stems, seeds, and fruits have been investigated. These studies demonstrated that *C. moschata* has extensive bioactivities, such as hepatoprotection (Makni et al., 2008), antidiabetes (Jiang, 2011), anticancer (Zhang et al., 2012), and antiobesity properties (Lee et al., 2010). *C. moschata* has numerous traditional medicinal uses in South and Central America. Seeds are toasted and eaten to kill worms and other intestinal parasites and used as a diuretic; a preparation from the flowers has been used to treat measles and smallpox. It is a cooling astringent fruit, which is a useful traditional remedy for treating irritable bowel syndrome, purifies blood, increases appetite, and cures leprosy, fatigue, and muscle cramping. Fruits and seeds are used in Ayurveda to treat hemorrhage of pulmonary organs, rheumatism, and urinary disease. There is a pharmacological validation of the ethnomedicinal claims regarding the usefulness of this plant as a drug against ulcers (Govindani et al., 2012).

29.4 OTHER SQUASHES

29.4.1 ACORN SQUASH

Easily found in supermarkets. As the name suggests, this winter squash is small and round-shaped like an acorn. It is one of my favorite baking squashes; it is easy to slice into halves and fill with butter. A small acorn squash weighs from 1 to 3 lb and has sweet, slightly fibrous flesh. Its distinct ribs run the length of its hard, blackish-green or golden-yellow skin. In addition to the dark-green acorn, there are now golden and multicolored varieties.

29.4.2 AMBERCUP SQUASH

Ambercup squash is a relative of the buttercup squash that resembles a small pumpkin with orange skin. The bright-orange flesh has a dry, sweet taste. Peel it, cube the flesh, roast it, and serve it like cut-up sweet potatoes. It has an extraordinarily long storage life.

29.4.3 AUTUMN CUP SQUASH

This is a hybrid, semibush, buttercup-/kabocha-type, dark-green squash. It has a rich-flavored flesh and produces high yields. Its fruit size is 6 in, with a weight of about 2–3 lb. The flesh is yellow/orange that is stringless, dry, and sweet.

29.4.4 BANANA SQUASH

As regards shape and skin color, this winter squash is reminiscent of a banana. It grows up to 2 ft in length and about 6 in. in diameter. Its bright-orange, finely textured flesh is sweet. Banana squash is often cut into smaller pieces.

29.4.5 BUTTERNUT SQUASH

This kind of squash is easily found in supermarkets. The fruit is beige-colored and shaped like a vase or a bell. This is a more watery squash and tastes somewhat similar to sweet potatoes. It has a bulbous end and pale, creamy skin, with a choice, fine-textured, deep-orange flesh and a sweet, nutty flavor. Some people say it is like butterscotch. It weighs from 2 to 5 lb. The oranger the color, the riper, drier, and sweeter the squash. Butternut is a common squash used in making soup because it tends not to be stringy.

29.4.6 BUTTERCUP SQUASH

Buttercup squash is part of the turban squash family (hard shells with turban-like shapes) and are a popular variety of winter squash. This squash has a dark-green skin, sometimes accented with lighter-green streaks. It has a sweet and creamy orange flesh. This squash is much sweeter than other winter varieties. Buttercup squash can be baked, mashed, pureed, steamed, simmered, or stuffed and can replace sweet potatoes in most recipes.

29.4.7 CARNIVAL SQUASH

Carnival squash is cream-colored with orange spots or pale green with dark-green spots in vertical stripes. Carnival squash have hard, thick skins, and only the flesh is eaten. It is sometimes labeled as a type of acorn squash. The delicious yellow meat is reminiscent of sweet potatoes and butternut squash and can be baked or steamed and then combined with butter and fresh herbs. This is also great in soups.

29.4.8 DELICATA SQUASH

Delicata squash is also called peanut squash and Bohemian squash. This is one of the tastier winter squashes, with a creamy pulp that tastes a bit like corn and sweet potatoes. The size may range from 5 to 10 in. in length. The squash can be baked or steamed. The thin skin is also edible. The delicata squash is actually an heirloom variety and a fairly recent reentry into the culinary world. It was originally introduced by the Peter Henderson Company, New York, in 1894, and was popular through the 1920s. Then it fell into obscurity for about 75 years, possibly because of its thinner, more tender skin, which is not suited for transportation over thousands of miles and storage over months.

29.4.9 FAIRYTALE PUMPKIN SQUASH

Its French name is Musquee de Provence. The fruits are flattened like a cheese but each rib makes a deep convolution. The fairytale pumpkin is a very unique eating and ornamental pumpkin. It is thick but tender, and the deep-orange flesh is very flavored, sweet, thick, and firm. It is a 115–125 day pumpkin and takes a long time to turn to its cheese color. The distinctive coach-like shape and warm russet color make it also perfect for fall decorating.

This pumpkin is usually used for baking. Cut it into pieces and bake in the oven.

29.4.10 GOLD NUGGET SQUASH

This a variety of winter squash which is sometimes referred to as an Oriental pumpkin that has the appearance of a small pumpkin in shape and color. It ranges in size from 1 to 3 lb. Golden nugget squashes are small, weighing on average about 1 lb. Both the skin and the flesh are orange. Gold nugget squash may be cooked whole or split lengthwise (removing seeds). Pierce whole squash in several places and bake halved squash hollow side up.

29.4.11 HUBBARD SQUASH

The extrahard skins make them one of the best keeping winter squashes. These are very large and irregularly shaped, with a skin that is quite "warted" and irregular. They range from big to enormous, have a blue/gray skin, and taper at the ends. Like all winter squashes, they have an inedible skin; large, fully developed seeds that must be scooped out; and a dense flesh. Hubbard squash is often sold in pieces because it can grow to very large sizes. The yellow flesh of these tends to be very moist and longer cooking times in the oven are needed. They are generally peeled and boiled, cut up and roasted, or cut small and steamed.

29.4.12 KABOCHA SQUASH

Kabocha squash is also known as Ebisu, Delica, Hoka, Hokkaido, or Japanese pumpkin. Kabocha is the generic Japanese word for squash, but it refers most commonly to a squash of the buttercup type. This squash has a green, bluish-gray, or a deep-orange skin. The flesh is deep yellow.

Kabocha squash may be cooked whole or split lengthwise (removing seeds). It has a rich, sweet flavor and is often dry and flaky when cooked. It is used in any dish in which buttercup squash would work.

29.4.13 SPAGHETTI SQUASH

Spaghetti squash is also called vegetable spaghetti, vegetable marrow, or noodle squash. A small, watermelon-shaped variety ranges in size from 2 to 5 lb or more. It has a golden-yellow flesh, oval rind, and a mild, nutlike flavor.

The yellowiest spaghetti squash will be the ripest and best to eat. Those that are nearly white are not very ripe. Although it may seem counterintuitive, larger spaghetti squash are more flavorful than smaller ones.

29.4.14 SWEET DUMPLING SQUASH

This small, mildly sweet-tasting squash resembles a miniature pumpkin with its top pushed in. It has cream-colored skin with green specks. Weighing only about 7 oz, it has sweet and tender orange flesh and is a great size for stuffing and baking as individual servings. Sweet dumplings are tiny but great for roasting and presenting whole.

29.4.15 TURBAN SQUASH

It is named for its shape. Turban squash has colors that vary from bright orange to green or white. It has golden-yellow flesh, and its taste is reminiscent of hazelnut. It has a bulblike cap swelling from its blossom end and comes in bizarre shapes with extravagant coloration that makes them popular as harvest ornamentals. It is popular for centerpieces, and its top can be sliced off so it can be hollowed and filled with soup. As a larger variety of the buttercup squash, the turban squash has a bright-orange-red rind. Its flesh and storage ability are comparable to the buttercup squash. This is used in recipes for making pie or sugar pumpkin.

29.5 BOTTLE GOURD

The bottle gourd (*Lagenaria siceraria*) is a domesticated member of the Cucurbitaceae family (cucumber, melon, and squash family). It is synonymous to white-flowered gourd and calabash gourd (*L. siceraria* (Molina) Standl.) and is one of the important cucurbits grown throughout the world for its tender fruits (Arvind et al., 2011). In addition to its use as cooked vegetables, its fruits have secondary uses as well, like the large, strong, hard-shelled, and buoyant fruits have long been used as containers for water and food, musical instruments (drums and flutes), and fishing floats. A total of six species have been recognized as belonging to the genus *Lagenaria* or white-flowered gourds. One is the domesticated monoecious species *L. siceraria*, while five of them are wild, perennial, dioecious forms from Africa and Madagascar. The basic haploid chromosome number in the genus is 11 ($2n = 22$). Bottle gourd was domesticated in Asia and at the same time indigenous to Africa (Whitaker and Davis, 1962b). Tropical Africa remains the primary gene pool for this species (Singh, 1990).

Along with several wild perennial *Lagenaria* species, the bottle gourd has long been believed to be indigenous to Africa. However, until the recent discovery and morphological and genetic characterization of a wild population of *L. siceraria* in Zimbabwe, the bottle gourd had only ever been well documented as a domesticated plant (Decker-Walters et al., 2004). Although native to Africa, it had reached Asia and the Americas 9000–8000 years ago, possibly as a wild species whose fruits had floated across the sea (experiments have shown that domesticated bottle gourds contain still-viable seeds even after floating in seawater for more than 7 months) (Whitaker and Carter, 1954). The bottle gourd had a broad New World distribution 8000 years ago. Independent domestications from wild populations are believed to have occurred in both the Old World and New World (a variety of plants and animals were independently domesticated in multiple parts of the world 5,000–10,000 years ago). A range of data suggest that the bottle gourd was present in the Americas as a domesticated plant by 10,000 BP, which would make it among the earliest domesticated species in the New World. Comparisons of DNA sequences from archaeological bottle gourd specimens and modern Asian and African landraces identify Asia as the source of its introduction to the New World. Erickson et al. (2005) suggested that this "utility species" (along with another such species, the domestic dog) was domesticated long before any food crops or livestock species and that both were brought to the Americas by Paleo-Indian populations as they colonized the New World. Clarke et al. (2006) developed chloroplast and nuclear markers to investigate the origins of bottle gourds in Polynesia and suggested that their work also has implications for understanding the complex history of domestication and dispersal of the species as a whole. It is grown in rainy season as well as in summer season.

L. siceraria is generally cultivated in all tropical parts of the world including India and few African countries. *L. siceraria* is reported to have $2n = 2X = 22$ chromosomes. It is an annual, vigorous climbing species, monoecious, and highly cross-pollinated having wide genetic variability across the globe. Yetisir et al. (2008) vividly summarized the collection and characterization of bottle gourd germplasm (162 accessions) collected in 15 surveys from southern Turkey. Mladenovic et al. (2012) studied the landraces of bottle gourd from the Balkan Peninsula and compared them to those from Africa, Asia, and America. They observed that reduction in trait variation in bottle gourds was due to the preference for certain shapes and sizes of the fruit. India has a rich genetic variability in terms of fruit size and shape (Sivaraj and Pandravada, 2005; Peter et al., 2007) and is one of the centers of diversity according to de Candolle (as quoted in Chadha and Lal, 1993). However, very little work was done to document the variability for various traits in germplasm in India. Mathew et al. (2000) studied 28 accessions from different parts of India for the variability and concluded that there exists a potential source of gene sanctuary for bottle gourd that can be harnessed.

29.5.1 USES

Bottle gourd is characterized by differently shaped fruits that can be used as utensils or decorative ornaments, while younger juicy fruits are edible and nutritious (Berenji, 1992, 1999, 2000). Prasad and Prasad (1979) have created unique bottle gourd varieties in India, primarily for human consumption. It has been used in varied and specific ways in cultures of different nations. Scientists believe that, of all currently known plants, bottle gourd is the only species that had been used worldwide in prehistoric times.

Bottle gourd variability has been studied by many authors, including Heiser (1979), Decker-Walters et al. (2001), Morimoto and Mvere (2004), Morimoto et al. (2005), and Achigan Dako et al. (2008). In 2005, studies in India demonstrated significant variability of their regional collection. Sivaraj and Pandravada (2005) cite that 54 analyzed genotypes showed variability in quantitative and qualitative characteristics. Owing to the essential role of bottle guard in cultures of many nations, it is not surprising that the fruits have also been used in religious and ceremonial rituals. After the discovery of terracotta utensils, the use of bottle gourd in cookware declined but did not completely cease, as the thick outer skin of these plants is still used for this purpose, particularly by indigenous peoples of tropical and subtropical regions. In Serbia, bottle gourd had significant presence in rural life, primarily household and agricultural uses. More recently, its role in vegetable cultivation as a rootstock for watermelon grafting and in art as a decorative ornament is becoming more prominent. Even though in Serbia it is rare to find large fields under bottle gourd crop, this is one of the most popular species in India and some other parts of the world (Sivaraj and Pandravada, 2005). Mladenovic et al. (2011) revealed significant variability in quantitative and qualitative bottle gourd properties, which can be further used in cross breeding and producing fruits of variable characteristics.

29.5.2 BOTANY

A bottle gourd vine is a quick-growing annual with a hairy stem, long forked tendrils, and a musky odor. It is characterized by stems that are prostrate or climbing in growth habit, angular, ribbed, thick, brittle, and softly hairy, up to 5 m long, and when the stems are cut, they exude no sap. Many forms of the bottle gourd have been cultivated for specific purposes, and the sizes of the vines, leaves, and flowers, as well as the sizes and shapes of the fruits, vary greatly. Leaves are simple, up to 400 mm long and 400 mm broad, with short and soft hairs. The leaves are broadly egg, kidney, or heart-shaped in outline with undivided, angular, or faintly 3–7 lobes. The lobes are round and margins shallowly toothed. Leaf stalks are up to 300 mm long, thick, often hollow, and densely hairy. It has two small, lateral glands inserted at the leaf base. Tendrils split in two (Decker et al., 2004).

Flowers stalked are solitary, and the female flower stalks are shorter than male flowers. It is monoecious in habit, that is, male and female flowers are present on the same plant. Flowers have petals that are five in number, crisped in texture, and cream or white with darker veins in color, with pale yellow at the base; these are obovate, up to 45 mm long. Flowers open in the evenings and soon show wilting. The fruit may be smooth, knobby, or ridged. Some are only 3 in. long, while others may be more than 3 ft long. Shapes vary from globe to dish, bottle, dumbbell, club, crookneck, or coiled. These are green in color when immature but yellowish or pale brown when mature. The pulp dries out completely on ripening, leaving a thick, hard, hollow shell with almost nothing inside except the seeds. Seeds are many in number and embedded in a spongy pulp. These are 7–20 mm long and compressed, with two flat facial ridges in some variants (Decker et al., 2004).

29.5.3 Medicinal Value

It is packed with nutrition and recommended for high blood pressure and urinary disorders. Bottle gourd is called lauki in Hindi. It has very useful medicinal properties. It is a good source of carbohydrates, vitamins A and C, and minerals. It helps in maintaining blood pressure and cholesterol, helps controlling diabetes, reduces fats, and is also useful in weight loss. It is a rich source of vitamins, minerals, and fiber. It enhances growth of WBC and increases immunity. The benefits found in bottle gourd are listed as follows (MakG, 2013):

29.5.3.1 Provides More Nutrients and Very Low Calories and Fat

In case you are on a low-calorie diet, bottle gourd is the best option to choose. Ninety-six percent of the gourd is composed of water. Bottle gourd has lots of dietary fiber and offers just 15 cal/100 g and just 0.1 g fat in 100 g. It offers vitamin C plus some B vitamins and trace elements as well as minerals like iron, sodium, and potassium.

29.5.3.2 Aids in Digestion

Bottle gourd offers a great deal of equally soluble as well as insoluble fiber and water. It will help in digestion and enables in digestion-related difficulties like bowel problems, unwanted gas, as well as acidity. The insoluble fiber also helps in conditions such as piles.

Bottle gourd, being a good method of obtaining fiber content and simple to break down, greatly assists in healing digestion-associated problems like constipation, flatulence, and also piles.

29.5.3.3 Gives Cooling Effect

Bottle gourd has cooling as well as soothing capability because of its excellent water content. It really is completely helpful throughout summers and also specifically to those who work underneath the hot sun since it has the capacity to prevent heat strokes and also helps recover the water lost because of sweating. It may be ingested in the form of juice or cooked like a vegetable.

29.5.3.4 Cures Urinary Disorders

Bottle gourd has a diuretic impact on the entire body. Therefore, it will help in eliminating the extra water retained in the body as swelling as well as bloating. Make a glass of fresh bottle gourd juice and also include a teaspoon of lime juice and consume. This particular alkaline combination assists in dilution of acidic urine as well as in reduction of burning sensation within the urinary system due to a higher acid content in urine.

29.5.3.5 Helps in Weight Loss

Bottle gourd offers outstanding nutrition along with minimal calories and fat; therefore, it is best in losing weight. It may also help in reducing the cholesterol within the blood. It really is loaded with fiber and water content. Drink fresh bottle gourd juice, the very first thing in the morning, before eating anything; this will help ward off food cravings.

29.5.3.6 Prevents Premature Graying of Hair

Historical Ayurveda states that consuming fresh bottle gourd juice each morning regularly can be quite helpful for the reduction as well as management of prematurely graying hair.

29.5.3.7 Provides Benefits for Skin

Bottle gourd in the natural form may be used in inner cleaning the skin. It will help in managing oil release on the face and helps keep acne breakouts as well as pimples away. It will help control different types of skin ailment.

29.5.3.8 Aids in Sleeping Better and Treats Insomnia

Sesame oil whenever combined with bottle gourd juice helps to heal difficulties of insomnia. Eating cooked leaves of the gourd helps in cooling down the brain and enables curing sleep-related issues and helps in sleeping better.

29.5.3.9 Revitalizes and Replaces Lost Fluids

People struggling with diarrhea, high-grade fever, or any other health problems that are related to serious sweating as well as water losses need to have a glass of bottle gourd juice. It will help in replenishing lost body water and also reduces extreme thirst in diabetics. It treats exhaustion and also revitalizes.

29.5.3.10 Balances Liver Function

This particular vegetable is excellent for controlling liver function. It is usually suggested by Ayurvedic doctors once the liver is swollen and cannot effectively process food for optimum nutrition and also intake.

29.5.3.11 Supplies Vitamins and Minerals

Bottle gourd is a superb source of minerals and vitamins that are invaluable for the child's growth. Parents must ensure that kids, especially babies, have an all-round as well as a balance diet as far as possible so that no insufficiencies can occur. Vitamin and mineral deficiencies may result in illnesses. In this case, feeding your kids bottle gourd is an excellent beginning.

This vegetable is excellent for balancing liver function for most people. It is often recommended by Ayurvedic physicians when the liver is inflamed and cannot efficiently process food for maximum nutrition and assimilation. Bottle gourd juice is used in treatment of stomach acidity, indigestions, and ulcers. A glass of lauki juice with a little salt added to it prevents excessive loss of sodium, quenches thirst, and helps in preventing fatigue, keeping you refreshed in summer. Juice from its leaves is good for jaundice (Rao et al., 2009).

29.6 GOURDS

29.6.1 *Luffa* ssp.

The genus *Luffa* is distributed in the tropical regions of world (Chakravarty, 1982). It is a small Old World genus that comprises of nine species in the world. Of these, five species, namely, *L. echinata* Roxb., *L. acutangula*, *L. aegyptiaca*, *L. graveolens* Roxb., and *L. hermaphrodita* Singh

(1990) Bhandari and two debatable species, *L. tuberose* and *L. umbellata*, exist in India (Krishna et al., 2013). These are considered as New World species (Pandey et al., 2008). The New World species of *Luffa* are differentiated from the Old World (*L. echinata* Roxb., *L. acutangula*, *L. aegyptiaca*, and *L. graveolens* Roxb.) and three species from the New World are *L. guinguefida* (Hook. and Arn.) Seem., *L. operculata* (L.) Cogn., and *L. astorii* Svens (Pandey et al., 2008). Of these seven species, three—*L. acutangula*, *L. cylindrica*, and *L. hermaphrodita*—are edible species and cultivated in tropical and subtropical climates, and the remaining four species occur wild, mostly confined to northwestern/eastern Himalayas, northeastern plains, east coast, and peninsular tract (Arora and Nayar, 1984).

According to Heiser and Schilling (1990), *Luffa* consists of seven species: four well-differentiated species of the Old World tropics and three rather similar species of the Neotropics. The seven species are divided in two groups:

1. *L. aegyptiaca* (Syn. *L. cylindrica*), *L. acutangula*, *L. echinata*, and *L. graveolens*
2. *L. quinquefida*, *L. operculata*, and *L. astorii*

The genus has a long history of cultivation in tropical Asia and Africa and the probable center of origin and the primary center of diversity for *Luffa* is India (Sirohi et al., 2005). The species of *Luffa* are distributed in tropical regions of the world. *L. acutangula* occurs in India, Myanmar, Sri Lanka, Malaysia, Indonesia, China, and Africa. *L. graveolens* Roxb is found in India, Australia, and Indonesia. *L. hermaphrodita* (Singh et al.) Bhandari is endemic to India. *L. tuberose* is found in India and tropical Africa (Chakravarty, 1982). *L. acutangula* and *L. aegyptiaca* (sponge gourd) are minor cucurbitaceous vegetables cultivated in the lower plains (Gopalan et al., 1993; Chandra, 1995).

Sponge gourd (*Luffa cylindrica* Roem Syn. *Luffa aegyptiaca*) is one of the important members of this group. Sponge gourd has been cultivated for centuries in the Middle East and India, China, Japan, and Malaysia (Porterfield, 1955). Sponge gourd is native to tropical Asia, probably India and Southeast Asia.

All species have 26 chromosomes ($2n = 26$) (Dutt and Roy, 1990; Heiser and Schilling, 1990). The early spread of the genus *Luffa* was in the New and Old World, but both cultivated species originated in India (Heiser and Schilling, 1990). Morphological variability was evident in the cultivated species of *Luffa*. Diversity analysis based on the morphology of the plant parts, presence of flavonoids, and breeding systems classified *L. acutangula* and *L. aegyptiaca* into a single clad (Schilling and Heiser, 1981). They sequenced 51 accessions of *Luffa*, representing the geographic range of the genus and as much as possible topotypical or type material. Phylogenies from four noncoding plastid regions and the nuclear ribosomal DNA spacer region showed that eight clades of specimens had geographical–morphological coherence. Based on phylogenetic studies, *L. graveolens* was found more distantly related to *L. acutangula* than to *L. echinata* (Dutt and Roy, 1976). Cruz et al. (1997) observed more morphological variation in *L. aegyptiaca* than *L. acutangula*, but total seed profiles showed reverse results.

29.6.1.1 Uses

When the fruit of *Luffa* becomes old and dry, the endocarp becomes a persistent fibrous vascular network, which is used in various ways. A major use is as a sponge for washing and scrubbing utensils as well as the human body. It is also used for the manufacture of hats, insoles of shoes, car wipers, pot holders, table mats, doormats and bath mats, sandals, and gloves. The fiber has also been used for its shock- and sound-absorbing properties, for instance, in helmets and armored vehicles and as a filter in engines. In Ghana, the dry fiber is used to filter water and palm wine. In Central Africa, it is used to brush clothes. Fungal biosorbents immobilized on *L. cylindrica* sponges have been used for the biosorption of heavy metals from olive oil mill wastewater and other wastewaters (Iqbal and Edyvean, 2004, 2005; Ahmadi et al., 2006a,b). The young fruit is eaten fresh or cooked as a vegetable, but it has to be picked before the fibrous vascular bundles harden and before the purging compounds develop. In India and China, a type of curry is prepared with the fruit, which is peeled, sliced, and fried. In Japan, the fruits are eaten fresh or sliced and dried to be eaten later. The leaves are also eaten as a vegetable. The roasted seeds are edible and contain edible oil. The oil has been used in the United States in soap manufacture. In cosmetics, *L. cylindrica* seed oil is used for its antifungal, anti-inflammatory, and antitumor properties. Because it prevents synthesis of certain proteins, it is also considered toxic to skin cancer cells. It is high in the essential amino acid, arginine. The seeds have laxative properties, when eaten, due to their high oil content. The kernels provide essential, valuable, and useful minerals needed for good body development. Over half of the seed is oil. The bitter and toxic seedcake is unsuitable as feed for cattle but can be used as fertilizer, given that it is rich in nitrogen and phosphorus. *L. cylindrica* fiber has a high water absorption capacity, making it suitable as an absorbent, for instance, to decolor aqueous effluents. Another potential use is the reinforcement of resin matrix composite materials, but in this case a barrier layer, for instance, of glass fibers, is necessary to avoid water absorption from the environment. The sponge has a high degree of porosity, high specific pore volume, and stable physical properties and is nontoxic and biodegradable. These properties make it suitable as support matrix to plant, algal, bacterial, and yeast cells. The fiber strands contain 50%–62% α-cellulose, 20%–28% hemicellulose, and 10%–12% lignin. The raw leaves contain per 100 g edible portion: moisture 94.0 g, energy 58 kJ (14 kcal), protein 1.6 g, fat 0.1 g, carbohydrates (including fiber) 2.7 g, ash 1.6 g, Ca 330 mg, and P 33 mg (Leung et al., 1968). Leaf extracts, containing saponins, alkaloids, and cardiac glycosides, have shown antibacterial activity against *Bacillus subtilis*, *Escherichia coli*, *Staphylococcus aureus*, and *Salmonella typhi*. Aqueous extracts of the leaves have shown in vitro oxytocic activity.

The immature fruit contains per 100 g edible portion: moisture 94.0 g, energy 88 kJ (21 kcal), protein 0.6 g, fat 0.2 g, carbohydrates 4.9 g, Ca 16 mg, P 24 mg, Fe 0.6 mg, vitamin A 235 IU, thiamin 0.04 mg, riboflavin 0.02 mg, niacin 0.3 mg, and ascorbic acid 7 mg. The fruit also contains saponins. Fruits of wild forms are bitter and poisonous. Various antioxidant compounds have been isolated from the fruit. Lucyosides isolated from the fruit have shown antitussive activity. Ethanolic extracts of the fruit have shown antibacterial and antifungal activity.

29.6.1.2 Medicinal Value

It has long been used as a medicinal herb to treat asthma, intestinal worms, sinusitis (Chakravarty, 1990; Schultes, 1990), edema, pharyngitis, and rhinitis (Khare, 2007). Leaves are used in amenorrhea, decayed teeth, parasitic affections, skin diseases (Porterfield, 1955), chronic bronchitis (Khare, 2007) pain, inflammation, carbuncles, and abscesses (Perry, 1980). Flowers are effective in treating migraine (Khare, 2007; Sutharshana, 2013). Stem is used in respiratory complaints (Porterfield, 1955) and fruits in hemorrhage from bowels or bladder, hernia, hemorrhoids, jaundice, menorrhagia, scarlet fever (Porterfield, 1955), bronchitis, hematuria, leprosy, splenopathy, and syphilis (Prajapati et al., 2003).

Phytochemically, leaves contain flavonoids (Schilling and Heiser, 1981), saponins (Liang et al., 1993, 1996), and triterpenes (Nauking Institute of Materia Medica, 1980), whereas in fruits, ascorbic acid, anthocyanins, flavonoids (Bor et al., 2006), and triterpenoid saponins (Partap et al., 2012) are present. The flowers are rich in flavonoids (Schilling and Heiser, 1981), and carotenoids, flavonoids, and oleanolic acid were found in the peel (Kao et al., 2012), whereas polypeptides are reported in seeds (Abirami et al., 2011). Pharmacologically, antitussive, antiasthmatic, cardiac stimulant, hepatoprotective, and hypolipidemic properties (Partap et al., 2012) and analgesic (Velmurugan et al., 2011), anti-inflammatory (Muthumani et al., 2010; Abirami et al., 2011; Khan et al., 2013), and antiemetic activities (Khan et al., 2013) are reported.

29.6.1.3 Botany

It is monoecious, annual, climbing, or trailing herb that grows up to 15 m long. The stem is five angled and finely hairy, having 2–6-fid tendrils. Leaves are alternate and simple and stipules are absent. The margins of the leaves are palmately 3–7-lobed with triangular or ovate lobes, cordate at the base, lobes acute or subacute and apiculate at the apex sinuate dentate, scabrous. Male inflorescence racemose, peduncle 7–32 cm long, finely hairy, whereas the female flowers are solitary. Flowers unisexual, regular, 5-merous, 5–10 cm in diameter; petals free, entire, broadly obovate, 2–4.5 cm long, deep yellow in color. Male flowers on bracteate pedicels 3–13 mm long, receptacle tube obconic below, expanded above, 3–8 mm long, with triangular lobes 8–12 mm long, sepals ovate, 8–14 mm long, stamens three or five, free, inserted on the receptacle tube, connectives broad. Female flowers on pedicel 1.5–14.5 cm long, receptacle tube shortly cylindrical and 2.5–6 mm long, with ovate lobes c. 1 cm long, sepals ovate-lanceolate or lanceolate, 8–16 mm long, ovary inferior, stigmas three, two-lobed. The fruit has a shape of an ellipsoid or cylindrical capsule up to 60(–90) cm × 10(–12) cm, is beaked and not prominently ribbed, is brownish in color, is dehiscent by an apical operculum, and is glabrous and many-seeded. Seeds lenticular, broadly elliptical in outline, compressed, 10–15 mm × 6–11 mm × 2–3 mm, smooth, dull black, with a narrow, membranous winglike border. Seedling with epigeal germination; cotyledons ovate, c. 5 cm long.

REFERENCES

Abirami, M. S., Indhumathy, R., Sashikala, D. G., Satheesh, K. D., Sudarvoli, M., and Nandini, R. (2011). Evaluation of the wound healing and anti-inflammatory activity of whole plant of *Luffa cylindrica* (Linn.) in rats. *Pharmacologyonline* 3: 281–285.

Achigan-Dako, E. G., Fuchs, J., Ahanchede, A., and Blattne, F. R. (2008). Flow cytometric analysis in *Lagenaria siceraria* (Cucurbitaceae) indicates correlation of genome size with usage types and growing elevation. *Plant Syst. Evol.* 276: 9–19.

Achigan-Dako, G. E., Fanou, N., Kouke, A., Avohou, H., Vodouhe, S. R., and Ahanchede, A. (2006). Evaluation agronomique de trois especes de Egusi (Cucurbitaceae) utilisees dans l'alimentation au Benin et elaboration d'un modele de prediction du rendement. *Biotechnol. Agron. Soc. Environ. (Base)* 10: 121–129.

Ahamed, K. U., Akhter, B., Islam, M. R., Ara, N., and Humauan, M. R. (2011). An assessment of morphology and yield characteristics of pumpkin (*Cucurbita moschata*) genotypes in northern Bangladesh. *Trop. Agric. Res. Extens.* 14(1): 2011.

Ahmadi, M., Vahabzadeh, F., Bonakdarpour, B., and Mehranian, M. (2006a). Empirical modeling of olive oil mill wastewater treatment using Luffa-immobilized *Phanerochaete chrysosporium*. *Process Biochem.* 41(2): 1148–1154.

Ahmadi, M., Vahabzadeh, F., Bonakdarpour, B., Mehranian, M., and Mofarrah, E. (2006b). Phenolic removal in olive oil mill wastewater using loofah immobilized *Phanerochaete chrysosporium*. *World J. Microbiol. Biotechnol.* 22: 119–127.

Ajmone-Marsan, P., Livini, C., Messmer, M. M., Melchinger, A. E., and Motto, M. (1992). Cluster analysis of RFLP data from related maize inbred lines of the BSSS and LSC heterotic groups and comparison with pedigree data. *Euphytica* 60: 139–148.

Arora, R. K. and Nayar, E. R. (1984). *Wild Relatives of Crop Plants in India*. NBPGR Science Monograph No. 7. National Bureau of Plant Genetic Resources, New Delhi, India, 90pp.

Arvind, K., Singh, B., Mukesh, K., and Naresh, R. K. (2011). Genetic variability, heritability and genetic advance for yield and its components in bottle gourd (*Lagenaria siceraria* M.). *Ann. Hortic.* 4: 101–103.

Baranek, M., Stift, G., Vollmann, J., and Lelley, T. (2000). Genetic diversity within and between the species *Cucurbita pepo*, *C. moschata* and *C. maxima* as revealed by RAPD markers. *Cucurbit Genet. Coop. Rep.* 23: 73–77.

Bar-Nun, N. and Mayer, A. M. (1990). Cucurbitacins protect cucumber tissue against *Botrytis cinerea*. *Phytochemistry* 29: 787–791.

Berenji, J. (1992). Tikve. *Bilten za hmelj, sirak i lekovito bilje*. 23–24: 86–89.

Berenji, J. (1999). Tikve–hrana, lek i ukras. *Zbornik radova Instituta za ratarstvoi povrtarstvo, Novi Sad*. 13: 63–75.

Berenji, J. (2000). Vrg, *Lagenaria siceraria* (Molina) Standl. *Zbornik radova Instituta za ratarstvo i povr-tarstvo, Novi Sad.* 33: 279–289.

Bor, J. Y., Chen, H. Y., and Yen, G. C. (2006). Evaluation of antioxidant activity and inhibitory effect on nitric oxide production of some common vegetables. *J. Agric. Food Chem.* 54: 1680–1686.

Chadha, M. L. and Lal, T. (1993). Improvement of cucurbits. In: Chadha, K. L. and Kalloo, G. (eds.), *Advances in Horticulture.* Malhotra Publishing House, New Delhi, India, pp. 151–155.

Chakravarty, H. L. (1982). *Fascicles of Flora of India, Fascicle II, Cucurbitaceae.* Botanical Survey of India, Howrah, India, 73pp.

Chakravarty, H. L. (1990). Cucurbits of India and their role in the development of vegetable crops. In: Bates, D. M., Robinson, R. W., and Jeffrey, C. (eds.), *Biology and Utilization of the Cucurbitaceae.* Cornell University Press, Ithaca, NY, pp. 325–334.

Chambliss, O. L. and Jones, C. M. (1966a). Cucurbitacins: Specific insect attractants in *Cucurbitaceae. Science* 153: 1392–1393.

Chandra, U. (1995). Distribution, domestication and genetic diversity of *Luffa* gourd in Indian subcontinent. *Indian J. Plant Genet. Resour.* 8: 189–196.

Chung, H. D., Youn, S. J., and Choi, Y. J. (1998). Ecological and morphological characteristics of the Korean native squash (*Cucurbita moschata*). *J. Kor. Soc. Hortic. Sci.* 39(4): 377–384.

Clarke, A. C., Burtenshaw, M. K., McLenachan, P. A., Erickson, D. A., and Penny, D. (2006). Reconstructing the origins and dispersal of the polynesian bottle gourd (*Lagenaria siceraria*). *Mol. Biol. Evol.* 23(5): 893–900.

Cruz, V. M. V., Tolentino, M. I. S., Altoveros, N. C., Villavicencio, M. L. H., Siopongco, L. B., DelaVina, A. C., and Laude, R. P. (1997). Correlations among accessions of Southeast Asian *Luffa* genetic resources and variability estimated by morphological and biochemical methods. *Philipp. J. Crop Sci.* 22(3): 13–40.

Decker, D. S. (1988). Origin(s), evolution, and systematics of *Cucurbita pepo* (Cucurbitaceae). *Econ. Bot.* 42(1): 4–15,

Decker-Walters, D. S., Staub, J., Lopez-Sese, A., and Nakata, E. (2001). Diversity in land races and cultivars of bottle gourd (*Lagenaria siceraria,* Cucurbitaceae) as assessed by random amplified polymorphic DNA. *Genet. Resour. Crop Evol.* 48: 369–380.

Decker-Walters, D. S., Walters, T. W., Poluszny, U., and Kevan, P.G. (1990). Genealogy and gene flow among annual domesticated species of *Cucurbita. Can. J. Bot.* 68: 782–789.

Decker-Walters, D. S., Wilkins-Ellert, M., Chung, S. M., and Staub, J. E. (2004). Discovery and genetic assessment of wild bottle gourd [*Lagenaria siceraria* (Mol.) Standley; Cucurbitaceae] from Zimbabwe. *Econ. Bot.* 58(4): 501–508.

Dutt, B. and Roy, R. P. (1976). Cytogenetic studies in an experimental amphidiploid in *Luffa. Caryologia* 29: 16.

Erickson, D. L., Smith, B. D., Clarke, A. C., Sandweiss, D. H., and Tuross, N. (2005). An Asian origin for a 10,000-year-old domesticated plant in the Americas. *Proc. Natl. Acad. Sci. USA* 102(51): 18315–18320.

Esquinas-Alcazar, J. T. and Gulick, P. J. (1983). *Genetic Resources of Cucurbitaceae.* International Board for Plant Genetics Resources, Rome, Italy.

Esteras, C., Diez, M. J., Pico, B., Sifres, A., Valcarcel, J. V., and Nuez, F. (2008). Diversity of Spanish landra-ces of *Cucumis sativus* and *Cucurbita* ssp. In: Pitrat, M. (ed.), *Proceeding of the IXth Eucarpia Meeting on Genetics and Breeding of Cucurbitaceae,* INRA, Avignon, France, May 21–24, 2008, pp. 67–76.

Ferriol, M., Pico, B., Cordova, P. F., and Nuez, F. (2004a). Molecular diversity of a germplasm collection of Squash (*Cucurbita moschata*) determined by SRAP and AFLP Markers. *Crop Sci.* 44: 653–664.

Ferriol, M., Picó, B., and Nuez, B. (2003). Genetic diversity of some accessions of *Cucurbita maxima* from Spain using RAPD and SBAP markers. *Genet. Resour. Crop Evol.* 50: 227–238.

Ferriol, M., Pico, B., and Nuez, F. (2004b). Morphological and molecular diversity of a collection of *Cucurbita maxima* landraces. *J. Am. Soc. Hortic. Sci.* 129(1): 60–69.

Ferriol, M., Pico, B., and Nuez, F. (2013). Genetic diversity of a germplasm collection of *Cucurbita pepo* using SRAP and AFLP markers. *Theor. Appl. Genet.* 107(2): 271–282.

Franco, J., Crossa, J., Ribaut, J. M., Betran, J., Warburton, M. L., and Khairallah, M. (2001). A method for combining molecular markers and phenotypic attributes for classifying plant genotypes. *Theor. Appl. Genet.* 103: 944–952.

Franco, J., Crossa, J., Villaseñor, J., Taba, S., and Eberhart, S. A. (1997). Classifying Mexican maize accessions using hierarchical and density search methods. *Crop Sci.* 37: 972–980.

Franco, J., Crossa, J., Villaseñor, J., Taba, S., and Eberhart, S. A. (1998). Classifying genetic resources by categorical and continuous variables. *Crop Sci.* 38: 1688–1696.

Fu, C. L., Shi, H., and Li, Q. H. (2006). A review on pharmacological activities and utilization technologies of pumpkin. *Plant Foods Hum. Nutr.* 61: 73–80.

Gopalan, C., Ramasastri, B. V., and Balasubramanian, S. C. (1993). *Nutritive Value of Indian Foods*, 2nd edn. National Institute of Nutrition, Indian Council of Medical Research, Hyderabad, India.

Govindani, H., Dey, A., Deb, L., Rout, S. P., Parial, S. D., and Jain, A. (2012). Protective role of methanolic and aqueous extracts of *Cucurbita moschata* Linn. fruits in inflammation and drug induced gastric ulcer in Wister rats. *Int. J. PharmTech Res.* 4(4): 1758–1765.

Gwanama, C., Labuschagne, M. T., and Botha, A. M. (2000). Analysis of genetic variation in *Cucurbita moschata* by random amplified polymorphic DNA (RAPD) markers. *Euphytica* 113: 19–24.

Heiser, B. C. (1979). *The Gourd Book*. University of Oklahoma Press, Norman, OK.

Heiser, C. B. and Schilling, E. E. (1990). The genus *Luffa*: A problem in phytogeography. In: Bates, D. M., Robinson, R. W., and Jeffrey, C (eds.), *Biology and Utilization of the Cucurbitaceae*. Cornell University Press, Ithaca, NY, pp. 120–133.

Iqbal, M. and Edyvean, R. G. J. (2004). Biosorption of lead, copper and zinc ions on loofa sponge immobilized biomass of *Phanerochaete chrysosporium*. *Miner. Eng.* 17: 217.

Iqbal, M. and Edyvean, R. G. J. (2005). Loofa sponge immobilized fungal biosorbent: A robust system for cadmium and other dissolved metal removal from aqueous solution, *Chemosphere* 61: 510; *Biochim. Biophys. Acta* 740: 52.

Jeffrey, D. (1990). An outline classification of the *Cucurbitaceae*. In: Bates, D. M., Robinson, R. W., and Jeffrey, C. (eds.), *Biology and Utilization of the Cucurbitaceae*. Cornell University, Ithaca, NY, pp. 449–463.

Jiang, Z. and Du, Q. (2011). Glucose-lowering activity of novel tetrasaccharide glyceroglycolipids from the fruits of *Cucurbita moschata*. *Bioorg. Med. Chem. Lett.* 21: 1001–1003.

Kao, T. H., Huang, C. W., and Chen, B. H. (2012). Functional components in *Luffa cylindrica* and their effects on anti-inflammation of macrophage cells. *Food Chem.* 135: 386–395.

Khan, K. W., Ahmed, S. W., Ahmed, S., and Hasan, M. M. (2013). Antiemetic and anti-inflammatory activity of leaves and flower extracts of *Luffa cylindrica* (L.) Roem. *J. Ethnobiol. Trad. Med. Photon* 118: 258–263.

Khare, C. P. (2007). *Indian Medicinal Plants: An Illustrated Dictionary*. Springer Science+Business Media, New York, pp. 384–385.

Kostalova, Z., Hromadkova, Z., and Ebringerova, A. (2009). Chemical evaluation of seeded fruit biomass of oil pumpkin (*Cucurbita pepo* L. var. *styriaca*). *Chem. Papers* 63: 406–413.

Lee, J., Kim, D., Choi, J., Choi, H., Ryu, J. H., Jeong, J., Park, E. J., Kim, S. H., and Kim, S. (2012). Dehydrodiconiferyl alcohol isolated from *Cucurbita moschata* shows anti-adipogenic and anti-lipogenic effects in 3T3-L1 cells and primary mouse embryonic fibroblasts. *J. Biol. Chem.* 287: 8839–8851.

Lee, J. M., Kubota, C., Tsao, S. J., HoyosEchevarria, P., Morra, L., and Oda, M. (2010). Current status of vegetable grafting: Diffusion grafting techniques, automation. *Sci. Hortic.* 127: 93–105.

Leung, W. T. W., Busson, F., and Jardin, C. (1968). *Food Composition Table for Use in Africa*. FAO, Rome, Italy, 306pp.

Lewis W. H., Elvin-Lewis, M. P. F., and Walter, H. (1997). *Medical Botany*. Wiley, New York, 291pp.

Liang, L., Liu, C. Y., Li, G. Y., Lu, L. E., and Cai, Y. C. (1996). Studies on the chemical components from leaves of *Luffa cylindrica* Roem. *Yaoxue Xuebao* 31: 122–125.

Liang, L., Lu, L. E., and Cai, Y. C. (1993). Chemical components from leaves of *Luffa cylindrica*. *Yaoxue Xuebao* 28: 836–839.

Lira, R., Andres, T. C., and Nee, M. (1995). *Cucurbita* L. In: Lira, R. (ed.), *Estudios Taxonómicos y Ecogeográficos de las Cucurbitaceae Latinoamericanas de Importancia Económica: Cucurbita, Sechium, Sicana y Cyclanthera. Systematic and Ecogeographic Studies on Crop Genepools 9*, International Plant Genetic Resources Institute, Rome, Italy, pp. 1–115.

Lu, A. and Jeffrey, C. (2011). Flora of China, Vol. 19. 1. Cucurbitaceae: 35. Cucurbita. Web. http://www.efloras.org/florataxon.aspx?flora_id=2&taxon_id=200022621.

MakG. (2013). Health benefits of bottle gourd. www.healthbenefitstimes.com.

Makni, M., Fetoui, H., Gargouri, N. K., Garoui, E. M., Jaber, H., and Makni, J. (2008). Hypolipidemic and hepatoprotective effects of flax and pumpkin seed mixture rich in ω-3 and ω-6 fatty acids in hypercholesterolemic rats. *Food Chem. Toxicol.* 46: 3714–3720.

Mathew, A., Markose, B. L., Rajan, S., and Peter, K. V. (2000). Genetic variability in bottle gourd, *Lagenaria siceraria* (Molina) Standley. *Cucurbits Genet. Coop. Rep.* 23: 78–79.

Mladenovic, E., Berenji, J., Ognjanov, V., Kraljevic-Balalic, M., Ljubojevic, M., and Cukanovic, J. (2011). Conservation and morphological characterization of bottle gourd for ornamental use. In *46th Croatian and 6th International Symposium on Agriculture*, Opatija, Croatia, pp. 550–553.

Mladenovic, E., Berenji, J., Ognjanov, V, Ljubojević, M., and Čukanović, J. (2012). Genetic variability of bottle gourd *Lagenaria siceraria* (Mol.) Standley and its morphological characterization by multivariate analysis. *Arch. Biol. Sci.* 64: 573–583.

Morimoto, Y., Maundu, P., Fujimaki, H., and Morishima, H. (2005). Diversity of landraces of the white-flowered gourd (*Lagenaria siceraria*) and its wild relatives in Kenya: Fruit and seed morphology. *Genet. Resour. Crop Evol.* 52: 737–747.

Morimoto, Y. and Mvere, B. (2004). *Lagenaria siceraria* (Molina) Standley. In: Grubben, G. J. H. and Denton, O. A. (eds.), *Plant Resources of Tropical Africa 2. Vegetables*. PROTA Foundation, Wageningen, the Netherlands, pp. 353–358.

Muthumani, P., Meera, R., Mary, S., Jeenamathew, D. P., and Kameswari, B. (2010). Phytochemical screening and anti inflammatory, bronchodilator and antimicrobial activities of the seeds of *Luffa cylindrica*. *Res. J. Pharm. Biol. Chem. Sci.* 1: 11–22.

Nauking Institute of Materia Medica. (1980). A note on chemical study on the vines (leaves) of *Luffa cylendrica* (L.) Roem. *Chung Tsao Yao.* 11: 55–64.

Nee, M. (1990). The domestication of *Cucurbita* (Cucurbitaceae). *Econ. Bot.* 44(3 suppl.): 56–68.

Norrman, R. and Haarberg, J. (1980). *Nature and Language: A Semiotic Study of Cucurbits in Literature*. Routledge and Kegan Paul, London, U.K.

Pandey, S., Kumar, S., Choudhary, B. R., Yadav, D. S., and Rai, M. (2008). Component analysis in pumpkin (*Cucurbita moschata* Duch. ex Poir.). *Veg. Sci.* 35(1): 35–37.

Paris, H. S. (1989). Historical records, origins, and development of the edible cultivar groups of *C. pepo* (Cucurbitaceae). *Econ. Bot.* 43(4): 423–443.

Partap, S., Kumar, A. S., Neeraj, K., and Jha, K. K. (2012). *Luffa cylindrica*: An important medicinal plant. *J. Nat. Prod. Plant Resour.* 2: 127–134.

Perry, L. M. (1980). *Medicinal Plants of East and Southeast Asia: Attributed Properties and Uses*. The MIT Press, Cambridge, MA, 116pp.

Peter, K. V., Sadhan Kumar, P. G., and George, T. E. (2007). Underutilized vegetables. In: Peter, K. V. (ed.), *Underutilized and Underexploited Horticultural Crops*. New India Publishing Agency, New Delhi, India, pp. 233–246.

Porterfield Jr, W. M. (1955). Loofah: The sponge gourd. *Econ. Bot.* 9: 211–223.

Prajapati, N. D., Purohit, S. S., Sharma, A. K., and Kumar, T. (2003). *A Handbook of Medicinal Plants: A Complete Source Book*. Agrobios, Jodhpur, India, p. 324.

Prakash, K., Pandey, A., Radhamany, J., and Bisht, I. (2013). Morphological variability in cultivated and wild species of *Luffa* in India. *Genet. Resour. Crop Evol.* 60: 2319–2329.

Prasad, R. and Prasad, A. (1979). A note on the heritability and genetic advance in bottle gourd (*Lagenaria siceraria* (Mol.) Standl.). *Ind. J. Hortic.* 36: 446–448.

Rios, H., Fernandez, A., and Batista, O. (1997). Cuban pumpkin genetic variability under low input conditions. *Cucurbit Genet. Coop. Rep.* 20: 48–49.

Robinson, R. W. and Decker-Walters, D. S. (1997). *Cucurbits*. CAB International, New York, 226pp. (Crop Production Science in Horticulture No. 6).

Roubik, D. W. (1995). *Pollination of Cultivated Plants in the Tropics*. FAO Agricultural Services Bulletin No. 118, Rome, Italy.

Russel, J. R., Fuller, J. D., Macaulay, M., Hatz, B. G., Jahoor, A., Powell, W., and Waugh, R. (1997). Direct comparison of levels of genetic variation among barley accessions detected by RFLPs, AFLPs, SSRs, and RAPDs. *Theor. Appl. Genet.* 95: 714–722.

Schilling, E. E. and Heiser, C. (1981). Flavonoids and systematics of *Luffa*. *Biochem. System. Ecol.* 9: 263–265.

Schultes, R. E. (1990). Biodynamic cucurbits in the New World tropics. In: Bates, D. M., Robinson, R. W., and Jeffrey, C. (eds.), *Biology and Utilization of the Cucurbitaceae*. Cornell University, Ithaca, NY.

Singh, A. K. (1990). Cytogenetics and evolution in the Cucurbitaceae. In: Bates, D. M., Robinson, R. W., and Jeffrey, C. (eds.), *Biology and Utilization of the Cucurbitaceae*. Cornell University, Ithaca, NY, pp. 10–28, 485pp.

Sirohi, P. S., Munshi, A. D., Kumar, G., and Behera, T. K. (2005). Cucurbits. In: Dhillon, B. S., Tyagi, R. K., Saxena, S., and Randhawa, G. J. (eds.), *Plant Genetic Resources: Horticultural Crops*. Narosa Publishing House, New Delhi, India, pp. 34–58.

Sivaraj, N. and Pandravada, S. R. (2005). Morphological diversity for fruit characters in bottle gourd (*Lagenaria siceraria* (Mol.) Standl.) germplasm from tribal pockets of Telangana region of Andhra Pradesh, India. *Asian-Agric. Hist. J.* 9: 305–310.

Sutharshana, V. (2013). Protective role of *Luffa cylindrica*. *J. Pharm. Sci. Res.* 5(9): 184–186.

Tapley, W. T., Enzie, W. D., and Van Eseltine, G. P. (1937). *The Vegetables of New York: The Cucurbits*. State of New York, Education Department, Albany, NY.

Velmurugan, V., Shiny, G., and Surya, S. P. (2011). Phytochemical and biological screening of *Luffa cylindrica* Linn. fruit. *Int. J. PharmTech Res.* 3(3): 1582–1585.

Wessel-Beaver, L. (1998). Pumpkin breeding with a tropical twist. *Cucurbit Netw. News* 5: 2–3.

Whitaker, T. W. (1990). Cucurbits of potential economic importance. In: Bates, D. M., Robinson, R. W., and Jeffrey, C. (eds.), *Biology and Utilization of the Cucurbitaceae*. Cornell University Press, Ithaca, NY, pp. 318–324.

Whitaker, T. W. and Bemis, W. P. (1975). Origin and evolution of the cultivated *Cucurbita*. *Bull. Torrey Bot. Club* 102: 362–368.

Whitaker, T. W. and Carter, G. F. (1954). Oceanic drift of gourds-experimental observations. *Am. J. Bot.* 41(9): 697–700.

Whitaker, T. W. and Davis, G. N. (1962a). *Cucurbits*. Interscience Publishers Inc., New York.

Whitaker, T. W. and Davis, G. N. (1962b). *Cucurbits: Botany, Cultivation and Utilization*. Inter Science, New York, 250pp.

Wu, J., Chang, Z., Wu, Q., Zhan, H., and Xie, S. (2011). Molecular diversity of Chinese *Cucurbita moschata* germplasm collections detected by AFLP markers. *Sci. Hortic.* 128: 7–13.

Yetisir, H., Sakar, M., and Serce, S. (2008). Collection and morphological characterization of *Lagenaria siceraria* germplasm form the Mediterranean region of Turkey. *Genet. Resour. Crop Evol.* 55: 1257–1266.

Youn, S. J. and Chung, H. D. (1998). Genetic relationship among the local varieties of the Korean native squashes (*Cucurbita moschata*) using RAPD technique. *J. Kor. Soc. Hortic. Sci.* 35: 429–437.

Younis, Y. M. H., Ghirmay, S., and Al-Shihry, S. S. (2000). African *Cucurbita pepo* L.: Properties of seed and variability in fatty acid composition of seed oil. *Phytochemistry* 54: 71–75.

Zhang, Q., Yu, E., and Medina, A. (2012). Development of advanced interspecific-bridge lines among *Cucurbita pepo, C. maxima*, and *C. Moschata*. *Hortic. Sci.* 47(4): 452–458.

30 Snake Gourd
Taxonomy, Botany, Cultural Practices, Harvesting, Major Diseases, and Pests

A.V.V. Koundinya and M.K. Pandit

CONTENTS

30.1 INTRODUCTION

Snake gourd (*Trichosanthes cucumerina* var. *anguina*), also known as snake tomato, viper gourd, and long tomato, is a commonly grown tropical or subtropical annual, climbing, herbaceous vine, raised for its strikingly long fruit, used as a vegetable, medicine, and a lesser known use, crafting didgeridoos. The name snake gourd might have been given due to its long, slender, snake-like fruits. Immature fruits of snake gourd are used for culinary purposes; occasionally, shoots and tender leaves are also used as vegetables. Snake gourd fruits and seeds are nutritive, and they contain essential nutrients and vitamins in appreciable amounts (Table 30.1). It is suitable for diabetic patients due to its low calories and high water content. It can be used against constipation as it is rich in fiber content. Fruits are also rich in flavonoids, carotenoids, and phenolic acids, which make the plant pharmacologically and therapeutically active (Prabha et al. 2010). *T. curcumineria* is used as an abortifacient, anthelmintic, stomachic, refrigerant, purgative, laxative, hydragogue, hemagglutinant, emetic, cathartic, in the treatment for malaria and bronchitis (Nadkani 2002; Madhava et al. 2008), and as an antioxidant (Adebooye 2008). It has a promising place in the Ayurvedic and Siddha system of medicine due to its various medicinal values like antidiabetic, hepatoprotective, cytotoxic, anti-inflammatory, and larvicidal effects (Prabha et al. 2010; Sandhya et al. 2010). The seeds are found to have high crude fat content; hence, its oil is investigated and found suitable for use in biodiesel production (Adesina and Amoo 2014). In Nigeria, snake gourd is substituted for solanaceous tomato not only because of its sweet taste, aroma, and deep red endocarp pulp when fully ripe, which prevents the fruit pulp from turning sour as quickly as tomato paste (Adebooye and Oloyede 2006), but also due to its nutraceutical properties (Adebooye 2008) and medicinal importance (Chuku et al. 2008).

TABLE 30.1
Proximate Composition of Snake Gourd Fruit and Seed

	Content	
Parameters	Fruit (Ojiako and Igwe 2008)	Seed (Adesina and Amoo 2014)
Moisture	94.6%	5.21%
Ash	2.50	2.93%
Fiber	0.8 g/100 g	—
Protein	0.50 g/100 g	28.59%
Carbohydrate	3.30 g/100 g	10.60%
Energy	18 kcal	—
Fat	0.3 g/100 g	51.53%
Mineral	0.5 g/100 g	—
Calcium	50 mg/100 g	1.43%
Iron	1.10 mg/100 g	2.028%
Phosphorous	20 mg/100 g	135
Sodium	25.4 mg/100 g	5.06%
Potassium	34 mg/100 g	1.10
Magnesium	53 mg/100 g	0.758%
Zinc	—	0.082%
Copper	0.11 mg/100 g	0.068%
Vitamin A	160 IU	—
Thiamin	0.04 mg/100 g	—
Riboflavin	0.06 mg/100 g	—
Niacin	0.30 mg/100 g	—
Vitamin C	5.0 mg/100 g	—
Oxalic acid	34 mg/100 g	—

Cultivation of snake gourd during off season, when tomato prices are high, is economical and fetches good income to the farmer. The good food value of this plant is an indicator that its cultivation and utilization should be promoted. To promote its cultivation, there is the need to develop agronomic practice packages that farmers would need (Oloyede and Adebooye 2005).

30.2 TAXONOMY

Snake gourd, botanically *Trichosanthes cucumerina* L. var. *anguina* (L.) Haines, a variant of *T. cucumerina*, belongs to the family Cucurbitaceae, subfamily Cucubitoideae, tribe Trichosantheae C Jaffr., and subtribe Tricosanthnae Pax. (Jeffrey 1980). *Trichosanthes* is the largest genera in the family Cucurbitaceae. The genus *Trichosanthes* has been classified by several workers into different subgenera and sections based on flower, fruit, and seed characters, but none of them is adopted widely. The difficulty involved in classifying the genus is lack of recent monographic revision of the genus and all the prior classifications are based on regional studies (Bharathi et al. 2013). Dioecious sex expression and nighttime anthesis are hindrances for taxonomic study of these species (Duyfjes and Pruesapan 2004). Pollen grain and seed characters are useful and are studied to understand the diversity and characterize the species of *Trichosanthes* (Rugayah and de Wilde 1999). There are about 44 species in the genus *Trichosanthes*, out of which 22 are present in India as per Chakravarthy (1982). But, in India, 26 taxa are present, including four intraspecific species (Pradheep et al. 2010). The genus *Trichosanthes* is divided into two sections, namely *Eutrichosanthes* with 23 species and *Pseudotrichosanthes* with 3 species only by Kundu (1942). All the species in the genus *Trichosanthes* are dioecious except *T. cucumerina* L. var. *anguina* (L.) Haines and *T. cucumerina* L., which are monoecious. Few cultivated species such as *T. cucumerina* L. var. *anguina* (L.) Haines. (snake gourd), *T. dioica* Roxb. (pointed gourd), and *T. cucumeroides* (Ser.) Maxim. (Japanese snake gourd) are reported. *T. celebica*, *T. ovigera*, and *T. villosa* are grown as minor vegetables. *T. celebica* leaves are used as a vegetable, and *T. ovigera* is cultivated in China and Japan. Its fruits are edible, and starch is extracted from tubers. *T. villosa* grows wild in Java, the Philippines, Thailand, and Indo-China; its boiled fruits are eaten and leaves are used for covering the body to reduce fever and pain of swollen legs (Gildemacher et al. 1993). Another two species, *T. kirilowii* and *T. rosthornii*, are grown for their medicinal value (Bharathi et al. 2013). Some important wild relatives of snake gourd are *T. bracteata* (syn. *T. palmeta*), *T. nervifolia*, *T. wallichiana* (syn. *T. multiloba*), *T. cordata*, *T. japonica*, and *T. shikokiana* (Varghese 1973).

30.3 CHROMOSOME NUMBER

Snake gourd and a majority of the species are diploids (2n = 2x = 22) with the basic chromosome number of x = 11. Polyploidy is also reported in the genus *Trichosanthes*. A natural auto-triploid (2n = 33) is reported in *T. dioica* (Singh 1979). Tetraploids (2n = 44) in the genus are *T. bracteata* (Thakur 1973) and *T. lepiniana* (Huang et al. 1994). Some accessions of *T. kirilowii* are found as octaploid (2n = 88) based on cytological analyses by Qian et al. (2012).

30.4 ORIGIN

Snake gourd is known to be originated and first domesticated in India or in the Indo-Malayan region (Robinson and Walters 1997). Maximum species diversity is found in northeast India, followed by southern India (Pradheep et al. 2010). Snake gourd might have originated directly from the wild *T. cucumerina*, which have small, bitter fruits and is grown wild in India or perhaps through the intermediary species *T. lobata* (Singh and Roy 1979). There is another perception that *T. lobata* is a cross between *T. cucumerina* and *T. anguina*, which is supported by the intermediate fruit size of interspecific hybrid between *T. cucumerina* and *T. anguina* similar to *T. lobata* (Mishra 1966). It is distributed throughout the humid tropical or subtropical parts of the world starting from Malaysia to Northern Australia in one direction and through China and Japan in another direction (Sarker et al. 1987). It is also grown in Latin and South America and some parts of Africa (Robinson and Walters 1997). It is widely grown in countries like India, Java, and Mauritius (Miniraj et al. 1993).

30.5 BOTANY

The stem is viny, long, and slender, furrowed with branched tendrils (Figure 30.1). Leaves are prominently lobed, deeply cordate at base, and densely pubescent (Robinson and Walters 1997; Pandey 2008). Variations in leaf shape in snake gourd can be seen in Figure 30.2. Snake gourd is monoecious, bearing male and female flowers separately on the same plant. A tendency toward hermaphrodism is also observed (Miniraj et al. 1993). Flowers are small, fringed, white in color, and borne single or raceme in the leaf axils. Male flowers are borne in lower nodes either solitary or raceme. Corolla fimbriate, having long, densely pubescent peduncles of 10–20 cm length, bear 8–15 flowers near the apex (Pandey 2008). Male flowers contain three stamens and anthers are connate. Female flowers solitarily appear after one week of the appearance of male flowers (Miniraj et al. 1993). They are sessile, having long and narrow ovary and penta-partite stigma. Fruits are long, sometimes up to 4 m length, slender, serpent-like, and twisted (Robinson and Walters 1997; Pandey 2008). Fruits are green/pale green/white/green and white striped in color at immature stage (Figure 30.3) and turns bright orange when ripe. The flesh is white and fibrous. Brown color seeds are present in the flesh (Robinson and Walters 1997).

30.6 FLORAL BIOLOGY

Snake gourd is a cross-pollinated crop mainly due to dicliny and pollination is assisted by insects, mainly the bees. Anthesis begins in the snake gourd during the evening from 5:30 to 7:45 p.m. and continues up to 9:30 p.m. Anther dehiscence takes place before anthesis and completes within 1 h. Pollen grains remain viable from 10 h before anther dehiscence to 46–49 h after anther dehiscence. Stigma is receptive from 12 h before to 12 h after anthesis (Deshpande et al. 1980).

FIGURE 30.1 Snake gourd plant trained over trellis.

FIGURE 30.2 Variation for leaf shape in snake gourd.

FIGURE 30.3 (**See color insert.**) Short-fruited-type snake gourd.

30.7 SOIL AND CLIMATE REQUIREMENT

Snake gourd is a warm-season annual cucurbit that grows well in summer and rainy seasons. It is an important summer vegetable and requires a temperature range of 30°C–35°C for better growth and fruit development; a temperature below 20°C restricts its growth. Short days promote the growth of snake gourd. It grows well on sandy, loamy soils rich in organic matter and having provisions for drainage. Water stagnation should be avoided as it cannot tolerate waterlogging conditions. Soil should be neutral in reaction, with an optimum pH range from 6.0 to 7.0 (Bharathi et al. 2013).

30.8 SEASON

In North Indian plains, it is grown during summer season by sowing January to February and rainy season by sowing June to July. In addition, a third crop can be grown in south India where winter is mild and frost-free. In Nigeria, the crop grown during April to July is considered as being early and the crop from August to November is considered as being late (Oloyede and Adebooye 2005). Avoid rainy season sowing in areas with high humidity and rainfall as the crop is highly susceptible to downy mildew disease. Lower fruit length and girth is observed during summer (January to April) rather than rainy season (June to September) and winter (October to January) in Kerala, a southern Indian state (Celine et al. 2010). In Nigeria, the crop grown from April to July has a significantly higher number of leaves, vine growth, and marketable fruits than the crop grown from August to November. Aborted flowers and cull fruits are more in the latter crop. Ascorbic acid content is more in the earlier crop while ether extract, crude fiber, and total sugar are more in the latter crop (Oloyede and Adebooye 2005).

30.9 PROPAGATION AND CULTURAL PRACTICES

Snake gourd is propagated sexually through seeds. In order to raise a crop of 1 ha, 5–6 kg seeds will suffice. Heavy seeds obtained from large fruits produce heavy seedlings (Devadas et al. 1998). Germination will be completed within 7–8 days. Quick germination can be ensured by soaking the seeds one night before sowing. Soaking of the snake gourd seeds in cow dung slurry @ 1 kg/1 kg of seeds for 1/2 h before sowing is an indigenous practice that helps in early germination and withstanding the drought conditions (Reddy 2010; Sridhar et al. 2013). Mixing of seeds with ash at the time of sowing is another indigenous practice to ensure good growth (Reddy 2010). Seed treatment with *Trichoderma viride* @ 4 g/kg seeds or *Pseudomonas fluorescens* @ 10 g/kg seeds or carbendazim @ 2 g/kg seeds reduces soilborne fungal diseases. Presowing electric current treatment of seed has been recognized as an innovative tool for yield improvement. Improvement of yield in snake gourd was achieved when seeds were treated with 200 mA of electrical stimulus for 3 min (Bera et al. 2006). Low temperatures cause slow germination. Seedlings can be raised in plug or pro-trays and can be transplanted to the field at the two-leaf stage.

Tissue culture can also be employed to produce disease-free, good quality planting material. *In vitro* propagation in *Trichosanthes cucumerina* is studied by Pillai et al. (2008) and Prabha et al. (2010). Shoot tips and nodal cuttings are found as good explants for micro propagation. Callus initiation occurs within 20 days when MS medium is supplemented with BAP (0.5 mg/L) and NAA (0.5 mg/L). Maximum direct shoot formation of 84% is observed when MS medium is supplemented with BA (1 mg/L) and NAA (1 mg/L). Highest direct root induction of 87.6% is observed in 10 days when MS medium is supplemented with BAP (1 mg/L) and IBA (1 mg/L) (Prabha et al. 2010). The highest culture response (97%) and multiplication rate of 1:7, that is, seven shoots per explant for next subculture in 4 weeks coupled with rooting, is achieved with nodal explants in MS medium supplemented with 0.46 μM kinetin and 2.46 μM indole-3-butyric acid (Pillai et al. 2008).

30.9.1 SOWING

The seeds can be sown in raised beds, ridges, or pits. For sowing in raised beds or ridges, the land is to be ploughed thoroughly and leveled. Raised beds of 250 × 100 × 15 cm size are prepared and seeds are sown on the edges. Ridges are made at 150–200 cm spacing and seeds are sown on the ridges at 100 cm spacing. Pit sowing is preferred where training systems are fixed permanently. Pits of 45 × 45 × 45 cm are dug and seeds are placed in the pits. Generally, 5–6 seeds are sown and are thinned to 3–4 seedlings.

30.9.2 Manures and Fertilizers Applications

Snake gourds respond well to fertilizer application, but excessive use of nitrogen promotes more stem growth at the expense of fruit production (Bharathi et al. 2013). At the time of land preparation, 10–15 t FYM is incorporated in case of raised beds and ridges. In pit sowing, fill ¾ of the pits with FYM and soil before sowing. Besides, 40–60 kg N, 30–40 kg P, and 30–40 kg K are applied for proper growth and development (More and Shinde 2001a). However, higher doses of nitrogen (105–140 kg/ha) significantly increase the growth and yield attributes, quality of fruits, and water-use efficiency of the crop, but reduce the shelf life of fruits (Syriac 1998). Nitrogen is applied in split doses such as one-half as basal doses and the remaining half at the time of fruit set. Maximum fruit yield in snake gourd is realized when 105 kg/ha N is applied along with the spraying of 200 ppm ethephon and drip irrigation at 5 mm CPE (Syriac and Pillai 2001).

30.9.3 Plant Growth Regulators

Snake gourd is a monoecious plant bearing male and female flowers on the same plant. Spraying of growth regulators can induce the production of more female flowers, thereby changing the ratio of male and female flowers. Ethephon or ethrel is the most commonly used growth regulator to alter sex expression of snake gourd. Foliar application of ethephon at 250 ppm changes the male-to-female ratio and increases fruit yield (Cantliffe 1976). Application of ethephon at the 2–4 leaf stage reduces the days to first female flower and increases the number of female flowers (Kohinoor and Mian 2005). Ethephon at higher concentrations 100 and 200 ppm favorably influences the growth and yield components and also improves the water-use efficiency of the crop (Syriac 1998). Fruit yield and yield components like the number of female flowers, fruit number/plant, fruit size, and fruit weight can be increased by the application of ethrel at 150 ppm (Ramaswamy et al. 1976). Cycocel is another growth regulator used to alter sex expression of snake gourd. Cycocel at 1500 ppm causes minimum ratio of male to female flowers, whereas earliest node for first female flower as well as lesser days to flower are observed when cycocel is sprayed at 1000 ppm (Ahmad and Gupta 1981). Fruit drop can be reduced by spraying auxins. The application of planofix reduces fruit drop and increases fruit size but inhibits the development of white stripes (Ramaswamy et al. 1976). Seed treatment with GA_3 induces first tendril emergence at lower nodes and seedling height, thereby promoting rapid growth (Sardar and Mukherjee 1990).

30.9.4 Training

Training is the most important cultural operation in snake gourd. The fruits of snake gourd are tender and soft. If fruits touch the ground, it may lead to spoilage due to soilborne microorganisms. Initially, the seedlings are supported with bamboo stakes. When they grow a sufficient height about 2 m, they are trained over trellis made up of bamboo or galvanized iron wire. From these trellis, the fruits grow longitudinally downward. Single staking followed by bower system of training (Figure 30.1) is found best for longer and maximum number of fruits and higher yield. Generally, straight fruits with no curve at the end are preferred in the market. Hence, sometimes, farmers tie a stone or a weight at the distal or blossom end of the fruit when they are 1/2 ft long to allow straight growth of the fruit without coiling (More and Shinde 2001a; Reddy 2010). Generally, it is followed in long-fruited types; short-fruited types do not require such training (Bharathi et al. 2013).

30.9.5 Intercropping

Intercropping indicates growing of two or more crops simultaneously on the same piece of land with a definite row pattern. Snake gourd can be grown as an intercrop in coconut orchards during monsoon or rainy season (Hegde 1997). Intercropping with legumes helps greater biological nitrogen fixation in soil. The highest land equivalent ratio and net returns are observed when snake gourd is intercropped with black gram plants in a 1:9 row pattern (Thanunathan et al. 2001).

30.9.6 IRRIGATION

Slight irrigation is done immediately after sowing to ensure proper germination. Plants are irrigated through the basins made around each plant of 30 cm radius. Flowering and fruit set are the critical stages of irrigation. Adequate soil moisture is maintained during these stages. In summer season crop, successive irrigations are given at 6–7 days interval depending upon soil moisture content. In rainy season crop, avoid irrigations when there is ample rainfall and make provisions for drainage to avoid water stagnation in the basins. Drip irrigation is also found effective, especially in areas with water scarcity. Frequent drip irrigation with 5 mm cumulative pan evaporation (CPE) positively influences the number of fruits per plant, fruit weight, and girth of fruits (Syriac 1998). Flooding and water logging should be avoided to reduce the incidence of *Fusarium* wilt and root rot (Bharathi et al. 2013).

30.9.7 WEEDING

Weed growth is more during rainy season crop than summer season crop. About 2–3 hand weeding and light hoeing can be done in the initial stages of vine growth. Organic or black polythene mulches can be useful in controlling weeds, besides conserving soil moisture.

30.10 HARVESTING

Snake gourd varieties have wide variations in fruit length, diameter, and skin color (light green, dark green, and green stripes). Depending upon varietal characteristics, young, tender, immature fruits are harvested when they attain 1/4 or 1/3 of their full size (More and Shinde 2001a). Edible quality can be determined by piercing the thumb nail into the fruit. If thumb nail penetrates readily, the fruit is edible, if not it is overmature (Bautista and Mabesa 1977). Maturity and delayed harvesting cause fiber development, loss in weight, and a hard rind. Harvesting commences from 2 to 3 months of sowing or 2–3 weeks of anthesis. Fruits are harvested in regular pickings at 6–7 days interval. The number of pickings may vary from 7 to 8 depending upon varietal yielding ability.

30.10.1 YIELD

A single vine can produce 20–40 fruits within 2–3 months. The average yield ranges from 18 to 25 t/ha (Bharathi et al. 2013).

30.11 POSTHARVEST TECHNOLOGY

As the fruits are soft and tender, they have less shelf life. They can be stored up to 2 days at room temperature. Shelf life can be extended up to 2 weeks at 16°C and 85%–90% relative humidity (Robinson and Walters 1997). Dipping of snake gourd fruits in wax emulsion (6%–12%) + sodium phenyl phenoate for 30–60 s prolong the shelf life up to 14 days at room temperature (Hazra et al. 2009).

30.12 RIVER BED CULTIVATION

Growing of vegetables, especially cucurbits, in river beds during winter season promotes their off-season cultivation. It can be treated as a kind of vegetable forcing wherein the cucurbits are grown under subnormal conditions, literally on sand, during winter months from November to February (Singh 2012). In Kerala (India), the river beds replenished with fresh silt after the floods recede are used by the farmers for cultivation of snake gourd crop (Pandey 2008). Trenches are dug at 50–60 cm width and 60–90 cm depth after cessation of monsoon and recession of flood from

October to November. In the northwest region of India, where winter temperatures are low, trenches are surrounded by planting of grass stubbles, which prevent shifting of sand dunes and protect from cold waves. Due to winter, sprouted seeds are sown in the trench. Before sowing, trenches are to be applied with FYM. Initially, watering should be done by pots until roots reach the water regime. Application of 30–60 g urea per pit at the time of thinning will be useful. After 30–40 days of sowing, top dressing of 40 g urea is usually done in two split doses. When the plants grow, grass is to be spread on which the plants will spread. Fruits are ready for harvesting by the month of March to April, which will fetch a high price in the market (Singh 2012).

30.13 MAJOR DISEASES

30.13.1 DOWNY MILDEW

Casual organism: *Pseudoperonospora cubensis.*

Symptoms: Yellow color water–soaked spots appear on the upper surface of leaf. These spots do not expand beyond the veins; hence, they are angular in shape. Purple or grayish spores appear on the lower surface corresponding to the yellow spots on the upper surface of the leaf. Eventually, the leaf dries and under severe conditions, defoliation occurs (More and Shinde 2001b; Hazra and Som 2006; Bharathi et al. 2013). The disease spreads rapidly under high humid conditions especially when coupled with moderate to high temperatures of 20°C–22°C (Bharathi et al. 2013).

Control: Growing of resistant varieties successfully eliminates the disease. Hand removal and burning of infected leaves is useful in the initial stages. Spraying of 0.2% dithane M-45 or difolaton at 10 days interval at least three times is effective. Using 0.3% copper oxychloride is also found effective in controlling the disease (More and Shinde 2001b; Hazra and Som 2006).

30.13.2 ANTHRACNOSE

Casual organism: *Colletotrichum lagenarium.*

Symptoms: Light brown to dark brown color spots along with a raised rim around the spots appear on the leaf surface. The spots coalesce and cause shriveled or scorching appearance on leaves.

Control: Clean cultivation and proper rotation with noncucurbitaceous crops minimizes the initial inoculum load. As it is a seed-borne disease, soaking of seeds in 0.1% HgCl$_2$ for 5 min reduces surface infection. Spraying of systemic fungicides like 0.1% bavistin or benomyl, 0.2% dithane M-45, or 0.3% difolaton four times at 10 days interval effectively controls the disease (Hazra and Som 2006).

30.13.3 FUSARIUM WILT

Casual organism: *Fusarium oxysporum* f. sp. *niveum.*

Symptoms: Fungus blocks the vascular tissues mainly xylem, thereby causing yellowing followed by wilting of leaves due to lack of water supply, and vascular bundles turn yellow or brown.

Control: Since it is a soilborne disease, drenching with capton (1.5 g/L) and bavistin (2 g/L) around the root zone followed by the application of *Trichoderma* (100 g/pit) is recommended for disease control (Bharathi et al. 2013). Growing resistant varieties and seed treatment with thiram @ 2.5 g/kg of seed or 0.2% zirum for 30 min is effective (Hazra and Som 2006). Hot water treatment at 55°C for 15 min helps in eliminating the seed-borne infection (More and Shinde 2001b).

30.13.4 VIRAL DISEASES

Casual organism: Cucumber mosaic virus (CMV), Tobamovirus, cucumber green mottle mosaic virus (CuGMMV), and papaya ring spot virus (PRSV).

Vectors: Sap, seed, and aphids.

Symptoms: Snake gourd is infected by several viruses. Mosaic viruses cause yellow-color chlorotic mosaic-like symptoms on the leaf surface followed by distortion and stunting of plants (Hazra and Som 2006; Bharathi et al. 2013). Tobamovirus causing mild chlorotic spots, severe mosaic and chlorosis, in snake gourd is reported in the Tirupati region of Andhra Pradesh, India, by Subbaiah and Gopal (1997). The first record of CuGMMV and a serologically PRSV-related virus in snake gourd is reported in Sri Lanka by Ariyaratne et al. (2005). Tobamovirus and CuGMMV are transmitted by seeds. Leaf mosaic with distortion, reduction of internode length, and fruit distortion were the prominent symptoms of the CuGMMV disease (Ariyaratne et al. 2005). PRSV is reported on snake gourd plants in Coimbatore, Tamilnadu (India), by Kumar et al. (2014). PRSV-attacked leaves of snake gourd show symptoms of mosaic, blistering vein thickening and leaf distortion (Kumar et al. 2014).

Control: Hand collection and destruction of infected plants, clean cultivation, and control of vectors with insecticides like 0.05% monochrotophos, 0.05% dimethoate, or 0.02% methyl parathion. Spraying of the crop with 0.75% mineral oil emulsion in water at weekly intervals checks the feeding of aphids and reduces the spread of the virus (Hazra and Som 2006).

30.14 MAJOR PESTS

30.14.1 SNAKE GOURD SEMILOOPER

Scientific name: *Anadevidia peponis* (Figure 30.4).

Marks of identification: Light green color larva of 30–40 mm length with white longitudinal stripes.

Nature of damage: Larva feeds on green tissue of leaves, thereby reducing the photosynthetic area.

Control: Hand picking and destruction of larvae is an organic approach. Spraying of 0.1% diflubenzuron effectively kills the eggs, larvae, and pupae (Mule and Patil 2000). Neem oil is found to be highly effective as an ovicide followed by neem cake extract and 1.0% malathion (Thenmozhi and Kingsly 2004).

30.14.2 FRUIT FLY

Scientific name: *Bactrocera cucurbitae*.

Marks of identification: Adult flies are ferruginous-brown in color with three parallel lemon yellow stripes on the thorax and hyaline wings. Maggots are white in color and triangular in shape.

Nature of damage: Female flies lay eggs 2–4 mm deep in ripe and green fruits by puncturing the pericarp. Eggs are also laid on flower buds and stems of host plants. Maggots will develop from eggs, and they make burrows by feeding on the flesh, leaving the excreta inside the fruit, which causes a foul smell. The fruits attacked in the early stages fail to develop properly and drop or rot on the plants (Dhillon et al. 2005). Full-grown maggots come out by making a hole on the surface of fruits for pupation. The holes may lead to secondary infestation by fungi.

FIGURE 30.4   (**See color insert.**) Snake gourd semilooper.

Control: Since the maggots damage the fruit internally, it is difficult to control the pest with insecticides. Integrated management of this pest is effective, which is described later (Dhillon et al. 2005).

1. Field sanitation and hand removal of all damaged fruits and destroying them.
2. Bagging of fruits with paper bags at 2–3 days interval.
3. Cue-lure traps or parapheromone lures attract and eliminate adult male flies in reproduction.
4. Protein baits contain corn protein, and toxicants like Spinosad are effective.
5. Utilization of biocontrol agents like *Opius flatcheri*, which is a parasitoid of *B. cucurbitae*.
6. Development cultivation of fruit fly-resistant varieties in areas of severe infestation.
7. Insecticides such as malathion (0.5%), dicholorovas (0.2%), neem oil (1.2%), and neem cake (4.0%) can be effective against the fruit fly.

30.14.3   Red Pumpkin Beetle

Scientific name: *Aulacophora fevicolis*, *A. cincta*, and *A. intermedia*.

Marks of identification: Adult beetles are 6–8 mm long and shiny yellowish-red to yellowish-brown in color.

Nature of damage: Eggs are laid in the soil. Grubs feed on root tissue and cause direct damage to the newly developed seedlings. Adult beetles feed voraciously on the foliage, flower buds, and flowers and make irregular holes on the foliage (Narayanan and Batra 1960). Damage is more severe and the crop needs to be resown if it is attacked at the seedling or 2–4 leaf stage.

Control: Clean cultivation and hand picking and destroying of adults and grubs are admissible in the initial stages. Incorporate 10% carbaryl in pits before sowing the seeds to destroy grubs and pupae. Dusting the crop with 5% carbaryl or spraying 0.2% carbaryl or 0.04% malathion is effective (Hazra and Som 2006). Some traditional practices followed in eastern India to control red pumpkin beetles are spraying of 1:10 (w/v) grinded river black soil in water, spraying of 1:10 (w/v) red chilli powder, spraying of 1:15 (w/v) cow dung suspension, and dusting of 100% fly ash powder (Nath and Ray 2012).

REFERENCES

Adebooye, O. C. 2008. Phyto-constituents and anti-oxidant activity of the pulp of snake tomato (*Trichosanthes cucumerina* L.). *African Journal of Traditional, Complementary and Alternative Medicines*, 5(2):173–179.

Adebooye, O. C. and Oloyede, F. M. 2006. Responses of fruit yield and quality of *Trichosanthes cucumerina* landraces to phosphorus rate. *Journal of Vegetable Science*, 12(1):5–19.

Adesina, A. O. and Amoo, I. A. 2014. Chemical composition and biodiesel production from snake gourd (*Tricosanthes cucumerina* L.) seeds. *International Journal of Scientific Research*, 4(2):1–10.

Ahmad, I. and Gupta, P. K. 1981. Effect of cyocel on sex expression in three members of cucurbitaceae. *Indian Journal of Horticulture*, 38:100–102.

Ariyaratne, I., Weeraratne, W. A. P. G., and Ranatunge, R. K. R. 2005. Identification of a new mosaic virus disease of snake gourd in Sri Lanka. *Annals of the Sri Lanka Department of Agriculture*, 7:13–21.

Bautista, O. K. and Mabesa, R. C. 1977. *Vegetable Production*. Los Banos, Philippines: University of the Philippines.

Bera, A. K., Pati, M. K., and Ghanti, P. 2006. Effect of pre-sown electrical stimulus of seed on growth and yield of ridge gourd (*Luffa acutangula* Roxb.) and snake gourd (*Trichosanthes anguina* L.). *Indian Journal of Plant Physiology*, 11(3):291–294.

Bharathi, L. K., Behera, T. K., Sureja, A. K., Jhon, K. J., and Wehner, T. C. 2013. Snake gourd and pointed gourd: Botany and horticulture. *Horticultural Reviews*, 41:457–495.

Cantliffe, D. J. 1976. Improved fruit set on cucumber by plant growth regulator sprays. *Proceedings of the Florida State Horticulture Society*, 89:9496.

Celine, V. A., Seeja, G., and Gokulapalan, C. 2010. Evaluation of snake gourd genotypes for different seasons in the humid tropics. *Indian Journal of Horticulture*, 67(Special Issue):185–188.

Chakravarthy, H. L. 1982. *Fascicles of Flora of India-II Cucurbitaceae*, 136pp. Howrah, India: Botanical Survey of India.

Chuku, E. C., Ogbonn, D. N., Onuegbu, B. A., and Adeleke, M. T. V. 2008. Comparative studies on the fungi and bio-chemical characteristics of snake gourd (*Trichosanthes cucumerina* L.) and toamto (*Lycopersicon esculentus* Mill.) in Rivers state, Nigeria. *Journal of Applied Sciences*, 8(1):168–172.

Deshpande, A. A., Bankapur, V. M., and Vankatasubbiah, K. A. 1980. Studies on floral biology of snake gourd (*Trichosanthes anguina* L.) and Ash gourd (*Benincasa hispida* Thumb Cogn.). *Mysore Journal of Agricultural Science*, 14:8–10.

Devadas, V. S., Rani, T. G., Kuriakose, K. J., Seena, P. G. and Gopalakrishnan, T. R. 1998. Effect of fruit grading on seed quality in snake gourd (*Trichosanthes anguina* L.). *Horticultural Journal*, 11(1):103–108.

Dhillon, M. K., Singh, R., Naresh, J. S., and Sharma, H. C. 2005. The melon fruit fly, *Bactrocera cucurbitae*: A review of its biology and management. *The Journal of Insect Science*, 5:40.

Duyfjes, B. E. E. and Pruesapan, K. 2004. The genus *Trichosanthes* (Cucurbitaceae) in Thai land. *Thai Forest Bulletin (Botany)*, 32:786–109.

Gildemacher, B. H., Jansen, G. J., and Chayamarit, K. 1993. *Trichosanthes* L. In *Plant Resources of South -East Asia No. 8. Vegetables*, eds. J. S. Siemonsma and P. Kasem, pp. 271–274. Wageningen, the Netherlands: Pudoc Scientific Publishers.

Hazra, P. and Som, M. G. 2006. *Vegetable Science*, 491pp. Ludhiana, India: Kalyani Publishers.

Hazra, P., Chattopadhyay, A., Thapa, U., Pandit, M. K., and Dutta, S. 2009. Vegetable Science. In *Basics of Horticulture*, ed. K. V. Peter, pp. 101–279. New Delhi, India: New India Publishing Agency.

Hegde, M. R. 1997. Performance of vegetable varieties under coconut (*Cocos nucifera*) shade. *Indian Journal of Agronomy*, 42(3):535–539.

Huang, L. Q., Yue, C. and Li, M. Y. C. 1994. Chromosome numbers of 8 species in *Trichosanthes*. *Acta Botanica Yunnanica*, 16(1):95–96.

Jeffrey, C. 1980. A review of the Cucurbitaceae. *Botanical Journal of Linnean Society*, 81:233–247.

Kohinoor, H. and Mian, M. A. K. 2005. Effect of ethrel on flower production and fruit yield of snake gourd (*Trichosanthes anguina* L.). *Bulletin of Institute of Tropical Agriculture, Kyushu University*, 28(2):59–67.

Kumar, S., Sankarlingam, A., and Rabindran, R. 2014. Characterization and confirmation of papaya ringspot virus-W strain infecting *Trichosanthes cucumerina* at Tamil Nadu. *India Journal of Plant Pathology and Microbiology*, 5(2):1–4.

Kundu, B. C. 1942. A revsion of Indian species of *Hodgsonia* and *Trichosanthes*. *Journal of Bombay Natural History Society*, 43:362–368.

Madhava, K. C., Sivaji, K. and Tulasi, K. R. 2008. *Flowering Plants of Chitoor Dist A.P.*, 141pp. Tirupati, India: Students Offset Printers.

Miniraj, N., Prasanna, K. P. and Peter, K. V. 1993. Snake gourd. In *Genetic Improvement in Vegetable Crops*, eds. G. Kalloo and B. O. Bergh, pp. 259–264. New York: Pergamon Press.

Mishra, A. R. 1966. Cytogenetic investigation in Cucurbitaceae. PhD dissertation, Patna University, Patna, India.

More, T. A. and Shinde, K. G. 2001a. Snake gourd. In *Vegetables Tuber Crops and Spices*, eds. S. Thumbraj and N. Singh, pp. 289–292, 315–319. New Delhi, India: Directorate of Information and Publication in Agriculture, Indian Council of Agricultural Research.

More, T. A. and Shinde, K. G. 2001b. Diseases and pests of Cucurbits. In *Vegetables Tuber Crops and Spices*, eds. S. Thumbraj and N. Singh, pp. 315–319. New Delhi, India: Directorate of Information and Publication in Agriculture, Indian Council of Agricultural Research.

Mule, R. S. and Patil, R. S. 2000. Efficacy of diflubenzuron against snake gourd semilooper. *Journal of Maharashtra Agricultural Universities*, 25(1):27–31.

Nadkani, K. M. 2002. *Indian Material Medica*, 2nd edn., Vol. 1, pp. 1235–1236. Mumbai, India: Popular Prakashan.

Narayanan, E. S. and Batra, H. N. 1960. Red pumpkin beetle and their control. Indian Council of Agricultural Research, New Delhi, India, Vol. 2, pp. 1–68.

Nath, D. and Ray, D. C. 2012. Traditional management of red pumpkin beetle, *Raphidopala foveicollis* Lucas in Cachar district, Assam. *Indian Journal of Traditional Knowledge*, 11(2):346–350.

Ojiako, O. A. and Igwe, C. U. 2008. The nutritive, anti nutritive and hepatotoxic properties of *T. anguina* (snake tomato) fruits from Nigeria. *Pakistan Journal of Nutrition*, 7(1):85–89.

Oloyede, F. M. and Adebooye, O. C. 2005. Effect of season on growth, fruit yield and nutrient profile of two landraces of *Tricosanthes cucumerina* L. *African Journal of Biotechnology*, 4(10):1040–1044.

Pandey, A. K. 2008. *Underutilized Vegetable Crops*, 366pp. New Delhi, India: Satish Serial Publishing House.

Pillai, G. S., Martin, G., Raghu, A. V., Lyric, P. S., Balachandran, I., and Ravindran, P. N. 2008. Optimizing culture conditions for in vitro propagation of *Trichosanthes cucumerina* L.: An important medicinal plant. *Journal of Herbs, Spices and Medicinal Plants*, 14(1–2):13–28.

Prabha, A. L., Nandagopalan, V., Piramila, H. M. and Prabhakaran, D. S. 2010. Standardization of *in vitro* studies on direct and indirect organogenesis of *Trichosanthes cucumerina*. In *Proceedings of Sixth International Plant Tissue Culture and Biotechnology Conference: Role of Biotechnology in Food Security and Climate Change*, eds. A. S. Islam, M. M. Haque, R. H. Sarker, and M. I. Hoque, pp. 73–77. Bangladesh Association of Plant Tissue Culture Biotechnology, Dhaka, Bangladesh.

Pradheep, K., Singh, P. K., Srivastava, R., and Dhariwal, O. P. 2010. Trichosanthes-collection and conservation, *ICAR News*, 16(1):4.

Qian, C. L., HuaQi, X., Yang, J. H., and Zhang, M. F. 2012. Molecular phylogeny of Chinese snake gourd (*Trichosanthes kirilowii* Maxim) based on cytological and AFLP analyses. *International Journal of Cytology, Cytosystematics and Cytogenetics*, 65(3):216–222.

Ramaswamy, N. C., Govindaswamy, V., and Ramanujam, C. 1976. Effect of ethrel and planofix on flowering and yield of snake gourd (*Trichosanthes anguina* L.). *Annamalai Agricultural University Annual Research*, 6:187189.

Reddy, R. 2010. Traditional practices in agriculture, p. 151. http://www.angoc.org/wp-content/uploads/2010/07/06/traditional-practices-in-agriculture/Traditional-Practices-in-Agriculture-FULL.pdf (accessed November 7, 2014).

Robinson, R. W. and Walters, D. S. D. 1997. *Cucurbits*, 226pp. Wallingford, U.K.: CAB International.

Rugayah, E. A. and de Wilde, W. J. O. 1999. Conspectus of *Trichosanthes* (Cucurbitaceae) in Malaysia. *Reinwarditia*, 11:227–280.

Sandhya, S., Vinod, K. R., Chandra Sekhar, J., Aradhana, R., and Nath, V. S. 2010. An updated review on *Tricosanthes cucumerina* L. *International Journal of Pharmaceutical Sciences Review and Research*, 1(2):56–60.

Sarker, D. D. E., Dutta, K. B., and Sen, R. 1987. Cytomorphology of some wild and cultivated members of Trichosanthes, *Cytologia*, 52:405.

Sardar, A. K. and Mukherjee, K. K. 1990. Effects of EMS, colchicine and GA$_3$ on tendril emergence in snake gourd (*Trichosanthes anguina* L.). *Phytomorphology*, 40(1–2):175–177.

Singh, A. K. 1979. Cucurbitaceae and polyploidy. *Cytologia*, 44:897–905.

Singh, A. K. and Roy, R. P. 1979. An analysis of interspecific hybrids in *Trichosanthes* L. *Cryologia*, 32:329.

Singh, P. K. 2012. Cucurbits cultivation under Diara-Lands. *Asian Journal of Agriculture and Rural Development*, 2(2):248–252.

Sridhar, S., Ashok Kumar, S., Thooyavathy, R. A., and Vijayalakshmi, K. 2013. *Seed Treatment Techniques*, pp. 1–14. Chennai, India: Centre for Indian Knowledge Systems (CIKS) Seed Node of the Revitalising Rainfed Agriculture Network.

Subbaiah, K. V. and Gopal, D. V. R. S. 1997. Characterization and identification of a tobamovirus infecting snake gourd (*Trichosanthes anguina* L.) in Andhra Pradesh. *Indian Journal of Virology*, 13(2):153–157.

Syriac, E. K. 1998. Nutrient-growth regulator interaction in snake gourd (*Trichosanthes anguina* L.) under drip irrigation system. PhD dissertation, Kerala Agricultural University, Thrissur, India.

Syriac, E. K. and Pillai, G. R. 2001. Effect of nitrogen, ethephon and drip irrigation on the yield and yield attributes of snake gourd (*Trichosanthes anguina* L.). *Annals of Agricultural Research*, 22(4):508–513.

Thakur, G. K. 1973. A natural tetraploid in the genus *Trichosanthes* from Bihar. *Proceedings of the 60th Indain Science Congress*, 324pp., Chandigarh, India.

Thanunathan, K., Imayavaramban, V., Kalyanasundaram, S., Kandasamy, S., and Singaravel, R. 2001. Intercropping of black gram (*Phaseolus mungo*) in snake gourd (*Trichosanthes anguina*). *Research on Crops*, 2(1):34–36.

Thenmozhi, A. and Kingsly, S. 2004. Ovicidal effect of neem on snake gourd pest, *Plusia peponis* (Lepidoptera: Noctuidae). *Journal of Experimental Zoology*, 7(2):325–328

Varghese, B. M. 1973. Studies and cytology and evolution of South Indian Cucurbitaceae. PhD dissertation, Kerala Agricultural University, Thrissur, India.

31 Snake Gourd
Nutritional Values, Medicinal Properties and Health Benefits, Cultivable Cultivars, Cultural Practices, Diseases and Pests, and Crop Improvement

Rakesh Kr. Dubey, Vikas Singh,
Garima Upadhyay, and A.K. Pandey

CONTENTS

31.1 INTRODUCTION

Taxonomy

Kingdom:	Plantae
Division:	Angiosperms
Class:	Eudicots
Order:	Cucurbitales
Tribe:	Trichosantheae
Sub tribe:	Trichosanthinae
Family:	Cucurbitaceae
Sub family:	Cucurbitoidae
Genus:	*Trichosanthes*
Species:	*cucumerina*

Snake gourd is widely cultivated in tropical regions and used as a vegetable. Snake gourd has high economic importance among vegetables in India and is commonly grown in southeast Asia, Indo-Malaysia, China, Japan, and northern Australia as well as parts of Latin America, tropical Africa, and Mauritius (Miniraj et al., 1993; Bharathi et al., 2013). Twenty-four species of *Trichosanthes* are found in India, primarily along the Malabar coast of the Western ghats, in the low and medium elevation zones of the Eastern ghats, and northeastern region. Snake gourd is an annual herbaceous plant with trailing vines. Flowers are fragrant with white, delicately fringed, hair-like petals and fruits are 30–150 cm long, narrow, cylindrical, slender, and pointed at both ends. Snake gourds have two basic shapes: the first one has a long, narrow fruit that is tapered from top to bottom and the other type has the same width from top to bottom. At the widest part, the fruit diameter ranges from 4 to 10 cm and surface color is greenish-white and usually stripped (Bharathi et al., 2013). The fruits at maturity become orange-red, more fibrous, and extremely bitter in taste. It is grown commercially in Kerala, Tamil Nadu, Karnataka, Telangna, and Andhra Pradesh, and sporadically in eastern Uttar Pradesh, Bihar, West Bengal, Odisha, Assam, Manipur, Nagaland, Meghalaya, Sikkim, and Arunachal Pradesh in India. Biodiesel from snake gourd oil was found to have a moderate kinematic viscosity but relatively high flash point when compared with the standards outlined in ASTM PS 121 standards for biodiesel (Adesina and Amoo, 2013).

31.2 VERNACULAR NAME

Serpent cucumber, viper gourd, *chichinda*

31.3 COMPOSITION AND HUMAN NUTRITION

Immature fruits of snake gourd are used as vegetables in southeast Asian countries, while in the eastern part of Nigeria, the ripe pulp is consumed (Bharathi et al., 2013). The red pulp can be used to improve food appearance as it can be blended and used to produce a paste for stew, which tastes like and serves the role of tomato (Enwere, 1998). The fruits have poor nutrient content when compared to other common vegetables but do have vitamins C and A (Ojiako and Igwe, 2008) as well as proteins, fat, fiber, carbohydrates, and vitamin E. The total phenolics and flavonoids contents are 46.8% and 78.0%, respectively (Adebooye, 2008). The predominant mineral elements per 100 g are potassium (122 mg) and phosphorus (135 mg) (Bharathi et al., 2013). The proximate analysis and nutritive composition of snake gourd is presented in Table 31.1.

TABLE 31.1
Proximate Analysis and Nutritive Composition of Snake Gourd (per 100 g)

Nutrient Component	Content	Nutrient Component	Content
Moisture (g)	94.6	Iron (mg)	1.1
Protein (g)	0.5	Phosphorus (mg)	20
Fat (g)	0.3	Vitamin A (IU)	160.0
Minerals (g)	0.5	Thiamin (mg)	0.04
Carbohydrates (g)	3.3	Riboflavin (mg)	0.06
Fiber (g)	0.8	Niacin (mg)	0.3
Calcium (mg)	50.0	Energy (kcal)	18
Magnesium (mg)	53.0	Sodium (mg)	25.4
Potassium (mg)	34.0	Copper (mg)	0.11
Sulfur (mg)	35.0	Vitamin C (mg)	5.0
Chlorine (mg)	21.0	Oxalic acid (mg)	34.0

Source: Adapted from Ojiako, O.A. and Igwe, C.U., *Pak. J. Nutr.*, 7(1), 85, 2008.

31.4 MEDICINAL PROPERTIES AND HEALTH BENEFITS

Snake gourd is mentioned in earliest Ayurvedic texts for its medicinal properties. *Trichosanthes* species are also used in Chinese herbal medicine (Dou and Li, 2004) and Thai traditional medicine (Kanchanapoom and Yamasaki, 2002). *Trichosanthes cucumerina* has a prominent place in alternative systems of Indian medicine due to its various pharmacological activities involving antidiabetic, hepatoprotective, hypoglycemic, cytotoxic, anti-inflammatory, antifertility, and larvicidal effects (Sandhya et al., 2010). Hot water extract of *T. cucumerina* showed significant protection against ethanol- or indomethacin-induced gastric damage, increasing the protective mucus layer and decreasing the acidity of the gastric juice and antihistamine activity (Arawala et al., 2010; Bharathi et al., 2013). The important medicinal properties of snake gourd could be summarized as follows: infusion of snake gourd leaves is useful in the treatment of jaundice, if given in 30–60 g doses with a decoction of coriander seeds thrice daily. The juice of the fresh leaves is beneficial in heart disorders like palpitation and pain in the heart on physical exertion; it should be taken in doses of 1–2 tablespoon thrice daily. A decoction of snake gourd is effective in bilious fevers, as a febrifuge or thirst reliever and laxative; its efficacy increases provided it is given with chirata and honey. In obstinate cases of fevers, a combined infusion of this plant and coriander is more beneficial. About 30 g of each should be infused in water overnight. The strained liquid should be given in two doses the next day. A decoction of the leaves with the addition of coriander is also useful in bilious fever. The leaf juice is used to induce vomiting. The latter is also applied locally as a liniment in cases of liver congestion. In remittent fevers, it is applied over the whole body. The root of snake gourd serves as a purgative and tonic. Whereas its juice is a strong purgative, an infusion of the dried fruit is a mild purgative. It also aids in digestion. Its leaves are useful as an emetic and purgative in children suffering from constipation. A teaspoon of the fresh juice can be given early in the morning to the ailing children. The leaf juice is useful in the treatment of alopecia, a disease of the scalp resulting in complete or partial baldness. Snake gourd is rich in water content and has cooling effect on the body and helps in handling the summer heat (Nadkani, 2002).

31.5 USES

The immature fruits are used as salad and also cooked as vegetable in various ways. The fruits have an unpleasant odor that disappears after cooking. Occasionally, shoots and tender leaves are also used as cooked vegetable. The fruit increases appetite, acts as a tonic and stomachic, and cures biliousness. The root and seeds are anthelmintic and also used for the treatment of diarrhea, bronchitis, and fever (Singh and Bahadur, 2008).

31.6 ORIGIN AND DISTRIBUTION

Trichosanthes is the largest genus in the Cucurbitaceae family with 91 species (de Borer and Thulin, 2012). The wild species of *Trichosanthes* are restricted to southern and eastern Asia, tropical Australia, and Fiji, while India or Indo-Malayan region is considered the center of origin (De Candolle, 1882). The genus has its center of diversity in southeast Asia from India eastward to Taiwan, the Philippines, and Japan and southeastward to Australia, Fiji, and Pacific Islands (de Wilde and Dufjes, 2010). Of the 91 species, 24 have been reported from India (Chakravarty, 1959), 17 from Thailand (Duyfjes and Pruesapan, 2004), 18 from Malaysia (Backer and Bakhuizen-van den Brink, 1963), 4 from northern Australia, and 4 from Japan (Bharathi et al., 2013). Snake gourd occurs in the wild form in India, southeast Asia, and tropical Australia, but India is thought to be its place of origin. In India, maximum diversity in snake gourd can be seen in eastern and western peninsular region and northeast region.

31.7 BOTANY

Snake gourd is a monoecious annual with nontuberous roots. Hermaphrodite flowers have been observed but were nonfunctional (Singh, 1953). Flowers open from the bud stage in 8–16 days, with anthesis starting in early evening between 5:00 and 7:00 p.m. The stigma remains receptive from 12 h before anthesis to 12 h after anthesis, but maximum receptivity remains at the time of anthesis (Deshpande et al., 1980; Singh et al., 1989). The floral habit (ratio of staminate to pistillate flowers) in the monoecious species of snake gourd varies from 25:1 to 225:1 (Singh, 1953). The major pollinators are bees (*Apis florae* and *Apis dorsata*) and beetles (*Conpophilus* sp.) (Bharathi et al., 2013). Fruits are orange in color when ripe and pulpy red at maturity. The plant has hairy, angular, six to seven lobed leaves that emit a fetid odor when bruised.

31.8 TAXONOMY

Trichosanthes genus is in the subtribe Trichosanthinae, tribe Trichosantheae, subfamily Cucurbitoidae of the family Cucurbitaceae (Kocyan et al., 2007). Yueh and Cheng (1980) have subdivided *Trichosanthes* on the basis of the male bract, fruit pulp, and seed characters into subgenus Cucumeroides (with two sections: Cucumeroides and Tetragonosperma) and subgenus *Trichosanthes* (with five sections: Foliobracteola, Involucraria, Pedatae, Trichosanthes, and Truncata). Generally pollen grains are 3(4)-porate, 3(4)-colporate, psilate, perforate, (micro) retic, regulate, and verrucate (Jeffery, 1990).

31.9 CYTOGENETICS

The basic chromosome number of *Trichosanthes* is x = 11 with many diploid and some polyploid species. Unlike the majority of taxa of the Cucurbitaceae, the chromosomes of *Trichosanthes* are large enough to allow karyotypic analysis. The metcentric to submetacentric chromosomes range from 5.74 to 1.48 μm (Bharathi et al., 2013). The sporophytic chromosome number of *T. cucumerina* is 2n = 22 (Singh and Roy, 1979a) with 11 bivalents at meiosis (Datta and Basu, 1978). Cultivated snake gourd (*T. cucumerina* var. *anguina*) might have arisen directly from *T. cucumerina* var. cucumerina (Singh, 1990).

31.10 CLIMATE AND SOIL

Snake gourd prefers a tropical warm and humid climate for better plant growth and fruit development. A temperature range of 30°C–35°C, with a minimum threshold limit of 20°C, is ideal for its successful cultivation. A temperature below 20°C restricts plant growth and a temperature above 35°C is harmful for its growth, flowering, and fruiting. Plants are susceptible to frost. The most ideal soil for the cultivation of snake gourd is sandy loam or loam that is well drained and rich in organic matter. The soil pH 6.0–7.0 is most ideal for its successful cultivation. Plants do not tolerate dry soil and require a good moisture reserve in the soil. The field is prepared to a fine tilth by repeated ploughing and planking. Plants are sensitive to waterlogged condition, and heavy soils are unsuitable for snake gourd production (Singh and Bahadur, 2008).

31.11 CULTIVABLE VARIETIES

31.11.1 PLR (SG) 1

PLR (SG) 1 is developed by Tamil Nadu Agricultural University (TNAU), Coimbatore, India, through pure line selection from white long type and is suitable for cultivation under irrigated conditions only. Its fruit has excellent cooking quality due to less fiber and high flesh content. This variety can be cultivated during June to September, November to March, and April to May, with a yield potential of 350–400 q/ha.

31.11.2   PLR (SG) 2

PLR (SG) 2, developed by TNAU, Coimbatore, India, has an excellent cooking quality due to less fiber content. Fruits are plumpy, fleshy with an attractive white color and preferred in local markets and supermarkets. An average single fruit weighs 600 g. The short fruit enables easy handling and long distance transport, with a yield potential of 300–400 q/ha.

31.11.3   MANUSHRI

Manushri is an early, high yielding variety developed by Kerala Agricultural University, Vellanikkara, India, through a selection from local collection that yields uniform, medium length (60–70 cm), attractive, white fruits with green stripes at the pedicel end and is suitable for growing in the warm, humid tropics. On average, each fruit weighs between 600 and 700 g, with a yield potential of 500–600 q/ha.

31.11.4   PKM 1

PKM 1 is a mutant induced from H 375 variety developed by TNAU, Coimbatore, India, and is suitable for growing year round. The vines grow vigorously. Fruit color is dark green with white stripes on the outer side and light green inside, with a mean fruit weight of 700 g. The fruits are extra long (150–200 cm). Its average yield potential is 250–300 q/ha in a crop duration of 150 days.

31.11.5   MDU 1

MDU 1, an F_1 hybrid developed by TNAU, Coimbatore, India, between parentage of Panripudal and Selection-1, is an early flowering type (84 days) with a sex ratio of 1:38. It produces 13 fruits per vine weighing 7.15 kg, with an average yield of 317.5 q/ha in a crop duration of 145 days. The fruits are of medium length (66.94 cm), with white stripes on a green background. On average, each fruit weighs 55 g. The fruits are fairly rich in vitamin A (44.4 mg/100 g) and very low in fiber content (0.6%).

31.11.6   CO 1

CO 1 is an early maturing variety with moderately spreading habit, developed by TNAU, Coimbatore, India. The first fruit becomes ready for harvesting about 70 days after sowing. Fruits are 160–180 cm long, dark green with white stripes and light-green flesh and are of good cooking quality. A single fruit weighs 500–750 g. Each plant per vine bears 10–12 fruits, with an average fruit yield of 180–200 q/ha in a crop duration of 135–140 days.

31.11.7   CO 2

CO 2 is developed by TNAU, Coimbatore, India, through selection and is suitable for high density planting. Fruits are short (30 cm) and light greenish-white in color. It gives an average yield of 350–400 q/ha in a crop duration of 110–120 days.

31.11.8   CO 4

CO 4 is an early maturing variety developed by TNAU, Coimbatore, India. The first marketable fruit may be harvested about 70 days after sowing. Fruits are 160–190 cm long, dark green with white stripes and light-green flesh. Each plant bears 10–12 fruits that weigh about 4–5 kg, with a yield potential of 300 q/ha.

31.11.9   TA 19

TA 19 is developed by Kerala Agricultural University, Vellanikkara, India. At the immature stage, fruits are about 60 cm long and light green with white stripes at the stylar end. It is ready for harvest 63–70 days after sowing. The average weight of a single fruit is 600 g, with a yield potential of 350–400 q/ha.

31.11.10   KONKAN SWETA

Konkan Sweta is developed by Konkan Krishi Vidyapeeth, Dapoli, Maharastra, India. The fruits are of medium length (90–100 cm) and white in color. Its fruit is fleshy if harvested timely or otherwise becomes hollow. Average fruit yield is 150–200 q/ha in a crop duration of 120–130 days.

31.11.11   BABY

Baby is a small-fruited variety developed through selection. The fruits are small (30–40 cm long and 22 cm diameter), elongated, and uniformly white in color. Its single fruit weighs about 474 g, and yield potential is about 580 q/ha in 8–10 pickings. First picking starts about 55 days after sowing, and total crop duration ranges between 120 and 140 days.

31.11.12   KAUMUDI

Kaumudi is developed by Kerala Agricultural University, Vellanikkara, India, through selection. Fruits are uniformly white, with an acute tip. Average fruit length is about 100 cm and breadth about 9.0 cm, with average fruit weight 1.3 kg. The yield potential is about 500 q/ha in a harvesting period of 140–150 days.

31.11.13   R K-325

R K-325 is an F_1 hybrid developed by R K Seed Farms (R), Delhi, India. The variety is early in maturing, with high fruiting. First fruit picking starts 60–65 days after sowing. Fruit is greenish white in color. Fruit length varies from 35 to 40 cm, with a diameter of 4.5 cm. Average fruit weight varies from 125 to 150 g. Average fruit yield varies between 600 and 650 q/ha, and the variety is suitable for long transportation.

31.12   SEED AND SOWING

About 4–6 kg seeds are enough for sowing 1 ha area; however, seed sowing in pits at a spacing of 2.0 × 2.5 m requires about 1.5 kg seeds/ha. Seeds collected from larger fruits produce heavier seedlings and vigorous plants. The seed, before sowing in the field, should be treated with *Trichoderma viride* @ 4.0 g or *Pseudomonas fluorescens* @ 10.0 g or carbendazim @ 2.0 g/kg of seeds. In south India, it is grown during both spring–summer and rainy seasons. The spring–summer crop is sown from December to January and the rainy season crop from June to July. In northern India, generally one crop is taken during the rainy season, which is sown from June to July. It is also planted from October to November where winters are very mild. Normally, the seeds will germinate and emerge in 1 week, unless the soil is too dry or the weather is too cold. Seedlings can be raised in the nursery and transplanted to the field at the two true-leaf stage. Zhang et al. (2000) tested different cytokinins for bud induction in vitro and reported that a mixture of benzyladenine and IAA was the best combination.

31.13   SOWING AND PLANTING

The followings are the important sowing/planting methods applied in raising a healthy crop. The details are as follows.

31.13.1   FLAT BED OR SHALLOW PIT METHOD

Shallow pits of $60 \times 60 \times 45$ cm size are dug and left open for 2 weeks before sowing seeds. The pits are filled with a mixture of soil, compost (4.0 kg/pit), and recommended dose of nitrogen, phosphorus, and potassium. Carbofuran @ 1.5 g/pit should also be thoroughly mixed in the pit soil before sowing the seed. After filling the pits, two to three seeds per pit should be sown at 2–3 cm depth. Spacing of 1.5–2.0 m row-to-row and 75–100 cm plant-to-plant is advocated for sowing/planting of snake gourd.

31.13.2   SOWING OF SEEDS ON RAISED BEDS

In this method of sowing, channels of 40–50 cm width are prepared manually or mechanically, keeping 2.0–2.5 m distance between two channels. Seeds are sown on both the edges of the channel at a spacing of 1 m. Usually, per hill, two to three seeds are sown to have optimum plant population. The vines are allowed to spread over the raised portion between the channels. In this way, around 3500–4500 plants can be accommodated in 1 ha area.

31.14   PLANT POPULATION DENSITY

In snake gourd, population density varies from 3,500 to 10,000 plants/ha, depending upon varieties and methods of cultivation. Adopting suitable training practices about 8,000–10,000 plants can be accommodated in an area of 1 ha (Singh and Bahadur, 2008).

31.15   NUTRIENT MANAGEMENT

Snake gourd responds well to manures and fertilizer, but too much nitrogenous fertilizer leads to excessive vine growth at the expense of fruit production. During field preparation, about 20–25 t/ha of well-rotten farmyard manure should be incorporated in the soil, supplemented with N 50, P 30, and K 20 kg/ha. The full amount of phosphorus and potassium and half of nitrogen are applied as basal dressing and the remaining amount of nitrogen is top dressed when the plants start bearing fruits. Syriac and Pillai (2001) reported maximum fruit yield using 105 kg nitrogen per hectare, 200 ppm Ethephan, and drip irrigation at 5 mm cumulative pan evaporation.

31.16   IRRIGATION MANAGEMENT

For initial growth, the crop should be irrigated at an interval of 5–6 days. Snake gourd in its entire crop duration requires a steady supply of moisture but is sensitive to excessive water application or water logging conditions (Bharathi et al., 2013). The first irrigation may be given immediately after sowing, if moisture in the field at the time of sowing is not sufficient for proper germination of seeds. At the time of fruit set and fruit development, sufficient moisture should be maintained, and there should not be any dry spell during this period. Flooding and soil water logging should be avoided to reduce the incidence of *Fusarium* wilt and root rots.

31.17 WEED MANAGEMENT

Weed control is important, particularly during the initial stages of plant growth. Use of paddy straw or black polythene as mulch will reduce heat and moisture stress as well as weed incidence (Bharathi et al., 2013). Two weeding by hand or machine, supplemented with preemergence application of 1.0 kg/ha pendimethalin or alachlor or preplant incorporation of 1.0 kg/ha fluchloralin, effectively controls weeds. However, these herbicides control the weeds in the early growth stages. Later, if weeds pose a problem, one manual weeding may be done to protect against the weeds.

31.18 TRAINING OF VINES

Snake gourd fruits are very long, slender, and soft; thus, training is an essential practice to allow the fruits to grow downward and to keep them straight. Vines are trailed on a high trellis to avoid fruit twisting or coiling. The bower system of training is most commonly used in snake gourd. When seedlings start producing tendrils, they are staked to thin bamboo poles using string or banana fibers to enable the vines to spread on a trellis. Maximum fruit length, diameter, number of fruits per plant, and yield have been obtained by training to a single stake followed by the bower system. Vine training over bower system is considered the best for snake gourd. Overhead trellises are made at about 2.0 m height. A small weight can be tied to the bottom end of the developing fruit to make them grow straight. The vines dry up at the cessation of fruiting during October to November. Vines can then be pruned to leave 30–35 cm of vine above the soil (Singh and Bahadur, 2008) (Figure 31.1).

31.19 USE OF PLANT GROWTH REGULATORS

The proportion of female flowers can be increased either by pruning the vines or by applying Ethrel @ 150 ppm. Application of ethephan to seedlings at the two- to four-leaf stage reduces the days to first pistillate flowering and increases the number of pistillate flowers (Kohinoor and Mian, 2005). Application of 0.1%–0.2% potassium naphthenate on 18-day-old plants can increase the number of fruits per plant (Singh and Bahadur, 2008).

FIGURE 31.1 **(See color insert.)** Snake gourd fruit, flower, and twig.

31.20   HARVEST AND POSTHARVEST MANAGEMENT

The harvest of snake gourd starts 70–80 days after seed sowing. The fruits can be harvested as early as 15–20 days after anthesis, and fruits attained a length of 30–100 cm depending on the cultivars and season. Flowering and harvesting usually continue for 1–2 months. Harvesting of fruits should be done at an interval of 5–6 days, with a total of seven to eight harvests per cropping season (Bharathi et al., 2013). Most often, the fruits are harvested when still green and tender since fruits at full maturity become lighter in weight, fibrous, and slightly hard, which are not usually preferred by consumers in the market. The snake gourd fruits have a short shelf life; thus, it should be sent to the market without delay. The fruits need to be packed in baskets or in other containers immediately after harvesting to protect them from excessive moisture loss and injuries during transport. Fruits can be stored for 10–14 days at 16°C–17°C temperature and 85%–90% relative humidity.

31.21   YIELD

A single plant of a traditional cultivar yields about 6–10 fruits and of an improved cultivar up to 50 fruits. Single fruit weight varies from 300 g to 1.0 kg, and average fruit yield varies from 200 to 300 q/ha depending on the cultivar, season, and packaging practices adopted.

31.22   PESTS AND DISEASES

31.22.1   Red Pumpkin Beetle (*Aulacophora foveicollis*)

The beetle feeds on cotyledons and foliage. Adults feed on cotyledons. Grubs develop in soil and, occasionally, feed on roots, causing wilting of plants. The adult beetles emerge from the soil and make typical shot holes on the foliage. The grubs also damage underground portions of plants by boring into the roots, stems, and sometimes into the fruits touching the soil.

31.22.1.1   Management
- Summer deep ploughing is recommended to expose the grubs present in the soil.
- Spraying of carbaryl (Sevin) two to three times @ 0.1%–0.2% or Rogor @ 0.1% near the base of the plants just after germination is recommended for its control.
- Spraying of 0.01% *neem* seed kernels extract and 0.4% *neem* oil can effectively control the beetles.

31.22.2   Fruit Fly (*Bactrocera cucurbitae*)

Snake gourd yield is drastically reduced if the crop is attacked by fruit fly in the early stage. Among cucurbits, snake gourd is the most preferred crop by fruit fly, as this pest completes incubation, larva, and pupa stages within 3 weeks. An adult female lays 20–30 white, cigarette-shaped eggs upon tender fruits. After hatching, the white maggots feed inside the fruits, which causes deformity, rotting, and premature dropping of the fruits.

31.22.2.1   Management
- Collect and destroy the infested plant parts and fruits.
- Expose the pupae by deep ploughing in hot summer months.
- Spray the crop with cypermethrin @ 0.025% to control the undersurface population of flies.
- Application of carbaryl @ 0.1% at the tender fruit stage gives excellent control of fruit fly.
- Use 20 × 15 cm poly-bags fishmeal traps with bait comprising 5.0 g of fishmeal + 1.0 mL of dichlorvos in cotton @ 50 traps/ha.

- Bait spray with malathion (50 EC) 20.0 mL + yeast hydroxylate (10.0 g) + 20.0 L water + 500.0 g molasses gives effective control of fruit fly.
- Foliar spraying of fenthion @ 0.05% with 5.0% jaggery at fruit formation is also recommended.

31.22.3 SEMILOOPER (*ANADEVIDIA PEPONIS*)

The caterpillars of this insect damage the young leaves. An adult female lays about 150 eggs singly during the nighttime on the upper- and lower-leaf surface. The life cycle is completed within 20–31 days. Its adults survive only for 5–10 days. The sex ratio (male to female) is usually 1:1.27.

31.22.3.1 Management

- Foliar spray of 0.05% quinalphos, monocrotophos, or endosulfan and/or 0.03% dimethoate is highly effective against this pest.
- Foliar spray of malathion 50 EC (1 mL/L) or methyl demeton 25 EC (1 mL/L) or fenthion 100 EC (1 mL/L) also controls the semilooper effectively.

31.22.4 LEAF MINER (*LIRIOMYZA TRIFOLII*)

Leaf miner infestation takes place during the months of March and April. The eggs hatch within 2–3 days and larvae scrap the chlorophyll and leaf tissues, feeding between lower and upper epidermis and making zigzag tunnels. The larvae make mines in leaves, especially in mature leaves.

31.22.4.1 Management

- Pluck and destroy the old leaves severely infested by leaf miners.
- Spray the crop with 4% neem seed kernel extract at weekly intervals.
- Foliar spray of dimethoate (30 EC) @ 1 mL/L is effective to control this pest.

31.22.5 ROOT KNOT NEMATODES (*MELOIDOGYNE INCOGNITA, MELOIDOGYNE HAPLA,* AND *MELOIDOGYNE JAVANICA*)

These long, slender, and microscopic organisms attack the root system. Infective larvae enter the roots, form galls, and affect the growth system adversely, and consequently, crop yield is reduced.

31.22.5.1 Management

- Apply 2–4 kg a.i./ha carbofuran 3G, 10 kg a.i./ha phorate, or 2–4 kg a.i./ha aldicarb during field preparation.
- Soil fumigation with Nemagon or DD mixture 200–400 L/ha is very effective in controlling nematodes.
- Soil application of neem cake @ 80–100 kg/ha is also advantageous.

31.23 DISEASES

31.23.1 DOWNY MILDEW (*PSEUDOPERONOSPORA CUBENSIS*)

Symptoms appear as numerous small, irregular, yellow lesions surrounded by green tissues scattered all over the leaves. In due course, lesions grow in size and coalesce with each other. Old lesions become necrotic, and they are clearly demarcated by slightly yellowish areas. Severely infected leaves roll upward with a brownish tinge that produces a blighted appearance. Grayish-black downy fungal growth is observed on the undersurface of the leaf. The crop should be grown with wide spacing on well-drained soil.

31.23.1.1 Management
- Field sanitation by burning crop debris helps in reducing the inoculums.
- Protective spray of mancozeb (0.25%) at 7 days interval gives good control.
- In severe cases, one spray of metalaxyl + mancozeb @ 0.2% may also be done.

31.23.2 ANTHRACNOSE (*COLLETOTRICHUM LAGENARIUM*)

The disease infects all the aboveground parts of the crop. Yellowish, water-soaked areas, which coalesce with each other on enlarging and turn brown to black, appear on leaves. The necrotic portion dries and shatters. Elongated, water-soaked, sunken lesions also appear on the stem. Light yellow to brown discoloration of the stem lesions takes place due to abundant sporulation. Sunken, dark brown to black lesions that vary in size also appear on fruits, depending upon the age of the crop plants and weather conditions.

31.23.2.1 Management
- Field sanitation by burning of crop debris reduces the primary inoculums.
- Always collect seed from healthy fruits and disease-free fields.
- Seeds must be treated with carbendazim or captan @ 2.5 g/kg of seed.
- Foliar spray of chlorothalonil 0.2% gives good control of the disease, but spraying must be started before the infection occurs.
- Foliar spray of fungicides such as Benomyl, Bavistin, or Thiophanate-M (0.1%) reduces disease incidence.

31.23.3 FUSARIUM ROOT ROT (*FUSARIUM SOLANI* F. SP. *CUCURBITAE*)

The symptoms include vascular browning, gummosis, and tyloses in xylem vessels of mature plants, and subsequently, the whole plant becomes wilted. The pathogen is both seed- and soilborne.

31.23.3.1 Management
- Follow long crop rotation and clean cultivation.
- Use only disease-free, healthy seed.
- Grow resistant varieties in disease-infested areas.
- Avoid root injury during intercultural operations.

31.23.4 POWDERY MILDEW (*SPHAEROTHECA FULIGINEA*)

The infection appears first on the upper side of the leaves and stems as white to dull white powdery growth, which quickly covers most of the leaf surface and leads to serious reduction in photosynthetic rate. In due course, all the aboveground parts are infected. Finally, the lesions turn brown and necrotic. The affected leaves become yellowish and dry and get defoliated. The fruits do not develop properly and are sometimes covered with white, powdery masses.

31.23.4.1 Management
- The fungicidal spray penconazole (0.05%) can give very good control of the disease.
- Spray the crop with Karathane 1.5 mL/L, carbendazim 0.1%, Calixin 0.1%, or Sulfex 2.5 g/L of water.

31.23.5 MOSAIC

Mosaic caused by cucumber mosaic virus (CMV) is also prevalent in snake gourd. Symptoms include a mosaic with dark-green raised blisters on the leaf lamina, reduced leaf size, shortened internodes, and retarded growth. Infected plants flower sparingly and set only few fruits. The virus is sap and insect (*Aphis gossypii* and *Aphis craccivora*) transmissible. *A. craccivora* acquires this virus with 5 min acquisition feeding and transmits it within 5 min feeding on healthy plants. A minimum of five aphids is required for its transmission. The vector could not retain the virus for long periods, the relationship being nonpersistent. The virus was identified as a strain of *Cucumis* virus-1 (CMV).

31.23.5.1 Management of Mosaic

The most effective management of viral diseases demands integration of management practices such as avoidance of sources of infection, control of vectors, modification of cultural practices, and resistance of host plant. Rouging and destroying of virus-infected plants are effective measures to restrict the spread of mosaic disease (Singh and Bahadur, 2008).

31.24 CROP IMPROVEMENT

A multiple allelic series controls fruit skin color, which segregates in monohybrid ratio in the F_2 generation. Deep green is dominant over green, yellow, and white; green is dominant over yellow and white; and yellow is dominant over white (Sardar and Mukharjee, 1987). Kumaresan et al. (2006) reported that days to first staminate flower opening, days to first pistillate flower opening, main vine length, number of fruit per vine, fruit length, diameter, fruit weight, and yield per vine were controlled by both additive and nonadditive gene action. The southern peninsular tract of India was identified as the priority region for collection of snake gourd germplasm distribution, variability, extent of gene erosion, and a survey of existing collections. Wide variability of snake gourd exists in South India, and institutions like Tamil Nadu Agricultural University, Coimbatore and Kerala Agricultural University have made a diverse collection of germplasm. N. I. Vavilov Research Institute of Plant Industry (VIR), St. Petersburg, the Russian Federation, collected 15 genotypes of *T. cucumerina* var. *anguina* (Bharathi et al., 2013). Variability exists for earliness, size of fruit, fruit/plant, and yield. Local strains have been selected by several institutions and many cultivars have been developed using pure line selection, including CO-1, CO-2, TA-19, and Baby. Natarajan et al. (1985) reported a hybrid snake gourd MDU 1. Mutation breeding holds promise in snake gourd improvement. The cultivar PKM-1 was developed through induced mutation (Pillai et al., 1979). The breeding goals of snake gourd could be summarized as vigorous and highly branched plants, earliness (lower node of appearance of first pistillate flower), highly pistillate to staminate flower ratio, high fruit quality and appearance including green or white color depending on consumer preference, and a range of fruit sizes such as short (30–35 cm), medium (60–70 cm), and long (160–180 cm). Fruit should be thick, heavy, nonfibrous, and tender at the marketable stage. There should be resistance against insects (fruit fly) and diseases (fruit rot and mosaic virus) (Bharathi et al., 2013).

REFERENCES

Adebooye, O.C. 2008. Phytoconstituents and antioxidant activity of the pulp of snake tomato (*Tricosanthes cucumerina*). *African Journal of Traditional, Complementary and Alternative Medicines* 5(2): 173–179.

Adesina, A.O. and Amoo, I.A. 2013. Chemical composition and biodiesel production from snake gourd (*Trichosanthes cucumerina*) seeds. *International Journal of Science and Research* 2(1): 41–48.

Arawala, L.D.A.M., Thabrew, M.I., and Arambewela, L.S.R. 2010. Gastroprotective activity of *Trichosanthes cucumerina* in rats. *Journal of Ethnopharmacology* 127(3): 750–754.

Backer, C.A. and Bakhuizen-van den Brink, Jr. R.C. 1963. *Flora of Java*, Vols. 1–3. Wolters-Noordhoff, Groningen, the Netherlands.

Bharathi, L.K., Bahera, T.K., Sureja, A.K., John, K.J., and Wehner, T.C. 2013. Snake gourd and pointed gourd: Botany and horticulture In: J. Janick, ed. *Horticultural Reviews*. John Wiley & Sons, Inc., Vol. 41, pp. 457–495.

Chakravarty, H.L. 1959. *Monograph on Indian Cucurbitaceae*. Records of the Botanical Survey of India, Vol. 17, pp. 1–234.

Datta, S.K. and Basu, R.K. 1978. Cytomorphological, biochemical and palynological studies in *Triehosonthes anguina* and *T. cucumerina*. *Cytologia* 43: 107–117.

de Boer, H.J. and Thulin, M. 2012. Synopsis of *Trichosanthes* (Cucurbitaceae) based on recent molecular phylogenetic data. *Phytokeys* 12: 23–33.

De Candolle, A. 1882. *Origin of Cultivated Plants*. Hafner Publishing Co., New York.

de Wilde, W.J.J.O. and Duyfjes, B.E.E. 2010. Cucurbitaceae. In: H.P. Nooteboom, ed. *Flora Malesiana*. Foundation Flara Malesiana, Leiden, the Netherlands, Vol. 19, pp. 1–342.

Deshpande, A.A., Bankapur, V.M., and Venkatasubbaiah, K.A. 1980. Studies on floral biology of snake gourd (*Trichosanthes anguina* L.) and ash gourd [*Benincasa hispida* (Thunb.) Cogn.]. *Mysore Journal of Agricultural Science* 14: 8–10.

Dou, C.M. and Li, J.C. 2004. Effect of extracts of *Trichosanthes* root tubers on HepA-H cells and HeLa cells. *World Journal of Gastroenterology* 10(14): 2091–2094.

Duyfjes, B.E.E. and Pruesapan, K. 2004. The genus *Trichosanthes* L. Thailand. *Thai Forest Bulletin (Bol)* 32: 76–109.

Enwere, N.J. 1998. *Foods of Plant Origin*. Afroorbis Publications, Nsukka, Nigeria, pp. 153–168.

Jeffrey, C. 1990. An outline classification of the Cucurbitaceae. In: D.M. Bates, R.W. Robinson, and C. Jeffrey, eds. *Biology and Utilization of the Cucurbitaceae*. Cornell University Press, Ithaca, NY, pp. 449–463.

Kanchanapoorn, T.K.R. and Yamasaki, K. 2002. Cucurbitane, hexanorcucurbitane and octanorcucurbitane glycosides from fruits of *Trichosanthes tricuspidata*. *Phytochemistry* 59: 215–228.

Kocyan, A., Zhang, L.B., Schaefer, H., and Renner, S.S. 2007. A multi-locus chloroplast phylogeny for the Cucurbitaceae and its implications for character evolution and classification. *Molecular Phylogenetics and Evolution* 44(2): 553–577.

Kohinoor, H. and Mian, M.A.K. 2005. Effect of ethrel on flower production and fruit yield of snakegourd (*Trichosanthes anguina*). *Bulletin of the Institute of Tropical Agriculture Kyush University* 28(2): 59–67.

Kumaresan, G.R., Makesh, S., and Ramaswamy, N. 2006. Combining ability studies for yield and its components in snake gourd. *Crop Research* 31(1): 103–106.

Miniraj, N., Prasanna, K.P., and Peter, K.V. 1993. Snake gourd. In: G. Kalloo and B.D. Bergh, eds. *Genetic Improvement of Vegetable Crops*. Pergamon Press, Oxford, U.K., pp. 259–264.

Nadkani, K.M. 2002. *Indian Material Medica*, 2nd edn. Popular Prakashan, Mumbai, India, Vol. 1, pp. 1235–1236.

Natarajan, S., Shanrnugavelu, E.G., Nambisan, K.M.P., and Krishnan, M. 1985. New varieties of horticultural crops released by Tamil Nadu Agricultural University, Coimbatore during 1985. 4. MDU 1 snake gourd. *South Indian Horticulture* 33: 454.

Ojiako, O.A. and Igwe, C.U. 2008. The nutritive, anti-nutritive and hepatotoxic properties of *Trichosanthes anguina* (snake tomato) fruits from Nigeria. *Pakistan Journal of Nutrition* 7(1): 85–89.

Pillai, O.A.A., Velukutty, B., and Jeyapal, R. 1979. PKM: 1, a new snake gourd (*Trichosanthes anguina* L.). *South Indian Horticulture* 27: 371–382.

Sandhya, S., Vinod, K.R., Chandrasekhar, J., Aradhana, R., and Nath, V.S. 2010. An updated review on *Trichosanthes cucumerina* L. *International Journal of Pharmaceutical Sciences Review* 1(2): 56–60.

Sardar, A.K. and Mukherjee, K.K. 1987. Inheritance of fruit coat colour in *Trichosanthes anguina* L. *Theoretical and Applied Genetics* 74: 171–172.

Singh, A.K. and Roy, R.P. 1979a. Cytological studies in *Trichosanthes* L. *Journal of cytology and genetics*. 14: 50–57.

Singh, A.K., Singh, R.D., and Singh, J.P. 1989. Studies on floral biology in pointed gourd (*Trichosanthes dioica* Roxb.). *Vegetable Science* 16(2): 185–190.

Singh, K.P. and Bahadur, A. 2008. Snake gourd. In: M.K. Rana, ed. *Scientific Cultivation of Vegetables*. Kalyani Publishers, New Delhi, India, pp. 247–255.

Singh, R.N. 1953. Studies in the sex expression, sex ratio and floral abnormalities in the genus *Trichosanthes* L. *Indian Journal of Horticulture* 10(3): 98–106.

Syriac, E.K. and Pillai, G.R. 2001. Effect of nitrogen, Ethephon and drip irrigation on the yield and yield attributes of Snake gourd (*Trichosanthes anguina* L.). *Annals of Agricultural Research* 22(4): 508–513.

Yueh, C.H. and Cheng, C.Y. 1980. The Chinese medicinal species of the genus *Trichosanthes* L. *Acta Phytotoxonomica Sinica* 1: 333–352.

Zhang, L.H., Cheng, H.W., Chi, and Xue, W.X. 2000. Bud induction of serpent gourd (*Trichosanthes anguina* L.) in vitro. *Cucurbit Genetics Cooperative Report* 23: 80–82.

Index

Printed in the United States
by Baker & Taylor Publisher Services